Elementary Algebra

Mark D. Turner
Charles P. McKeague

xyztextbooks

Elementary Algebra

Mark D. Turner
Charles P. McKeague

Publisher: XYZ Textbooks

Project Manager: Katherine Heistand Shields

Composition: Katherine Heistand Shields, Matthew Hoy

Sales: Amy Jacobs, Richard Jones, Bruce Spears

Cover Design: Kyle Schoenberger

Printed in the United States of America

ISBN-13: 978-1-63098-063-4 / ISBN-10: 1-63098-063-3

For product information and technology assistance, contact us at
XYZ Textbooks, 1-877-745-3499

For permission to use material from this text or product,
e-mail: **info@mathtv.com**

XYZ Textbooks
1339 Marsh Street
San Luis Obispo, CA 93401
USA

For your course and learning solutions, visit www.xyztextbooks.com

Brief Contents

1 The Basics 1

2 Linear Equations and Inequalities 91

3 Linear Equations and Inequalities in Two Variables 181

4 Systems of Linear Equations 265

5 Exponents and Polynomials 309

6 Factoring 383

7 Rational Expressions 445

8 Roots and Radical Expressions 521

9 Quadratic Equations 585

Contents

1 The Basics 1

1.1 Variables, Symbols, and the Order of Operations 3
1.2 The Real Numbers 15
1.3 Addition of Real Numbers 29
1.4 Subtraction of Real Numbers 39
1.5 Multiplication and Division of Real Numbers 47
1.6 Fractions 59
1.7 Properties of Real Numbers 77
Chapter 1 Summary 87
Chapter 1 Test 90

2 Linear Equations and Inequalities 91

2.1 Simplifying Expressions 93
2.2 Addition Property of Equality 103
2.3 Multiplication Property of Equality 111
2.4 Solving Linear Equations 119
2.5 Formulas and Percents 129
2.6 Applications 141
2.7 More Applications 155
2.8 Linear Inequalities 167
Chapter 2 Summary 177
Chapter 2 Test 180

3 Linear Equations and Inequalities in Two Variables 181

3.1 Paired Data and Graphing Ordered Pairs 183

3.2 Graphing Linear Equations in Two Variables 195

3.3 More On Graphing: Intercepts 209

3.4 The Slope of a Line 219

3.5 Slope-Intercept Form 233

3.6 Point-Slope Form 243

3.7 Linear Inequalities in Two Variables 251

Chapter 3 Summary 259

Chapter 3 Test 262

4 Systems of Linear Equations 265

4.1 Solving Linear Systems by Graphing 267

4.2 The Elimination Method 277

4.3 The Substitution Method 287

4.4 Applications 295

Chapter 4 Summary 305

Chapter 4 Test 307

5 Exponents and Polynomials 309

5.1 Multiplication with Exponents and Scientific Notation 311

5.2 Division with Exponents 321

5.3 Operations with Monomials 335

5.4 Addition and Subtraction of Polynomials 345

5.5 Multiplication with Polynomials 353

5.6 Binomial Squares and Other Special Products 361

5.7 Division with Polynomials 369

Chapter 5 Summary 379

Chapter 5 Test 381

6 Factoring 383

6.1 The Greatest Common Factor and Factoring by Grouping 385

6.2 Factoring Trinomials 395

6.3 More on Factoring Trinomials 401

6.4 Special Factoring Patterns 409

6.5 Factoring: A General Review 417

6.6 Solving Equations by Factoring 423

6.7 Applications 431

Chapter 6 Summary 441

Chapter 6 Test 444

7 Rational Expressions 445

7.1 Reducing Rational Expressions to Lowest Terms 447

7.2 Multiplication and Division of Rational Expressions 459

7.3 Addition and Subtraction of Rational Expressions 469

7.4 Complex Fractions 479

7.5 Equations Involving Rational Expressions 487

7.6 Proportions 493

7.7 Applications 499

7.8 Variation 509

Chapter 7 Summary 517

Chapter 7 Test 520

8 Roots and Radical Expressions 521

8.1 Definitions and Common Roots 523

8.2 Simplified Form and Properties of Radicals 535

8.3 Addition and Subtraction of Radical Expressions 547

8.4 Multiplication of Radicals 555

8.5 Division of Radicals 563

8.6 Equations Involving Radicals 573

Chapter 8 Summary 581

Chapter 8 Test 584

9 Quadratic Equations 585

9.1 Square Root Property 587

9.2 Completing the Square 597

9.3 The Quadratic Formula 605

9.4 Complex Numbers 613

9.5 Complex Solutions to Quadratic Equations 619

9.6 Graphing Parabolas 625

Chapter 9 Summary 635

Chapter 9 Test 637

Answers A-1

Index I-1

Preface: A Note to Instructors

Elementary Algebra

by Mark D. Turner and Charles P. McKeague

Description

Elementary Algebra fits the traditional, one-semester, elementary algebra course. The prerequisite is Basic Mathematics or Prealgebra.

The normal sequence of topics is enriched with modeling applications, study skills, and other built-in features such as the "Spotlight on Success" (as detailed below). Encourage your students to make regular use of these features—you'll find that they will gradually build a foundation of successful studying practices that can benefit them in all their future courses.

This textbook is more than just the book itself. We built this book to be the hub of a math "toolbox" of sorts. While most students still prefer to use the printed book in their studies, the eBook extends the reach of the book, giving students and instructors access to a wide array of supporting tools:

- The eBook includes free access to the Elementary Algebra eBook, plus over 20 eBooks covering 8 math courses—great for remediation.
- It also includes free access to 10,000 MathTV videos, with 3-4 tutorials for every single example.
- Plus all of the accompanying worksheets and digital supplements for the book are found in the eBook.
- Additionally, QR code technology connects the printed textbook directly to the digital resources through the students' smartphones and tablets.
- The associated Matched Problems Worksheets gives you the opportunity to flip your classroom, without the need to create new materials yourself.

Textbook Features

Every section has been written so that it can be discussed in a single class session. The clean layout and conversational style used by the authors make it easy for students to read, and important information, such as definitions and properties, are highlighted so that they can easily be located and referenced by students.

In addition, the following features provide both instructors and students a vast array of resources which can be used to enrich the learning environment and promote student success.

Chapter Introductions We begin each chapter with a brief application that involves some of the concepts to be presented later in the chapter. These "glimpses" of what is to come help motivate students by showing them an interesting, real-world scenario that requires the knowledge and skills they will be attaining. Acting as a mini-theme for the chapter, students will encounter examples and problems in the following sections that directly relate to this application.

Learning Objectives Every section now begins with a short set of learning objectives, listing the specific, measurable knowledge and skills students will be acquiring in that section. Objectives help students identify and focus on the important concepts in each section and increase the likelihood of their success by having established and clear goals. For instructors, the objectives can help in organizing class lessons and learning activities, and in creating student assessments.

Getting Ready for Class It's always a challenge to get students to read, especially in math. To encourage them, we place four key questions at the end of every section, under the heading "Getting Ready for Class." If students have read the section, they should be able to answer each of these with ease. Even a minimal attempt to answer these questions can enhance the students' in-class experience.

We've heard some innovative strategies from early adopters of the textbook. Instructors in one department use these questions as a student's "ticket" to class. They collect the answers as students come in the door. A simple action that can start the class off right!

Getting Ready for the Next Section When students finish a section, they can feel a great sense of accomplishment. We want to help maintain their momentum as much as possible, which is why we offer a brief preview of the next stop in their mathematical journey. A small set of "Getting Ready for the Next Section" problems appear near the end of every problem set. Many are the exact same problems that students will see when they read through the next section of the text. Students who consistently work these problems will be much better prepared for class.

Learning Objectives Assessments Every problem set concludes with a short section of multiple-choice questions that can be used by students or instructors to measure the extent to which students have met the learning objectives for that section. These questions can easily be adapted for use with clicker technology, or used as a quick exit quiz at the close of each class session, to verify that students understand the main points from the lesson.

Paying Attention to Instructions Students tend to get stuck making the same mistakes over and over again, developing bad habits. The more time you spend teaching, the more you can start to anticipate these problems. For example, we know that students don't always pay attention to instructions when they are doing their homework, and it can get them into trouble on tests and quizzes.

The only way to fix this is to address it head-on. Our strategy was to build problems into the text itself, under the heading "Paying Attention to Instructions," that challenge students to carefully read the instructions for each problem. Small nuances make a big difference in mathematics, and we hope that this feature will help you demonstrate to your students just how important this is.

"How To" Segments Many sections include a "How To" segment that outlines the steps in the method or process used to solve certain types of problems. These summaries help students internalize the particular problem solving strategy introduced in that section.

Spotlight on Success Each student has a unique approach to learning. A one-size-fits-all strategy doesn't work, which is why our MathTV videos have always used multiple tutors for each example. We want to give every student the chance to succeed!

The "Spotlight on Success" feature is designed in the same spirit. Scattered throughout the text, this feature offers students a variety of strategies from many different sources. Some are from peers, others are from instructors. Many of the spotlights feature the same peer tutors that students will see on the MathTV videos that accompany the text.

Real-Data Application Problems Students do better in math when they see its application to real-world problems. That's why we've included as many applied problems as we can. Many times the charts and graphics in the text look like the types of charts and graphics students see in the media.

Not only does this help them make the connection, but it can give them a greater mathematical "sense," and even make the concepts a little easier to understand.

Facts from Geometry Anything we can do to help solidify abstract concepts is a good thing. Geometry can give students another view of the problem, and potentially, another avenue for understanding algebraic concepts. Students who are visual learners will love this feature.

Chapter Summaries and Tests Every chapter concludes with a Chapter Summary and Chapter Test. The chapter summary lists the main properties and definitions found in the chapter, with examples given in the margin. The chapter test provides a representative sample of the various types of problems students have encountered in the chapter. These features are valuable assets to students in preparing for exams or refreshing their skills with previously learned concepts.

Connecting Print and Digital We want students to get the most out of their course materials, which is why we think of the textbook as a "toolbox" for students. QR codes are integrated throughout the textbook, to connect the printed version to the digital assets. As they read their printed book, support material is quickly and conveniently available via one scan of the accompanying QR code. No hunting and searching and scrolling—a direct link to additional support.

Supplements for the Instructor

Please contact your sales representative—or see xyztextbooks.com/instructors for more info.

MathTV.com Every example in every XYZ Textbook is worked on video by multiple student instructors. Students benefit from seeing multiple approaches, and gain confidence in learning from their peers. The MathTV library contains over 10,000 videos, from basic math through calculus. It's great for learning the material at hand, and for remediation too.

Complete Solutions Manual Available online, this manual contains complete solutions to all the exercises in the problem sets.

Printable Test Items Choose from a bank of pre-created tests for most of our textbooks.

Supplements for the Student

MathTV.com Students have access to math instruction 24 hours a day, seven days a week. Assistance with any problem or subject is never more than a few clicks away.

XYZ eBook This textbook is available online for both instructors and students. Tightly integrated with MathTV.com, students can read the book and watch videos of the author and peer instructors explaining each example. Access to the online book is available free with the purchase of a new book.

QR Codes QR Codes connect print to digital! Scan the QR codes located inside printed textbooks* with your mobile device, and it will take you directly to the accompanying MathTV video.

*QR codes are not available in every textbook.

Student Solutions Manual Contains complete solutions to all the odd-numbered exercises in the text. Available for purchase separately.

A Note from Charles P. McKeague

I am extremely pleased to have Mark Turner as my co-author on this book. We have worked together for years—Mark is the co-author of the last four editions of my trigonometry book. He is an award-winning teacher and, as you will see, an excellent writer. He is an innovative instructor, bringing many new ideas into the classroom—and he gives regular presentations to share his knowledge and strategies with other instructors.

There are many things I like about working with Mark. First, and maybe most important, is that the integration of Mark's writing style with mine is seamless. When I read over our material, I often cannot tell which one of us wrote it. Mark also has an attention to detail and an eye for consistency that very few authors have. I know this is a better book because of Mark's contributions to it.

Preface: A Note to Students

We want you to succeed.

Welcome to the XYZ Textbooks/MathTV community. We want you to succeed in this course. As you will see as you progress through this book, and access the other tools we have for you, we are different than other publishers. Here's how:

Our Authors Are Real Teachers The best textbooks are written by the best teachers, period. We select our authors based on one factor only: Can they teach? All of our authors have won awards for teaching. The result is the best instruction you can get in written form, produced by award-winning, experienced instructors.

Innovative Products The foundation of our products is the textbook, which is the hub for all the resources you will need to do well in your math course. These include eBooks, videos, worksheets, and a variety of ways to access these resources, from QR codes built into our books, to our MathTV Mobile site.

Peer Tutors Math can be difficult. Sometimes your class time and textbook are not enough. We understand that, which is why we created MathTV, providing you with a set of instructional videos by students just like you, who have found a way to master the material you are studying. You'll get to see how your peers solve each problem, sometimes offering a different view from how an instructor solves the problem, and other times, solving the problem in the same way your instructor does, giving you confidence that that is the way the problem should be solved. It also means that you can get help anytime (not just during office hours).

Fair Prices We're small, independently owned and independently run. Why does that matter to you? Because we do not have the overhead and expenses that the larger publishers have. Yes, we want to be a profitable business, but we believe that we can keep our prices reasonable and still give you everything you need to be successful. Also, we want you to use this book, and the best way to make sure that happens is to make it affordable.

Unlimited Access When you purchase one of our products, we give you access to all of our products. Why? Because everything you need to know about math is not contained in one book. Suppose you need to review a topic from a math course you completed previously? No problem. Suppose you want to see an alternate approach? It's all yours. As a member of our XYZ Textbooks/MathTV community, you have access to everything we produce, including all our eBooks.

We know you can do it.

We Believe in You We have seen students with all varieties of backgrounds and levels in mathematics do well in the courses we supply books and materials for. In fact, we have never run across a student that could not be successful in algebra. And that carries over to you: We believe in you; we believe you can be successful in whatever math class you are taking. Our job is to supply you with the tools you can use to attain success, and you supply the drive and ambition.

We Know College Can Be Difficult It is not always the material you are studying that makes college difficult. We know that many of you are working, some part time, some full time. We know many of you have families to support, or look after. We understand that your time can be limited. We take all this to heart when we create the materials we think you will need. For example, we make our videos available on your smart phone, tablet, and on the Internet. That way, no matter where you are, you will have access to help when you get stuck on a problem.

We Believe in What We Do We know you will see the value in the things we have created. That's why the first chapter in every one of our eBooks is free, and so are all the resources that come with it. We want you to try us out for free. See what you think. We wouldn't do that if we didn't believe in what we do here.

Here are some strategies to help.

I often find my students getting frustrated and asking themselves the question, "Why can't I understand this stuff the first time through?" The answer is, "You're not expected to."

Learning a topic in mathematics isn't always accomplished the first time through the material. If you don't understand a topic the first time you see it, that's perfectly normal.

Stick with it. Understanding mathematics takes time. You may find that you need to read over new material a number of times before you can begin to work problems. The process of understanding requires reading the book, studying the examples, working problems, and getting your questions answered.

How to Be Successful in Mathematics

1. If you are in a lecture class, be sure to attend all class sessions on time. You simply will not know exactly what went on in class unless you were there. Missing class and then expecting to find out what went on from someone else is not a good strategy. Make the time to be there—and to be attentive.
2. Read the book. It is best to read the section that will be covered in class beforehand. It's OK if you don't fully understand everything you read! Reading in advance at least gives you a sense of what will be discussed, which puts you in a good position when you get to class.

3. Work problems every day and check your answers. The secret to success in mathematics is working problems. The more problems you work, the better you will perform. It's really that simple. The answers to the odd-numbered problems are given in the back of the book. When you have finished an assignment, be sure to compare your answers with those in the book. If you have made a mistake, find out what it is, and try to correct it.
4. Do it on your own. Having someone else show you how to work a problem is not the same as working the problem yourself. It is absolutely OK to get help when you are stuck. As a matter of fact, it is a good idea. Just be sure you do the work yourself. After all, when it's test time, it's all you! Get confident in every problem type, and you will do well.

5. Review every day. After you have finished the problems your instructor has assigned, take another 15 minutes and review a section you have already completed. This simple trick works wonders. The more you review, the longer you will retain the material you have learned. Since math topics build upon one another, this will help you throughout the term.
6. Don't expect to understand every new topic the first time you see it. Sometimes it will come easy and sometimes it won't. Don't beat yourself up over it—that's just the way things are in mathematics. It's perfectly normal. Expecting to understand each new topic the first time you see it can lead to disappointment and frustration. The process of understanding takes time and practice. It requires that you read the book, work problems, and get your questions answered.

7. Spend as much time as it takes for you to master the material. What's the exact amount of time you need to spend on mathematics to master it? There's no way to know except to do it. You will find out as you go what is or isn't enough time for you. Some sections may take less time, and some may take more. If you end up spending 2 or more hours on each section, OK. Then that's how much time it takes; trying to get by with less will not work.
8. Relax. It's probably not as difficult as you think. You might get stuck at points. That's OK, everyone does. Take a break if you need to. Seek some outside help. Watch a MathTV video of the problem. There is a solution, and you *will* find it—even if it takes a while.

1

The Basics

Chapter Outline

1.1 Variables, Symbols, and the Order of Operations

1.2 The Real Numbers

1.3 Addition of Real Numbers

1.4 Subtraction of Real Numbers

1.5 Multiplication and Division of Real Numbers

1.6 Fractions

1.7 Properties of Real Numbers

iStockphoto.com © Rawpixel Ltd

Much of what we do in mathematics is concerned with recognizing patterns. If you recognize the patterns in the following two sequences, then you can easily extend each sequence.

Sequence of odd numbers = 1, 3, 5, 7, 9,...

Sequence of squares = 1, 4, 9, 16, 25,...

Once we have classified groups of numbers as to the characteristics they share, we sometimes discover that a relationship exists between the groups. Although it may not be obvious at first, there is a relationship that exists between the two sequences shown. The introduction to *The Book of Squares*, written in 1225 by the mathematician known as Fibonacci, begins this way:

"I thought about the origin of all square numbers and discovered that they arise out of the increasing sequence of odd numbers."

The relationship that Fibonacci refers to is shown visually here.

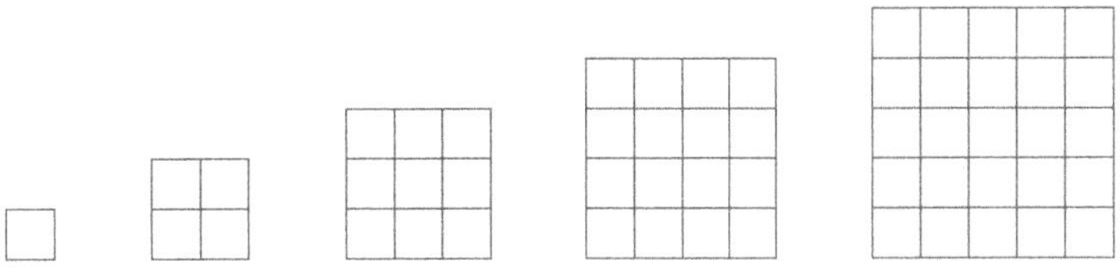

Many times we describe a relationship or pattern in a number of different ways. Here we have a visual description of a relationship. In this chapter we will work on describing relationships numerically and verbally (in writing).

Success Skills

Some of the students enrolled in my college algebra classes develop difficulties early in the course. Their difficulties are not associated with their ability to learn mathematics; they all have the potential to pass the course. Students who get off to a poor start do so because they have not developed the study skills necessary to be successful in algebra. Here is a list of things you can do to begin to develop effective study skills.

1. **Put Yourself on a Schedule** The general rule is that you spend 2 hours on homework for every hour you are in class. Make a schedule for yourself in which you set aside 2 hours each day to work on algebra. Once you make the schedule, stick to it. Don't just complete your assignments and stop. Use all the time you have set aside. If you complete an assignment and have time left over, read the next section in the book, and then work more problems.

2. **Find Your Mistakes and Correct Them** There is more to studying algebra than just working problems. You must always check your answers with the answers in the back of the book. When you have made a mistake, find out what it is and correct it. Making mistakes is part of the process of learning mathematics. In the prologue to *The Book of Squares*, Leonardo Fibonacci (ca. 1170–ca. 1250) had this to say about the content of his book:

   > I have come to request indulgence if in any place it contains something more or less than right or necessary; for to remember everything and be mistaken in nothing is divine rather than human . . .

   Fibonacci knew, as you know, that human beings make mistakes. You cannot learn algebra without making mistakes.

3. **Gather Information on Available Resources** You need to anticipate that you will need extra help sometime during the course. One resource is your instructor; you need to know your instructor's office hours and where the office is located. Another resource is the math lab or study center, if they are available at your school. It also helps to have the phone numbers of other students in the class, in case you miss class. You want to anticipate that you will need these resources, so now is the time to gather them together.

Variables, Symbols, and the Order of Operations

1.1

Learning Objectives

In this section, we will learn how to:

1. Translate a phrase into symbols.
2. Evaluate an exponent.
3. Simplify arithmetic expressions using the order of operations.
4. Interpret a bar chart.
5. Find the perimeter and area for squares, rectangles, and triangles.

Introduction

Suppose you have a checking account that costs you \$15 a month, plus \$0.05 for each check you write. If you write 10 checks in a month, then the monthly charge for your checking account will be

$$15 + 10(0.05)$$

Do you add 15 and 10 first and then multiply by 0.05? Or do you multiply 10 and 0.05 first and then add 15? If you don't know the answer to this question, you will after you have read through this section.

iStockphoto.com/©Jeffrey Smith

Variables

The difference between arithmetic and algebra is the use of variables in expressions. In algebra, we use a *variable* to represent an unknown number.

Consider our previous discussion regarding the monthly charge for a checking account. The expression

$$15 + 10(0.05)$$

is arithmetic because all of the values are known. However, the usefulness of this expression is very limited because it only gives us the monthly charge if ten checks were written. If we let the letter n represent the number of checks written, then the expression becomes

$$15 + n(0.05)$$

This is an algebraic expression, and it can be used to find the monthly charge for *any* number of checks we may write. The letter n is called a variable because the value it takes on may vary.

We use the variables a and b in the following lists so that the relationships shown there are true for all numbers that we will encounter in this book. By using variables, the following statements are general statements about all numbers, rather than specific statements about only a few numbers.

Symbols

First, we consider the symbols used to represent the four basic operations: addition, subtraction, multiplication, and division.

OPERATION SYMBOLS

Addition:	$a + b$	The *sum* of a and b
Subtraction:	$a - b$	The *difference* of a and b
Multiplication:	$a \cdot b, (a)(b), a(b), (a)b, ab$	The *product* of a and b
Division:	$a \div b, a/b, \frac{a}{b}, b\overline{)a}$	The *quotient* of a and b

Note: In the past you may have used the notation 3×5 to denote multiplication. In algebra it is best to avoid this notation if possible, because the multiplication symbol $\times$ can be confused with the variable x when written by hand.

When we encounter the word *sum*, the implied operation is addition. To find the sum of two numbers, we simply add them. *Difference* implies subtraction, *product* implies multiplication, and *quotient* implies division. Notice also that there is more than one way to write the product or quotient of two numbers.

GROUPING SYMBOLS

Parentheses ( ) and brackets [ ] are the symbols used for grouping numbers together. Occasionally, braces { } are also used for grouping, although they are usually reserved for set notation, as we shall see.

The following examples illustrate the relationship between the operation symbols and grouping symbols and the English language.

VIDEO EXAMPLES

SECTION 1.1

EXAMPLES For each phrase, write an equivalent expression in symbols.

	Phrase	Equivalent Expression
1.	The sum of 4 and 1	$4 + 1$
2.	The difference of 8 and 1	$8 - 1$
3.	Twice the sum of 3 and 4	$2(3 + 4)$
4.	The difference of the product of 3 times x and 15	$3x - 15$
5.	The product of 3 and the difference of x and 15	$3(x - 15)$
6.	The quotient of y and 2	$\frac{y}{2}$

Exponents

The last type of notation we need to discuss is the notation that allows us to write repeated multiplications in a more compact form—*exponents*. In the expression 2^3, the 2 is called the *base* and the 3 is called the *exponent*. The exponent 3 tells us the number of times the base appears in the product; that is,

$$2^3 = 2 \cdot 2 \cdot 2 = 8$$

The expression 2^3 is said to be in exponential form, whereas $2 \cdot 2 \cdot 2$ is said to be in expanded form. Here are some additional examples of expressions involving exponents.

EXAMPLE 7 Evaluate each exponent.

a. 5^2 **b.** 2^5 **c.** 10^3

SOLUTION

a. $5^2 = 5 \cdot 5 = 25$ Base 5, exponent 2

b. $2^5 = 2 \cdot 2 \cdot 2 \cdot 2 \cdot 2 = 32$ Base 2, exponent 5

c. $10^3 = 10 \cdot 10 \cdot 10 = 1{,}000$ Base 10, exponent 3

Notation and Vocabulary Here is how we read expressions containing exponents.

Mathematical Expression	Written Equivalent
5^2	five to the second power
5^3	five to the third power
5^4	five to the fourth power
5^5	five to the fifth power
5^6	five to the sixth power

We have a shorthand vocabulary for second and third powers because the area of a square with a side of 5 is 5^2, and the volume of a cube with a side of 5 is 5^3.

5^2 can be read "five squared." 5^3 can be read "five cubed."

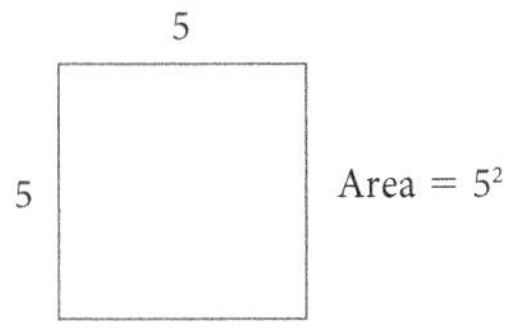

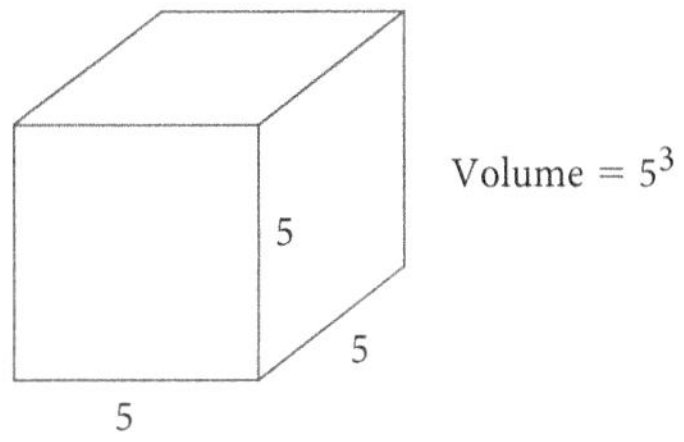

Order of Operations

The operation and grouping symbols are to mathematics what punctuation symbols are to English. Consider the following sentence:

Paul said John is tall.

It can have two different meanings, depending on how it is punctuated.

1. "Paul," said John, "is tall."
2. Paul said, "John is tall."

Let's take a look at a similar situation in mathematics. Consider the following mathematical statement:

$$5 + 2 \cdot 7$$

If we add the 5 and 2 first and then multiply by 7, we get an answer of 49. However, if we multiply the 2 and the 7 first and then add 5, we are left with 19. We have a problem that seems to have two different answers, depending on whether we add first or multiply first. We would like to avoid this type of situation. Every problem like $5 + 2 \cdot 7$ should have only one answer. Therefore, we will use the following rule for the order of operations.

RULE *Order of Operations*

When evaluating a mathematical expression, we will perform the operations in the following order, beginning with the expression in the innermost parentheses or brackets first and working our way out.

1. Simplify all numbers with exponents, working from left to right if more than one of these expressions is present.
2. Then do all multiplications and divisions left to right.
3. Perform all additions and subtractions left to right.

EXAMPLE 8 Simplify each expression following the order of operations.

a. $5 + 8 \cdot 2$ **b.** $12 \div 4 \cdot 2$

c. $2[5 + 2(6 + 3 \cdot 4)]$ **d.** $10 + 12 \div 4 + 2 \cdot 3$

e. $2^4 + 3^3 \div 9 - 4^2$

SOLUTION

a. $5 + 8 \cdot 2 = 5 + 16$ Multiply $8 \cdot 2$ first

$= 21$

b. $12 \div 4 \cdot 2 = 3 \cdot 2 = 6$ Work left to right

c. $2[5 + 2(6 + 3 \cdot 4)] = 2[5 + 2(6 + 12)]$ Simplify within the innermost grouping symbols first

$= 2[5 + 2(18)]$

$= 2[5 + 36]$ Next, simplify inside the brackets

$= 2[41]$

$= 82$ Multiply

d. $10 + 12 \div 4 + 2 \cdot 3 = 10 + 3 + 6$ Multiply and divide left to right

$= 19$ Add left to right

e. $2^4 + 3^3 \div 9 - 4^2 = 16 + 27 \div 9 - 16$ Simplify numbers with exponents

$= 16 + 3 - 16$ Then, divide

$= 19 - 16$ Finally, add and subtract left to right

$= 3$

Reading Tables and Bar Charts

The following table shows the average amount of caffeine in a number of beverages. The diagram in Figure 1 is a *bar chart*. It is a visual presentation of the information in the table. The table gives information in numerical form, whereas the chart gives the same information in a geometric way. In mathematics, it is important to be able to move back and forth between the two forms.

Caffeine Content of Hot Drinks

Drink (6-ounce cup)	Caffeine (milligrams)
Brewed coffee	100
Instant coffee	70
Tea	50
Cocoa	5
Decaffeinated coffee	4

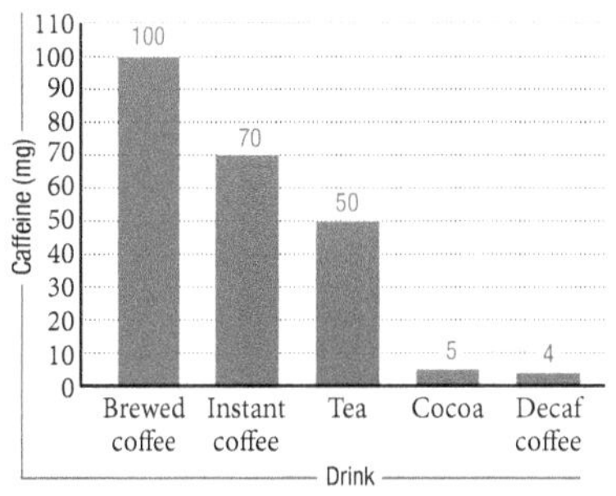

FIGURE 1

EXAMPLE 9 Referring to the table and Figure 1, suppose you have 3 cups of brewed coffee, 1 cup of tea, and 2 cups of decaf in one day. Write an expression that will give the total amount of caffeine in these six drinks, and then simplify the expression.

SOLUTION From the table or the bar chart, we find the number of milligrams of caffeine in each drink; then we write an expression for the total amount of caffeine:

$$3(100) + 50 + 2(4)$$

Using the rule for order of operations, we get 358 total milligrams of caffeine.

Geometry

We conclude this section by looking at some formulas for the area and perimeter of several geometric figures.

FACTS FROM GEOMETRY *Formulas for Area and Perimeter*

A square, rectangle, and triangle are shown in the following figures. Note that we have labeled the dimensions of each with variables. The formulas for the perimeter and area of each object are given in terms of its dimensions.

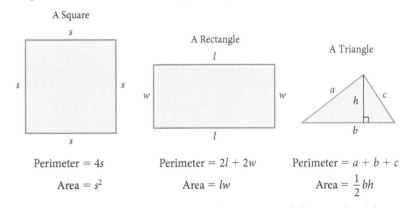

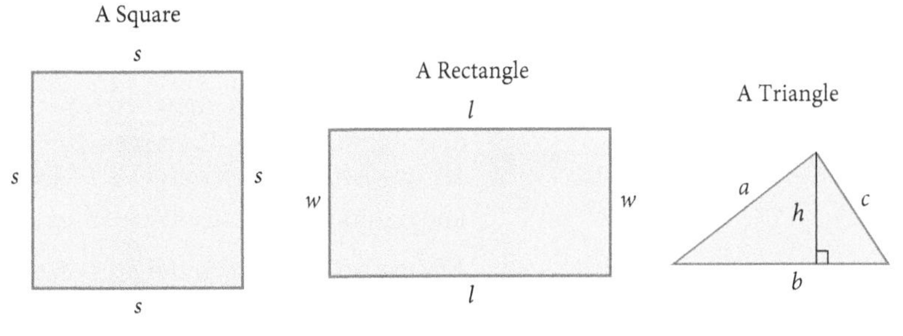

A Square	A Rectangle	A Triangle
Perimeter $= 4s$	Perimeter $= 2l + 2w$	Perimeter $= a + b + c$
Area $= s^2$	Area $= lw$	Area $= \frac{1}{2}bh$

The formula for perimeter gives us the distance around the outside of the object along its sides, whereas the formula for area gives us a measure of the amount of surface the object has.

Note: The vertical line labeled h in the triangle is its height, or altitude. It extends from the top of the triangle down to the base, meeting the base at an angle of 90°. The altitude of a triangle is always perpendicular to the base. The small square shown where the altitude meets the base is used to indicate that the angle formed is 90°.

EXAMPLE 10 Find the perimeter and area of each figure.

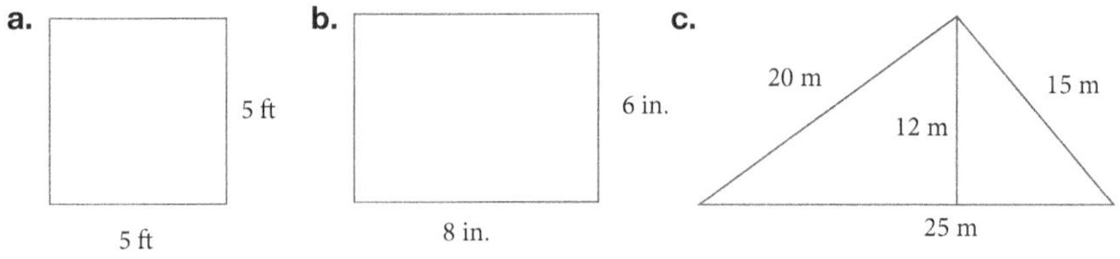

SOLUTION We use the preceding formulas to find the perimeter and the area. In each case, the units for perimeter are linear units, whereas the units for area are square units.

a. Perimeter $= 4s = 4 \cdot 5$ feet $= 20$ feet

Area $= s^2 = (5 \text{ feet})^2 = 25$ square feet

b. Perimeter $= 2l + 2w = 2(8 \text{ inches}) + 2(6 \text{ inches}) = 28$ inches

Area $= lw = (8 \text{ inches})(6 \text{ inches}) = 48$ square inches

c. Perimeter $= a + b + c = (20 \text{ meters}) + (25 \text{ meters}) + (15 \text{ meters})$

$= 60$ meters

Area $= \frac{1}{2}bh = \frac{1}{2}(25 \text{ meters})(12 \text{ meters}) = 150$ square meters

Getting Ready for Class

Each section of the book will end with some problems and questions like the ones below. They are for you to answer after you have read through the section but before you go to class. All of them require that you give written responses in complete sentences. Writing about mathematics is a valuable exercise. If you write with the intention of explaining and communicating what you know to someone else, you will find that you understand the topic you are writing about even better than you did before you started writing. As with all problems in this course, you want to approach these writing exercises with a positive point of view. You will get better at giving written responses to questions as you progress through the course. Even if you never feel comfortable writing about mathematics, just the process of attempting to do so will increase your understanding and ability in mathematics.

After reading through the preceding section, respond in your own words and in complete sentences.

A. What is a variable?

B. Explain the relationship between an exponent and its base.

C. Write the first step in the order of operations.

D. How could you place grouping symbols in the expression $20 - 8 \cdot 2$ so that the value of the expression is 24 instead of 4?

Problem Set 1.1

For each phrase below, write an equivalent expression in symbols.

1. The sum of x and 5.
2. The difference of x and 4.
3. The product of 5 and y.
4. The product of 8 and y.
5. The product of 5 and the difference of y and 16.
6. The product of 3 and the sum of y and 6.
7. The quotient of x and 3.
8. The quotient of x and the difference of x and 4.

Evaluate each exponent.

9. 3^2 **10.** 4^2 **11.** 7^2 **12.** 9^2 **13.** 2^3 **14.** 3^3
15. 4^3 **16.** 5^3 **17.** 2^4 **18.** 3^4 **19.** 10^2 **20.** 10^4
21. 11^2 **22.** 111^2

Use the order of operations to simplify each expression as much as possible.

23. $2 \cdot 3 + 5$ **24.** $8 \cdot 7 + 1$ **25.** $2(3 + 5)$ **26.** $8(7 + 1)$
27. $5 + 2 \cdot 6$ **28.** $8 + 9 \cdot 4$ **29.** $(5 + 2) \cdot 6$ **30.** $(8 + 9) \cdot 4$
31. $5 \cdot 4 + 5 \cdot 2$ **32.** $6 \cdot 8 + 6 \cdot 3$ **33.** $5(4 + 2)$ **34.** $6(8 + 3)$
35. $8 + 2(5 + 3)$ **36.** $7 + 3(8 - 2)$ **37.** $(8 + 2)(5 + 3)$ **38.** $(7 + 3)(8 - 2)$
39. $20 + 2(8 - 5) + 1$ **40.** $10 + 3(7 + 1) + 2$
41. $5 + 2(3 \cdot 4 - 1) + 8$ **42.** $11 - 2(5 \cdot 3 - 10) + 2$
43. $8 + 10 \div 2$ **44.** $16 - 8 \div 4$
45. $4 + 8 \div 4 - 2$ **46.** $6 + 9 \div 3 + 2$
47. $3 + 12 \div 3 + 6 \cdot 5$ **48.** $18 + 6 \div 2 + 3 \cdot 4$
49. $3 \cdot 8 + 10 \div 2 + 4 \cdot 2$ **50.** $5 \cdot 9 + 10 \div 2 + 3 \cdot 3$
51. $(5 + 3)(5 - 3)$ **52.** $(7 + 2)(7 - 2)$ **53.** $5^2 - 3^2$ **54.** $7^2 - 2^2$
55. $(4 + 5)^2$ **56.** $(6 + 3)^2$ **57.** $4^2 + 5^2$ **58.** $6^2 + 3^2$
59. $3 \cdot 10^2 + 4 \cdot 10 + 5$ **60.** $6 \cdot 10^2 + 5 \cdot 10 + 4$
61. $2 \cdot 10^3 + 3 \cdot 10^2 + 4 \cdot 10 + 5$ **62.** $5 \cdot 10^3 + 6 \cdot 10^2 + 7 \cdot 10 + 8$
63. $10 - 2(4 \cdot 5 - 16)$ **64.** $15 - 5(3 \cdot 2 - 4)$
65. $4[7 + 3(2 \cdot 9 - 8)]$ **66.** $5[10 + 2(3 \cdot 6 - 10)]$
67. $5(7 - 3) + 8(6 - 4)$ **68.** $3(10 - 4) + 6(12 - 10)$
69. $3(4 \cdot 5 - 12) + 6(7 \cdot 6 - 40)$ **70.** $6(8 \cdot 3 - 4) + 5(7 \cdot 3 - 1)$
71. $3^4 + 4^2 \div 2^3 - 5^2$ **72.** $2^5 + 6^2 \div 2^2 - 3^2$
73. $5^2 + 3^4 \div 9^2 + 6^2$ **74.** $6^2 + 2^5 \div 4^2 + 7^2$

Simplify each expression.

75. $20 \div 2 \cdot 10$ **76.** $40 \div 4 \cdot 5$ **77.** $24 \div 8 \cdot 3$ **78.** $24 \div 4 \cdot 6$
79. $36 \div 6 \cdot 3$ **80.** $36 \div 9 \cdot 2$ **81.** $48 \div 12 \cdot 2$ **82.** $48 \div 8 \cdot 3$
83. $16 - 8 + 4$ **84.** $16 - 8 + 8$ **85.** $24 - 14 + 8$ **86.** $24 - 16 + 6$
87. $36 - 6 + 12$ **88.** $36 - 9 + 20$ **89.** $48 - 12 + 17$ **90.** $48 - 13 + 15$

Find the perimeter and area of each figure.

91.

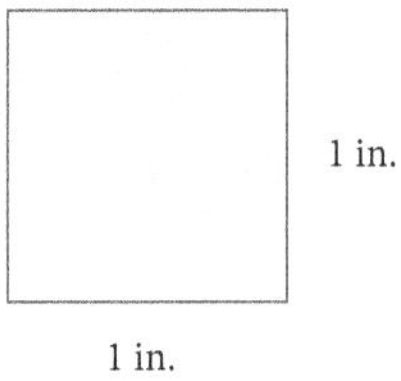
1 in.
1 in.

92.

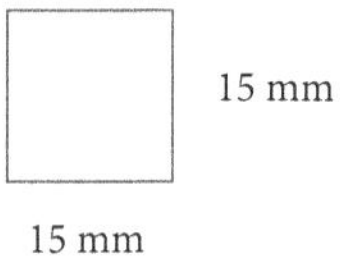
15 mm
15 mm

93.

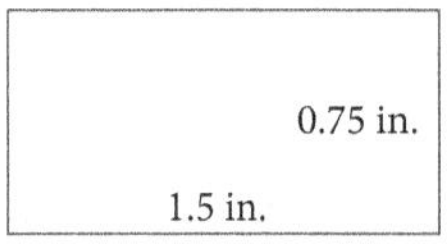
0.75 in.
1.5 in.

94.

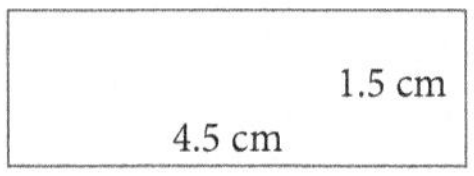
1.5 cm
4.5 cm

95.

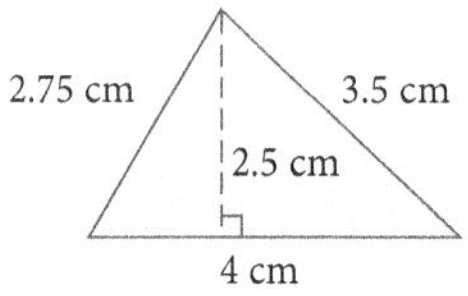
2.75 cm
3.5 cm
2.5 cm
4 cm

96.

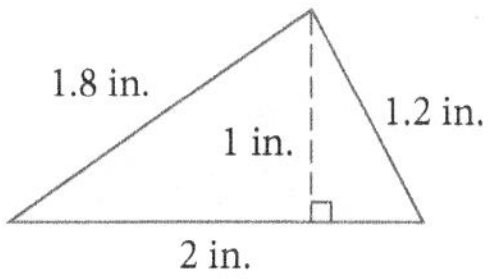
1.8 in.
1.2 in.
1 in.
2 in.

Applying the Concepts

Food Labels In 1993 the government standardized the way in which nutrition information was presented on the labels of most packaged food products. The standardized food label shown here is from a package of cookies. Use the information on the label to answer the following questions.

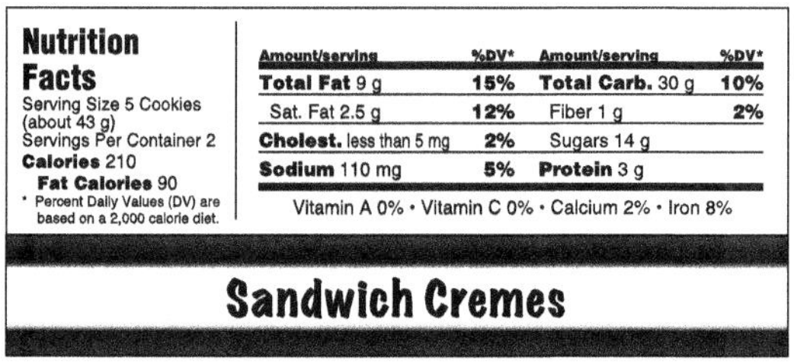
Nutrition Facts
Serving Size 5 Cookies (about 43 g)
Servings Per Container 2
Calories 210
Fat Calories 90
* Percent Daily Values (DV) are based on a 2,000 calorie diet.
Amount/serving %DV* Amount/serving %DV*
Total Fat 9 g 15% Total Carb. 30 g 10%
Sat. Fat 2.5 g 12% Fiber 1 g 2%
Cholest. less than 5 mg 2% Sugars 14 g
Sodium 110 mg 5% Protein 3 g
Vitamin A 0% • Vitamin C 0% • Calcium 2% • Iron 8%
Sandwich Cremes

97. How many cookies are in the package?

98. If you paid $2.25 for the package of cookies, how much did each cookie cost?

99. If the "calories" category stands for calories per serving, how many calories would you consume by eating the whole package of cookies?

100. Suppose that, while swimming, you burn 11 calories each minute. If you swim for 20 minutes, will you burn enough calories to cancel out the calories you added by eating 5 cookies?

Food Labels The food label shown here was taken from a bag of corn chips. Use the information to answer the following questions.

101. Approximately how many chips are in the bag?

102. If the bag of chips costs $1.69, approximately how much does one serving of chips cost?

103. The table toward the bottom of the label gives the recommended amount of total fat that should be consumed by a person eating 2,000 calories per day and by a person eating 2,500 calories per day. Use the numbers in the table to estimate the recommended fat intake for a person eating 3,000 calories per day.

104. Deidre burns 256 calories per hour by trotting on her horse at a constant rate. How long must she ride to burn the calories consumed by eating four servings of these chips?

Nutrition Facts

Serving Size 1 oz. (28 g/About 32 chips)
Servings Per Container 7

Amount Per Serving

Calories 160	Calories from Fat 90
	%?Daily Value*
Total Fat 10 g	**15%**
Saturated Fat 1.5 g	**8%**
Cholesterol 0 mg	**0%**
Sodium 160 mg	**7%**
Total Carbohydrate 15 g	**5%**
Dietary Fiber 1 g	**4%**
Sugars 0 g	
Protein 2 g	

Vitamin A 0% • Vitamin C 0%
Calcium 2% • Iron 0%

* Percent Daily Values are based on a 2,000 calorie diet. Your daily values may be higher or lower depending on your calorie needs:

	Calories:	2,000	2,500
Total Fat	Less than	65 g	80 g
Sat Fat	Less than	20 g	25 g
Cholesterol	Less than	300 mg	300 mg
Sodium	Less than	2,400 mg	2,400 mg
Total Carbohydrate		300 g	375 g
Dietary Fiber		25 g	30 g

Calories per gram:
Fat 9 • Carbohydrate 4 • Protein 4

105. Reading Charts The following bar chart gives the amount of caffeine in five different soft drinks. How much caffeine is in each of the following?

a. A 6-pack of Jolt **b.** 2 Cokes plus 3 Tabs

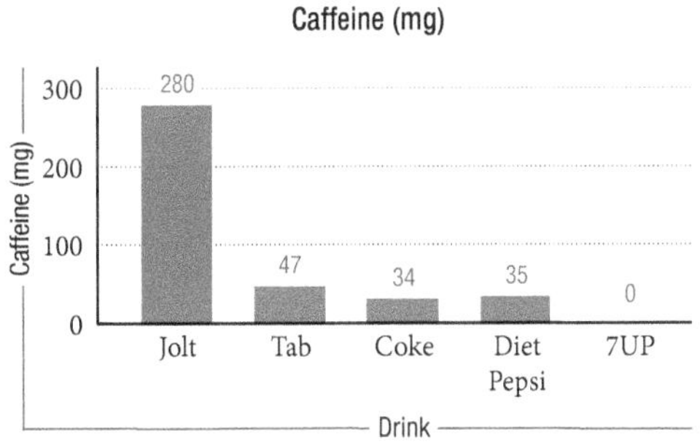

106. Reading Charts The following bar chart gives the amount of caffeine in five different nonprescription drugs. How much caffeine is in each of the following?

a. A box of 12 Excedrin **b.** 1 Dexatrim plus 4 Excedrin

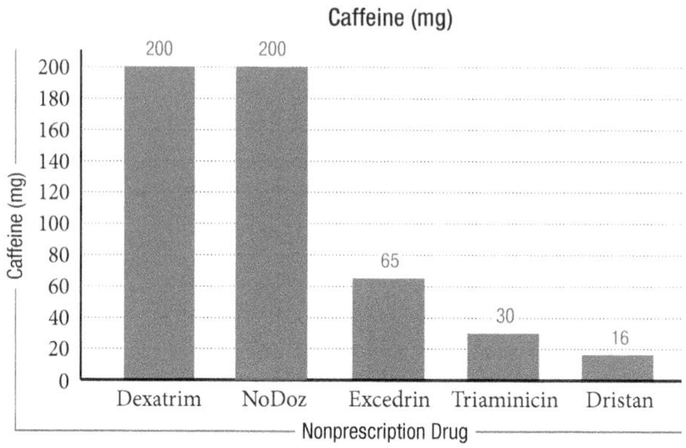

107. Reading Tables and Charts The following bar chart gives the number of calories burned by a 150-pound person during 1 hour of various exercises. The accompanying table should display the same information. Use the bar chart to complete the table.

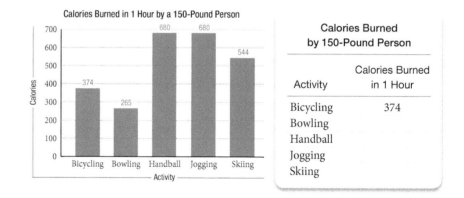

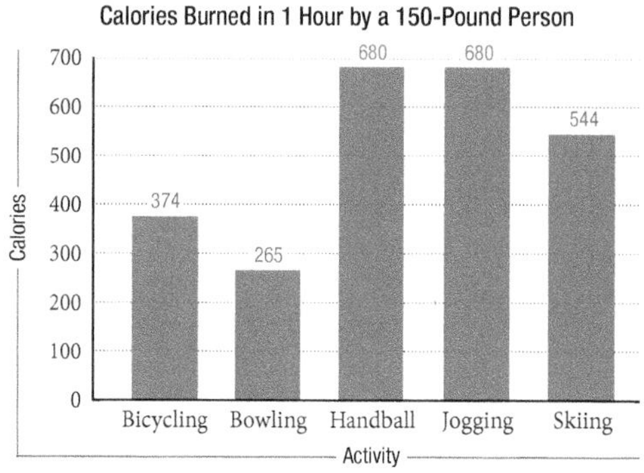

Calories Burned by 150-Pound Person

Activity	Calories Burned in 1 Hour
Bicycling	374
Bowling	
Handball	
Jogging	
Skiing	

108. Reading Tables and Charts The following table and bar chart give the number of calories consumed by eating some popular fast foods. The accompanying table should display the same information. Use the bar chart to complete the table.

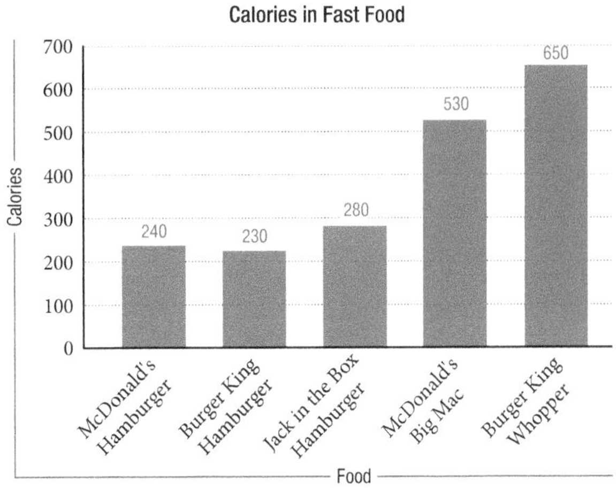

Calories In Fast Food

Food	Calories
McDonald's Hamburger	240
Burger King Hamburger	
Jack in the Box Hamburger	
McDonald's Big Mac	
Burger King Whopper	

109. Geometry Find the area and perimeter of an 8.5-by-11-inch piece of notebook paper.

110. Geometry Find the area and perimeter of an 8.5-by-5.5-inch piece of paper.

Learning Objectives Assessment

The following problems can be used to help assess if you have successfully met the learning objectives for this section.

111. Translate into symbols: The product of 4 and the difference of x and 9.

a. $4(9 - x)$ **b.** $9 - 4x$ **c.** $4x - 9$ **d.** $4(x - 9)$

112. Evaluate: 8^2.

a. 16 **b.** 256 **c.** 64 **d.** 512

113. Use the order of operations to simplify $6 + 4(3 \cdot 5 - 8) - 10$.

a. 24 **b.** 60 **c.** 132 **d.** 7

114. Referring to Figure 1 of this section, how much more caffeine will you have by drinking a cup of instant coffee instead of a cup of tea?

a. 70 mg **b.** 50 mg **c.** 120 mg **d.** 20 mg

115. Find the perimeter of an 8x10 inch photo.

a. 18 in. **b.** 36 in. **c.** 80 in^2 **d.** 40 in^2

The Real Numbers 1.2

Learning Objectives

In this section, we will learn how to:

1. Use the number line to represent real numbers.
2. Classify subsets of real numbers.
3. Find the absolute value of a number.
4. Find the opposite of a number.

Introduction

The bar chart shown here gives the record low temperature, in degrees Fahrenheit, for each month of the year in the city of Jackson, Wyoming. Notice that some of these temperatures are represented by negative numbers.

iStockPhoto.com/©Nicholas Belton

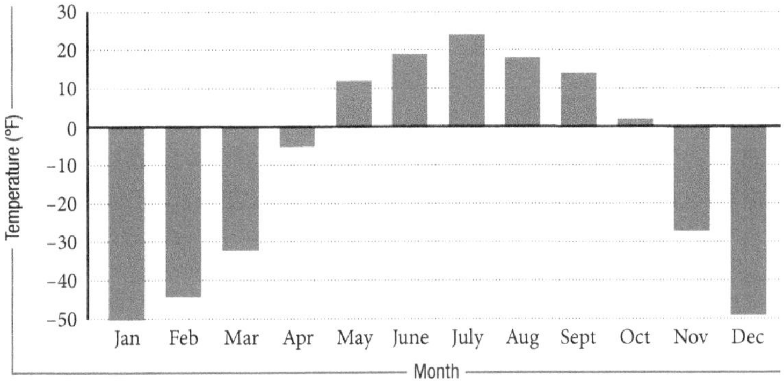

FIGURE 1

In this section we start our work with negative numbers. To represent negative numbers in algebra, we use what is called the *real number line.* Here is how we construct a real number line: We first draw a straight line and label a convenient point on the line with 0. Then we mark off equally spaced distances in both directions from 0. Label the points to the right of 0 with the numbers 1, 2, 3,…(the dots mean "and so on"). The points to the left of 0 we label in order, $-1, -2, -3, \ldots$. Here is what it looks like.

Note: If there is no sign ($+$ or $-$) in front of a number, the number is assumed to be positive ($+$).

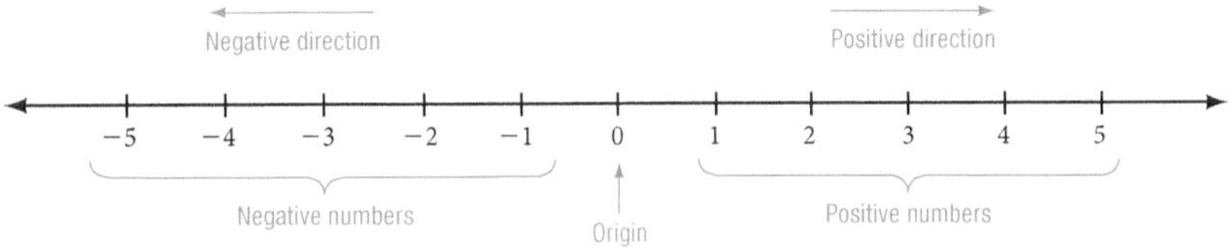

The numbers increase in value going from left to right. If we "move" to the right, we are moving in the positive direction. If we move to the left, we are moving in the negative direction.

VIDEO EXAMPLES

SECTION 1.2

EXAMPLE 1 Locate and label the points on the real number line associated with the numbers -3.5, $-1\frac{1}{4}$, $\frac{1}{2}$, $\frac{3}{4}$, 2.5.

SOLUTION We draw a real number line from -4 to 4 and label the points in question.

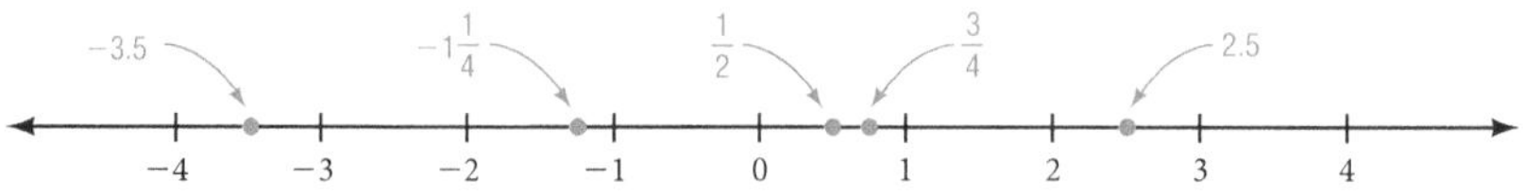

Note: There are other numbers on the number line that you may not be as familiar with. They are irrational numbers such as π, $\sqrt{2}$, $\sqrt{3}$. We will introduce these numbers later in the section.

DEFINITION *coordinate*

The number associated with a point on the real number line is called the ***coordinate*** of that point.

In the preceding example, the numbers $\frac{1}{2}$, $\frac{3}{4}$, 2.5, -3.5, and $-1\frac{1}{4}$ are the coordinates of the points they represent.

DEFINITION *real numbers*

The numbers that can be represented with points on the real number line are called ***real numbers***.

Real numbers include whole numbers, fractions, decimals, and other numbers that are not as familiar to us as these.

When we compare two numbers on the number line, the number on the left is always less than the number on the right. For instance, -3 is less than -1 because it is to the left of -1 on the number line. When making comparisons, we use the following symbols.

COMPARISON SYMBOLS

Equality:	$a = b$	a is equal to b
	$a \neq b$	a is not equal to b
Inequality:	$a < b$	a is less than b (a lies to the left of b on the number line)
	$a > b$	a is greater than b (a lies to the right of b on the number line)
	$a \geq b$	a is greater than or equal to b
	$a \leq b$	a is less than or equal to b

Subsets of the Real Numbers

The numbers that make up the set of real numbers can be classified as *counting numbers, whole numbers, integers, rational numbers,* and *irrational numbers;* each is said to be a *subset* of the real numbers.

DEFINITION *subset*

Set A is called a ***subset*** of set B if set A is contained in set B; that is, if each and every element in set A is also a member of set B.

Note: In previous math classes, you may have see the term "natural numbers." Natural numbers and counting numbers are the same subset of the real numbers.

Here is a detailed description of the major subsets of the real numbers.

The *counting numbers* are the numbers with which we count. They are the numbers 1, 2, 3, and so on. The notation we use to specify a group of numbers like this is *set notation.* We use the symbols { and } to enclose the members of the set.

Counting numbers = {1, 2, 3,...}

EXAMPLE 2 Which of the numbers in the following set are not counting numbers?

$$\left\{-3, 0, \frac{1}{2}, 1, 1.5, 3\right\}$$

SOLUTION The numbers $-3, 0, \frac{1}{2}$, and 1.5 are not counting numbers.

The *whole numbers* include the counting numbers and the number 0.

Whole numbers = {0, 1, 2, . . . }

The set of *integers* includes the whole numbers and the opposites of all the counting numbers. (Later in this section, we give a formal definition of the word *opposite* as it is used in mathematics.)

Integers = $\{\ldots, -3, -2, -1, 0, 1, 2, 3, \ldots\}$

When we refer to positive integers, we are referring to the numbers 1, 2, 3,.... Likewise, the negative integers are $-1, -2, -3, \ldots$. The number 0 is neither positive nor negative.

EXAMPLE 3 Which of the numbers in the following set are not integers?

$$\left\{-5, -1.75, 0, \frac{2}{3}, 1, \pi, 3\right\}$$

SOLUTION The only numbers in the set that are not integers are -1.75, $\frac{2}{3}$, and π.

The set of *rational numbers* is the set of numbers commonly called "fractions" together with the integers. The set of rational numbers is difficult to list in the same way we have listed the other sets, so we will use a different kind of notation:

$$\textbf{Rational numbers} = \left\{\frac{a}{b} \mid a \text{ and } b \text{ are integers } (b \neq 0)\right\}$$

This notation is read "The set of elements $\frac{a}{b}$ such that a and b are integers (and b is not 0)." If a number can be put in the form $\frac{a}{b}$, where a and b are both from the set of integers, then it is called a rational number.

EXAMPLE 4 Show why each of the numbers in the following set is a rational number.

$$\left\{-3, -\frac{2}{3}, 0, 0.333\ldots, 0.75\right\}$$

SOLUTION The number -3 is a rational number because it can be written as the ratio of -3 to 1; that is,

$$-3 = \frac{-3}{1}$$

Similarly, the number $-\frac{2}{3}$ can be thought of as the ratio of -2 to 3, whereas the number 0 can be thought of as the ratio of 0 to 1.

Any repeating decimal, such as 0.333 . . . (the dots indicate that the 3s repeat forever), can be written as the ratio of two integers. In this case 0.333 . . . is the same as the fraction $\frac{1}{3}$.

Finally, any decimal that terminates after a certain number of digits can be written as the ratio of two integers. The number 0.75 is equal to the fraction $\frac{3}{4}$ and is therefore a rational number.

Still other numbers exist, each of which is associated with a point on the real number line, that cannot be written as the ratio of two integers. In decimal form they never terminate and never repeat a sequence of digits indefinitely. They are called *irrational numbers* (because they are not rational):

Irrational numbers = {nonrational numbers; nonrepeating, nonterminating decimals}

One irrational number you have probably seen before is π. It is not 3.14. Rather, 3.14 is an approximation to π. It cannot be written as a terminating decimal number. Other representations for irrational numbers are $\sqrt{2}$, $\sqrt{3}$, $\sqrt{5}$, $\sqrt{6}$, and, in general, the square root of any number that is not itself a perfect square. (If you are not familiar with square roots, you will be after Chapter 8.)

To summarize, the set of real numbers is the set of numbers that are either rational or irrational; that is, a real number is either rational or irrational.

Real numbers = {all rational numbers and all irrational numbers}

Fractions on the Number Line

As we mentioned previously, rational numbers can also be thought of as fractions. Here is the formal definition of a fraction.

DEFINITION *fraction*

If a and b are real numbers, then the expression

$$\frac{a}{b} \qquad b \neq 0$$

is called a ***fraction***. The top number a is called the ***numerator***, and the bottom number b is called the ***denominator***. The restriction $b \neq 0$ keeps us from writing an expression that is undefined. (As you will see, division by zero is not allowed.)

The number line can be used to visualize fractions. The denominator indicates the number of equal parts in the interval from 0 to 1 on the number line. The numerator indicates how many of those parts we have. If we take that part of the number line from 0 to 1 and divide it into three equal parts, we say that we have divided it into thirds (Figure 2). Each of the three segments is $\frac{1}{3}$ (one third) of the whole segment from 0 to 1.

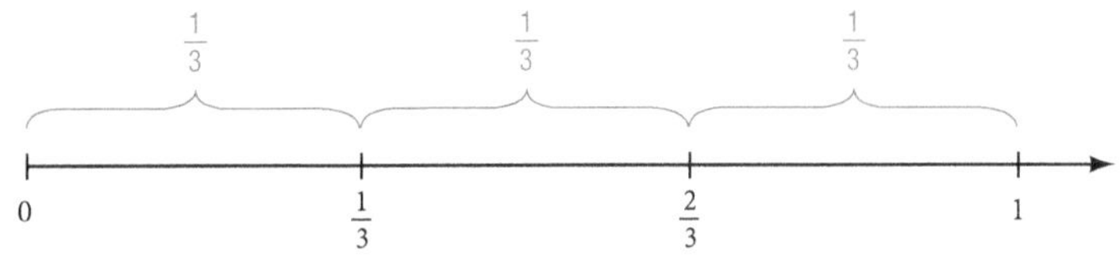

FIGURE 2

Two of these smaller segments together are $\frac{2}{3}$ (two thirds) of the whole segment. And three of them would be $\frac{3}{3}$ (three thirds), or the whole segment.

Let's do the same thing again with six equal divisions of the segment from 0 to 1 (Figure 3). In this case we say each of the smaller segments has a length of $\frac{1}{6}$ (one sixth).

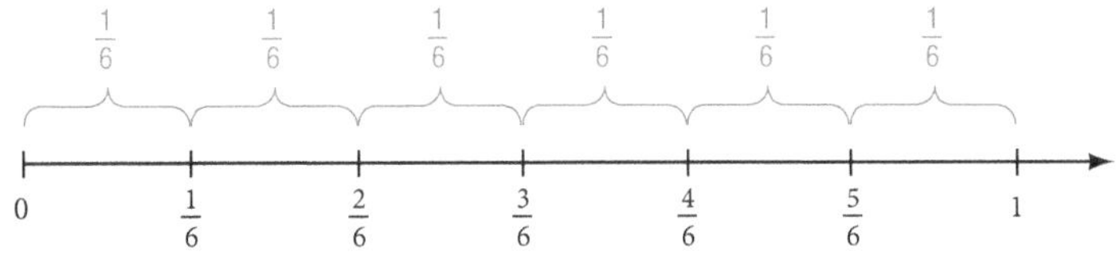

FIGURE 3

The same point we labeled with $\frac{1}{3}$ in Figure 2 is now labeled with $\frac{2}{6}$. Likewise, the point we labeled earlier with $\frac{2}{3}$ is now labeled $\frac{4}{6}$. It must be true then that

$$\frac{2}{6} = \frac{1}{3} \qquad \text{and} \qquad \frac{4}{6} = \frac{2}{3}$$

Actually, there are many fractions that name the same point as $\frac{1}{3}$. If we were to divide the segment between 0 and 1 into 12 equal parts, 4 of these 12 equal parts $\left(\frac{4}{12}\right)$ would be the same as $\frac{2}{6}$ or $\frac{1}{3}$; that is,

$$\frac{4}{12} = \frac{2}{6} = \frac{1}{3}$$

Even though these three fractions look different, each names the same point on the number line, as shown in Figure 4. All three fractions have the same value because they all represent the same number.

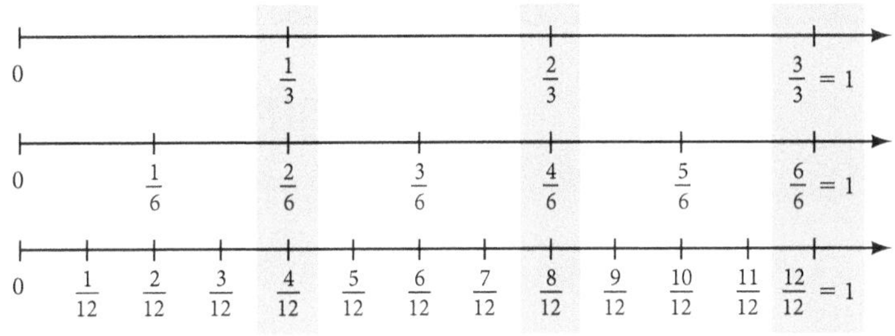

FIGURE 4

DEFINITION *equivalent*

Fractions that represent the same number are said to be ***equivalent***. Equivalent fractions may look different, but they must have the same value.

It is apparent that every fraction has many different representations, each of which is equivalent to the original fraction. The next two properties give us a way of changing the terms of a fraction without changing its value.

PROPERTY 1

Multiplying the numerator and denominator of a fraction by the same nonzero number never changes the value of the fraction.

PROPERTY 2

Dividing the numerator and denominator of a fraction by the same nonzero number never changes the value of the fraction.

We can use Property 2 to reduce a fraction to *lowest terms* by dividing the largest common factor that appears in both the numerator and denominator.

EXAMPLE 5 Reduce $\frac{18}{30}$ to lowest terms.

SOLUTION The largest number that divides evenly into 18 and 30 is 6, so 6 is the largest common factor of the numerator and denominator.

$$\frac{18}{30} = \frac{6(3)}{6(5)}$$

$$= \frac{3}{5} \qquad \text{Divide the numerator and denominator by 6}$$

We say that $\frac{3}{5}$ is in lowest terms because 3 and 5 have no common factor other than 1.

Absolute Values and Opposites

Representing numbers on the number line lets us give each number two important properties: a direction from zero and a distance from zero. The direction from zero is represented by the sign in front of the number. (A number without a sign is understood to be positive.) The distance from zero is called the *absolute value* of the number, as the following definition indicates.

DEFINITION *absolute value*

The ***absolute value*** of a real number is its distance from zero on the number line. If x represents a real number, then the absolute value of x is written $|x|$.

EXAMPLE 6 Write each expression without absolute value symbols.

a. $|5|$ **b.** $|-5|$ **c.** $\left|-\frac{1}{2}\right|$

SOLUTION

a. $|5| = 5$ The number 5 is 5 units from zero

b. $|-5| = 5$ The number -5 is 5 units from zero

c. $\left|-\frac{1}{2}\right| = \frac{1}{2}$ The number $-\frac{1}{2}$ is $\frac{1}{2}$ units from zero

The absolute value of a number is never negative. It is the distance the number is from zero without regard to which direction it is from zero. When working with the absolute value of sums and differences, we must simplify the expression inside the absolute value symbols first and then find the absolute value of the simplified expression.

EXAMPLE 7 Simplify each expression.

a. $|8 - 3|$ **b.** $|3 \cdot 2^3 + 2 \cdot 3^2|$ **c.** $|9 - 2| - |8 - 6|$

SOLUTION

a. $|8 - 3| = |5| = 5$

b. $|3 \cdot 2^3 + 2 \cdot 3^2| = |3 \cdot 8 + 2 \cdot 9| = |24 + 18| = |42| = 42$

c. $|9 - 2| - |8 - 6| = |7| - |2| = 7 - 2 = 5$

Another important concept associated with numbers on the number line is that of opposites. Here is the definition.

DEFINITION ***opposites***

Numbers the same distance from zero but in opposite directions from zero are called ***opposites.*** If a is any real number, then we denote the opposite of a by $-a$.

EXAMPLE 8 Give the opposite of each number.

a. 5 **b.** -3 **c.** $\frac{1}{4}$ **d.** -2.3

SOLUTION

	Number	Opposite	
a.	5	$-(5) = 5$	5 and -5 are opposites
b.	-3	$-(-3) = 3$	-3 and 3 are opposites
c.	$\frac{1}{4}$	$-\left(\frac{1}{4}\right) = -\frac{1}{4}$	$\frac{1}{4}$ and $-\frac{1}{4}$ are opposites
d.	-2.3	$-(-2.3) = 2.3$	-2.3 and 2.3 are opposites

Each negative number is the opposite of some positive number, and each positive number is the opposite of some negative number. The opposite of a negative number is a positive number. In symbols, if a represents a positive number, then

$$-(-a) = a$$

Opposites always have the same absolute value. And, when you add any two opposites, the result is always zero:

$$a + (-a) = 0$$

Getting Ready for Class

After reading through the preceding section, respond in your own words and in complete sentences.

A. What is a real number?

B. What is a whole number?

C. Is every integer also a rational number? Explain.

D. How do you find the opposite of a number?

Draw a number line that extends from -5 to $+5$. Label the points with the following coordinates.

1. 5 **2.** -2 **3.** -4 **4.** -3

5. 1.5 **6.** -1.5 **7.** $\frac{9}{4}$ **8.** $\frac{8}{3}$

Given the numbers in the set $\{-3, -2.5, 0, 1, \frac{3}{2}, \sqrt{15}\}$:

9. List all the whole numbers.

10. List all the integers.

11. List all the rational numbers.

12. List all the irrational numbers.

13. List all the real numbers.

Given the numbers in the set $\{-10, -8, -0.333\ldots, -2, 9, \frac{25}{3}, \pi\}$:

14. List all the whole numbers.

15. List all the integers.

16. List all the rational numbers.

17. List all the irrational numbers.

18. List all the real numbers.

Identify the following statements as either true or false.

19. Every whole number is also an integer.

20. The set of whole numbers is a subset of the set of integers.

21. A number can be both rational and irrational.

22. The set of rational numbers and the set of irrational numbers have some elements in common.

23. Some whole numbers are also negative integers.

24. Every rational number is also a real number.

25. All integers are also rational numbers.

26. The set of integers is a subset of the set of rational numbers.

For each of the following numbers, give the opposite and the absolute value. (Assume all variables are nonzero.)

27. 10 **28.** 8 **29.** $\frac{3}{4}$ **30.** $\frac{5}{7}$ **31.** $\frac{11}{2}$ **32.** $\frac{16}{3}$

33. -3 **34.** -5 **35.** $-\frac{2}{5}$ **36.** $-\frac{3}{8}$ **37.** x **38.** a

Place one of the symbols $<$ or $>$ between each of the following to make the resulting statement true.

39. $-5 \square -3$ **40.** $-8 \square -1$ **41.** $-3 \square -7$ **42.** $-6 \square 5$

43. $|-4| \square -|-4|$ **44.** $3 \square -|-3|$ **45.** $7 \square -|-7|$ **46.** $-7 \square |-7|$

Simplify each expression.

47. $|8 - 2|$

48. $|6 - 1|$

49. $|5 \cdot 2^3 - 2 \cdot 3^2|$

50. $|2 \cdot 10^2 + 3 \cdot 10|$

51. $|7 - 2| - |4 - 2|$

52. $|10 - 3| - |4 - 1|$

53. $10 - |7 - 2(5 - 3)|$

54. $12 - |9 - 3(7 - 5)|$

55. $15 - |8 - 2(3 \cdot 4 - 9)| - 10$

56. $25 - |9 - 3(4 \cdot 5 - 18)| - 20$

Applying the Concepts

57. **Football Yardage** A football team gains 6 yards on one play and then loses 8 yards on the next play. To what number on the number line does a loss of 8 yards correspond? The total yards gained or lost on the two plays corresponds to what negative number?
58. **Checking Account Balance** A woman has a balance of \$20 in her checking account. If she writes a check for \$30, what negative number can be used to represent the new balance in her checking account?

Temperature In the United States, temperature is measured on the Fahrenheit temperature scale. On this scale, water boils at 212 degrees and freezes at 32 degrees. To denote a temperature of 32 degrees on the Fahrenheit scale, we write

32°F, which is read "32 degrees Fahrenheit"

Use this information for Problems 59 and 60.

59. **Temperature and Altitude** Marilyn is flying from Seattle to San Francisco on a Boeing 737 jet. When the plane reaches an altitude of 35,000 feet, the temperature outside the plane is 64 degrees below zero Fahrenheit. Represent the temperature with a negative number. If the temperature outside the plane gets warmer by 10 degrees, what will the new temperature be?
60. **Temperature Change** At 10:00 in the morning in White Bear Lake, Minnesota, John notices the temperature outside is 10 degrees below zero Fahrenheit. Write the temperature as a negative number. An hour later it has warmed up by 6 degrees. What is the temperature at 11:00 that morning?

Wind Chill The table below is a table of wind chill temperatures. The left column gives the air temperature, and the first row is wind speed in miles per hour. The numbers within the table indicate how cold the weather will feel. For example, if the thermometer reads 30°F and the wind is blowing at 15 miles per hour, the wind chill temperature is 9°F. Use Table 1 to answer Problems 61 and 62.

Wind Chill Temperatures

Air Temperature (°F)	Wind Speed (mph)				
	10	15	20	25	30
30°	16°	9°	4°	1°	−2°
25°	10°	2°	−3°	−7°	−10°
20°	3°	−5°	−10°	−15°	−18°
15°	−3°	−11°	−17°	−22°	−25°
10°	−9°	−18°	−24°	−29°	−33°
5°	−15°	−25°	−31°	−36°	−41°
0°	−22°	−31°	−39°	−44°	−49°
−5°	−27°	−38°	−46°	−51°	−56°

TABLE 1

61. **Reading Tables** Find the wind chill temperature if the thermometer reads 20°F and the wind is blowing at 25 miles per hour.

62. **Reading Tables** Which will feel colder: a day with an air temperature of 10°F with a 25-mile-per-hour wind, or a day with an air temperature of 25° F and a 10-mile-per-hour wind?

63. **Scuba Diving** Steve is scuba diving near his home in Maui. At one point he is 100 feet below the surface. Represent this number with a negative number. If he descends another 5 feet, what negative number will represent his new position?

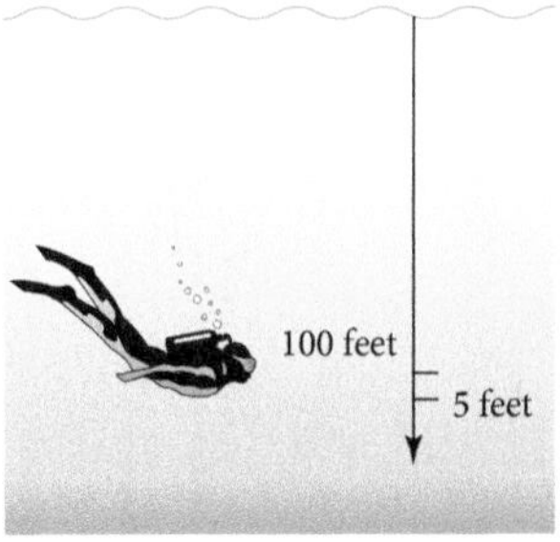

64. **Reading a Chart** The chart shows yields for certificates of deposit during one week in 2006. Write a mathematical statement using one of the symbols < or > to compare the following:

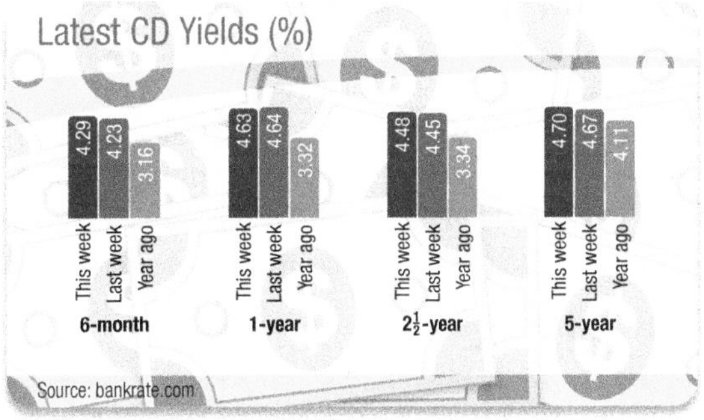

a. 6 month yield a year ago to 1 year yield last week

b. $2\frac{1}{2}$ year yield this week to 5 year yield a year ago

c. 5 year yield last week to 6 month yield this week

Calories and Exercise The table here gives the amount of energy expended per hour for various activities for a person weighing 120, 150, or 180 pounds. Use the table to answer questions 65–68.

Energy Expended from Exercising

	Calories per Hour		
Activity	120 lb	150 lb	180 lb
Bicycling	299	374	449
Bowling	212	265	318
Handball	544	680	816
Horseback trotting	278	347	416
Jazzercise	272	340	408
Jogging	544	680	816
Skiing (downhill)	435	544	653

65. Suppose you weigh 120 pounds. How many calories will you burn if you play handball for 2 hours and then ride your bicycle for an hour?

66. How many calories are burned by a person weighing 150 pounds who jogs for $\frac{1}{2}$ hour and then goes bicycling for 2 hours?

67. Two people go skiing. One weighs 180 pounds and the other weighs 120 pounds. If they ski for 3 hours, how many more calories are burned by the person weighing 180 pounds?

68. Two people spend 3 hours bowling. If one weighs 120 pounds and the other weighs 150 pounds, how many more calories are burned during the evening by the person weighing 150 pounds?

69. Use the chart shown here to answer the following questions.

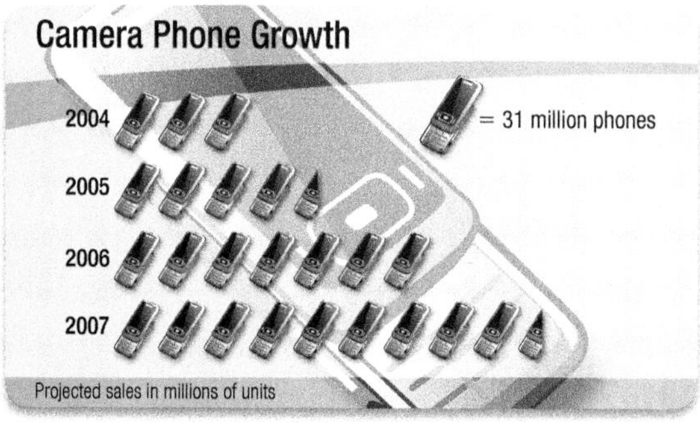

a. How many millions of camera phones were sold in 2004?

b. True or false? The chart shows sales in 2005 to be more than 155 million camera phones.

c. True or false? The chart shows sales in 2007 to be less than 310 million camera phones.

70. Improving Your Quantitative Literacy Quantitative literacy is a subject discussed by many people involved in teaching mathematics. The person they are concerned with when they discuss it is you. We are going to work at improving your quantitative literacy, but before we do that we should answer the question, What is quantitative literacy? Lynn Arthur Steen, a noted mathematics educator, has stated that quantitative literacy is "the capacity to deal effectively with the quantitative aspects of life."

a. Give a definition for the word *quantitative*.

b. Give a definition for the word *literacy*.

c. Are there situations that occur in your life that you find distasteful, or that you try to avoid, because they involve numbers and mathematics? If so, list some of them here. (For example, some people find the process of buying a car particularly difficult because they feel that the numbers and details of the financing are beyond them.)

Learning Objectives Assessment

The following problems can be used to help assess if you have successfully met the learning objectives for this section.

71. What is the coordinate of the point shown on the number line below?

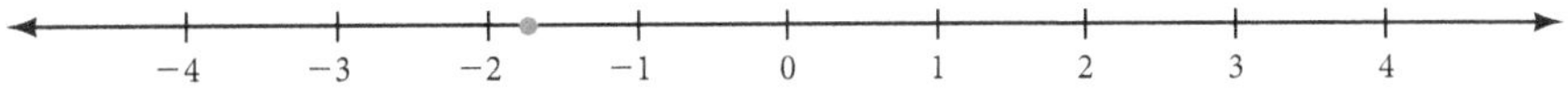

a. 1.75 **b.** 2.5 **c.** $-1\frac{3}{4}$ **d.** $-2\frac{1}{4}$

72. Given the numbers in the set $\left\{-4.5, 2, 0, 3\frac{1}{3}, -1, \sqrt{2}\right\}$ list all the integers.

a. 2, 0 **b.** 2, 0, −1

c. $-4.5, 2, 0, 3\frac{1}{3}, -1$ **d.** All of them

73. Find $|-9|$.

a. 9 **b.** −9 **c.** 0 **d.** $-\frac{1}{9}$

74. Find the opposite of −11.

a. 11 **b.** −11 **c.** 0 **d.** $-\frac{1}{11}$

Addition of Real Numbers 1.3

Learning Objectives

In this section, we will learn how to:

1. Add real numbers.
2. Translate a phrase involving addition into a mathematical expresssion.
3. Find the next number in an arithmetic sequence.

Introduction

Suppose that you are playing a friendly game of poker with some friends, and you lose \$3 on the first hand and \$4 on the second hand. If you represent winning with positive numbers and losing with negative numbers, how can you translate this situation into symbols? Because you lost \$3 and \$4 for a total of \$7, one way to represent this situation is with addition of negative numbers:

iStockPhoto.com/©Alexander Fairfull

$$(-\$3) + (-\$4) = -\$7$$

From this equation, we see that the sum of two negative numbers is a negative number. To generalize addition with positive and negative numbers, we use the number line.

Because real numbers have both a distance from zero (absolute value) and a direction from zero (sign), we can think of addition of two numbers in terms of distance and direction from zero.

Let's look at a problem for which we know the answer. Suppose we want to add the numbers 3 and 4. The problem is written $3 + 4$. To put it on the number line, we read the problem as follows:

1. The 3 tells us to "start at the origin and move 3 units in the positive direction."
2. The $+$ sign is read "and then move."
3. The 4 means "4 units in the positive direction."

To summarize, $3 + 4$ means to start at the origin, move 3 units in the positive direction, and then move 4 units in the positive direction.

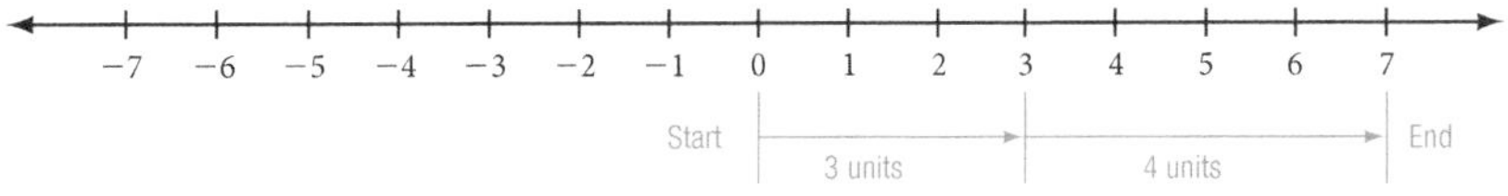

We end up at 7, which is the answer to our problem: $3 + 4 = 7$.

Let's try other combinations of positive and negative 3 and 4 on the number line.

VIDEO EXAMPLES

SECTION 1.3

EXAMPLE 1 Add $3 + (-4)$.

SOLUTION Starting at the origin, move 3 units in the positive direction and then 4 units in the negative direction.

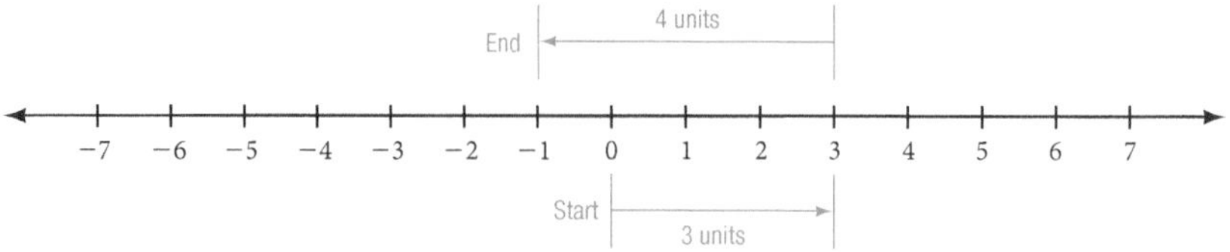

We end up at -1; therefore, $3 + (-4) = -1$.

EXAMPLE 2 Add $-3 + 4$.

SOLUTION Starting at the origin, move 3 units in the negative direction and then 4 units in the positive direction.

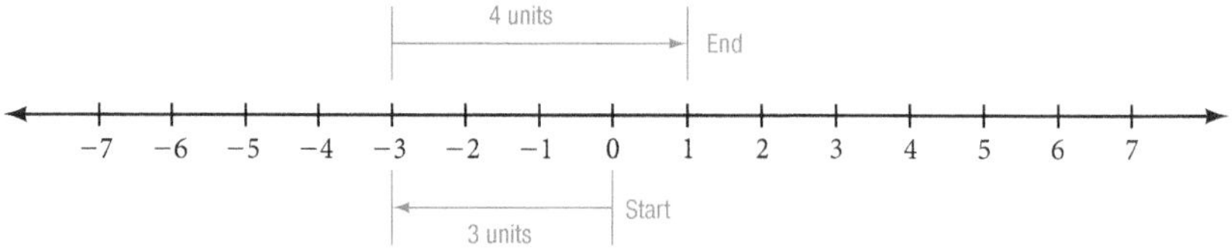

We end up at $+1$; therefore, $-3 + 4 = 1$.

EXAMPLE 3 Add $-3 + (-4)$.

SOLUTION Starting at the origin, move 3 units in the negative direction and then 4 units in the negative direction.

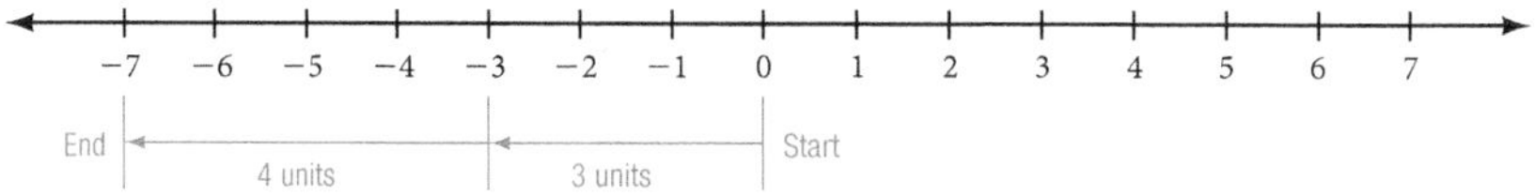

We end up at -7; therefore, $-3 + (-4) = -7$.

Here is a summary of what we have just completed:

$$3 + 4 = 7$$
$$3 + (-4) = -1$$
$$-3 + 4 = 1$$
$$-3 + (-4) = -7$$

Let's do four more problems on the number line and then summarize our results into a rule we can use to add any two real numbers.

EXAMPLE 4 Show that $5 + 7 = 12$.

SOLUTION

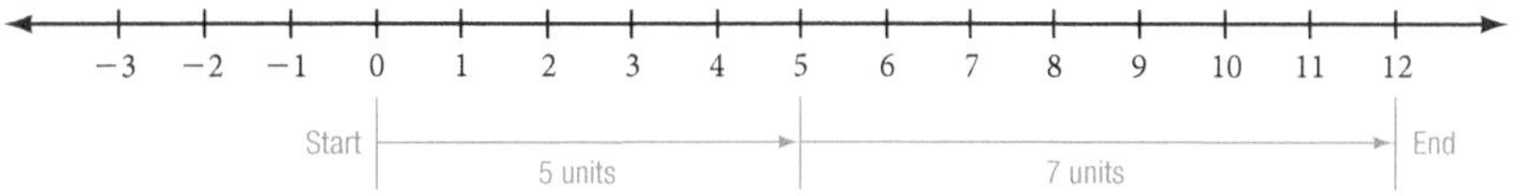

EXAMPLE 5 Show that $5 + (-7) = -2$.

SOLUTION

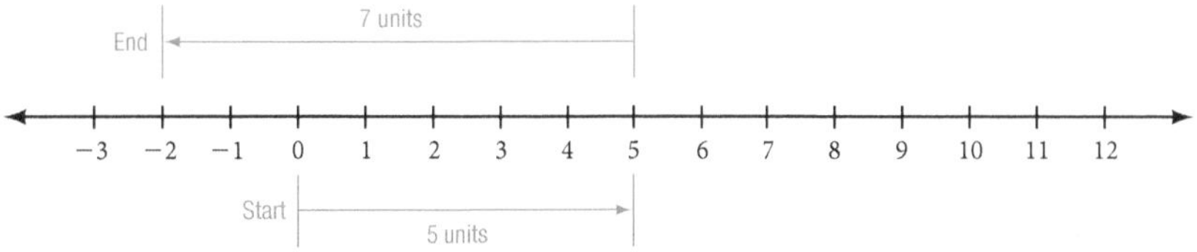

EXAMPLE 6 Show that $-5 + 7 = 2$.

SOLUTION

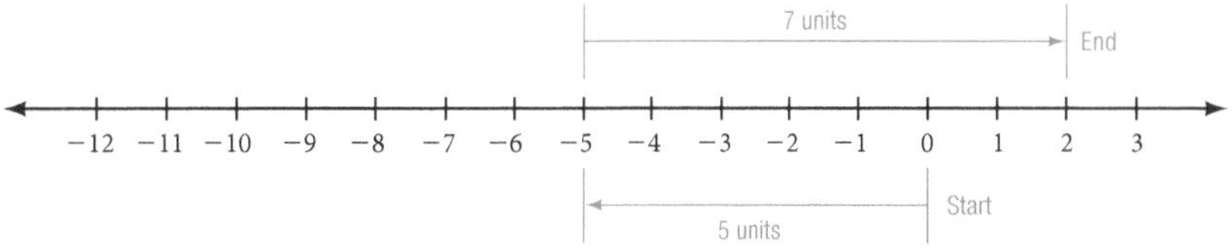

EXAMPLE 7 Show that $-5 + (-7) = -12$.

SOLUTION

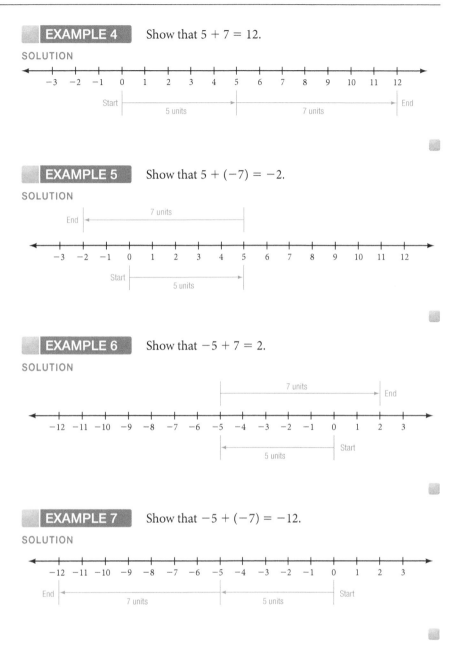

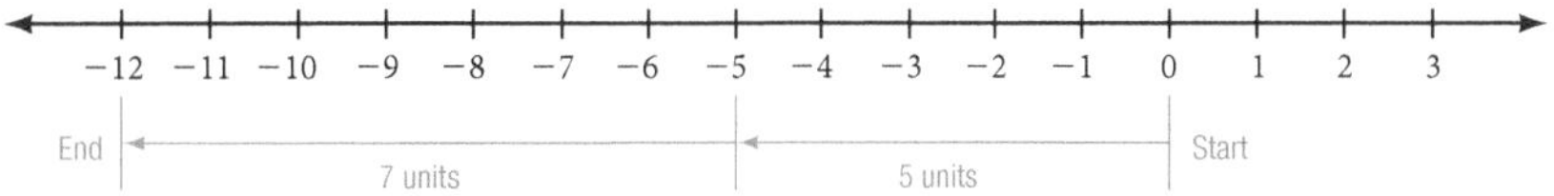

If we look closely at the results of the preceding addition problems, we can see that they support (or justify) the following rule.

Note: This rule is what we have been working towards. The rule is very important. Be sure that you understand it and can use it. The problems we have done up to this point have been done simply to justify this rule. Now that we have the rule, we no longer need to do our addition problems on the number line.

RULE

To add two real numbers with

1. The *same* sign: Simply add their absolute values and use the common sign. (If both numbers are positive, the answer is positive. If both numbers are negative, the answer is negative.)
2. *Different* signs: Subtract the smaller absolute value from the larger. The answer will have the sign of the number with the larger absolute value.

This rule covers all possible combinations of addition with real numbers. You must memorize it. After you have worked a number of problems, it will seem almost automatic.

EXAMPLE 8 Add all combinations of positive and negative 10 and 13.

SOLUTION Rather than work these problems on the number line, we use the rule for adding positive and negative numbers to obtain our answers:

$$10 + 13 = 23$$
$$10 + (-13) = -3$$
$$-10 + 13 = 3$$
$$-10 + (-13) = -23$$

EXAMPLE 9 Add all possible combinations of positive and negative 12 and 17.

SOLUTION Applying the rule for adding positive and negative numbers, we have

$$12 + 17 = 29$$
$$12 + (-17) = -5$$
$$-12 + 17 = 5$$
$$-12 + (-17) = -29$$

EXAMPLE 10 Add $-3 + 2 + (-4)$.

SOLUTION Applying the rule for order of operations, we add left to right:

$$\begin{aligned} -3 + 2 + (-4) &= -1 + (-4) \\ &= -5 \end{aligned}$$

EXAMPLE 11 Add $-8 + [2 + (-5)] + (-1)$.

SOLUTION Adding inside the brackets first and then left to right, we have

$$\begin{aligned} -8 + [2 + (-5)] + (-1) &= -8 + (-3) + (-1) \\ &= -11 + (-1) \\ &= -12 \end{aligned}$$

EXAMPLE 12 Simplify $-10 + 2(-8 + 11) + (-4)$.

SOLUTION First, we simplify inside the parentheses. Then, we multiply. Finally, we add left to right:

$$\begin{aligned} -10 + 2(-8 + 11) + (-4) &= -10 + 2(3) + (-4) \\ &= -10 + 6 + (-4) \\ &= -4 + (-4) \\ &= -8 \end{aligned}$$

Translating Words into Symbols

The following table shows a list of common phrases that are used to indicate a sum of two numbers, along with the corresponding mathematical expression.

Phrase	Equivalent Expression
the sum of a and b	$a + b$
a added to b	$b + a$
a increased by b	$a + b$
a plus b	$a + b$
a more than b	$b + a$
the total of a and b	$a + b$
a exceeded by b	$a + b$

EXAMPLE 13 Write the mathematical expression that is equivalent to the phrase, "5 more than the sum of -2 and x."

SOLUTION This phrase involves two additions. We break the translation down into steps.

"5 more than"	$\underline{\quad ? \quad} + 5$
"the sum of"	$(\underline{\quad ? \quad} + \underline{\quad ? \quad}) + 5$
"-2 and x"	$(-2 + x) + 5$

Arithmetic Sequences

A *sequence* is simply a list of numbers that have been placed in a particular order. In mathematics, most sequences follow some kind of pattern. When we notice a pattern to a sequence of numbers and then use the pattern to extend the sequence, we are using what is called *inductive reasoning*.

The pattern in a sequence of numbers is easy to identify when each number in the sequence comes from the preceding number by adding the same amount each time. This leads us to our next level of classification, in which we classify groups of sequences with a common characteristic.

DEFINITION *arithmetic sequence*

An ***arithmetic sequence*** is a sequence of numbers in which each number (after the first number) comes from adding the same amount to the number before it.

Here is an example of an arithmetic sequence:

$$2, 5, 8, 11, \ldots$$

Each number is obtained by adding 3 to the number before it.

EXAMPLE 14 Each sequence below is an arithmetic sequence. Find the next two numbers in each sequence.

a. 7, 10, 13,... **b.** 9.5, 10, 10.5,... **c.** 5, 0, −5,...

SOLUTION Because we know that each sequence is arithmetic, we know to look for the number that is added to each term to produce the next consecutive term.

a. 7, 10, 13,...: Each term is found by adding 3 to the term before it. Therefore, the next two terms will be 16 and 19.

b. 9.5, 10, 10.5,...: Each term comes from adding 0.5 to the term before it. Therefore, the next two terms will be 11 and 11.5.

c. 5, 0, −5,...: Each term comes from adding −5 to the term before it. Therefore, the next two terms will be $-5 + (-5) = -10$ and $-10 + (-5) = -15$.

Getting Ready for Class

After reading through the preceding section, respond in your own words and in complete sentences.

A. Explain how you would add 3 and −5 on the number line.

B. How do you add two negative numbers?

C. Give three different phrases that could be used to represent the expression $a + b$.

D. What is an arithmetic sequence?

Problem Set 1.3

1. Add all combinations of positive and negative 3 and 5. (Look back to Examples 8 and 9.)
2. Add all combinations of positive and negative 6 and 4.
3. Add all combinations of positive and negative 15 and 20.
4. Add all combinations of positive and negative 18 and 12.

Work the following problems. You may want to begin by doing a few on the number line.

5. $6 + (-3)$	**6.** $7 + (-8)$	**7.** $13 + (-20)$	**8.** $15 + (-25)$
9. $18 + (-32)$	**10.** $6 + (-9)$	**11.** $-6 + 3$	**12.** $-8 + 7$
13. $-30 + 5$	**14.** $-18 + 6$	**15.** $-6 + (-6)$	**16.** $-5 + (-5)$
17. $-9 + (-10)$	**18.** $-8 + (-6)$	**19.** $-10 + (-15)$	**20.** $-18 + (-30)$

Work the following problems using the rule for addition of real numbers. You may want to refer back to the rule for order of operations.

21. $5 + (-6) + (-7)$
22. $6 + (-8) + (-10)$
23. $-7 + 8 + (-5)$
24. $-6 + 9 + (-3)$
25. $5 + [6 + (-2)] + (-3)$
26. $10 + [8 + (-5)] + (-20)$
27. $[6 + (-2)] + [3 + (-1)]$
28. $[18 + (-5)] + [9 + (-10)]$
29. $20 + (-6) + [3 + (-9)]$
30. $18 + (-2) + [9 + (-13)]$
31. $-3 + (-2) + [5 + (-4)]$
32. $-6 + (-5) + [-4 + (-1)]$
33. $(-9 + 2) + [5 + (-8)] + (-4)$
34. $(-7 + 3) + [9 + (-6)] + (-5)$
35. $[-6 + (-4)] + [7 + (-5)] + (-9)$
36. $[-8 + (-1)] + [8 + (-6)] + (-6)$
37. $(-6 + 9) + (-5) + (-4 + 3) + 7$
38. $(-10 + 4) + (-3) + (-3 + 8) + 6$

The problems that follow involve some multiplication. Be sure that you work inside the parentheses first, then multiply, and finally, add left to right.

39. $-5 + 2(-3 + 7)$
40. $-3 + 4(-2 + 7)$
41. $9 + 3(-8 + 10)$
42. $4 + 5(-2 + 6)$
43. $-10 + 2(-6 + 8) + (-2)$
44. $-20 + 3(-7 + 10) + (-4)$
45. $2(-4 + 7) + 3(-6 + 8)$
46. $5(-2 + 5) + 7(-1 + 6)$

Recall that the word sum indicates addition. Write the numerical expression that is equivalent to each of the following phrases and then simplify.

47. The sum of 5 and 9
48. The sum of 6 and -3
49. Four added to the sum of -7 and -5
50. Six added to the sum of -9 and 1
51. The sum of -2 and -3 increased by 10
52. The sum of -4 and -12 increased by 2

Write the algebraic expression that is equivalent to each of the following phrases. Use x to represent the unknown number.

53. The sum of 4 and a number
54. Five more than a number
55. The total of -8 and a number
56. A number added to -6
57. The sum of a number and -2, increased by 3
58. Seven more than the sum of a number and -1

Answer the following questions.

59. What number do you add to -8 to get -5?
60. What number do you add to 10 to get 4?
61. The sum of what number and -6 is -9?
62. The sum of what number and -12 is 8?

Each sequence below is an arithmetic sequence. In each case, find the next two numbers in the sequence.

63. 3, 8, 13, 18, ...
64. 1, 5, 9, 13, ...
65. 10, 15, 20, 25, ...
66. 10, 16, 22, 28, ...
67. 20, 15, 10, 5, ...
68. 24, 20, 16, 12, ...
69. 6, 0, -6, ...
70. 1, 0, -1, ...
71. 8, 4, 0, ...
72. 5, 2, -1, ...
73. Is the sequence of odd numbers an arithmetic sequence?
74. Is the sequence of squares an arithmetic sequence?

Applying the Concepts

75. Temperature Change The temperature at noon is 12 degrees below 0 Fahrenheit. By 1:00 it has risen 4 degrees. Write an expression using the numbers -12 and 4 to describe this situation.

76. Stock Value On Monday a certain stock gains 2 points. On Tuesday it loses 3 points. Write an expression using positive and negative numbers with addition to describe this situation and then simplify.

77. Gambling On three consecutive hands of draw poker a gambler wins \$10, loses \$6, and then loses another \$8. Write an expression using positive and negative numbers and addition to describe this situation and then simplify.

iStockPhoto.com/©Alexander Fairfull

78. Number Problem You know from your past experience with numbers that subtracting 5 from 8 results in 3 ($8 - 5 = 3$). What addition problem that starts with the number 8 gives the same result?

79. Checkbook Balance Suppose that you balance your checkbook and find that you are overdrawn by \$30; that is, your balance is $-\$30$. Then you go to the bank and deposit \$40. Translate this situation into an addition problem, the answer to which gives the new balance in your checkbook.

80. Checkbook Balance The balance in your checkbook is $-\$25$. If you make a deposit of \$75, and then write a check for \$18, what is the new balance?

Profit, Revenue, and Costs In business, the difference of revenue and cost is profit, or $P = R - C$, where P is profit, R is revenue, and C is costs. The bar charts below show the costs and revenue for the Baby Steps Shoe Company for a recent 5-year period. Use this information to answer the questions below.

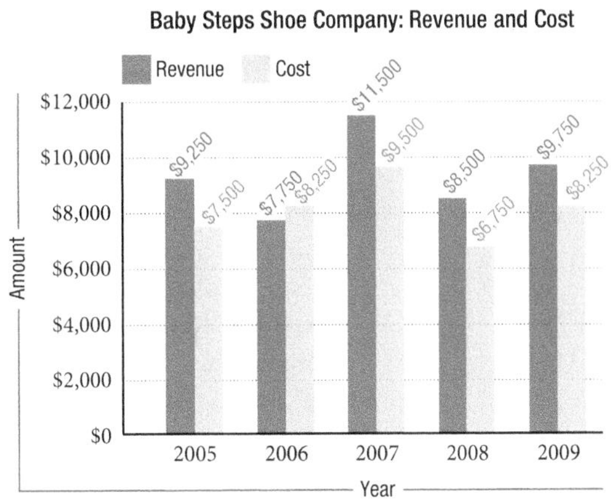

81. What was the profit for the year 2007?

82. In what year was the largest increase in costs from the previous year? How much was the increase?

83. What was the only year the company had a loss? (Profit is negative.) How much was the loss?

Learning Objectives Assessment

The following problems can be used to help assess if you have successfully met the learning objectives for this section.

84. Simplify: $12 + (-17)$.

a. 5 **b.** -5 **c.** 29 **d.** -29

85. Translate the phrase "four more than -3" into an equivalent numerical expression.

a. $4 > -3$ **b.** $4 - 3$ **c.** $(-3) + 4$ **d.** $4(-3)$

86. Find the next number in the arithmetic sequence 50, 44, 38, 32, ...

a. 26 **b.** 28 **c.** 24 **d.** 22

SPOTLIGHT ON SUCCESS *Student Instructor Cynthia*

Each time we face our fear, we gain strength, courage, and confidence in the doing.
—Unknown

I must admit, when it comes to math, it takes me longer to learn the material compared to other students. Because of that, I was afraid to ask questions, especially when it seemed like everyone else understood what was going on. Because I wasn't getting my questions answered, my quiz and exam scores were only getting worse. I realized that I was already paying a lot to go to college and that I couldn't afford to keep doing poorly on my exams. I learned how to overcome my fear of asking questions by studying the material before class, and working on extra problem sets until I was confident enough that at least I understood the main concepts. By preparing myself beforehand, I would often end up answering the question myself. Even when that wasn't the case, the professor knew that I tried to answer the question on my own. If you want to be successful, but you are afraid to ask a question, try putting in a little extra time working on problems before you ask your instructor for help. I think you will find, like I did, that it's not as bad as you imagined it, and you will have overcome an obstacle that was in the way of your success.

Subtraction of Real Numbers 1.4

Learning Objectives

In this section, we will learn how to:

1. Subtract real numbers.
2. Translate a phrase involving subtraction into a mathematical expression.

Introduction

Suppose that the temperature at noon is 20° Fahrenheit and 12 hours later, at midnight, it has dropped to −15° Fahrenheit. What is the difference between the temperature at noon and the temperature at midnight? Intuitively, we know the difference in the two temperatures is 35°. We also know that the word difference indicates subtraction. The difference between 20 and −15 is written

$$20 - (-15)$$

It must be true that $20 - (-15) = 35$. In this section we will see how our definition for subtraction confirms that this last statement is in fact correct.

In the previous section we spent some time developing the rule for addition of real numbers. Because we want to make as few rules as possible, we can define subtraction in terms of addition. By doing so, we can then use the rule for addition to solve our subtraction problems.

RULE

To subtract one real number from another, simply add its opposite.

Algebraically, the rule is written like this: If a and b represent two real numbers, then it is always true that

$$\underbrace{a - b}_{\text{To subtract } b} = \underbrace{a + (-b)}_{\text{add the opposite of } b}$$

This is how subtraction is defined in algebra. This definition of subtraction will not conflict with what you already know about subtraction, but it will allow you to do subtraction using negative numbers.

VIDEO EXAMPLES

SECTION 1.4

EXAMPLE 1 Subtract all possible combinations of positive and negative 7 and 2.

SOLUTION

$$\left.\begin{aligned} 7 - 2 &= 7 + (-2) = 5 \\ -7 - 2 &= -7 + (-2) = -9 \end{aligned}\right\} \quad \text{Subtracting 2 is the same as adding } -2$$

$$\left.\begin{aligned} 7 - (-2) &= 7 + 2 = 9 \\ -7 - (-2) &= -7 + 2 = -5 \end{aligned}\right\} \quad \text{Subtracting } -2 \text{ is the same as adding } 2$$

Notice that each subtraction problem is first changed to an addition problem. The rule for addition is then used to arrive at the answer.

We have defined subtraction in terms of addition, and we still obtain answers consistent with the answers we are used to getting with subtraction. Moreover, we now can do subtraction problems involving both positive and negative numbers.

As you proceed through the following examples and the problem set, you will begin to notice shortcuts you can use in working the problems. You will not always have to change subtraction to addition of the opposite to be able to get answers quickly. Use all the shortcuts you wish as long as you consistently get the correct answers.

EXAMPLE 2 Subtract all combinations of positive and negative 8 and 13.

SOLUTION

$$8 - 13 = 8 + (-13) = -5$$
$$-8 - 13 = -8 + (-13) = -21$$
Subtracting +13 is the same as adding −13

$$8 - (-13) = 8 + 13 = 21$$
$$-8 - (-13) = -8 + 13 = 5$$
Subtracting −13 is the same as adding +13

EXAMPLE 3 Simplify each expression as much as possible.

a. $7 + (-3) - 5$ **b.** $8 - (-2) - 6$ **c.** $-2 - (-3 + 1) - 5$

SOLUTION

a. $7 + (-3) - 5 = 7 + (-3) + (-5)$ Begin by changing all subtractions to additions
$= 4 + (-5)$
$= -1$ Then add left to right

b. $8 - (-2) - 6 = 8 + 2 + (-6)$ Begin by changing all subtractions to additions
$= 10 + (-6)$
$= 4$ Then add left to right

c. $-2 - (-3 + 1) - 5 = -2 - (-2) - 5$ Do what is in the parentheses first
$= -2 + 2 + (-5)$
$= -5$

The next two examples involve multiplication and exponents as well as subtraction. Remember, according to the order of operations, we evaluate the numbers containing exponents and multiply before we subtract.

EXAMPLE 4 Simplify $2 \cdot 5 - 3 \cdot 8 - 4 \cdot 9$.

SOLUTION First, we multiply left to right, and then we subtract:

$$\begin{aligned} 2 \cdot 5 - 3 \cdot 8 - 4 \cdot 9 &= 10 - 24 - 36 \\ &= -14 - 36 \\ &= -50 \end{aligned}$$

EXAMPLE 5 Simplify $3 \cdot 2^3 - 2 \cdot 4^2$.

SOLUTION We begin by evaluating each number that contains an exponent. Then we multiply before we subtract:

$$\begin{aligned} 3 \cdot 2^3 - 2 \cdot 4^2 &= 3 \cdot 8 - 2 \cdot 16 \\ &= 24 - 32 \\ &= -8 \end{aligned}$$

EXAMPLE 6 Subtract 7 from -3.

SOLUTION First, we write the problem in terms of subtraction. We then change to addition of the opposite:

$$\begin{aligned} -3 - 7 &= -3 + (-7) \\ &= -10 \end{aligned}$$

EXAMPLE 7 Subtract -5 from 2.

SOLUTION Subtracting -5 is the same as adding $+5$:

$$\begin{aligned} 2 - (-5) &= 2 + 5 \\ &= 7 \end{aligned}$$

EXAMPLE 8 Find the difference of 9 and 2.

SOLUTION Written in symbols, the problem looks like this:

$$9 - 2 = 7$$

The difference of 9 and 2 is 7.

EXAMPLE 9 Find the difference of 3 and -5.

SOLUTION Subtracting -5 from 3 we have

$$\begin{aligned} 3 - (-5) &= 3 + 5 \\ &= 8 \end{aligned}$$

Translating Words into Symbols

The following table shows a list of common phrases that are used to indicate a subtraction of two numbers, along with the corresponding mathematical expression.

Phrase	Equivalent Expression
the difference of a and b	$a - b$
a subtracted from b	$b - a$
a decreased by b	$a - b$
a minus b	$a - b$
a less than b	$b - a$
a reduced by b	$a - b$
a fewer than b	$b - a$

EXAMPLE 10 Write the mathematical expression that is equivalent to the phrase, "9 less than the difference of x and -4."

SOLUTION This phrase involves two subtractions. Here is the translation, broken down into steps.

"9 less than"	$\underline{\ \ ?\ \ } - 9$
"the difference of"	$[\underline{\ \ ?\ \ } - \underline{\ \ ?\ \ }] - 9$
"x and -4"	$[x - (-4)] - 9$

Subtracting and Taking Away

For some people taking algebra for the first time, subtraction of positive and negative numbers can be a problem. These people may believe that $-5 - 9$ should be -4 or 4, not -14. If this is happening to you, you probably are thinking of subtraction in terms of taking one number away from another. Thinking of subtraction in this way works well with positive numbers if you always subtract the smaller number from the larger. In algebra, however, we encounter many situations other than this. The definition of subtraction, that $a - b = a + (-b)$, clearly indicates the correct way to use subtraction; that is, when working subtraction problems, you should think "addition of the opposite," not "take one number away from another." To be successful in algebra, you need to apply properties and definitions exactly as they are presented here.

Getting Ready for Class

After reading through the preceding section, respond in your own words and in complete sentences.

A. Why do we define subtraction in terms of addition?

B. Write the definition for $a - b$.

C. Explain in words how you would subtract 3 from -7.

D. Explain how it is possible to subtract two negative numbers and get a positive result.

Problem Set 1.4

The following problems are intended to give you practice with subtraction of positive and negative numbers. Remember, in algebra subtraction is not taking one number away from another. Instead, subtracting a number is equivalent to adding its opposite.

Subtract.

1. $5 - 8$ **2.** $6 - 7$ **3.** $3 - 9$ **4.** $2 - 7$

5. $5 - 5$ **6.** $8 - 8$ **7.** $-8 - 2$ **8.** $-6 - 3$

9. $-4 - 12$ **10.** $-3 - 15$ **11.** $-6 - 6$ **12.** $-3 - 3$

13. $-8 - (-1)$ **14.** $-6 - (-2)$ **15.** $15 - (-20)$ **16.** $20 - (-5)$

17. $-4 - (-4)$ **18.** $-5 - (-5)$

Simplify each expression by following the order of operations.

19. $3 - 2 - 5$ **20.** $4 - 8 - 6$ **21.** $9 - 2 - 3$

22. $8 - 7 - 12$ **23.** $-6 - 8 - 10$ **24.** $-5 - 7 - 9$

25. $-22 + 4 - 10$ **26.** $-13 + 6 - 5$ **27.** $10 - (-20) - 5$

28. $15 - (-3) - 20$ **29.** $8 - (2 - 3) - 5$ **30.** $10 - (4 - 6) - 8$

31. $7 - (3 - 9) - 6$ **32.** $4 - (3 - 7) - 8$ **33.** $5 - (-8 - 6) - 2$

34. $4 - (-3 - 2) - 1$ **35.** $-(5 - 7) - (2 - 8)$ **36.** $-(4 - 8) - (2 - 5)$

37. $-(3 - 10) - (6 - 3)$ **38.** $-(3 - 7) - (1 - 2)$ **39.** $16 - [(4 - 5) - 1]$

40. $15 - [(4 - 2) - 3]$ **41.** $5 - [(2 - 3) - 4]$ **42.** $6 - [(4 - 1) - 9]$

43. $21 - [-(3 - 4) - 2] - 5$ **44.** $30 - [-(10 - 5) - 15] - 25$

The following problems involve multiplication and exponents. Use the order of operations to simplify each expression as much as possible.

45. $2 \cdot 8 - 3 \cdot 5$ **46.** $3 \cdot 4 - 6 \cdot 7$ **47.** $3 \cdot 5 - 2 \cdot 7$

48. $6 \cdot 10 - 5 \cdot 20$ **49.** $5 \cdot 9 - 2 \cdot 3 - 6 \cdot 2$ **50.** $4 \cdot 3 - 7 \cdot 1 - 9 \cdot 4$

51. $3 \cdot 8 - 2 \cdot 4 - 6 \cdot 7$ **52.** $5 \cdot 9 - 3 \cdot 8 - 4 \cdot 5$ **53.** $2 \cdot 3^2 - 5 \cdot 2^2$

54. $3 \cdot 7^2 - 2 \cdot 8^2$ **55.** $4 \cdot 3^3 - 5 \cdot 2^3$ **56.** $3 \cdot 6^2 - 2 \cdot 3^2 - 8 \cdot 6^2$

Rewrite each of the following phrases as an equivalent expression in symbols, and then simplify.

57. Subtract 4 from -7. **58.** Subtract 5 from -19.

59. Subtract -8 from 12. **60.** Subtract -2 from 10.

61. Subtract -7 from -5. **62.** Subtract -9 from -3.

63. Subtract 17 from the sum of 4 and -5.

64. Subtract -6 from the sum of 6 and -3.

Recall that the word *difference* indicates subtraction. The difference of a and b is $a - b$, in that order. Write a numerical expression that is equivalent to each of the following phrases, and then simplify.

65. The difference of 8 and 5. **66.** The difference of 5 and 8.

67. The difference of -8 and 5. **68.** The difference of -5 and 8.

69. The difference of 8 and -5. **70.** The difference of 5 and -8.

Write the algebraic expression that is equivalent to each of the following phrases. Use x to represent the unknown number.

71. A number decreased by 6

72. A number minus -2

73. The difference of -4 and a number

74. Three fewer than a number

75. The sum of a number and 12, decreased by 5

76. The difference of a number and 8, increased by 1

Answer the following questions.

77. What number do you subtract from 8 to get -2?

78. What number do you subtract from 1 to get -5?

79. What number do you subtract from 8 to get 10?

80. What number do you subtract from 1 to get 5?

Applying the Concepts

81. Savings Account Balance A man with $1,500 in a savings account makes a withdrawal of $730. Write an expression using subtraction that describes this situation.

First Bank
Account No. 12345

Date	Withdrawals	Deposits	Balance
1/1/16			1,500
2/2/16	730		

82. Checkbook Balance Bob has $98 in his checking account when he writes a check for $65 and then another check for $53. Write a subtraction problem that gives the new balance in Bob's checkbook. What is his new balance?

83. Gambling A man who has lost $35 playing roulette in Las Vegas wins $15 playing blackjack. He then loses $20 playing the wheel of fortune. Write an expression using the numbers -35, 15, and 20 to describe this situation and then simplify it.

84. Altitude Change An airplane flying at 10,000 feet lowers its altitude by 1,500 feet to avoid other air traffic. Then it increases its altitude by 3,000 feet to clear a mountain range. Write an expression that describes this situation and then simplify it.

85. Temperature Change The temperature inside a weather probe is 73°F before reentry. During reentry the temperature inside the probe increases 10°. On landing it drops 8°F. Write an expression using the numbers 73, 10, and 8 to describe this situation. What is the temperature inside the probe on landing?

86. Temperature Change The temperature at noon is 23°F. Six hours later it has dropped 19°F, and by midnight it has dropped another 10°F. Write a subtraction problem that gives the temperature at midnight. What is the temperature at midnight?

87. Depreciation Stacey buys a used car for \$4,500. With each year that passes, the car drops \$550 in value. Write a sequence of numbers that gives the value of the car at the beginning of each of the first five years she owns it. Can this sequence be considered an arithmetic sequence?

88. Depreciation Wade buys a computer system for \$6,575. Each year after that he finds that the system is worth \$1,250 less than it was the year before. Write a sequence of numbers that gives the value of the computer system at the beginning of each of the first four years he owns it. Can this sequence be considered an arithmetic sequence?

89. Grass Growth The bar chart below shows the growth of a certain species of grass over a period of 10 days.

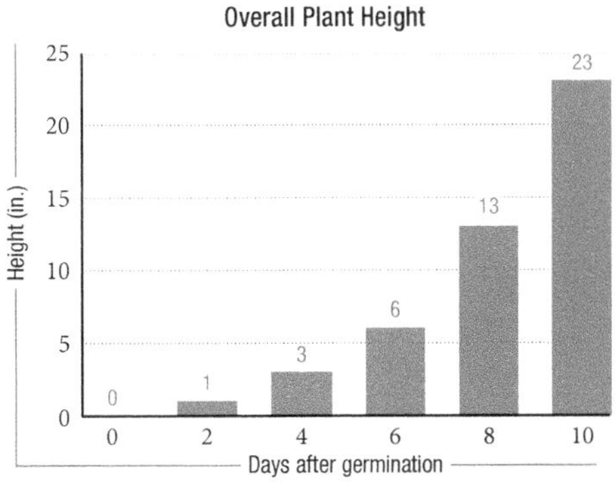

Day	Plant Height (inches)
0	0
2	
4	
6	
	13
10	

a. Use the chart to fill in the missing entries in the table.
b. How much higher is the grass after 8 days than after 2 days?

90. Wireless Phone Costs The bar chart below shows the cost of wireless phone use over a recent six-year period.

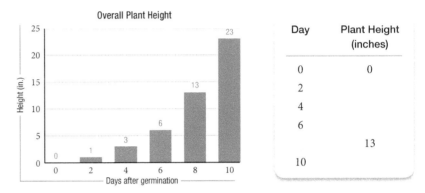

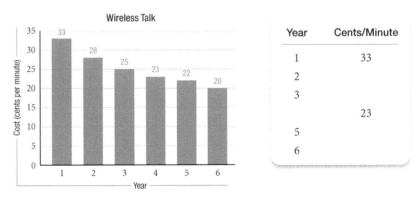

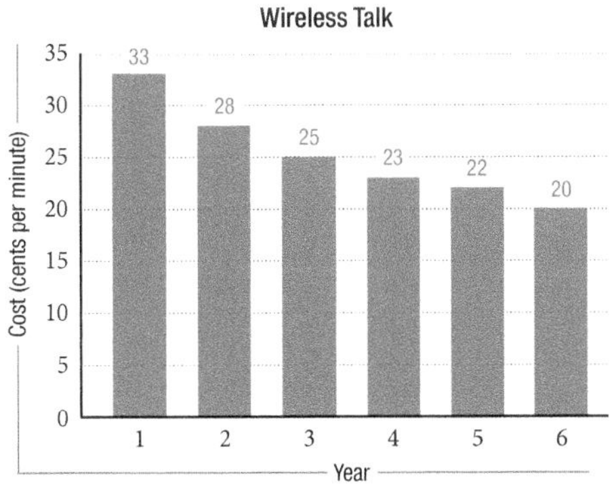

Year	Cents/Minute
1	33
2	
3	
	23
5	
6	

a. Use the chart to fill in the missing entries in the table.
b. What is the difference in cost between Year 1 and Year 2?

91. Triathlon Project Use the chart to answer the following questions.

a. Do you think the numbers in the chart have been rounded? If so, to which place were they rounded?

b. How many more participants were there in 2004 than in 2000?

c. If the trend from 2000 to 2004 continued, how many participants were there in 2008?

Learning Objectives Assessment

The following problems can be used to help assess if you have successfully met the learning objectives for this section.

92. Simplify: $-12-(-17)$.

a. 29 **b.** -29 **c.** -5 **d.** 5

93. Translate the phrase "four less than -9" into an equivalent numerical expression.

a. $-9-4$ **b.** $-9+4$ **c.** $4<-9$ **d.** $4-(-9)$

Multiplication and Division of Real Numbers 1.5

Learning Objectives

In this section, we will learn how to:

1. Multiply real numbers.
2. Divide real numbers.
3. Translate a phrase involving multiplication or division into a mathematical expression.
4. Find the next number in a geometric sequence.

Introduction

Suppose that you own 5 shares of a stock and the price per share drops \$3. How much money have you lost? Intuitively, we know the loss is \$15. Because it is a loss, we can express it as −\$15. To describe this situation with numbers, we would write

5 shares — each lose \$3 — for a total of \$15

$$5(-3) = -15$$

Reasoning in this manner, we conclude that the product of a positive number with a negative number is a negative number. Let's look at multiplication in more detail.

Multiplication

From our experience with counting numbers, we know that multiplication is simply repeated addition; that is, $3(5) = 5 + 5 + 5$. We will use this fact, along with our knowledge of negative numbers, to develop the rule for multiplication of any two real numbers. The following example illustrates multiplication with all of the possible combinations of positive and negative numbers.

VIDEO EXAMPLES

SECTION 1.5

EXAMPLE 1 Multiply.

a. $3(5)$ **b.** $3(-5)$ **c.** $-3(5)$ **d.** $-3(-5)$

SOLUTION

a. Two positives: $3(5) = 5 + 5 + 5$

$= 15$ Positive answer

b. One positive: $3(-5) = -5 + (-5) + (-5)$

$= -15$ Negative answer

c. One negative: $-3(5) = 5(-3)$ Commutative property

$= -3 + (-3) + (-3) + (-3) + (-3)$

$= -15$ Negative answer

d. Two negatives: $-3(-5) = ?$

With two negatives, $-3(-5)$, it is not possible to work the problem in terms of repeated addition. (It doesn't "make sense" to write -5 down a -3 number of times.) The answer is probably $+15$ (that's just a guess), but we need some justification for saying so. We will solve a different problem and in doing so get the answer to the problem $(-3)(-5)$.

Note: You may have to read the explanation for Example 1(d) several times before you understand it completely. The purpose of the explanation in Example 1(d) is simply to justify the fact that the product of two negative numbers is a positive number. If you have no trouble believing that, then it is not so important that you understand everything in the explanation.

Here is a problem to which we know the answer. We will work it two different ways.

$$-3[5 + (-5)] = -3(0) = 0$$

The answer is zero. We also can work the problem using the distributive property.

$$-3[5 + (-5)] = -3(5) + (-3)(-5) \quad \text{Distributive property}$$
$$= -15 + ?$$

Because the answer to the problem is 0, our ? must be +15. (What else could we add to −15 to get 0? Only +15.)

Here is a summary of the results we have obtained from the first four examples:

Original Numbers Have		The Answer is
the same sign	$3(5) = 15$	positive
different signs	$3(-5) = -15$	negative
different signs	$-3(5) = -15$	negative
the same sign	$-3(-5) = 15$	positive

By examining Example 1 and the preceding table, we can use the information there to write the following rule. This rule tells us how to multiply any two real numbers.

RULE

To multiply any two real numbers, simply multiply their absolute values. The sign of the answer is

1. *Positive* if both numbers have the same sign (both + or both −).
2. *Negative* if the numbers have opposite signs (one +, the other −).

The following example illustrates how we use the preceding rule to multiply real numbers.

EXAMPLE 2 Multiply.

a. $-8(-3) = 24$
b. $-10(-5) = 50$
c. $-4(-7) = 28$

If the two numbers in the product have the same sign, the answer is positive

d. $5(-7) = -35$
e. $-4(8) = -32$
f. $-6(10) = -60$

If the two numbers in the product have different signs, the answer is negative

Note: Students have trouble with the expression $-8(-3)$ because they want to subtract rather than multiply. Because we are very precise with the notation we use in algebra, the expression $-8(-3)$ has only one meaning—multiplication. A subtraction problem that uses the same numbers is $-8 - 3$. Compare the two following lists.

All Multiplication	No Multiplication
$5(4)$	$5 + 4$
$-5(4)$	$-5 + 4$
$5(-4)$	$5 - 4$
$-5(-4)$	$-5 - 4$

In the following example, we combine the order of operations with the rule for multiplication to simplify expressions. Remember, the order of operations specifies that we are to work inside the parentheses first and then simplify numbers containing exponents. After this, we multiply and divide, left to right. The last step is to add and subtract, left to right.

EXAMPLE 3 Simplify as much as possible.

a. $-5(-3)(-4)$ **b.** $4(-3) + 6(-5) - 10$

c. $(-2)^3$ **d.** $-3(-2)^3 - 5(-4)^2$

e. $6 - 4(7 - 2)$

SOLUTION

a. $-5(-3)(-4) = 15(-4)$

$= -60$

b. $4(-3) + 6(-5) - 10 = -12 + (-30) - 10$ Multiply

$= -42 - 10$ Add

$= -52$ Subtract

c. $(-2)^3 = (-2)(-2)(-2)$ Definition of exponents

$= -8$ Multiply, left to right

d. $-3(-2)^3 - 5(-4)^2 = -3(-8) - 5(16)$ Exponents first

$= 24 - 80$ Multiply

$= -56$ Subtract

e. $6 - 4(7 - 2) = 6 - 4(5)$ Inside parentheses first

$= 6 - 20$ Multiply

$= -14$ Subtract

Division

Suppose you and four friends bought equal shares of an investment for a total of \$15,000 and then sold it later for only \$13,000. How much did each person lose? Because the total amount of money lost can be represented by −\$2,000, and there are 5 people with equal shares, we can represent each person's loss with division:

$$\frac{-\$2{,}000}{5} = -\$400$$

From this discussion it seems reasonable to say that a negative number divided by a positive number is a negative number. Notice we can rephrase this result in terms of multiplication.

Because $\quad -400(5) = -2000$

we have $\quad -\frac{2000}{5} = -400$

Whether multiplying or dividing two numbers of unlike signs, the result is a negative number.

Following similar reasoning,

$$\text{Because} \quad 400(-5) = -2000$$

$$\text{we have} \quad \frac{-2000}{-5} = 400$$

The quotient of two negative numbers must be positive.

Because every division problem can be written as a multiplication problem, the rule for division of two real numbers is based upon the rule for multiplication.

RULE

To divide any two nonzero real numbers, simply divide their absolute values. The sign of the answer is

1. ***Positive*** if both numbers have the same sign (both + or both −).
2. ***Negative*** if the numbers have opposite signs (one +, the other −).

EXAMPLE 4 Divide.

a. $\frac{6}{2}$ **b.** $\frac{6}{-2}$ **c.** $\frac{-6}{2}$ **d.** $\frac{-6}{-2}$

SOLUTION

a. $\frac{6}{2} = 3$ The quotient of two positives is positive

b. $\frac{6}{-2} = -3$

c. $\frac{-6}{2} = -3$

(b and c) The quotient of a positive and a negative is a negative

d. $\frac{-6}{-2} = 3$ The quotient of two negatives is positive

Here are some more examples. If the original numbers have the same signs, the answer will be positive. If the original numbers have different signs, the answer will be negative.

EXAMPLE 5 Divide.

a. $\frac{12}{6} = 2$ Like signs give a positive answer

b. $\frac{12}{-6} = -2$ Unlike signs give a negative answer

c. $\frac{-12}{6} = -2$ Unlike signs give a negative answer

d. $\frac{-12}{-6} = 2$ Like signs give a positive answer

e. $\frac{15}{-3} = -5$ Unlike signs give a negative answer

f. $\frac{-40}{-5} = 8$ Like signs give a positive answer

g. $\frac{-14}{2} = -7$ Unlike signs give a negative answer

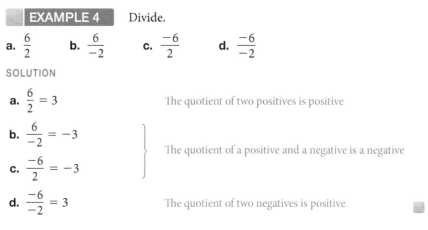

From the examples we have done so far, we can make the following generalization about quotients that contain negative signs:

If a and b are numbers and b is not equal to 0, then

$$\frac{-a}{b} = \frac{a}{-b} = -\frac{a}{b} \quad \text{and} \quad \frac{-a}{-b} = \frac{a}{b}$$

The last step in each of the following examples involves reducing a fraction to lowest terms. Recall that to reduce a fraction to lowest terms, we divide the numerator and denominator by the largest number that divides each of them exactly.

Note It is important to remember that we cannot divide by 0. Dividing any number, positive or negative, by zero gives an *undefined* expression for an answer.

EXAMPLE 6 Simplify as much as possible.

a. $\frac{-4(5)}{6}$ **b.** $\frac{30}{-4-5}$

SOLUTION

a. $\frac{-4(5)}{6} = \frac{-20}{6}$ Simplify numerator

$= -\frac{10}{3}$ Reduce to lowest terms by dividing numerator and denominator by 2

b. $\frac{30}{-4-5} = \frac{30}{-9}$ Simplify denominator

$= -\frac{10}{3}$ Reduce to lowest terms by dividing numerator and denominator by 3

Note When there is one negative sign in a fraction (either in the numerator or denominator, but not both), it is customary to write the final answer with the negative sign before the entire fraction. In other words,

$$\frac{-10}{3} = \frac{10}{-3} = -\frac{10}{3}$$

In the examples that follow, the numerators and denominators contain expressions that are somewhat more complicated than those we have seen thus far. To apply the rule for order of operations to these examples, we treat fraction bars the same way we treat grouping symbols; that is, fraction bars separate numerators and denominators so that each will be simplified separately.

EXAMPLE 7 Simplify.

a. $\frac{2(-3)+4}{12}$ **b.** $\frac{5(-4)+6(-1)}{2(3)-4(1)}$

SOLUTION

a. $\frac{2(-3)+4}{12} = \frac{-6+4}{12}$ In the numerator, we multiply before we add

$= \frac{-2}{12}$ Addition

$= -\frac{1}{6}$ Reduce to lowest terms by dividing the numerator and the denominator by 2

b. $\frac{5(-4)+6(-1)}{2(3)-4(1)} = \frac{-20+(-6)}{6-4}$ Multiplication before addition

$= \frac{-26}{2}$ Simplify numerator and denominator

$= -13$ Divide -26 by 2

We must be careful when we are working with expressions such as $(-5)^2$ and -5^2 that we include the negative sign with the base only when parentheses indicate we are to do so.

Unless there are parentheses to indicate otherwise, we consider the base to be only the number directly below and to the left of the exponent. If we want to include a negative sign with the base, we must use parentheses.

To simplify a more complicated expression, we follow the same rule. For example,

$$-7^2 - 3^2 = -49 - 9$$ The bases are 7 and 3; the sign between the two terms is a subtraction sign

For another example,

$$5^3 - 3^4 = 125 - 81$$ We simplify exponents first, then subtract

EXAMPLE 8 Simplify.

a. $\dfrac{5^2 - 3^2}{-5 + 3}$ **b.** $\dfrac{(3 + 2)^2}{-3^2 - 2^2}$

SOLUTION

a. $\dfrac{5^2 - 3^2}{-5 + 3} = \dfrac{25 - 9}{-2}$ Simplify numerator and denominator separately

$$= \frac{16}{-2}$$

$$= -8$$

b. $\dfrac{(3 + 2)^2}{-3^2 - 2^2} = \dfrac{5^2}{-9 - 4}$ Simplify numerator and denominator separately

$$= \frac{25}{-13}$$

$$= -\frac{25}{13}$$

Translating Words into Symbols

The following table shows a list of common phrases that are used to indicate a product or quotient of two numbers, along with the corresponding mathematical expression.

Phrase	Equivalent Expression
the product of a and b	$a \cdot b$
a times b	$a \cdot b$
a multiplied by b	$a \cdot b$
twice b	$2b$
the quotient of a and b	$\frac{a}{b}$
the ratio of a to b	$\frac{a}{b}$
a divided by b	$\frac{a}{b}$
a into b	$\frac{b}{a}$

EXAMPLE 9 Write the mathematical expression that is equivalent to the phrase, "The quotient of two and one less than the product of x and y."

SOLUTION This phrase involves a division, a multiplication and a subtraction. Here are the steps.

"The quotient of 2 and" $\dfrac{2}{?}$

"one less than" $\quad \dfrac{2}{\underline{?} - 1}$

"the product of x and y" $\quad \dfrac{2}{x \cdot y - 1}$

Division with the Number 0

As we discussed previously, for every division problem there is an associated multiplication problem involving the same numbers. We can use this relationship between division and multiplication to clarify division involving the number 0.

First, dividing 0 by a number other than 0 is allowed and always results in 0. To see this, consider dividing 0 by 5. We know the answer is 0 because of the relationship between multiplication and division. This is how we write it:

$$\frac{0}{5} = 0 \qquad \text{because} \qquad 0 = 0(5)$$

However, dividing a nonzero number by 0 is not allowed in the real numbers. Suppose we were attempting to divide 5 by 0. We don't know if there is an answer to this problem, but if there is, let's say the answer is a number that we can represent with the letter n. If 5 divided by 0 is a number n, then

$$\frac{5}{0} = n \qquad \text{and} \qquad 5 = n(0)$$

This is impossible, because no matter what number n is, when we multiply it by 0 the answer must be 0. It can never be 5. In algebra, we say expressions like $\frac{5}{0}$ are undefined because there is no answer to them; that is, division by 0 is not allowed in the real numbers.

The only other possibility for division involving the number 0 is 0 divided by 0. We will treat problems like $\frac{0}{0}$ as if they were undefined also.

Application

EXAMPLE 10 Figure 1 gives the calories that are burned in 1 hour for a variety of exercise by a person weighing 150 pounds. Figure 2 gives the calories that are consumed by eating some popular fast foods. Find the net change in calories for a 150-pound person playing handball for 2 hours and then eating a Whopper.

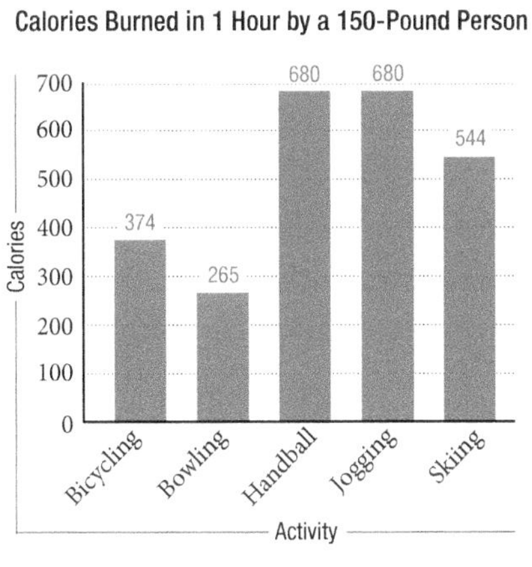

FIGURE 1

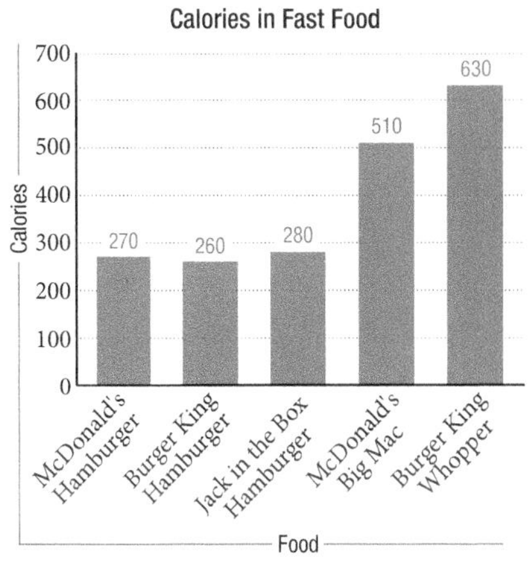

FIGURE 2

SOLUTION The net change in calories will be the difference of the calories gained from eating and the calories lost from exercise.

$$\text{Net change in calories} = 630 - 2(680) = -730 \text{ calories}$$

Geometric Sequences

DEFINITION *geometric sequence*

A ***geometric sequence*** is a sequence of numbers in which each number (after the first number) comes from the number before it by multiplying by the same amount each time.

For example, the sequence

$$2, 6, 18, 54, \ldots$$

is a geometric sequence because each number is obtained by multiplying the number before it by 3.

EXAMPLE 11 Each sequence below is a geometric sequence. Find the next number in each sequence.

a. 5, 10, 20,… **b.** 3, -15, 75,… **c.** 8, 4, 2,…

SOLUTION Because each sequence is a geometric sequence, we know that each term is obtained from the previous term by multiplying by the same number each time.

a. 5, 10, 20,…: Starting with 5, each number is obtained from the previous number by multiplying by 2 each time. The next number will be $20 \cdot 2 = 40$.

b. 3, -15, 75,…: The sequence starts with 3. After that, each number is obtained by multiplying by -5 each time. The next number will be $75(-5) = -375$.

c. 8, 4, 2,…: This sequence starts with 8. Multiplying each number in the sequence by 0.5 produces the next number in the sequence. To extend the sequence, we multiply 2 by 0.5: $2(0.5) = 1$. The next number in the sequence is 1.

Getting Ready for Class

After reading through the preceding section, respond in your own words and in complete sentences.

A. How do you multiply two negative numbers?

B. How do you divide two numbers with different signs?

C. Why is division by 0 not allowed with real numbers?

D. What is a geometric sequence?

Problem Set 1.5

Use the rule for multiplying real numbers to find each of the following products.

1. $7(-6)$ **2.** $8(-4)$ **3.** $-8(2)$ **4.** $-16(3)$

5. $-3(-1)$ **6.** $-7(-1)$ **7.** $-11(-11)$ **8.** $-12(-12)$

9. $-3(2)(-1)$ **10.** $-2(3)(-4)$ **11.** $-3(-4)(-5)$ **12.** $-5(-6)(-7)$

13. $-2(-4)(-3)(-1)$ **14.** $-1(-3)(-2)(-1)$

15. $(-7)^2$ **16.** $(-8)^2$ **17.** $(-3)^3$ **18.** $(-2)^4$

Find the following quotients (divide).

19. $\frac{8}{-4}$ **20.** $\frac{10}{-5}$ **21.** $\frac{-48}{16}$ **22.** $\frac{-32}{4}$

23. $\frac{-7}{21}$ **24.** $\frac{-25}{100}$ **25.** $\frac{-39}{-13}$ **26.** $\frac{-18}{-6}$

27. $\frac{-6}{-42}$ **28.** $\frac{-4}{-28}$ **29.** $\frac{0}{-32}$ **30.** $\frac{0}{17}$

The following problems review all four operations with positive and negative numbers. Perform the indicated operations.

31. $-3 + 12$ **32.** $5 + (-10)$ **33.** $-3 - 12$ **34.** $5 - (-10)$

35. $-3(12)$ **36.** $5(-10)$ **37.** $-3 \div 12$ **38.** $5 \div (-10)$

The following problems involve more than one operation. Use the order of operations to simplify each expression as much as possible.

39. $\frac{3(-2)}{-10}$ **40.** $\frac{4(-3)}{24}$ **41.** $\frac{-5(-5)}{-15}$

42. $\frac{-7(-3)}{-35}$ **43.** $\frac{-8(-7)}{-28}$ **44.** $\frac{-3(-9)}{-6}$

45. $-2(2 - 5)$ **46.** $-3(3 - 7)$ **47.** $-5(8 - 10)$

48. $-4(6 - 12)$ **49.** $(4 - 7)(6 - 9)$ **50.** $(3 - 10)(2 - 6)$

51. $(-3 - 2)(-5 - 4)$ **52.** $(-3 - 6)(-2 - 8)$ **53.** $-3(-6) + 4(-1)$

54. $-4(-5) + 8(-2)$ **55.** $2(3) - 3(-4) + 4(-5)$ **56.** $5(4) - 2(-1) + 5(6)$

57. $\frac{27}{4 - 13}$ **58.** $\frac{27}{13 - 4}$ **59.** $\frac{20 - 6}{5 - 5}$

60. $\frac{10 - 12}{3 - 3}$ **61.** $\frac{-3 + 9}{2 \cdot 5 - 10}$ **62.** $\frac{2 + 8}{2 \cdot 4 - 8}$

63. $\frac{15(-5) - 25}{2(-10)}$ **64.** $\frac{10(-3) - 20}{5(-2)}$ **65.** $\frac{27 - 2(-4)}{-3(5)}$

66. $\frac{20 - 5(-3)}{10(-3)}$ **67.** $\frac{12 - 6(-2)}{12(-2)}$ **68.** $\frac{3(-4) + 5(-6)}{10 - 6}$

69. $4(-3)^2 + 5(-6)^2$

70. $2(-5)^2 + 4(-3)^2$

71. $7(-2)^3 - 2(-3)^3$

72. $10(-2)^3 - 5(-2)^4$

73. $6 - 4(8 - 2)$

74. $7 - 2(6 - 3)$

75. $9 - 4(3 - 8)$

76. $8 - 5(2 - 7)$

77. $-4(3 - 8) - 6(2 - 5)$

78. $-8(2 - 7) - 9(3 - 5)$

79. $\dfrac{5^2 - 2^2}{-5 + 2}$

80. $\dfrac{7^2 - 4^2}{-7 + 4}$

81. $\dfrac{8^2 - 2^2}{8^2 + 2^2}$

82. $\dfrac{4^2 - 6^2}{4^2 + 6^2}$

83. $\dfrac{(5 + 3)^2}{-5^2 - 3^2}$

84. $\dfrac{(7 + 2)^2}{-7^2 - 2^2}$

85. $\dfrac{(8 - 4)^2}{8^2 - 4^2}$

86. $\dfrac{(6 - 2)^2}{6^2 - 2^2}$

87. $7 - 2[-6 - 4(-3)]$

88. $6 - 3[-5 - 3(-1)]$

89. $7 - 3[2(-4 - 4) - 3(-1 - 1)]$

90. $5 - 3[7(-2 - 2) - 3(-3 + 1)]$

91. $8 - 6[-2(-3 - 1) + 4(-2 - 3)]$

92. $4 - 2[-3(-1 + 8) + 5(-5 + 7)]$

93. $\dfrac{-4 \cdot 3^2 - 5 \cdot 2^2}{-8(7)}$

94. $\dfrac{-2 \cdot 5^2 + 3 \cdot 2^3}{-3(13)}$

95. $\dfrac{3 \cdot 10^2 + 4 \cdot 10 + 5}{345}$

96. $\dfrac{5 \cdot 10^2 + 6 \cdot 10 + 7}{567}$

97. $\dfrac{7 - [(2 - 3) - 4]}{-1 - 2 - 3}$

98. $\dfrac{2 - [(3 - 5) - 8]}{-3 - 4 - 5}$

99. $\dfrac{6(-4) - 2(5 - 8)}{-6 - 3 - 5}$

100. $\dfrac{3(-4) - 5(9 - 11)}{-9 - 2 - 3}$

101. $\dfrac{3(-5 - 3) + 4(7 - 9)}{5(-2) + 3(-4)}$

102. $\dfrac{-2(6 - 10) - 3(8 - 5)}{6(-3) - 6(-2)}$

103. $\dfrac{|3 - 9|}{3 - 9}$

104. $\dfrac{|4 - 7|}{4 - 7}$

105. Simplify each expression.

a. $20 \div 4 \cdot 5$

b. $-20 \div 4 \cdot 5$

c. $20 \div (-4) \cdot 5$

d. $20 \div 4(-5)$

e. $-20 \div 4(-5)$

106. Simplify each expression.

a. $32 \div 8 \cdot 4$

b. $-32 \div 8 \cdot 4$

c. $32 \div (-8) \cdot 4$

d. $32 \div 8(-4)$

e. $-32 \div 8(-4)$

Answer the following questions.

107. Five added to the product of 3 and -10 is what number?

108. If the product of -8 and -2 is decreased by 4, what number results?

109. What number results if 8 is subtracted from the product of -9 and 2?

110. What number results if -8 is subtracted from the product of -9 and 2?

111. What is the quotient of -12 and -4?

112. The quotient of -4 and -12 is what number?

113. What number do we divide by -5 to get 2?

114. What number do we divide by -3 to get 4?

115. Twenty-seven divided by what number is -9?

116. Fifteen divided by what number is -3?

117. If the quotient of -20 and 4 is decreased by 3, what number results?

118. If -4 is added to the quotient of 24 and -8, what number results?

Here are some problems you will see later in the book. Simplify.

119. $3(x - 5) + 4$

120. $5(x - 3) + 2$

121. $2(3) - 4 - 3(-4)$

122. $2(3) + 4(5) - 5(2)$

Each of the following is a geometric sequence. In each case, find the next number in the sequence.

123. $1, 2, 4, \ldots$

124. $1, 5, 25, \ldots$

125. $10, -20, 40, \ldots$

126. $10, -30, 90, \ldots$

127. $3, -6, 12, \ldots$

128. $-3, 6, -12, \ldots$

Applying the Concepts

129. Temperature Change The temperature is 25°F at 5:00 in the afternoon. If the temperature drops 6°F every hour after that, what is the temperature at 9:00 in the evening?

130. Investment Value Suppose you purchase \$500 worth of a mutual fund and find that the value of your purchase doubles every 2 years. Write a sequence of numbers that gives the value of your purchase every 2 years for the first 10 years you own it. Is this sequence a geometric sequence?

131. Investment Suppose that you and 3 friends bought equal shares of an investment for a total of \$15,000 and then sold it later for only \$13,600. How much did each person lose?

132. Investment If 8 people invest \$500 each in a stamp collection and after a year the collection is worth \$3,800, how much did each person lose?

133. Temperature Change Suppose that the temperature outside is dropping at a constant rate. If the temperature is 75°F at noon and drops to 61°F by 4:00 in the afternoon, by how much did the temperature change each hour?

134. Temperature Change In a chemistry class, a thermometer is placed in a beaker of hot water. The initial temperature of the water is 165°F. After 10 minutes the water has cooled to 72°F. If the water temperature drops at a constant rate, by how much does the water temperature change each minute?

135. Internet Mailing Lists A company sells products on the Internet through an email list. They predict that they sell one \$50 product for every 25 people on their mailing list.

a. What is their projected revenue if their list contains 10,000 email addresses?

b. What is their projected revenue if their list contains 25,000 email addresses?

c. They can purchase a list of 5,000 email addresses for \$5,000. Is this a wise purchase?

136. Internet Mailing Lists A new band has a following on the Internet. They sell their CDs through an email list. They predict that they sell one $15 CD for every 10 people on their mailing list.

a. What is their projected revenue if their list contains 5,000 email addresses?

b. What is their projected revenue if their list contains 20,000 email addresses?

c. If they need to make $45,000, how many people do they need on their email list?

137. Reading Charts Refer to the bar charts below to find the net change in calories for a 150-pound person who bowls for 3 hours and then eats 2 Whoppers.

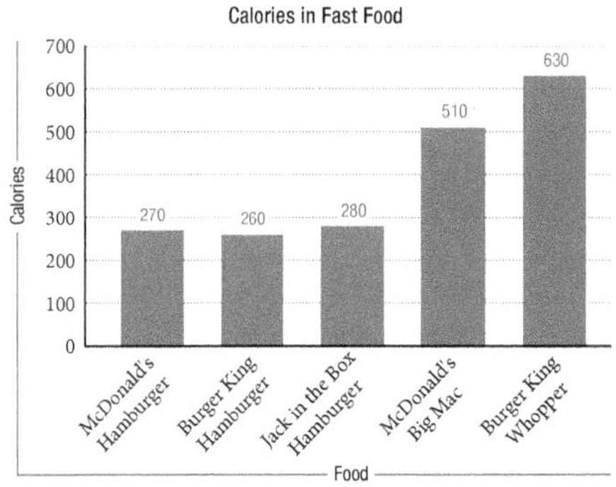

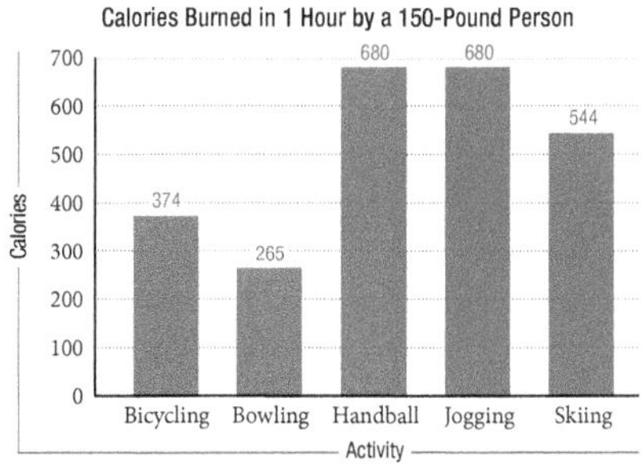

Learning Objectives Assessment

The following problems can be used to help assess if you have successfully met the learning objectives for this section.

138. Multiply: $-9(-5)$.

a. -45 **b.** 45 **c.** 14 **d.** -14

139. Divide: $\frac{-18}{0}$.

a. Undefined **b.** 1 **c.** 0 **d.** -18

140. Translate into symbols: the quotient of 3 and the sum of x and 1.

a. $3(x + 1)$ **b.** $\frac{3}{x} + 1$ **c.** $\frac{3}{x + 1}$ **d.** $\frac{x + 1}{3}$

141. Find the next number in the geometric sequence $2, -6, 18, -54, \ldots$.

a. 108 **b.** 162 **c.** -162 **d.** -108

Fractions 1.6

Learning Objectives

In this section, we will learn how to:

1. Factor a number into a product of primes.
2. Reduce a fraction to lowest terms.
3. Find the reciprocal of a number.
4. Multiply and divide fractions.
5. Add and subtract fractions.

Introduction

In Section 1.2, we introduced the concept of a fraction, and learned how to obtain an equivalent fraction using multiplication or division. Now we will continue our study of fractions by learning how to multiply, divide, add, or subtract them. But before we do so, we need to consider some things involving prime numbers.

Prime Numbers and Factoring

The following diagram shows the relationship between multiplication and factoring:

Multiplication

Factors → $3 \cdot 4 = 12$ ← Product

Factoring

When we read the problem from left to right, we say the product of 3 and 4 is 12. Or we multiply 3 and 4 to get 12. When we read the problem in the other direction, from right to left, we say we have *factored* 12 into 3 times 4, or 3 and 4 are *factors* of 12.

The number 12 can be factored still further:

$$\begin{aligned} 12 &= 4 \cdot 3 \\ &= 2 \cdot 2 \cdot 3 \\ &= 2^2 \cdot 3 \end{aligned}$$

The numbers 2 and 3 are called *prime factors* of 12 because neither of them can be factored any further.

def DEFINITION *factor*

If a and b represent integers, then a is said to be a ***factor*** (or divisor) of b if a divides b evenly; that is, if a divides b with no remainder.

def DEFINITION *prime number*

A ***prime number*** is any positive integer larger than 1 whose only positive factors (divisors) are itself and 1.

Note: The number 15 is not a prime number because it has factors of 3 and 5; that is, $15 = 3 \cdot 5$. When a whole number larger than 1 is not prime, it is said to be *composite*.

Here is a list of the first few prime numbers.

Prime numbers = {2, 3, 5, 7, 11, 13, 17, 19, 23, 29, 31, 37, 41, . . .}

When a number is not prime, we can factor it into the product of prime numbers. To factor a number into the product of primes, we simply factor it until it cannot be factored further.

VIDEO EXAMPLES

SECTION 1.6

EXAMPLE 1 Factor the number 60 into the product of prime numbers.

SOLUTION We begin by writing 60 as the product of any two positive integers whose product is 60, like 6 and 10:

$$60 = 6 \cdot 10$$

We then factor these numbers:

$$\begin{aligned} 60 &= 6 \cdot 10 \\ &= (2 \cdot 3) \cdot (2 \cdot 5) \\ &= 2 \cdot 2 \cdot 3 \cdot 5 \\ &= 2^2 \cdot 3 \cdot 5 \end{aligned}$$

Note: It is customary to write the prime factors in order from smallest to largest.

EXAMPLE 2 Factor the number 630 into the product of primes.

SOLUTION Let's begin by writing 630 as the product of 63 and 10:

$$\begin{aligned} 630 &= 63 \cdot 10 \\ &= (7 \cdot 9) \cdot (2 \cdot 5) \\ &= 7 \cdot 3 \cdot 3 \cdot 2 \cdot 5 \\ &= 2 \cdot 3^2 \cdot 5 \cdot 7 \end{aligned}$$

It makes no difference which two numbers we start with, as long as their product is 630. We will always get the same result because a number has only one set of prime factors.

$$\begin{aligned} 630 &= 18 \cdot 35 \\ &= 3 \cdot 6 \cdot 5 \cdot 7 \\ &= 3 \cdot 2 \cdot 3 \cdot 5 \cdot 7 \\ &= 2 \cdot 3^2 \cdot 5 \cdot 7 \end{aligned}$$

Note: There are some "tricks" to finding the divisors of a number. For instance, if a number ends in 0 or 5, then it is divisible by 5. If a number ends in an even number (0, 2, 4, 6, or 8), then it is divisible by 2. A number is divisible by 3 if the sum of its digits is divisible by 3. For example, 921 is divisible by 3 because the sum of its digits is $9 + 2 + 1 = 12$, which is divisible by 3.

When we have factored a number into the product of its prime factors, we not only know what prime numbers divide the original number, but we also know all of the other numbers that divide it as well. For instance, if we were to factor 210 into its prime factors, we would have $210 = 2 \cdot 3 \cdot 5 \cdot 7$, which means that 2, 3, 5, and 7 divide 210, as well as any combination of products of 2, 3, 5, and 7. That is, because 3 and 7 divide 210, then so does their product 21. Because 3, 5, and 7 each divide 210, then so does their product 105.

21 divides 210

$$210 = 2 \cdot 3 \cdot 5 \cdot 7$$

105 divides 210

Reducing Fractions and Equivalent Fractions

Although there are many ways in which factoring is used in arithmetic and algebra, one simple application is in reducing fractions to lowest terms.

Recall that we reduce fractions to lowest terms by dividing the numerator and denominator by the same number. We can use the prime factorization of numbers to help us reduce fractions with large numerators and denominators.

EXAMPLE 3 Reduce $\frac{210}{231}$ to lowest terms.

SOLUTION First we factor 210 and 231 into the product of prime factors. Then we reduce to lowest terms by dividing the numerator and denominator by any factors they have in common.

$$\frac{210}{231} = \frac{2 \cdot 3 \cdot 5 \cdot 7}{3 \cdot 7 \cdot 11}$$ Factor the numerator and denominator completely

$$= \frac{2 \cdot 3 \cdot 5 \cdot 7}{3 \cdot 7 \cdot 11}$$ Divide the numerator and denominator by $3 \cdot 7$

$$= \frac{2 \cdot 5}{11}$$

$$= \frac{10}{11}$$

Note: The small lines we have drawn through the factors that are common to the numerator and denominator are used to indicate that we have divided the numerator and denominator by those factors.

EXAMPLE 4 Write $\frac{3}{4}$ as an equivalent fraction with denominator 20.

SOLUTION The denominator of the original fraction is 4. The fraction we are trying to find must have a denominator of 20. We know that if we multiply 4 by 5, we get 20. Property 1 for fractions indicates that we are free to multiply the denominator by 5 as long as we do the same to the numerator.

$$\frac{3}{4} = \frac{3 \cdot 5}{4 \cdot 5} = \frac{15}{20}$$

The fraction $\frac{15}{20}$ is equivalent to the fraction $\frac{3}{4}$. We can accomplish the same result by multiplying or original fraction by the number 1 in the form $\frac{5}{5}$.

$$\frac{3}{4} = \frac{3}{4} \cdot \frac{5}{5} = \frac{15}{20}$$

Reciprocals and Multiplication with Fractions

The next concept we want to cover in this section is the concept of reciprocals. Understanding reciprocals requires some knowledge of multiplication with fractions. To multiply two fractions, we simply multiply numerators and multiply denominators.

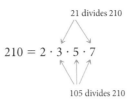

EXAMPLE 5 Multiply $\frac{3}{4} \cdot \frac{5}{7}$.

SOLUTION The product of the numerators is 15, and the product of the denominators is 28:

$$\frac{3}{4} \cdot \frac{5}{7} = \frac{3 \cdot 5}{4 \cdot 7} = \frac{15}{28}$$

EXAMPLE 6 Multiply $7\left(\frac{1}{3}\right)$.

SOLUTION The number 7 can be thought of as the fraction $\frac{7}{1}$:

$$7\left(\frac{1}{3}\right) = \frac{7}{1}\left(\frac{1}{3}\right) = \frac{7 \cdot 1}{1 \cdot 3} = \frac{7}{3}$$

EXAMPLE 7 Expand and multiply $\left(\frac{2}{3}\right)^3$.

SOLUTION Using the definition of exponents from the previous section, we have

$$\left(\frac{2}{3}\right)^3 = \frac{2}{3} \cdot \frac{2}{3} \cdot \frac{2}{3} = \frac{8}{27}$$

We can apply the rule for multiplication of positive and negative numbers to fractions in the same way we apply it to other numbers. We multiply absolute values: The product is positive if both fractions have the same sign and negative if they have different signs. Here is an example.

EXAMPLE 8 Multiply.

a. $-\frac{3}{4}\left(\frac{5}{7}\right)$ **b.** $-6\left(\frac{1}{2}\right)$ **c.** $-\frac{2}{3}\left(-\frac{3}{2}\right)$

SOLUTION

a. $-\frac{3}{4}\left(\frac{5}{7}\right) = -\frac{3 \cdot 5}{4 \cdot 7}$ Different signs give a negative answer

$= -\frac{15}{28}$

b. $-6\left(\frac{1}{2}\right) = -\frac{6}{1}\left(\frac{1}{2}\right)$ Different signs give a negative answer

$= -\frac{6}{2}$

$= -3$

c. $-\frac{2}{3}\left(-\frac{3}{2}\right) = \frac{2 \cdot 3}{3 \cdot 2}$ Same signs give a positive answer

$= \frac{6}{6}$

$= 1$

We are now ready for the definition of reciprocals.

DEFINITION *reciprocals*

Two numbers whose product is 1 are called ***reciprocals***.

EXAMPLE 9 Give the reciprocal of each number.

a. 5 **b.** -2 **c.** $\frac{1}{3}$ **d.** $-\frac{3}{4}$

SOLUTION

	Number	Reciprocal	
a.	5	$\frac{1}{5}$	Because $5\left(\frac{1}{5}\right) = \frac{5}{1}\left(\frac{1}{5}\right) = \frac{5}{5} = 1$
b.	-2	$-\frac{1}{2}$	Because $-2\left(-\frac{1}{2}\right) = -\frac{2}{1}\left(-\frac{1}{2}\right) = \frac{2}{2} = 1$
c.	$\frac{1}{3}$	3	Because $\frac{1}{3}(3) = \frac{1}{3}\left(\frac{3}{1}\right) = \frac{3}{3} = 1$
d.	$-\frac{3}{4}$	$-\frac{4}{3}$	Because $-\frac{3}{4}\left(-\frac{4}{3}\right) = \frac{12}{12} = 1$

We know that division by the number 2 is the same as multiplication by $\frac{1}{2}$; that is, 6 divided by 2 is 3, which is the same as 6 times $\frac{1}{2}$. Similarly, dividing a number by 5 gives the same result as multiplying by $\frac{1}{5}$. We can extend this idea to all real numbers with the following rule.

RULE

If a and b represent any two real numbers (b cannot be 0), then it is always true that

$$a \div b = \frac{a}{b} = a\left(\frac{1}{b}\right)$$

Division by a number is the same as multiplication by its reciprocal.

Division with Fractions

We can apply the definition of division to fractions. Because dividing by a fraction is equivalent to multiplying by its reciprocal, we can divide a number by the fraction $\frac{3}{4}$ by multiplying it by the reciprocal of $\frac{3}{4}$, which is $\frac{4}{3}$. For example,

$$\frac{2}{5} \div \frac{3}{4} = \frac{2}{5} \cdot \frac{4}{3} = \frac{8}{15}$$

You may have learned this rule in previous math classes. In some math classes, multiplication by the reciprocal is referred to as "inverting the divisor and multiplying." No matter how you say it, division by any number (except 0) is always equivalent to multiplication by its reciprocal. Here are additional examples that involve division by fractions.

EXAMPLE 10 Divide.

a. $\frac{2}{3} \div \frac{5}{7}$ **b.** $-\frac{3}{4} \div \frac{7}{9}$ **c.** $8 \div \left(-\frac{4}{5}\right)$

SOLUTION

a. $\frac{2}{3} \div \frac{5}{7} = \frac{2}{3} \cdot \frac{7}{5}$ Rewrite as multiplication by the reciprocal

$= \frac{14}{15}$ Multiply

b. $-\frac{3}{4} \div \frac{7}{9} = -\frac{3}{4} \cdot \frac{9}{7}$ Rewrite as multiplication by the reciprocal

$= -\frac{27}{28}$ Multiply

c. $8 \div \left(-\frac{4}{5}\right) = \frac{8}{1}\left(-\frac{5}{4}\right)$ Rewrite as multiplication by the reciprocal

$= -\frac{40}{4}$ Multiply

$= -10$ Divide 40 by 4

Adding and Subtracting Fractions

You may recall from previous math classes that to add two fractions with the same denominator, you simply add their numerators and put the result over the common denominator:

$$\frac{3}{4} + \frac{2}{4} = \frac{3+2}{4} = \frac{5}{4}$$

We will justify this process in the next section when we introduce the distributive property. For now, we simply state it as a fact. The reason we add numerators but do not add denominators is that we must follow the distributive property.

In symbols we have the following.

RULE ***Addition and Subtraction with Fractions***

If a, b, and c are integers and c is not equal to 0, then

$$\frac{a}{c} + \frac{b}{c} = \frac{a+b}{c}$$

This rule holds for subtraction as well; that is,

$$\frac{a}{c} - \frac{b}{c} = \frac{a-b}{c}$$

EXAMPLE 11 Find the sum or difference. Reduce all answers to lowest terms. (Assume all variables represent nonzero numbers.)

a. $\frac{3}{8} + \frac{1}{8}$ **b.** $\frac{a+5}{8} - \frac{3}{8}$ **c.** $\frac{9}{x} - \frac{3}{x}$ **d.** $\frac{3}{7} + \frac{2}{7} - \frac{9}{7}$

SOLUTION

a. $\frac{3}{8} + \frac{1}{8} = \frac{3+1}{8}$ Add numerators; keep the same denominator

$= \frac{4}{8}$ The sum of 3 and 1 is 4

$= \frac{1}{2}$ Reduce to lowest terms

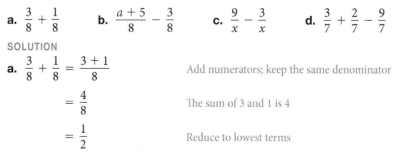

b. $\frac{a+5}{8} - \frac{3}{8} = \frac{a+5-3}{8}$ Combine numerators; keep the same denominator

$= \frac{a+2}{8}$

c. $\frac{9}{x} - \frac{3}{x} = \frac{9-3}{x}$ Subtract numerators; keep the same denominator

$= \frac{6}{x}$ The difference of 9 and 3 is 6

d. $\frac{3}{7} + \frac{2}{7} - \frac{9}{7} = \frac{3+2-9}{7}$

$= \frac{-4}{7}$

$= -\frac{4}{7}$ Unlike signs give a negative answer

As Example 11 indicates, addition and subtraction are simple, straightforward processes when all the fractions have the same denominator. We will now turn our attention to the process of adding fractions that have different denominators. To get started, we need the following definition.

DEFINITION *least common denominator (LCD)*

The ***least common denominator (LCD)*** for a set of denominators is the smallest number that is exactly divisible by each denominator. (Note that in some books the least common denominator is also called the least common multiple.)

In other words, all the denominators of the fractions involved in a problem must divide into the least common denominator exactly; that is, they divide it without giving a remainder.

EXAMPLE 12 Find the LCD for the fractions $\frac{5}{12}$ and $\frac{7}{18}$.

Note: The ability to find least common denominators is very important in mathematics. The discussion here is a detailed explanation of how to do it.

SOLUTION The least common denominator for the denominators 12 and 18 must be the smallest number divisible by both 12 and 18. We can factor 12 and 18 completely and then build the LCD from these factors. Factoring 12 and 18 completely gives us

$$12 = 2 \cdot 2 \cdot 3 \qquad 18 = 2 \cdot 3 \cdot 3$$

Now, if 12 is going to divide the LCD exactly, then the LCD must have factors of $2 \cdot 2 \cdot 3$. If 18 is to divide it exactly, it must have factors of $2 \cdot 3 \cdot 3$. We don't need to repeat the factors that 12 and 18 have in common:

12 divides the LCD

$$12 = 2 \cdot 2 \cdot 3 \qquad \text{LCD} = 2 \cdot 2 \cdot 3 \cdot 3$$
$$18 = 2 \cdot 3 \cdot 3$$

18 divides the LCD

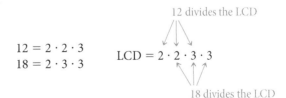

In other words, first we write down the factors of 12, then we attach the factors of 18 that do not already appear as factors of 12. We start with $2 \cdot 2 \cdot 3$ because those are the factors of 12. Then we look at the first factor of 18. It is 2. Because 2 already appears in the expression $2 \cdot 2 \cdot 3$, we don't need to attach another one. Next, we look at the factors $3 \cdot 3$. The expression $2 \cdot 2 \cdot 3$ has one 3. For it to contain the expression $3 \cdot 3$, we attach another 3. The final expression, our LCD, is $2 \cdot 2 \cdot 3 \cdot 3$.

The LCD for 12 and 18 is 36. It is the smallest number that is divisible by both 12 and 18; 12 divides it exactly three times, and 18 divides it exactly two times.

We can use the results of Example 12 to find the sum of the fractions $\frac{5}{12}$ and $\frac{7}{18}$.

EXAMPLE 13 Add $\frac{5}{12} + \frac{7}{18}$.

SOLUTION We can add fractions only when they have the same denominators. In Example 12 we found the LCD for $\frac{5}{12}$ and $\frac{7}{18}$ to be 36. We change $\frac{5}{12}$ and $\frac{7}{18}$ to equivalent fractions that each have 36 for a denominator.

$$\frac{5}{12} = \frac{5}{12} \cdot \frac{3}{3} = \frac{15}{36}$$

$$\frac{7}{18} = \frac{7}{18} \cdot \frac{2}{2} = \frac{14}{36}$$

The fraction $\frac{15}{36}$ is equivalent to $\frac{5}{12}$, because it was obtained by multiplying both the numerator and denominator by 3. Likewise, $\frac{14}{36}$ is equivalent to $\frac{7}{18}$ because it was obtained by multiplying the numerator and denominator by 2. All we have left to do is to add numerators:

$$\frac{15}{36} + \frac{14}{36} = \frac{29}{36}$$

The sum of $\frac{5}{12}$ and $\frac{7}{18}$ is the fraction $\frac{29}{36}$. Let's write the complete problem again step-by-step.

$$\frac{5}{12} + \frac{7}{18} = \frac{5}{12} \cdot \frac{3}{3} + \frac{7}{18} \cdot \frac{2}{2}$$ Rewrite each fraction as an equivalent fraction with denominator 36

$$= \frac{15}{36} + \frac{14}{36}$$

$$= \frac{29}{36}$$ Add numerators; keep the common denominator

EXAMPLE 14 Find the LCD for $\frac{3}{4}$ and $\frac{1}{6}$.

SOLUTION We factor 4 and 6 into products of prime factors and build the LCD from these factors:

$$\left.\begin{array}{l} 4 = 2 \cdot 2 \\ 6 = 2 \cdot 3 \end{array}\right\} \quad \text{LCD} = 2 \cdot 2 \cdot 3 = 12$$

The LCD is 12. Both denominators divide it exactly; 4 divides 12 exactly three times, and 6 divides 12 exactly two times.

EXAMPLE 15 Add $\frac{3}{4} + \frac{1}{6}$.

SOLUTION In Example 14 we found that the LCD for these two fractions is 12. We begin by changing $\frac{3}{4}$ and $\frac{1}{6}$ to equivalent fractions with denominator 12:

$$\frac{3}{4} = \frac{3}{4} \cdot \frac{3}{3} = \frac{9}{12}$$

$$\frac{1}{6} = \frac{1}{6} \cdot \frac{2}{2} = \frac{2}{12}$$

The fraction $\frac{9}{12}$ is equal to the fraction $\frac{3}{4}$ because it was obtained by multiplying the numerator and denominator of $\frac{3}{4}$ by 3. Likewise, $\frac{2}{12}$ is equivalent to $\frac{1}{6}$ because it was obtained by multiplying the numerator and denominator of $\frac{1}{6}$ by 2. To complete the problem, we add numerators:

$$\frac{9}{12} + \frac{2}{12} = \frac{11}{12}$$

The sum of $\frac{3}{4}$ and $\frac{1}{6}$ is $\frac{11}{12}$. Here is how the complete problem looks:

$$\frac{3}{4} + \frac{1}{6} = \frac{3}{4} \cdot \frac{3}{3} + \frac{1}{6} \cdot \frac{2}{2}$$ Rewrite each fraction as an equivalent fraction with denominator 12

$$= \frac{9}{12} + \frac{2}{12}$$

$$= \frac{11}{12}$$ Add numerators; keep the same denominator

EXAMPLE 16 Subtract $\frac{7}{15} - \frac{3}{10}$.

SOLUTION Let's factor 15 and 10 completely and use these factors to build the LCD:

15 divides the LCD

$$\left.\begin{array}{l} 15 = 3 \cdot 5 \\ 10 = 2 \cdot 5 \end{array}\right\} \text{LCD} = 2 \cdot 3 \cdot 5 = 30$$

10 divides the LCD

Changing to equivalent fractions and subtracting, we have

$$\frac{7}{15} - \frac{3}{10} = \frac{7}{15} \cdot \frac{2}{2} - \frac{3}{10} \cdot \frac{3}{3}$$ Rewrite as equivalent fractions with the LCD for denominator

$$= \frac{14}{30} - \frac{9}{30}$$

$$= \frac{5}{30}$$ Subtract numerators; keep the LCD

$$= \frac{1}{6}$$ Reduce to lowest terms

As a summary of what we have done so far and as a guide to working other problems, we will now list the steps involved in adding and subtracting fractions with different denominators.

HOW TO *Add or Subtract Any Two Fractions*

Step 1: Factor each denominator completely and use the factors to build the LCD. (Remember, the LCD is the smallest number divisible by each of the denominators in the problem.)

Step 2: Rewrite each fraction as an equivalent fraction that has the LCD for its denominator.

Step 3: Add or subtract the numerators of the fractions produced in step 2. This is the numerator of the sum or difference. The denominator of the sum or difference is the LCD.

Step 4: Reduce the fraction produced in step 3 to lowest terms if it is not already in lowest terms.

The idea behind adding or subtracting fractions is really very simple. We can add or subtract only fractions that have the same denominators. If the fractions we are trying to add or subtract do not have the same denominators, we rewrite each of them as an equivalent fraction with the LCD for a denominator.

Here are some further examples of sums and differences of fractions.

EXAMPLE 17 Add $\frac{1}{6} + \frac{1}{8} + \frac{1}{4}$.

SOLUTION We begin by factoring the denominators completely and building the LCD from the factors that result:

8 divides the LCD

$$\begin{aligned} 6 &= 2 \cdot 3 \\ 8 &= 2 \cdot 2 \cdot 2 \qquad \text{LCD} = 2 \cdot 2 \cdot 2 \cdot 3 = 24 \\ 4 &= 2 \cdot 2 \end{aligned}$$

4 divides the LCD 6 divides the LCD

We then change to equivalent fractions and add as usual:

$$\begin{aligned} \frac{1}{6} + \frac{1}{8} + \frac{1}{4} &= \frac{1}{6} \cdot \frac{4}{4} + \frac{1}{8} \cdot \frac{3}{3} + \frac{1}{4} \cdot \frac{6}{6} \\ &= \frac{4}{24} + \frac{3}{24} + \frac{6}{24} \\ &= \frac{13}{24} \end{aligned}$$

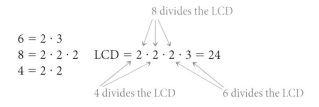

EXAMPLE 18 Subtract $3 - \frac{5}{6}$.

SOLUTION The denominators are 1 $\left(\text{because } 3 = \frac{3}{1}\right)$ and 6. The smallest number divisible by both 1 and 6 is 6.

$$\begin{aligned} 3 - \frac{5}{6} &= \frac{3}{1} - \frac{5}{6} \\ &= \frac{3}{1} \cdot \frac{6}{6} - \frac{5}{6} \\ &= \frac{18}{6} - \frac{5}{6} \\ &= \frac{13}{6} \end{aligned}$$

EXAMPLE 19 Find the next number in each sequence.

a. $\frac{1}{2}, 0, -\frac{1}{2}, \ldots$ **b.** $\frac{1}{2}, 1, \frac{3}{2}, \ldots$ **c.** $\frac{1}{2}, \frac{1}{4}, \frac{1}{8}, \ldots$

SOLUTION

a. $\frac{1}{2}, 0, -\frac{1}{2}, \ldots$: Adding $-\frac{1}{2}$ to each term produces the next term. The fourth term will be $-\frac{1}{2} + \left(-\frac{1}{2}\right) = -1$. This is an arithmetic sequence.

b. $\frac{1}{2}, 1, \frac{3}{2}, \ldots$: Each term comes from the term before it by adding $\frac{1}{2}$. The fourth term will be $\frac{3}{2} + \frac{1}{2} = 2$. This sequence is also an arithmetic sequence.

c. $\frac{1}{2}, \frac{1}{4}, \frac{1}{8}, \ldots$: This is a geometric sequence in which each term comes from the term before it by multiplying by $\frac{1}{2}$ each time. The next term will be $\frac{1}{8} \cdot \frac{1}{2} = \frac{1}{16}$.

Getting Ready for Class

After reading through the preceding section, respond in your own words and in complete sentences.

A. What is the reciprocal of a number?

B. How do we divide fractions?

C. What is a least common denominator?

D. What is the first step in adding two fractions that have different denominators?

Problem Set 1.6

Label each of the following numbers as prime or composite. If a number is composite, then factor it completely.

1. 48 **2.** 72 **3.** 37 **4.** 23
5. 1,023 **6.** 543

Factor the following into the product of primes. When the number has been factored completely, write its prime factors from smallest to largest.

7. 144 **8.** 288 **9.** 38 **10.** 63
11. 105 **12.** 210 **13.** 180 **14.** 900
15. 385 **16.** 1,925 **17.** 121 **18.** 546
19. 420 **20.** 598 **21.** 620 **22.** 2,310

Reduce each fraction to lowest terms by first factoring the numerator and denominator into the product of prime factors and then dividing out any factors they have in common.

23. $\frac{105}{165}$ **24.** $\frac{165}{385}$ **25.** $\frac{525}{735}$ **26.** $\frac{550}{735}$ **27.** $\frac{385}{455}$ **28.** $\frac{385}{735}$

29. $\frac{322}{345}$ **30.** $\frac{266}{285}$ **31.** $\frac{205}{369}$ **32.** $\frac{111}{185}$ **33.** $\frac{215}{344}$ **34.** $\frac{279}{310}$

35. Simplify the expression $3 \cdot 8 + 3 \cdot 7 + 3 \cdot 5$, and then factor the result into the product of primes. (Notice one of the factors of the answer is 3.)

36. Simplify the expression $5 \cdot 4 + 5 \cdot 9 + 5 \cdot 3$, and then factor the result into the product of primes.

Write each of the following fractions as an equivalent fraction with denominator 24.

37. $\frac{3}{4}$ **38.** $\frac{5}{6}$ **39.** $\frac{1}{2}$ **40.** $\frac{1}{8}$ **41.** $\frac{5}{8}$ **42.** $\frac{7}{12}$

Write each fraction as an equivalent fraction with denominator 60.

43. $\frac{3}{5}$ **44.** $\frac{5}{12}$ **45.** $\frac{11}{30}$ **46.** $\frac{9}{10}$

Place one of the symbols $<$ or $>$ between each of the following to make the resulting statement true.

47. $-\frac{3}{4} \square -\frac{1}{4}$ **48.** $-\frac{2}{3} \square -\frac{1}{3}$ **49.** $-\frac{3}{2} \square -\frac{3}{4}$ **50.** $-\frac{8}{3} \square -\frac{17}{3}$

Multiply the following.

51. $\frac{2}{3} \cdot \frac{4}{5}$ **52.** $\frac{1}{4} \cdot \frac{3}{5}$ **53.** $\frac{1}{2}(3)$ **54.** $\frac{1}{3}(2)$

55. $\frac{1}{4}(5)$ **56.** $\frac{1}{5}(4)$ **57.** $\frac{4}{3} \cdot \frac{3}{4}$ **58.** $\frac{5}{7} \cdot \frac{7}{5}$

59. $6\left(\frac{1}{6}\right)$ **60.** $8\left(\frac{1}{8}\right)$ **61.** $3 \cdot \frac{1}{3}$ **62.** $4 \cdot \frac{1}{4}$

Expand and multiply.

63. $\left(\frac{3}{4}\right)^2$ **64.** $\left(\frac{5}{6}\right)^2$ **65.** $\left(\frac{2}{3}\right)^3$ **66.** $\left(\frac{1}{2}\right)^3$ **67.** $\left(\frac{1}{10}\right)^4$ **68.** $\left(\frac{1}{10}\right)^5$

Multiply the following.

69. $-\frac{2}{3}\cdot\frac{5}{7}$ **70.** $-\frac{6}{5}\cdot\frac{2}{7}$ **71.** $-8\left(\frac{1}{2}\right)$ **72.** $-12\left(\frac{1}{3}\right)$

73. $-\frac{3}{4}\left(-\frac{4}{3}\right)$ **74.** $-\frac{5}{8}\left(-\frac{8}{5}\right)$ **75.** $\left(-\frac{3}{4}\right)^2$ **76.** $\left(-\frac{2}{5}\right)^2$

77. $-\frac{1}{3}(-3x)$ **78.** $-\frac{1}{5}(-5x)$

Divide and reduce all answers to lowest terms.

79. $\frac{4}{5}\div\frac{3}{4}$ **80.** $\frac{6}{8}\div\frac{3}{4}$ **81.** $-\frac{5}{6}\div\left(-\frac{5}{8}\right)$ **82.** $-\frac{7}{9}\div\left(-\frac{1}{6}\right)$

83. $\frac{10}{13}\div\left(-\frac{5}{4}\right)$ **84.** $\frac{5}{12}\div\left(-\frac{10}{3}\right)$ **85.** $-\frac{5}{6}\div\frac{5}{6}$ **86.** $-\frac{8}{9}\div\frac{8}{9}$

87. $-\frac{3}{4}\div\left(-\frac{3}{4}\right)$ **88.** $-\frac{6}{7}\div\left(-\frac{6}{7}\right)$

89. Simplify each expression.

a. $8\div\frac{4}{5}$ **b.** $8\div\frac{4}{5}-10$ **c.** $(-10)8\div\frac{4}{5}$ **d.** $8\div\left(-\frac{4}{5}\right)-10$

90. Simplify each expression.

a. $10\div\frac{5}{6}$ **b.** $10\div\frac{5}{6}-12$ **c.** $(-12)10\div\frac{5}{6}$ **d.** $10\div\left(-\frac{5}{6}\right)-12$

Find the following sums and differences, and reduce to lowest terms. Assume all variables represent nonzero numbers.

91. $\frac{3}{6}+\frac{1}{6}$ **92.** $\frac{2}{5}+\frac{3}{5}$ **93.** $\frac{3}{8}-\frac{5}{8}$ **94.** $\frac{1}{7}-\frac{6}{7}$

95. $-\frac{1}{4}+\frac{3}{4}$ **96.** $-\frac{4}{9}+\frac{7}{9}$ **97.** $\frac{x}{3}-\frac{1}{3}$ **98.** $\frac{x}{8}-\frac{1}{8}$

99. $\frac{1}{4}+\frac{2}{4}+\frac{3}{4}$ **100.** $\frac{2}{5}+\frac{3}{5}+\frac{4}{5}$ **101.** $\frac{x+7}{2}-\frac{1}{2}$ **102.** $\frac{x+5}{4}-\frac{3}{4}$

103. $\frac{1}{10}-\frac{3}{10}-\frac{4}{10}$ **104.** $\frac{3}{20}-\frac{1}{20}-\frac{4}{20}$ **105.** $\frac{1}{a}+\frac{4}{a}+\frac{5}{a}$ **106.** $\frac{5}{a}+\frac{4}{a}+\frac{3}{a}$

107.

First Number a	Second Number b	The Sum of a and b $a+b$
$\frac{1}{2}$	$\frac{1}{3}$	
$\frac{1}{3}$	$\frac{1}{4}$	
$\frac{1}{4}$	$\frac{1}{5}$	
$\frac{1}{5}$	$\frac{1}{6}$	

108.

First Number a	Second Number b	The Sum of a and b $a+b$
1	$\frac{1}{2}$	
1	$\frac{1}{3}$	
1	$\frac{1}{4}$	
1	$\frac{1}{5}$	

109.

First Number a	Second Number b	The Sum of a and b $a+b$
$\frac{1}{12}$	$\frac{1}{2}$	
$\frac{1}{12}$	$\frac{1}{3}$	
$\frac{1}{12}$	$\frac{1}{4}$	
$\frac{1}{12}$	$\frac{1}{6}$	

110.

First Number a	Second Number b	The Sum of a and b $a+b$
$\frac{1}{8}$	$\frac{1}{2}$	
$\frac{1}{8}$	$\frac{1}{4}$	
$\frac{1}{8}$	$\frac{1}{16}$	
$\frac{1}{8}$	$\frac{1}{24}$	

Find the LCD for each of the following; then use the methods developed in this section to add and subtract as indicated.

111. $\frac{4}{9}+\frac{1}{3}$ **112.** $\frac{1}{2}+\frac{1}{4}$ **113.** $2+\frac{1}{3}$ **114.** $3+\frac{1}{2}$

115. $-\frac{3}{4}+1$ **116.** $-\frac{3}{4}+2$ **117.** $\frac{1}{2}+\frac{2}{3}$ **118.** $\frac{2}{3}+\frac{1}{4}$

119. $\frac{5}{12}-\left(-\frac{3}{8}\right)$ **120.** $\frac{9}{16}-\left(-\frac{7}{12}\right)$ **121.** $-\frac{1}{20}+\frac{8}{30}$ **122.** $-\frac{1}{30}+\frac{9}{40}$

123. $\frac{17}{30}+\frac{11}{42}$ **124.** $\frac{19}{42}+\frac{13}{70}$ **125.** $\frac{25}{84}+\frac{41}{90}$ **126.** $\frac{23}{70}+\frac{29}{84}$

127. $\frac{13}{126}-\frac{13}{180}$ **128.** $\frac{17}{84}-\frac{17}{90}$ **129.** $\frac{3}{4}+\frac{1}{8}+\frac{5}{6}$ **130.** $\frac{3}{8}+\frac{2}{5}+\frac{1}{4}$

131. $\frac{1}{2}+\frac{1}{3}+\frac{1}{4}+\frac{1}{6}$ **132.** $\frac{1}{8}+\frac{1}{4}+\frac{1}{5}+\frac{1}{10}$

133. $1-\frac{5}{2}$ **134.** $1-\frac{5}{3}$ **135.** $1+\frac{1}{2}$ **136.** $1+\frac{2}{3}$

137. Find the sum of $\frac{3}{7}$, 2, and $\frac{1}{9}$. **138.** Find the sum of 6, $\frac{6}{11}$, and 11.

139. Give the difference of $\frac{7}{8}$ and $\frac{1}{4}$. **140.** Give the difference of $\frac{9}{10}$ and $\frac{1}{100}$.

Find the fourth term in each sequence.

141. $\frac{1}{3}, 0, -\frac{1}{3}, \ldots$

142. $\frac{2}{3}, 0, -\frac{2}{3}, \ldots$

143. $\frac{1}{3}, 1, \frac{5}{3}, \ldots$

144. $1, \frac{3}{2}, 2, \ldots$

145. $1, \frac{1}{5}, \frac{1}{25}, \ldots$

146. $1, -\frac{1}{2}, \frac{1}{4}, \ldots$

Find the perimeter of each figure.

147.

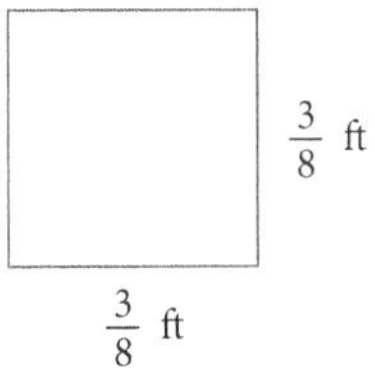

148.

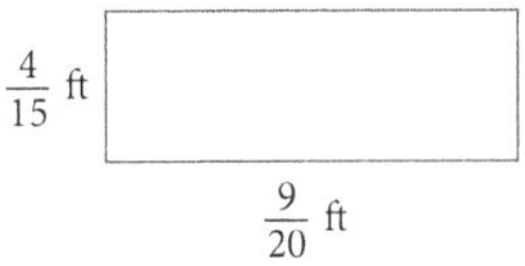

149.

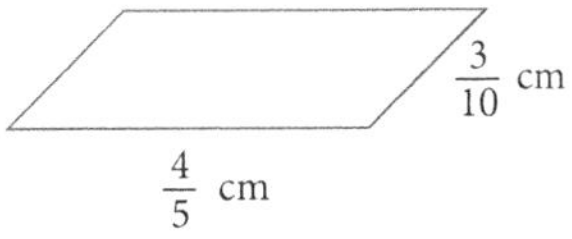

150.

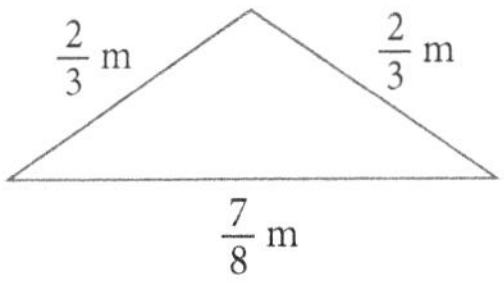

Applying the Concepts

151. Sewing If $\frac{6}{7}$ yard of material is needed to make a blanket, how many blankets can be made from 12 yards of material?

152. Manufacturing A clothing manufacturer is making scarves that require $\frac{3}{8}$ yard for material each. How many can be made from 27 yards of material?

153. Capacity Suppose a bag of candy holds exactly $\frac{1}{4}$ pound of candy. How many of these bags can be filled from 12 pounds of candy?

154. Capacity A certain size bottle holds exactly $\frac{4}{5}$ pint of liquid. How many of these bottles can be filled from a 20-pint container?

155. Cooking A man is making cookies from a recipe that calls for $\frac{3}{4}$ teaspoon of oil. If the only measuring spoon he can find is a $\frac{1}{8}$ teaspoon, how many of these will he have to fill with oil in order to have a total of $\frac{3}{4}$ teaspoon of oil?

156. Cooking A cake recipe calls for $\frac{1}{2}$ cup of sugar. If the only measuring cup available is a $\frac{1}{8}$ cup, how many of these will have to be filled with sugar to make a total of $\frac{1}{2}$ cup of sugar?

157. Cartons of Milk If a small carton of milk holds exactly $\frac{1}{2}$ pint, how many of the $\frac{1}{2}$-pint cartons can be filled from a 14-pint container?

158. Pieces of Pipe How many pieces of pipe that are $\frac{2}{3}$ foot long must be laid together to make a pipe 16 feet long?

159. Capacity One carton of milk contains $\frac{1}{2}$ pint while another contains 4 pints. How much milk is contained in both cartons?

160. Baking A recipe calls for $\frac{2}{3}$ cup of flour and $\frac{3}{4}$ cup of sugar. What is the total amount of flour and sugar called for in the recipe?

161. Budget A family decides that they can spend $\frac{5}{8}$ of their monthly income on house payments. If their monthly income is \$2,120, how much can they spend for house payments?

162. Savings A family saves $\frac{3}{16}$ of their income each month. If their monthly income is \$1,264, how much do they save each month?

Reading a Pie Chart The pie chart below shows how the students at one of the universities in California are distributed among the different schools at the university. Use the information in the pie chart to answer questions 167 and 168.

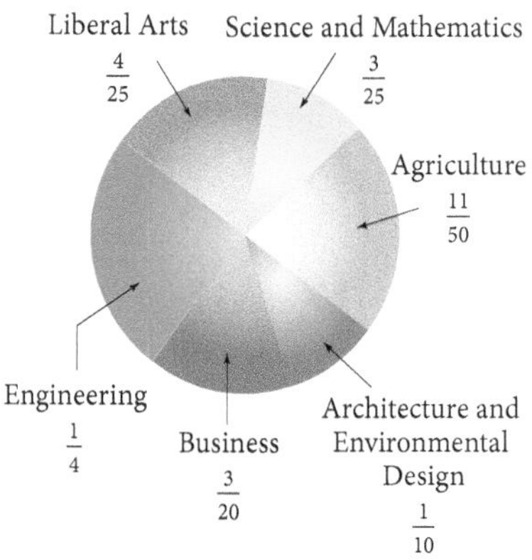

163. If the students in the Schools of Engineering and Business are combined, what fraction results?

164. What fraction of the university's students are enrolled in the Schools of Agriculture, Engineering, and Business combined?

165. Final Exam Grades The table below gives the fraction of students in a class of 40 that received grades of A, B, or C on the final exam. Fill in all the missing parts of the table.

Grade	Number of Students	Fraction of Students
A		$\frac{1}{8}$
B		$\frac{1}{5}$
C		$\frac{1}{2}$
below C		
Total	40	1

166. Flu During a flu epidemic a company with 200 employees has $\frac{1}{10}$ of their employees call in sick on Monday and another $\frac{3}{10}$ call in sick on Tuesday. What is the total number of employees calling in sick during this 2-day period?

167. Subdivision A 6-acre piece of land is subdivided into $\frac{3}{5}$-acre lots. How many lots are there?

168. Cutting Wood A 12-foot piece of wood is cut into shelves. If each shelf is $\frac{3}{4}$ foot in length, how many shelves are there?

Learning Objectives Assessment

The following problems can be used to help assess if you have successfully met the learning objectives for this section.

169. Factor 132 into a product of primes.

a. $2 \cdot 3^2 \cdot 7$ **b.** $2^2 \cdot 3 \cdot 11$ **c.** $2 \cdot 66$ **d.** $2 \cdot 3 \cdot 23$

170. Reduce $\frac{126}{168}$ to lowest terms.

a. $\frac{21}{28}$ **b.** $\frac{63}{84}$ **c.** $\frac{4}{7}$ **d.** $\frac{3}{4}$

171. Find the reciprocal of $-\frac{6}{7}$.

a. $\frac{6}{7}$ **b.** $-\frac{6}{7}$ **c.** $-\frac{7}{6}$ **d.** $\frac{7}{6}$

172. Divide: $\frac{2}{15} \div \left(-\frac{4}{3}\right)$.

a. $-\frac{1}{10}$ **b.** $-\frac{8}{45}$ **c.** $\frac{1}{10}$ **d.** $-\frac{2}{5}$

173. Add: $\frac{2}{5} + \frac{4}{3}$.

a. $\frac{26}{15}$ **b.** $\frac{8}{15}$ **c.** $\frac{3}{4}$ **d.** $\frac{22}{15}$

SPOTLIGHT ON SUCCESS *Napa Valley College*

You may think that all your mathematics instructors started their college math sequence with precalculus or calculus, but that is not always the case. Diane Van Deusen, a full time mathematics instructor at Napa Valley College in Napa, California, started her career in mathematics in the same class you are taking. Here is part of her story from her website:

Dear Student,

Welcome to elementary algebra! Since we will be spending a significant amount of time together this semester, I thought I should introduce myself to you, and tell you how I ended up with a career in education.

I was not encouraged to attend college after high school, and in fact, had no interest in "more school". Consequently, I didn't end up taking a college class until I was 31 years old! Before returning to and while attending college, I worked locally in the restaurant business as a waitress and bartender and in catering. In fact, I sometimes wait tables a few nights a week during my summer breaks.

When I first came back to school, at Napa Valley College (NVC), I thought I might like to enter the nursing program but soon found out nursing was not for me. As I started working on general education requirements, I took elementary algebra and was surprised to learn that I really loved mathematics, even though I had failed 8th grade algebra! As I continued to appreciate and value my own education, I decided to become a teacher so that I could support other people seeking education goals. After earning my AA degree from NVC, I transferred to Sonoma State where I earned my bachelor's degree in mathematics with a concentration in statistics. Finally, I attended Cal State Hayward to earn my master's degree in applied statistics. It took me ten years in all to do this.

I feel that having been a returning student while a single, working parent, also an EOPS and Financial Aid recipient, I fully understand the complexity of the life of a community college student. If at any time you have questions about the college, the class or just need someone to talk to, my door is open.

I sincerely hope that my classroom will provide a positive and satisfying learning experience for you.

Diane Van Deusen

Elementary algebra is a great place to start your journey into college mathematics. You can start here and go as far as you want in mathematics. Who knows, you may end up teaching mathematics one day, just like Diane Van Deusen.

Properties of Real Numbers 1.7

Learning Objectives

In this section, we will learn how to:

1. Rewrite expressions using the properties of real numbers.
2. Simplify expressions using the associative properties of addition and multiplication.
3. Simplify expressions using the distributive property.

Introduction

In this section we will list all the facts (properties) that you know from past experience are true about numbers in general. We will give each property a name so we can refer to it later in this book. Mathematics is very much like a game. The game involves numbers. The rules of the game are the properties and rules we are developing in this chapter. The goal of the game is to extend the basic rules to as many situations as possible.

You know from past experience with numbers that it makes no difference in which order you add two numbers; that is, $3 + 5$ is the same as $5 + 3$. This fact about numbers is called the *commutative property of addition*. We say addition is a commutative operation. Changing the order of the numbers does not change the answer.

There is one other basic operation that is commutative. Because $3(5)$ is the same as $5(3)$, we say multiplication is a commutative operation. Changing the order of the two numbers you are multiplying does not change the answer.

For all properties listed in this section, a, b, and c represent real numbers.

PROPERTY *Commutative Property of Addition*

In symbols: $a + b = b + a$

In words: Changing the ***order*** of the numbers in a sum will not change the result.

PROPERTY *Commutative Property of Multiplication*

In symbols: $a \cdot b = b \cdot a$

In words: Changing the ***order*** of the numbers in a product will not change the result.

VIDEO EXAMPLES

SECTION 1.7

For example, the statement $5 + 8 = 8 + 5$ is an example of the commutative property of addition and the statement $2 \cdot y = y \cdot 2$ is an example of the commutative property of multiplication.

Note: At this point, some students are confused by the expression $x + 8$; they feel that there is more to do, but they don't know what. At this point, there isn't any more that can be done with $x + 8$ unless we know what x is. So $x + 8$ is as far as we can go with this problem.

EXAMPLE 1 Simplify: $5 + x + 3$.

SOLUTION The expression $5 + x + 3$ can be simplified using the commutative property of addition:

$$5 + x + 3 = x + 5 + 3 \quad \text{Commutative property of addition}$$
$$= x + 8 \quad \text{Addition}$$

The other two basic operations, subtraction and division, are not commutative. The order in which we subtract or divide two numbers makes a difference in the answer.

Another property of numbers that you have used many times has to do with grouping. You know that when we add three numbers it makes no difference which two we add first. When adding $3 + 5 + 7$, we can add the 3 and 5 first and then the 7, or we can add the 5 and 7 first and then the 3. Mathematically, it looks like this: $(3 + 5) + 7 = 3 + (5 + 7)$. This property is true of multiplication as well. Operations that behave in this manner are called *associative* operations. The answer will not change when we change the association (or grouping) of the numbers.

PROPERTY *Associative Property of Addition*

In symbols: $a + (b + c) = (a + b) + c$

In words: Changing the ***grouping*** of the numbers in a sum will not change the result.

Note: Subtraction and division are not associative operations. Using the numbers 24, 12, 2, can you show why they are not?

PROPERTY *Associative Property of Multiplication*

In symbols: $a(bc) = (ab)c$

In words: Changing the ***grouping*** of the numbers in a product will not change the result.

The following examples illustrate how the associative properties can be used to simplify expressions that involve both real numbers and variables.

EXAMPLE 2 Simplify.

a. $4 + (5 + x)$ **b.** $-5(2x)$ **c.** $6(-5y)$

SOLUTION

a. $4 + (5 + x) = (4 + 5) + x$ Associative property of addition

$= 9 + x$ Addition

b. $-5(2x) = (-5 \cdot 2)x$ Associative property of multiplication

$= -10x$ Multiplication

c. $6(-5y) = [6(-5)]y$ Associative property

$= -30y$ Multiplication

EXAMPLE 3 Simplify.

a. $\frac{1}{5}(5x)$ **b.** $-2\left(-\frac{1}{2}x\right)$ **c.** $12\left(\frac{2}{3}x\right)$

SOLUTION

a. $\frac{1}{5}(5x) = \left(\frac{1}{5} \cdot 5\right)x$ Associative property of multiplication

$= 1x$ Multiplication

$= x$

b. $-2\left(-\frac{1}{2}x\right) = \left[(-2)\left(-\frac{1}{2}\right)\right]x$ Associative property of multiplication

$= 1x$ Multiplication

$= x$

c. $12\left(\frac{2}{3}x\right) = \left(12 \cdot \frac{2}{3}\right)x$ Associative property of multiplication

$= 8x$ Multiplication

The associative and commutative properties apply to problems that are either all multiplication or all addition. There is a third basic property that involves both addition and multiplication. It is called the *distributive property* and looks like this.

PROPERTY *Distributive Property*

In symbols: $a(b + c) = ab + ac$

In words: Multiplication ***distributes*** over addition.

Note: Because subtraction is defined in terms of addition, it is also true that the distributive property applies to subtraction as well as addition; that is, $a(b - c) = ab - ac$ for any three real numbers a, b, and c.

You will see as we progress through the book that the distributive property is used very frequently in algebra. We can give a visual justification to the distributive property by finding the areas of rectangles. Figure 1 shows a large rectangle that is made up of two smaller rectangles. We can find the area of the large rectangle two different ways.

Method 1

We can calculate the area of the large rectangle directly by finding its length and width. The width is 5 inches, and the length is (3 + 4) inches.

$$\begin{aligned}\text{Area of large rectangle} &= 5(3 + 4)\\ &= 5(7)\\ &= 35 \text{ square inches}\end{aligned}$$

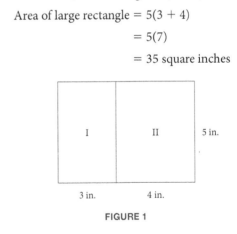

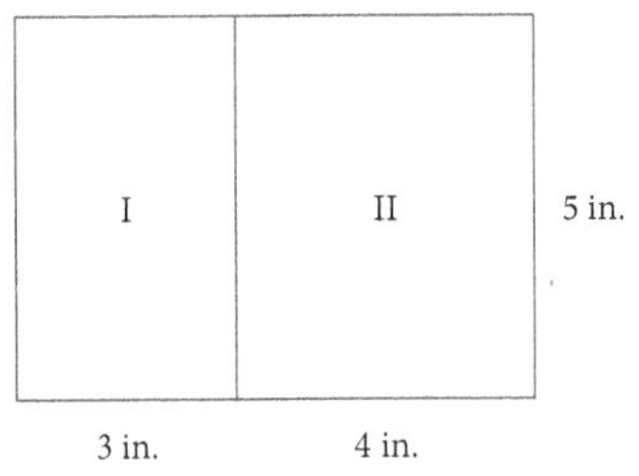

FIGURE 1

Method 2

Because the area of the large rectangle is the sum of the areas of the two smaller rectangles, we find the area of each small rectangle and then add to find the area of the large rectangle.

$$\begin{aligned}\text{Area of large rectangle} &= \text{Area of rectangle I} + \text{Area of rectangle II}\\ &= 5(3) + 5(4)\\ &= 15 + 20\\ &= 35 \text{ square inches}\end{aligned}$$

In both cases the result is 35 square inches. Because the results are the same, the two original expressions must be equal. Stated mathematically, $5(3 + 4) = 5(3) + 5(4)$. We can either add the 3 and 4 first and then multiply that sum by 5, or we can multiply the 3 and the 4 separately by 5 and then add the products. In either case we get the same answer.

Here are some examples that illustrate how we use the distributive property.

EXAMPLE 4 Apply the distributive property to each expression, and then simplify the result.

a. $2(x + 3)$ **b.** $5(2x - 8)$ **c.** $-5(2x + 4y)$ **d.** $-4(3x - 5) - 8$

SOLUTION

a. $2(x + 3) = 2(x) + 2(3)$ Distributive property

$= 2x + 6$ Multiplication

b. $5(2x - 8) = 5(2x) - 5(8)$ Distributive property

$= 10x - 40$ Multiplication

Notice in part *b* that multiplication distributes over subtraction as well as addition.

c. $-5(2x + 4y) = -5(2x) + (-5)(4y)$ Distributive property

$= -10x + (-20y)$ Multiplication

$= -10x - 20y$

d. $-4(3x - 5) - 8 = -4(3x) - (-4)(5) - 8$ Distributive property

$= -12x - (-20) - 8$ Multiplication

$= -12x + 20 - 8$ Definition of subtraction

$= -12x + 12$ Subtraction

Next we have some expressions to simplify that involve fractions.

EXAMPLE 5 Apply the distributive property to each expression, and then simplify the result.

a. $-\frac{1}{3}(2x - 6)$ **b.** $3\left(\frac{1}{3}x + 5\right)$ **c.** $a\left(1 + \frac{1}{a}\right)$ **d.** $12\left(\frac{2}{3}x + \frac{1}{2}y\right)$

SOLUTION

a. $-\frac{1}{3}(2x - 6) = -\frac{1}{3}(2x) - \left(-\frac{1}{3}\right)(6)$ Distributive property

$= -\frac{2}{3}x - (-2)$ Multiplication

$= -\frac{2}{3}x + 2$

b. $3\left(\frac{1}{3}x + 5\right) = 3 \cdot \frac{1}{3}x + 3 \cdot 5$ Distributive property

$= x + 15$ Multiplication

c. $a\left(1 + \frac{1}{a}\right) = a \cdot 1 + a \cdot \frac{1}{a}$ Distributive property

$= a + 1$ Multiplication

d. $12\left(\frac{2}{3}x + \frac{1}{2}y\right) = 12 \cdot \frac{2}{3}x + 12 \cdot \frac{1}{2}y$ Distributive property

$= 8x + 6y$ Multiplication

In the next example, we use the rule for division, along with properties of real numbers, to simplify expressions.

EXAMPLE 6 Simplify each expression.

a. $10\left(\frac{x}{2}\right)$ **b.** $a\left(\frac{3}{a} - 4\right)$

SOLUTION

a. $10\left(\frac{x}{2}\right) = 10\left(\frac{1}{2}x\right)$ Rule for division

$= \left(10 \cdot \frac{1}{2}\right)x$ Associative property of multiplication

$= 5x$ Multiplication

b. $a\left(\frac{3}{a} - 4\right) = a \cdot \frac{3}{a} - a \cdot 4$ Distributive property

$= 3 - 4a$ Multiplication

If you recall from the previous section, to add two fractions with the same denominator, we add their numerators and put the result over the common denominator:

$$\frac{3}{4} + \frac{2}{4} = \frac{3+2}{4} = \frac{5}{4}$$

We can now justify why this process works. The reason we add numerators but do not add denominators is that we must follow the distributive property. To see this, you first have to recall that $\frac{3}{4}$ can be written as $3 \cdot \frac{1}{4}$, and $\frac{2}{4}$ can be written as $2 \cdot \frac{1}{4}$ (dividing by 4 is equivalent to multiplying by $\frac{1}{4}$). Here is the addition problem again, this time showing the use of the distributive property:

$$\frac{3}{4} + \frac{2}{4} = 3 \cdot \frac{1}{4} + 2 \cdot \frac{1}{4}$$

$$= (3 + 2) \cdot \frac{1}{4} \quad \text{Distributive property}$$

$$= 5 \cdot \frac{1}{4}$$

$$= \frac{5}{4}$$

What we have here is the sum of the numerators placed over the *common denominator.*

Note: Most people who have done any work with adding fractions know that you add fractions that have the same denominator by adding their numerators but not their denominators. However, most people don't know why this works. The reason why we add numerators but not denominators is because of the distributive property. That is what the discussion at the right is all about. If you really want to understand addition of fractions, pay close attention to this discussion.

Special Numbers

In addition to the three properties mentioned so far, we want to include in our list two special numbers that have unique properties. They are the numbers zero and one.

PROPERTY *Additive Identity Property*

There exists a unique number 0 such that

In symbols: $a + 0 = a$ and $0 + a = a$

PROPERTY *Multiplicative Identity Property*

There exists a unique number 1 such that
In symbols: $a(1) = a$ and $(1)a = a$

PROPERTY *Additive Inverse Property*

In symbols: $a + (-a) = 0$
In words: Opposites add to 0.

PROPERTY *Multiplicative Inverse Property*

For every real number a, except 0, there exists a unique real number $\frac{1}{a}$ such that

In symbols: $a\left(\frac{1}{a}\right) = 1$

In words: Reciprocals multiply to 1.

Of all the basic properties listed, the commutative, associative, and distributive properties are the ones we will use most often. They are important because they will be used as justifications or reasons for many of the things we will do.

The following example illustrates how we use the preceding properties. Each sub-example contains an algebraic expression that has been changed in some way. The property that justifies the change is written to the right.

EXAMPLE 7 State the property that justifies the given statement.

a. $x + 5 = 5 + x$	Commutative property of addition
b. $(2 + x) + y = 2 + (x + y)$	Associative property of addition
c. $6(x + 3) = 6x + 18$	Distributive property
d. $2 + (-2) = 0$	Additive inverse property
e. $3\left(\frac{1}{3}\right) = 1$	Multiplicative inverse property
f. $(2 + 0) + 3 = 2 + 3$	Additive identity property
g. $(2 + 3) + 4 = 3 + (2 + 4)$	Commutative and associative properties of addition
h. $(x + 2) + y = (x + y) + 2$	Commutative and associative properties of addition

As a final note on the properties of real numbers, we should mention that although some of the properties are stated for only two or three real numbers, they hold for as many numbers as needed. For example, the distributive property holds for expressions like $3(x + y + z + 5 + 2)$; that is,

$$3(x + y + z + 5 + 2) = 3x + 3y + 3z + 15 + 6$$

It is not important how many numbers are contained in the sum, only that it is a sum. Multiplication, you see, distributes over addition, whether there are two numbers in the sum or 200.

Getting Ready for Class

After reading through the preceding section, respond in your own words and in complete sentences.

A. What is the commutative property of addition?

B. Do you know from your experience with numbers that the commutative property of addition is true? Explain why.

C. Write the commutative property of multiplication in symbols and words.

D. How do you rewrite expressions using the distributive property?

Problem Set 1.7

State the property or properties that justify the following.

1. $3 + 2 = 2 + 3$
2. $5 + 0 = 5$
3. $4\left(\frac{1}{4}\right) = 1$
4. $10(0.1) = 1$
5. $4 + x = x + 4$
6. $3(x - 10) = 3x - 30$
7. $2(y + 8) = 2y + 16$
8. $3 + (4 + 5) = (3 + 4) + 5$
9. $(3 + 1) + 2 = 1 + (3 + 2)$
10. $(5 + 2) + 9 = (2 + 5) + 9$
11. $(8 + 9) + 10 = (8 + 10) + 9$
12. $(7 + 6) + 5 = (5 + 6) + 7$
13. $3(x + 2) = 3(2 + x)$
14. $2(7y) = (7 \cdot 2)y$
15. $x(3y) = 3(xy)$
16. $a(5b) = 5(ab)$
17. $4(xy) = 4(yx)$
18. $3[2 + (-2)] = 3(0)$
19. $8[7 + (-7)] = 8(0)$
20. $7(1) = 7$

Each of the following problems has a mistake in it. Correct the right-hand side.

21. $3(x + 2) = 3x + 2$
22. $5(4 + x) = 4 + 5x$
23. $9(a + b) = 9a + b$
24. $2(y + 1) = 2y + 1$
25. $3(0) = 3$
26. $5\left(\frac{1}{5}\right) = 5$
27. $3 + (-3) = 1$
28. $8(0) = 8$
29. $10(1) = 0$
30. $3 \cdot \frac{1}{3} = 0$

Use the associative property to rewrite each of the following expressions, and then simplify the result. (See Examples 2 and 3.)

31. $4 + (2 + x)$
32. $5 + (6 + x)$
33. $(x + 2) + 7$
34. $(x + 8) + 2$
35. $3(5x)$
36. $5(3x)$
37. $-9(6y)$
38. $-6(9y)$
39. $\frac{1}{2}(3a)$
40. $\frac{1}{3}(2a)$
41. $-\frac{1}{3}(3x)$
42. $-\frac{1}{4}(4x)$
43. $\frac{1}{2}(2y)$
44. $\frac{1}{7}(7y)$
45. $-\frac{3}{4}\left(\frac{4}{3}x\right)$
46. $-\frac{3}{2}\left(\frac{2}{3}x\right)$
47. $-\frac{6}{5}\left(-\frac{5}{6}a\right)$
48. $-\frac{2}{5}\left(-\frac{5}{2}a\right)$

Apply the distributive property to each of the following expressions. Simplify when possible.

49. $8(x + 2)$
50. $5(x + 3)$
51. $8(x - 2)$
52. $5(x - 3)$
53. $4(y + 1)$
54. $4(y - 1)$
55. $3(6x + 5)$
56. $3(5x + 6)$
57. $-2(3a + 7)$
58. $-5(3a + 2)$
59. $-9(6y - 8)$
60. $-2(7y - 4)$
61. $\frac{1}{3}(3x + 6)$
62. $\frac{1}{2}(2x + 4)$
63. $6(2x + 3y)$
64. $8(3x + 2y)$
65. $4(3a - 2b)$
66. $5(4a - 8b)$
67. $\frac{1}{2}(6x + 4y)$
68. $\frac{1}{3}(6x + 9y)$
69. $-4(a + 2)$
70. $-7(a + 6)$
71. $-\frac{1}{2}(3x - 6)$
72. $-\frac{1}{4}(2x - 4)$

73. $10\left(\frac{x}{2} + \frac{3}{5}\right)$ **74.** $6\left(\frac{x}{3} + \frac{5}{2}\right)$ **75.** $15\left(\frac{x}{5} - \frac{4}{3}\right)$ **76.** $6\left(\frac{x}{3} - \frac{1}{2}\right)$

77. $x\left(\frac{3}{x} + 1\right)$ **78.** $x\left(\frac{4}{x} + 3\right)$ **79.** $-21\left(\frac{x}{7} - \frac{y}{3}\right)$ **80.** $-36\left(\frac{x}{4} - \frac{y}{9}\right)$

81. $a\left(\frac{3}{a} - \frac{2}{a}\right)$ **82.** $a\left(\frac{7}{a} + \frac{1}{a}\right)$ **83.** $4(a + 4) + 9$ **84.** $6(a + 2) + 8$

85. $2(3x + 5) + 2$ **86.** $7(2x + 1) + 3$

87. $7(2x + 4) + 10$ **88.** $3(5x + 6) + 20$

89. $-3(2x - 5) - 7$ **90.** $-4(3x - 1) - 8$

91. $-5(3x + 4) - 10$ **92.** $-3(4x + 5) - 20$

Here are some problems you will see later in the book. Simplify.

93. $\left(\frac{1}{2} \cdot 18\right)^2$ **94.** $\left[\frac{1}{2}(-10)\right]^2$ **95.** $\left(\frac{1}{2} \cdot 3\right)^2$

96. $\left(\frac{1}{2} \cdot 5\right)^2$ **97.** $\frac{1}{2}(4x + 2)$ **98.** $\frac{1}{3}(6x + 3)$

99. $\frac{3}{4}(8x - 4)$ **100.** $\frac{2}{5}(5x + 10)$ **101.** $\frac{5}{6}(6x + 12)$

102. $\frac{2}{3}(9x - 3)$ **103.** $10\left(\frac{3}{5}x + \frac{1}{2}\right)$ **104.** $8\left(\frac{1}{4}x - \frac{5}{8}\right)$

105. $15\left(\frac{1}{3}x + \frac{2}{5}\right)$ **106.** $12\left(\frac{1}{12}m + \frac{1}{6}\right)$ **107.** $12\left(\frac{1}{2}m - \frac{5}{12}\right)$

108. $8\left(\frac{1}{8} + \frac{1}{2}m\right)$ **109.** $21\left(\frac{1}{3} + \frac{1}{7}x\right)$ **110.** $6\left(\frac{3}{2}y + \frac{1}{3}\right)$

111. $6\left(\frac{1}{2}x - \frac{1}{3}y\right)$ **112.** $12\left(\frac{1}{4}x + \frac{2}{3}y\right)$ **113.** $-\frac{1}{3}(-2x + 6)$

114. $-\frac{1}{2}(-2x + 6)$ **115.** $8\left(-\frac{1}{4}x + \frac{1}{8}y\right)$ **116.** $9\left(-\frac{1}{9}x + \frac{1}{3}y\right)$

117. $0.09(x + 2{,}000)$ **118.** $0.04(x + 7{,}000)$ **119.** $0.12(x + 500)$

120. $0.06(x + 800)$ **121.** $a\left(1 + \frac{1}{a}\right)$ **122.** $a\left(1 - \frac{1}{a}\right)$

123. $a\left(\frac{1}{a} - 1\right)$ **124.** $a\left(\frac{1}{a} + 1\right)$

Applying the Concepts

125. Getting Dressed While getting dressed for work, a man puts on his socks and puts on his shoes. Are the two statements "put on your socks" and "put on your shoes" commutative? That is, will changing the order of the events always produce the same result?

126. Skydiving A skydiver flying over the jump area is about to do two things: jump out of the plane and pull the rip cord. Are the two events "jump out of the plane" and "pull the rip cord" commutative?

127. Division Give an example that shows that division is not a commutative operation; that is, find two numbers for which changing the order of division gives two different answers.

128. Subtraction Simplify the expression $10 - (5 - 2)$ and the expression $(10 - 5) - 2$ to show that subtraction is not an associative operation.

129. Hours Worked Carlo works as a waiter. He works double shifts 4 days a week. The lunch shift is 2 hours and the dinner shift is 3 hours. Find the total number of hours he works per week using the numbers 2, 3, and 4. Do the calculation two different ways so that the results give further justification for the distributive property.

Learning Objectives Assessment

The following problems can be used to help assess if you have successfully met the learning objectives for this section.

130. Which property of real numbers justifies $(2 + x) + 5 = 5 + (2 + x)$?

a. Associative property of addition **b.** Commutative property of addition
c. Distributive property **d.** Additive inverse property

131. Use the associative property to simplify $-4(8x)$.

a. $-32x$ **b.** $4x$ **c.** $-32 - 4x$ **d.** $8x - 4$

132. Simplify $-3(2x - 5)$ using the distributive property.

a. $-6x - 5$ **b.** $-6x + 15$ **c.** $-6x - 15$ **d.** $2x - 8$

Chapter 1 Summary

Note: We will use the margins in the chapter summaries to give examples that correspond to the topic being reviewed whenever it is appropriate.

The number(s) in brackets next to each heading indicates the section(s) in which that topic is discussed.

Exponents [1.1]

Exponents are notation used to indicate repeated multiplication. In the expression 3^4, 3 is the *base* and 4 is the *exponent.*

$$3^4 = 3 \cdot 3 \cdot 3 \cdot 3 = 81$$

EXAMPLES

1. $2^5 = 2 \cdot 2 \cdot 2 \cdot 2 \cdot 2 = 32$
$5^2 = 5 \cdot 5 = 25$
$10^3 = 10 \cdot 10 \cdot 10 = 1{,}000$
$1^4 = 1 \cdot 1 \cdot 1 \cdot 1 = 1$

Order of Operations [1.1]

When evaluating a mathematical expression, we will perform the operations in the following order, beginning with the expression in the innermost parentheses or brackets and working our way out.

1. Simplify all numbers with exponents, working from left to right if more than one of these numbers is present.
2. Then do all multiplications and divisions left to right.
3. Finally, perform all additions and subtractions left to right.

2. $10 + (2 \cdot 3^2 - 4 \cdot 2)$
$= 10 + (2 \cdot 9 - 4 \cdot 2)$
$= 10 + (18 - 8)$
$= 10 + 10$
$= 20$

Comparison Symbols [1.2]

$a = b$	a is equal to b.
$a \neq b$	a is not equal to b.
$a < b$	a is less than b.
$a \not< b$	a is not less than b.
$a > b$	a is greater than b.
$a \not> b$	a is not greater than b.
$a \geq b$	a is greater than or equal to b.
$a \leq b$	a is less than or equal to b.

Subsets of the Real Numbers [1.2]

Counting numbers:	$\{1, 2, 3, \ldots\}$
Whole numbers:	$\{0, 1, 2, 3, \ldots\}$
Integers:	$\{\ldots, -3, -2, -1, 0, 1, 2, 3, \ldots\}$
Rational numbers:	{all numbers that can be expressed as the ratio of two integers}
Irrational numbers:	{all numbers on the number line that cannot be expressed as the ratio of two integers}
Real numbers:	{all numbers that are either rational or irrational}

3. a. 7 and 100 are counting numbers, but 0 and -2 are not.

b. 0 and 241 are whole numbers, but -4 and $\frac{1}{2}$ are not.

c. -15, 0, and 20 are integers.

d. -4, $-\frac{1}{2}$, 0.75, and 0.666 . . . are rational numbers.

e. $-\pi$, $\sqrt{3}$, and π are irrational numbers.

f. All the numbers listed above are real numbers.

Absolute Value [1.2]

4. $|5| = 5$
$|-5| = 5$

The *absolute value* of a real number is its distance from zero on the real number line. Absolute value is never negative.

Opposites [1.2]

5. The numbers 3 and -3 are opposites; their sum is 0:
$3 + (-3) = 0$

Any two real numbers the same distance from zero on the number line but in opposite directions from zero are called *opposites*. Opposites always add to zero.

Addition of Real Numbers [1.3]

6. Add all combinations of positive and negative 10 and 13.

$$10 + 13 = 23$$
$$10 + (-13) = -3$$
$$-10 + 13 = 3$$
$$-10 + (-13) = -23$$

To add two real numbers with

1. The same sign: Simply add their absolute values and use the common sign.
2. Different signs: Subtract the smaller absolute value from the larger absolute value. The answer has the same sign as the number with the larger absolute value.

Subtraction of Real Numbers [1.4]

7. Subtracting 2 is the same as adding -2:
$7 - 2 = 7 + (-2) = 5$

To subtract one number from another, simply add the opposite of the number you are subtracting; that is, if a and b represent real numbers, then

$$a - b = a + (-b)$$

Multiplication of Real Numbers [1.5]

8.

$$3(5) = 15$$
$$3(-5) = -15$$
$$-3(5) = -15$$
$$-3(-5) = 15$$

To multiply two real numbers, simply multiply their absolute values. Like signs give a positive answer. Unlike signs give a negative answer.

Division of Real Numbers [1.5]

9. $-\frac{6}{2} = -3$
$\frac{-6}{-2} = 3$

To divide two real numbers, simply divide their absolute values. Like signs give a positive answer. Unlike signs give a negative answer.

Factoring [1.6]

10. The number 150 can be factored into the product of prime numbers:

$$150 = 15 \cdot 10$$
$$= (3 \cdot 5)(2 \cdot 5)$$
$$= 2 \cdot 3 \cdot 5^2$$

Factoring is the reverse of multiplication.

Multiplication

$$\text{Factors} \longrightarrow 3 \cdot 5 = 15 \longleftarrow \text{Product}$$

Factoring

Least Common Denominator (LCD) [1.6]

11. The LCD for $\frac{5}{12}$ and $\frac{7}{18}$ is 36.

The *least common denominator* (LCD) for a set of denominators is the smallest number that is exactly divisible by each denominator.

Reciprocals [1.6]

12. The numbers 2 and $\frac{1}{2}$ are reciprocals; their product is 1:

$$2\left(\frac{1}{2}\right) = 1$$

Any two real numbers whose product is 1 are called *reciprocals*. Every real number has a reciprocal except 0.

Addition and Subtraction of Fractions [1.6]

13. $$\frac{5}{12} + \frac{7}{18} = \frac{5}{12} \cdot \frac{3}{3} + \frac{7}{18} \cdot \frac{2}{2} = \frac{15}{36} + \frac{14}{36} = \frac{29}{36}$$

To add (or subtract) two fractions with a common denominator, add (or subtract) numerators and use the common denominator.

$$\frac{a}{c} + \frac{b}{c} = \frac{a+b}{c} \quad \text{and} \quad \frac{a}{c} - \frac{b}{c} = \frac{a-b}{c}$$

Properties of Real Numbers [1.7]

	For Addition	*For Multiplication*
Commutative:	$a + b = b + a$	$a \cdot b = b \cdot a$
Associative:	$a + (b + c) = (a + b) + c$	$a \cdot (b \cdot c) = (a \cdot b) \cdot c$
Identity:	$a + 0 = a$	$a \cdot 1 = a$
Inverse:	$a + (-a) = 0$	$a\left(\frac{1}{a}\right) = 1$
Distributive:	$a(b + c) = ab + ac$	

COMMON MISTAKE

1. Interpreting absolute value as changing the sign of the number inside the absolute value symbols. $|-5| = +5, |+5| = -5$. (The first expression is correct; the second one is not.) To avoid this mistake, remember: Absolute value is a distance and distance is always measured in positive units.
2. Using the phrase "two negatives make a positive." This works only with multiplication and division. With addition, two negative numbers produce a negative answer. It is best not to use the phrase "two negatives make a positive" at all.

Chapter 1 Test

Evaluate each expression. [1.1]

1. 12^2

2. 4^3

Simplify using the order of operations. [1.1, 1.3, 1.4, 1.5]

3. $10 + 2(7 - 3) - 4^2$

4. $15 + 24 \div 6 - 3^2$

From the set of numbers $\{-3, -\frac{1}{2}, 2, \sqrt{5}, \pi\}$ list all the elements that are in the following sets. [1.2]

5. Integers

6. Rational numbers

Write an expression in symbols that is equivalent to each English phrase, and then simplify it.

7. The sum of 6 and -9 [1.3]

8. The difference of -5 and -12 [1.4]

9. The product of 6 and -7 [1.5]

10. The quotient of 32 and -8 [1.5]

Find the next number in each sequence. [1.3, 1.5]

11. $-3, 1, 5, 9, \ldots$

12. $81, -27, 9, -3, \ldots$

Simplify the following: [1.5]

13. $-2(3) - 7$

14. $2(3)^3 - 4(-2)^4$

15. $9 + 4(2 - 6)$

16. $5 - 3[-2(1 + 4) + 3(-3)]$

17. $\dfrac{-4(3) + 5(-2)}{-5 - 6}$

18. $\dfrac{4(3 - 5) - 2(-6 + 8)}{4(-2) + 10}$

Factor into the product of primes. [1.6]

19. 660

20. 4,725

Combine. [1.6]

21. $\dfrac{5}{24} + \dfrac{9}{36}$

22. $\dfrac{5}{y} + \dfrac{6}{y}$

Match each expression below with the letter of the property that justifies it. [1.7]

23. $4(2y) = (4 \cdot 2)y$

24. $5(x - 3) = 5x - 15$

25. $4 + x = x + 4$

26. $(a + 5) - 2 = a + (5 - 2)$

a. Commutative property of addition
b. Commutative property of multiplication
c. Associative property of addition
d. Associative property of multiplication
e. Distributive property

Apply the associative property, and then simplify. [1.7]

27. $5 + (7 + 3x)$

28. $3(-5y)$

Multiply by applying the distributive property. [1.7]

29. $-5(2x - 3)$

30. $\dfrac{1}{3}(6x + 12)$

2

Linear Equations and Inequalities

Chapter Outline

2.1 Simplifying Expressions

2.2 Addition Property of Equality

2.3 Multiplication Property of Equality

2.4 Solving Linear Equations

2.5 Formulas and Percents

2.6 Applications

2.7 More Applications

2.8 Linear Inequalities

One year, I flew to Europe for vacation. From time to time, the video screens on the plane displayed statistics about the flight. At one point during the flight, the temperature outside the plane was −60°F. When I returned home, I did some research and found that the relationship between temperature T and altitude A can be described with the formula

$$T = -0.0035A + 70$$

when the temperature on the ground is 70°F.

To find the temperature at an altitude of 20,000 feet, we can evaluate the expression when $A = 20{,}000$:

When	$A = 20{,}000$
the formula	$T = -0.0035A + 70$
becomes	$T = -0.0035(20{,}000) + 70$
	$= -70 + 70$
	$= 0$

At an altitude of 20,000 feet, the temperature is 0°F.

Given the temperature, we can find the altitude by solving the equation for A. For example, if the temperature is −60°F, we can solve

$$-60 = -0.0035A + 70$$

to obtain $A \approx 37{,}143$ feet.

In this chapter, we will learn how to evaluate formulas and solve equations similar to the one above.

Success Skills

If you have successfully completed Chapter 1, then you have made a good start at developing the study skills necessary to succeed in all math classes. Here is the list of study skills for this chapter.

1. **Imitate Success** Your work should look like the work you see in this book and the work your instructor shows. The steps shown in solving problems in this book were written by someone who has been successful in mathematics. The same is true of your instructor. Your work should imitate the work of people who have been successful in mathematics.

2. **List Difficult Problems** Begin to make lists of problems that give you the most difficulty. These are problems in which you are repeatedly making mistakes.

3. **Begin to Develop Confidence with Word Problems** It seems that the major difference between those people who are good at working word problems and those who are not is confidence. The people with confidence know that no matter how long it takes them, they eventually will be able to solve the problem. Those without confidence begin by saying to themselves, "I'll never be able to work this problem." Are you like that? If you are, what you need to do is put your old ideas about you and word problems aside for a while and make a decision to be successful. Sometimes that's all it takes. Instead of telling yourself that you can't do word problems, that you don't like them, or that they're not good for anything anyway, decide to do whatever it takes to master them.

Simplifying Expressions 2.1

Learning Objectives

In this section, we will learn how to:

1. Combine similar terms.
2. Simplify expressions containing parentheses.
3. Find the value of an expression.
4. Find the first few terms of a sequence.

Introduction

If a cellular phone company charges \$35 per month plus \$0.25 for each minute, or fraction of a minute, that you use one of their cellular phones, then the amount of your monthly bill is given by the expression $35 + 0.25t$. To find the amount you will pay for using that phone 30 minutes in one month, you substitute 30 for t and simplify the resulting expression. This process is one of the topics we will study in this section.

The expression $35 + 0.25t$ contains two terms. For our immediate purposes, a *term* is a number or a number and one or more variables multiplied together. For example, the number 5 is a term, as are the expressions $3x$, $-7y$, and $15xy$. The numerical factor in a term is called the *coefficient*. Here are some examples of terms and their coefficients.

Term	Coefficient	
$3x$	3	
$-7xy + 4$	-7	
$\frac{x}{2}$	$\frac{1}{2}$	Because $\frac{x}{2} = \frac{1}{2} \cdot x$
x^2	1	Because $x^2 = 1 \cdot x^2$
9	9	

If the term is a number only and does not contain any variables, then it is called a *constant term*, and the coefficient is the number itself. For instance, in the above examples, 9 is a constant term.

As you will see in the next few sections, the first step in solving an equation is to simplify both sides as much as possible. In the first part of this section, we will practice simplifying expressions by combining what are called *similar* (or like) terms.

DEFINITION *similar terms*

Two or more terms with the same variable part are called ***similar (or like) terms.***

The terms $3x$ and $4x$ are similar because their variable parts are identical. Likewise, the terms $18y$, $-10y$, and $6y$ are similar terms. Here are some more examples of like terms and unlike terms.

Like Terms	Unlike Terms
$x, -\frac{1}{2}x$	$x, -\frac{2}{x}$
$5x^2, -8x^2$	$5x, -8x^2$
$3xy, 11xy$	$3x^2y, 11xy^2$
$-\frac{1}{4}, \frac{5}{6}$	$-\frac{x}{4}, \frac{y}{6}$

Notice that similar terms can only differ in their coefficients.

To simplify an algebraic expression, we simply reduce the number of terms in the expression. We accomplish this by applying the distributive property along with our knowledge of addition and subtraction of positive and negative real numbers. The following examples illustrate the procedure.

VIDEO EXAMPLES

SECTION 2.1

EXAMPLE 1 Simplify by combining similar terms.

a. $3x + 4x$ **b.** $7a - 10a$ **c.** $18y - 10y + 6y$

SOLUTION We combine similar terms by applying the distributive property.

a. $3x + 4x = (3 + 4)x$ Distributive property

$= 7x$ Add 3 and 4

b. $7a - 10a = (7 - 10)a$ Distributive property

$= -3a$ Add 7 and −10

c. $18y - 10y + 6y = (18 - 10 + 6)y$ Distributive property

$= 14y$ Add 18, −10, and 6

When the expression we intend to simplify is more complicated, we use the commutative and associative properties first.

EXAMPLE 2 Simplify each expression.

a. $3x + 5 + 2x - 3$ **b.** $4a - 7 - 2a + 3$ **c.** $5x + 8 - x - 6$

SOLUTION We combine similar terms by applying the commutative and associative properties first, and then the distributive property.

a. $3x + 5 + 2x - 3 = 3x + 2x + 5 - 3d$ Commutative property

$= (3x + 2x) + (5 - 3)$ Associative property

$= (3 + 2)x + (5 - 3)$ Distributive property

$= 5x + 2$ Add

b. $4a - 7 - 2a + 3 = (4a - 2a) + (-7 + 3)$ Commutative and associative properties

$= (4 - 2)a + (-7 + 3)$ Distributive property

$= 2a - 4$ Add

c. $5x + 8 - x - 6 = (5x - x) + (8 - 6)$ Commutative and associative properties

$= (5 - 1)x + (8 - 6)$ Distributive property

$= 4x + 2$ Add

Notice that in each case the result has fewer terms than the original expression. Because there are fewer terms, the resulting expression is said to be simpler than the original expression.

Simplifying Expressions Containing Parentheses

If an expression contains parentheses, it is often necessary to apply the distributive property to remove the parentheses. This step will change the product in the expression into a sum or difference of individual terms, which will then allow us to combine any similar terms.

EXAMPLE 3 Simplify the expression: $5(2x - 8) - 3$.

SOLUTION We begin by distributing the 5 across $2x - 8$. We then combine similar terms:

$$5(2x - 8) - 3 = 10x - 40 - 3 \quad \text{Distributive property}$$
$$= 10x - 43$$

EXAMPLE 4 Simplify: $7 - 3(2y + 1)$.

SOLUTION By the rule for order of operations, we must multiply before we add or subtract. For that reason, it would be incorrect to subtract 3 from 7 first. Instead, we multiply -3 and $2y + 1$ to remove the parentheses and then combine similar terms:

$$7 - 3(2y + 1) = 7 - 6y - 3 \quad \text{Distributive property}$$
$$= -6y + 4$$

EXAMPLE 5 Simplify: $5(x - 2) - (3x + 4)$.

SOLUTION We begin by applying the distributive property to remove the parentheses. The expression $-(3x + 4)$ can be thought of as $-1(3x + 4)$. Thinking of it in this way allows us to apply the distributive property:

$$-1(3x + 4) = -1(3x) + (-1)(4)$$
$$= -3x - 4$$

The complete solution looks like this:

$$5(x - 2) - (3x + 4) = 5x - 10 - 3x - 4 \quad \text{Distributive property}$$
$$= 2x - 14 \quad \text{Combine similar terms}$$

As you can see from the explanation in Example 5, we use the distributive property to simplify expressions in which parentheses are preceded by a negative sign. In general we can write

$$\begin{aligned}-(a+b) &= -1(a+b)\\ &= -a+(-b)\\ &= -a-b\end{aligned}$$

The negative sign outside the parentheses ends up changing the sign of each term within the parentheses. In words, we say "the opposite of a sum is the sum of the opposites."

The Value of an Expression

An expression like $3x + 2$ has a certain value depending on what number we assign to x. For instance, when x is 4, $3x + 2$ becomes $3(4) + 2$, or 14. When x is -8, $3x + 2$ becomes $3(-8) + 2$, or -22. The value of an expression is found by replacing the variable with a given number.

EXAMPLE 6 Find the value of the following expressions by replacing the variable with the given number.

Expression	The Variable	Value of the Expression
a. $3x - 1$	$x = 2$	$3(2) - 1 = 6 - 1$ $= 5$
b. $7a + 4$	$a = -3$	$7(-3) + 4 = -21 + 4$ $= -17$
c. $2x - 3 + 4x$	$x = -1$	$2(-1) - 3 + 4(-1) = -2 - 3 + (-4)$ $= -9$
d. $2x - 5 - 8x$	$x = 5$	$2(5) - 5 - 8(5) = 10 - 5 - 40$ $= -35$
e. $y^2 - 6y + 9$	$y = 4$	$4^2 - 6(4) + 9 = 16 - 24 + 9$ $= 1$

Simplifying an expression should not change its value; that is, if an expression has a certain value when x is 5, then it will always have that value no matter how much it has been simplified as long as x is 5. If we were to simplify the expression in Example 6d first, it would look like

$$2x - 5 - 8x = -6x - 5$$

When x is 5, the simplified expression $-6x - 5$ is

$$-6(5) - 5 = -30 - 5 = -35$$

It has the same value as the original expression when x is 5.

We also can find the value of an expression that contains two variables if we know the values for both variables.

EXAMPLE 7 Find the value of the expression $2x - 3y + 4$ when x is -5 and y is 6.

SOLUTION Substituting -5 for x and 6 for y, the expression becomes

$$\begin{aligned} 2(-5) - 3(6) + 4 &= -10 - 18 + 4 \\ &= -28 + 4 \\ &= -24 \end{aligned}$$

EXAMPLE 8 Find the value of the expression $x^2 - 2xy + y^2$ when x is 3 and y is -4.

SOLUTION Replacing each x in the expression with the number 3 and each y in the expression with the number -4 gives us

$$\begin{aligned} 3^2 - 2(3)(-4) + (-4)^2 &= 9 - 2(3)(-4) + 16 \\ &= 9 - (-24) + 16 \\ &= 33 + 16 \\ &= 49 \end{aligned}$$

Sequences

As the next example indicates, when we substitute the counting numbers, in order, into an algebraic expression, we form a sequence of numbers. To review, recall that the sequence of counting numbers (also called the sequence of positive integers) is

$$\text{Counting numbers} = 1, 2, 3, \ldots$$

EXAMPLE 9 Substitute 1, 2, 3, and 4 for n in the expression $2n - 1$.

SOLUTION Substituting as indicated, we have

When $n = 1$, $2n - 1 = 2 \cdot 1 - 1 = 1$

When $n = 2$, $2n - 1 = 2 \cdot 2 - 1 = 3$

When $n = 3$, $2n - 1 = 2 \cdot 3 - 1 = 5$

When $n = 4$, $2n - 1 = 2 \cdot 4 - 1 = 7$

As you can see, substituting the first four counting numbers into the expression $2n - 1$ produces the first four terms in the sequence of odd numbers.

The next example is similar to Example 9 but uses tables to display the information.

EXAMPLE 10 Fill in the tables below to find the sequences formed by substituting the first four counting numbers into the expressions $2n$ and n^2.

a.

n	1	2	3	4
$2n$				

b.

n	1	2	3	4
n^2				

SOLUTION Proceeding as we did in the previous example, we substitute the numbers 1, 2, 3, and 4 into the given expressions.

a. When $n = 1, 2n = 2 \cdot 1 = 2$

When $n = 2, 2n = 2 \cdot 2 = 4$

When $n = 3, 2n = 2 \cdot 3 = 6$

When $n = 4, 2n = 2 \cdot 4 = 8$

As you can see, the expression $2n$ produces the sequence of even numbers when n is replaced by the counting numbers. Placing these results into our first table gives us

n	1	2	3	4
$2n$	2	4	6	8

b. The expression n^2 produces the sequence of squares when n is replaced by 1, 2, 3, and 4. In table form we have

n	1	2	3	4
n^2	1	4	9	16

Getting Ready for Class

After reading through the preceding section, respond in your own words and in complete sentences.

A. What are similar terms?

B. Explain how the distributive property is used to combine similar terms.

C. What is wrong with writing $3x + 4x = 7x^2$?

D. Explain how you would find the value of $5x + 3$ when x is 6.

Simplify the following expressions.

1. $3x - 6x$ **2.** $7x - 5x$ **3.** $-2a + a$

4. $3a - a$ **5.** $7x + 3x + 2x$ **6.** $8x - 2x - x$

7. $3a - 2a + 5a$ **8.** $7a - a + 2a$ **9.** $4x - 3 + 2x$

10. $5x + 6 - 3x$ **11.** $3a + 4a + 5$ **12.** $6a + 7a + 8$

13. $2x - 3 + 3x - 2$ **14.** $6x + 5 - 2x + 3$ **15.** $3a - 1 + a + 3$

16. $-a + 2 + 8a - 7$ **17.** $-4x + 8 - 5x - 10$ **18.** $-9x - 1 + x - 4$

19. $7a + 3 + 2a + 3a$ **20.** $8a - 2 + a + 5a$ **21.** $5(2x - 1) + 4$

22. $2(4x - 3) + 2$ **23.** $7(3y + 2) - 8$ **24.** $6(4y + 2) - 7$

25. $-3(2x - 1) + 5$ **26.** $-4(3x - 2) - 6$ **27.** $5 - 2(a + 1)$

28. $7 - 8(2a + 3)$ **29.** $6 - 4(x - 5)$ **30.** $12 - 3(4x - 2)$

31. $-9 - 4(2 - y) + 1$ **32.** $-10 - 3(2 - y) + 3$ **33.** $-6 + 2(2 - 3x) + 1$

34. $-7 - 4(3 - x) + 1$ **35.** $(4x - 7) - (2x + 5)$ **36.** $(7x - 3) - (4x + 2)$

37. $8(2a + 4) - (6a - 1)$ **38.** $9(3a + 5) - (8a - 7)$ **39.** $3(x - 2) + (x - 3)$

40. $2(2x + 1) - (x + 4)$ **41.** $4(2y - 8) - (y + 7)$ **42.** $5(y - 3) - (y - 4)$

43. $-9(2x + 1) - (x + 5)$ **44.** $-3(3x - 2) - (2x + 3)$

Evaluate the following expressions when x is 2. (Find the value of the expressions if x is 2.)

45. $3x - 1$ **46.** $4x + 3$ **47.** $-2x - 5$ **48.** $-3x + 6$

49. $x^2 - 8x + 16$ **50.** $x^2 - 10x + 25$ **51.** $(x - 4)^2$ **52.** $(x - 5)^2$

Evaluate the following expressions when x is -5. Then simplify the expression, and check to see that it has the same value for $x = -5$.

53. $7x - 4 - x - 3$ **54.** $3x + 4 + 7x - 6$

55. $5(2x + 1) + 4$ **56.** $2(3x - 10) + 5$

Evaluate the following expressions when x is -3 and y is 5.

57. $x^2 - 2xy + y^2$ **58.** $x^2 + 2xy + y^2$ **59.** $(x - y)^2$

60. $(x + y)^2$ **61.** $x^2 + 6xy + 9y^2$ **62.** $x^2 + 10xy + 25y^2$

63. $(x + 3y)^2$ **64.** $(x + 5y)^2$

Find the value of $12x - 3$ for each of the following values of x.

65. $\frac{1}{2}$ **66.** $\frac{1}{3}$ **67.** $\frac{1}{4}$ **68.** $\frac{1}{6}$

69. $\frac{3}{2}$ **70.** $\frac{2}{3}$ **71.** $\frac{3}{4}$ **72.** $\frac{5}{6}$

73. Fill in the tables below to find the sequences formed by substituting the first four counting numbers into the expressions $3n$ and n^3.

a.

n	1	2	3	4
$3n$				

b.

n	1	2	3	4
n^3				

74. Fill in the tables below to find the sequences formed by substituting the first four counting numbers into the expressions $2n - 1$ and $2n + 1$.

a.

n	1	2	3	4
$2n - 1$				

b.

n	1	2	3	4
$2n + 1$				

Find the sequences formed by substituting the first four counting numbers, in order, into the following expressions.

75. $3n - 2$ **76.** $2n - 3$ **77.** $n^2 - 2n + 1$ **78.** $(n - 1)^2$

Here are some problems you will see later in the book. Simplify.

79. $7 - 3(2y + 1)$

80. $4(3x - 2) - (6x - 5)$

81. $0.08x + 0.09x$

82. $0.04x + 0.05x$

83. $(x + y) + (x - y)$

84. $(-12x - 20y) + (25x + 20y)$

85. $3x + 2(x - 2)$

86. $2(x - 2) + 3(5x)$

87. $4(x + 1) + 3(x - 3)$

88. $5(x + 1) + 3(x - 1)$

89. $x + (x + 3)(-3)$

90. $x - 2(x + 2)$

91. $3(4x - 2) - (5x - 8)$

92. $2(5x - 3) - (2x - 4)$

93. $-(3x + 1) - (4x - 7)$

94. $-(6x + 2) - (8x - 3)$

95. $(x + 3y) + 3(2x - y)$

96. $(2x - y) - 2(x + 3y)$

97. $3(2x + 3y) - 2(3x + 5y)$

98. $5(2x + 3y) - 3(3x + 5y)$

99. $-6\left(\frac{1}{2}x - \frac{1}{3}y\right) + 12\left(\frac{1}{4}x + \frac{2}{3}y\right)$

100. $6\left(\frac{1}{3}x + \frac{1}{2}y\right) - 4\left(x + \frac{3}{4}y\right)$

101. $0.08x + 0.09(x + 2{,}000)$

102. $0.06x + 0.04(x + 7{,}000)$

103. $0.10x + 0.12(x + 500)$

104. $0.08x + 0.06(x + 800)$

Find the value of $b^2 - 4ac$ for the given values of a, b, and c. (You will see these problems later in the book.)

105. $a = 1, b = -5, c = -6$

106. $a = 1, b = -6, c = 7$

107. $a = 2, b = 4, c = -3$

108. $a = 3, b = 4, c = -2$

Applying the Concepts

109. Temperature and Altitude If the temperature on the ground is 70°F, then the temperature at A feet above the ground can be found from the expression $-0.0035A + 70$. Find the temperature at the following altitudes.

a. 8,000 feet **b.** 12,000 feet **c.** 24,000 feet

110. Perimeter of a Rectangle The expression $2l + 2w$ gives the perimeter of a rectangle with length l and width w. Find the perimeter of the rectangles with the following lengths and widths.

a. Length = 8 meters
Width = 5 meters

b. Length = 10 feet
Width = 3 feet

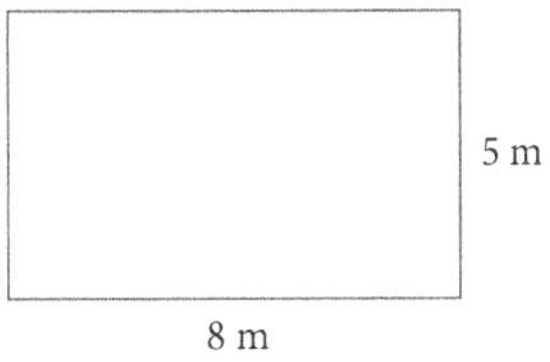

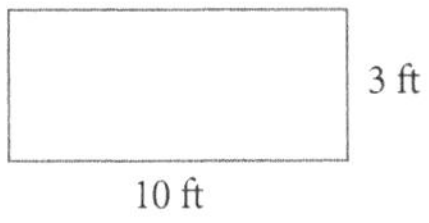

111. Cellular Phone Rates A cellular phone company charges \$35 per month plus \$0.25 for each minute, or fraction of a minute, that you use one of their cellular phones. The expression $35 + 0.25t$ gives the amount of money you will pay for using one of their phones for t minutes a month. Find the monthly bill for using one of their phones.

a. 10 minutes in a month **b.** 20 minutes in a month

c. 30 minutes in a month

112. Cost of Bottled Water A water bottling company charges \$7.00 per month for their water dispenser and \$1.10 for each gallon of water delivered. If you have g gallons of water delivered in a month, then the expression $7 + 1.1g$ gives the amount of your bill for that month. Find the monthly bill for each of the following deliveries.

a. 10 gallons **b.** 20 gallons **c.** 30 gallons

Learning Objectives Assessment

The following problems can be used to help assess if you have successfully met the learning objectives for this section.

113. Simplify: $7x + 4 - 2x - 1$.

a. $8x$ **b.** $8 + x$ **c.** $5x + 3$ **d.** $9x + 5$

114. Simplify: $9 - 3(4 - x) + 2x$.

a. $5x - 3$ **b.** $-x - 3$ **c.** $-4x + 24$ **d.** $x - 3$

115. Evaluate: $x^2 - 5x - 4$ when x is -2.

a. 2 **b.** -10 **c.** 10 **d.** 18

116. Find the sequence formed by substituting the first three counting numbers into the expression $3n + 2$.

a. 1, 2, 3 **b.** 5, 8, 11
c. 5, 10, 15 **d.** 33, 34, 35

Getting Ready for the Next Section

These are problems that you must be able to work in order to understand the material in the next section. The problems below are exactly the type of problems you will see in the explanations and examples in the next section.

Simplify.

117. $17 - 5$

118. $12 + (-2)$

119. $2 - 5$

120. $25 - 20$

121. $-2.4 + (-7.3)$

122. $8.1 + 2.7$

123. $-\frac{1}{2} + \left(-\frac{3}{4}\right)$

124. $-\frac{1}{6} + \left(-\frac{2}{3}\right)$

125. $4(2 \cdot 9 - 3) - 7$

126. $5(3 \cdot 45 - 4) - 14 \cdot 45$

127. $4(2a - 3) - 7a$

128. $5(3a - 4) - 14a$

129. Find the value of $2x - 3$ when x is 5.

130. Find the value of $3x + 4$ when x is -2.

Addition Property of Equality 2.2

Learning Objectives

In this section, we will learn how to:

1. Identify a solution to an equation.
2. Use the addition property of equality to solve a linear equation.
3. Solve linear equations involving grouping symbols.

Introduction

When light comes into contact with any object, it is reflected, absorbed, and transmitted, as shown below.

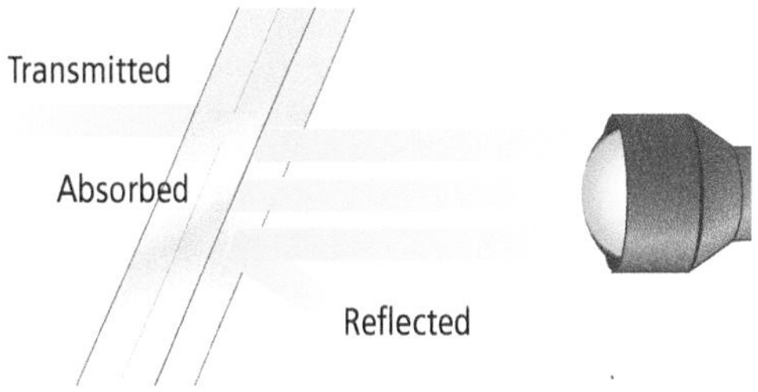

For a certain type of glass, 88% of the light hitting the glass is transmitted through to the other side, whereas 6% of the light is absorbed into the glass. To find the percent of light that is reflected by the glass, we can solve the equation

$$88 + R + 6 = 100$$

Solving equations of this type is what we study in this section.

Equations and Solutions

An *equation* is a mathematical statement that two quantities are the same, or equal. As comparisons, equations can be true or false. For instance, the equation $5 + 3 = 11$ is false, but the equation $5 + 6 = 11$ is true.

If the equation contains a variable, then the statement may be either true or false, depending on the value being used to replace the variable. For example, the equation

$$88 + R + 6 = 100$$

from the introduction to this section is true if $R = 6$, but will be false for any other value of R. To solve an equation we must find all replacements for the variable that make the equation a true statement.

DEFINITION *solution set*

The ***solution set*** for an equation is the set of all numbers that when used in place of the variable make the equation a true statement.

For example, the equation $x + 2 = 5$ has the solution set $\{3\}$ because when x is 3 the equation becomes the true statement $3 + 2 = 5$, or $5 = 5$. We sometimes say that $x = 3$ *satisfies* the equation $x + 2 = 5$. When giving the solution to this equation, we can say the solution is $x = 3$, or we can say the solution set is $\{3\}$.

VIDEO EXAMPLES

SECTION 2.2

Note We can use a question mark over the equal signs to show that we don't know yet whether the two sides of the equation are equal.

EXAMPLE 1 Is 5 a solution to $2x - 3 = 7$?

SOLUTION We substitute 5 for x in the equation, and then simplify to see if a true statement results. A true statement means we have a solution; a false statement indicates the number we are using is not a solution.

When	$x = 5$	
the equation	$2x - 3 = 7$	
becomes	$2(5) - 3 \stackrel{?}{=} 7$	
	$10 - 3 \stackrel{?}{=} 7$	
	$7 = 7$	A true statement

Because $x = 5$ turns the equation into the true statement $7 = 7$, we know 5 is a solution to the equation.

EXAMPLE 2 Is -2 a solution to $8 = 3x + 4$?

SOLUTION Substituting -2 for x in the equation, we have

$$8 \stackrel{?}{=} 3(-2) + 4$$

$$8 \stackrel{?}{=} -6 + 4$$

$$8 = -2 \qquad \text{A false statement; } 8 \neq -2$$

Substituting -2 for x in the equation produces a false statement. Therefore, $x = -2$ is not a solution to the equation.

The important thing about an equation is its solution set. Therefore, we make the following definition to classify together all equations with the same solution set.

DEFINITION ***equivalent equations***

Two or more equations with the same solution set are said to be ***equivalent equations.***

Equivalent equations may look different but must have the same solution set.

EXAMPLE 3

a. $x + 2 = 5$ and $x = 3$ are equivalent equations because both have solution set $\{3\}$.

b. $a - 4 = 3$, $a - 2 = 5$, and $a = 7$ are equivalent equations because they all have solution set $\{7\}$.

c. $y + 3 = 4$, $y - 8 = -7$, and $y = 1$ are equivalent equations because they all have solution set $\{1\}$.

The Addition Property of Equality

If two numbers are equal and we increase (or decrease) both of them by the same amount, the resulting quantities are also equal. We can apply this concept to equations. Adding the same amount to both sides of an equation always produces an equivalent equation—one with the same solution set. This fact about equations is called the *addition property of equality* and can be stated more formally as follows.

Note We will use this property many times in the future. Be sure you understand it completely by the time you finish this section.

PROPERTY *Addition Property of Equality*

For any three algebraic expressions A, B, and C,

if	$A = B$
then	$A + C = B + C$

In words: Adding the same quantity to both sides of an equation will not change the solution set.

This property is just as simple as it seems. We can add any amount to both sides of an equation and always be sure we have not changed the solution set.

Consider the equation $x + 6 = 5$. We want to solve this equation for the value of x that makes it a true statement. We want to end up with x on one side of the equal sign and a number on the other side. Because we want x by itself, we will add -6 to both sides:

Note Recall that the additive inverse property tells us that a number and its opposite add to zero. This is precisely why we chose to add -6 to both sides.

$$x + 6 + (-6) = 5 + (-6) \quad \text{Addition property of equality}$$
$$x + 0 = -1 \quad \text{Add}$$
$$x = -1$$

All three equations say the same thing about x. They all say that x is -1. All three equations are equivalent. The last one is just easier to read.

Here are some further examples of how the addition property of equality can be used to solve equations.

EXAMPLE 4 Solve the equation $x - 5 = 12$ for x.

SOLUTION Because we want x alone on the left side, we choose to add 5 to both sides:

$$x - 5 + 5 = 12 + 5 \quad \text{Addition property of equality}$$
$$x + 0 = 17$$
$$x = 17$$

If we want to write our solution using set notation, we would write $\{17\}$ for the solution set.

To check our solution to Example 4, we substitute 17 for x in the original equation:

When	$x = 17$	
the equation	$x - 5 = 12$	
becomes	$17 - 5 \stackrel{?}{=} 12$	
	$12 = 12$	A true statement

As you can see, our solution checks. The purpose for checking a solution to an equation is to catch any mistakes we may have made in the process of solving the equation.

EXAMPLE 5 Solve for a: $a + \frac{3}{4} = -\frac{1}{2}$.

SOLUTION Because we want a by itself on the left side of the equal sign, we add the opposite of $\frac{3}{4}$ to each side of the equation.

$$a + \frac{3}{4} + \left(-\frac{3}{4}\right) = -\frac{1}{2} + \left(-\frac{3}{4}\right) \quad \text{Addition property of equality}$$

$$a + 0 = -\frac{1}{2} \cdot \frac{2}{2} + \left(-\frac{3}{4}\right) \quad \text{LCD on the right side is 4}$$

$$a = -\frac{2}{4} + \left(-\frac{3}{4}\right) \quad \tfrac{2}{4} \text{ is equivalent to } \tfrac{1}{2}$$

$$a = -\frac{5}{4} \quad \text{Add fractions}$$

The solution is $a = -\frac{5}{4}$. To check our result, we replace a with $-\frac{5}{4}$ in the original equation. The left side then becomes $-\frac{5}{4} + \frac{3}{4}$, which reduces to $-\frac{1}{2}$, so our solution checks.

EXAMPLE 6 Solve for x: $7.3 + x = -2.4$.

SOLUTION Again, we want to isolate x, so we add the opposite of 7.3 to both sides:

$$7.3 + (-7.3) + x = -2.4 + (-7.3) \quad \text{Addition property of equality}$$

$$0 + x = -9.7$$

$$x = -9.7$$

The solution set is $\{-9.7\}$.

Sometimes it is necessary to simplify each side of an equation before using the addition property of equality. The reason we simplify both sides first is that we want as few terms as possible on each side of the equation before we use the addition property of equality. The following examples illustrate this procedure.

EXAMPLE 7 Solve for x: $-x + 2 + 2x = 7 + 5$.

SOLUTION We begin by combining similar terms on each side of the equation. Then we use the addition property to solve the simplified equation.

$$x + 2 = 12 \quad \text{Simplify both sides first}$$

$$x + 2 + (-2) = 12 + (-2) \quad \text{Addition property of equality}$$

$$x + 0 = 10$$

$$x = 10$$

The next example illustrates how we solve an equation involving grouping symbols.

EXAMPLE 8 Solve: $4(2a - 3) - 7a = 2 - 5$.

SOLUTION We must begin by applying the distributive property to separate terms on the left side of the equation. Following that, we combine similar terms and then apply the addition property of equality.

$$4(2a - 3) - 7a = 2 - 5 \quad \text{Original equation}$$

$$8a - 12 - 7a = 2 - 5 \quad \text{Distributive property}$$

$$a - 12 = -3 \quad \text{Simplify each side}$$

$$a - 12 + 12 = -3 + 12 \quad \text{Add 12 to each side}$$

$$a = 9 \quad \text{Add}$$

To check our solution, we replace a with 9 in the original equation.

$$4(2 \cdot 9 - 3) - 7 \cdot 9 \stackrel{?}{=} 2 - 5$$
$$4(15) - 63 \stackrel{?}{=} -3$$
$$60 - 63 \stackrel{?}{=} -3$$
$$-3 = -3 \qquad \text{A true statement}$$

Note Again, we place a question mark over the equal sign because we don't know yet whether the expressions on the left and right side of the equal sign will be equal.

We can also add a term involving a variable to both sides of an equation.

EXAMPLE 9 Solve: $3x - 5 = 2x + 7$.

SOLUTION We can solve this equation in two steps. First, we add $-2x$ to both sides of the equation. When this has been done, x appears on the left side only. Second, we add 5 to both sides:

$$3x + (-2x) - 5 = 2x + (-2x) + 7 \qquad \text{Add } -2x \text{ to both sides}$$
$$x - 5 = 7 \qquad \text{Simplify each side}$$
$$x - 5 + 5 = 7 + 5 \qquad \text{Add 5 to both sides}$$
$$x = 12 \qquad \text{Simplify each side}$$

Note In my experience teaching algebra, I find that students make fewer mistakes if they think in terms of addition rather than subtraction. So, you are probably better off if you continue to use the addition property just the way we have used it in the examples in this section. But, if you are curious as to whether you can subtract the same number from both sides of an equation, the answer is yes.

PROPERTY *A Note on Subtraction*

Although the addition property of equality is stated for addition only, we can subtract the same number from both sides of an equation as well. Because subtraction is defined as addition of the opposite, subtracting the same quantity from both sides of an equation does not change the solution.

$$x + 2 = 12 \qquad \text{Original equation}$$
$$x + 2 - 2 = 12 - 2 \qquad \text{Subtract 2 from each side}$$
$$x = 10 \qquad \text{Subtract}$$

Getting Ready for Class

After reading through the preceding section, respond in your own words and in complete sentences.

A. What is a solution to an equation?

B. What are equivalent equations?

C. Explain in words the addition property of equality.

D. How do you check a solution to an equation?

Problem Set 2.2

1. Is $x = 4$ a solution to the equation $3x - 5 = 7$?
2. Is $x = -3$ a solution to the equation $2x + 1 = 5$?
3. Is $y = -2$ a solution to the equation $3y - 4(y + 6) + 2 = 8$?
4. Is $y = 6$ a solution to the equation $7y - 5(y - 1) - 3 = 14$?

For each of the following equations determine whether each given value is a solution to the equation.

5. $2m + 3 = m - 5$
   a. -1 b. -8
6. $3x + 7 = 4x - 6$
   a. 13 b. 1
7. $2x + \frac{5}{3} = x - \frac{1}{2}$
   a. $-\frac{13}{6}$ b. $\frac{7}{6}$
8. $5a - \frac{1}{2} = 4a + \frac{3}{4}$
   a. $-\frac{3}{2}$ b. $\frac{5}{4}$

Solve the following equations.

9. $x - 3 = 8$
10. $x - 2 = 7$
11. $x + 2 = 6$
12. $x + 5 = 4$
13. $a + \frac{1}{2} = -\frac{1}{4}$
14. $a + \frac{1}{3} = -\frac{5}{6}$
15. $x + 2.3 = -3.5$
16. $x + 7.9 = 23.4$
17. $y + 11 = -6$
18. $y - 3 = -1$
19. $x - \frac{5}{8} = -\frac{3}{4}$
20. $x - \frac{2}{5} = -\frac{1}{10}$
21. $m - 6 = -10$
22. $m - 10 = -6$
23. $6.9 + x = 3.3$
24. $7.5 + x = 2.2$
25. $5 = a + 4$
26. $12 = a - 3$
27. $-\frac{5}{9} = x - \frac{2}{5}$
28. $-\frac{7}{8} = x - \frac{4}{5}$

Simplify both sides of the following equations as much as possible, and then solve.

29. $4x + 2 - 3x = 4 + 1$
30. $5x + 2 - 4x = 7 - 3$
31. $8a - \frac{1}{2} - 7a = \frac{3}{4} + \frac{1}{8}$
32. $9a - \frac{4}{5} - 8a = \frac{3}{10} - \frac{1}{5}$
33. $-3 - 4x + 5x = 18$
34. $10 - 3x + 4x = 20$
35. $-11x + 2 + 10x + 2x = 9$
36. $-10x + 5 - 4x + 15x = 0$
37. $-2.5 + 4.8 = 8x - 1.2 - 7x$
38. $-4.8 + 6.3 = 7x - 2.7 - 6x$
39. $2y - 10 + 3y - 4y = 18 - 6$
40. $15 - 21 = 8x + 3x - 10x$

The following equations contain parentheses. Apply the distributive property to remove the parentheses, then simplify each side before using the addition property of equality.

41. $2(x + 3) - x = 4$

42. $5(x + 1) - 4x = 2$

43. $-3(x - 4) + 4x = 3 - 7$

44. $-2(x - 5) + 3x = 4 - 9$

45. $5(2a + 1) - 9a = 8 - 6$

46. $4(2a - 1) - 7a = 9 - 5$

47. $-(x + 3) + 2x - 1 = 6$

48. $-(x - 7) + 2x - 8 = 4$

49. $4y - 3(y - 6) + 2 = 8$

50. $7y - 6(y - 1) + 3 = 9$

51. $-3(2m - 9) + 7(m - 4) = 12 - 9$

52. $-5(m - 3) + 2(3m + 1) = 15 - 8$

Solve the following equations by the method used in Example 9 in this section. Check each solution in the original equation.

53. $4x = 3x + 2$

54. $6x = 5x - 4$

55. $8a = 7a - 5$

56. $9a = 8a - 3$

57. $2x = 3x + 1$

58. $4x = 3x + 5$

59. $3y + 4 = 2y + 1$

60. $5y + 6 = 4y + 2$

61. $2m - 3 = m + 5$

62. $8m - 1 = 7m - 3$

63. $4x - 7 = 5x + 1$

64. $3x - 7 = 4x - 6$

65. $5x - \frac{2}{3} = 4x + \frac{4}{3}$

66. $3x - \frac{5}{4} = 2x + \frac{1}{4}$

67. $8a - 7.1 = 7a + 3.9$

68. $10a - 4.3 = 9a + 4.7$

69. $11y - 2.9 = 12y + 2.9$

70. $20y + 9.9 = 21y - 9.9$

Applying the Concepts

71. **Light** When light comes into contact with any object, it is reflected, absorbed, and transmitted, as shown in the following figure. If T represents the percent of light transmitted, R the percent of light reflected, and A the percent of light absorbed by a surface, then the equation $T + R + A = 100$ shows one way these quantities are related.

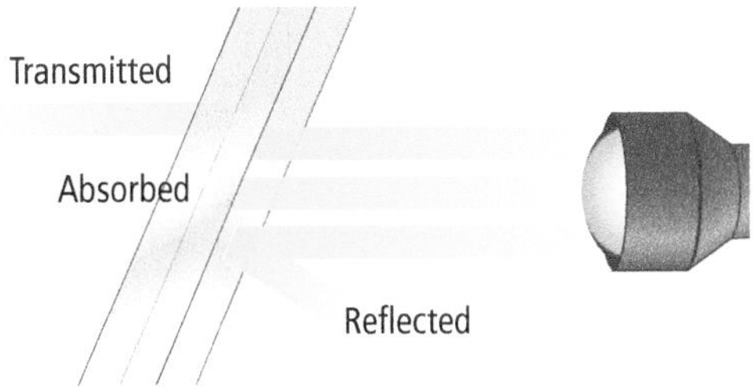

a. For glass, $T = 88$ and $A = 6$, meaning that 88% of the light hitting the glass is transmitted and 6% is absorbed. Substitute $T = 88$ and $A = 6$ into the equation $T + R + A = 100$ and solve for R to find the percent of light that is reflected.

b. For flat black paint, $A = 95$ and no light is transmitted, meaning that $T = 0$. What percent of light is reflected by flat black paint?

c. A pure white surface can reflect 98% of light, so $R = 98$. If no light is transmitted, what percent of light is absorbed by the pure white surface?

d. Typically, shiny gray metals reflect 70–80% of light. Suppose a thick sheet of aluminum absorbs 25% of light. What percent of light is reflected by this shiny gray metal? (Assume no light is transmitted.)

72. Geometry The three angles shown in the triangle at the front of the tent in the following figure add up to 180°. Use this fact to write an equation containing x, and then solve the equation to find the number of degrees in the angle at the top of the triangle.

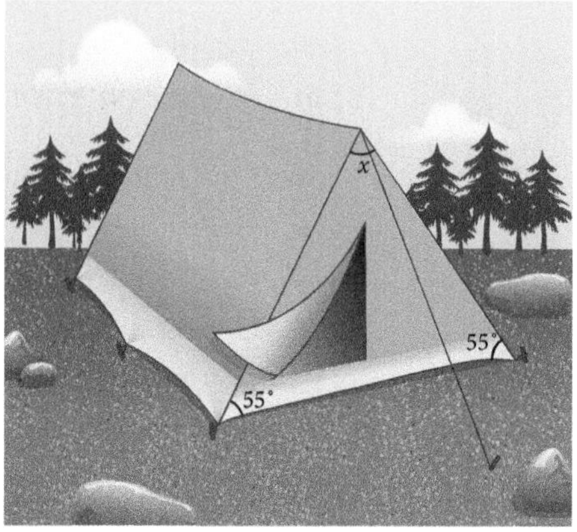

Learning Objectives Assessment

The following problems can be used to help assess if you have successfully met the learning objectives for this section.

73. Which of the following is a solution to the equation $5x - 6 = 4x + 3$?

a. 9 **b.** -3 **c.** 3 **d.** -9

74. Solve: $9 + x = 27$.

a. 36 **b.** 18 **c.** 3 **d.** -18

75. Solve: $-4(x - 2) + 5x = 3 - 9$.

a. -4 **b.** 2 **c.** -14 **d.** 7

Getting Ready for the Next Section

To understand all of the explanations and examples in the next section you must be able to work the problems below.

Simplify.

76. $\frac{3}{2}\left(\frac{2}{3}y\right)$ **77.** $-\frac{5}{2}\left(-\frac{2}{5}y\right)$ **78.** $\frac{1}{5}(5x)$ **79.** $-\frac{1}{4}(-4a)$

80. $\frac{1}{5}(30)$ **81.** $-\frac{1}{4}(24)$ **82.** $\frac{3}{2}(4)$ **83.** $\frac{1}{26}(13)$

84. $12\left(-\frac{3}{4}\right)$ **85.** $12\left(\frac{1}{2}\right)$ **86.** $\frac{3}{2}\left(-\frac{5}{4}\right)$ **87.** $\frac{5}{3}\left(-\frac{6}{5}\right)$

88. $13 + (-5)$ **89.** $-13 + (-5)$ **90.** $-\frac{3}{4} + \left(-\frac{1}{2}\right)$ **91.** $-\frac{7}{10} + \left(-\frac{1}{2}\right)$

92. $7x + (-4x)$ **93.** $5x + (-2x)$

Multiplication Property of Equality 2.3

Learning Objectives

In this section, we will learn how to:

1. Use the multiplication property of equality to solve linear equations.
2. Solve linear equations using both the addition and multiplication properties of equality.

Introduction

We all have to pay taxes. According to Figure 1, people have been paying taxes for quite a long time.

FIGURE 1 *Collection of taxes, ca. 3000 B.C. Clerks and scribes appear at the right, with pen and papyrus, and officials and taxpayers appear at the left.*

Suppose 21% of your monthly pay is withheld for federal income taxes and another 8% is withheld for Social Security, state income tax, and other miscellaneous items, leaving you with \$987.50 a month in take-home pay. The amount you earned before the deductions were removed from your check, your gross income G, is given by the equation

$$G - 0.21G - 0.08G = 987.5$$

In this section, we will learn how to solve equations of this type.

In the previous section, we found that adding the same number to both sides of an equation never changed the solution set. The same idea holds for multiplication by numbers other than zero. We can multiply both sides of an equation by the same nonzero number and always be sure we have not changed the solution set. (The reason we cannot multiply both sides by zero will become apparent later.) This fact about equations is called the *multiplication property of equality*, which can be stated formally as follows.

Note This property is also used many times throughout the book. Make every effort to understand it completely.

PROPERTY *Multiplication Property of Equality*

For any three algebraic expressions A, B, and C, where $C \neq 0$,

if $\quad A = B$

then $\quad AC = BC$

In words: Multiplying both sides of an equation by the same nonzero number will not change the solution set.

VIDEO EXAMPLES

SECTION 2.3

EXAMPLE 1 Solve: $5x = 30$.

SOLUTION We have $5x$ on the left side but would like to have just x. To isolate the variable x, we need to change the coefficient from a 5 to a 1. The multiplicative inverse property gives us a way to do just that. We choose to multiply both sides by $\frac{1}{5}$ because it is the reciprocal of 5, and $\left(\frac{1}{5}\right)(5) = 1$. Here is the process:

$$
\begin{aligned}
5x &= 30 \\
\frac{1}{5}(5x) &= \frac{1}{5}(30) && \text{Multiplication property of equality} \\
\left(\frac{1}{5}\cdot 5\right)x &= \frac{1}{5}(30) && \text{Associative property of multiplication} \\
1x &= 6 && \text{Multiply} \\
x &= 6
\end{aligned}
$$

We can see from Example 1 that multiplication by any number except zero will not change the solution set. If, however, we were to multiply both sides by zero, we may get an equation that is not equivalent. Consider the equation

$$3 = -2$$

This equation is clearly false. However, if we multiply both sides by 0, we obtain

$$0(3) = 0(-2)$$

$$0 = 0$$

which is true. A false statement has been turned into a true one. The two are not equivalent. This is the only restriction of the multiplication property of equality. We are free to multiply both sides of an equation by any number except zero.

PROPERTY ***A Note on Division***

Because ***division*** is defined as multiplication by the reciprocal, multiplying both sides of an equation by the same number is equivalent to dividing both sides of the equation by the reciprocal of that number; that is, multiplying each side of an equation by $\frac{1}{5}$ and dividing each side of the equation by 5 are equivalent operations. If we were to solve the equation $5x = 30$ from Example 1 using division instead of multiplication, the steps would look like this:

$$
\begin{aligned}
5x &= 30 && \text{Original equation} \\
\frac{5x}{5} &= \frac{30}{5} && \text{Divide each side by 5} \\
x &= 6
\end{aligned}
$$

Using division instead of multiplication on a problem like this may save you some writing. However, with multiplication, it is easier to explain "why" we end up with just one x on the left side of the equation. (The "why" has to do with the associative property of multiplication.) My suggestion is that you continue to use multiplication to solve equations like this one until you understand the process completely. Then, if you find it more convenient, you can use division instead of multiplication.

Here are some more examples that use the multiplication property of equality. We use the multiplication property of equality to isolate the variable by changing the coefficient of the variable into a 1.

EXAMPLE 2 Solve for a: $-4a = 24$.

SOLUTION Because we want a alone on the left side, we choose to multiply both sides by $-\frac{1}{4}$:

$$-\frac{1}{4}(-4a) = -\frac{1}{4}(24) \quad \text{Multiplication property of equality}$$

$$\left[-\frac{1}{4}(-4)\right]a = -\frac{1}{4}(24) \quad \text{Associative property}$$

$$a = -6 \quad \text{Multiply}$$

Note It is always a good idea to check your solution in the original equation. For Example 2, we would replace a with -6 and get $(-4)(-6) \overset{?}{=} 24$, or $24 = 24$, which is a true statement.

EXAMPLE 3 Solve for t: $-\frac{t}{3} = 5$.

SOLUTION Because division by 3 is the same as multiplication by $\frac{1}{3}$, we can write $-\frac{t}{3}$ as $-\frac{1}{3}t$. To solve the equation, we multiply each side by the reciprocal of $-\frac{1}{3}$, which is -3.

$$-\frac{t}{3} = 5 \quad \text{Original equation}$$

$$-\frac{1}{3}t = 5 \quad \text{Dividing by 3 is equivalent to multiplying by } \tfrac{1}{3}$$

$$-3\left(-\frac{1}{3}t\right) = -3(5) \quad \text{Multiply each side by } -3$$

$$t = -15 \quad \text{Multiply}$$

We say the solution is $t = -15$ and the solution set is $\{-15\}$.

EXAMPLE 4 Solve: $\frac{2}{3}y = 4$.

SOLUTION We can multiply both sides by $\frac{3}{2}$ and have $1y$ on the left side:

$$\frac{3}{2}\left(\frac{2}{3}y\right) = \frac{3}{2}(4) \quad \text{Multiplication property of equality}$$

$$\left(\frac{3}{2} \cdot \frac{2}{3}\right)y = \frac{3}{2}(4) \quad \text{Associative property}$$

$$y = 6 \quad \text{Simplify } \tfrac{3}{2}(4) = \tfrac{3}{2}\left(\tfrac{4}{1}\right) = \tfrac{12}{2} = 6$$

Note Notice in Examples 2 through 4 that if the variable is being multiplied by a number like -4 or $\frac{2}{3}$, we always multiply by the number's reciprocal, $-\frac{1}{4}$ or $\frac{3}{2}$, to end up with just the variable on one side of the equation.

EXAMPLE 5 Solve: $5 + 8 = 10x + 20x - 4x$.

SOLUTION Our first step will be to simplify each side of the equation:

$$13 = 26x \quad \text{Simplify both sides first}$$

$$\frac{1}{26}(13) = \frac{1}{26}(26x) \quad \text{Multiplication property of equality}$$

$$\frac{13}{26} = x \quad \text{Multiply}$$

$$\frac{1}{2} = x \quad \text{Reduce to lowest terms}$$

Note It makes no difference on which side of the equal sign x ends up. Most people prefer to have x on the left side because we read from left to right, and it seems to sound better to say x is 6 rather than 6 is x. Both expressions, however, have exactly the same meaning.

In the next four examples, we will use both the addition property of equality and the multiplication property of equality. We use the addition property of equality first in order to get a single term containing the variable on one side of the equation. Then we use the multiplication property of equality to isolate the variable itself by changing the coefficient into a 1.

EXAMPLE 6 Solve for x: $6x + 5 = -13$.

SOLUTION We begin by adding -5 to both sides of the equation in order to isolate the term containing the variable:

$$6x + 5 + (-5) = -13 + (-5) \quad \text{Add } -5 \text{ to both sides}$$

$$6x = -18 \quad \text{Simplify}$$

Now we isolate x itself using the multiplication property of equality:

$$\frac{1}{6}(6x) = \frac{1}{6}(-18) \quad \text{Multiply both sides by } \frac{1}{6}$$

$$x = -3$$

EXAMPLE 7 Solve for x: $5x = 2x + 12$.

SOLUTION We begin by adding $-2x$ to both sides of the equation:

$$5x + (-2x) = 2x + (-2x) + 12 \quad \text{Add } -2x \text{ to both sides}$$

$$3x = 12 \quad \text{Simplify}$$

$$\frac{1}{3}(3x) = \frac{1}{3}(12) \quad \text{Multiply both sides by } \frac{1}{3}$$

$$x = 4 \quad \text{Simplify}$$

So $x = 4$ is the solution to our equation. Or we can write $\{4\}$ as the solution set.

Note Notice that in Example 7 we used the addition property of equality first to combine all the terms containing x on the left side of the equation. Once this had been done, we used the multiplication property to isolate x on the left side.

EXAMPLE 8 Solve for x: $3x - 4 = -2x + 6$.

SOLUTION We begin by adding $2x$ to both sides:

$$3x + 2x - 4 = -2x + 2x + 6 \quad \text{Add } 2x \text{ to both sides}$$

$$5x - 4 = 6 \quad \text{Simplify}$$

Now we add 4 to both sides:

$$5x - 4 + 4 = 6 + 4 \quad \text{Add 4 to both sides}$$

$$5x = 10 \quad \text{Simplify}$$

$$\frac{1}{5}(5x) = \frac{1}{5}(10) \quad \text{Multiply by } \frac{1}{5}$$

$$x = 2 \quad \text{Simplify}$$

The next example involves fractions. You will see that the properties we use to solve equations containing fractions are the same as the properties we used to solve the previous equations.

EXAMPLE 9 Solve: $\frac{2}{3}x + \frac{1}{2} = -\frac{3}{4}$.

SOLUTION We can solve this equation by applying our properties and working with the fractions.

$$\frac{2}{3}x + \frac{1}{2} + \left(-\frac{1}{2}\right) = -\frac{3}{4} + \left(-\frac{1}{2}\right) \quad \text{Add } -\tfrac{1}{2} \text{ to each side}$$

$$\frac{2}{3}x = -\frac{5}{4} \quad \text{Note that } -\tfrac{3}{4} + \left(-\tfrac{1}{2}\right) = -\tfrac{3}{4} + \left(-\tfrac{2}{4}\right)$$

$$\frac{3}{2}\left(\frac{2}{3}x\right) = \frac{3}{2}\left(-\frac{5}{4}\right) \quad \text{Multiply each side by } \tfrac{3}{2}$$

$$x = -\frac{15}{8}$$

As we saw in the previous section, when grouping symbols appear in the equation, we use the distributive property to separate terms. Then we can proceed with the addition and multiplication properties of equality.

EXAMPLE 10 Solve: $2(x + 3) = 10$.

SOLUTION To begin, we apply the distributive property to the left side of the equation to separate terms:

$$2x + 6 = 10 \quad \text{Distributive property}$$

$$2x + 6 + (-6) = 10 + (-6) \quad \text{Addition property of equality}$$

$$2x = 4$$

$$\frac{1}{2}(2x) = \frac{1}{2}(4) \quad \text{Multiply each side by } \tfrac{1}{2}$$

$$x = 2 \quad \text{The solution is 2}$$

Getting Ready for Class

After reading through the preceding section, respond in your own words and in complete sentences.

A. Explain in words the multiplication property of equality.

B. Explain in words how you would solve the equation $3x = 7$ using the multiplication property of equality.

C. Why is it okay to divide both sides of an equation by the same nonzero number?

D. In solving the equation $2x + 5 = 9$, which property of equality do we use first? Why?

Problem Set 2.3

Solve the following equations. Be sure to show your work.

1. $5x = 10$ **2.** $6x = 12$ **3.** $7a = 28$ **4.** $4a = 36$

5. $-8x = 4$ **6.** $-6x = 2$ **7.** $8m = -16$ **8.** $5m = -25$

9. $-3x = -9$ **10.** $-9x = -36$ **11.** $-7y = -28$ **12.** $-15y = -30$

13. $2x = 0$ **14.** $7x = 0$ **15.** $-5x = 0$ **16.** $-3x = 0$

17. $\frac{x}{3} = 2$ **18.** $\frac{x}{4} = 3$ **19.** $-\frac{m}{5} = 10$ **20.** $-\frac{m}{7} = 1$

21. $-\frac{x}{2} = -\frac{3}{4}$ **22.** $-\frac{x}{3} = \frac{5}{6}$ **23.** $\frac{2}{3}a = 8$ **24.** $\frac{3}{4}a = 6$

25. $-\frac{3}{5}x = \frac{9}{5}$ **26.** $-\frac{2}{5}x = \frac{6}{15}$ **27.** $-\frac{5}{8}y = -20$ **28.** $-\frac{7}{2}y = -14$

Simplify both sides as much as possible, and then solve.

29. $-4x - 2x + 3x = 24$ **30.** $7x - 5x + 8x = 20$

31. $4x + 8x - 2x = 15 - 10$ **32.** $5x + 4x + 3x = 4 + 8$

33. $-3 - 5 = 3x + 5x - 10x$ **34.** $10 - 16 = 12x - 6x - 3x$

35. $18 - 13 = \frac{1}{2}a + \frac{3}{4}a - \frac{5}{8}a$ **36.** $20 - 14 = \frac{1}{3}a + \frac{5}{6}a - \frac{2}{3}a$

Solve the following equations by multiplying both sides by -1.

37. $-x = 4$ **38.** $-x = -3$ **39.** $-x = -4$ **40.** $-x = 3$

41. $15 = -a$ **42.** $-15 = -a$ **43.** $-y = \frac{1}{2}$ **44.** $-y = -\frac{3}{4}$

Solve each of the following equations using the method shown in Examples 6–9 in this section.

45. $3x - 2 = 7$ **46.** $2x - 3 = 9$ **47.** $2a + 1 = 3$

48. $5a - 3 = 7$ **49.** $\frac{1}{8} + \frac{1}{2}x = \frac{1}{4}$ **50.** $\frac{1}{3} + \frac{1}{7}x = -\frac{8}{21}$

51. $6x = 2x - 12$ **52.** $8x = 3x - 10$ **53.** $2y = -4y + 18$

54. $3y = -2y - 15$ **55.** $-7x = -3x - 8$ **56.** $-5x = -2x - 12$

57. $8x + 4 = 2x - 5$ **58.** $5x + 6 = 3x - 6$ **59.** $6m - 3 = m + 2$

60. $6m - 5 = m + 5$ **61.** $9y + 2 = 6y - 4$ **62.** $6y + 14 = 2y - 2$

Solve each of the following equations using the method shown in Example 10 in this section.

63. $2(x + 3) = 12$ **64.** $3(x - 2) = 6$ **65.** $6(x - 1) = -18$

66. $4(x + 5) = 16$ **67.** $2(4a + 1) = -6$ **68.** $3(2a - 4) = 12$

69. $14 = 2(5x - 3)$ **70.** $-25 = 5(3x + 4)$ **71.** $-2(3y + 5) = 14$

72. $-3(2y - 4) = -6$ **73.** $-5(2a + 4) = 0$ **74.** $-3(3a - 6) = 0$

75. $1 = \frac{1}{2}(4x + 2)$ **76.** $1 = \frac{1}{3}(6x + 3)$ **77.** $3(t - 4) + 5 = -4$

78. $5(t - 1) + 6 = -9$ **79.** $4(2x + 1) - 7 = 1$ **80.** $6(3y + 2) - 8 = -2$

81. Solve each equation.

a. $2x = 3$ **b.** $2 + x = 3$ **c.** $2x + 3 = 0$

d. $2x + 3 = -5$ **e.** $2x + 3 = 7x - 5$

82. Solve each equation.

a. $5t = 10$ **b.** $5 + t = 10$ **c.** $5t + 10 = 0$

d. $5t + 10 = 12$ **e.** $5t + 10 = 8t + 12$

Applying the Concepts

83. Break-Even Point Movie theaters pay a certain price for the movies that you and I see. Suppose a theater pays \$1,500 for each showing of a popular movie. If they charge \$7.50 for each ticket they sell, then the equation $7.5x = 1{,}500$ gives the number of tickets they must sell to equal the \$1,500 cost of showing the movie. This number is called the break-even point. Solve the equation for x to find the break-even point.

84. Basketball Laura plays basketball for her community college. In one game she scored 13 points total, with a combination of free throws, field goals, and three-pointers. Each free throw is worth 1 point, each field goal is 2 points, and each three-pointer is worth 3 points. If she made 1 free throw and 3 field goals, then solving the equation

$$1 + 3(2) + 3x = 13$$

will give us the number of three-pointers she made. Solve the equation to find the number of three-point shots Laura made.

85. Taxes Suppose 21% of your monthly pay is withheld for federal income taxes and another 8% is withheld for Social Security, state income tax, and other miscellaneous items. If you are left with \$987.50 a month in take-home pay, then the amount you earned before the deductions were removed from your check is given by the equation

$$G - 0.21G - 0.08G = 987.5$$

Solve this equation to find your gross income.

86. Rhind Papyrus The *Rhind Papyrus* is an ancient document that contains mathematical riddles. One problem asks the reader to find a quantity such that when it is added to one-fourth of itself the sum is 15. The equation that describes this situation is

$$x + \frac{1}{4}x = 15$$

Solve this equation.

Learning Objectives Assessment

The following problems can be used to help assess if you have successfully met the learning objectives for this section.

87. Solve: $3x - 8x = 20$.

a. 15 **b.** 25 **c.** 4 **d.** -4

88. Solve: $4a - 5 = 8$.

a. $\frac{13}{4}$ **b.** 7 **c.** $\frac{3}{4}$ **d.** -3

Getting Ready for the Next Section

To understand all of the explanations and examples in the next section you must be able to work the problems below.

Solve each equation.

89. $2x = 4$ **90.** $3x = 24$ **91.** $30 = 5x$

92. $0 = 5x$ **93.** $0.17x = 510$ **94.** $0.1x = 400$

Apply the distributive property and then simplify if possible.

95. $3(x - 5) + 4$ **96.** $5(x - 3) + 2$ **97.** $0.09(x + 2{,}000)$

98. $0.04(x + 7{,}000)$ **99.** $7 - 3(2y + 1)$ **100.** $4 - 2(3y + 1)$

101. $3(2x - 5) - (2x - 4)$ **102.** $4(3x - 2) - (6x - 5)$

Simplify.

103. $10x + (-5x)$ **104.** $12x + (-7x)$ **105.** $0.08x + 0.09x$ **106.** $0.06x + 0.04x$

Solving Linear Equations 2.4

Learning Objectives

In this section, we will learn how to:

1. Solve a linear equation containing grouping symbols.
2. Solve a linear equation containing fractions.
3. Identify an identity or a contradiction.

Introduction

We will now use the material we have developed in the first three sections of this chapter to build a method for solving any linear equation.

DEFINITION *linear equation*

A ***linear equation*** in one variable is any equation that can be put in the form $ax + b = 0$, where a and b are real numbers and a is not zero.

Each of the equations we will solve in this section is a *linear equation in one variable*. The general method of solving linear equations is actually very simple. It is based on the properties we developed in Chapter 1 and on two very simple properties we developed in Sections 2.3 and 2.4. We can add any number to both sides of the equation and multiply both sides by any nonzero number. The equation may change in form, but the solution set will not.

The steps we use to solve a linear equation in one variable are listed here. The overall goal is to isolate the variable on one side of the equation.

Note You may have some previous experience solving equations. Even so, you should solve the equations in this section using the method developed here. Your work should look like the examples in the text. If you have learned shortcuts or a different method of solving equations somewhere else, you can always go back to them later. What is important now is that you are able to solve equations by the methods shown here.

HOW TO *Solve Linear Equations in One Variable*

Step 1a: Use the distributive property to separate terms, if necessary.

1b: If fractions are present, consider multiplying both sides by the LCD to eliminate the fractions. If decimals are present, consider multiplying both sides by a power of 10 to clear the equation of decimals.

1c: Combine similar terms on each side of the equation.

Step 2: Use the addition property of equality to get all variable terms on one side of the equation and all constant terms on the other side. A ***variable term*** is a term that contains the variable (for example, $5x$). A ***constant term*** is a term that does not contain the variable (the number 3, for example).

Step 3: Use the multiplication property of equality to change the coefficient of the variable term into 1.

Step 4: Check your solution in the original equation to be sure that you have not made a mistake in the solution process.

As you will see as you work through the examples in this section, it is not always necessary to use all four steps when solving equations. The number of steps used depends on the equation.

The examples that follow show a variety of equations and their solutions. When you have finished this section and worked the problems in the problem set, the steps in the solution process should be a description of how you operate when solving equations. That is, you want to work enough problems so that the Strategy for Solving Linear Equations is second nature to you.

Equations Containing Grouping Symbols

VIDEO EXAMPLES

SECTION 2.4

EXAMPLE 1 Solve for x: $3(x - 5) + 4 = 13$.

SOLUTION Our first step will be to apply the distributive property to the left side of the equation:

Step 1a: $3x - 15 + 4 = 13$ — Distributive property

Step 1c: $3x - 11 = 13$ — Simplify the left side

Step 2:
$$\begin{cases} 3x - 11 + 11 = 13 + 11 & \text{Add 11 to both sides} \\ 3x = 24 \end{cases}$$

Step 3:
$$\begin{cases} \frac{1}{3}(3x) = \frac{1}{3}(24) & \text{Multiply both sides by } \frac{1}{3} \\ x = 8 & \text{The solution is 8} \end{cases}$$

Step 4: When $x = 8$

the equation $3(x - 5) + 4 = 13$

becomes $3(8 - 5) + 4 \stackrel{?}{=} 13$

$3(3) + 4 \stackrel{?}{=} 13$

$9 + 4 \stackrel{?}{=} 13$

$13 = 13$ — A true statement

EXAMPLE 2 Solve: $5(x - 3) + 2 = 5(2x - 8) - 3$.

SOLUTION In this case, we apply the distributive property on each side of the equation:

Step 1a: $5x - 15 + 2 = 10x - 40 - 3$ — Distributive property

Step 1c: $5x - 13 = 10x - 43$ — Simplify each side

Step 2:
$$\begin{cases} 5x + (-5x) - 13 = 10x + (-5x) - 43 & \text{Add } -5x \text{ to both sides} \\ -13 = 5x - 43 \\ -13 + 43 = 5x - 43 + 43 & \text{Add 43 to both sides} \\ 30 = 5x \end{cases}$$

Step 3:
$$\begin{cases} \frac{1}{5}(30) = \frac{1}{5}(5x) & \text{Multiply both sides by } \frac{1}{5} \\ 6 = x & \text{The solution is 6} \end{cases}$$

Step 4: Replacing x with 6 in the original equation, we have

$$5(6-3)+2 \stackrel{?}{=} 5(2\cdot 6-8)-3$$
$$5(3)+2 \stackrel{?}{=} 5(12-8)-3$$
$$5(3)+2 \stackrel{?}{=} 5(4)-3$$
$$15+2 \stackrel{?}{=} 20-3$$
$$17 = 17 \qquad \text{A true statement}$$

EXAMPLE 3 Solve: $7-3(2y+1)=16$.

SOLUTION We begin by multiplying -3 times the sum of $2y$ and 1:

Step 1a: $7-6y-3=16$ — Distributive property

Step 1c: $-6y+4=16$ — Simplify the left side

Step 2: $-6y+4+(-4)=16+(-4)$ — Add -4 to both sides
$-6y=12$

Step 3: $-\frac{1}{6}(-6y)=-\frac{1}{6}(12)$ — Multiply both sides by $-\frac{1}{6}$
$y=-2$

There are two things to notice about the example that follows: first, the distributive property is used to remove parentheses that are preceded by a negative sign, and second, the addition property and the multiplication property are not shown in as much detail as in the previous examples.

EXAMPLE 4 Solve: $3(2x-5)-(2x-4)=6-(4x+5)$.

SOLUTION When we apply the distributive property to remove the grouping symbols and separate terms, we have to be careful with the signs. Remember, we can think of $-(2x-4)$ as $-1(2x-4)$, so that

$$-(2x-4)=-1(2x-4)=-2x+4$$

It is not uncommon for students to make a mistake with this type of simplification and write the result as $-2x-4$, which is incorrect. Here is the complete solution to our equation:

$3(2x-5)-(2x-4)=6-(4x+5)$	Original equation
$6x-15-2x+4=6-4x-5$	Distributive property
$4x-11=-4x+1$	Simplify each side
$8x-11=1$	Add $4x$ to each side
$8x=12$	Add 11 to each side
$x=\frac{12}{8}$	Multiply each side by $\frac{1}{8}$
$x=\frac{3}{2}$	Reduce to lowest terms

The solution, $\frac{3}{2}$, checks when replacing x in the original equation.

Equations Containing Fractions

If the equation contains fractions, we can use the multiplication property of equality to clear fractions from the equation by multiplying both sides by the LCD. The next examples illustrate how this is done.

EXAMPLE 5 Solve: $\frac{2}{3}x + \frac{1}{2} = -\frac{3}{4}$.

SOLUTION The LCD of all fractions in the equation is 12. We multiply both sides of the equation by 12 in order to eliminate fractions from the equation.

Step 1b:

$$12\left(\frac{2}{3}x + \frac{1}{2}\right) = 12\left(-\frac{3}{4}\right) \quad \text{Multiply each side by the LCD 12}$$

$$12\left(\frac{2}{3}x\right) + 12\left(\frac{1}{2}\right) = 12\left(-\frac{3}{4}\right) \quad \text{Distributive property on the left side}$$

$$8x + 6 = -9 \quad \text{Multiply}$$

Step 2:

$$8x = -15 \quad \text{Add } -6 \text{ to each side}$$

Step 3:

$$x = -\frac{15}{8} \quad \text{Multiply each side by } \tfrac{1}{8}$$

As you can see, we obtain the same solution as we did in Example 9 of the previous section where we worked with fractions.

> *Note* Our original equation has denominators of 3, 2, and 4. The LCD for these three denominators is 12, and it has the property that all three denominators will divide it evenly. Therefore, if we multiply both sides of our equation by 12, each denominator will divide into 12 and we will be left with an equation that does not contain any denominators other than 1.

EXAMPLE 6 Solve: $\frac{1}{3}(x - 4) + 2x = \frac{3}{5}(4x + 1)$

SOLUTION This equation contains both grouping symbols and fractions. Following our strategy for solving linear equations, we use the distributive property first to separate terms, and then use the LCD to eliminate fractions.

Step 1a:

$$\frac{1}{3}x - \frac{4}{3} + 2x = \frac{12}{5}x + \frac{3}{5} \quad \text{Distributive property}$$

Step 1b:

$$15\left(\frac{1}{3}x - \frac{4}{3} + 2x\right) = 15\left(\frac{12}{5}x + \frac{3}{5}\right) \quad \text{Multiply both sides by the LCD 15}$$

$$15\left(\frac{1}{3}x\right) - 15\left(\frac{4}{3}\right) + 15(2x) = 15\left(\frac{12}{5}x\right) + 15\left(\frac{3}{5}\right) \quad \text{Distributive property}$$

$$5x - 20 + 30x = 36x + 9 \quad \text{Multiply}$$

Step 1c:

$$35x - 20 = 36x + 9 \quad \text{Simplify the left side}$$

Step 2:

$$35x + (-35x) - 20 = 36x + (-35x) + 9 \quad \text{Add } -35x \text{ to both sides}$$

$$-20 = x + 9$$

$$-20 + (-9) = x + 9 + (-9) \quad \text{Add } -9 \text{ to both sides}$$

$$-29 = x$$

If we substitute $x = -29$ into the original equation, we get $-69 = -69$, which is a true statement; so the solution set is $\{-29\}$.

Equations Containing Decimals

Now we will consider equations that contain one or more decimals.

EXAMPLE 7 Solve the equation $0.08x + 0.09(x + 2{,}000) = 690$.

SOLUTION We can solve the equation in its original form by working with the decimals, or we can eliminate the decimals first by using the multiplication property of equality and solving the resulting equation. Both methods follow.

Method 1

Working with the decimals.

$$0.08x + 0.09(x + 2{,}000) = 690 \quad \text{Original equation}$$

Step 1a: $$0.08x + 0.09x + 0.09(2{,}000) = 690 \quad \text{Distributive property}$$

Step 1c: $$0.17x + 180 = 690 \quad \text{Simplify the left side}$$

Step 2: $$\begin{cases} 0.17x + 180 + (-180) = 690 + (-180) & \text{Add } -180 \text{ to each side} \\ 0.17x = 510 \end{cases}$$

Step 3: $$\begin{cases} \dfrac{0.17x}{0.17} = \dfrac{510}{0.17} & \text{Divide each side by } 0.17 \\ x = 3{,}000 \end{cases}$$

Note that we divided each side of the equation by 0.17 to obtain the solution. This is still an application of the multiplication property of equality because dividing by 0.17 is equivalent to multiplying by $\frac{1}{0.17}$.

Method 2

Eliminating the decimals in the beginning.

$$0.08x + 0.09(x + 2{,}000) = 690 \quad \text{Original equation}$$

Step 1a: $$0.08x + 0.09x + 180 = 690 \quad \text{Distributive property}$$

Step 1b: $$\begin{cases} 100(0.08x + 0.09x + 180) = 100(690) & \text{Multiply both sides by } 100 \\ 8x + 9x + 18{,}000 = 69{,}000 \end{cases}$$

Step 1c: $$17x + 18{,}000 = 69{,}000 \quad \text{Simplify the left side}$$

Step 2: $$17x = 51{,}000 \quad \text{Add } -18{,}000 \text{ to each side}$$

Step 3: $$\begin{cases} \dfrac{17x}{17} = \dfrac{51{,}000}{17} & \text{Divide each side by } 17 \\ x = 3{,}000 \end{cases}$$

Substituting 3,000 for x in the original equation, we have

Step 4: $$\begin{cases} 0.08(3{,}000) + 0.09(3{,}000 + 2{,}000) \stackrel{?}{=} 690 \\ 0.08(3{,}000) + 0.09(5{,}000) \stackrel{?}{=} 690 \\ 240 + 450 \stackrel{?}{=} 690 \\ 690 = 690 & \text{A true statement} \end{cases}$$

Identities and Contradictions

We conclude this section by considering some special cases. First, we offer the following definitions.

DEFINITION *identity*

An ***identity*** is an equation that is true when the variable is replaced by any real number for which the equation is defined.

With linear equations, the solution set for an identity will be the set of all real numbers. The opposite extreme is an equation that is false for any value of the variable.

DEFINITION *contradiction*

A ***contradiction*** is an equation that has no solutions. It will be a false statement for any replacement of the variable.

The solution set for a contradiction is the empty set, { }, which can also be expressed using the null set symbol, $\varnothing$.

EXAMPLE 8 Solve: $9a - 2(3a + 5) = 3(a - 1) - 7$

SOLUTION We begin by applying the distributive property to both sides of the equation.

Step 1a: $9a - 6a - 10 = 3a - 3 - 7$ Distributive property

Step 1c: $3a - 10 = 3a - 10$ Simplify

Notice that the left and right sides of the equation are identical. If we replace a with any real number, the result will be a true statement. For example,

If $a = 0$ we have
$$3(0) - 10 = 3(0) - 10$$
$$0 - 10 = 0 - 10$$
$$-10 = -10$$

If $a = 5$ we have
$$3(5) - 10 = 3(5) - 10$$
$$15 - 10 = 15 - 10$$
$$5 = 5$$

This equation is an identity. The solution set is the set of all real numbers.

EXAMPLE 9 Solve: $\frac{1}{6}(5x + 3) - \frac{1}{2}x = \frac{1}{3}(x - 5)$

SOLUTION First we distribute, and then eliminate fractions.

Step 1a: $\frac{5}{6}x + \frac{1}{2} - \frac{1}{2}x = \frac{1}{3}x - \frac{5}{3}$ Distributive property

Step 1b:
$$12\left(\frac{5}{6}x + \frac{1}{2} - \frac{1}{2}x\right) = 12\left(\frac{1}{3}x - \frac{5}{3}\right)$$ Multiply both sides by LCD 12
$$10x + 6 - 6x = 4x - 20$$

Step 1c: $4x + 6 = 4x - 20$ Simplify

Step 2:
$$4x + (-4x) + 6 = 4x + (-4x) - 20$$ Add $-4x$ to both sides
$$6 = -20$$

This statement will never be true. It will be false no matter what value we replace x with. The equation is a contradiction, and therefore has no solutions. The solution set is $\varnothing$.

Getting Ready for Class

After reading through the preceding section, respond in your own words and in complete sentences.

A. What is the first step in solving a linear equation containing parentheses?

B. If an equation contains fractions, what can you do to eliminate the fractions?

C. If an equation contains decimals, what can you do to eliminate the decimals?

D. What is an identity? How can you recognize if an equation is an identity?

Problem Set 2.4

Solve each equation.

1. $x + (2x - 1) = 2$
2. $x + (5x + 2) = 20$
3. $15 = 3(x - 1)$
4. $12 = 4(x - 5)$
5. $6 - 5(2a - 3) = 1$
6. $-8 - 2(3 - a) = 0$
7. $x - (3x + 5) = -3$
8. $x - (4x - 1) = 7$
9. $7(2y - 1) - 6y = -1$
10. $4(4y - 3) + 2y = 3$
11. $5x - 8(2x - 5) = 7$
12. $3x + 4(8x - 15) = 10$
13. $4x - (-4x + 1) = 5$
14. $-2x - (4x - 8) = -1$
15. $3(x - 3) + 2(2x) = 5$
16. $2(x - 2) + 3(5x) = 30$
17. $3x + 2(x - 2) = 6$
18. $5x - (x - 5) = 25$
19. $5x + 10(x + 8) = 245$
20. $5x + 10(x + 7) = 175$
21. $x + (x + 3)(-3) = x - 3$
22. $x - 2(x + 2) = x - 2$
23. $5(y + 2) = 4(y + 1)$
24. $3(y - 3) = 2(y - 2)$
25. $50(x - 5) = 30(x + 5)$
26. $34(x - 2) = 26(x + 2)$
27. $5(x + 2) + 3(x - 1) = -9$
28. $4(x + 1) + 3(x - 3) = 2$
29. $-2(3y + 1) = 3(1 - 6y) - 9$
30. $-5(4y - 3) = 2(1 - 8y) + 11$
31. $2(t - 3) + 3(t - 2) = 28$
32. $-3(t - 5) - 2(2t + 1) = -8$
33. $5x + 10(x + 3) + 25(x + 5) = 435$
34. $5(x + 3) + 10x + 25(x + 7) = 390$
35. $5(x - 2) - (3x + 4) = 3(6x - 8) + 10$
36. $3(x - 1) - (4x - 5) = 2(5x - 1) - 7$
37. $2(5x - 3) - (2x - 4) = 5 - (6x + 1)$
38. $3(4x - 2) - (5x - 8) = 8 - (2x + 3)$
39. $-(3x + 1) - (4x - 7) = 4 - (3x + 2)$
40. $-(6x + 2) - (8x - 3) = 8 - (5x + 1)$
41. $\frac{x}{2} + 4 = \frac{x}{5} - 1$
42. $\frac{x}{3} - 2 = \frac{x}{4} + 6$
43. $\frac{x}{7} - \frac{5}{21} = \frac{x}{3} + \frac{2}{7}$
44. $\frac{7}{6} - \frac{x}{2} = \frac{11}{12} + \frac{x}{3}$
45. $\frac{4}{9}x + \frac{5}{3} = 2 + \frac{2}{3}x$
46. $5 - \frac{3}{8}x = \frac{1}{2} + \frac{1}{4}x$
47. $\frac{1}{2}(x - 3) = \frac{1}{4}(x + 1)$
48. $\frac{1}{3}(x - 4) = \frac{1}{2}(x - 6)$
49. $\frac{3}{4}(8x - 4) + 3 = \frac{2}{5}(5x + 10) - 1$
50. $\frac{5}{6}(6x + 12) + 1 = \frac{2}{3}(9x - 3) + 5$
51. $0.5x + 0.2(18 - x) = 5.4$
52. $0.1x + 0.5(40 - x) = 32$
53. $0.06x + 0.08(100 - x) = 6.5$
54. $0.05x + 0.07(100 - x) = 6.2$

55. $0.2x - 0.5 = 0.5 - 0.2(2x - 13)$

56. $0.4x - 0.1 = 0.7 - 0.3(6 - 2x)$

57. $0.08x + 0.09(x + 2{,}000) = 860$

58. $0.11x + 0.12(x + 4{,}000) = 940$

59. $0.10x + 0.12(x + 500) = 214$

60. $0.08x + 0.06(x + 800) = 104$

61. $-0.7(2x - 7) = 0.3(11 - 4x)$

62. $-0.3(2x - 5) = 0.7(3 - x)$

63. $0.2x + 0.5(12 - x) = 3.6$

64. $0.3x + 0.6(25 - x) = 12$

Solve each equation and write the solution set. State if the equation is an identity or a contradiction.

65. $2x + 4 = 2x - 4$

66. $2x + 4 = 4 + 2x$

67. $3x - 5 = 5 - 3x$

68. $6x + 7 = 7 - 6x$

69. $4(3x + 2) - x = 7x + 2(2x + 4)$

70. $2(5x - 1) + 6 = 4(4x + 3) - 6x$

71. $\frac{4}{3} + \frac{1}{2}x - \frac{1}{6} = \frac{1}{3}x + 1 + \frac{1}{6}x$

72. $\frac{2}{3}x + \frac{1}{4} - \frac{1}{6}x = \frac{1}{3} + \frac{1}{2}x - \frac{1}{12}$

The next two problems are intended to give you practice reading, and paying attention to, the instructions that accompany the problems you are working. Working these problems is an excellent way to get ready for a test or a quiz.

73. Work each problem according to the instructions given.

a. Solve: $4x - 5 = 0$.

b. Solve: $4x - 5 = 25$.

c. Add: $(4x - 5) + (2x + 25)$.

d. Solve: $4x - 5 = 2x + 25$.

e. Multiply: $4(x - 5)$.

f. Solve: $4(x - 5) = 2x + 25$.

74. Work each problem according to the instructions given.

a. Solve: $3x + 6 = 0$.

b. Solve: $3x + 6 = 4$.

c. Add: $(3x + 6) + (7x + 4)$.

d. Solve: $3x + 6 = 7x + 4$.

e. Multiply: $3(x + 6)$.

f. Solve: $3(x + 6) = 7x + 4$.

Learning Objectives Assessment

The following problems can be used to help assess if you have successfully met the learning objectives for this section.

75. Solve: $4(2x - 1) + 9 = 7 - (5x + 3)$.

a. $-\frac{1}{13}$ **b.** 0 **c.** -2 **d.** $-\frac{8}{13}$

76. Solve: $\frac{x}{8} + \frac{1}{3} - x = \frac{x}{12} + 2 - \frac{5}{6}$.

a. $-\frac{10}{11}$ **b.** $\frac{13}{19}$ **c.** $-\frac{20}{23}$ **d.** $\varnothing$

77. Solve: $0.05x + 0.04(x - 300) = 312$.

a. 373 **b.** 3,480 **c.** 3,600 **d.** 168

78. Solve: $3 + 9(x - 4) = 7x + 2(x - 15)$

a. $\varnothing$ **b.** 0 **c.** $\frac{31}{6}$ **d.** All real numbers

Getting Ready for the Next Section

To understand all of the explanations and examples in the next section you must be able to work the problems below.

Solve each equation.

79. $40 = 2x + 12$ **80.** $80 = 2x + 12$ **81.** $12 + 2y = 6$ **82.** $3x + 18 = 6$

83. $24x = 6$ **84.** $45 = 0.75x$ **85.** $70 = x \cdot 210$ **86.** $15 = x \cdot 80$

Apply the distributive property.

87. $\frac{1}{2}(-3x + 6)$ **88.** $-\frac{1}{4}(-5x + 20)$

Formulas and Percents 2.5

Learning Objectives

In this section, we will learn how to:

1. Solve a formula for a given variable.
2. Solve basic percent problems.

Introduction

In this section, we continue solving equations by working with formulas and percents. To begin, here is the definition of a *formula*.

DEFINITION *formula*

In mathematics, a ***formula*** is an equation that contains more than one variable.

The equation $P = 2l + 2w$, which tells us how to find the perimeter of a rectangle, is an example of a formula.

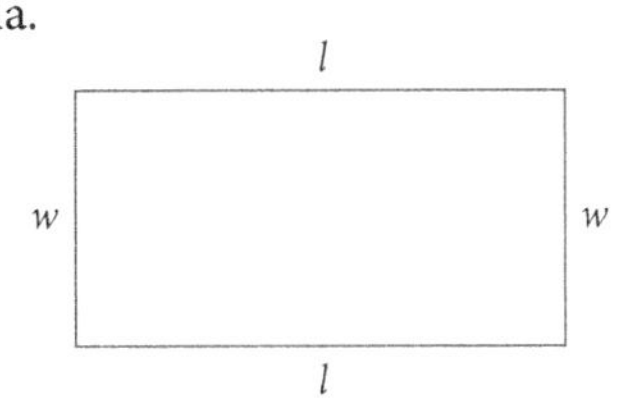

To begin our work with formulas, we will consider some examples in which we are given numerical replacements for all but one of the variables.

VIDEO EXAMPLES

SECTION 2.5

EXAMPLE 1 The perimeter P of a rectangular livestock pen is 40 feet. If the width w is 6 feet, find the length.

SOLUTION First, we substitute 40 for P and 6 for w in the formula $P = 2l + 2w$. Then we solve for l:

$$\begin{array}{lll} \text{When} & P = 40 \text{ and } w = 6 & \\ \text{the formula} & P = 2l + 2w & \\ \text{becomes} & 40 = 2l + 2(6) & \\ \text{or} & 40 = 2l + 12 & \text{Multiply 2 and 6} \\ & 28 = 2l & \text{Add } -12 \text{ to each side} \\ & 14 = l & \text{Multiply each side by } \tfrac{1}{2} \end{array}$$

To summarize our results, if a rectangular pen has a perimeter of 40 feet and a width of 6 feet, then the length must be 14 feet.

EXAMPLE 2 Find y when $x = 4$ in the formula $3x + 2y = 6$.

SOLUTION We substitute 4 for x in the formula and then solve for y:

When	$x = 4$	
the formula	$3x + 2y = 6$	
becomes	$3(4) + 2y = 6$	
or	$12 + 2y = 6$	Multiply 3 and 4
	$2y = -6$	Add -12 to each side
	$y = -3$	Multiply each side by $\frac{1}{2}$

In the next examples, we will solve a formula for one of its variables without being given numerical replacements for the other variables.

Consider the formula for the area of a triangle:

$A = \frac{1}{2}bh$

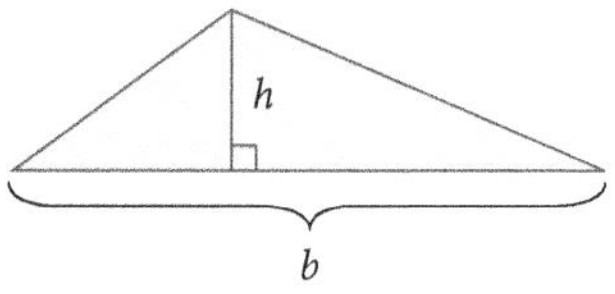

where A = area, b = length of the base, and h = height of the triangle.

Suppose we want to solve this formula for h. What we must do is isolate the variable h on one side of the equal sign. We begin by multiplying both sides by 2 to eliminate the fraction:

$$2 \cdot A = 2 \cdot \frac{1}{2}bh$$

$$2A = bh$$

Then we divide both sides by b:

$$\frac{2A}{b} = \frac{bh}{b}$$

$$h = \frac{2A}{b}$$

The original formula $A = \frac{1}{2}bh$ and the final formula $h = \frac{2A}{b}$ both give the same relationship among A, b, and h. The first one has been solved for A and the second one has been solved for h.

RULE ***Solving a formula for a variable***

To solve a formula for one of its variables, we must isolate that variable on either side of the equal sign. All other variables and constants will appear on the other side.

EXAMPLE 3 Solve $3x + 2y = 6$ for y.

SOLUTION To solve for y, we must isolate y on the left side of the equation. To begin, we use the addition property of equality to add $-3x$ to each side:

$$3x + 2y = 6 \quad \text{Original formula}$$
$$3x + (-3x) + 2y = (-3x) + 6 \quad \text{Add } -3x \text{ to each side}$$
$$2y = -3x + 6 \quad \text{Simplify the left side}$$
$$\frac{1}{2}(2y) = \frac{1}{2}(-3x + 6) \quad \text{Multiply each side by } \frac{1}{2}$$
$$y = -\frac{3}{2}x + 3 \quad \text{Multiplication}$$

EXAMPLE 4 Solve $h = vt - 16t^2$ for v.

SOLUTION Let's begin by interchanging the left and right sides of the equation. That way, the variable we are solving for, v, will be on the left side.

$$vt - 16t^2 = h \quad \text{Exchange sides}$$
$$vt - 16t^2 + 16t^2 = h + 16t^2 \quad \text{Add } 16t^2 \text{ to each side}$$
$$vt = h + 16t^2$$
$$\frac{vt}{t} = \frac{h + 16t^2}{t} \quad \text{Divide each side by } t$$
$$v = \frac{h + 16t^2}{t}$$

We know we are finished because we have isolated the variable we are solving for on the left side of the equation and it does not appear on the other side.

EXAMPLE 5 Solve for y: $\frac{y-1}{x} = \frac{3}{2}$.

SOLUTION Although we will do more extensive work with formulas of this form later in the book, we need to know how to solve this particular formula for y in order to understand some things in the next chapter. We begin by multiplying each side of the formula by x. Doing so will simplify the left side of the equation, and make the rest of the solution process simple.

$$\frac{y-1}{x} = \frac{3}{2} \quad \text{Original formula}$$
$$x \cdot \frac{y-1}{x} = \frac{3}{2} \cdot x \quad \text{Multiply each side by } x$$
$$y - 1 = \frac{3}{2}x \quad \text{Simplify each side}$$
$$y = \frac{3}{2}x + 1 \quad \text{Add 1 to each side}$$

This is our solution. If we look back to the first step, we can justify our result on the left side of the equation this way: Dividing by x is equivalent to multiplying by its reciprocal $\frac{1}{x}$. Here is what it looks like when written out completely:

$$x \cdot \frac{y-1}{x} = x\left(\frac{1}{x}\right)(y-1) = 1(y-1) = (y-1)$$

Basic Percent Problems

The next examples in this section show how basic percent problems can be translated directly into equations. To understand these examples, you must recall that *percent* means "per hundred"; that is, 75% is the same as $\frac{75}{100}$, 0.75, and, in reduced fraction form, $\frac{3}{4}$. Likewise, the decimal 0.25 is equivalent to 25%. To change a decimal to a percent, we move the decimal point two places to the right and write the % symbol. To change from a percent to a decimal, we drop the % symbol and move the decimal point two places to the left. The table that follows gives some of the most commonly used fractions and decimals and their equivalent percents.

Note You are probably familiar with the repeating decimals 0.333... and 0.666... for $\frac{1}{3}$ and $\frac{2}{3}$, respectively. The decimals we have written, $0.33\frac{1}{3}$ and $0.66\frac{2}{3}$, are equivalent to those repeating decimals, but are more helpful when converting to percents.

Fraction	Decimal	Percent
$\frac{1}{2}$	0.5	50%
$\frac{1}{4}$	0.25	25%
$\frac{3}{4}$	0.75	75%
$\frac{1}{3}$	$0.33\frac{1}{3}$	$33\frac{1}{3}\%$
$\frac{2}{3}$	$0.66\frac{2}{3}$	$66\frac{2}{3}\%$
$\frac{1}{5}$	0.2	20%
$\frac{2}{5}$	0.4	40%

EXAMPLE 6 What number is 25% of 60?

SOLUTION To solve a problem like this, we let x represent the number in question (that is, the number we are looking for). Then we translate the sentence directly into an equation by using an equal sign for the word "is" and multiplication for the word "of." Here is how it is done:

What number is 25% of 60?

$$x = 0.25 \cdot 60$$
$$x = 15$$

Notice that we must write 25% as a decimal in order to do the arithmetic in the problem.

The number 15 is 25% of 60.

EXAMPLE 7 What percent of 24 is 6?

SOLUTION Translating this sentence into an equation, as we did in Example 6, we have

What percent of 24 is 6?

$$x \cdot 24 = 6$$
or
$$24x = 6$$

Next, we multiply each side by $\frac{1}{24}$. (This is the same as dividing each side by 24.)

$$\begin{aligned} \frac{1}{24}(24x) &= \frac{1}{24}(6) \\ x &= \frac{6}{24} \\ &= \frac{1}{4} \\ &= 0.25, \text{ or } 25\% \end{aligned}$$

25% of 24 is 6, or in other words, the number 6 is 25% of 24.

EXAMPLE 8 45 is 75% of what number?

SOLUTION Again, we translate the sentence directly:

45 is 75% of what number?

$$45 = 0.75 \cdot x$$

Next we multiply each side by $\frac{1}{0.75}$ (which is the same as dividing each side by 0.75):

$$\begin{aligned} \frac{1}{0.75}(45) &= \frac{1}{0.75}(0.75x) \\ \frac{45}{0.75} &= x \\ 60 &= x \end{aligned}$$

The number 45 is 75% of 60.

EXAMPLE 9 At one time, the American Dietetic Association (ADA) recommended eating foods in which the calories from fat are less than 30% of the total calories. The nutrition labels from two kinds of granola bars are shown in Figure 1. For each bar, what percent of the total calories come from fat?

BAR I

Nutrition Facts
Serving Size 2 bars (47g)
Servings Per Container 6

Amount Per Serving	
Calories	210
Calories from Fat	**70**
	% Daily Value*
Total Fat 8g	**12%**
Saturated Fat 1g	**5%**
Cholesterol 0mg	**0%**
Sodium 150mg	**6%**
Total Carbohydrate 32g	**11%**
Dietary Fiber 2g	**10%**
Sugars 12g	
Protein 4g	

* Percent Daily Values are based on a 2,000 calorie diet. Your daily values may be higher or lower depending on your calorie needs.

BAR II

Nutrition Facts
Serving Size 1 bar (21g)
Servings Per Container 8

Amount Per Serving	
Calories	80
Calories from Fat	**15**
	% Daily Value*
Total Fat 1.5g	**2%**
Saturated Fat 0g	**0%**
Cholesterol 0mg	**0%**
Sodium 60mg	**3%**
Total Carbohydrate 16g	**5%**
Dietary Fiber 1g	**4%**
Sugars 5g	
Protein 2g	

* Percent Daily Values are based on a 2,000 calorie diet. Your daily values may be higher or lower depending on your calorie needs.

FIGURE 1

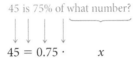

SOLUTION The information needed to solve this problem is located towards the top of each label. Each serving of Bar I contains 210 calories, of which 70 calories come from fat. To find the percent of total calories that come from fat, we must answer this question:

70 is what percent of 210?

For Bar II, one serving contains 80 calories, of which 15 calories come from fat. To find the percent of total calories that come from fat, we must answer this question:

15 is what percent of 80?

Translating each equation into symbols, we have

70 is what percent of 210?	15 is what percent of 80?
$70 = x \cdot 210$	$15 = x \cdot 80$
$x = \frac{70}{210}$	$x = \frac{15}{80}$
$x = 0.33$ to the nearest hundredth	$x = 0.19$ to the nearest hundredth
$x = 33\%$	$x = 19\%$

Comparing the two bars, 33% of the calories in Bar I are fat calories, whereas 19% of the calories in Bar II are fat calories. According to the ADA, Bar II is the healthier choice.

Applying the Concepts

As we mentioned in Chapter 1, in the U.S. system, temperature is measured on the Fahrenheit scale. In the metric system, temperature is measured on the Celsius scale. On the Celsius scale, water boils at 100 degrees and freezes at 0 degrees. To denote a temperature of 100 degrees on the Celsius scale, we write

100°C, which is read "100 degrees Celsius"

Table 1 is intended to give you an intuitive idea of the relationship between the two temperature scales. Table 2 gives the formulas, in both symbols and words, that are used to convert between the two scales.

TABLE 1

	Temperature	
Situation	**Fahrenheit**	**Celsius**
Water freezes	32°F	0°C
Room temperature	68°F	20°C
Normal body temperature	98.6°F	37°C
Water boils	212°F	100°C

TABLE 2

To Convert from	Formula in Symbols	Formula in Words
Fahrenheit to Celsius	$C = \frac{5}{9}(F - 32)$	Subtract 32, then multiply by $\frac{5}{9}$
Celsius to Fahrenheit	$F = \frac{9}{5}C + 32$	Multiply by $\frac{9}{5}$, then add 32

EXAMPLE 10 Mr. McKeague traveled to Buenos Aires with a group of friends. It was a hot day when they arrived. One of the bank kiosks indicated the temperature was 25°C. Someone asked what that would be on the Fahrenheit scale (the scale they were familiar with), and Budd, one of his friends said, "just multiply by 2 and add 30."

©Nikada/iStockPhoto.com

a. What was the temperature in °F according to Budd's approximation?

b. What is the actual temperature in °F?

c. Why does Budd's estimate work?

d. Write a formula for Budd's estimate.

SOLUTION

a. According to Budd, we multiply by 2 and add 30, so

$$2 \cdot 25 + 30 = 50 + 30 = 80°\text{F}$$

b. Using the formula $F = \frac{9}{5}C + 32$, with $C = 25$, we have

$$F = \frac{9}{5}(25) + 32 = 45 + 32 = 77°\text{F}$$

c. Budd's estimate works because $\frac{9}{5}$ is approximately 2 and 30 is close to 32.

d. In symbols, Budd's estimate is $F = 2 \cdot C + 30$.

Getting Ready for Class

After reading through the preceding section, respond in your own words and in complete sentences.

A. What is a solution to an equation?

B. What is a formula?

C. How do you solve a formula for one of its variables?

D. What does percent mean?

Problem Set 2.5

Use the formula $P = 2l + 2w$ to find the length l of a rectangular lot if

1. The width w is 50 feet and the perimeter P is 300 feet.
2. The width w is 75 feet and the perimeter P is 300 feet.

Use the formula $2x + 3y = 6$ to find y if

3. $x = 3$
4. $x = -2$
5. $x = 0$
6. $x = -3$

Use the formula $2x - 5y = 20$ to find x if

7. $y = 2$
8. $y = -4$
9. $y = 0$
10. $y = -6$

Use the equation $y = (x + 1)^2 - 3$ to find the value of y when

11. $x = -2$
12. $x = -1$
13. $x = 1$
14. $x = 2$

15. Use the formula $y = \dfrac{20}{x}$ to find y when

    a. $x = 10$
    b. $x = 5$

16. Use the formula $y = 2x^2$ to find y when

    a. $x = 5$
    b. $x = -6$

17. Use the formula $y = Kx$ to find K if

    a. $y = 15$ and $x = 3$
    b. $y = 72$ and $x = 4$

18. Use the formula $y = Kx^2$ to find K if

    a. $y = 32$ and $x = 4$
    b. $y = 45$ and $x = 3$

Solve each of the following formulas for the indicated variable.

19. $A = lw$ for l
20. $d = rt$ for r
21. $V = lwh$ for h
22. $PV = nRT$ for P
23. $P = a + b + c$ for a
24. $P = a + b + c$ for b
25. $x - 3y = -1$ for x
26. $x + 3y = 2$ for x
27. $-3x + y = 6$ for y
28. $2x + y = -17$ for y
29. $2x + 3y = 6$ for y
30. $4x + 5y = 20$ for y
31. $y - 3 = -2(x + 4)$ for y
32. $y + 5 = 2(x + 2)$ for y
33. $y - 3 = -\frac{2}{3}(x + 3)$ for y
34. $y - 1 = -\frac{1}{2}(x + 4)$ for y
35. $P = 2l + 2w$ for w
36. $P = 2l + 2w$ for l
37. $h = vt + 16t^2$ for v
38. $h = vt - 16t^2$ for v
39. $A = \pi r^2 + 2\pi rh$ for h
40. $A = 2\pi r^2 + 2\pi rh$ for h

41. Solve for y.

    a. $\dfrac{y - 1}{x} = \dfrac{3}{5}$
    b. $\dfrac{y - 2}{x} = \dfrac{1}{2}$
    c. $\dfrac{y - 3}{x} = 4$

42. Solve for y.

a. $\frac{y+1}{x} = -\frac{3}{5}$ b. $\frac{y+2}{x} = -\frac{1}{2}$ c. $\frac{y+3}{x} = -4$

Solve each formula for y.

43. $\frac{x}{7} - \frac{y}{3} = 1$

44. $\frac{x}{5} - \frac{y}{9} = 1$

45. $-\frac{1}{4}x + \frac{1}{8}y = 1$

46. $-\frac{1}{9}x + \frac{1}{3}y = 1$

Translate each of the following questions into an equation, and then solve that equation.

47. What number is 25% of 40?

48. What number is 75% of 40?

49. What number is 12% of 2,000?

50. What number is 9% of 3,000?

51. What percent of 28 is 7?

52. What percent of 28 is 21?

53. What percent of 40 is 14?

54. What percent of 20 is 14?

55. 32 is 50% of what number?

56. 16 is 50% of what number?

57. 240 is 12% of what number?

58. 360 is 12% of what number?

Applications

In the introduction to this chapter, we discussed the formula

$$T = -0.0035A + 70$$

which describes the relationship between the temperature T and altitude A when the temperature on the ground is 70°F. Use this formula for Problems 59 through 64.

59. Evaluate the formula when $A = 10{,}000$. What is the practical meaning of your answer?

60. Evaluate the formula when $A = 30{,}000$. What is the practical meaning of your answer?

61. Find A if $T = 56$.

62. Find A if $T = -42$.

63. Solve the formula for the variable A.

64. Use your result from Problem 63 to find A if $T = -7$.

65. Let $F = 212$ in the formula $C = \frac{5}{9}(F - 32)$, and solve for C. Does the value of C agree with the information in Table 1?

66. Let $C = 100$ in the formula $F = \frac{9}{5}C + 32$, and solve for F. Does the value of F agree with the information in Table 1?

67. Let $F = 68$ in the formula $C = \frac{5}{9}(F - 32)$, and solve for C. Does the value of C agree with the information in Table 1?

68. Let $C = 37$ in the formula $F = \frac{9}{5}C + 32$, and solve for F. Does the value of F agree with the information in Table 1?

69. Solve the formula $F = \frac{9}{5}C + 32$ for C.

70. Solve the formula $C = \frac{5}{9}(F - 32)$ for F.

71. How far off is Budd's estimate when the temperature is 30°C? (See Example 10)

72. How far off is Budd's estimate when the temperature is 0°C? (See Example 10)

Circumference The circumference of a circle is given by the formula $C = 2\pi r$, where r is the radius of the circle.

73. Find the circumference if the radius is 7 meters and π is $\frac{22}{7}$.

74. Find the circumference if the radius is 28 meters and π is $\frac{22}{7}$.

75. Find the radius if the circumference is 9.42 inches and π is 3.14.

76. Find the radius if the circumference is 12.56 inches and π is 3.14.

Volume The volume of a cylinder is given by the formula $V = \pi r^2 h$, where r is the radius and h is the height of the cylinder.

77. Find the volume if the height is 2 centimeters, the radius is 3 centimeters, and π is 3.14.

78. Find the volume if the height is 3 centimeters, the radius is 2 centimeters, and π is 3.14.

79. Find the height if the volume is 42 cubic feet, the radius is $\frac{7}{22}$ feet, and π is $\frac{22}{7}$.

80. Find the height if the volume is 84 cubic inches, the radius is $\frac{7}{11}$ inches, and π is $\frac{22}{7}$.

Nutrition Labels The nutrition label in Figure 2 is from a quart of vanilla ice cream. The label in Figure 3 is from a pint of vanilla frozen yogurt. Use the information on these labels for problems 81–84. Round your answers to the nearest tenth of a percent.

Nutrition Facts
Serving Size 1/2 cup (65g)
Servings 8

Amount/Serving	
Calories 150	Calories from Fat 90
	% Daily Value*
Total Fat 10g	**16%**
Saturated Fat 6g	**32%**
Cholesterol 35mg	**12%**
Sodium 30mg	**1%**
Total Carbohydrate 14g	**5%**
Dietary Fiber 0g	**0%**
Sugars 11g	
Protein 2g	
Vitamin A 6% •	Vitamin C 0%
Calcium 6% •	Iron 0%

* Percent Daily Values are based on a 2,000 calorie diet.

FIGURE 2 *Vanilla ice cream*

Nutrition Facts
Serving Size 1/2 cup (98g)
Servings Per Container 4

Amount Per Serving	
Calories 160	Calories from Fat 25
	% Daily Value*
Total Fat 2.5g	**4%**
Saturated Fat 1.5g	**7%**
Cholesterol 45mg	**15%**
Sodium 55mg	**2%**
Total Carbohydrate 26g	**9%**
Dietary Fiber 0g	**0%**
Sugars 19g	
Protein 8g	
Vitamin A 0% •	Vitamin C 0%
Calcium 25% •	Iron 0%

* Percent Daily Values are based on a 2,000 calorie diet.

FIGURE 3 *Vanilla frozen yogurt*

81. What percent of the calories in one serving of the vanilla ice cream are fat calories?

82. What percent of the calories in one serving of the frozen yogurt are fat calories?

83. One serving of frozen yogurt is 98 grams, of which 26 grams are carbohydrates. What percent of one serving are carbohydrates?

84. One serving of vanilla ice cream is 65 grams. What percent of one serving is sugar?

Learning Objectives Assessment

The following problems can be used to help assess if you have successfully met the learning objectives for this section.

85. Solve the formula $2x - 3y = 6$ for y.

a. $y = \frac{2}{3}x - 2$ **b.** $y = 2 - \frac{2}{3}x$

c. $y = -2x - 2$ **d.** $y = 2x - 2$

86. 18 is what percent of 60?

a. 0.3% **b.** 30% **c.** 3.3% **d.** 10.8%

Getting Ready for the Next Section

To understand all of the explanations and examples in the next section, you must be able to work the problems below.

Write an equivalent expression in English. Include the words *sum* and *difference* when possible.

87. $4 + 1$ **88.** $7 + 3$ **89.** $6 - 2$ **90.** $8 - 1$

91. $x - 15$ **92.** $2x + 3$ **93.** $4(x - 3)$ **94.** $2(2x - 5)$

For each of the following phrases, write an equivalent mathematical expression.

95. Twice the sum of 6 and 3

96. Four added to the product of 5 and -1

97. The sum of twice 5 and 3

98. Twice the difference of 8 and 2

99. The sum of a number and five

100. The difference of ten and a number

101. Five times the sum of a number and seven

102. Five times the difference of twice a number and six

SPOTLIGHT ON SUCCESS *University of North Alabama*

Pride is a personal commitment.
It is an attitude which separates excellence from mediocrity.
—William Blake

University of NORTH ALABAMA 1830

The University of North Alabama places its Pride Rock, a 60-pound granite stone engraved with a lion's paw print, behind the north end zone at all home football games. The rock reminds current Lion players of the proud athletic traditions that have been established at the school, and to take pride in their efforts on the field.

Photo courtesy UNA

The same idea holds true for your work in your math class. Take pride in it. When you turn in an assignment, it should be accurate and easy for the instructor to read. It shows that you care about your progress in the course and that you take pride in your work. The work that you turn in to your instructor is a reflection of you. As the quote from William Blake indicates, pride is a personal commitment; a decision that you make, yourself. And once you make that commitment to take pride in the work you do in your math class, you have directed yourself toward excellence, and away from mediocrity.

Applications 2.6

Learning Objectives

In this section, we will learn how to:

1. Solve number problems.
2. Solve age problems.
3. Solve geometry problems.
4. Solve coin problems.

Introduction

As you begin reading through the examples in this section, you may find yourself asking why some of these problems seem so contrived. The title of the section is "Applications," but many of the problems here don't seem to have much to do with "real life." You are right about that. Consider the following quote from *A Tour of the Calculus*, by David Berlinski:

> "The examples offered by elementary algebra are often uninspiring if only because no one wishes to really know which numbers correspond to the unknowns, the unknowns in word problems referring always to a strangely meditative farmer standing forlornly on that illustrated textbook hill of his, wondering in a way that suggests nothing of the power of mathematics how many turnips he might grow if he had two tons of fertilizer. ... But as always in any great art, the matter and the method need not necessarily be the same at first, the matter trivial (farmers and their fields), but the method, in the case of elementary algebra, suggesting in incremental steps the power of a system of equations adequate to nothing less than the description of the world."

The problems in this section may be contrived, but the strategy we will learn in solving these problems is very powerful, and it is easiest to master this strategy by beginning with some simple (if uninspiring) problems.

To begin this section, we list the steps used in solving application problems. We call this strategy the *Blueprint for Problem Solving*. It is an outline that will overlay the solution process we use on all application problems.

HOW TO *Use the Blueprint for Problem Solving*

Step 1: ***Read*** the problem, and then mentally ***list*** the items that are known and the items that are unknown.

Step 2: ***Assign a variable*** to one of the unknown items. (In most cases this will amount to letting x represent the item that is asked for in the problem.) Then ***translate*** the other ***information*** in the problem to expressions involving the variable.

Step 3: ***Reread*** the problem, and then ***write an equation***, using the items and variables listed in steps 1 and 2, that describes the situation. We call this equation a ***mathematical model*** for the problem.

Step 4: ***Solve the equation*** found in step 3.

Step 5: ***Write*** your ***answer*** using a complete sentence.

Step 6: ***Reread*** the problem, and ***check*** your solution with the original words in the problem.

There are a number of substeps within each of the steps in our blueprint. For instance, with steps 1 and 2 it is always a good idea to draw a diagram or picture if it helps visualize the relationship between the items in the problem. In other cases, a table helps organize the information. As you gain more experience using the blueprint to solve application problems, you will find additional techniques that expand the blueprint.

To help with problems of the type shown next in Example 1, here are some additional common English words and phrases used to indicate equality.

English	Algebra
a is *b*	$a = b$
a was *b*	$a = b$
a is equal to *b*	$a = b$
a equals *b*	$a = b$
a represents *b*	$a = b$
a results in *b*	$a = b$

Number Problems

VIDEO EXAMPLES

SECTION 2.6

EXAMPLE 1 The sum of twice a number and three is seven. Find the number.

SOLUTION Using the Blueprint for Problem Solving as an outline, we solve the problem as follows:

Step 1: ***Read*** the problem, and then mentally ***list*** the items that are known and the items that are unknown.

Known items: The numbers 3 and 7

Unknown items: The number in question

Step 2: ***Assign a variable*** to one of the unknown items. Then ***translate*** the other ***information*** in the problem to expressions involving the variable.

Let x represent the number asked for in the problem, then "The sum of twice a number and three" translates to $2x + 3$.

Step 3: ***Reread*** the problem, and then ***write an equation,*** using the items and variables listed in steps 1 and 2, that describes the situation.

With all word problems, the word *is* translates to $=$.

$$\underbrace{\text{The sum of twice } x \text{ and } 3}_{2x + 3} \ \underset{=}{\text{is}} \ \underset{7}{7}$$

$$2x + 3 = 7$$

Step 4: ***Solve the equation*** found in step 3.

$$2x + 3 = 7$$

$$2x + 3 + (-3) = 7 + (-3)$$

$$2x = 4$$

$$\frac{1}{2}(2x) = \frac{1}{2}(4)$$

$$x = 2$$

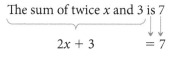

Step 5: ***Write*** your ***answer*** using a complete sentence.

The number is 2.

Step 6: ***Reread*** the problem, and ***check*** your solution with the original words in the problem.

The sum of twice 2 and 3 is 7; a true statement.

You may find some examples and problems in this section that you can solve without using algebra or our blueprint. It is very important that you solve these problems using the methods we are showing here. The purpose behind these problems is to give you experience using the blueprint as a guide to solving problems written in words. Your answers are much less important than the work that you show to obtain your answer. You will be able to condense the steps in the blueprint later in the course. For now, though, you need to show your work in the same detail that we are showing in the examples in this section.

EXAMPLE 2 One number is three more than twice another; their sum is eighteen. Find the numbers.

SOLUTION

Step 1: **Read and list.**

Known items: Two numbers that add to 18. One is 3 more than twice the other.

Unknown items: The numbers in question.

Step 2: **Assign a variable, and translate information.**

Let x be the first number. The other is $2x + 3$.

Step 3: **Reread, and write an equation.**

$$\underbrace{\text{Their sum}}_{x + (2x + 3)} \quad \underset{=}{\text{is}} \quad \underset{18}{18}$$

$$x + (2x + 3) = 18$$

Step 4: **Solve the equation.**

$$x + (2x + 3) = 18$$
$$3x + 3 = 18$$
$$3x + 3 + (-3) = 18 + (-3)$$
$$3x = 15$$
$$x = 5$$

Step 5: **Write the answer.**

The first number is 5. The other is $2 \cdot 5 + 3 = 13$.

Step 6: **Reread, and check.**

The sum of 5 and 13 is 18, and 13 is 3 more than twice 5.

Age Problem

Remember as you read through the steps in the solutions to the examples in this section that step 1 is done mentally. Read the problem, and then mentally list the items that you know and the items that you don't know. The purpose of step 1 is to give you direction as you begin to work application problems. Finding the solution to an application problem is a process; it doesn't happen all at once. The first step is to read the problem with a purpose in mind. That purpose is to mentally note the items that are known and the items that are unknown.

EXAMPLE 3 Bill is 6 years older than Tom. Three years ago Bill's age was four times Tom's age. Find the age of each boy now.

SOLUTION Applying the Blueprint for Problem Solving, we have

Step 1: **Read and list.**

Known items: Bill is 6 years older than Tom. Three years ago Bill's age was four times Tom's age.

Unknown items: Bill's age and Tom's age

Step 2: **Assign a variable, and translate information.**
Let x represent Tom's age now. That makes Bill $x + 6$ years old now. A table like the one shown here can help organize the information in an age problem. Notice how we placed the x in the box that corresponds to Tom's age now.

	Three Years Ago	Now
Bill		$x + 6$
Tom		x

If Tom is x years old now, 3 years ago he was $x - 3$ years old. If Bill is $x + 6$ years old now, 3 years ago he was $x + 6 - 3 = x + 3$ years old. We use this information to fill in the remaining blanks in the table.

	Three Years Ago	Now
Bill	$x + 3$	$x + 6$
Tom	$x - 3$	x

Step 3: **Reread, and write an equation.**
Reading the problem again, we see that

$$\underbrace{\text{Three years ago Bill's age}}_{x+3}\ \underset{=}{\text{was}}\ \underset{4}{\text{four}}\ \underset{\cdot}{\text{times}}\ \underbrace{\text{Tom's age}}_{(x-3)}$$

Step 4: **Solve the equation.**

$$x + 3 = 4(x - 3)$$
$$x + 3 = 4x - 12$$
$$x + (-x) + 3 = 4x + (-x) - 12$$
$$3 = 3x - 12$$
$$3 + 12 = 3x - 12 + 12$$
$$15 = 3x$$
$$x = 5$$

Step 5: **Write the answer.**
Tom is 5 years old. Bill is 11 years old.

Step 6: **Reread, and check.**
If Tom is 5 and Bill is 11, then Bill is 6 years older than Tom. Three years ago Tom was 2 and Bill was 8. At that time, Bill's age was four times Tom's age. As you can see, the answers check with the original problem.

Geometry Problem

To understand Example 4 completely, you need to recall from Chapter 1 that the perimeter of a rectangle is the sum of the lengths of the sides. The formula for the perimeter is $P = 2l + 2w$.

EXAMPLE 4 The length of a rectangle is 5 inches more than twice the width. The perimeter is 34 inches. Find the length and width.

SOLUTION When working problems that involve geometric figures, a sketch of the figure helps organize and visualize the problem.

Step 1: **Read and list.**
Known items: The figure is a rectangle. The length is 5 inches more than twice the width. The perimeter is 34 inches.
Unknown items: The length and the width

Step 2: **Assign a variable, and translate information.**
Because the length is given in terms of the width (the length is 5 more than twice the width), we let x be the width of the rectangle. The length is 5 more than twice the width, so it must be $2x + 5$. The diagram below is a visual description of the relationships we have listed so far.

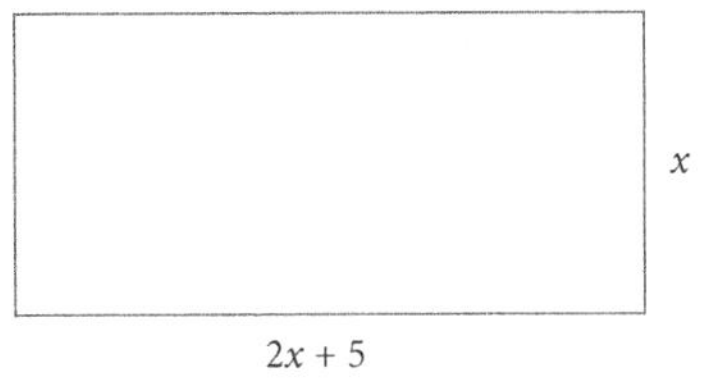

Step 3: **Reread, and write an equation.**
The equation that describes the situation is

$$2(\text{length}) + 2(\text{width}) = \text{perimeter}$$
$$2(2x+5) + 2x = 34$$

Step 4: **Solve the equation.**

$$\begin{aligned} 2(2x+5)+2x &= 34 && \text{Original equation} \\ 4x+10+2x &= 34 && \text{Distributive property} \\ 6x+10 &= 34 && \text{Add } 4x \text{ and } 2x \\ 6x &= 24 && \text{Add } -10 \text{ to each side} \\ x &= 4 && \text{Divide each side by 6} \end{aligned}$$

Step 5: **Write the answer.**
The width x is 4 inches. The length is $2x + 5 = 2(4) + 5 = 13$ inches.

Step 6: **Reread, and check.**
If the length is 13 and the width is 4, then the perimeter must be $2(13) + 2(4) = 26 + 8 = 34$, which checks with the original problem.

Coin Problem

EXAMPLE 5 Jennifer has \$2.45 in dimes and nickels. If she has 8 more dimes than nickels, how many of each coin does she have?

SOLUTION

Step 1: **Read and list.**
Known items: The type of coins, the total value of the coins, and that there are 8 more dimes than nickels.
Unknown items: The number of nickels and the number of dimes

Step 2: **Assign a variable, and translate information.**
If we let x be the number of nickels, then $x + 8$ is the number of dimes. Because the value of each nickel is 5 cents, the amount of money in nickels is $5x$. Similarly, because each dime is worth 10 cents, the amount of money in dimes is $10(x + 8)$. Here is a table that summarizes the information we have so far:

	Nickels	Dimes
Number	x	$x + 8$
Value (in cents)	$5x$	$10(x + 8)$

Step 3: **Reread, and write an equation.**
Because the total value of all the coins is 245 cents, the equation that describes this situation is

$$\text{Amount of money in nickels} + \text{Amount of money in dimes} = \text{Total amount of money}$$
$$5x + 10(x+8) = 245$$

Step 4: **Solve the equation.**
To solve the equation, we apply the distributive property first.

$$
\begin{aligned}
5x + 10x + 80 &= 245 && \text{Distributive property} \\
15x + 80 &= 245 && \text{Add } 5x \text{ and } 10x \\
15x &= 165 && \text{Add } -80 \text{ to each side} \\
x &= 11 && \text{Divide each side by } 15
\end{aligned}
$$

Step 5: **Write the answer.**
The number of nickels is $x = 11$.
The number of dimes is $x + 8 = 11 + 8 = 19$.

Step 6: **Reread, and check.**
To check our results

11 nickels are worth $5(11) = 55$ cents
19 dimes are worth $10(19) = 190$ cents

The total value is 245 cents = \$2.45

When you begin working the problems in the problem set that follows, there are a few things to remember. The first is that you may have to read the problems a number of times before you begin to see how to solve them. The second thing to remember is that word problems are not always solved correctly the first time you try them. Sometimes it takes a few attempts and some wrong answers before you can set up and solve these problems correctly.

Getting Ready for Class

After reading through the preceding section, respond in your own words and in complete sentences.

A. What good does it do you to solve application problems even when they don't have much to do with real life?

B. What is the first step in the Blueprint for Problem Solving?

C. What is the last thing you do when solving an application problem?

D. With coin problems, why do we multiply the number of coins by the value of each coin?

Problem Set 2.6

For Problems 1-8, translate each sentence into an equivalent equation in symbols.

1. The sum of x and 5 is 14.
2. The difference of x and 4 is 8.
3. The quotient of x and 3 is equal to the sum of x and 2.
4. The quotient of x and 2 is equal to the difference of x and 4.
5. Twice the difference of a number and nine, when increased by five, results in eleven.
6. Three times the sum of a number and six, when decreased by four, gives a result of twenty.
7. Half the sum of a number and 5 is three times the difference of the number and 5.
8. One-third a number added to one-half the number equals two less than the number.

Solve the following word problems. Follow the steps given in the Blueprint for Problem Solving.

Number Problems

9. The sum of a number and five is thirteen. Find the number.
10. The difference of ten and a number is negative eight. Find the number.
11. The sum of twice a number and four is fourteen. Find the number.
12. The difference of four times a number and eight is sixteen. Find the number.
13. Five times the sum of a number and seven is thirty. Find the number.
14. Five times the difference of twice a number and six is negative twenty. Find the number.
15. One number is two more than another. Their sum is eight. Find both numbers.
16. One number is three less than another. Their sum is fifteen. Find the numbers.
17. One number is four less than three times another. If their sum is increased by five, the result is twenty-five. Find the numbers.
18. One number is five more than twice another. If their sum is decreased by ten, the result is twenty-two. Find the numbers.

Age Problems

19. Shelly is 3 years older than Michele. Four years ago, the sum of their ages was 67. Find the age of each person now.

	Four Years Ago	Now
Shelly	$x - 1$	$x + 3$
Michele	$x - 4$	x

20. Cary is 9 years older than Dan. In 7 years, the sum of their ages will be 93. Find the age of each man now. (Begin by filling in the table.)

	Now	In Seven Years
Cary	$x + 9$	
Dan	x	$x + 7$

21. Cody is twice as old as Evan. Three years ago, the sum of their ages was 27. Find the age of each boy now.

	Three Years Ago	Now
Cody		
Evan	$x - 3$	x

22. Justin is 2 years older than Ethan. In 9 years, the sum of their ages will be 30. Find the age of each boy now.

	Now	In Nine Years
Justin		
Ethan	x	

23. Fred is 4 years older than Barney. Five years ago, the sum of their ages was 48. How old are they now?

	Five Years Ago	Now
Fred		
Barney		x

24. Tim is 5 years older than JoAnn. Six years from now, the sum of their ages will be 79. How old are they now?

	Now	Six Years From Now
Tim		
JoAnn	x	

25. Jack is twice as old as Lacy. In 3 years, the sum of their ages will be 54. How old are they now?

26. John is 4 times as old as Martha. Five years ago, the sum of their ages was 50. How old are they now?

27. Pat is 20 years older than his son Patrick. In 2 years, Pat will be twice as old as Patrick. How old are they now?

28. Diane is 23 years older than her daughter Amy. In 6 years, Diane will be twice as old as Amy. How old are they now?

Geometry Problems

29. The perimeter of a square is 36 inches. Find the length of one side.

30. The perimeter of a square is 44 centimeters. Find the length of one side.

31. The perimeter of a square is 60 feet. Find the length of one side.

32. The perimeter of a square is 84 meters. Find the length of one side.

33. One side of a triangle is three times the shortest side. The third side is 7 feet more than the shortest side. The perimeter is 62 feet. Find all three sides.

34. One side of a triangle is half the longest side. The third side is 10 meters less than the longest side. The perimeter is 45 meters. Find all three sides.

35. One side of a triangle is half the longest side. The third side is 12 feet less than the longest side. The perimeter is 53 feet. Find all three sides.

36. One side of a triangle is 6 meters more than twice the shortest side. The third side is 9 meters more than the shortest side. The perimeter is 75 meters. Find all three sides.

37. The length of a rectangle is 5 inches more than the width. The perimeter is 34 inches. Find the length and width.

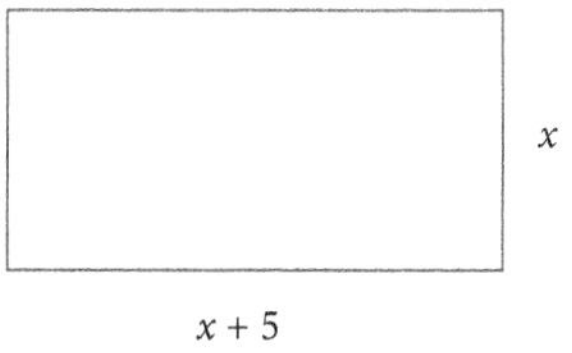

38. The width of a rectangle is 3 feet less than the length. The perimeter is 10 feet. Find the length and width.

39. The length of a rectangle is 7 inches more than twice the width. The perimeter is 68 inches. Find the length and width.

40. The length of a rectangle is 4 inches more than three times the width. The perimeter is 72 inches. Find the length and width.

41. The length of a rectangle is 6 feet more than three times the width. The perimeter is 36 feet. Find the length and width.

42. The length of a rectangle is 3 feet less than twice the width. The perimeter is 54 feet. Find the length and width.

Coin Problems

43. Marissa has \$4.40 in quarters and dimes. If she has 5 more quarters than dimes, how many of each coin does she have?

	Dimes	**Quarters**
Number	x	$x + 5$
Value (in cents)	$10(x)$	$25(x + 5)$

44. Kendra has \$2.75 in dimes and nickels. If she has twice as many dimes as nickels, how many of each coin does she have?

	Nickels	Dimes
Number	x	$2x$
Value (in cents)	$5(x)$	

45. Tanner has \$4.35 in nickels and quarters. If he has 15 more nickels than quarters, how many of each coin does he have?

	Nickels	Quarters
Number	$x + 15$	x
Value (in cents)		

46. Connor has \$9.00 in dimes and quarters. If he has twice as many quarters as dimes, how many of each coin does he have?

	Dimes	Quarters
Number	x	$2x$
Value (in cents)		

47. Sue has \$2.10 in dimes and nickels. If she has 9 more dimes than nickels, how many of each coin does she have?

48. Mike has \$1.55 in dimes and nickels. If he has 7 more nickels than dimes, how many of each coin does he have?

49. Katie has a collection of nickels, dimes, and quarters with a total value of \$4.35. There are 3 more dimes than nickels and 5 more quarters than nickels. How many of each coin is in her collection? (*Hint:* Let x be the number of nickels.)

	Nickels	Dimes	Quarters
Number	x		
Value			

50. Mary Jo has \$3.90 worth of nickels, dimes, and quarters. The number of nickels is 3 more than the number of dimes. The number of quarters is 7 more than the number of dimes. How many of each coin does she have? (*Hint:* Let x be the number of dimes.)

	Nickels	Dimes	Quarters
Number		x	
Value			

51. Cory has a collection of nickels, dimes, and quarters with a total value of \$2.55. There are 6 more dimes than nickels and twice as many quarters as nickels. How many of each coin is in her collection?

	Nickels	Dimes	Quarters
Number	x		
Value			

52. Kelly has a collection of nickels, dimes, and quarters with a total value of \$7.40. There are four more nickels than dimes and twice as many quarters as nickels. How many of each coin is in her collection?

	Nickels	Dimes	Quarters
Number		x	
Value			

Learning Objectives Assessment

The following problems can be used to help assess if you have successfully met the learning objectives for this section.

For each of the following, which equation is an appropriate model for the given problem?

53. One number is three more than another. Their sum is twenty. Find both numbers.

a. $x + 3 = 20$ **b.** $x + 3 = x + 20$

c. $x + (x + 3) = 20$ **d.** $x = (x + 3) + 20$

54. Allison is 8 years older than Kaitlin. Five years ago, Allison was three times as old as Kaitlin. Find Kaitlin's age now.

a. $x + 3 = 3(x - 5)$ **b.** $x + 8 = 3x - 5$

c. $x - 5 = 3(x + 8)$ **d.** $3(x + 8) - 5 = x$

55. The length of a rectangle is 2 meters less than three times the width. The perimeter is 36 meters. Find the width.

a. $3x - 2 = 36$ **b.** $(3x - 2) + x = 36$

c. $3(x - 2) + x = 36$ **d.** $2(3x - 2) + 2x = 36$

56. Valerie has \$2.75 in quarters and dimes. If she has four more quarters than dimes, how many dimes does she have?

a. $10(x + 4) + 25x = 275$ **b.** $x + (x + 4) = 275$

c. $10x + 25(x + 4) = 2.75$ **d.** $10x + 25(x + 4) = 275$

Getting Ready for the Next Section

To understand all of the explanations and examples in the next section you must be able to work the problems below.

Simplify the following expressions.

57. $x + 2x + 2x$

58. $x + 2x + 3x$

59. $x + 0.075x$

60. $x + 0.065x$

61. $0.09(x + 2{,}000)$

62. $0.06(x + 1{,}500)$

Solve the following equations.

63. $0.02x + 0.06(x + 1{,}500) = 570$

64. $0.08x + 0.09(x + 2{,}000) = 690$

65. $x + 2x + 3x = 180$

66. $2x + 3x + 5x = 180$

SPOTLIGHT ON SUCCESS *Student Instructor Lauren*

There are a lot of word problems in algebra and many of them involve topics that I don't know much about. I am better off solving these problems if I know something about the subject. So, I try to find something I can relate to. For instance, an example may involve the amount of fuel used by a pilot in a jet airplane engine. In my mind, I'd change the subject to something more familiar, like the mileage I'd be getting in my car and the amount spent on fuel, driving from my hometown to my college. Changing these problems to more familiar topics makes math much more interesting and gives me a better chance of getting the problem right. It also helps me to understand how greatly math affects and influences me in my everyday life. We really do use math more than we would like to admit—budgeting our income, purchasing gasoline, planning a day of shopping with friends—almost everything we do is related to math. So the best advice I can give with word problems is to learn how to associate the problem with something familiar to you.

You should know that I have always enjoyed math. I like working out problems and love the challenges of solving equations like individual puzzles. Although there are more interesting subjects to me, and I don't plan on pursuing a career in math or teaching, I do think it's an important subject that will help you in any profession.

More Applications 2.7

Learning Objectives

In this section, we will learn how to:

1. Solve consecutive integer problems.
2. Solve interest problems.
3. Solve mixture problems.
4. Solve problems involving triangles.

Introduction

Now that you have worked through a number of application problems using our blueprint, you probably have noticed that step 3, in which we write an equation that describes the situation, is the key step. Anyone with experience solving application problems will tell you that there will be times when your first attempt at writing a model results in the wrong equation. Remember, mistakes are part of the process of learning to do things correctly. Many times the correct equation will become obvious after you have written an equation that is partially wrong. In any case it is better to write an equation that is partially wrong and be actively involved with the problem than to write nothing at all. Application problems, like other problems in algebra, are not always solved correctly the first time.

Consecutive Integers

Our first example involves *consecutive integers*. When we ask for consecutive integers, we mean integers that are next to each other on the number line, like

$$5 \text{ and } 6, \qquad 13 \text{ and } 14, \qquad \text{or } -4 \text{ and } -3$$

In the dictionary, consecutive is defined as following one another in uninterrupted order. If we ask for consecutive odd integers, then we mean odd integers that follow one another on the number line. For example,

$$3 \text{ and } 5, \qquad 11 \text{ and } 13, \qquad \text{and } -9 \text{ and } -7$$

are consecutive odd integers. As you can see, to get from one odd integer to the next consecutive odd integer we add 2.

If we are asked to find two consecutive integers and we let x equal the first integer, the next one must be $x + 1$, because consecutive integers always differ by 1. Likewise, if we are asked to find two consecutive odd or even integers, and we let x equal the first integer, then the next one will be $x + 2$ because consecutive even or odd integers always differ by 2. The following table summarizes this information.

In Words	Using Algebra	Example
Two consecutive integers	$x, x + 1$	The sum of two consecutive integers is 15. $x + (x + 1) = 15$ or $7 + 8 = 15$
Three consecutive integers	$x, x + 1, x + 2$	The sum of three consecutive integers is 24. $x + (x + 1) + (x + 2) = 24$ or $7 + 8 + 9 = 24$
Two consecutive odd integers	$x, x + 2$	The sum of two consecutive odd integers is 16. $x + (x + 2) = 16$ or $7 + 9 = 16$
Two consecutive even integers	$x, x + 2$	The sum of two consecutive even integers is 18. $x + (x + 2) = 18$ or $8 + 10 = 18$

VIDEO EXAMPLES

SECTION 2.7

EXAMPLE 1 The sum of two consecutive odd integers is 28. Find the two integers.

SOLUTION

Step 1: **Read and list.**
Known items: Two consecutive odd integers. Their sum is equal to 28.
Unknown items: The numbers in question.

Step 2: **Assign a variable, and translate information.**
If we let x be the first of the two consecutive odd integers, then $x + 2$ is the next consecutive one.

Step 3: **Reread, and write an equation.**
Their sum is 28.

$$x + (x + 2) = 28$$

Step 4: **Solve the equation.**

$$2x + 2 = 28 \quad \text{Simplify the left side}$$
$$2x = 26 \quad \text{Add } -2 \text{ to each side}$$
$$x = 13 \quad \text{Multiply each side by } \tfrac{1}{2}$$

Step 5: **Write the answer.**
The first of the two integers is 13. The second of the two integers will be two more than the first, which is 15.

Step 6: **Reread, and check.**
Suppose the first integer is 13. The next consecutive odd integer is 15. The sum of 15 and 13 is 28.

Interest

If a person invests an amount of money P, called the *principal*, in an account that has an *annual interest rate* r, then the *interest* the person earns after one year is given by the formula

$$\text{Interest} = \text{Rate(Principal)}$$
$$I = rP$$

When using this formula, we must remember to express the interest rate r as a decimal.

EXAMPLE 2 Suppose you invest a certain amount of money in an account that earns 8% in annual interest. At the same time, you invest \$2,000 more than that in an account that pays 9% in annual interest. If the total interest from both accounts at the end of the year is \$690, how much is invested in each account?

SOLUTION

Step 1: **Read and list.**

Known items: The interest rates, the total interest earned, and how much more is invested at 9%

Unknown items: The amounts invested in each account

Step 2: **Assign a variable, and translate information.**

Let x be the amount of money invested at 8%. From this, $x + 2{,}000$ is the amount of money invested at 9%. The interest earned on x dollars invested at 8% is $0.08x$. The interest earned on $x + 2{,}000$ dollars invested at 9% is $0.09(x + 2{,}000)$.

Here is a table that summarizes this information:

	Dollars Invested at 8%	Dollars Invested at 9%
Number of	x	$x + 2{,}000$
Interest on	$0.08x$	$0.09(x + 2{,}000)$

Step 3: **Reread, and write an equation.**

Because the total amount of interest earned from both accounts is \$690, the equation that describes the situation is

$$\text{Interest earned at 8\%} + \text{Interest earned at 9\%} = \text{Total interest earned}$$

$$0.08x + 0.09(x + 2{,}000) = 690$$

Step 4: **Solve the equation.**

$$0.08x + 0.09(x + 2{,}000) = 690$$

$$0.08x + 0.09x + 180 = 690 \quad \text{Distributive property}$$

$$0.17x + 180 = 690 \quad \text{Add } 0.08x \text{ and } 0.09x$$

$$0.17x = 510 \quad \text{Add } -180 \text{ to each side}$$

$$x = 3{,}000 \quad \text{Divide each side by } 0.17$$

Step 5: **Write the answer:**

The amount of money invested at 8% is \$3,000, whereas the amount of money invested at 9% is $x + 2{,}000 = 3{,}000 + 2{,}000 = \$5{,}000$.

Step 6: **Reread, and check.**

The interest at 8% is 8% of $3{,}000 = 0.08(3{,}000) = \$240$

The interest at 9% is 9% of $5{,}000 = 0.09(5{,}000) = \$450$

The total interest is \$690

Mixture Problem

Another type of application that involves working with percents is creating mixtures of various solutions. In this context, a solution is a liquid that contains two different substances.

For example, the liquid in the cooling system for most automobiles is a solution containing both water and antifreeze. If the cooling system holds 20 quarts of a 60% antifreeze solution, then

$$60\% \text{ of } 20 \text{ quarts} = 0.60(20 \text{ qt}) = 12 \text{ qt is antifreeze, and}$$
$$40\% \text{ of } 20 \text{ quarts} = 0.40(20 \text{ qt}) = 8 \text{ qt is water}$$

EXAMPLE 3 How much of a 30% alcohol solution and 60% alcohol solution must be mixed to get 15 gallons of 50% alcohol solution?

SOLUTION

Step 1: **Read and list.**

Known items: There are two solutions that together must total 15 gallons. 30% of one of the solutions is alcohol and the rest is water, whereas the other solution is 60% alcohol and the other is 40% water. The mixture must be 50% alcohol.

Unknown items: The gallons of each individual solution we need.

Step 2: **Assign a variable, and translate information.**

Let x represent the number of gallons of the 30% alcohol solution. Because a total of 15 gallons of solution is required, $15 - x$ is the number of gallons of 60% alcohol solution that must be used. The amount of alcohol in the 30% solution is $0.30x$. The amount of alcohol in the 60% solution is $0.60(15 - x)$. The amount of alcohol in the mixture will be $0.50(15) = 7.5$ gallons.

Here is a table that summarizes this information.

	30% Solution	60% Solution	Mixture
Number of Gallons	x	$15 - x$	15
Gallons of alcohol	$0.30x$	$0.60(15 - x)$	$0.50(15)$

Step 3: **Reread, and write an equation.**

Because the total amount of alcohol from the two solutions must add up to the amount of alcohol in the mixture, the equation that describes the situation is

Gallons alcohol in 30% solution	+	Gallons alcohol in 60% solution	=	Gallons alcohol in mixture
$0.30x$	+	$0.60(15 - x)$	=	$0.50(15)$

Step 4: **Solve the equation.**

$$0.30x + 0.60(15 - x) = 0.50(15)$$

$$0.30x + 9 - 0.60x = 7.5 \quad \text{Distributive property}$$

$$9 - 0.30x = 7.5 \quad \text{Simplify left side}$$

$$-0.30x = -1.5 \quad \text{Add } -9 \text{ to each side}$$

$$x = 5 \quad \text{Divide each side by } -0.30$$

Step 5: **Write the answer.**
5 gallons of the 30% alcohol solution should be mixed with $15 - 5 = 10$ gallons of the 60% alcohol solution.

Step 6: **Reread, and check.**
$5 + 10 = 15$ gallons, and $0.30(5) + 0.60(10) = 1.5 + 6 = 7.5$ gallons, which is 50% of 15 gallons.

Triangles

FACTS FROM GEOMETRY *Labeling Triangles and the Sum of the Angles in a Triangle*

One way to label the important parts of a triangle is to label the vertices with capital letters and the sides with small letters, as shown in Figure 1.

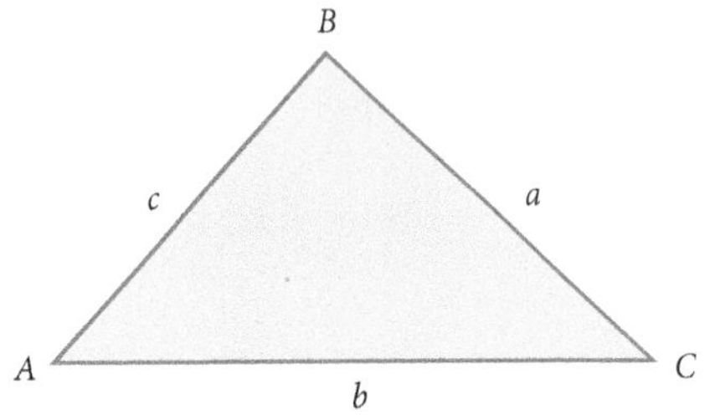

FIGURE 1

In Figure 1, notice that side a is opposite vertex A, side b is opposite vertex B, and side c is opposite vertex C. Also, because each vertex is the vertex of one of the angles of the triangle, we refer to the three interior angles as A, B, and C.

In any triangle, the sum of the interior angles is 180°. For the triangle shown in Figure 1, the relationship is written

$$A + B + C = 180°$$

EXAMPLE 4 The angles in a triangle are such that one angle is twice the smallest angle, whereas the third angle is three times as large as the smallest angle. Find the measure of all three angles.

SOLUTION

Step 1: **Read and list.**
Known items: The sum of all three angles is 180°, one angle is twice the smallest angle, the largest angle is three times the smallest angle.
Unknown items: The measure of each angle

Step 2: **Assign a variable, and translate information.**
Let x be the smallest angle, then $2x$ will be the measure of another angle and $3x$ will be the measure of the largest angle.

Step 3: **Reread, and write an equation.**
When working with geometric objects, drawing a generic diagram sometimes will help us visualize what it is that we are asked to find. In Figure 2, we draw a triangle with angles A, B, and C.

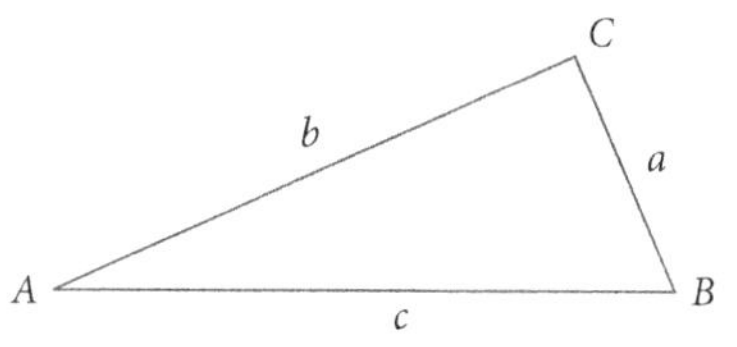

FIGURE 2

We can let the value of $A = x$, the value of $B = 2x$, and the value of $C = 3x$. We know that the sum of angles A, B, and C will be 180°, so our equation becomes

$$x + 2x + 3x = 180°$$

Step 4: **Solve the equation.**

$$\begin{aligned} x + 2x + 3x &= 180° \\ 6x &= 180° \\ x &= 30° \end{aligned}$$

Step 5: **Write the answer.**
The smallest angle A measures 30°
Angle B measures $2x$, or $2(30°) = 60°$
Angle C measures $3x$, or $3(30°) = 90°$

Step 6: **Reread, and check.**
The angles must add to 180°.

$$\begin{aligned} A + B + C &= 180° \\ 30° + 60° + 90° &\stackrel{?}{=} 180° \\ 180° &= 180° \quad \text{Our answers check} \end{aligned}$$

Getting Ready for Class

After reading through the preceding section, respond in your own words and in complete sentences.

A. If x is an integer, what would the next smaller consecutive integer be?

B. In Example 3, explain why the equation contains the product 0.50(15).

C. How do we label triangles?

D. What rule is always true about the three angles in a triangle?

Problem Set 2.7

Consecutive Integer Problems

1. The sum of two consecutive integers is 11. Find the numbers.
2. The sum of two consecutive integers is 15. Find the numbers.
3. The sum of two consecutive integers is -9. Find the numbers.
4. The sum of two consecutive integers is -21. Find the numbers.
5. The sum of two consecutive odd integers is 28. Find the numbers.
6. The sum of two consecutive odd integers is 44. Find the numbers.
7. The sum of two consecutive even integers is 106. Find the numbers.
8. The sum of two consecutive even integers is 66. Find the numbers.
9. The sum of two consecutive even integers is -30. Find the numbers.
10. The sum of two consecutive odd integers is -76. Find the numbers.
11. The sum of three consecutive odd integers is 57. Find the numbers.
12. The sum of three consecutive odd integers is -51. Find the numbers.
13. The sum of three consecutive even integers is 132. Find the numbers.
14. The sum of three consecutive even integers is -108. Find the numbers.

Interest Problems

15. Suppose you invest money in two accounts. One of the accounts pays 8% annual interest, whereas the other pays 9% annual interest. If you have \$2,000 more invested at 9% than you have invested at 8%, how much do you have invested in each account if the total amount of interest you earn in a year is \$860? (Begin by completing the following table.)

	Dollars Invested at 8%	Dollars Invested at 9%
Number of	x	
Interest on		

16. Suppose you invest a certain amount of money in an account that pays 11% interest annually, and \$4,000 more than that in an account that pays 12% annually. How much money do you have in each account if the total interest for a year is \$940?

	Dollars Invested at 11%	Dollars Invested at 12%
Number of	x	
Interest on		

17. Tyler has two savings accounts that his grandparents opened for him. The two accounts pay 10% and 12% in annual interest; there is \$500 more in the account that pays 12% than there is in the other account. If the total interest for a year is \$214, how much money does he have in each account?

18. Travis has a savings account that his parents opened for him. It pays 6% annual interest. His uncle also opened an account for him, but it pays 8% annual interest. If there is \$800 more in the account that pays 6%, and the total interest from both accounts is \$104, how much money is in each of the accounts?

19. A stockbroker has money in three accounts. The interest rates on the three accounts are 8%, 9%, and 10%. If she has twice as much money invested at 9% as she has invested at 8%, three times as much at 10% as she has at 8%, and the total interest for the year is \$280, how much is invested at each rate? (*Hint:* Let $x =$ the amount invested at 8%.)

20. An accountant has money in three accounts that pay 9%, 10%, and 11% in annual interest. He has twice as much invested at 9% as he does at 10% and three times as much invested at 11% as he does at 10%. If the total interest from the three accounts is \$610 for the year, how much is invested at each rate? (*Hint:* Let $x =$ the amount invested at 10%.)

Mixture Problems

21. How many liters of 50% alcohol solution and 10% alcohol solution must be mixed to obtain 24 liters of 30% alcohol solution?

	50% Solution	10% Solution	Mixture
Number of Liters	x		24
Liters of Alcohol			

22. How many liters of 25% alcohol solution and 50% alcohol solution must be mixed to obtain 30 liters of 40% alcohol solution?

	25% Solution	50% Solution	Mixture
Number of Liters	x		30
Liters of Alcohol			

23. A mixture of 8% disinfectant solution is to be made from 10% and 5% disinfectant solutions. How much of each solution should be used if 25 gallons of 8% solution are needed?

24. How much 70% antifreeze solution and 40% antifreeze solution should be combined to give 30 gallons of 50% antifreeze solution?

25. Coffee beans worth \$9.50 per pound are to be mixed with coffee beans worth \$12.00 per pound to make 40 pounds of beans worth \$10.00 per pound. How many pounds of each type of coffee bean should be used?

26. Peanuts worth \$6.00 per pound are to be combined with cashews worth \$11.00 per pound to make 60 pounds of a blend worth \$8.00 per pound. How many pounds of each type of nut should be used?

Triangle Problems

27. Two angles in a triangle are equal and their sum is equal to the third angle in the triangle. What are the measures of each of the three interior angles?

28. One angle in a triangle measures twice the smallest angle, whereas the largest angle is six times the smallest angle. Find the measures of all three angles.

29. The smallest angle in a triangle is $\frac{1}{5}$ as large as the largest angle. The third angle is twice the smallest angle. Find the three angles.

30. One angle in a triangle is half the largest angle but three times the smallest. Find all three angles.

31. One angle of a triangle measures 20° more than the smallest, while a third angle is twice the smallest. Find the measure of each angle.

32. One angle of a triangle measures 50° more than the smallest, while a third angle is three times the smallest. Find the measure of each angle.

Miscellaneous Problems

33. Ticket Prices Miguel is selling tickets to a barbecue. Adult tickets cost \$6.00 and children's tickets cost \$4.00. He sells six more children's tickets than adult tickets. The total amount of money he collects is \$184. How many adult tickets and how many children's tickets did he sell?

	Adult	Child
Number	x	$x + 6$
Income	$6(x)$	$4(x + 6)$

34. Working Two Jobs Maggie has a job working in an office for \$10 an hour and another job driving a tractor for \$12 an hour. One week she works in the office twice as long as she drives the tractor. Her total income for that week is \$416. How many hours did she spend at each job?

Job	Office	Tractor
Hours Worked	$2x$	x
Wages Earned	$10(2x)$	$12x$

35. **Phone Bill** The cost of a long-distance phone call is \$0.41 for the first minute and \$0.32 for each additional minute. If the total charge for a long-distance call is \$5.21, how many minutes was the call? Hint: Let $x =$ the number of additional minutes. After you solve for x, you will need to add 1 minute to your answer. Can you see why?

36. **Phone Bill** Danny, who is 1 year old, is playing with the telephone when he accidentally presses one of the buttons his mother has programmed to dial her friend Sue's number. Sue answers the phone and realizes Danny is on the other end. She talks to Danny, trying to get him to hang up. The cost for a call is \$0.23 for the first minute and \$0.14 for every minute after that. If the total charge for the call is \$3.73, how long did it take Sue to convince Danny to hang up the phone?

37. **Hourly Wages** JoAnn works in the publicity office at the state university. She is paid \$12 an hour for the first 35 hours she works each week and \$18 an hour for every hour after that. If she makes \$492 one week, how many hours did she work?

38. **Hourly Wages** Diane had a part-time job that paid her \$6.50 an hour. During one week she worked 26 hours and was paid \$178.10. She realized when she saw her check that she had been given a raise. How much per hour was that raise?

39. **Office Numbers** Professors Wong and Gil have offices in the mathematics building at Miami Dade College. Their office numbers are consecutive odd integers with a sum of 14,660. What are the office numbers of these two professors?

40. **Cell Phone Numbers** Diana and Tom buy two cell phones. The phone numbers assigned to each are consecutive integers with a sum of 11,109,295. If the smaller number is Diana's, what are their phone numbers?

41. **Age** Marissa and Kendra are 2 years apart in age. Their ages are two consecutive even integers. Kendra is the younger of the two. If Marissa's age is added to twice Kendra's age, the result is 26. How old is each girl?

42. **Age** Justin's and Ethan's ages form two consecutive odd integers. What is the difference of their ages?

43. **Arrival Time** Jeff and Carla Cole are driving separately from San Luis Obispo, California, to the north shore of Lake Tahoe, a distance of 425 miles. Jeff leaves San Luis Obispo at 11:00 a.m. and averages 55 miles per hour on the drive, Carla leaves later, at 1:00 p.m. but averages 65 miles per hour. Which person arrives in Lake Tahoe first?

44. **Piano Lessons** Tyler is taking piano lessons. Because he doesn't practice as often as his parents would like him to, he has to pay for part of the lessons himself. His parents pay him \$0.50 to do the laundry and \$1.25 to mow the lawn. In one month, he does the laundry 6 more times than he mows the lawn. If his parents pay him \$13.50 that month, how many times did he mow the lawn?

At one time, the Texas Junior College Teachers Association annual conference was held in Austin. At that time a taxi ride in Austin was \$1.25 for the first $\frac{1}{5}$ of a mile and \$0.25 for each additional $\frac{1}{5}$ of a mile. Use this information for Problems 47 and 48.

45. Cost of a Taxi Ride If the distance from one of the convention hotels to the airport is 7.5 miles, how much will it cost to take a taxi from that hotel to the airport?

46. Cost of a Taxi Ride Suppose the distance from one of the hotels to one of the western dance clubs in Austin is 12.4 miles. If the fare meter in the taxi gives the charge for that trip as \$16.50, is the meter working correctly?

47. Geometry The width and length of a rectangle are consecutive even integers. The perimeter is 44 meters. Find the width and length.

48. Geometry The width and length of a rectangle are consecutive odd integers. The perimeter is 128 meters. Find the width and length.

49. Geometry The angles of a triangle are three consecutive integers. Find the measure of each angle.

50. Geometry The angles of a triangle are three consecutive even integers. Find the measure of each angle.

Ike and Nancy Lara give western dance lessons at the Elk's Lodge on Sunday nights. The lessons cost \$3.00 for members of the lodge and \$5.00 for nonmembers. Half of the money collected for the lesson is paid to Ike and Nancy. The Elk's Lodge keeps the other half. One Sunday night, Ike counts 36 people in the dance lesson. Use this information to work Problems 53 through 56.

51. Dance Lessons What is the least amount of money Ike and Nancy will make?

52. Dance Lessons What is the largest amount of money Ike and Nancy will make?

53. Dance Lessons At the end of the evening, the Elk's Lodge gives Ike and Nancy a check for \$80 to cover half of the receipts. Can this amount be correct?

54. Dance Lessons Besides the number of people in the dance lesson, what additional information does Ike need to know to always be sure he is being paid the correct amount?

Learning Objectives Assessment

The following problems can be used to help assess if you have successfully met the learning objectives for this section.

For each of the following, which equation is an appropriate model for the given problem?

55. The sum of two consecutive odd integers is 272. Find the numbers.

a. $x + (x + 2) = 272$ **b.** $2x = 272$

c. $x + (x + 1) = 272$ **d.** $x + (x - 1) = 272$

56. Suppose you invest some money in an account that pays 3% interest annually, and \$600 more than that amount in another account that pays 4% annually. If the total interest for the year is \$87, how much have you invested at 3%?

a. $3x + 4x + 600 = 87$ **b.** $0.03x + 0.04x + 600 = 87$

c. $3x + 4(x + 600) = 87$ **d.** $0.03x + 0.04(x + 600) = 87$

57. A mixture of 35% alcohol solution is to be made by combining 25% and 60% alcohol solutions. How much of the 25% solution should be used if 40 gallons of 35% solution are needed?

a. $0.25x + 0.60(40 - x) = 0.35(40)$ **b.** $0.25x + 0.60(40 - x) = 40$

c. $0.25x + 0.60x = 40$ **d.** $25x + 60x = 35(40)$

58. One angle in a triangle is 10° more than the smallest angle. A third angle is twice the smallest angle. Find all three angles.

a. $(x - 10) + 2x = 180$ **b.** $x + 2x - 10 = 90$

c. $x + (x + 10) + 2x = 180$ **d.** $x + (x + 10) + (x + 2) = 90$

Getting Ready for the Next Section

To understand all the explanations and examples in the next section you must be able to work the problems below.

Solve the following equations.

59. **a.** $x - 3 = 6$ **b.** $x + 3 = 6$ **c.** $-x - 3 = 6$ **d.** $-x + 3 = 6$

60. **a.** $x - 7 = 16$ **b.** $x + 7 = 16$ **c.** $-x - 7 = 16$ **d.** $-x + 7 = 16$

61. **a.** $\frac{x}{4} = -2$ **b.** $-\frac{x}{4} = -2$ **c.** $\frac{x}{4} = 2$ **d.** $-\frac{x}{4} = 2$

62. **a.** $3a = 15$ **b.** $3a = -15$ **c.** $-3a = 15$ **d.** $-3a = -15$

63. $2.5x - 3.48 = 4.9x + 2.07$

64. $2(1 - 3x) + 4 = 4x - 14$

65. $3(x - 4) = -2$

66. Solve for y: $2x - 3y = 6$

Linear Inequalities 2.8

Learning Objectives

In this section, we will learn how to:

1. Solve a linear inequality.
2. Graph a solution set for a linear inequality on the number line.

Introduction

Linear inequalities are solved by a method similar to the one used in solving linear equations. The overall goal is to isolate the variable. The only real differences between the methods are in the multiplication property for inequalities and in expressing the solution set.

An *inequality* differs from an equation only with respect to the comparison symbol between the two quantities being compared. In place of the equal sign, we use $<$ (less than), $\leq$ (less than or equal to), $>$ (greater than), or $\geq$ (greater than or equal to).

Solution Sets and Notation

We can use *set-builder notation* to write the solution set, and then visually represent the solution set by graphing it on a number line. If the inequality symbol includes equality ($\leq$, $\geq$) then we use a solid point when graphing the solution set. If the inequality symbol does not include equality ($<$, $>$) then we use an open circle when graphing the solution set.

Here are some examples of equations and inequalities showing the notation used to write and graph the solution set.

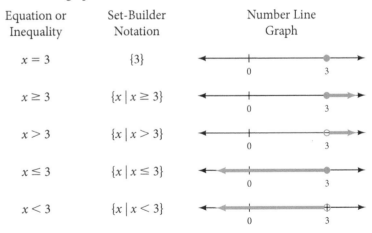

Equation or Inequality	Set-Builder Notation	Number Line Graph
$x = 3$	$\{3\}$	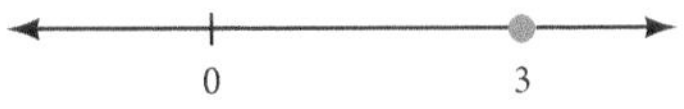
$x \geq 3$	$\{x \mid x \geq 3\}$	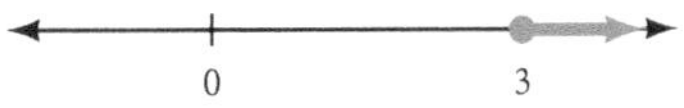
$x > 3$	$\{x \mid x > 3\}$	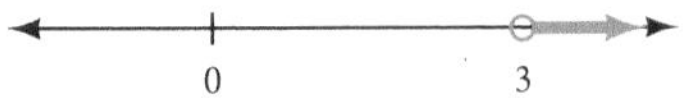
$x \leq 3$	$\{x \mid x \leq 3\}$	
$x < 3$	$\{x \mid x < 3\}$	

Solving Linear Inequalities

The addition property for inequalities is almost identical to the addition property for equality.

PROPERTY ***Addition Property for Inequalities***

For any three algebraic expressions A, B, and C,

if $\quad A < B$

then $\quad A + C < B + C$

In words: Adding the same quantity to both sides of an inequality will not change the solution set.

It makes no difference which inequality symbol we use to state the property. Adding the same amount to both sides always produces an inequality equivalent to the original inequality. Also, because subtraction can be thought of as addition of the opposite, this property holds for subtraction as well as addition.

VIDEO EXAMPLES

SECTION 2.8

EXAMPLE 1 Solve the inequality: $x + 5 < 7$.

SOLUTION To isolate x, we add -5 to both sides of the inequality.

$$x + 5 < 7$$

$$x + 5 + (-5) < 7 + (-5) \qquad \text{Addition property for inequalities}$$

$$x < 2$$

Using set-builder notation we can write the solution set as $\{x \mid x < 2\}$. We can go one step further here and graph the solution set. The solution set is all real numbers less than 2. To graph this set, we simply draw a straight line and label a convenient point 0 (zero) for reference. Then we label the 2 on the right side of zero and extend an arrow beginning at 2 and pointing to the left. We use an open circle at 2 because it is not included in the solution set. Here is the graph:

EXAMPLE 2 Solve: $x - 6 \le -3$.

SOLUTION Adding 6 to each side will isolate x on the left side.

$$x - 6 \le -3$$

$$x - 6 + 6 \le -3 + 6 \qquad \text{Add 6 to both sides}$$

$$x \le 3$$

The solution set is $\{x \mid x \le 3\}$. The graph of the solution set is

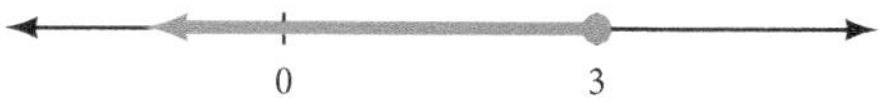

Notice that the point at the 3 is solid because 3 is included in the solution set. We always will use open circles on the graphs of solution sets with $<$ or $>$ and solid points on the graphs of solution sets with $\le$ or $\ge$.

To see the idea behind the multiplication property for inequalities, we will consider three true inequality statements and explore what happens when we multiply both sides by a positive number and then what happens when we multiply by a negative number.

Consider the following three true statements:

$$3 < 5 \qquad -3 < 5 \qquad -5 < -3$$

Now multiply both sides by the positive number 4:

$$4(3) < 4(5) \qquad 4(-3) < 4(5) \qquad 4(-5) < 4(-3)$$
$$12 < 20 \qquad -12 < 20 \qquad -20 < -12$$

Note This discussion is intended to show why the multiplication property for inequalities is written the way it is. You may want to look ahead to the property itself and then come back to this discussion if you are having trouble making sense out of it.

In each case, the inequality symbol in the result points in the same direction it did in the original inequality. We say the "sense" of the inequality doesn't change when we multiply both sides by a positive quantity.

Notice what happens when we go through the same process but multiply both sides by -4 instead of 4:

$$3 < 5 \qquad -3 < 5 \qquad -5 < -3$$

$$-4(3) > -4(5) \qquad -4(-3) > -4(5) \qquad -4(-5) > -4(-3)$$
$$-12 > -20 \qquad 12 > -20 \qquad 20 > 12$$

In each case, we have to change the direction in which the inequality symbol points to keep each statement true. Multiplying both sides of an inequality by a negative quantity always reverses the sense of the inequality. Our results are summarized in the multiplication property for inequalities.

Note Because division is defined in terms of multiplication, this property is also true for division. We can divide both sides of an inequality by any nonzero number we choose. If that number happens to be negative, we must also reverse the direction of the inequality symbol.

PROPERTY ***Multiplication Property for Inequalities***

For any three algebraic expressions A, B, and C, where $C \neq 0$,

if	$A < B$	
then	$AC < BC$	when C is positive
and	$AC > BC$	when C is negative

In words: Multiplying both sides of an inequality by a positive number does not change the solution set. When multiplying both sides of an inequality by a negative number, it is necessary to reverse the inequality symbol to produce an equivalent inequality.

We can multiply both sides of an inequality by any nonzero number we choose. If that number happens to be negative, we must also reverse the sense of the inequality.

EXAMPLE 3 Solve $3a < 15$ and graph the solution set.

SOLUTION We begin by multiplying each side by $\frac{1}{3}$. Because $\frac{1}{3}$ is a positive number, we do not reverse the direction of the inequality symbol:

$$3a < 15$$
$$\frac{1}{3}(3a) < \frac{1}{3}(15) \qquad \text{Multiply each side by } \frac{1}{3}$$
$$a < 5$$

The solution set is $\{a \mid a < 5\}$, and its graph is

EXAMPLE 4 Solve $-3a \le 18$, and graph the solution set.

SOLUTION We begin by multiplying both sides by $-\frac{1}{3}$. Because $-\frac{1}{3}$ is a negative number, we must reverse the direction of the inequality symbol at the same time that we multiply by $-\frac{1}{3}$.

$$-3a \le 18$$

$$-\frac{1}{3}(-3a) \ge -\frac{1}{3}(18) \quad \text{Multiply both sides by } -\tfrac{1}{3} \text{ and reverse the direction of the inequality symbol}$$

$$a \ge -6$$

The solution set is $\{a \mid a \ge -6\}$, and the graph of the solution set is

EXAMPLE 5 Solve $-\frac{x}{4} > 2$ and graph the solution set.

SOLUTION To isolate x, we multiply each side by -4. Because -4 is a negative number, we also must reverse the direction of the inequality symbol:

$$-\frac{x}{4} > 2$$

$$-4\left(-\frac{x}{4}\right) < -4(2) \quad \text{Multiply each side by } -4\text{, and reverse the direction of the inequality symbol}$$

$$x < -8$$

The solution set is $\{x \mid x < -8\}$, and its graph is

To solve more complicated inequalities, we use the following process.

HOW TO ***Solve Linear Inequalities in One Variable***

Step 1a: Use the distributive property to separate terms, if necessary.
1b: If fractions are present, consider multiplying both sides by the LCD to eliminate the fractions. If decimals are present, consider multiplying both sides by a power of 10 to clear the inequality of decimals.
1c: Combine similar terms on each side of the inequality.
Step 2: Use the addition property for inequalities to get all variable terms on one side of the inequality and all constant terms on the other side.
Step 3: Use the multiplication property for inequalities to get the variable by itself on one side of the inequality (change the coefficient of the variable term into 1).
Step 4: Write and graph the solution set.

EXAMPLE 6 Solve: $2.5x - 3.48 < -4.9x + 2.07$.

SOLUTION We have two methods we can use to solve this inequality. We can simply apply our properties to the inequality the way it is currently written and work with the decimal numbers, or we can eliminate the decimals to begin with and solve the resulting inequality.

Method 1 Working with the decimals.

$$\begin{aligned}
2.5x - 3.48 &< -4.9x + 2.07 && \text{Original inequality}\\
2.5x + 4.9x - 3.48 &< -4.9x + 4.9x + 2.07 && \text{Add } 4.9x \text{ to each side}\\
7.4x - 3.48 &< 2.07 &&\\
7.4x - 3.48 + 3.48 &< 2.07 + 3.48 && \text{Add 3.48 to each side}\\
7.4x &< 5.55 &&\\
\frac{7.4x}{7.4} &< \frac{5.55}{7.4} && \text{Divide each side by 7.4}\\
x &< 0.75 &&
\end{aligned}$$

Method 2 Eliminating the decimals in the beginning.

Because the greatest number of places to the right of the decimal point in any of the numbers is 2, we can multiply each side of the inequality by 100 and we will be left with an equivalent inequality that contains only integers.

$$\begin{aligned}
2.5x - 3.48 &< -4.9x + 2.07 && \text{Original inequality}\\
100(2.5x - 3.48) &< 100(-4.9x + 2.07) && \text{Multiply each side by 100}\\
100(2.5x) - 100(3.48) &< 100(-4.9x) + 100(2.07) && \text{Distributive property}\\
250x - 348 &< -490x + 207 && \text{Multiplication}\\
740x - 348 &< 207 && \text{Add } 490x \text{ to each side}\\
740x &< 555 && \text{Add 348 to each side}\\
\frac{740x}{740} &< \frac{555}{740} && \text{Divide each side by 740}\\
x &< 0.75 &&
\end{aligned}$$

The solution set by either method is $\{x \mid x < 0.75\}$. Here is the graph:

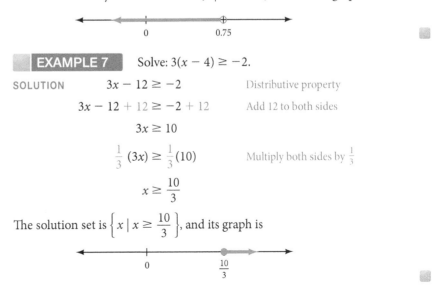

EXAMPLE 7 Solve: $3(x - 4) \geq -2$.

SOLUTION

$$\begin{aligned}
3x - 12 &\geq -2 && \text{Distributive property}\\
3x - 12 + 12 &\geq -2 + 12 && \text{Add 12 to both sides}\\
3x &\geq 10 &&\\
\frac{1}{3}(3x) &\geq \frac{1}{3}(10) && \text{Multiply both sides by } \tfrac{1}{3}\\
x &\geq \frac{10}{3} &&
\end{aligned}$$

The solution set is $\left\{x \mid x \geq \frac{10}{3}\right\}$, and its graph is

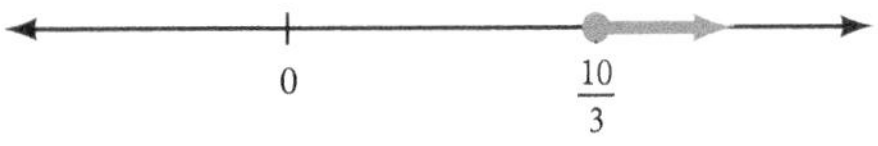

EXAMPLE 8 Solve $2(1 - 3x) + 4 < 4x - 14$ and graph the solution set.

SOLUTION

$2 - 6x + 4 < 4x - 14$	Distributive property
$-6x + 6 < 4x - 14$	Simplify
$-6x + 6 + (-6) < 4x - 14 + (-6)$	Add -6 to both sides
$-6x < 4x - 20$	
$-6x + (-4x) < 4x + (-4x) - 20$	Add $-4x$ to both sides
$-10x < -20$	
$\left(-\frac{1}{10}\right)(-10x) > \left(-\frac{1}{10}\right)(-20)$	Multiply by $-\frac{1}{10}$, reverse the direction of the inequality symbol
$x > 2$	

The solution set is $\{x \mid x > 2\}$. Here is the graph:

EXAMPLE 9 Solve $2x - 3y < 6$ for y.

SOLUTION We can solve this formula for y by first adding $-2x$ to each side and then multiplying each side by $-\frac{1}{3}$. When we multiply by $-\frac{1}{3}$ we must reverse the direction of the inequality symbol. Because this is a formula, we will not graph the solution set.

$2x - 3y < 6$	Original formula
$2x + (-2x) - 3y < (-2x) + 6$	Add $-2x$ to each side
$-3y < -2x + 6$	
$-\frac{1}{3}(-3y) > -\frac{1}{3}(-2x + 6)$	Multiply each side by $-\frac{1}{3}$
$y > \frac{2}{3}x - 2$	Distributive property

Application

Here are some common phrases used when describing inequalities and their equivalent mathematical expressions.

Phrase	Equivalent Expression
a is less than b	$a < b$
a is more than b	$a > b$
a is greater than b	$a > b$
a is at most b	$a \leq b$
a is no larger than b	$a \leq b$
a does not exceed b	$a \leq b$
a is at least b	$a \geq b$
a is no smaller than b	$a \geq b$

Our last example is similar to an example done earlier in this chapter. This time it involves an inequality instead of an equation.

We can modify our Blueprint for Problem Solving to solve application problems whose solutions depend on writing and then solving inequalities.

EXAMPLE 10 The sum of two consecutive odd integers is at most 28. What are the possibilities for the first of the two integers?

SOLUTION When we use the phrase "their sum is at most 28," we mean that their sum is less than or equal to 28.

Step 1: **Read and list.**
Known items: Two consecutive odd integers. Their sum is less than or equal to 28.
Unknown items: The numbers in question.

Step 2: **Assign a variable, and translate information.**
If we let x be the first of the two consecutive odd integers, then $x + 2$ is the next consecutive one.

Step 3: **Reread, and write an inequality.**
Their sum is at most 28.

$$x + (x + 2) \leq 28$$

Step 4: **Solve the inequality.**

$$2x + 2 \leq 28 \quad \text{Simplify the left side}$$

$$2x \leq 26 \quad \text{Add } -2 \text{ to each side}$$

$$x \leq 13 \quad \text{Multiply each side by } \tfrac{1}{2}$$

Step 5: **Write the answer.**
The first of the two integers must be an odd integer that is less than or equal to 13.

Step 6: **Reread, and check.**
Suppose the first integer is 13. The next consecutive odd integer is 15. The sum of 15 and 13 is 28. If the first odd integer is less than 13, the sum of it and the next consecutive odd integer will be less than 28.

Getting Ready for Class

After reading through the preceding section, respond in your own words and in complete sentences.

A. State the addition property for inequalities.

B. How is the multiplication property for inequalities different from the multiplication property of equality?

C. When do we reverse the direction of an inequality symbol?

D. Under what conditions do we not change the direction of the inequality symbol when we multiply both sides of an inequality by a number?

Problem Set 2.8

Solve the following inequalities using the addition property of inequalities. Graph each solution set.

1. $x - 5 < 7$ **2.** $x + 3 < -5$ **3.** $a - 4 \le 8$ **4.** $a + 3 \le 10$

5. $x - 4.3 > 8.7$ **6.** $x - 2.6 > 10.4$ **7.** $y + 6 \ge 10$ **8.** $y + 3 \ge 12$

9. $2 < x - 7$ **10.** $3 < x + 8$

Solve the following inequalities using the multiplication property of inequalities. If you multiply both sides by a negative number, be sure to reverse the direction of the inequality symbol. Graph the solution set.

11. $3x < 6$ **12.** $2x < 14$ **13.** $5a \le 25$ **14.** $4a \le 16$

15. $\frac{x}{3} > 5$ **16.** $\frac{x}{7} > 1$ **17.** $-2x > 6$ **18.** $-3x \ge 9$

19. $-3x \ge -18$ **20.** $-8x \ge -24$ **21.** $-\frac{x}{5} \le 10$ **22.** $-\frac{x}{9} \ge -1$

23. $-\frac{2}{3}y > 4$ **24.** $-\frac{3}{4}y > 6$

Solve the following inequalities.

25. $2x - 3 < 9$ **26.** $3x - 4 < 17$ **27.** $-\frac{1}{5}y - \frac{1}{3} \le \frac{2}{3}$

28. $-\frac{1}{6}y - \frac{1}{2} \le \frac{2}{3}$ **29.** $-7.2x + 1.8 > -19.8$ **30.** $-7.8x - 1.3 > 22.1$

31. $\frac{2}{3}x - 5 \le 7$ **32.** $\frac{3}{4}x - 8 \le 1$ **33.** $-\frac{2}{5}a - 3 > 5$

34. $-\frac{4}{5}a - 2 > 10$ **35.** $5 - \frac{3}{5}y > -10$ **36.** $4 - \frac{5}{6}y > -11$

37. $0.3(a + 1) \le 1.2$ **38.** $0.4(a - 2) \le 0.4$ **39.** $2(5 - 2x) \le -20$

40. $7(8 - 2x) > 28$ **41.** $3x - 5 > 8x$ **42.** $8x - 4 > 6x$

43. $\frac{1}{3}y - \frac{1}{2} \le \frac{5}{6}y + \frac{1}{2}$ **44.** $\frac{7}{6}y + \frac{4}{3} \le \frac{11}{6}y - \frac{7}{6}$

45. $-2.8x + 8.4 < -14x - 2.8$ **46.** $-7.2x - 2.4 < -2.4x + 12$

47. $3(m - 2) - 4 \ge 7m + 14$ **48.** $2(3m - 1) + 5 \ge 8m - 7$

49. $3 - 4(x - 2) \le -5x + 6$ **50.** $8 - 6(x - 3) \le -4x + 12$

Solve each of the following formulas for y.

51. $3x + 2y < 6$ **52.** $-3x + 2y < 6$ **53.** $2x - 5y > 10$

54. $-2x - 5y > 5$ **55.** $-3x + 7y \le 21$ **56.** $-7x + 3y \le 21$

57. $2x - 4y \ge -4$ **58.** $4x - 2y \ge -8$

The next two problems are intended to give you practice reading, and paying attention to, the instructions that accompany the problems you are working.

59. Work each problem according to the instructions given.

a. Evaluate when $x = 0$: $-5x + 3$. **b.** Solve: $-5x + 3 = -7$.

c. Is 0 a solution to $-5x + 3 < -7$? **d.** Solve: $-5x + 3 < -7$.

60. Work each problem according to the instructions given.

a. Evaluate when $x = 0$: $-2x - 5$. **b.** Solve: $-2x - 5 = 1$.

c. Is 0 a solution to $-2x - 5 > 1$? **d.** Solve: $-2x - 5 > 1$.

For each graph below, write an inequality whose solution is the graph.

61. 

62. 

63. 

64. 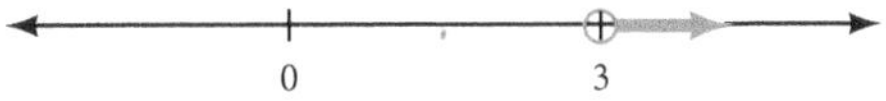

Applying the Concepts

65. Consecutive Integers The sum of two consecutive integers is at least 583. What are the possibilities for the first of the two integers?

66. Consecutive Integers The sum of two consecutive integers is at most 583. What are the possibilities for the first of the two integers?

67. Number Problems The sum of twice a number and six is less than ten. Find all solutions.

68. Number Problems Twice the difference of a number and three is greater than or equal to the number increased by five. Find all solutions.

69. Number Problems The product of a number and four is greater than the number minus eight. Find the solution set.

70. Number Problems The quotient of a number and five is less than the sum of seven and two. Find the solution set.

71. Geometry Problems The length of a rectangle is 3 times the width. If the perimeter is to be at least 48 meters, what are the possible values for the width? (If the perimeter is at least 48 meters, then it is greater than or equal to 48 meters.)

72. Geometry Problems The length of a rectangle is 3 more than twice the width. If the perimeter is to be at least 51 meters, what are the possible values for the width? (If the perimeter is at least 51 meters, then it is greater than or equal to 51 meters.)

73. Geometry Problems The numerical values of the three sides of a triangle are given by three consecutive even integers. If the perimeter is greater than 24 inches, what are the possibilities for the shortest side?

74. Geometry Problems The numerical values of the three sides of a triangle are given by three consecutive odd integers. If the perimeter is greater than 27 inches, what are the possibilities for the shortest side?

Learning Objectives Assessment

The following problems can be used to help assess if you have successfully met the learning objectives for this section.

75. Solve: $3(9 - 5x) < 72$.

a. $x < -3$ **b.** $x > -3$ **c.** $x < 3$ **d.** $x > 3$

76. Graph $x > -2$ on the number line.

a.
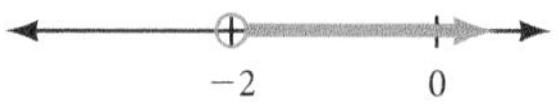

b.

c.
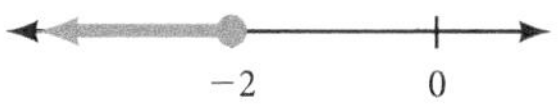

d.
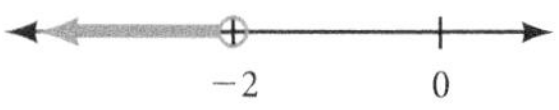

Maintaining Your Skills

The problems that follow review some of the more important skills you have learned in previous sections and chapters. You can consider the time you spend working these problems as time spent studying for exams.

Answer the following percent problems.

77. What number is 25% of 32?
78. What number is 15% of 75?
79. What number is 20% of 120?
80. What number is 125% of 300?
81. What percent of 36 is 9?
82. What percent of 16 is 9?
83. What percent of 50 is 5?
84. What percent of 140 is 35?
85. 16 is 20% of what number?
86. 6 is 3% of what number?
87. 8 is 2% of what number?
88. 70 is 175% of what number?

Simplify each expression.

89. $-|-5|$
90. $\left(-\frac{2}{3}\right)^3$
91. $-3 - 4(-2)$
92. $2^4 + 3^3 \div 9 - 4^2$
93. $5|3 - 8| - 6|2 - 5|$
94. $7 - 3(2 - 6)$
95. $5 - 2[-3(5 - 7) - 8]$
96. $\dfrac{5 + 3(7 - 2)}{2(-3) - 4}$

97. Find the difference of -3 and -9.

98. If you add -4 to the product of -3 and 5, what number results?

99. Apply the distributive property to $\frac{1}{2}(4x - 6)$.

100. Use the associative property to simplify $-6\left(\frac{1}{3}x\right)$.

For the set $\left\{-3, -\frac{4}{5}, 0, \frac{5}{8}, 2, \sqrt{5}\right\}$, which numbers are

101. Integers
102. Rational numbers

Chapter 2 Summary

EXAMPLES

Similar Terms [2.1]

1. The terms $2x$, $5x$, and $-7x$ are all similar because their variable parts are the same.

A term is a number or a number and one or more variables multiplied together. Similar terms are terms with the same variable part.

Simplifying Expressions [2.1]

2. Simplify: $3x + 4x$.

$$3x + 4x = (3 + 4)x$$
$$= 7x$$

In this chapter, we simplified expressions that contained variables by using the distributive property to combine similar terms.

Solution Set [2.2]

3. The solution set for the equation $x + 2 = 5$ is $\{3\}$ because $x = 3$ is the only real number that makes the equation true.

The solution set for an equation (or inequality) is all the numbers that, when used in place of the variable, make the equation (or inequality) a true statement.

Equivalent Equations [2.2]

4. The equations $a - 4 = 3$ and $a - 2 = 5$ are equivalent because both have solution set $\{7\}$.

Two equations are called equivalent if they have the same solution set.

Addition Property of Equality [2.2]

5. Solve: $x - 5 = 12$.

$$x - 5\ (+\ \mathbf{5}) = 12\ (+\ \mathbf{5})$$
$$x + 0 = 17$$
$$x = 17$$

When the same quantity is added to both sides of an equation, the solution set for the equation is unchanged. Adding the same amount to both sides of an equation produces an equivalent equation.

Multiplication Property of Equality [2.3]

6. Solve: $3x = 18$.

$$\mathbf{\frac{1}{3}}(3x) = \mathbf{\frac{1}{3}}(18)$$
$$x = 6$$

If both sides of an equation are multiplied by the same nonzero number, the solution set is unchanged. Multiplying both sides of an equation by a nonzero quantity produces an equivalent equation.

Strategy for Solving Linear Equations in One Variable [2.4]

7. Solve: $2(x + 3) = 10$.

$$2x + 6 = 10$$
$$2x + 6 + (-6) = 10 + (-6)$$
$$2x = 4$$
$$\frac{1}{2}(2x) = \frac{1}{2}(4)$$
$$x = 2$$

Step 1a: Use the distributive property to separate terms, if necessary.

1b: If fractions are present, consider multiplying both sides by the LCD to eliminate the fractions. If decimals are present, consider multiplying both sides by a power of 10 to clear the equation of decimals.

1c: Combine similar terms on each side of the equation.

Step 2: Use the addition property of equality to get all variable terms on one side of the equation and all constant terms on the other side. A variable term is a term that contains the variable (for example, $5x$). A constant term is a term that does not contain the variable (the number 3, for example).

Step 3: Use the multiplication property of equality to change the coefficient of the variable term into 1.

Step 4: Check your solution in the original equation to be sure that you have not made a mistake in the solution process.

Formulas [2.5]

8. Solving $P = 2l + 2w$ for l, we have

$$P - 2w = 2l$$
$$\frac{P - 2w}{2} = l$$

A formula is an equation with more than one variable. To solve a formula for one of its variables, we use the addition and multiplication properties of equality to move everything except the variable in question to one side of the equal sign so the variable in question is alone on the other side.

Blueprint for Problem Solving [2.6, 2.7]

Step 1: ***Read*** the problem, and then mentally ***list*** the items that are known and the items that are unknown.

Step 2: ***Assign a variable*** to one of the unknown items. (In most cases this will amount to letting x represent the item that is asked for in the problem.) Then ***translate*** the other ***information*** in the problem to expressions involving the variable.

Step 3: ***Reread*** the problem, and then ***write an equation,*** using the items and variables listed in steps 1 and 2, that describes the situation.

Step 4: ***Solve the equation*** found in step 3.

Step 5: ***Write*** your ***answer*** using a complete sentence.

Step 6: ***Reread*** the problem, and ***check*** your solution with the original words in the problem.

Addition Property for Inequalities [2.8]

Adding the same quantity to both sides of an inequality produces an equivalent inequality, one with the same solution set.

9. Solve: $x + 5 < 7$.

$$x + 5 + (\mathbf{-5}) < 7 + (\mathbf{-5})$$
$$x < 2$$

Multiplication Property for Inequalities [2.8]

Multiplying both sides of an inequality by a positive number never changes the solution set. If both sides are multiplied by a negative number, the direction of the inequality symbol must be reversed to produce an equivalent inequality.

10. Solve: $-3a \leq 18$.

$$-\frac{\mathbf{1}}{\mathbf{3}}(-3a) \geq -\frac{\mathbf{1}}{\mathbf{3}}(18)$$
$$a \geq -6$$

Strategy for Solving Linear Inequalities in One Variable [2.8]

Step 1a: Use the distributive property to separate terms, if necessary.

Step 1b: If fractions are present, consider multiplying both sides by the LCD to eliminate the fractions. If decimals are present, consider multiplying both sides by a power of 10 to clear the inequality of decimals.

Step 1c: Combine similar terms on each side of the inequality.

Step 2: Use the addition property for inequalities to get all variable terms on one side of the inequality and all constant terms on the other side.

Step 3: Use the multiplication property for inequalities to get the variable by itself on one side of the inequality. Remember to reverse the direction of the inequality symbol if both sides are multiplied by a negative number.

Step 4: Graph the solution set.

11. Solve: $3(x - 4) \geq -2$.

$$3x - 12 \geq -2$$
$$3x - 12 + \mathbf{12} \geq -2 + \mathbf{12}$$
$$3x \geq 10$$
$$\frac{\mathbf{1}}{\mathbf{3}}(3x) \geq \frac{\mathbf{1}}{\mathbf{3}}(10)$$
$$x \geq \frac{10}{3}$$

Chapter 2 Test

Simplify each of the following expressions. [2.1]

1. $5y - 3 - 6y + 4$

2. $3x - 4 + x + 3$

3. $4 - 2(y - 3) - 6$

4. $3(3x - 4) - 2(4x + 5)$

5. Find the value of $3x + 12 + 2x$ when $x = -3$. [2.1]

6. Find the value of $x^2 - 3xy + y^2$ when $x = -2$ and $y = -4$. [2.1]

7. Fill in the tables below to find the sequences formed by substituting the first four counting numbers into the expressions $(n + 2)^2$ and $n^2 + 2$. [2.1]

a.

n	$(n + 2)^2$
1	
2	
3	
4	

b.

n	$n^2 + 2$
1	
2	
3	
4	

Solve the following equations. [2.2, 2.3, 2.4]

8. $3x - 2 = 7$

9. $4y + 15 = y$

10. $\frac{1}{4}x - \frac{1}{12} = \frac{1}{3}x - \frac{1}{6}$

11. $-3(3 - 2x) - 7 = 8$

12. $3x - 9 = -6$

13. $0.05 + 0.07(100 - x) = 3.2$

14. $4(t - 3) + 2(t + 4) = 2t - 16$

15. $4x - 2(3x - 1) = 2x - 8$

For each of the following expressions, write an equivalent equation. [2.5]

16. What number is 40% of 56?

17. 720 is 24% of what number?

18. If $3x - 4y = 16$, find y when $x = 4$.

19. If $3x - 4y = 16$, find x when $y = 2$.

Solve each formula for the appropriate variable. [2.5]

20. Solve $2x + 6y = 12$ for y.

21. Solve $x^2 = v^2 + 2ad$ for a.

Solve each word problem. [2.6, 2.7]

22. Age Problem Paul is twice as old as Becca. Five years ago, the sum of their ages was 44. How old are they now?

23. Geometry The length of a rectangle is 5 less than 3 times the width. The perimeter is 150 centimeters. What are the length and width?

24. Coin Problem A man has a collection of dimes and nickels with a total value of \$1.70. If he has 8 more dimes than nickels, how many of each coin does he have?

25. Investing A woman has money in two accounts. One account pays 6% annual interest, whereas the other pays 12% annual interest. If she has \$500 more invested at 12% than she does at 6% and her total interest for a year is \$186, how much does she have in each account?

Solve each inequality, and graph the solution. [2.8]

26. $\frac{1}{2}x - 2 > 3$

27. $-6y \leq 24$

28. $0.3 - 0.2x < 1.1$

29. $3 - 2(n - 1) \geq 9$

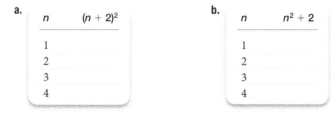

3 Linear Equations and Inequalities in Two Variables

Chapter Outline

3.1 Paired Data and Graphing Ordered Pairs

3.2 Graphing Linear Equations in Two Variables

3.3 More on Graphing: Intercepts

3.4 The Slope of a Line

3.5 Slope-Intercept Form

3.6 Point-Slope Form

3.7 Linear Inequalities in Two Variables

When light comes into contact with a surface that does not transmit light, then all the light that contacts the surface is either reflected off the surface or absorbed into the surface. If we let R represent the percentage of light reflected and w represent the percentage of light absorbed, then the relationship between these two variables can be written as

$$R + A = 100$$

which is a linear equation in two variables. The following table and graph show the same relationship as that described by the equation. The table is a numerical description; the graph is a visual description.

Reflected and Absorbed Light

Percent Reflected	Percent Absorbed
0	100
20	80
40	60
60	40
80	20
100	0

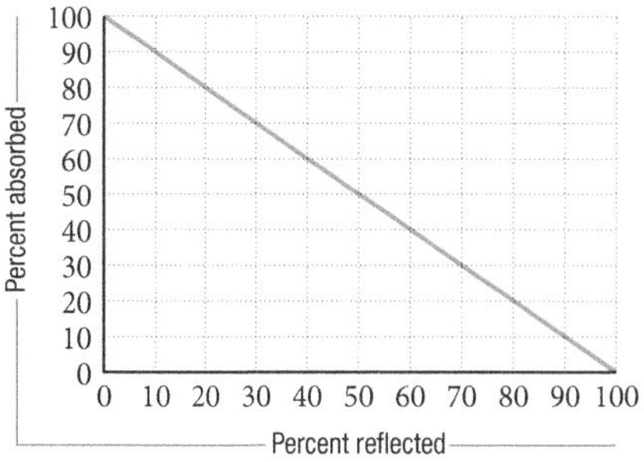

In this chapter, we learn how to build tables and draw graphs from linear equations in two variables.

Success Skills

Try to arrange your daily study habits so that you have very little studying to do the night before your next exam. The next two goals will help you achieve this.

1. **Review with the Exam in Mind** Each day you should review material that will be covered on the next exam. Your review should consist of working problems. Preferably, the problems you work should be problems from your list of difficult problems.

2. **Pay Attention to Instructions** Each of the following is a valid instruction with respect to the equation $y = 3x - 2$, and the result of applying the instructions will be different in each case:

Find x when y is 10.	(Section 2.5)
Solve for x.	(Section 2.5)
Graph the equation.	(Section 3.2)
Find the intercepts.	(Section 3.3)

There are many things to do with the equation $y = 3x - 2$. If you train yourself to pay attention to the instructions that accompany a problem as you work through the assigned problems, you will not find yourself confused about what to do with a problem when you see it on a test.

Paired Data and Graphing Ordered Pairs

3.1

Learning Objectives

In this section, we will learn how to:

1. Construct and interpret a scatter diagram.
2. Construct and interpret a line graph.
3. Graph an ordered pair on the rectangular coordinate system.

Introduction

This table and figure show the relationship between the table of values for the speed of a race car and the corresponding bar chart. In Figure 1, the horizontal line that shows the elapsed time in seconds is called the *horizontal axis*, and the vertical line that shows the speed in miles per hour is called the *vertical axis*.

The data in the table are called *paired data* because the information is organized so that each number in the first column is paired with a specific number in the second column. Each pair of numbers is associated with one of the solid bars in Figure 1. For example, the third bar in the bar chart is associated with the pair of numbers 3 seconds and 162.8 miles per hour. The first number, 3 seconds, is associated with the horizontal axis, and the second number, 162.8 miles per hour, is associated with the vertical axis.

Speed of a Race Car

Time in Seconds	Speed in Miles per Hour
0	0
1	72.7
2	129.9
3	162.8
4	192.2
5	212.4
6	228.1

Visually

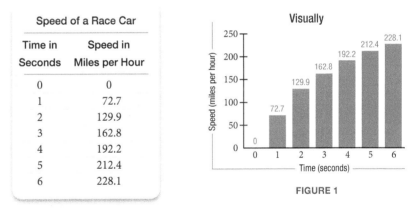

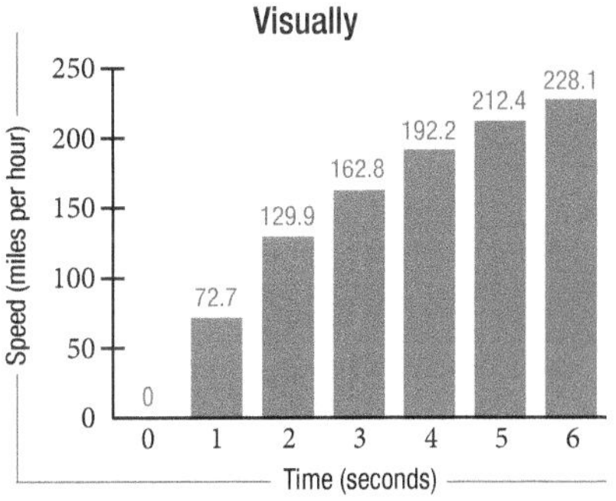

FIGURE 1

Scatter Diagrams and Line Graphs

The information in the table can be visualized with a scatter diagram and line graph as well. Figure 2 is a *scatter diagram* of the information in the table. We use dots instead of the bars shown in Figure 1 to show the speed of the race car at each second during the race. Figure 3 is called a *line graph*. It is constructed by taking the dots in Figure 2 and connecting each one to the next with a straight line.

Notice that we have labeled the axes in these two figures a little differently than we did with the bar chart by making the axes intersect at the number 0.

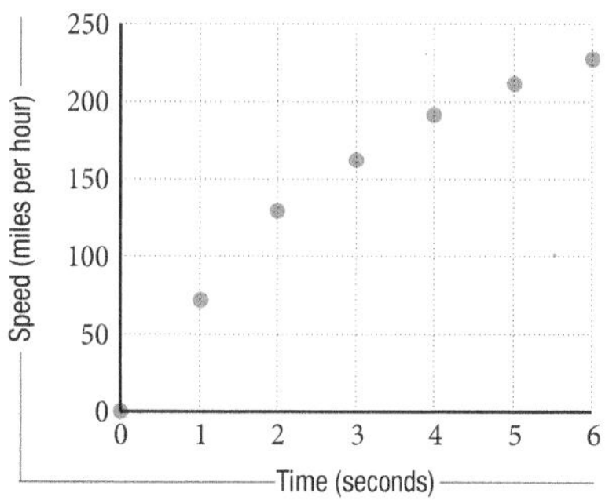

FIGURE 2

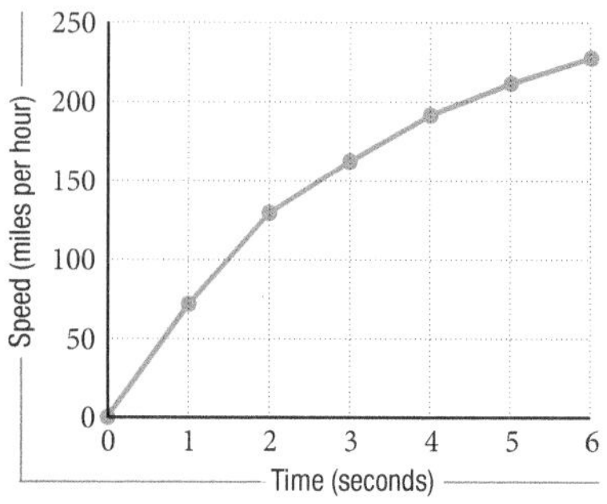

FIGURE 3

The number sequences we have worked with in the past can also be written as paired data by associating each number in the sequence with its position in the sequence. For instance, in the sequence of odd numbers

$$1, 3, 5, 7, 9, \ldots$$

the number 7 is the fourth number in the sequence. Its position is 4, and its value is 7. Here is the sequence of odd numbers written so that the position of each term is noted:

Position	1, 2, 3, 4, 5, . . .
Value	1, 3, 5, 7, 9, . . .

VIDEO EXAMPLES

SECTION 3.1

EXAMPLE 1 The tables below give the first five terms of the sequence of odd numbers and the sequence of squares as paired data. In each case construct a scatter diagram.

Odd Numbers

Position	Value
1	1
2	3
3	5
4	7
5	9

Squares

Position	Value
1	1
2	4
3	9
4	16
5	25

SOLUTION The two scatter diagrams are based on the data from these tables shown here. Notice how the dots in Figure 4 seem to line up in a straight line, whereas the dots in Figure 5 give the impression of a curve. We say the points in Figure 4 suggest a linear relationship between the two sets of data, whereas the points in Figure 5 suggest a nonlinear relationship.

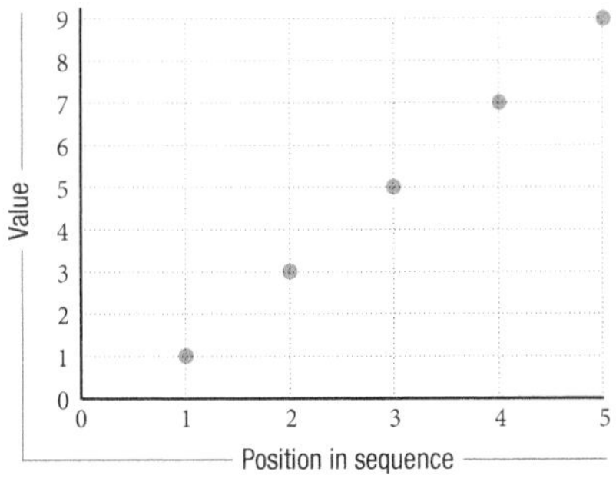

FIGURE 4

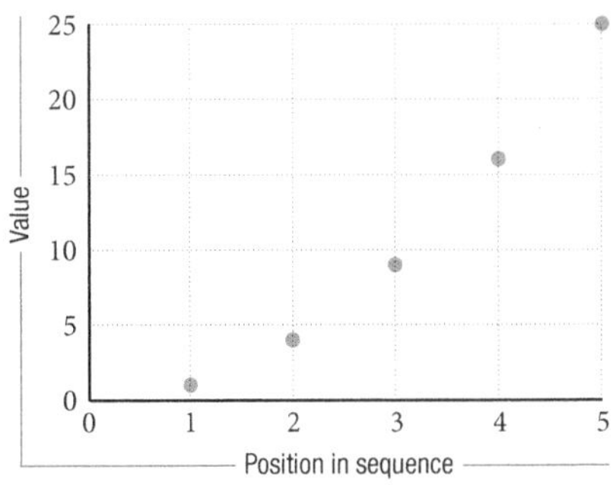

FIGURE 5

As you know, each dot in Figures 4 and 5 corresponds to a pair of numbers, one of which is associated with the horizontal axis and the other with the vertical axis. Paired data play a very important role in the equations we will solve in the next section. To prepare ourselves for those equations, we need to expand the concept of paired data to include negative numbers. At the same time, we want to standardize the position of the axes in the diagrams that we use to visualize paired data.

The Rectangular Coordinate System

DEFINITION *x-coordinate, y-coordinate*

A pair of numbers enclosed in parentheses and separated by a comma, such as $(-2, 1)$, is called an ***ordered pair*** of numbers. The first number in the pair is called the ***x-coordinate*** of the ordered pair; the second number is called the ***y-coordinate***. For the ordered pair $(-2, 1)$, the x-coordinate is -2 and the y-coordinate is 1.

Ordered pairs of numbers are important in the study of mathematics because they give us a way to visualize solutions to equations. To see the visual component of ordered pairs, we need the diagram shown in Figure 6. It is called the *rectangular coordinate system*.

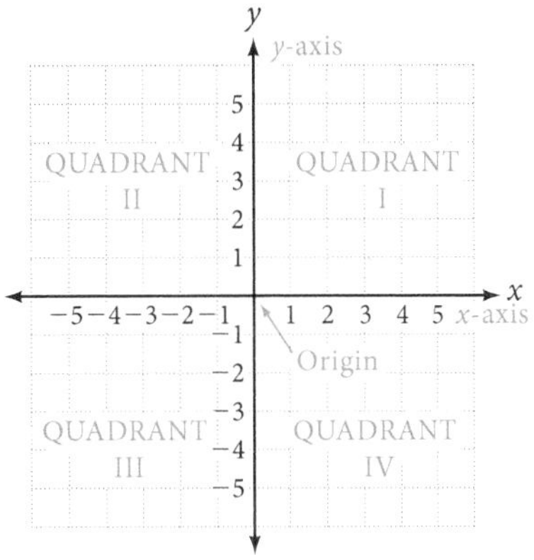

FIGURE 6

The rectangular coordinate system is built from two number lines oriented perpendicular to each other. The horizontal number line is exactly the same as our real number line and is called the *x-axis*. The vertical number line is also the same as our real number line with the positive direction up and the negative direction down. It is called the *y-axis*. The point where the two axes intersect is called the *origin*. As you can see from Figure 6, the axes divide the plane into four *quadrants*, which are numbered I through IV in a counterclockwise direction.

Graphing Ordered Pairs

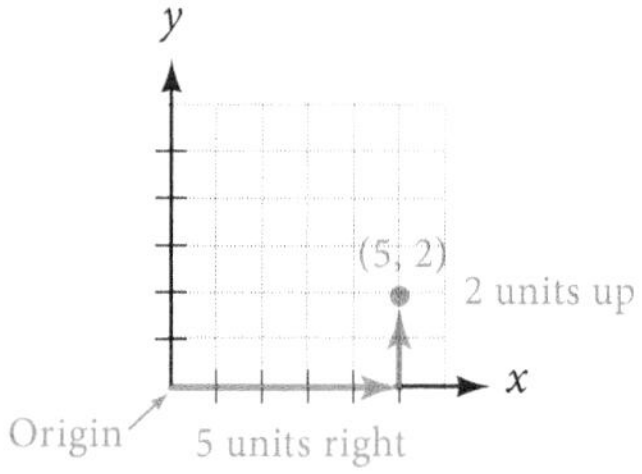

To graph the ordered pair (a, b), we start at the origin and move a units right or left (right if a is positive and left if a is negative). Then we move b units up or down (up if b is positive, down if b is negative). The point where we end up is the graph of the ordered pair (a, b). To graph the ordered pair (5, 2), we start at the origin and move 5 units to the right. Then, from that position, we move 2 units up.
Every ordered pair can be represented graphically as a unique point on the rectangular coordinate system.

EXAMPLE 2 Graph the ordered pairs (3, 4), (3, −4), (−3, 4), and (−3, −4). Also, identify which quadrant each point lies in.

SOLUTION

Note It is very important that you graph ordered pairs quickly and accurately. Remember, the first coordinate goes with the horizontal axis and the second coordinate goes with the vertical axis.

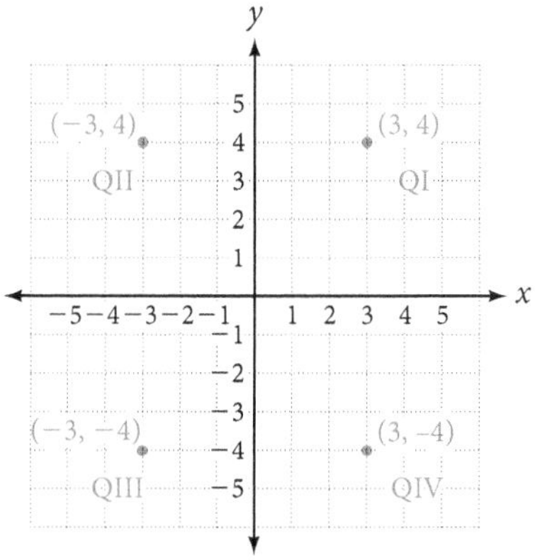

FIGURE 7

We can see in Figure 7 that when we graph ordered pairs, the x-coordinate corresponds to movement parallel to the x-axis (horizontal) and the y-coordinate corresponds to movement parallel to the y-axis (vertical).

The point (3, 4) lies in QI, (3, −4) lies in QIV, (−3, 4) lies in QII, and (−3, −4) lies in QIII. We have labeled each point with its corresponding quadrant in Figure 7.

EXAMPLE 3 Graph the ordered pairs $(-1, 3)$, $(2, 5)$, $(0, 0)$, $(0, -3)$, and $(4, 0)$. State which quadrant each point lies in.

SOLUTION See Figure 8.

Note If we do not label the axes of a coordinate system, we assume that each square is one unit long and one unit wide.

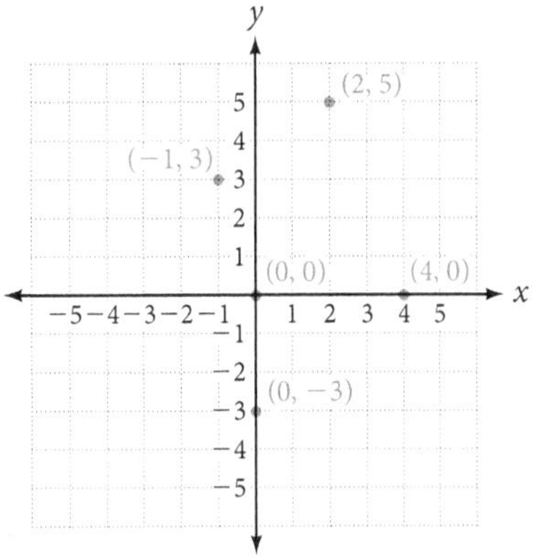

FIGURE 8

The point $(-1, 3)$ lies in QII, and $(2, 5)$ lies in QI. Because $(0, 0)$, $(0, -3)$, and $(4, 0)$ lie on one of the axes, these points are not in a particular quadrant.

Getting Ready for Class

After reading through the preceding section, respond in your own words and in complete sentences.

A. What is an ordered pair of numbers?

B. Explain in words how you would graph the ordered pair $(3, 4)$.

C. How do you construct a rectangular coordinate system?

D. Where is the origin on a rectangular coordinate system?

Problem Set 3.1

Graph each ordered pair on a rectangular coordinate system. Then indicate which quadrant, if any, the corresponding point lies in.

1. $(3, 2)$ **2.** $(3, -2)$ **3.** $(-3, 2)$ **4.** $(-3, -2)$

5. $(5, 1)$ **6.** $(5, -1)$ **7.** $(1, 5)$ **8.** $(1, -5)$

9. $(-1, 5)$ **10.** $(-1, -5)$ **11.** $\left(2, \frac{1}{2}\right)$ **12.** $\left(3, \frac{3}{2}\right)$

13. $\left(-4, -\frac{5}{2}\right)$ **14.** $\left(-5, -\frac{3}{2}\right)$ **15.** $(3, 0)$ **16.** $(-2, 0)$

17. $(0, 5)$ **18.** $(0, 0)$

Give the coordinates of each numbered point in the figure.

19–28.

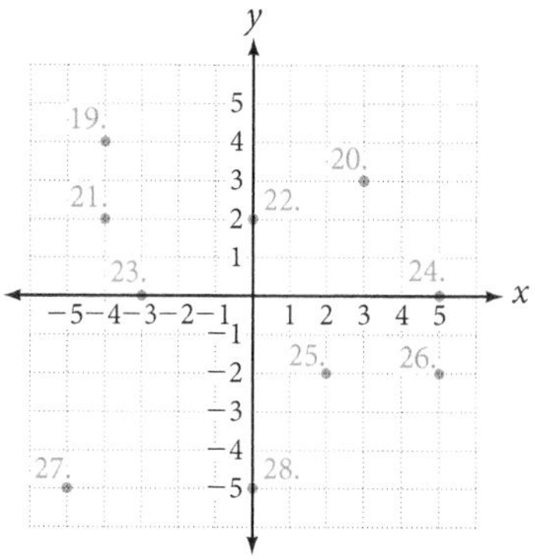

Graph the points $(4, 3)$ and $(-4, -1)$, and draw a straight line that passes through both of them. Then answer the following questions.

29. Does the graph of $(2, 2)$ lie on the line?

30. Does the graph of $(-2, 0)$ lie on the line?

31. Does the graph of $(0, -2)$ lie on the line?

32. Does the graph of $(-6, 2)$ lie on the line?

Graph the points $(-2, 4)$ and $(2, -4)$, and draw a straight line that passes through both of them. Then answer the following questions.

33. Does the graph of $(0, 0)$ lie on the line?

34. Does the graph of $(-1, 2)$ lie on the line?

35. Does the graph of $(2, -1)$ lie on the line?

36. Does the graph of $(1, -2)$ lie on the line?

Draw a straight line that passes through the points $(3, 4)$ and $(3, -4)$. Then answer the following questions.

37. Is the graph of $(3, 0)$ on this line?

38. Is the graph of $(0, 3)$ on this line?

39. Is there any point on this line with an x-coordinate other than 3?

40. If you extended the line, would it pass through a point with a y-coordinate of 10?

Draw a straight line that passes through the points (3, 4) and (−3, 4). Then answer the following questions.

41. Is the graph of (4, 0) on this line?

42. Is the graph of (0, 4) on this line?

43. Is there any point on this line with a y-coordinate other than 4?

44. If you extended the line, would it pass through a point with an x-coordinate of 10?

Applying the Concepts

45. Fibonacci Sequence The table below gives the first six terms of the Fibonacci sequence as paired data. Use the information in the table to construct a scatter diagram.

Fibonacci Sequence

Position	Value
1	1
2	1
3	2
4	3
5	5
6	8

46. Triangular Numbers The table below gives the first six terms of the sequence of triangular numbers as paired data. Use the information in the table to construct a scatter diagram.

Triangular Numbers

Position	Value
1	1
2	3
3	6
4	10
5	15
6	21

47. Non-Camera Phone Sales The table and bar chart show the sales of non-camera phones for the years 2006–2010. Use the information from the table and chart to construct a scatter diagram and a line graph.

Year	Sales (in Millions)
2006	300
2007	250
2008	175
2009	150
2010	125

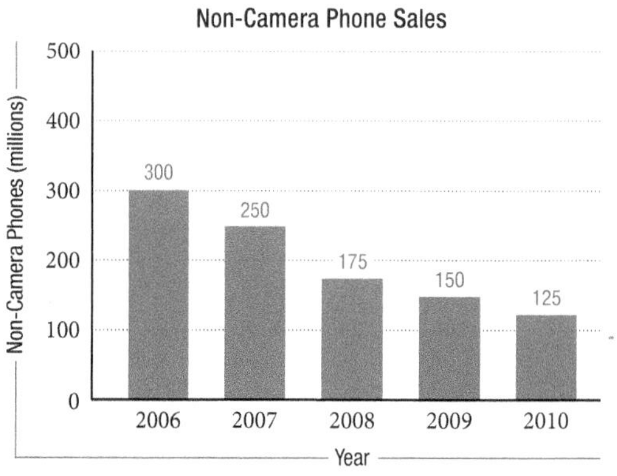

48. Camera Phone Sales The table and bar chart show the sales of camera phones from 2006 to 2010. Use the information from the table and chart to construct a scatter diagram and a line graph.

Year	Sales (in Millions)
2006	500
2007	650
2008	750
2009	875
2010	900

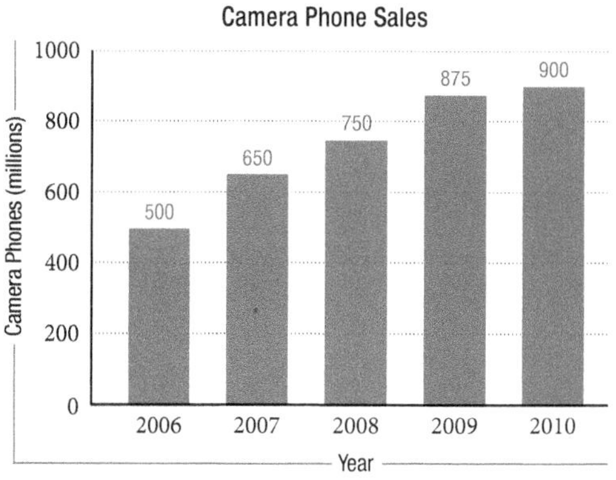

49. Hourly Wages Jane takes a job at the local Marcy's department store. Her job pays $8.00 per hour. The graph shows how much Jane earns for working from 0 to 40 hours in a week.

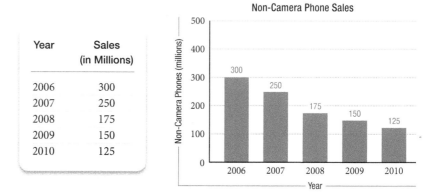

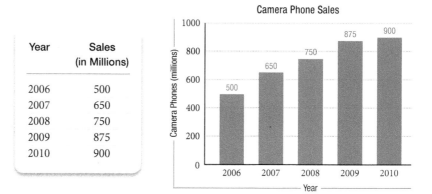

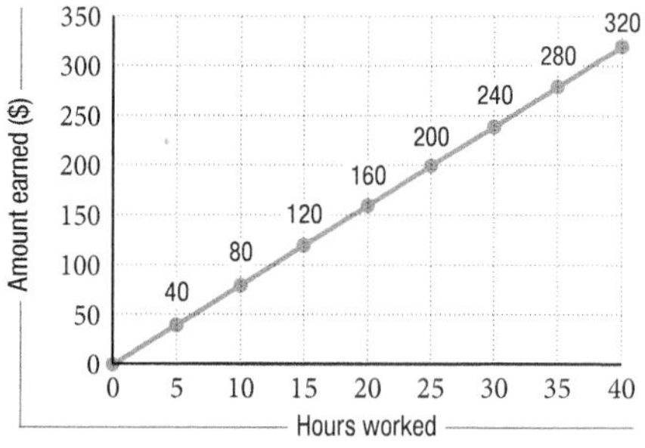

a. List three ordered pairs that lie on the line graph.

b. How much will she earn for working 40 hours?

c. If her check for one week is $240, how many hours did she work?

d. She works 35 hours one week, but her paycheck before deductions are subtracted out is for $260. Is this correct? Explain.

50. Hourly Wages Judy takes a job at Gigi's boutique. Her job pays \$9.00 per hour plus \$50 per week in commission. The graph shows how much Judy earns for working from 0 to 40 hours in a week.

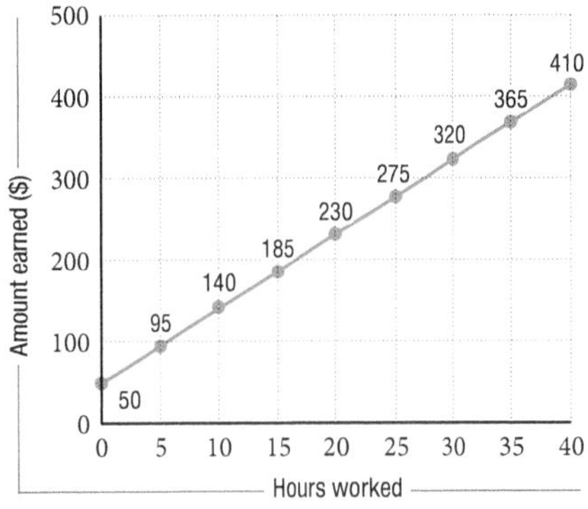

a. List three ordered pairs that lie on the line graph.

b. How much will she earn for working 40 hours?

c. If her check for one week is \$230, how many hours did she work?

d. She works 35 hours one week, but her paycheck before deductions are subtracted out is for \$365. Is this correct? Explain.

51. Kentucky Derby The line graph gives the monetary bets placed at the Kentucky Derby for specific years. If x represents the year in question and y represents the total wagering for that year, write five ordered pairs that describe the information in the table.

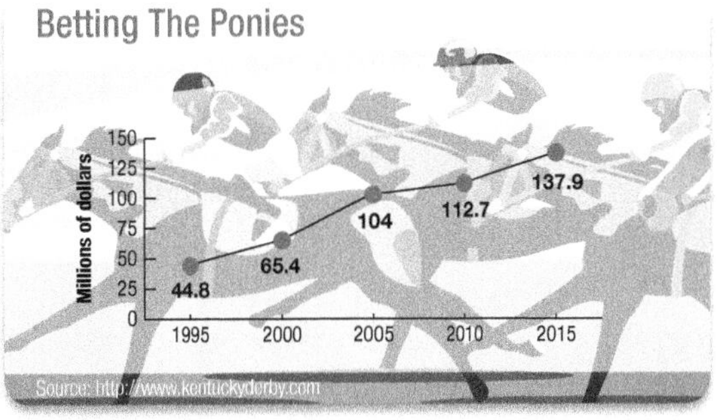

52. Health Care Costs Write 5 ordered pairs that lie on the curve shown below.

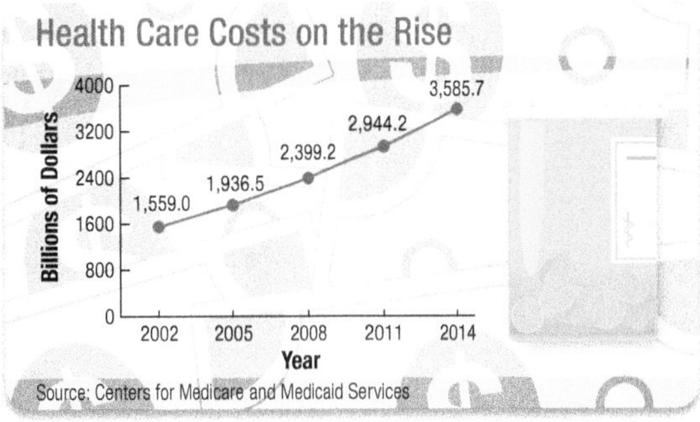

53. Rectangle $ABCD$ (Figure 11) has a length of 5 and a width of 3. Point D is the ordered pair (7, 2). Find points A, B, and C.

54. Rectangle $ABCD$ (Figure 12) has a length of 5 and a width of 3. Point D is the ordered pair (−1, 1). Find points A, B, and C.

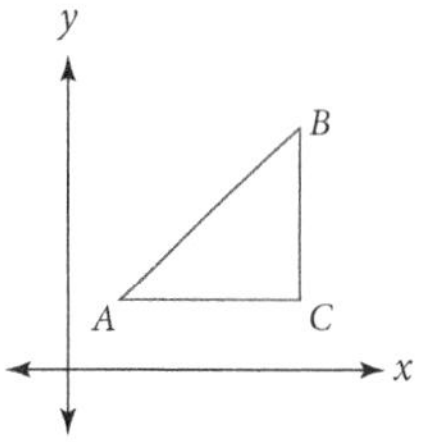

FIGURE 9

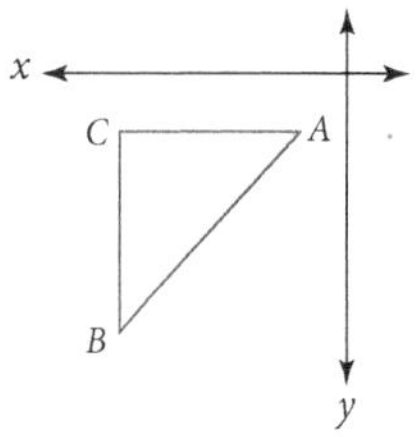

FIGURE 10

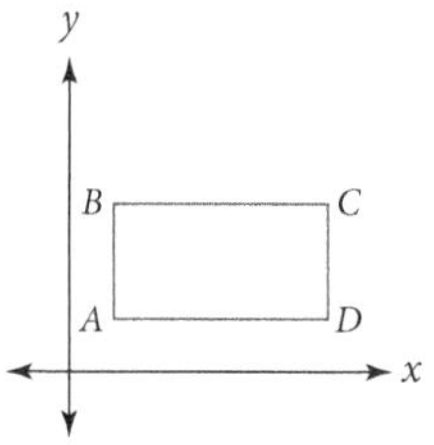

FIGURE 11

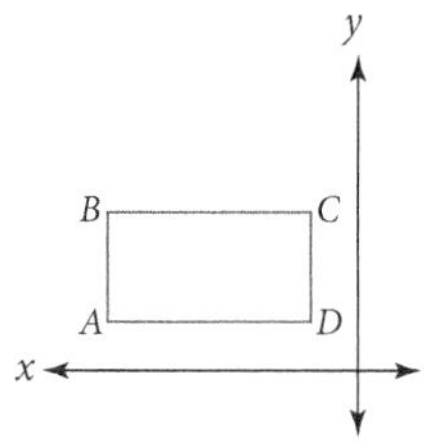

FIGURE 12

Learning Objectives Assessment

The following problems can be used to help assess if you have successfully met the learning objectives for this section.

55. Which of the following ordered pairs appears in the scatter diagram shown below?

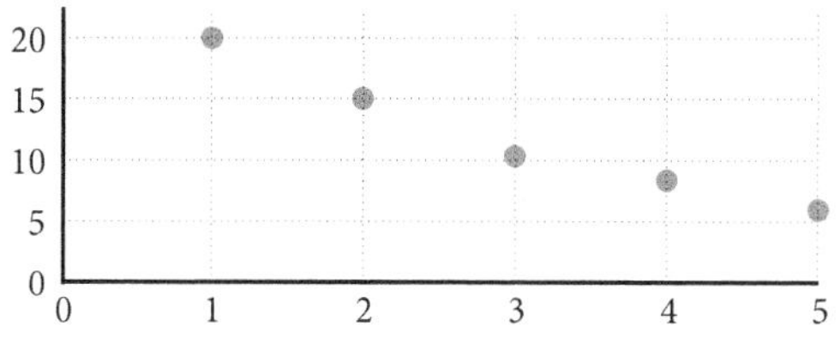

a. (15, 2) b. (10, 4) c. (3, 11) d. (1, 18)

56. The line graph gives the worldwide sales of Apple iPhones for specific years. How many iPhones were sold in 2013?

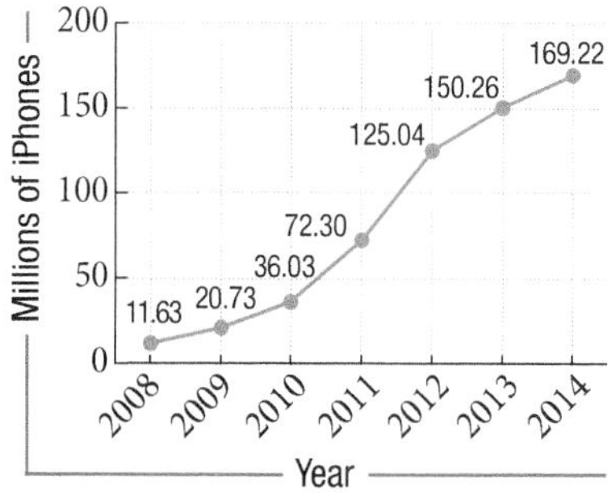

a. 125.04 b. 125.04 million
c. 150.26 million d. 150.26

57. Which of the following graphs correctly represents the ordered pair $(-3, 1)$?

a.

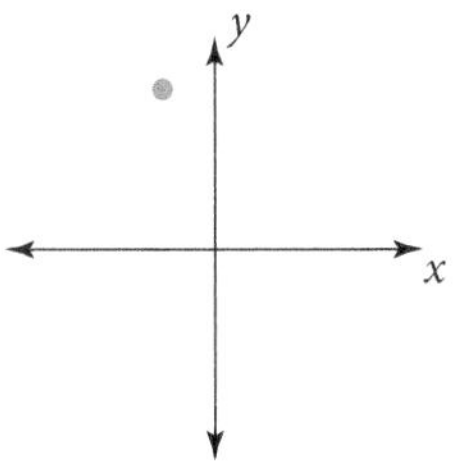

b.

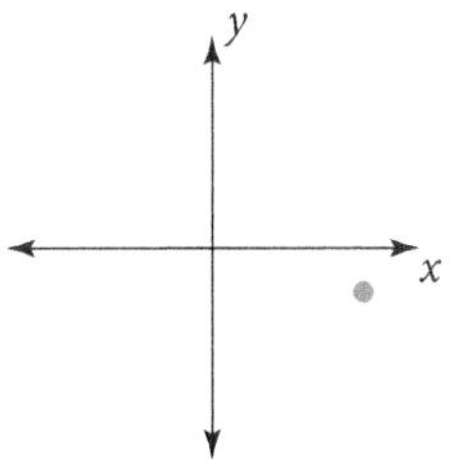

c.

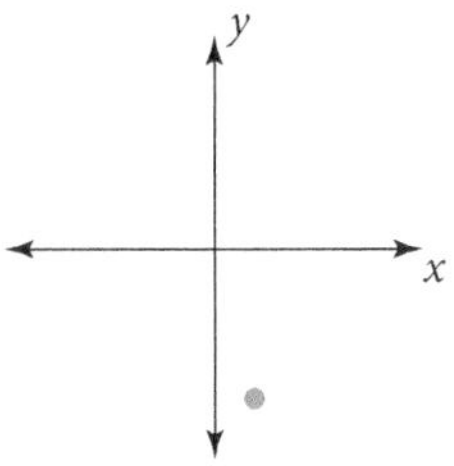

d.

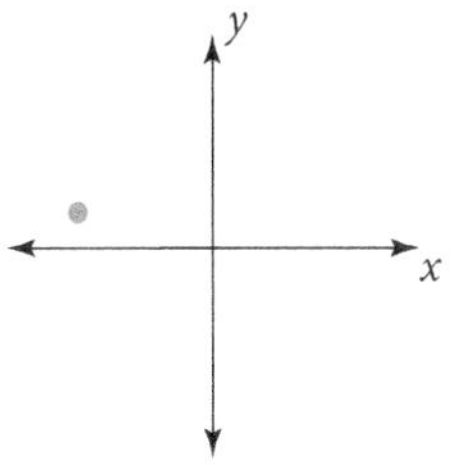

Getting Ready for the Next Section

58. Let $2x + 3y = 6$.

a. Find x if $y = 4$.
b. Find x if $y = -2$.
c. Find y if $x = 3$.
d. Find y if $x = 9$.

59. Let $2x - 5y = 20$.

a. Find x if $y = 0$.
b. Find x if $y = -6$.
c. Find y if $x = 0$.
d. Find y if $x = 5$.

60. Let $y = 2x - 1$.

a. Find x if $y = 7$.
b. Find x if $y = 3$.
c. Find y if $x = 0$.
d. Find y if $x = 5$.

61. Let $y = 3x - 2$.

a. Find x if $y = 4$.
b. Find x if $y = 3$.
c. Find y if $x = 2$.
d. Find y if $x = -3$.

62. Solve $5x + y = 4$ for y.

63. Solve $-3x + y = 5$ for y.

64. Solve $3x - 2y = 6$ for y.

65. Solve $2x - 3y = 6$ for y.

SPOTLIGHT ON SUCCESS *Student Instructor CJ*

We are what we repeatedly do.
Excellence, then, is not an act, but a habit.
—Aristotle

Something that has worked for me in college, in addition to completing the assigned homework, is working on some extra problems from each section. Working on these extra problems is a great habit to get into because it helps further your understanding of the material, and you see the many different types of problems that can arise. If you have completed every problem that your book offers, and you still don't feel confident that you have a full grasp of the material, look for more problems. Many problems can be found online or in other books. Your professors may even have some problems that they would suggest doing for extra practice. The biggest benefit to working all the problems in the course's assigned textbook is that often teachers will choose problems either straight from the book or ones similar to problems that were not assigned for tests. Doing this will ensure that you do your best in all your classes.

Graphing Linear Equations in Two Variables

3.2

Learning Objectives

In this section, we will learn how to:

1. Determine if an ordered pair is a solution to a linear equation in two variables.
2. Find solutions for a linear equation in two variables.
3. Graph a linear equation in two variables.
4. Graph horizontal and vertical lines.

Introduction

In this section we will begin to investigate equations in two variables. As you will see, equations in two variables have ordered pairs for solutions. We will use the rectangular coordinate system introduced in Section 3.1 to obtain a visual picture of *all* solutions to a linear equation in two variables. The process we use to obtain a visual picture of all solutions to an equation is called *graphing*. The picture itself is called the *graph* of the equation.

Let's begin this section by reviewing the relationship between equations in one variable and their solutions. If we solve the equation $3x - 2 = 10$, the solution is $x = 4$. If we graph this solution, we simply draw the real number line and place a point at the point whose coordinate is 4. The relationship between linear equations in one variable, their solutions, and the graphs of those solutions looks like this:

Equation	*Solution*	*Graph of Solution Set*
$3x - 2 = 10$	$x = 4$	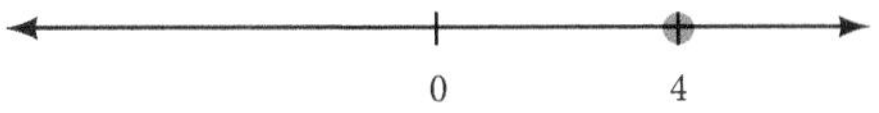

When the equation has one variable, the solution is a single number whose graph is a point on a line.

Note If this discussion seems a little long and confusing, you may want to look over some of the examples first and then come back and read this. Remember, it isn't always easy to read material in mathematics. What is important is that you understand what you are doing when you work problems. The reading is intended to assist you in understanding what you are doing. It is important to read everything in the book, but you don't always have to read it in the order it is written.

Solutions to Linear Equations in Two Variables

Consider the equation $2x + y = 3$. The first thing we notice is that there are two variables instead of one. Therefore, a solution to the equation $2x + y = 3$ will be not a single number but a pair of numbers, one for x and one for y, that makes the equation a true statement. One pair of numbers that works is $x = 2$ and $y = -1$, because when we substitute them for x and y in the equation we get a true statement.

$$2(2) + (-1) \stackrel{?}{=} 3$$
$$4 - 1 \stackrel{?}{=} 3$$
$$3 = 3 \quad \text{A true statement}$$

The pair of numbers $x = 2, y = -1$ is written as $(2, -1)$. As you know from Section 3.1, $(2, -1)$ is called an *ordered pair* because it is a pair of numbers written in a specific order. The first number is always associated with the variable x, and the second number is always associated with the variable y. We call the first number in the ordered pair the *x-coordinate* (or x component) and the second number the *y-coordinate* (or y component) of the ordered pair.

The ordered pair $(2, -1)$ is not the only solution. Another solution is $(0, 3)$ because when we substitute 0 for x and 3 for y we get

$$2(0) + 3 \stackrel{?}{=} 3$$
$$0 + 3 \stackrel{?}{=} 3$$
$$3 = 3 \qquad \text{A true statement}$$

As a matter of fact, for any number we want to use for x, there is another number we can use for y that will make the equation a true statement. There is an infinite number of ordered pairs that satisfy (are solutions to) the equation $2x + y = 3$; we have listed just a couple of them.

VIDEO EXAMPLES

SECTION 3.2

EXAMPLE 1 Which of the ordered pairs $(2, 3)$, $(1, 5)$, and $(-2, -4)$ are solutions to the equation $y = 3x + 2$?

SOLUTION If an ordered pair is a solution to the equation, then it must satisfy the equation; that is, when the coordinates are used in place of the variables in the equation, the equation becomes a true statement.

Try $(2, 3)$ in $y = 3x + 2$:

$$3 \stackrel{?}{=} 3(2) + 2$$
$$3 \stackrel{?}{=} 6 + 2$$
$$3 = 8 \qquad \text{A false statement}$$

Try $(1, 5)$ in $y = 3x + 2$:

$$5 \stackrel{?}{=} 3(1) + 2$$
$$5 \stackrel{?}{=} 3 + 2$$
$$5 = 5 \qquad \text{A true statement}$$

Try $(-2, -4)$ in $y = 3x + 2$:

$$-4 \stackrel{?}{=} 3(-2) + 2$$
$$-4 \stackrel{?}{=} -6 + 2$$
$$-4 = -4 \qquad \text{A true statement}$$

The ordered pairs $(1, 5)$ and $(-2, -4)$ are solutions to the equation $y = 3x + 2$, and $(2, 3)$ is not.

Now that we know how to determine if a given ordered pair is a solution to a linear equation in two variables, the next question is how can we *find* ordered pairs that will be solutions? The next examples illustrate how this is done.

EXAMPLE 2 Given the equation $2x + 3y = 6$, complete the following ordered pairs so they will be solutions to the equation: $(0,\ \ )$, $(\ \ , 1)$, $(3,\ \ )$.

SOLUTION To complete the ordered pair $(0,\ \ )$, we substitute 0 for x in the equation and then solve for y:

$$2(0) + 3y = 6$$
$$3y = 6$$
$$y = 2$$

The ordered pair is $(0, 2)$.

To complete the ordered pair (, 1), we substitute 1 for y in the equation and solve for x:

$$2x + 3(1) = 6$$
$$2x + 3 = 6$$
$$2x = 3$$
$$x = \frac{3}{2}$$

The ordered pair is $\left(\frac{3}{2}, 1\right)$.

To complete the ordered pair (3,), we substitute 3 for x in the equation and solve for y:

$$2(3) + 3y = 6$$
$$6 + 3y = 6$$
$$3y = 0$$
$$y = 0$$

The ordered pair is (3, 0).

Notice in each case that once we have substituted a number in place of one of the variables, the equation becomes a linear equation in one variable. We then use the method explained in Chapter 2 to solve for that variable.

EXAMPLE 3 Complete the following table for the equation $y = 2x - 1$.

x	y
0	
5	
	7
	3

SOLUTION When $x = 0$, we have

$$y = 2(0) - 1$$
$$y = 0 - 1$$
$$y = -1$$

When $x = 5$, we have

$$y = 2(5) - 1$$
$$y = 10 - 1$$
$$y = 9$$

When $y = 7$, we have

$$7 = 2x - 1$$
$$8 = 2x$$
$$4 = x$$

When $y = 3$, we have

$$3 = 2x - 1$$
$$4 = 2x$$
$$2 = x$$

The completed table is

x	y
0	-1
5	9
4	7
2	3

which means the ordered pairs $(0, -1)$, $(5, 9)$, $(4, 7)$, and $(2, 3)$ are among the solutions to the equation $y = 2x - 1$.

Graphing Linear Equations in Two Variables

We know from Section 3.1 that every ordered pair can be represented graphically as a point on the rectangular coordinate system. To graph a linear equation in two variables, we draw a picture that represents all solutions (ordered pairs) to the equation.

EXAMPLE 4 Graph the solution set for $x + y = 5$.

Note Our ability to graph an equation as we have done in Example 4 is due to the invention of the rectangular coordinate system. The French philosopher René Descartes (1595–1650) is the person usually credited with the invention of the rectangular coordinate system. As a philosopher, Descartes is responsible for the statement "I think, therefore I am." Until Descartes invented his coordinate system in 1637, algebra and geometry were treated as separate subjects. The rectangular coordinate system allows us to connect algebra and geometry by associating geometric shapes with algebraic equations. The study of the relationship between equations in algebra and their associated geometric figures is called *analytic geometry*. The rectangular coordinate system often is referred to as the *Cartesian coordinate system* in honor of Descartes.

SOLUTION We know from our previous work in this section that an infinite number of ordered pairs are solutions to the equation $x + y = 5$. We can't possibly list them all. What we can do is list a few of them and see if there is any pattern to their graphs.

Some ordered pairs that are solutions to $x + y = 5$ are $(0, 5)$, $(2, 3)$, $(3, 2)$, $(5, 0)$. The graph of each is shown in Figure 1.

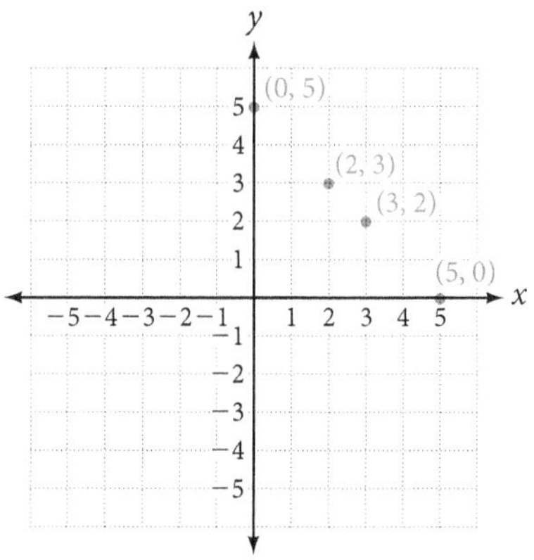

FIGURE 1

As you can see, all four points appear to lie on a line. If we were to continue graphing additional solutions, the gaps between points would start to fill in. Eventually we would have so many points that we would not be able to distinguish one from another, and the resulting image would appear to us as a continuous line.

So, by drawing a line through these points, we represent the entire solution set for the equation $x + y = 5$. Linear equations in two variables always have graphs that are lines. The graph of the solution set for $x + y = 5$ is shown in Figure 2.

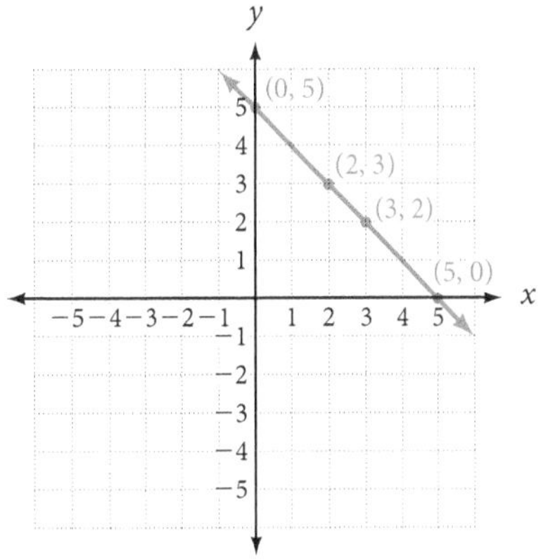

FIGURE 2

Every ordered pair that satisfies $x + y = 5$ has its graph on the line, and any point on the line has coordinates that satisfy the equation. So, there is a one-to-one correspondence between points on the line and solutions to the equation.

Here is the precise definition for a linear equation in two variables.

Note In standard form, we prefer that $a > 0$ and that any fractions have been cleared from the equation by multiplying both sides, if necessary, by the LCD.

DEFINITION *Linear Equation in Two Variables, Standard Form*

Any equation that can be put in the form $ax + by = c$, where a, b, and c are real numbers and a and b are not both 0, is called a ***linear equation in two variables***. The graph of any equation of this form is a straight line (that is why these equations are called "linear"). The form $ax + by = c$ is called ***standard form***.

To graph a linear equation in two variables, we simply graph its solution set; that is, we draw a line representing all the points whose coordinates satisfy the equation. Here are the steps to follow.

HOW TO *Graph a Linear Equation in Two Variables*

Step 1: Find any three ordered pairs that satisfy the equation. This can be done by using a convenient number for one variable and solving for the other variable.

Step 2: Graph the three ordered pairs found in step 1. Actually, we need only two points to graph a straight line. The third point serves as a check. If all three points do not line up, there is a mistake in our work.

Step 3: Draw a line through the three points graphed in step 2.

EXAMPLE 5 Graph the equation $y = 3x - 1$.

SOLUTION Because $y = 3x - 1$ can be put in the form $ax + by = c$, it is a linear equation in two variables. Hence, the graph of its solution set is a line. We can find some specific solutions by substituting numbers for x and then solving for the corresponding values of y. We are free to choose any numbers for x, so let's use 0, 2, and -1.

> *Note* It may seem that we have simply picked the numbers 0, 2, and -1 out of the air and used them for x. In fact we have done just that. Could we have used numbers other than these? The answer is yes, we can substitute any number for x; there will always be a value of y to go with it.

Let $x = 0$: $y = 3(0) - 1$
$y = 0 - 1$
$y = -1$

The ordered pair $(0, -1)$ is one solution.

Let $x = 2$: $y = 3(2) - 1$
$y = 6 - 1$
$y = 5$

The ordered pair $(2, 5)$ is a second solution.

Let $x = -1$: $y = 3(-1) - 1$
$y = -3 - 1$
$y = -4$

The ordered pair $(-1, -4)$ is a third solution.

In table form:

x	y
0	-1
2	5
-1	-4

Next, we graph the ordered pairs $(0, -1)$, $(2, 5)$, $(-1, -4)$ and draw a line through them.

The line we have drawn in Figure 3 is the graph of $y = 3x - 1$.

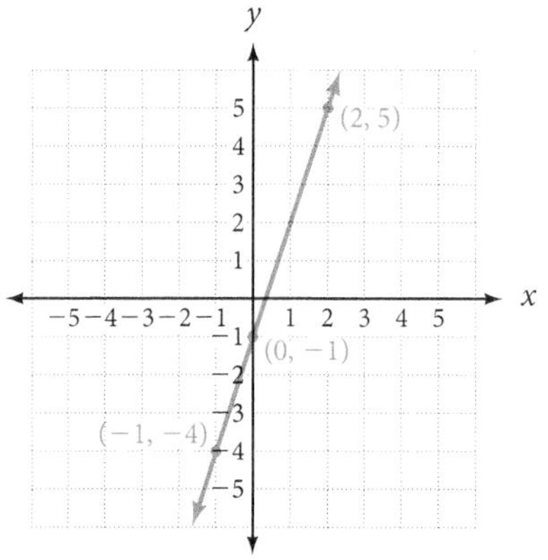

FIGURE 3

EXAMPLE 6 Graph the equation $3x - 2y = 6$.

SOLUTION It will be easier to find convenient values of x to use in the equation if we first solve the equation for y. To do so, we add $-3x$ to each side, and then we multiply each side by $-\frac{1}{2}$.

$$3x - 2y = 6 \quad \text{Original equation}$$

$$-2y = -3x + 6 \quad \text{Add } -3x \text{ to each side}$$

$$-\frac{1}{2}(-2y) = -\frac{1}{2}(-3x + 6) \quad \text{Multiply each side by } -\frac{1}{2}$$

$$y = \frac{3}{2}x - 3 \quad \text{Simplify each side}$$

Now, because each value of x will be multiplied by $\frac{3}{2}$, it will be to our advantage to choose values of x that are divisible by 2. That way, we will obtain values of y that do not contain fractions. This time, let's use 0, 2, and 4 for x.

When $x = 0$:
$$y = \frac{3}{2}(0) - 3$$
$$y = 0 - 3$$
$$y = -3$$

The ordered pair $(0, -3)$ is one solution.

When $x = 2$:
$$y = \frac{3}{2}(2) - 3$$
$$y = 3 - 3$$
$$y = 0$$

The ordered pair $(2, 0)$ is a second solution.

When $x = 4$:
$$y = \frac{3}{2}(4) - 3$$
$$y = 6 - 3$$
$$y = 3$$

The ordered pair $(4, 3)$ is a third solution.

Graphing the ordered pairs $(0, -3)$, $(2, 0)$, and $(4, 3)$ and drawing a line through them, we have the graph shown in Figure 4.

Note In Example 6 the values of x we used, 0, 2, and 4 are referred to as convenient values of x because they are easier to work with than some other numbers. For instance, if we let $x = 1$ in the equation $y = \frac{3}{2}x - 3$, we would have to add $\frac{3}{2}$ and -3 to find the corresponding value of y. Not only would the arithmetic be more difficult but also the ordered pair we obtained would have a fraction for its y-coordinate, making it more difficult to graph accurately.

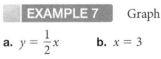

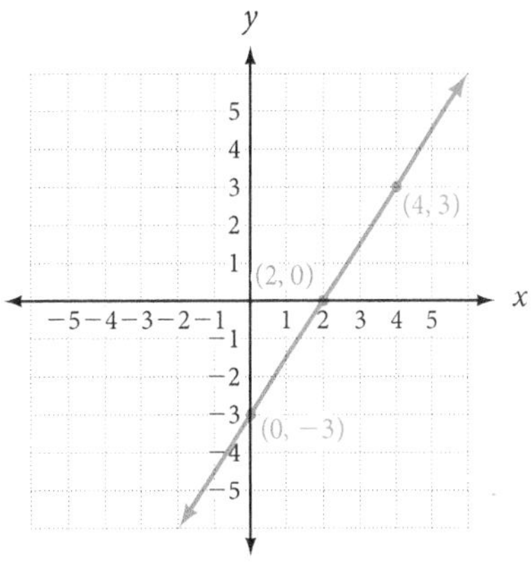

FIGURE 4

EXAMPLE 7 Graph each of the following lines.

a. $y = \frac{1}{2}x$ **b.** $x = 3$ **c.** $y = -2$

SOLUTION

a. The line $y = \frac{1}{2}x$ passes through the origin because $(0, 0)$ satisfies the equation. To sketch the graph we need at least one more point on the line. When x is 2, we obtain the point $(2, 1)$, and when x is -4, we obtain the point $(-4, -2)$.

The graph of $y = \frac{1}{2}x$ is shown in Figure 5.

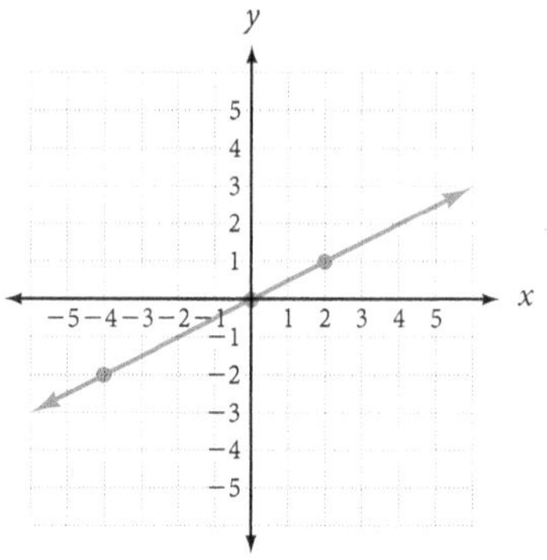

FIGURE 5

b. The line $x = 3$ is the set of all points whose x-coordinate is 3. The variable y does not appear in the equation, so the y-coordinate can be any number. Note that we can write our equation as a linear equation in two variables by writing it as $x + 0y = 3$. Because the product of 0 and y will always be 0, y can be any number.

For instance, if we use -4, 0, and 2 for y, then we obtain the solutions $(3, -4)$, $(3, 0)$, and $(3, 2)$.

The graph of $x = 3$ is shown in Figure 6. As you can see, the graph is a vertical line.

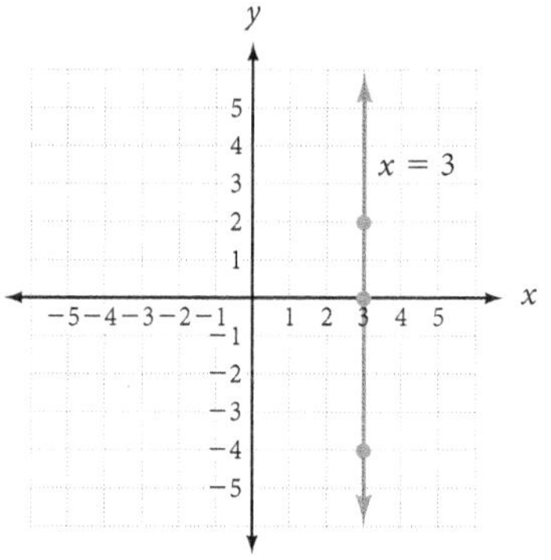

FIGURE 6

c. The line $y = -2$ is the set of all points whose y-coordinate is -2. The variable x does not appear in the equation, so the x-coordinate can be any number. Again, we can write our equation as a linear equation in two variables by writing it as $0x + y = -2$. Because the product of 0 and x will always be 0, x can be any number.

For instance, if we use -3, 0, and 1 for x, then we obtain the solutions $(-3, -2)$, $(0, -2)$, and $(1, -2)$.

The graph of $y = -2$ is shown in Figure 7. As you can see, the graph is a horizontal line.

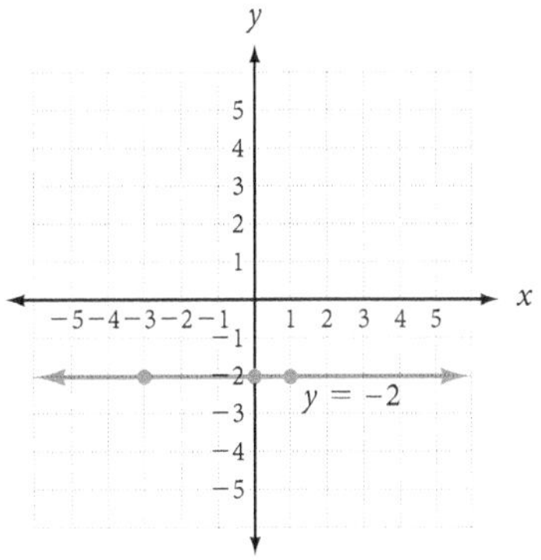

FIGURE 7

FACTS FROM GEOMETRY *Special Equations and Their Graphs*

For the equations below, m, a, and b are real numbers.

Through the Origin

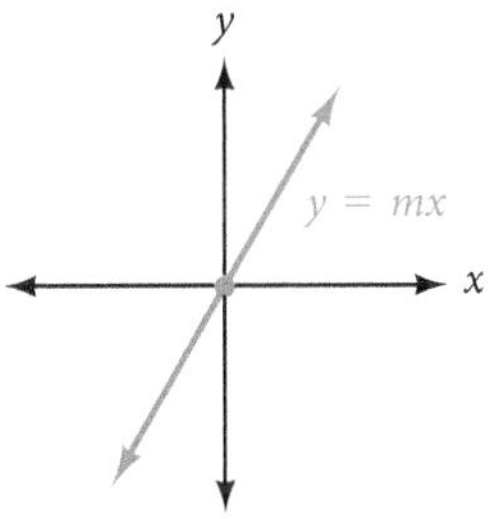

FIGURE 8 Any equation of the form $y = mx$ has a graph that passes through the origin.

Vertical Line

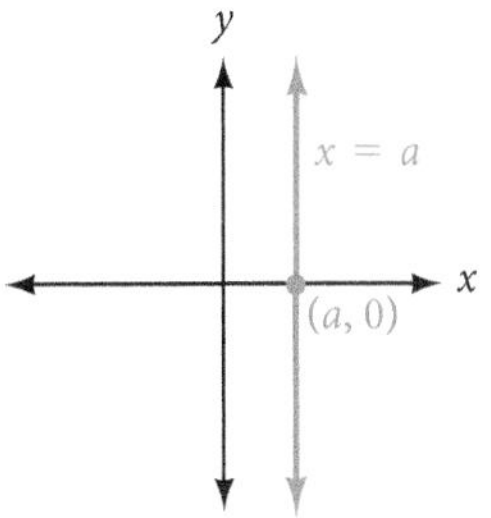

FIGURE 9 The graph of an equation of the form $x = a$ is a vertical line passing through the point $(a, 0)$ on the x-axis.

Horizontal Line

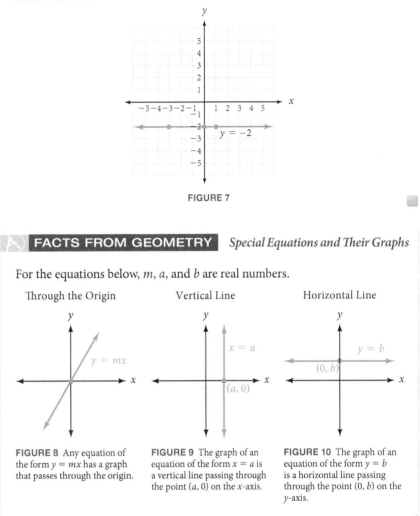

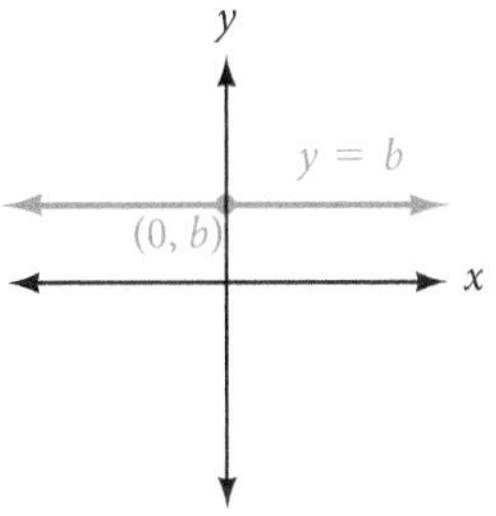

FIGURE 10 The graph of an equation of the form $y = b$ is a horizontal line passing through the point $(0, b)$ on the y-axis.

Getting Ready for Class

After reading through the preceding section, respond in your own words and in complete sentences.

A. How can you tell if an ordered pair is a solution to an equation?

B. How would you find a solution to $y = 3x - 5$?

C. Explain how you would go about graphing the line $x + y = 5$.

D. What kind of equations have horizontal lines for graphs?

Problem Set 3.2

For the following equations, tell which of the given ordered pairs are solutions.

1. $2x - 5y = 10$ $(2, 3), (0, -2), \left(\frac{5}{2}, 1\right)$

2. $3x + 7y = 21$ $(0, 3), (7, 0), (1, 2)$

3. $y = 7x - 2$ $(1, 5), (0, -2), (-2, -16)$

4. $y = -4x$ $(0, 0), (2, 4), (-3, 12)$

5. $x + y = 0$ $(1, 1), (2, -2), (3, 3)$

6. $x - y = 1$ $(0, 1), (0, -1), (1, 2)$

7. $x = 3$ $(3, 0), (3, -3), (5, 3)$

8. $y = -4$ $(3, -4), (-4, 4), (0, -4)$

For each equation, complete the given ordered pairs.

9. $2x + y = 6$ $(0,\ ), (\ , 0), (\ , -6)$

10. $3x - y = 5$ $(0,\ ), (1,\ ), (\ , 5)$

11. $3x + 4y = 12$ $(0,\ ), (\ , 0), (-4,\ )$

12. $5x - 5y = 20$ $(0,\ ), (\ , -2), (1,\ )$

13. $y = 4x - 3$ $(1,\ ), (\ , 0), (5,\ )$

14. $y = 3x - 5$ $(\ , 13), (0,\ ), (-2,\ )$

15. $y = 7x - 1$ $(2,\ ), (\ , 6), (0,\ )$

16. $y = 8x + 2$ $(3,\ ), (\ , 0), (\ , -6)$

17. $x = -5$ $(\ , 4), (\ , -3), (\ , 0)$

18. $y = 2$ $(5,\ ), (-8,\ ), \left(\frac{1}{2},\ \right)$

For each of the following equations, complete the given table.

19. $y = 3x$

x	y
1	
−3	
	12
	18

20. $y = -2x$

x	y
−4	
0	
	10
	12

21. $x + y = 5$

x	y
2	
3	
	0
	−4

22. $x - y = 8$

x	y
0	
4	
	−3
	−2

23. $2x - y = 4$

x	y
	0
	2
1	
−3	

24. $3x - y = 9$

x	y
	0
	−9
5	
−4	

25. $y = 6x - 1$

x	y
0	
	−7
−3	
	8

26. $y = 5x + 7$

x	y
0	
−2	
−4	
	−8

For the following equations, complete the given ordered pairs, and use the results to graph the solution set for the equation.

27. $x + y = 4$ $(0,\ ), (2,\ ), (\ , 0)$

28. $x - y = 3$ $(0,\ ), (2,\ ), (\ , 0)$

29. $y = 2x$ $(0,\ ), (-2,\ ), (2,\ )$

30. $y = \frac{1}{2}x$ $(0,\ ), (-2,\ ), (2,\ )$

31. $y = \frac{1}{3}x$ $(-3,\ ), (0,\ ), (3,\ )$

32. $y = 3x$ $(-2,\ ), (0,\ ), (2,\ )$

33. $y = 2x + 1$ $(0,\ ), (-1,\ ), (1,\ )$

34. $y = -2x + 1$ $(0,\ ), (-1,\ ), (1,\ )$

35. $y = 4$ $(0,\ ), (-1,\ ), (2,\ )$

36. $x = 3$ $(\ , -2), (\ , 0), (\ , 5)$

37. $y = \frac{1}{2}x + 3$ $(-2,\ ), (0,\ ), (2,\ )$

38. $y = \frac{1}{2}x - 3$ $(-2,\ ), (0,\ ), (2,\ )$

39. $y = -\frac{2}{3}x + 1$ $(-3,\ ), (0,\ ), (3,\ )$

40. $y = -\frac{2}{3}x - 1$ $(-3,\ ), (0,\ ), (3,\ )$

Solve each equation for y. Then, complete the given ordered pairs, and use them to graph the equation.

41. $2x + y = 3$ $(-1,\ ), (0,\ ), (1,\ )$

42. $3x + y = 2$ $(-1,\ ), (0,\ ), (1,\ )$

43. $3x + 2y = 6$ $(0,\ ), (2,\ ), (4,\ )$

44. $2x + 3y = 6$ $(0,\ ), (3,\ ), (6,\ )$

45. $-x + 2y = 6$ $(-2,\ ), (0,\ ), (2,\ )$

46. $-x + 3y = 6$ $(-3,\ ), (0,\ ), (3,\ )$

47. $4y = 2$ $(-4,\ ), (0,\ ), (4,\ )$

48. $2y = -5$ $(-4,\ ), (-2,\ ), (0,\ )$

Find three solutions to each of the following equations, and then graph the solution set.

49. $y = -\frac{1}{2}x$

50. $y = -2x$

51. $y = 3x - 1$

52. $y = -3x - 1$

53. $-2x + y = 1$

54. $-3x + y = 1$

55. $3x + 4y = 8$

56. $3x - 4y = 8$

57. $x = -2$

58. $y = 3$

59. $y = 2$

60. $x = -3$

Graph each equation.

61. $y = \frac{3}{4}x + 1$

62. $y = \frac{2}{3}x + 1$

63. $y = \frac{2}{3}x + \frac{2}{3}$

64. $y = -\frac{3}{4}x + \frac{3}{2}$

For each equation in each table below, indicate whether the graph is horizontal (H), or vertical (V), or whether it passes through the origin (O).

65.

Equation	H, V, and/or O
$x = 3$	
$y = 3$	
$y = 3x$	
$y = 0$	

66.

Equation	H, V, and/or O
$x = \frac{1}{2}$	
$y = \frac{1}{2}$	
$y = \frac{1}{2}x$	
$x = 0$	

67.

Equation	H, V, and/or O
$x = -\frac{3}{5}$	
$y = -\frac{3}{5}$	
$y = -\frac{3}{5}x$	
$x = 0$	

68.

Equation	H, V, and/or O
$x = -4$	
$y = -4$	
$y = -4x$	
$y = 0$	

69. Use the graph at the right to complete the table.

x	y
	-3
-2	
0	
	0
6	

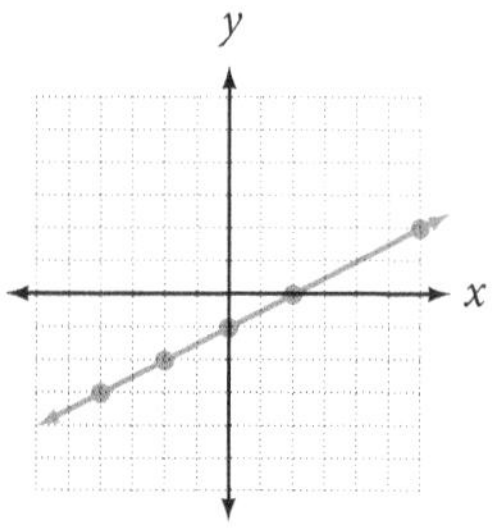

70. Use the graph at the right to complete the table. (*Hint:* Some parts have two answers.)

x	y
-3	6
	4
0	3
	1
6	

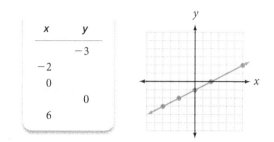

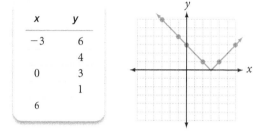

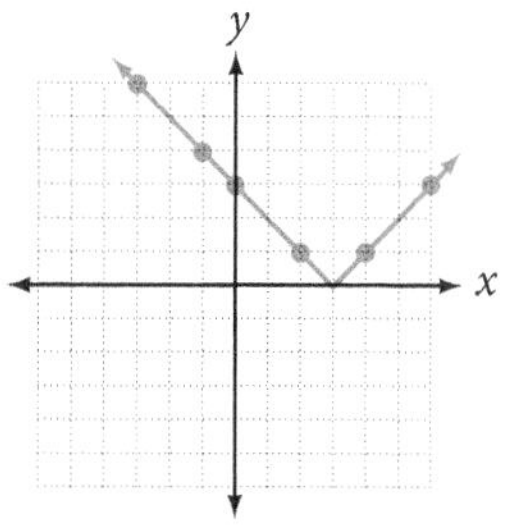

The next two problems are intended to give you practice reading, and paying attention to, the instructions that accompany the problems you are working. Working these problems is an excellent way to get ready for a test or a quiz.

71. Paying Attention to Instructions Work each problem according to the instructions given.

a. Solve: $2x + 5 = 10$

b. Find x when y is 0: $2x + 5y = 10$

c. Find y when x is 0: $2x + 5y = 10$

d. Graph: $2x + 5y = 10$

e. Solve for y: $2x + 5y = 10$

72. Paying Attention to Instructions Work each problem according to the instructions given.

a. Solve: $x - 2 = 6$

b. Find x when y is 0: $x - 2y = 6$

c. Find y when x is 0: $x - 2y = 6$

d. Graph: $x - 2y = 6$

e. Solve for y: $x - 2y = 6$

Applying the Concepts

73. Perimeter If the perimeter of a rectangle is 30 inches, then the relationship between the length l and the width w is given by the equation

$$2l + 2w = 30$$

What is the length when the width is 3 inches?

74. Perimeter The relationship between the perimeter P of a square and the length of its side s is given by the formula $P = 4s$. If each side of a square is 5 inches, what is the perimeter? If the perimeter of a square is 28 inches, how long is a side?

75. Hourly Wages Janai earns \$12 per hour working as a math tutor. We can express the amount she earns each week, y, for working x hours with the equation $y = 12x$. Indicate with a yes or no, which of the following could be one of Janai's paychecks. If you answer no, explain your answer.

a. \$60 for working five hours. **b.** \$100 for working nine hours

c. \$80 for working seven hours. **d.** \$168 for working 14 hours

76. Hourly Wages Erin earns \$15 per hour working as a graphic designer. We can express the amount she earns each week, y, for working x hours with the equation $y = 15x$. Indicate with a yes or no which of the following could be one of Erin's paychecks. If you answer no, explain your answer.

a. \$75 for working five hours. **b.** \$125 for working nine hours

c. \$90 for working six hours. **d.** \$500 for working 35 hours

77. Depreciation The equation $V = -45{,}000t + 600{,}000$, can be used to find the value, V, of a small crane at the end of t years, for $0 \le t \le 13$.

a. What is the value of the crane at the end of five years?

b. When is the crane worth \$330,000?

c. Is it true that the crane will be worth \$150,000 after nine years?

d. How much did the crane cost?

78. Depreciation The equation $V = -400t + 2{,}500$, can be used to find the value, V, of a notebook computer at the end of t years, for $0 \le t \le 6$.

a. What is the value of the notebook computer at the end of four years?

b. When is the notebook computer worth \$1,700?

c. Is it true that the notebook computer will be worth \$100 after five years?

d. How much did the notebook computer cost?

Learning Objectives Assessment

The following problems can be used to help assess if you have successfully met the learning objectives for this section.

79. Which ordered pair is a solution to the equation $3x - 4y = 8$?

a. $(0, 2)$ **b.** $(4, -1)$ **c.** $(-2, 1)$ **d.** $(-4, -5)$

80. Find an ordered pair with an x-coordinate of 2 that is a solution to $y = -3x + 5$.

a. $(1, 2)$ **b.** $(0, 2)$ **c.** $(2, -1)$ **d.** $(2, 11)$

81. Which equation has the graph shown in Figure 11?

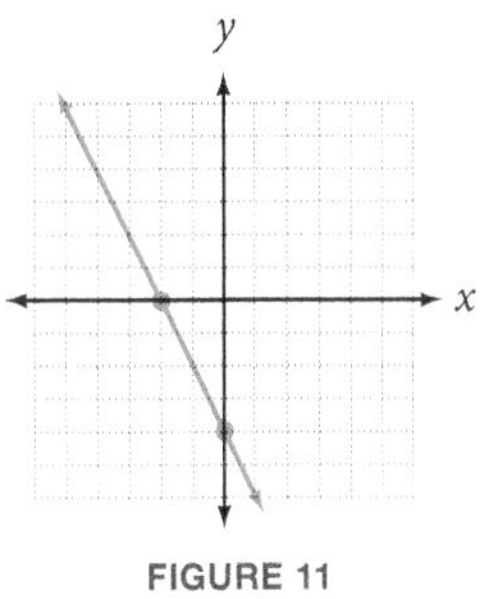

FIGURE 11

a. $-2x - 4y = 0$ **b.** $-2x - 4y = 1$

c. $2x + y = -4$ **d.** $2x - y = 4$

82. Graph the equation $x = 2$.

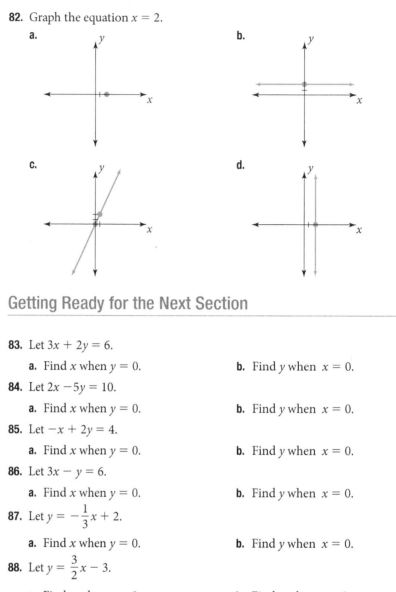

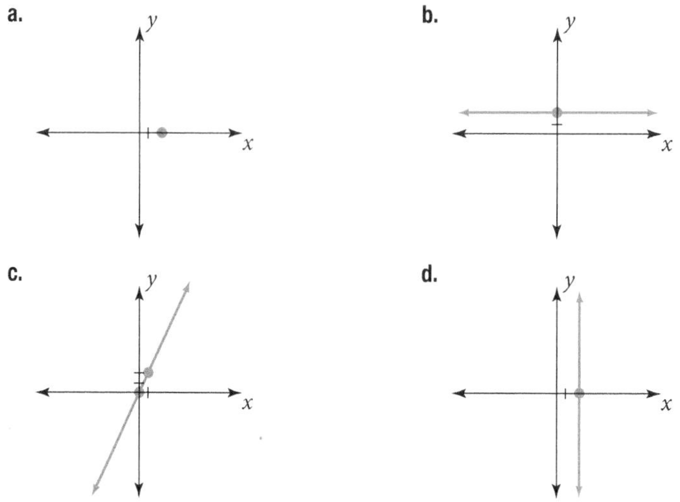

Getting Ready for the Next Section

83. Let $3x + 2y = 6$.

a. Find x when $y = 0$. **b.** Find y when $x = 0$.

84. Let $2x - 5y = 10$.

a. Find x when $y = 0$. **b.** Find y when $x = 0$.

85. Let $-x + 2y = 4$.

a. Find x when $y = 0$. **b.** Find y when $x = 0$.

86. Let $3x - y = 6$.

a. Find x when $y = 0$. **b.** Find y when $x = 0$.

87. Let $y = -\frac{1}{3}x + 2$.

a. Find x when $y = 0$. **b.** Find y when $x = 0$.

88. Let $y = \frac{3}{2}x - 3$.

a. Find x when $y = 0$. **b.** Find y when $x = 0$.

More on Graphing: Intercepts 3.3

Learning Objectives

In this section, we will learn how to:

1. Find the x-intercept and y-intercept for a line.
2. Use the intercepts to graph a line.

Introduction

In this section we continue our work with graphing lines by finding the points where a line crosses the axes of our coordinate system. To do so, we use the fact that any point on the x-axis has a y-coordinate of 0 and any point on the y-axis has an x-coordinate of 0. We begin with the following definition.

DEFINITION *x-intercept, y-intercept*

The ***x-intercept*** of a line is the x-coordinate of the point where the graph crosses the x-axis. The ***y-intercept*** is defined similarly. It is the y-coordinate of the point where the graph crosses the y-axis.

If the x-intercept is a, then the point $(a, 0)$ lies on the graph. This is true because any point on the x-axis has a y-coordinate of 0.

If the y-intercept is b, then the point $(0, b)$ lies on the graph. This is true because any point on the y-axis has an x-coordinate of 0.

Graphically, the relationship is shown in Figure 1.

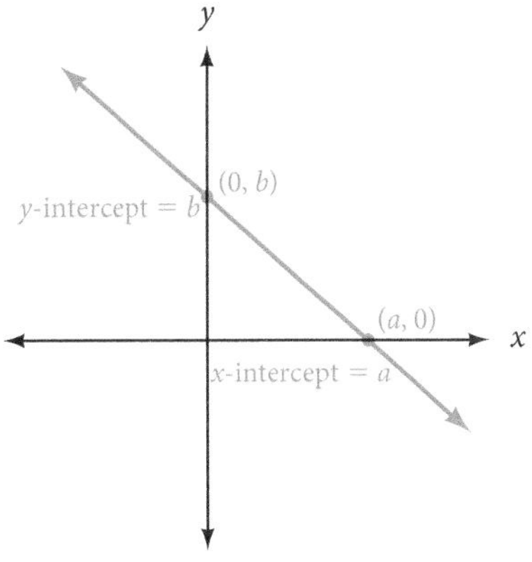

FIGURE 1

VIDEO EXAMPLES

SECTION 3.3

EXAMPLE 1 Find the x- and y-intercepts for $3x - 2y = 6$, and then use them to draw the graph.

SOLUTION To find where the graph crosses the x-axis, we let $y = 0$. (The y-coordinate of any point on the x-axis is 0.)

x-intercept:

When	$y = 0$	
the equation	$3x - 2y = 6$	
becomes	$3x - 2(0) = 6$	
	$3x - 0 = 6$	
	$x = 2$	Multiply each side by $\frac{1}{3}$

The graph crosses the x-axis at (2, 0), which means the x-intercept is 2.

y-intercept:

When	$x = 0$	
the equation	$3x - 2y = 6$	
becomes	$3(0) - 2y = 6$	
	$0 - 2y = 6$	
	$-2y = 6$	
	$y = -3$	Multiply each side by $-\frac{1}{2}$

The graph crosses the y-axis at (0, −3), which means the y-intercept is −3.

Plotting the x- and y-intercepts and then drawing a line through them, we have the graph of $3x - 2y = 6$, as shown in Figure 2.

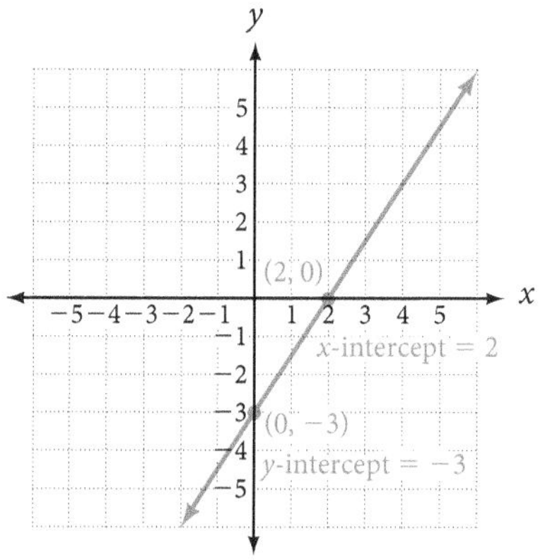

FIGURE 2

EXAMPLE 2 Graph $-x + 2y = 4$ by finding the intercepts and using them to draw the graph.

SOLUTION Again, we find the x-intercept by letting $y = 0$ in the equation and solving for x. Similarly, we find the y-intercept by letting $x = 0$ and solving for y.

***x*-intercept:**

When	$y = 0$
the equation	$-x + 2y = 4$
becomes	$-x + 2(0) = 4$
	$-x + 0 = 4$
	$-x = 4$
	$x = -4$ Multiply each side by -1

The x-intercept is -4, indicating that the point $(-4, 0)$ is on the graph of $-x + 2y = 4$.

***y*-intercept:**

When	$x = 0$
the equation	$-x + 2y = 4$
becomes	$-0 + 2y = 4$
	$2y = 4$
	$y = 2$ Multiply each side by $\frac{1}{2}$

The y-intercept is 2, indicating that the point (0, 2) is on the graph of $-x + 2y = 4$.

Plotting the intercepts and drawing a line through them, we have the graph of $-x + 2y = 4$, as shown in Figure 3.

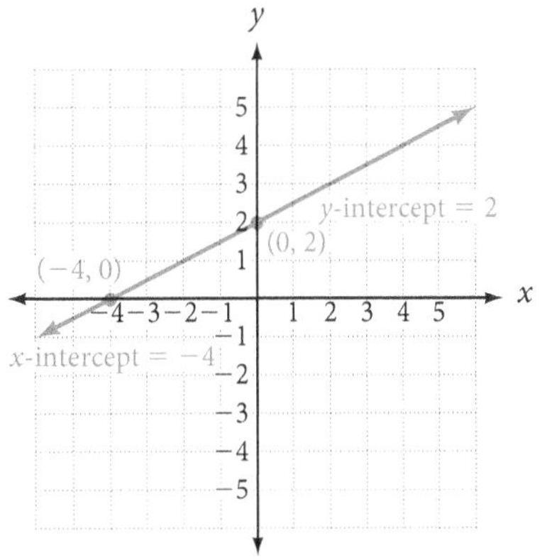

FIGURE 3

Graphing a line by finding the intercepts, as we have done in Examples 1 and 2, is an easy method of graphing if the equation is in standard form $ax + by = c$ and both the numbers a and b divide the number c evenly.

In our next example we use the intercepts to graph a line in which y is given in terms of x.

EXAMPLE 3 Use the intercepts for $y = -\frac{1}{3}x + 2$ to draw its graph.

SOLUTION Graph the line by finding the intercepts.

***x*-intercept:**

When	$y = 0$	
the equation	$y = -\frac{1}{3}x + 2$	
becomes	$0 = -\frac{1}{3}x + 2$	
	$-2 = -\frac{1}{3}x$	Add -2 to each side
	$6 = x$	Multiply each side by -3

The x-intercept is 6, which means the graph passes through the point (6, 0).

***y*-intercept:**

When	$x = 0$
the equation	$y = -\frac{1}{3}x + 2$
becomes	$y = -\frac{1}{3}(0) + 2$
	$y = 2$

The y-intercept is 2, which means the graph passes through the point (0, 2).

The graph of $y = -\frac{1}{3}x + 2$ is shown in Figure 4.

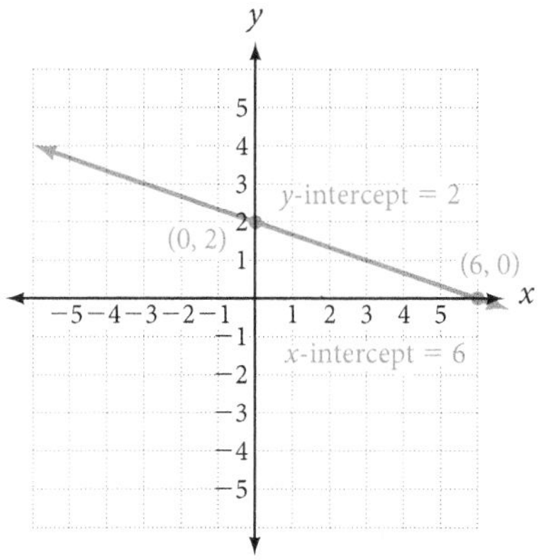

FIGURE 4

In the previous example, notice that the y-intercept was given by the constant term in the equation. If the equation is written with the variable y isolated, then this will always be true.

Consider the equation $y = ax + b$. To find the y-intercept, we substitute $x = 0$:

When	$x = 0$
the equation	$y = ax + b$
becomes	$y = a(0) + b$
	$y = b$

We summarize this fact in the following property.

PROPERTY *y-intercept*

The y-intercept of the line $y = ax + b$ is $y = b$. The graph of the line will cross the y-axis at the point $(0, b)$.

We will investigate this form of a linear equation in more detail in Section 3.5.

EXAMPLES

4. The y-intercept of $y = \frac{1}{2}x - 5$ is -5.

5. The y-intercept of $y = 2x + \frac{1}{3}$ is $\frac{1}{3}$.

6. The y-intercept of $y = 6$ is 6.

7. The y-intercept of $y = 4 - x$ is 4.

Getting Ready for Class

After reading through the preceding section, respond in your own words and in complete sentences.

A. What is the x-intercept for a graph?

B. What is the y-intercept for a graph?

C. How do we find the y-intercept for a line from the equation?

D. How do we graph a line using its intercepts?

Problem Set 3.3

Find the x- and y-intercepts for the following equations. Then use the intercepts to graph each equation.

1. $2x + y = 4$ **2.** $2x + y = 2$ **3.** $-x + y = 3$ **4.** $-x + y = 4$

5. $-x + 2y = 2$ **6.** $-x + 2y = 4$ **7.** $5x + 2y = 10$ **8.** $2x + 5y = 10$

9. $4x - 2y = 8$ **10.** $2x - 4y = 8$ **11.** $-4x + 5y = 20$ **12.** $-5x + 4y = 20$

13. $y = 2x - 6$ **14.** $y = 2x + 6$ **15.** $y = 2x + 2$ **16.** $y = -2x + 2$

17. $y = 2x - 1$ **18.** $y = -2x - 1$ **19.** $y = \frac{1}{2}x + 3$ **20.** $y = \frac{1}{2}x - 3$

21. $y = -\frac{1}{3}x - 2$ **22.** $y = -\frac{1}{3}x + 2$

For each of the following lines the x-intercept and the y-intercept are both 0, which means the graph of each will go through the origin, (0, 0). Graph each line by finding a point on each, other than the origin, and then drawing a line through that point and the origin.

23. $y = -2x$ **24.** $y = \frac{1}{2}x$ **25.** $y = -\frac{1}{3}x$ **26.** $y = -3x$

27. $y = \frac{2}{3}x$ **28.** $y = \frac{3}{2}x$

Complete each table.

29.

Equation	x-intercept	y-intercept
$3x + 4y = 12$		
$3x + 4y = 4$		
$3x + 4y = 3$		
$3x + 4y = 2$		

30.

Equation	x-intercept	y-intercept
$-2x + 3y = 6$		
$-2x + 3y = 3$		
$-2x + 3y = 2$		
$-2x + 3y = 1$		

31.

Equation	x-intercept	y-intercept
$x - 3y = 2$		
$y = \frac{1}{3}x - \frac{2}{3}$		
$x - 3y = 0$		
$y = \frac{1}{3}x$		

32.

Equation	x-intercept	y-intercept
$x - 2y = 1$		
$y = \frac{1}{2}x - \frac{1}{2}$		
$x - 2y = 0$		
$y = \frac{1}{2}x$		

Find the y-intercept for each of the following lines.

33. $y = 2x + 5$

34. $y = 3x - 1$

35. $y = \dfrac{1}{3}x - \dfrac{2}{3}$

36. $y = \dfrac{1}{4}x + \dfrac{5}{4}$

37. $y = 4 - x$

38. $y = 6 - 5x$

39. $y = \dfrac{7x + 1}{2}$

40. $y = \dfrac{2x + 3}{5}$

The next two problems are intended to give you practice reading, and paying attention to, the instructions that accompany the problems you are working. Working these problems is an excellent way to get ready for a test or a quiz.

41. Paying Attention to Instructions Work each problem according to the instructions given.

a. Solve: $2x - 3 = -3$

b. Find the x-intercept: $2x - 3y = -3$

c. Find y when x is 0: $2x - 3y = -3$

d. Graph: $2x - 3y = -3$

e. Solve for y: $2x - 3y = -3$

42. Paying Attention to Instructions Work each problem according to the instructions given.

a. Solve: $3x - 4 = -4$

b. Find the y-intercept: $3x - 4y = -4$

c. Find x when y is 0: $3x - 4y = -4$

d. Graph: $3x - 4y = -4$

e. Solve for y: $3x - 4y = -4$

From the graphs below, find the x- and y-intercepts for each line.

43.

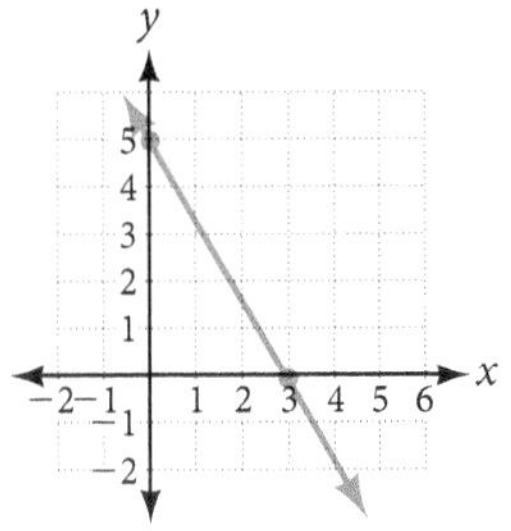

44.

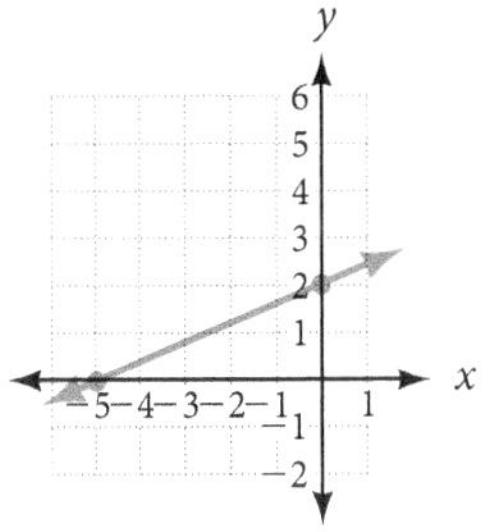

45.

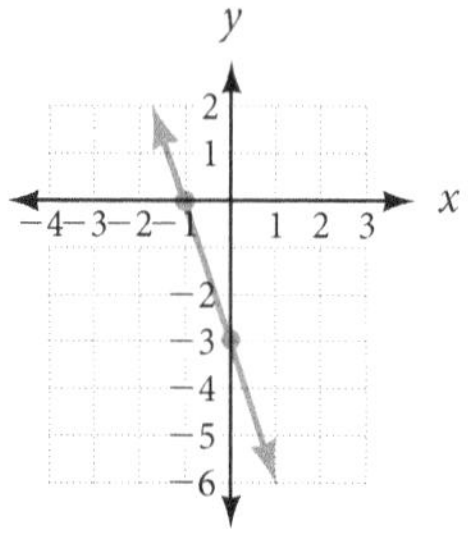

46.

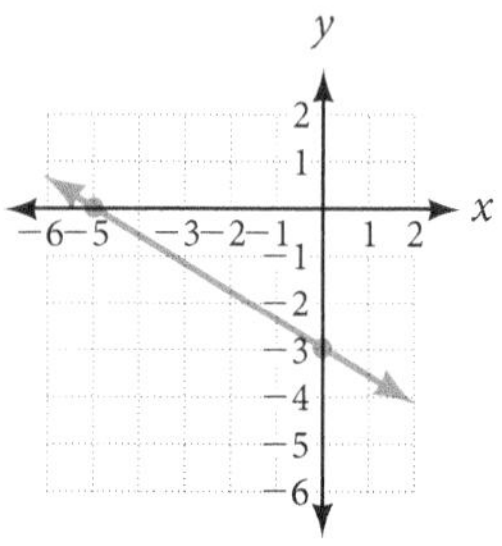

47. Graph the line that passes through the point $(-4, 4)$ and has an x-intercept of -2. What is the y-intercept of this line?

48. Graph the line that passes through the point $(-3, 4)$ and has a y-intercept of 3. What is the x-intercept of this line?

49. A line passes through the point $(1, 4)$ and has a y-intercept of 3. Graph the line and name its x-intercept.

50. A line passes through the point $(3, 4)$ and has an x-intercept of 1. Graph the line and name its y-intercept.

51. Graph the line that passes through the points $(-2, 5)$ and $(5, -2)$. What are the x- and y-intercepts for this line?

52. Graph the line that passes through the points $(5, 3)$ and $(-3, -5)$. What are the x- and y-intercepts for this line?

53. Use the graph at the right to complete the following table.

x	y
-2	
0	
	0
	-2

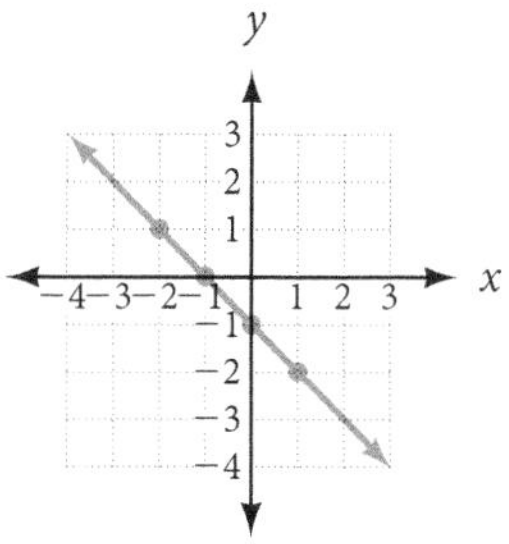

54. Use the graph at the right to complete the following table.

x	y
-2	
	0
	6

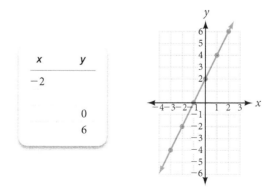

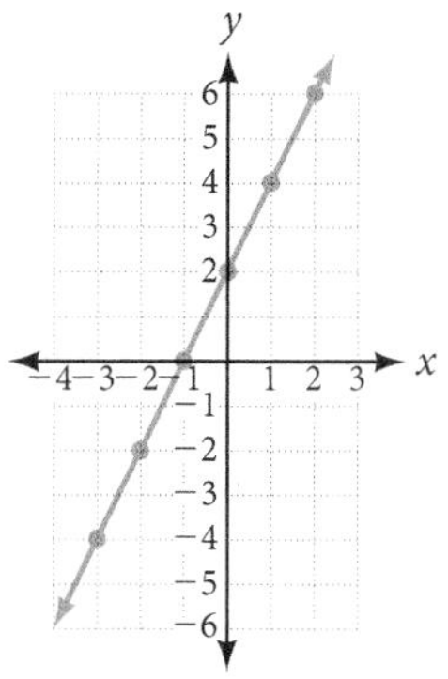

55. The vertical line $x = 3$ has only one intercept. Graph $x = 3$, and name its intercept. [Remember, ordered pairs (x, y) that are solutions to the equation $x = 3$ are ordered pairs with an x-coordinate of 3 and any y-coordinate.]

56. Graph the vertical line $x = -2$. Then name its intercept.

57. The horizontal line $y = 4$ has only one intercept. Graph $y = 4$, and name its intercept. [Ordered pairs (x, y) that are solutions to the equation $y = 4$ are ordered pairs with a y-coordinate of 4 and any x-coordinate.]

58. Graph the horizontal line $y = -3$. Then name its intercept.

Applying the Concepts

59. **Complementary Angles** The following diagram shows sunlight hitting the ground. Angle α (*alpha*) is called the angle of inclination, and angle θ (*theta*) is called the angle of incidence. As the sun moves across the sky, the values of these angles change. Assume that $\alpha + \theta = 90$, where both α and θ are in degrees. Graph this equation on a coordinate system where the horizontal axis is the α-axis and the vertical axis is the θ-axis. Find the intercepts first, and limit your graph to the first quadrant only.

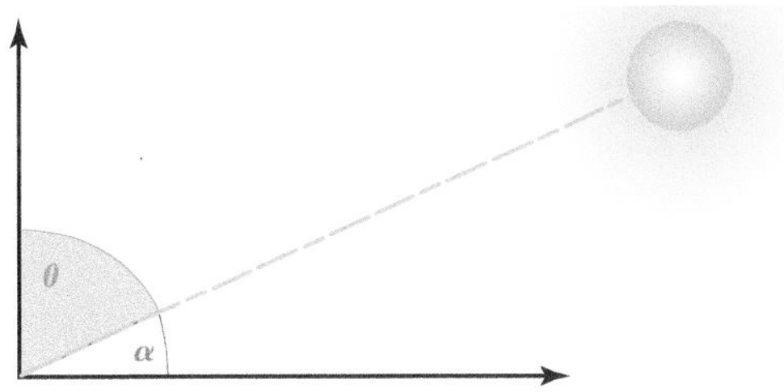

60. **Light** When light comes into contact with an impenetrable object, such as a thick piece of wood or metal, it is reflected or absorbed, but not transmitted, as shown in the following diagram. If we let R represent the percentage of light reflected and A the percentage of light absorbed by a surface, then the relationship between R and A is $R + A = 100$. Graph this equation on a coordinate system where the horizontal axis is the A-axis and the vertical axis is the R-axis. Find the intercepts first, and limit your graph to the first quadrant.

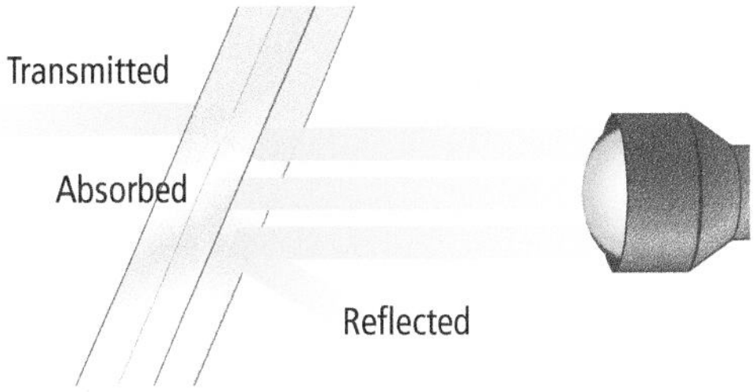

Learning Objectives Assessment

The following problems can be used to help assess if you have successfully met the learning objectives for this section.

61. Find the x-intercept of the line $3x - 4y = 8$.

a. $\frac{3}{8}$ **b.** -2 **c.** 3 **d.** $\frac{8}{3}$

62. Find the y-intercept of the line $y = 9x - 4$.

a. $\frac{4}{9}$ **b.** $\frac{9}{4}$ **c.** -4 **d.** 9

63. Graph the line having an x-intercept of -3 and a y-intercept of 2.

a.

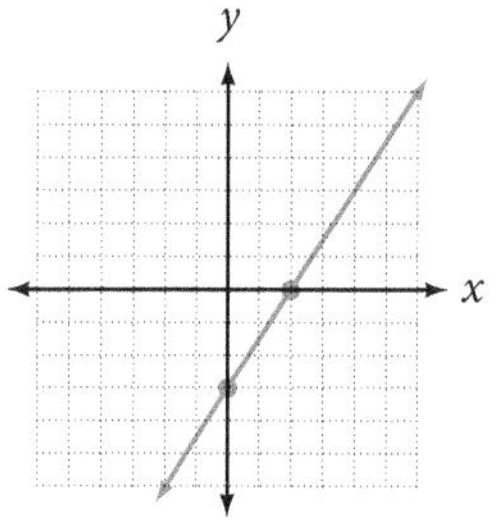

b.

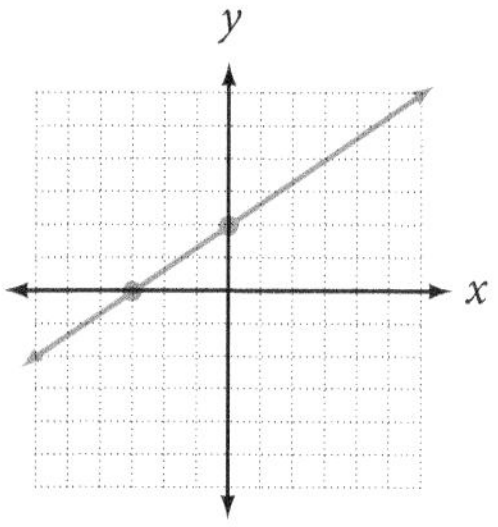

c.

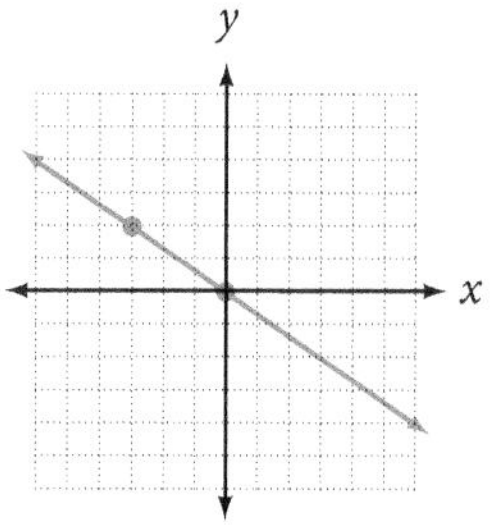

d.

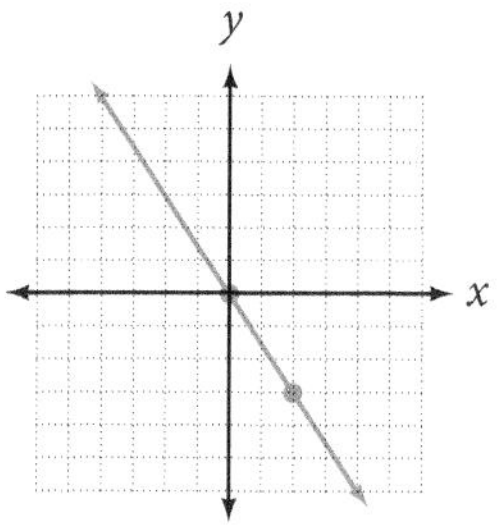

Getting Ready for the Next Section

64. Evaluate.

a. $\frac{5-2}{3-1}$ **b.** $\frac{2-5}{1-3}$

65. Evaluate.

a. $\frac{-4-1}{5-(-2)}$ **b.** $\frac{1+4}{-2-5}$

66. Evaluate the following expressions when $x = 3$, and $y = 5$.

a. $\frac{y-2}{x-1}$ **b.** $\frac{2-y}{1-x}$

67. Evaluate the following expressions when $x = -2$, and $y = 1$.

a. $\frac{-4-y}{5-x}$ **b.** $\frac{y+4}{x-5}$

The Slope of a Line 3.4

Learning Objectives

In this section, we will learn how to:

1. Use two ordered pairs to find the slope of a line.
2. Graph a line given a point and the slope.
3. Interpret slope as a rate of change.
4. Determine if two lines are parallel or perpendicular.

Introduction

In defining the slope of a line, we are looking for a number to associate with the line that does two things. First of all, we want the slope of a line to measure the "steepness" of the line; that is, in comparing two lines, the slope of the steeper line should have the larger absolute value. Second, we want a line that *rises* going from left to right to have a *positive* slope. We want a line that *falls* going from left to right to have a *negative* slope. (A line that neither rises nor falls going from left to right must, therefore, have 0 slope.) These are illustrated in Figure 1.

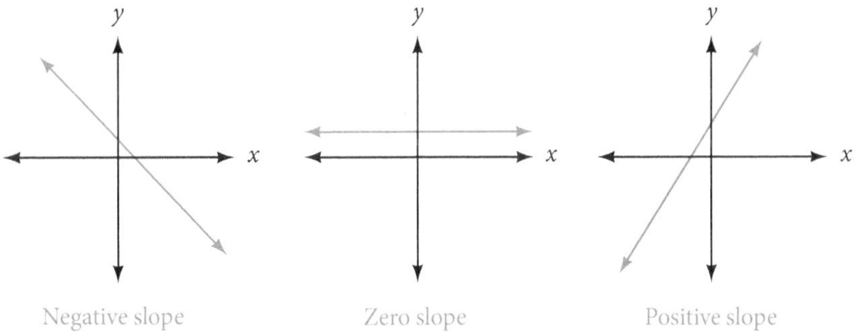

FIGURE 1

Slope

Suppose we know the coordinates of two points on a line. Because we are trying to develop a general formula for the slope of a line, we will use general points—call the two points $P_1(x_1, y_1)$ and $P_2(x_2, y_2)$. They represent the coordinates of any two different points on our line. We define the *slope* of our line to be the ratio of the vertical change to the horizontal change as we move from point (x_1, y_1) to point (x_2, y_2) on the line. (See Figure 2.)

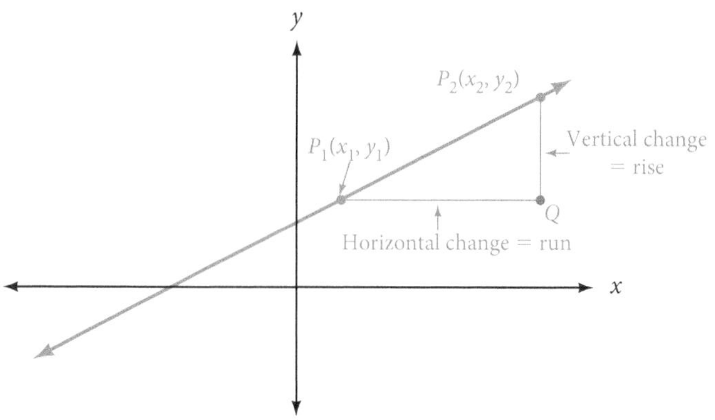

FIGURE 2

We call the vertical change the *rise* in the graph and the horizontal change the *run* in the graph. The slope, then, is

$$\text{Slope} = \frac{\text{vertical change}}{\text{horizontal change}} = \frac{\text{rise}}{\text{run}}$$

We would like to have a numerical value to associate with the rise in the graph and a numerical value to associate with the run in the graph. A quick study of Figure 2 shows that the coordinates of point Q must be (x_2, y_1), because Q is directly below point P_2 and right across from point P_1. We can draw our diagram again in the manner shown in Figure 3. It is apparent from this graph that the rise can be expressed as $(y_2 - y_1)$ and the run as $(x_2 - x_1)$.

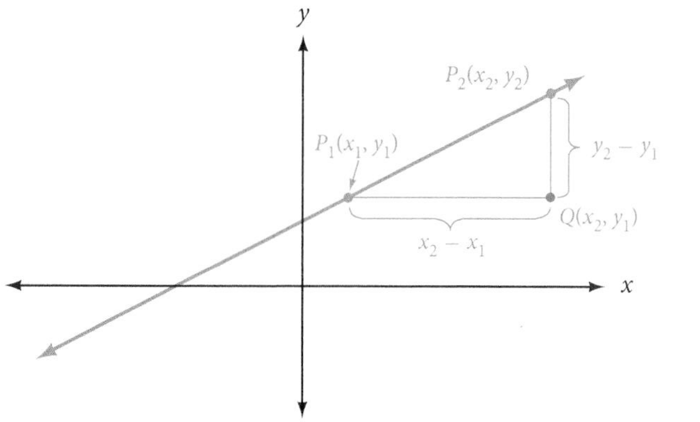

FIGURE 3

We usually denote the slope of a line by the letter m. Here is a formal definition of slope.

DEFINITION ***slope***

If points (x_1, y_1) and (x_2, y_2) are any two different points, then the ***slope*** of the line on which they lie is

$$\text{Slope} = m = \frac{\text{rise}}{\text{run}} = \frac{y_2 - y_1}{x_2 - x_1}$$

This definition of the *slope* of a line does just what we want it to do. If the line rises going from left to right, the slope will be positive. If the line falls from left to right, the slope will be negative. Also, the steeper the line, the larger absolute value the slope will have.

VIDEO EXAMPLES

SECTION 3.4

EXAMPLE 1 Find the slope of the line between the points (1, 2) and (3, 5).

SOLUTION We can let

$$(x_1, y_1) = (1, 2)$$

and

$$(x_2, y_2) = (3, 5)$$

then

$$m = \frac{y_2 - y_1}{x_2 - x_1} = \frac{5 - 2}{3 - 1} = \frac{3}{2}$$

The slope is $\frac{3}{2}$. For every vertical change of 3 units, there will be a corresponding horizontal change of 2 units. (See Figure 4.)

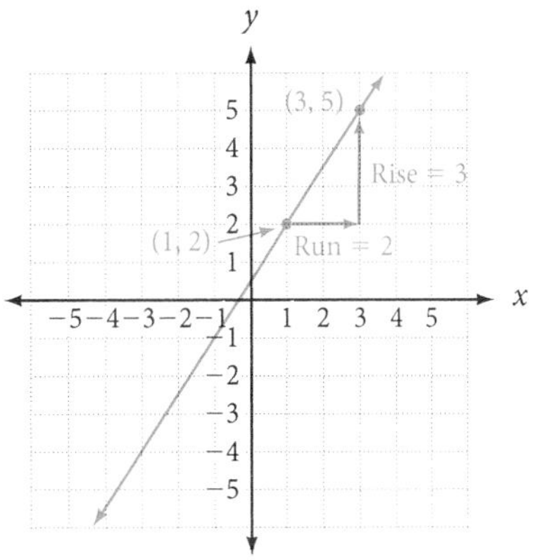

FIGURE 4

EXAMPLE 2 Find the slope of the line through (−2, 1) and (5, −4).

SOLUTION It makes no difference which ordered pair we call (x_1, y_1) and which we call (x_2, y_2).

$$\text{Slope} = m = \frac{y_2 - y_1}{x_2 - x_1} = \frac{-4 - 1}{5 - (-2)} = -\frac{5}{7}$$

The slope is $-\frac{5}{7}$. Every vertical change of −5 units (down 5 units) is accompanied by a horizontal change of 7 units (to the right 7 units). (See Figure 5.)

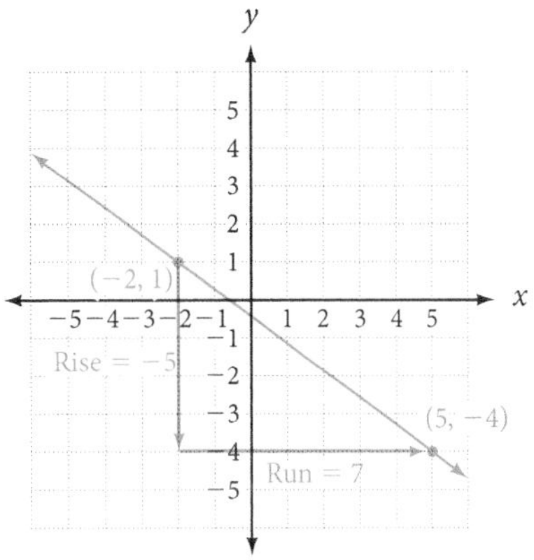

FIGURE 5

EXAMPLE 3 Find the slope of the line shown in Figure 6.

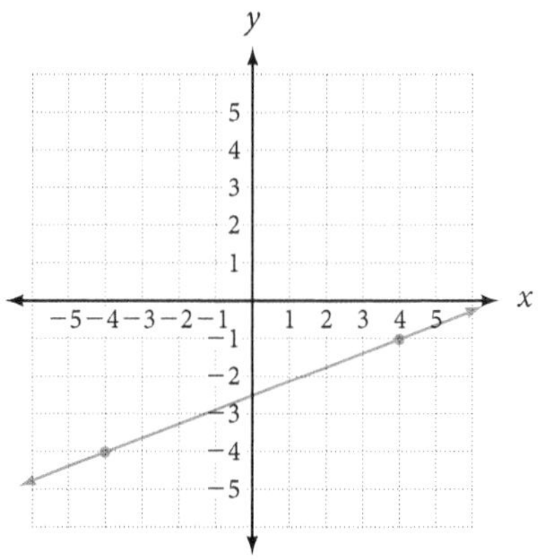

FIGURE 6

SOLUTION From Figure 6, we can see that the vertical change between the two points is 3 units and the horizontal change is 8 units (see Figure 7). Therefore, the slope of the line is

$$m = \frac{\text{rise}}{\text{run}} = \frac{3}{8}$$

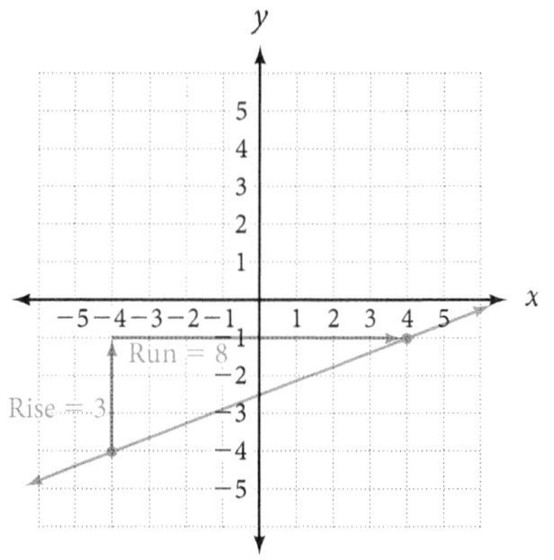

FIGURE 7

Graphing with Slope

EXAMPLE 4 Graph the line with slope $-\frac{3}{2}$ passing through the point (1, 1).

SOLUTION We begin by plotting the point (1, 1), as shown in Figure 8.

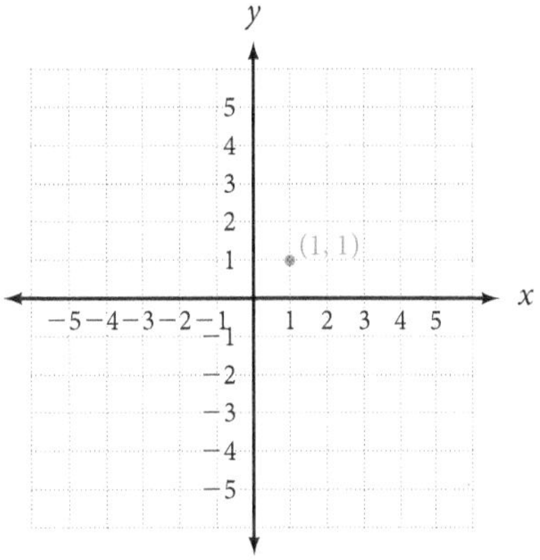

FIGURE 8

There are many lines that pass through the point shown in Figure 8, but only one of those lines has a slope of $-\frac{3}{2}$. The slope, $-\frac{3}{2}$, can be thought of as the rise in the graph divided by the run in the graph. Because the slope is negative, we can associate the negative with either the rise or the run (but not both).

If we think of the slope as $\frac{-3}{2}$, then the rise is -3 and the run is 2. Therefore, if we start at the point (1, 1) and move 3 units down (that's a rise of -3) and then 2 units to the right (a run of 2), we will be at another point on the graph. Figure 9 shows that the point we reach by doing so is the point $(3, -2)$, and shows the resulting line passing through both points.

Or, we can think of the slope as $\frac{3}{-2}$. Then the rise is 3 and the run is -2. If we start at the point (1, 1) and move 3 units up (a rise of 3) and then 2 units left (a run of -2), we will reach another point on the graph at $(-1, 4)$. Figure 10 shows this point and the resulting line.

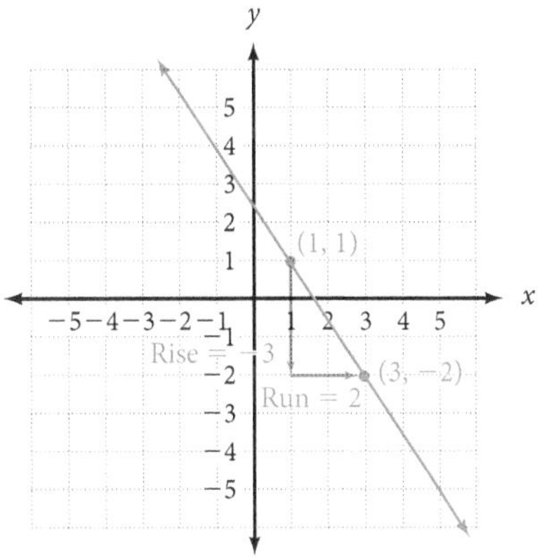

FIGURE 9

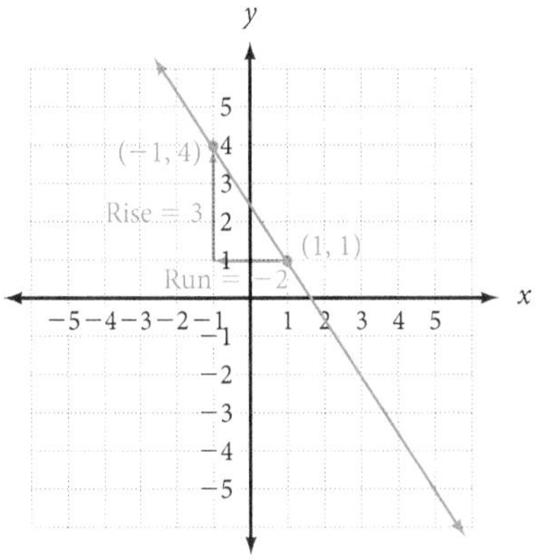

FIGURE 10

As you can see, the lines in Figures 9 and 10 are identical. It makes no difference whether we choose to associate the negative sign with the rise or the run. We will get the same line in either case.

EXAMPLE 5 Find the slope of the line containing $(3, -1)$ and $(3, 4)$.

SOLUTION Using the definition for slope, we have

$$m = \frac{y_2 - y_1}{x_2 - x_1} = \frac{4 - (-1)}{3 - 3} = \frac{5}{0}$$

The expression $\frac{5}{0}$ is undefined; that is, there is no real number to associate with it. In this case, we say the line *has no slope.*

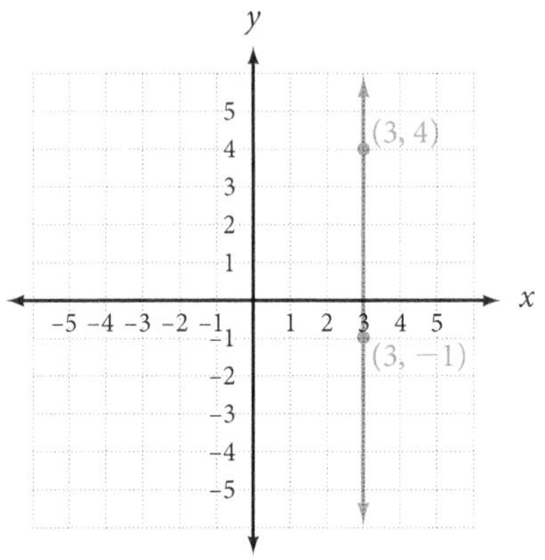

FIGURE 11

The graph of the line is shown in Figure 11. A line with no slope is a vertical line.

The following summary reminds us that all horizontal lines have equations of the form $y = b$ and slopes of 0. Because they cross the y-axis at $(0, b)$, the y-intercept is b; there is no x-intercept. Vertical lines have no slope and equations of the form $x = a$. Each will have an x-intercept at $(a, 0)$ and no y-intercept.

FACTS FROM GEOMETRY ***Special Equations and Their Graphs, Slopes, and Intercepts***

For the equations below, a and b are real numbers.

Vertical Line	**Horizontal Line**
Equation: $x = a$	Equation: $y = b$
No slope	Slope $= 0$
x-intercept $= a$	No x-intercept
No y-intercept	y-intercept $= b$

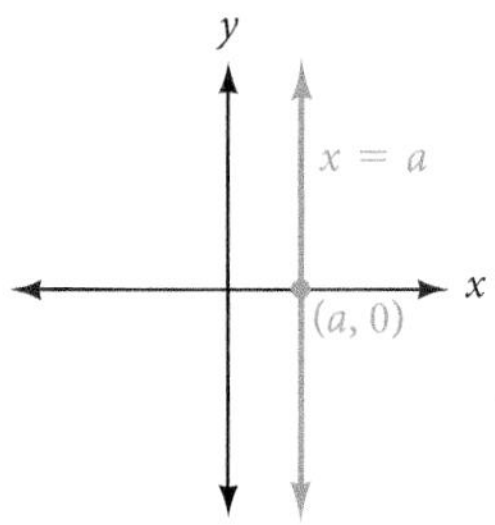

FIGURE 12

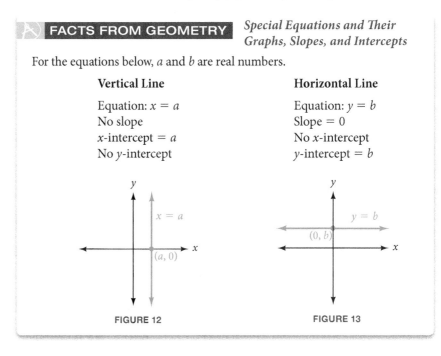

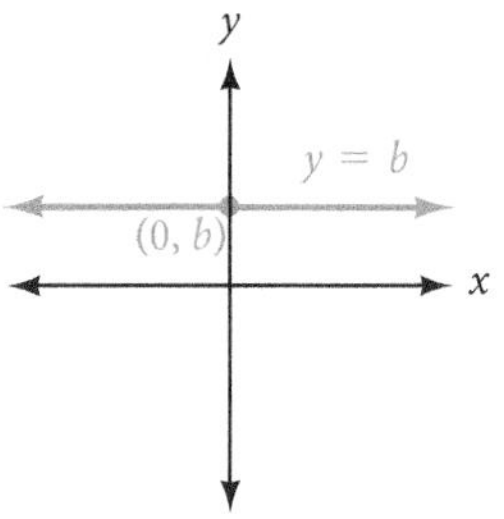

FIGURE 13

Slope as Rate of Change

If two quantities have a relationship that can be described by a linear equation, then we can interpret the slope of the line as a ***rate of change*** between these two quantities. Our next example illustrates how this is done.

EXAMPLE 6 The graph in Figure 14 shows the rise in U.S. retail e-commerce sales over a five-year period, which is approximately linear. Use the graph to find the slope of the line, and then interpret the slope as a rate of change.

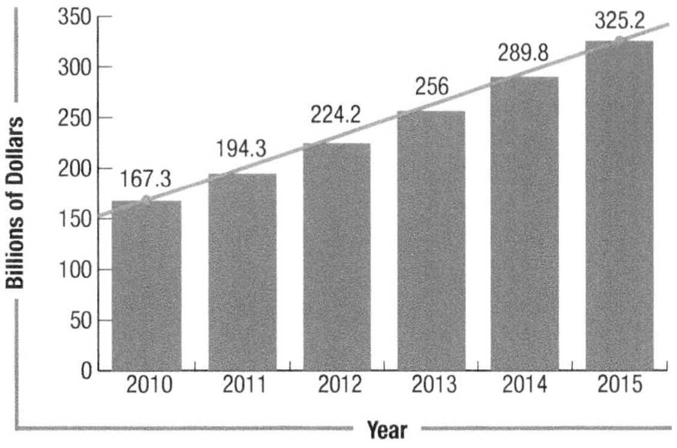

FIGURE 14

SOLUTION Using the points (2010, 167.3) and (2015, 325.2), we have

$$m = \frac{\text{rise}}{\text{run}} = \frac{325.2 - 167.3}{2015 - 2010} = \frac{157.9}{5} = 31.58$$

We see that U.S. retail e-commerce sales increased by \$157.9 billion dollars over the five-year period. This means that, on average, the U.S. retail e-commerce sales are increasing at a rate of 31.58 billion dollars per year during the five-year period between 2010 and 2015.

Parallel and Perpendicular Lines

We conclude our discussion of slope by considering parallel and perpendicular lines. Two distinct lines are *parallel* if they never intersect, and they are *perpendicular* if they intersect at a right angle. The following definition allows us to determine if two lines are parallel or perpendicular by looking at their slopes.

DEFINITION *parallel and perpendicular lines*

Two distinct, nonvertical lines are ***parallel*** if they have the same slope, or ***perpendicular*** if the product of their slopes is -1.

In the case of perpendicular lines, their slopes will have a product of -1 if they are both opposites and reciprocals of each other.

EXAMPLE 7 Determine if the lines passing through the given pairs of points are parallel, perpendicular, or neither: $(-1, -5)$ and $(2, 1)$, $(-2, 2)$ and $(6, -2)$.

SOLUTION We must find the slopes for both lines.

$$\text{Line 1:} \quad m_1 = \frac{1 - (-5)}{2 - (-1)} = \frac{6}{3} = 2$$

$$\text{Line 2:} \quad m_2 = \frac{-2 - 2}{6 - (-2)} = \frac{-4}{8} = -\frac{1}{2}$$

Because $m_1 \neq m_2$, the lines are not parallel. However, if we look at their product

$$m_1 m_2 = 2\left(-\frac{1}{2}\right) = -1$$

we see that the lines are perpendicular. Notice that $-\frac{1}{2}$ is the opposite and reciprocal of 2.

Getting Ready for Class

After reading through the preceding section, respond in your own words and in complete sentences.

A. What is the slope of a line?

B. Describe how to obtain the slope of a line if you know the coordinates of two points on the line.

C. Describe how you would graph a line from its slope and a point.

D. How do we know if two lines are parallel or perpendicular?

Problem Set 3.4

Find the slope of the line through the following pairs of points. Then plot each pair of points, draw a line through them, and indicate the rise and run in the graph in the same manner shown in Examples 1 and 2.

1. (2, 1), (4, 4) **2.** (3, 1), (5, 4) **3.** (1, 4), (5, 2)

4. (1, 3), (5, 2) **5.** (1, −3), (4, 2) **6.** (2, −3), (5, 2)

7. (−3, −2), (1, 3) **8.** (−3, −1), (1, 4) **9.** (−3, 2), (3, −2)

10. (−3, 3), (3, −1) **11.** (2, −5), (3, −2) **12.** (2, −4), (3, −1)

13. (4, −5), (4, 1) **14.** (−1, 4), (3, 4) **15.** (−4, −2), (−1, −2)

16. (−2, 0), (−2, 5) **17.** (0, 0) (−3, −1) **18.** (0, 0), (−3, 2)

Find the slope for each line.

19.

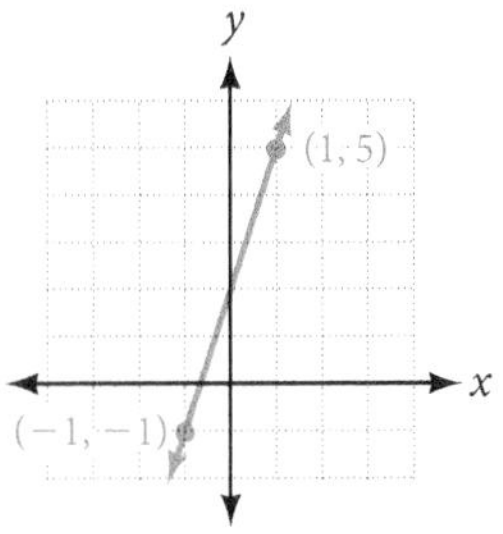

20.

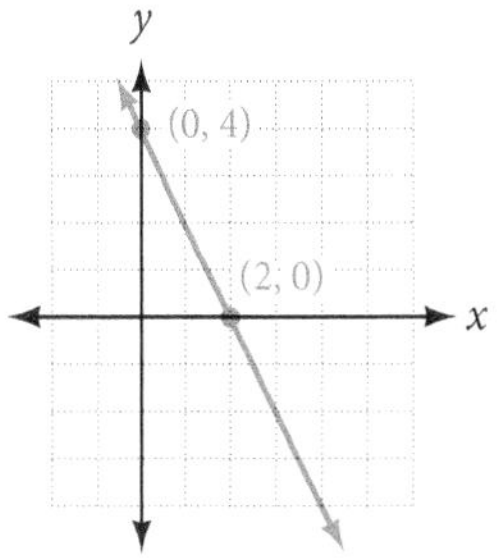

21.

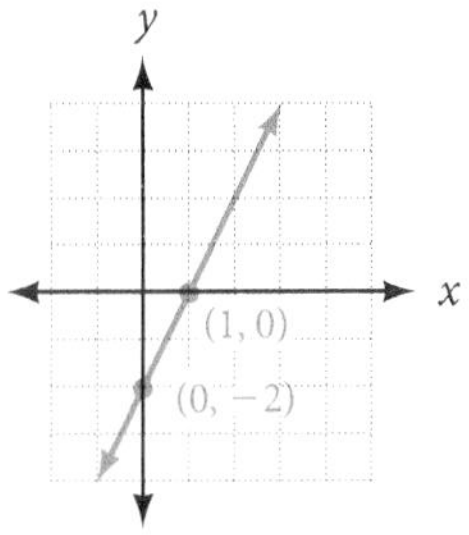

22.

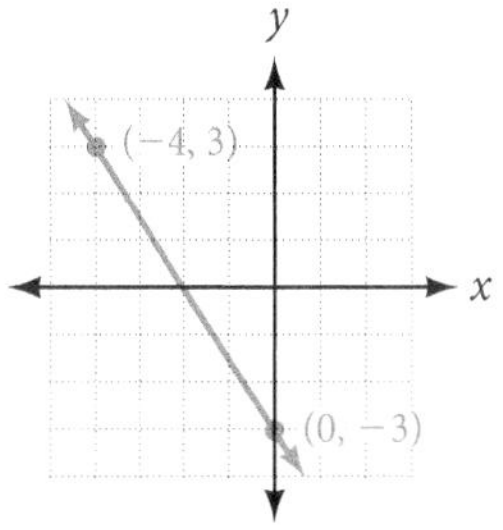

In each of the following problems, graph the line having the given slope and passing through the given point.

23. $m = 2$, (0, 1) **24.** $m = -2$, (0, 4) **25.** $m = \frac{2}{3}$, (1, 0)

26. $m = \frac{3}{4}$, (−2, 0) **27.** $m = \frac{3}{2}$, (−1, −3) **28.** $m = \frac{4}{3}$, (−2, 1)

29. $m = -\frac{4}{3}$, (1, 1) **30.** $m = -\frac{3}{5}$, (−3, −1) **31.** $m = 3$, (−1, −4)

32. $m = -3$, (−2, −2) **33.** $m = -\frac{1}{5}$, (0, 0) **34.** $m = \frac{1}{4}$, (0, 0)

35. Graph the line that has an x-intercept of 3 and a y-intercept of −2. What is the slope of this line?

36. Graph the line that has an x-intercept of 2 and a y-intercept of −3. What is the slope of this line?

37. Graph the line with x-intercept 4 and y-intercept 2. What is the slope of this line?

38. Graph the line with x-intercept -4 and y-intercept -2. What is the slope of this line?

39. Graph the line $y = 2x - 3$, then name the slope and y-intercept by looking at the graph.

40. Graph the line $y = -2x + 3$, then name the slope and y-intercept by looking at the graph.

41. Graph the line $y = \frac{1}{2}x + 1$, then name the slope and y-intercept by looking at the graph.

42. Graph the line $y = -\frac{1}{2}x - 2$, then name the slope and y-intercept by looking at the graph.

For each equation in each table, give the slope of the graph.

43.

Equation	Slope
$x = 3$	
$y = 3$	
$y = 3x$	

44.

Equation	Slope
$y = \frac{3}{2}$	
$x = \frac{3}{2}$	
$y = \frac{3}{2}x$	

45.

Equation	Slope
$y = -\frac{2}{3}$	
$x = -\frac{2}{3}$	
$y = -\frac{2}{3}x$	

46.

Equation	Slope
$x = -2$	
$y = -2$	
$y = -2x$	

If a line has the slope that is given, state the slope of a line that is (a) parallel to this line, and (b) perpendicular to this line.

47. $m = \frac{2}{3}$

48. $m = -\frac{3}{2}$

49. $m = -\frac{1}{4}$

50. $m = \frac{1}{5}$

51. $m = 2$

52. $m = -3$

53. $m = 0$

54. No slope

Determine if the lines passing through the given pairs of points are parallel, perpendicular or neither.

55. $(-3, 2)$ and $(6, 8)$, $(0, -1)$ and $(3, 1)$

56. $(-4, -4)$ and $(2, 5)$, $(-2, 0)$ and $(0, -3)$

57. $(-2, 6)$ and $(2, 4)$, $(-3, -5)$ and $(1, 3)$

58. $(0, -1)$ and $(4, 7)$, $(-4, -5)$ and $(-1, 1)$

59. $(-4, 0)$ and $(1, 5)$, $(0, -4)$ and $(2, -6)$

60. $(0, -2)$ and $(2, 6)$, $(-4, 1)$ and $(4, -1)$

61. Find y if the line through $(4, 2)$ and $(6, y)$ has a slope of 2.

62. Find y if the line through $(1, y)$ and $(7, 3)$ has a slope of 6.

Applying the Concepts

63. **Garbage Production** The table and completed line graph give the annual production of garbage in the United States for some specific years.
    a. Find the slope of each of the four line segments, A, B, C, and D.
    b. Interpret the slope of line segment A as a rate of change.

Year	Garbage (millions of tons)
1960	88
1970	121
1980	152
1990	205
2000	224

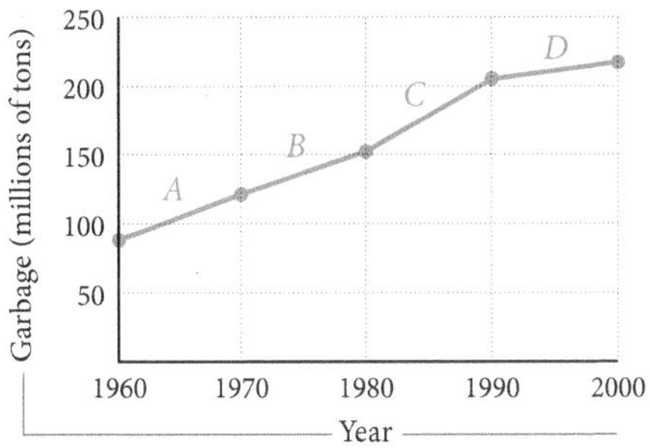

64. **Grass Height** The table and completed line graph give the growth of a certain plant species over time.
    a. Find the slopes of the line segments labeled A, B, and C.
    b. Interpret the slope of segment C as a rate of change.

Day	Plant Height
0	0
2	1
4	3
6	6
8	13
10	23

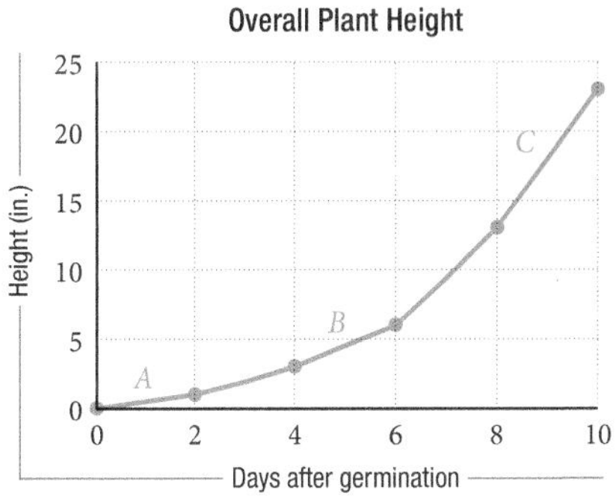

65. **Non-Camera Phone Sales** The table and line graph here each show the projected non-camera phone sales each year from 2006 to 2010.
    a. Find the slope of each of the three line segments, A, B, and C.
    b. Interpret the slope of segment B as a rate of change.

Year	Sales (in millions)
2006	300
2007	250
2008	175
2009	150
2010	125

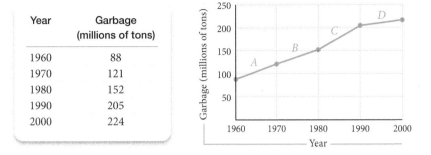

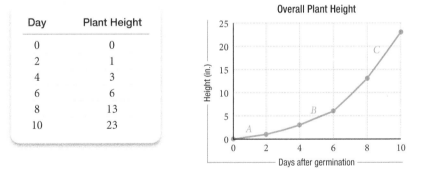

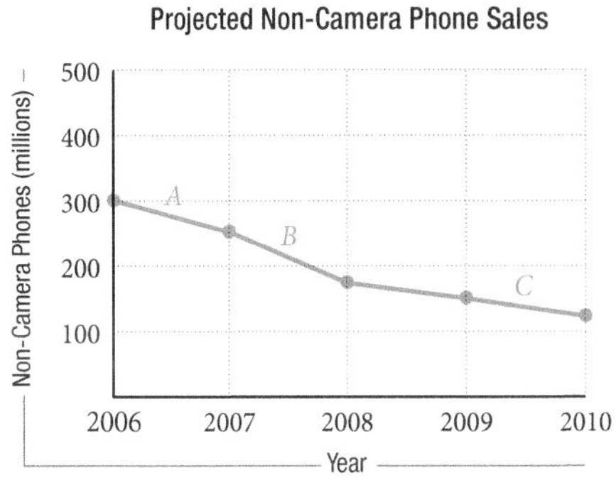

66. Camera Phone Sales The table and line graph here show the projected sales of camera phones from 2006 to 2010.

a. Find the slopes of line segments A, B, and C.

b. Interpret the slope of segment C as a rate of change.

Year	Sales (in millions)
2006	500
2007	650
2008	750
2009	875
2010	900

Projected Camera Phone Sales

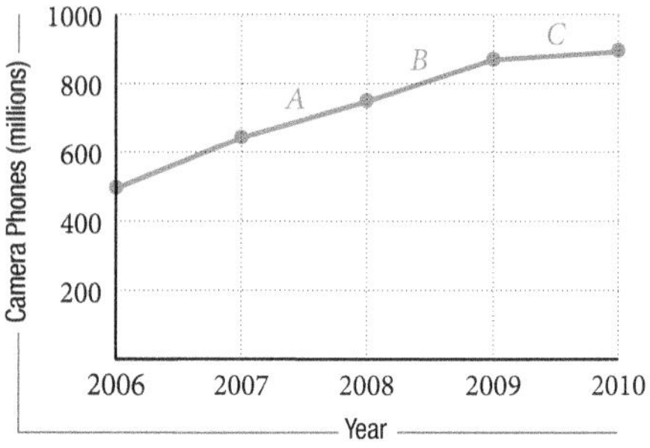

Learning Objectives Assessment

The following problems can be used to help assess if you have successfully met the learning objectives for this section.

67. Find the slope of the line passing through the points $(-3, 2)$ and $(1, -4)$.

a. $-\frac{3}{2}$ **b.** 1 **c.** $\frac{1}{2}$ **d.** $-\frac{2}{3}$

68. Sketch the graph of the line with slope 3 and passing through the point $(2, 1)$.

a.

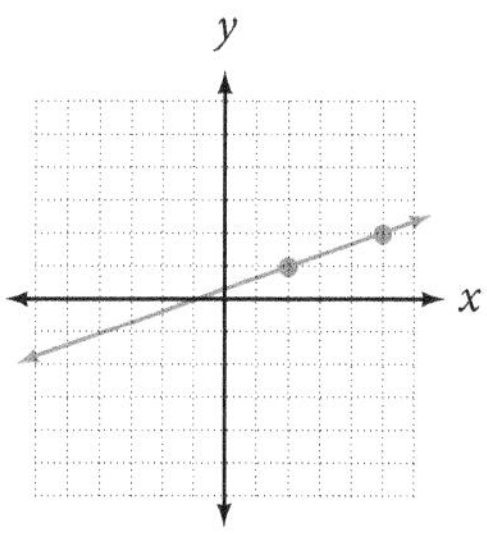

b.

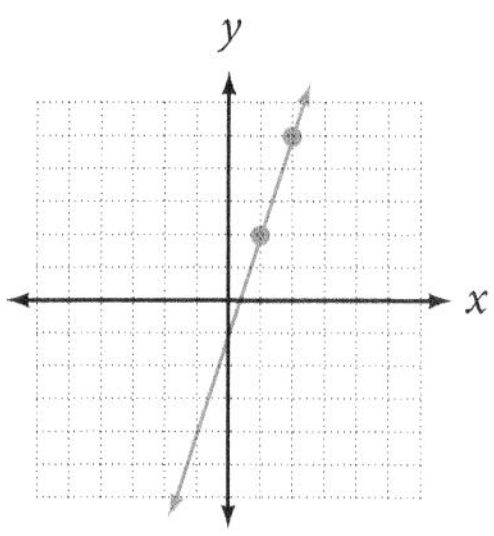

c.

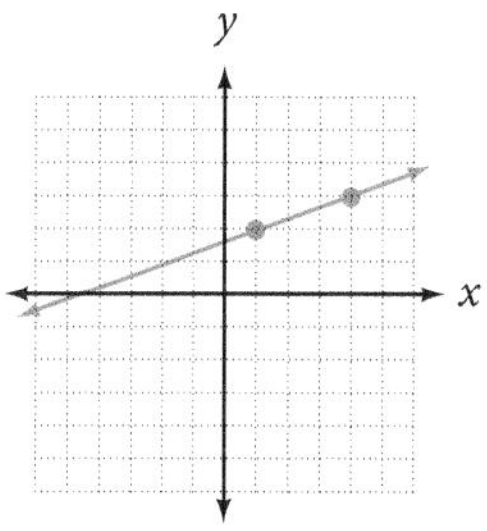

d.

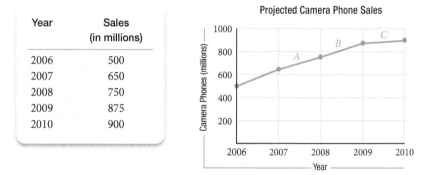

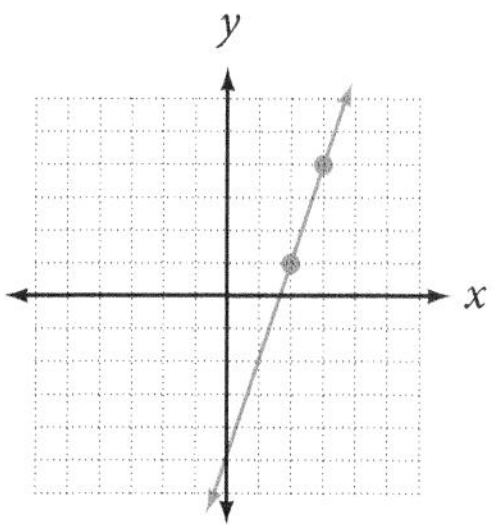

69. The table below shows the federal minimum wage from two past years. Find the slope of the line passing through these two points and then interpret the slope as a rate of change.

Year	Minimum Wage
1979	\$2.90
2009	\$7.25

a. The federal minimum wage increased at a rate of \$6.90 per year from 1979 to 2009.

b. The federal Minimum wage increased at a rate of \$4.35 per year from 1979 to 2009.

c. The federal minimum wage increased at a rate of 0.145 years per dollar from 1979 to 2009.

d. The federal minimum wage increased at a rate of 14.5 cents per year from 1979 to 2009.

70. Find the slope of a line that would be perpendicular to the line passing through $(-1, -3)$ and $(2, 3)$.

a. $-\frac{1}{2}$ **b.** 2 **c.** -2 **d.** $\frac{1}{2}$

Getting Ready for the Next Section

Solve each equation for y.

71. $-2x + y = 4$ **72.** $-4x + y = -2$ **73.** $2x + y = 3$

74. $3x + 2y = 6$ **75.** $4x - 5y = 20$ **76.** $-2x - 5y = 10$

SPOTLIGHT ON SUCCESS ***Student Instructor Julieta***

Success is no accident. It is hard work, perseverance, learning, studying, sacrifice, and most of all, love of what you are doing or learning to do.

—Pelé

Success really is no accident, nor is it something that happens overnight. Sure you may be sitting there wondering why you don't understand a certain lesson or topic, but you are not alone. There are many others who are sitting in your exact position. Throughout my first year in college (and more specifically in Calculus I) I learned that it is normal for any student to feel stumped every now and then. The students who do well are the ones who keep working, even when they are confused.

Pelé wasn't just born with all that legendary talent. It took dedication and hard work as well. Don't ever feel bad because there's something you don't understand—it's not worth it. Stick with it 100% and just keep working problems; I'm sure you'll be successful with whatever you set your mind to achieve in this course.

Slope-Intercept Form 3.5

Learning Objectives

In this section, we will learn how to:

1. Use slope-intercept form to find the slope and y-intercept of a line.
2. Use slope-intercept form to graph a line.
3. Use slope-intercept form to find the equation of a line.

Introduction

To this point in the chapter, most of the problems we have worked have used the equation of a line to find different types of information about the line. For instance, given the equation of a line, we can find points on the line, the graph of the line, the intercepts, and the slope of the line. In this section we reverse things somewhat and begin to move in the other direction; we will use information about a line, such as its slope and y-intercept, to find the equation of a line.

The Slope-Intercept Form of a Line

VIDEO EXAMPLES

SECTION 3.5

EXAMPLE 1 Find the slope and y-intercept of the line $y = \frac{3}{2}x + 1$.

SOLUTION From our work in Section 3.3, we know that the y-intercept is given by the constant term when an equation is in this form. Therefore, the y-intercept is 1 and the point (0, 1) lies on the line.

To find the slope, we need another point on the line. Using $x = 2$, we have

$$y = \frac{3}{2}(2) + 1 = 3 + 1 = 4$$

So (2, 4) is a second point on the line. Now we can find the slope.

$$m = \frac{y_2 - y_1}{x_2 - x_1} = \frac{4 - 1}{2 - 0} = \frac{3}{2}$$

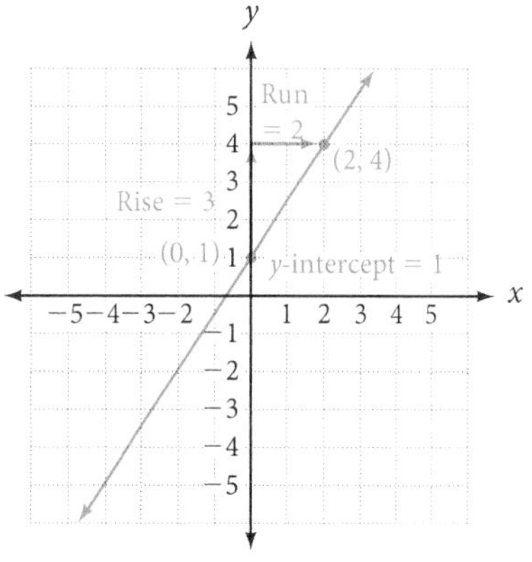

FIGURE 1

The graph of the line is shown in Figure 1.

What is interesting and useful about the equation from the previous example is that the coefficient of x is the slope of the line. It is no coincidence that it turned out this way. Whenever an equation has the form $y = mx + b$, the graph is always a line with slope m and y-intercept b.

To see that this is true in general, suppose we know that the slope of a line is m and the y-intercept is b. Because the y-intercept is b, then the point $(0, b)$ is on the line. If (x, y) is any other point on the line (see Figure 2), then we apply our slope formula to get

$$\frac{y-b}{x-0} = m \qquad \text{Slope} = \frac{\text{vertical change}}{\text{horizontal change}}$$

$$\frac{y-b}{x} = m \qquad x - 0 = x$$

$$y - b = mx \qquad \text{Multiply each side by } x$$

$$y = mx + b \qquad \text{Add } b \text{ to each side}$$

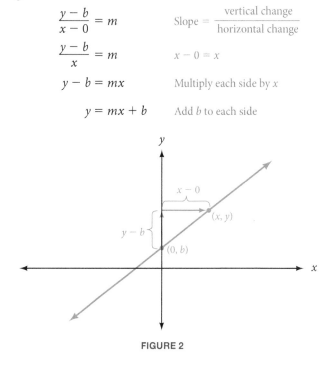

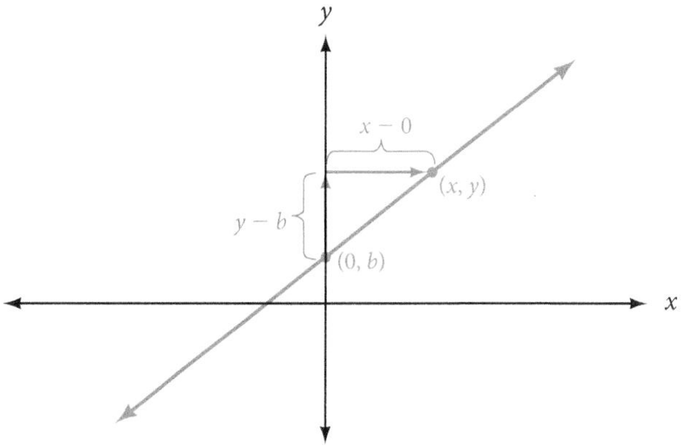

FIGURE 2

Here is a summary of what we have just found.

RULE ***Slope-Intercept Form of the Equation of a Line***

The equation of the line with slope m and y-intercept b is always given by

$$y = mx + b$$

Graphing with Slope-Intercept Form

EXAMPLE 2 Find the slope and y-intercept for $-2x + y = -4$. Then, use them to draw the graph.

SOLUTION To identify the slope and y-intercept from the equation, the equation must be in the form $y = mx + b$ (slope-intercept form). To write our equation in this form, we must solve the equation for y. To do so, we simply add $2x$ to each side of the equation.

$$-2x + y = -4 \qquad \text{Original equation}$$

$$y = 2x - 4 \qquad \text{Add } 2x \text{ to each side}$$

The equation is now in slope-intercept form, so the slope must be 2 and the y-intercept must be -4. The graph, therefore, crosses the y-axis at $(0, -4)$. Because the slope is 2, we can let the rise = 2 and the run = 1 and find a second point on the graph. The graph is shown in Figure 3.

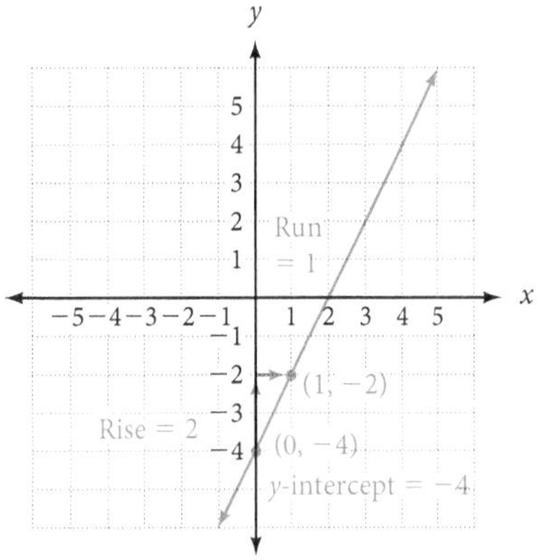

FIGURE 3

EXAMPLE 3 Find the slope and y-intercept for $3x - 2y = 6$. Then use them to draw the graph.

SOLUTION To find the slope and y-intercept from the equation, we can write the equation in the form $y = mx + b$. This means we must solve the equation $3x - 2y = 6$ for y.

$$3x - 2y = 6 \qquad \text{Original equation}$$

$$-2y = -3x + 6 \qquad \text{Add } -3x \text{ to each side}$$

$$-\frac{1}{2}(-2y) = -\frac{1}{2}(-3x + 6) \qquad \text{Multiply each side by } -\frac{1}{2}$$

$$y = \frac{3}{2}x - 3 \qquad \text{Simplify each side}$$

Now that the equation is written in slope-intercept form, we can identify the slope as $\frac{3}{2}$ and the y-intercept as -3. The graph is shown in Figure 4.

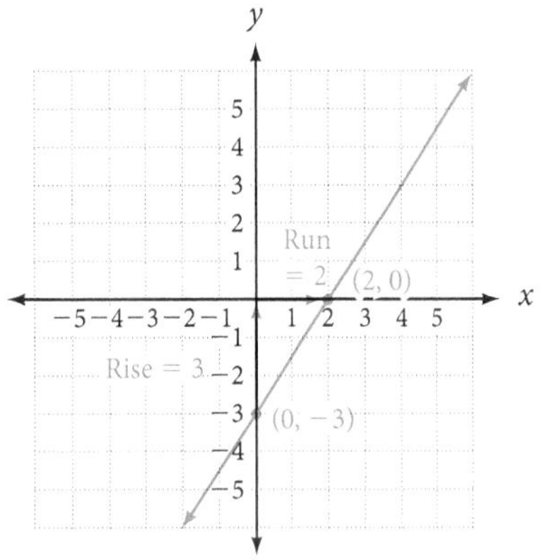

FIGURE 4

Before we move on to finding equations of lines, here is a summary of the methods we have discussed so far for sketching the graph of a line.

Methods of Graphing Lines

1. Substitute convenient values of x into the equation, and find the corresponding values of y. We used this method first for equations like $y = 2x - 3$. To use this method for equations that looked like $2x - 3y = 6$, we first solved them for y.
2. Find the x- and y-intercepts. This method works best for equations of the form $3x + 2y = 6$ where the numbers in front of x and y divide the constant term evenly.
3. Find the slope and y-intercept. This method works best when the equation has the form $y = mx + b$ and b is an integer.

Finding the Equation of a Line

EXAMPLE 4 Find the equation of the line with slope $-\frac{4}{3}$ and y-intercept 5. Then, graph the line.

SOLUTION Substituting $m = -\frac{4}{3}$ and $b = 5$ into the equation $y = mx + b$, we have

$$y = -\frac{4}{3}x + 5$$

Finding the equation from the slope and y-intercept is just that easy. If the slope is m and the y-intercept is b, then the equation is always $y = mx + b$.

Because the y-intercept is 5, the graph goes through the point (0, 5). To find a second point on the graph, we start at (0, 5) and move 4 units down (that's a rise of -4) and 3 units to the right (a run of 3). The point we reach is (3, 1). Drawing a line that passes through (0, 5) and (3, 1), we have the graph of our equation. (Note that we could also let the rise $= 4$ and the run $= -3$ and obtain the same graph.) The graph is shown in Figure 5.

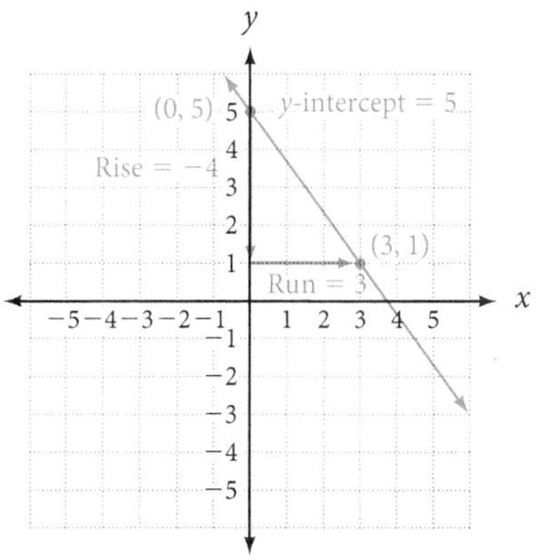

FIGURE 5

EXAMPLE 5 Find the equation of the line with slope -2 that passes through the point $(-4, 3)$. Write the answer in standard form.

SOLUTION We can find the equation using slope-intercept form $y = mx + b$. Because the point $(-4, 3)$ is not on the y-axis, it is not a y-intercept. We are not given the value of b. However we can substitute $x = -4$, $y = 3$, and $m = -2$ into $y = mx + b$ and solve for b.

Using $(x, y) = (-4, 3)$ and $m = -2$

in	$y = mx + b$	Slope-intercept form
gives us	$3 = -2(-4) + b$	Substitute values
	$3 = 8 + b$	Multiply
	$-5 = b$	Add -8 to both sides

So, the slope-intercept form of the line is $y = -2x - 5$. To write the equation in standard form, we simply add $2x$ to both sides to obtain $2x + y = -5$.

EXAMPLE 6 Find the equation of the line passing through the point $(2, 5)$ and perpendicular to the line $3x + 4y = 12$. Write the answer in slope-intercept form.

SOLUTION First, we must find the slope of the line $3x + 4y = 12$. To do so, we isolate y in order to write the equation in slope-intercept form.

$3x + 4y = 12$	Original equation
$4y = -3x + 12$	Add $-3x$ to both sides
$y = -\frac{3}{4}x + 3$	Multiply each side by $\frac{1}{4}$

The slope of this line is $-\frac{3}{4}$. Therefore, a line perpendicular to this line must have a slope of $\frac{4}{3}$. We can now find the equation of the line that was asked for using the process we followed in the previous example.

Using $(x, y) = (2, 5)$ and $m = \frac{4}{3}$

in	$y = mx + b$	Slope-intercept form
gives us	$5 = \frac{4}{3}(2) + b$	Substitute values
	$5 = \frac{8}{3} + b$	Multiply
	$5 - \frac{8}{3} = b$	Add $-\frac{8}{3}$ to both sides
	$\frac{7}{3} = b$	Simplify

In slope-intercept form, the desired line has the equation

$$y = \frac{4}{3}x + \frac{7}{3}$$

EXAMPLE 7 Find the equation of the horizontal line passing through the point $(-3, -1)$.

SOLUTION Any horizontal line has slope $m = 0$. To pass through $(-3, -1)$, the line must also cross the y-axis at $(0, -1)$. Therefore, $b = -1$. This gives us

$y = mx + b$	Slope-intercept form
$y = 0(x) + (-1)$	Substitute values
$y = -1$	Simplify

Getting Ready for Class

After reading through the preceding section, respond in your own words and in complete sentences.

A. What are m and b in the equation $y = mx + b$?

B. How would you find the slope and y-intercept for the line $3x - 2y = 6$?

C. How would you find the equation of a line given a point on the line and the slope?

D. In Example 6, explain why $b \neq 5$.

Problem Set 3.5

Find the slope and y-intercept for each of the following equations.

1. $y = 5x - 3$ **2.** $y = -4x + 1$ **3.** $y = -\frac{2}{3}x + \frac{7}{3}$

4. $y = \frac{5}{6}x - \frac{11}{6}$ **5.** $y = x + 9$ **6.** $y = -x - 8$

7. $y = \frac{x}{2} - \frac{5}{2}$ **8.** $y = \frac{x}{3} + \frac{4}{3}$ **9.** $y = \frac{1}{4} - 2x$

10. $y = \frac{2}{5} + 6x$ **11.** $y = 3x$ **12.** $y = 4x$

13. $y = -10$ **14.** $y = 7$

In each of the following problems, graph the line having the given slope m and y-intercept b.

15. $m = \frac{2}{3}, b = 1$ **16.** $m = \frac{3}{4}, b = -2$ **17.** $m = \frac{3}{2}, b = -3$

18. $m = \frac{4}{3}, b = 2$ **19.** $m = -\frac{4}{3}, b = 5$ **20.** $m = -\frac{3}{5}, b = 4$

21. $m = 3, b = -1$ **22.** $m = 3, b = -2$

Using the slope and y-intercept to graph each equation.

23. $y = -2x + 3$ **24.** $y = 2x - 5$ **25.** $y = \frac{3}{4}x + 2$

26. $y = -\frac{4}{3}x - 1$ **27.** $y = \frac{3}{2}x$ **28.** $y = -\frac{2}{5}x$

Find the slope and y-intercept for each of the following equations by writing them in the form $y = mx + b$. Then, graph each equation.

29. $-2x + y = 4$ **30.** $-2x + y = 2$ **31.** $3x + y = 3$

32. $3x + y = 6$ **33.** $3x + 2y = 6$ **34.** $2x + 3y = 6$

35. $4x - 5y = 20$ **36.** $2x - 5y = 10$ **37.** $-2x - 5y = 10$

38. $-4x + 5y = 20$

In each of the following problems, give the equation of the line with the given slope and y-intercept.

39. $m = \frac{2}{3}, b = 1$ **40.** $m = \frac{3}{4}, b = -2$ **41.** $m = \frac{3}{2}, b = -1$ **42.** $m = \frac{4}{3}, b = 2$

43. $m = -\frac{2}{3}, b = 3$ **44.** $m = -\frac{3}{5}, b = 4$ **45.** $m = 2, b = -4$ **46.** $m = -2, b = 4$

Find the slope and y-intercept for each line. Then write the equation of each line in slope-intercept form.

47.

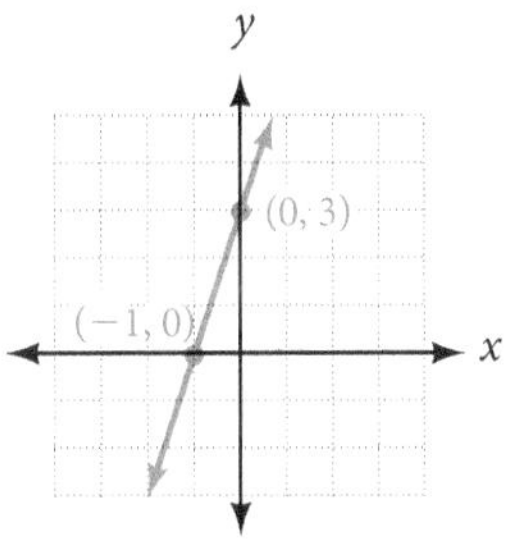

48.

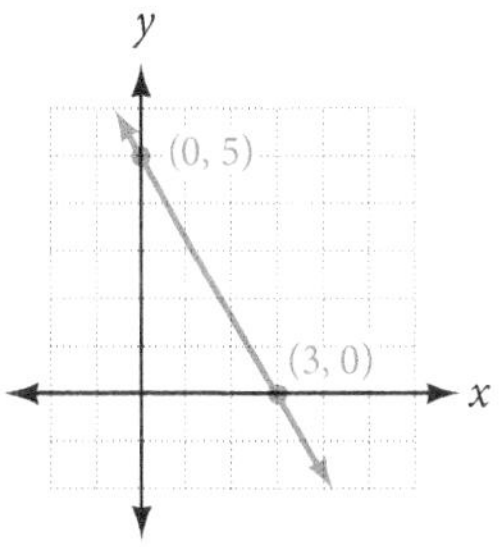

49.

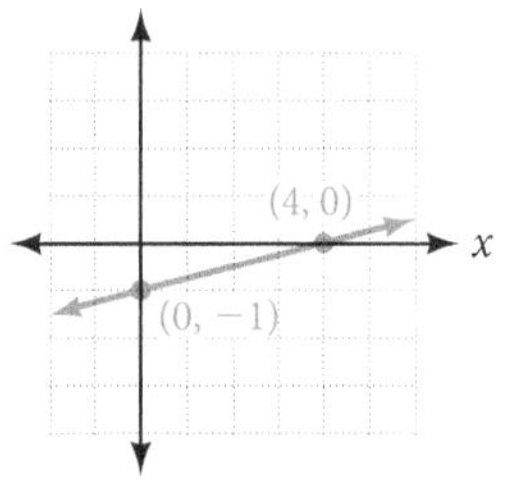

50.

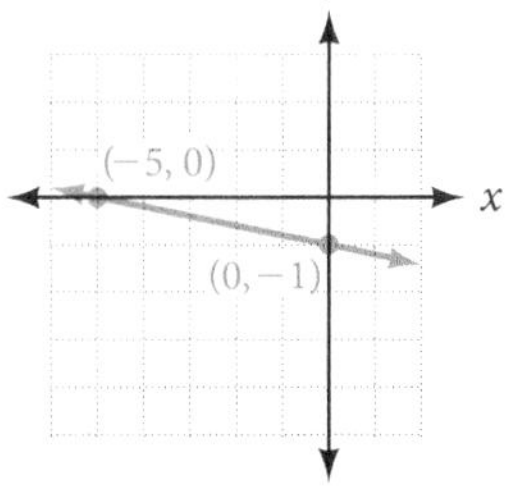

For the following problems, the slope and one point on a line are given. In each case use slope-intercept form to find the equation of the line. Then write your answer in standard form.

51. $(2, 3), m = 2$

52. $(-1, 5), m = 3$

53. $(1, -4), m = -\frac{1}{3}$

54. $(-2, -2), m = -\frac{1}{4}$

55. $(-2, 6), m = \frac{3}{5}$

56. $(0, 0), m = -\frac{5}{2}$

57. $(4, 0), m = -\frac{4}{3}$

58. $(0, -3), m = \frac{3}{4}$

Find the equation of the line passing through the given point and **a.** parallel to the given line, **b.** perpendicular to the given line. Write your answers in slope-intercept form.

59. $(1, 2); y = 3x + 4$

60. $(2, -3); y = -2x - 1$

61. $(-4, -1); x + 2y = 6$

62. $(-2, 5); x - 4y = 4$

63. $(3, 4); 2x + 3y = 6$

64. $(6, 0); 3x - 4y = 12$

65. Find the equation of the horizontal line passing through $(3, 2)$.

66. Find the equation of the horizontal line passing through $(-2, -3)$.

67. Find the equation of the line with zero slope passing through $(-1, 5)$.

68. Find the equation of the line with zero slope passing through the origin.

The next two problems are intended to give you practice reading, and paying attention to, the instructions that accompany the problems you are working. Working these problems is an excellent way to get ready for a test or a quiz.

69. Paying Attention to Instructions Work each problem according to the instructions given.

a. Solve: $-2x + 1 = 6$

b. Write in slope-intercept form: $-2x + y = 6$

c. Find the y-intercept: $-2x + y = 6$

d. Find the slope: $-2x + y = 6$

e. Graph: $-2x + y = 6$

70. Paying Attention to Instructions Work each problem according to the instructions given.

a. Solve: $x + 3 = -6$

b. Write in slope-intercept form: $x + 3y = -6$

c. Find the y-intercept: $x + 3y = -6$

d. Find the slope: $x + 3y = -6$

e. Graph: $x + 3y = -6$

Applying the Concepts

71. Value of a Copy Machine Cassandra buys a new color copier for her small business. It will cost $21,000 and will decrease in value each year. The graph below shows the value of the copier after the first 5 years of ownership.

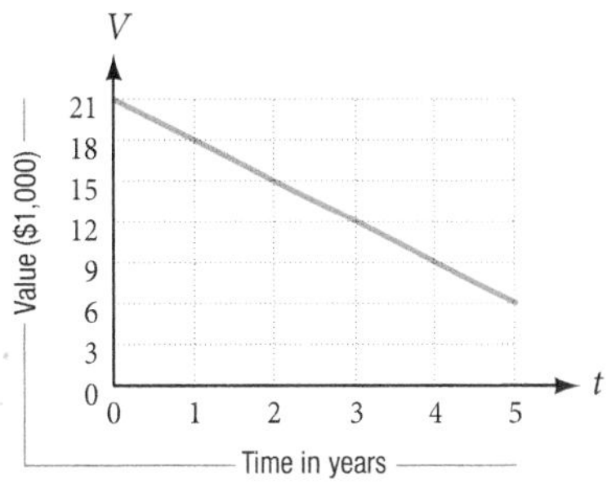

a. How much is the copier worth after 5 years?
b. After how many years is the copier worth $12,000?
c. Find the slope of this line.
d. By how many dollars per year is the copier decreasing in value?
e. Find the equation of this line where V is the value after t years.

72. Salesperson's Income Kevin starts a new job in sales next month. He will earn $1,000 per month plus a certain amount for each shirt he sells. The graph below shows the amount Kevin will earn per month based on how many shirts he sells.

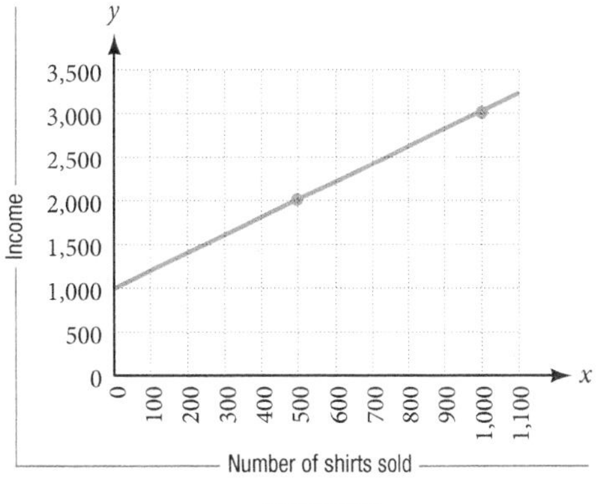

FIGURE 2

a. How much will he earn for selling 1,000 shirts?
b. How many shirts must he sell to earn $2,000 for a month?
c. Find the slope of this line.
d. How much money does Kevin earn for each shirt he sells?
e. Find the equation of this line where y is the amount he earns for selling x number of shirts.

Learning Objectives Assessment

The following problems can be used to help assess if you have successfully met the learning objectives for this section.

73. Find the slope of the line $y = -\frac{5}{6}x + \frac{13}{6}$.

a. -5 **b.** $\frac{13}{6}$ **c.** $-\frac{5}{6}$ **d.** 1

74. To graph $y = \frac{1}{2}x - 4$, we would

a. Plot a point at $(-4, 0)$, then rise 2 units and run 1 unit to locate a second point.

b. Plot a point at $(-4, 0)$, then rise 1 unit and run 2 units to locate a second point.

c. Plot a point at $(0, -4)$, then rise 2 units and run 1 unit to locate a second point.

d. Plot a point at $(0, -4)$, then rise 1 unit and run 2 units to locate a second point.

75. Use slope-intercept form to find the equation of the line with slope 3 and passing through $(-4, 2)$.

a. $y = 14x + 3$ **b.** $y = -10x + 3$

c. $y = 3x + 14$ **d.** $y = 3x - 10$

Getting Ready for the Next Section

Solve each equation for y.

76. $-y - 3 = -2(x + 4)$ **77.** $-y + 5 = 2(x + 2)$ **78.** $-y - 3 = -\frac{2}{3}(x + 3)$

79. $-y - 1 = -\frac{1}{2}(x + 4)$ **80.** $-\frac{y - 1}{x} = \frac{3}{2}$ **81.** $-\frac{y + 1}{x} = \frac{3}{2}$

Find the slope of the line through the following pairs of points.

82. $(-3, 3), (3, -1)$ **83.** $(-2, -4), (2, 1)$ **84.** $(0, 0), (4, 1)$

85. $(-3, 0), (0, -3)$ **86.** $(2, 1), (2, 5)$ **87.** $(1, -3), (3, -3)$

Point-Slope Form 3.6

Learning Objectives

In this section, we will learn how to:

1. Find the equation of a line using point-slope form.

Introduction

In the previous section, we saw how slope-intercept form could be used to graph a line or to find the equation of a line. Another useful form of the equation of a straight line is the point-slope form.

Let line l contain the point (x_1, y_1) and have slope m. If (x, y) is any other point on l, then by the definition of slope we have

$$\frac{y - y_1}{x - x_1} = m$$

Multiplying both sides by $(x - x_1)$ gives us

$$(x - x_1) \cdot \frac{y - y_1}{x - x_1} = m(x - x_1)$$

$$y - y_1 = m(x - x_1)$$

This last equation is known as the *point-slope form* of the equation of a straight line.

RULE ***Point-Slope Form of the Equation of a Line***

The equation of the line through (x_1, y_1) with slope m is given by

$$y - y_1 = m(x - x_1)$$

This form is used to find the equation of a line, either given one point on the line and the slope, or given two points on the line.

VIDEO EXAMPLES

SECTION 3.6

EXAMPLE 1 Find the equation of the line with slope -2 that contains the point $(-4, 3)$. Write the answer in standard form.

SOLUTION In Section 3.5 we solved this problem using slope-intercept form. Now we will see how it can be done using point-slope form.

Using	$(x_1, y_1) = (-4, 3)$ and $m = -2$	
in	$y - y_1 = m(x - x_1)$	Point-slope form
gives us	$y - 3 = -2(x + 4)$	Note: $x - (-4) = x + 4$
	$y - 3 = -2x - 8$	Multiply out right side
	$2x + y - 3 = -8$	Add $2x$ to each side
	$2x + y = -5$	Add 3 to each side

Figure 1 is the graph of the line that contains $(-4, 3)$ and has a slope of -2. Notice that the y-intercept on the graph matches that of the equation we found.

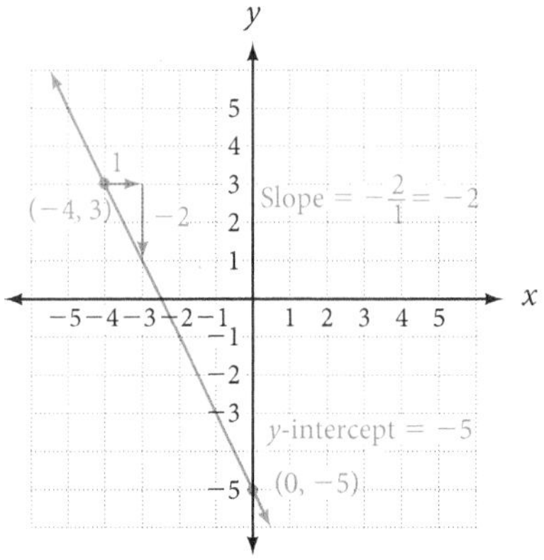

FIGURE 1

EXAMPLE 2 Find the equation of the line that passes through the points $(-3, 3)$ and $(3, -1)$. Write the answer in slope-intercept form.

SOLUTION We begin by finding the slope of the line:

$$m = \frac{3 - (-1)}{-3 - 3} = \frac{4}{-6} = -\frac{2}{3}$$

Using $(x_1, y_1) = (3, -1)$ and $m = -\frac{2}{3}$ in $y - y_1 = m(x - x_1)$ yields

$$y + 1 = -\frac{2}{3}(x - 3)$$

$$y + 1 = -\frac{2}{3}x + 2 \qquad \text{Multiply out right side}$$

$$y = -\frac{2}{3}x + 1 \qquad \text{Add } -1 \text{ to each side}$$

Figure 2 shows the graph of the line that passes through the points $(-3, 3)$ and $(3, -1)$. As you can see, the slope and y-intercept are $-\frac{2}{3}$ and 1, respectively.

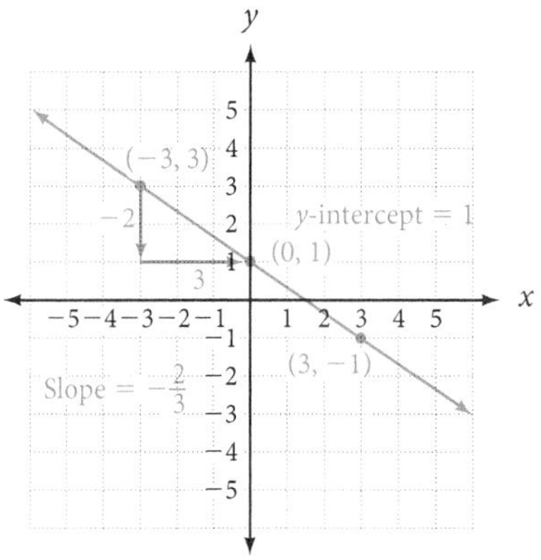

FIGURE 2

Note In Example 2 we could have used the point $(-3, 3)$ instead of $(3, -1)$ and obtained the same equation; that is, using $(x_1, y_1) = (-3, 3)$ and $m = -\frac{2}{3}$ in $y - y_1 = m(x - x_1)$ gives us

$$y - 3 = -\frac{2}{3}(x + 3)$$
$$y - 3 = -\frac{2}{3}x - 2$$
$$y = -\frac{2}{3}x + 1$$

which is the same result we obtained using $(3, -1)$.

EXAMPLE 3 Find the equation of the line passing through $(4, -5)$ with

a. zero slope **b.** no slope

SOLUTION

a. Using $(x_1, y_1) = (4, -5)$ and $m = 0$ in point-slope form gives us

$$y - (-5) = 0(x - 4)$$
$$y + 5 = 0$$
$$y = -5$$

This is the equation of a horizontal line. Recall that the equation of the horizontal line passing through (a, b) is $y = b$. We could have used this fact to get $y = -5$ with less effort, although point-slope form does lead us to the correct equation.

b. Because there is no slope, we cannot use point-slope form. We must recognize that a line having no slope is a vertical line. The equation of the vertical line passing through (a, b) is $x = a$. In this case, we get $x = 4$.

EXAMPLE 4 Find the equation of the line passing through the point $(-3, -2)$ and parallel to the line $3x - 5y = 10$. Write the answer in standard form.

SOLUTION First, we find the slope of the line $3x - 5y = 10$ by writing the equation in slope-intercept form.

$$3x - 5y = 10$$
$$-5y = -3x + 10 \quad \text{Add } -3x \text{ to both sides}$$
$$y = \frac{3}{5}x - 2 \quad \text{Multiply both sides by } -\frac{1}{5}$$

The slope of this line is $m = \frac{3}{5}$. Any line parallel to this line must have the same slope. Using $(x_1, y_1) = (-3, -2)$ and $m = \frac{3}{5}$ in point-slope form gives us

$$y - (-2) = \frac{3}{5}(x - (-3)) \quad \text{Substitute values}$$
$$y + 2 = \frac{3}{5}(x + 3) \quad \text{Simplify}$$
$$y + 2 = \frac{3}{5}x + \frac{9}{5} \quad \text{Distributive property}$$
$$5y + 10 = 3x + 9 \quad \text{Multiply both sides by 5 to clear fractions}$$
$$-3x + 5y + 10 = 9 \quad \text{Add } -3x \text{ to each side}$$
$$-3x + 5y = -1 \quad \text{Add } -10 \text{ to each side}$$
$$3x - 5y = 1 \quad \text{Multiply both sides by } -1$$

Figure 3 shows the graph of both lines.

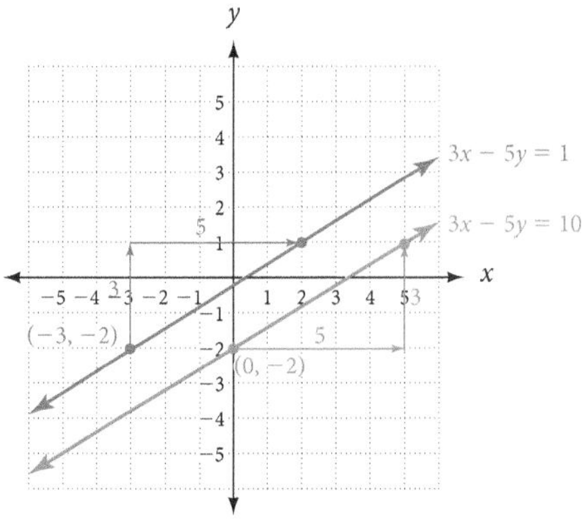

FIGURE 3

Here is a summary of the various equations for a line that we have introduced in this chapter.

Equations of Lines

$ax + by = c$	Standard form
$x = a$	Vertical line through (a, b)
$y = b$	Horizontal line through (a, b)
$y = mx + b$	Slope-intercept form of a line with slope m and y-intercept b.
$y - y_1 = m(x - x_1)$	Point-slope form of line passing through (x_1, y_1) with slope m.

Getting Ready for Class

After reading through the preceding section, respond in your own words and in complete sentences.

A. What is the point-slope form of the equation of a line?

B. What is the point-slope form most useful for?

C. How would you find the equation of a line from two points on the line?

D. Explain why point-slope form cannot be used to find the equation of a vertical line.

For each of the following problems, the slope and one point on a line are given. In each case use the point-slope form to find the equation of that line. Write your answers in standard form.

1. $(-2, -5), m = 2$

2. $(-1, -5), m = 2$

3. $(-4, 1), m = -\frac{1}{2}$

4. $(-2, 1), m = -\frac{1}{2}$

5. $(2, -3), m = \frac{3}{2}$

6. $(3, -4), m = \frac{4}{3}$

7. $(-1, 4), m = -3$

8. $(-2, 5), m = -3$

9. $(6, 0), m = -\frac{2}{3}$

10. $(-5, 0), m = \frac{3}{4}$

11. $(0, -1), m = \frac{1}{5}$

12. $(0, 4), m = -\frac{1}{3}$

13. $(5, 3), m = 0$

14. $(-3, 5), m = 0$

15. $(0, 0), m = \frac{3}{2}$

16. $(0, 0), m = -\frac{3}{4}$

Find the equation of the line that passes through each pair of points. Write your answers in slope-intercept form when possible.

17. $(-2, -4), (1, -1)$

18. $(2, 4), (-3, -1)$

19. $(-1, -5), (2, 1)$

20. $(-1, 6), (1, 2)$

21. $(-3, -2), (3, 6)$

22. $(-3, 6), (3, -2)$

23. $(-3, -1), (3, -5)$

24. $(-3, -5), (3, 1)$

25. $(2, 0), (0, 3)$

26. $(-4, 0), (0, 1)$

27. $(-1, 0), (-1, 5)$

28. $(3, -4), (3, 2)$

29. $(1, 1), (5, 1)$

30. $(-6, -3), (-2, -3)$

31. Find the equation of the line with x-intercept 3 and y-intercept 2.

32. Find the equation of the line with x-intercept 2 and y-intercept 3.

33. Find the equation of the line with x-intercept -2 and y-intercept -5.

34. Find the equation of the line with x-intercept -3 and y-intercept -5.

35. The equation of the vertical line that passes through the points $(3, -2)$ and $(3, 4)$ is either $x = 3$ or $y = 3$. Which one is it?

36. The equation of the horizontal line that passes through the points $(2, 3)$ and $(-1, 3)$ is either $x = 3$ or $y = 3$. Which one is it?

37. Find the equation of the line passing through the point $(-2, 3)$ with

a. zero slope **b.** no slope

38. Find the equation of the line passing through the point $(-1, -4)$ with

a. zero slope **b.** no slope

39. Find the equation of the line through $(6, 2)$ if the line is

a. horizontal **b.** vertical

40. Find the equation of the line through $(5, 0)$ if the line is

a. horizontal **b.** vertical

For the following problems, find the equation of the line passing through the given point and **a.** parallel to the given line, **b.** perpendicular to the given line. Write your answers in standard form.

41. $(3, -1), y = 2x - 5$

42. $(-1, 2), y = 3x + 4$

43. $(-4, -3), y = -\frac{1}{3}x + 1$

44. $(2, -6), y = \frac{1}{2}x - 3$

45. $(0, 5), 3x + 4y = 12$

46. $(0, -1), 4x + 3y = 15$

47. $(1, 0), 2x - 3y = 6$

48. $(3, 0), 2x - 5y = -10$

49. $(4, 2), y = -3$

50. $(-5, 1), x = 2$

Applying the Concepts

51. Temperature The following graph shows the relationship between temperature measured in degrees Fahrenheit (°F) and degrees Celsius (°C).

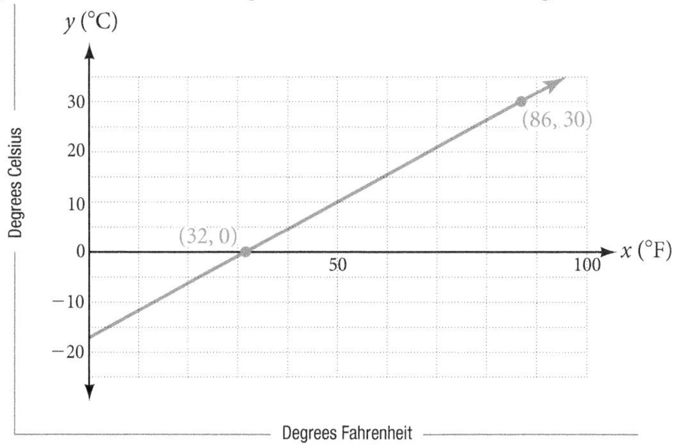

a. Water freezes at a temperature of 32°F. What is the freezing temperature of water in degrees Celsius?

b. Find the slope of this line.

c. Use the two points shown on the graph to find the equation of the line.

d. Water boils at a temperature of 212°F. What is the boiling point of water in degrees Celsius?

52. Pressure The following graph shows the relationship between depth below sea level, measured in feet, and the pressure at that depth in pounds per square inch (psi).

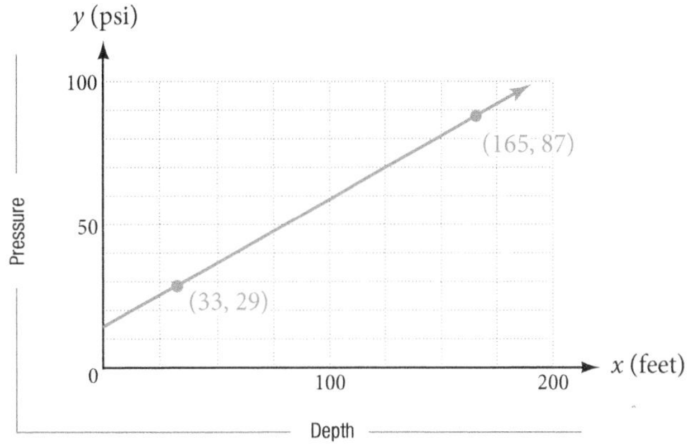

a. What is the pressure at a depth of 33 feet?
b. Find the slope of this line.
c. Use the two points shown on the graph to find the equation of the line.
d. What pressure would a diver experience at a depth of 100 feet?

Learning Objectives Assessment

The following problems can be used to help assess if you have successfully met the learning objectives for this section.

53. Find the equation of the line passing through $(2, -4)$ with slope $m = -3$.

a. $y = -3x - 4$ **b.** $y = -3x + 2$
c. $3x + y = -2$ **d.** $3x + y = -10$

54. Find the equation of the line passing through the points $(-2, 1)$ and $(2, -3)$.

a. $x - y = -3$ **b.** $x + y = -1$
c. $y = -x + 1$ **d.** $y = -x - 3$

Getting Ready for the Next Section

Graph each of the following lines.

55. $x + y = 4$ **56.** $x - y = -2$ **57.** $y = 2x - 3$ **58.** $y = 2x + 3$
59. $y = 2x$ **60.** $y = -2x$

Linear Inequalities in Two Variables

3.7

Learning Objectives

In this section, we will learn how to:

1. Graph the solution set for a linear inequality in two variables.

Introduction

A linear inequality in two variables is any expression that can be put in the form

$$ax + by < c$$

where a, b, and c are real numbers (a and b not both 0). The inequality symbol can be any of the following four: $<, \leq, >, \geq$.

Some examples of linear inequalities are

$$2x + 3y < 6 \qquad y \geq 2x + 1 \qquad x \leq 0 \qquad y > -4$$

Although not all of these inequalities have the form $ax + by < c$, each one can be put in that form.

The solution set for a linear inequality is an entire region of the coordinate plane. The boundary for the region is found by replacing the inequality symbol with an equal sign and graphing the resulting equation. If the inequality symbol used originally is $\leq$ or $\geq$, then the points on the boundary line satisfy the inequality and are part of the solution set. We indicate this by drawing the boundary as a solid line. However, if the original inequality symbol is $<$ or $>$, then the points on the boundary line do not satisfy the inequality and are not part of the solution set. In this case, we draw the boundary as a dashed line. (This is similar to our use of open or closed circles in Section 2.8.)

The boundary equation divides the coordinate plane into two regions. One of these regions is part of the solution set. To determine which region contains solutions, we choose any point not on the boundary and see if it satisfies the inequality.

Let's look at some examples.

VIDEO EXAMPLES

SECTION 3.7

EXAMPLE 1 Graph the solution set for $x + y \leq 4$.

SOLUTION The boundary for the solution set is the line $x + y = 4$. Because the statement $x + y \leq 4$ is true if either $x + y = 4$ or $x + y < 4$, the points on the boundary line are solutions, so we draw the boundary as a solid line.

The graph of the boundary is shown in Figure 1.

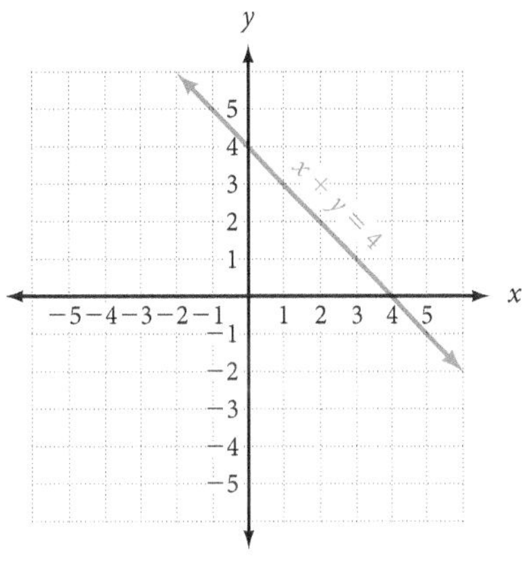

FIGURE 1

The boundary separates the coordinate plane into two regions: the region above the boundary and the region below the boundary. One of the regions contains the remaining solutions, which are the points that satisfy $x + y < 4$. To find the correct region, we simply choose any convenient point that is *not* on the boundary. We then substitute the coordinates of the point into the original inequality $x + y \leq 4$. If the point we choose satisfies the inequality, then it is a member of the solution set, and we can assume that all points on the same side of the boundary as the chosen point are also in the solution set. If the coordinates of our point do not satisfy the original inequality, then the solution set lies on the other side of the boundary.

In this example a convenient point not on the boundary is the origin. Substituting (0, 0) into $x + y \leq 4$ gives us

$$0 + 0 \overset{?}{\leq} 4$$

$$0 \leq 4 \qquad \text{A true statement}$$

Because the origin is a solution to the inequality $x + y \leq 4$, and the origin is below the boundary, all other points below the boundary are also solutions. We indicate this by shading the region of the coordinate plane below the boundary line.

The graph of $x + y \leq 4$ is shown in Figure 2. The solution set consists of all points on or below the line.

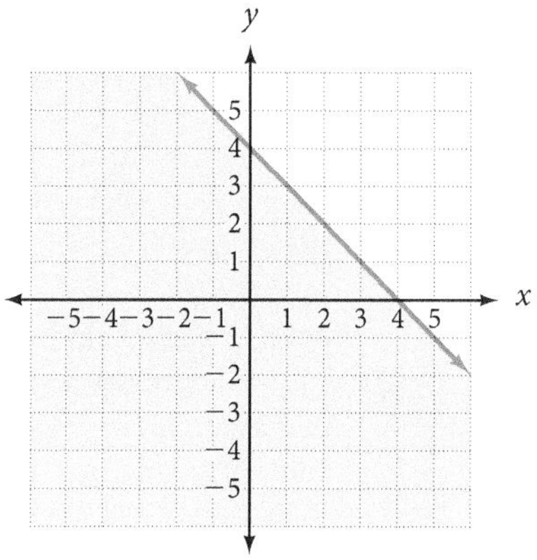

FIGURE 2

Here is a list of steps to follow when graphing the solution set for linear inequalities in two variables.

HOW TO ***Graph the Solution Set for Linear Inequalities in Two Variables***

Step 1: Replace the inequality symbol with an equal sign. The resulting equation represents the boundary for the solution set.

Step 2: Graph the boundary found in step 1 using a *solid line* if the boundary is included in the solution set (that is, if the original inequality symbol was either $\leq$ or $\geq$). Use a *dashed line* to graph the boundary if it is *not* included in the solution set. (It is not included if the original inequality was either $<$ or $>$).

Step 3: Choose any convenient point not on the boundary and substitute the coordinates into the *original* inequality. If the resulting statement is *true*, the solution set lies on the *same* side of the boundary as the chosen point. If the resulting statement is *false*, the solution set lies on the *opposite* side of the boundary.

Step 4: Shade the region on the side of the boundary line that contains solutions.

EXAMPLE 2 Graph the solution set for $y < 2x - 3$.

SOLUTION The boundary is the graph of $y = 2x - 3$. The boundary is not included because the original inequality symbol is $<$, which does not allow equality. We therefore use a dashed line to represent the boundary, as shown in Figure 3.

A convenient test point is again the origin. Using (0, 0) in $y < 2x - 3$, we have

$$0 \overset{?}{<} 2(0) - 3$$

$$0 < -3 \qquad \text{A false statement}$$

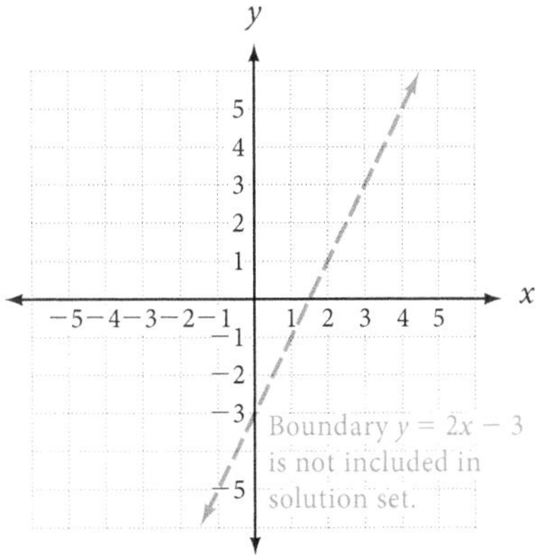

FIGURE 3

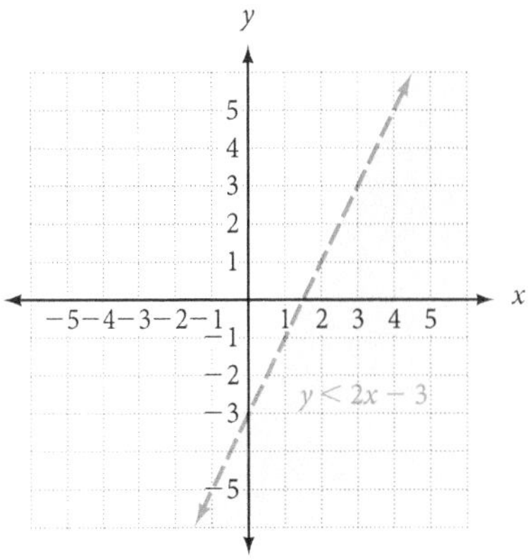

FIGURE 4

Because our test point gives us a false statement and it lies above the boundary, the solution set must lie on the other side of the boundary. Therefore, we shade the region below the boundary line as shown in Figure 4.

EXAMPLE 3 Graph the inequality $2x + 3y \leq 0$.

SOLUTION We begin by graphing the boundary $2x + 3y = 0$. The boundary is included in the solution set because the inequality symbol is $\leq$.

Because the origin (0, 0) lies on the boundary line, we cannot use it as a test point. We choose any other point not on the line, such as (1, 1). Using (1, 1) in $2x + 3y \leq 0$, we have

$$2(1) + 3(1) \overset{?}{\leq} 0$$

$$5 \leq 0 \qquad \text{A false statement}$$

Since (1, 1) does not satisfy the inequality and lies above the boundary, the solution set must be the region below the line. We therefore shade below the line, as shown in Figure 5.

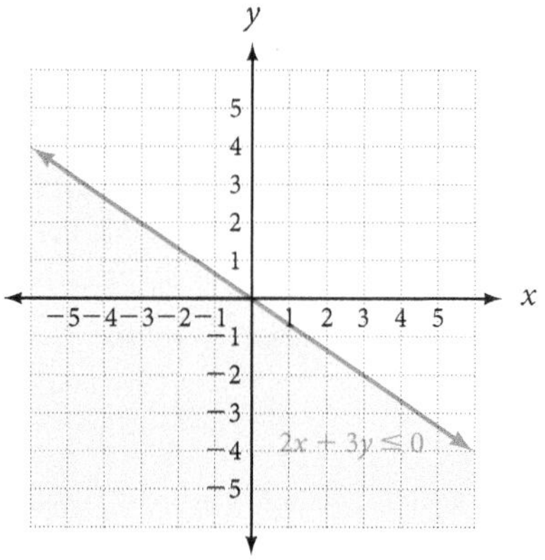

FIGURE 5

EXAMPLE 4 Graph the solution set for $x \leq 5$.

SOLUTION In this context, the inequality $x \leq 5$ can be interpreted to mean $x + 0y \leq 5$. It is still considered an inequality in two variables. The boundary is $x = 5$, which is a vertical line and is part of the solution set.

If we use (0, 0) as our test point, we see that it yields a true statement when its coordinates are substituted into $x + 0y \leq 5$. The solution set, therefore, lies to the left of the vertical line. Figure 6 shows the graph.

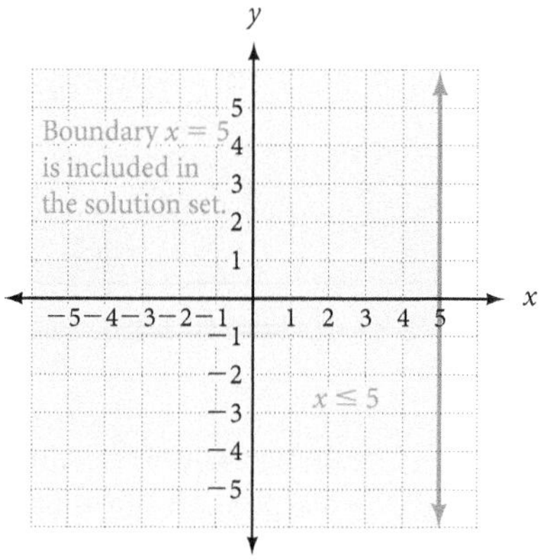

FIGURE 6

Getting Ready for Class

After reading through the preceding section, respond in your own words and in complete sentences.

A. When graphing a linear inequality in two variables, how do you find the equation of the boundary line?

B. What is the significance of a dashed line in the graph of an inequality?

C. When graphing a linear inequality in two variables, how do you know which side of the boundary line to shade?

D. Describe the set of ordered pairs that are solutions to $x + y < 6$.

Problem Set 3.7

Graph the solution set for the following linear inequalities.

1. $2x - 3y < 6$ **2.** $3x + 2y \geq 6$ **3.** $x - 2y \leq 4$ **4.** $2x + y > 4$

5. $x - y \leq 2$ **6.** $x - y \leq 1$ **7.** $3x - 4y \geq 12$ **8.** $4x + 3y < 12$

9. $5x - y \leq 5$ **10.** $4x + y > 4$ **11.** $2x + 6y \leq 12$ **12.** $x - 5y > 5$

13. $x \geq 1$ **14.** $x < 5$ **15.** $y > 2$ **16.** $y \leq -4$

17. $x + 3y \leq 0$ **18.** $3x - y > 0$ **19.** $2x + y > 3$ **20.** $5x + 2y < 2$

21. $y \leq 3x - 1$ **22.** $y \geq 3x + 2$ **23.** $y \leq -\frac{1}{2}x + 2$ **24.** $y < \frac{1}{3}x + 3$

The next two problems are intended to give you practice reading, and paying attention to, the instructions that accompany the problems you are working.

25. Paying Attention to Instructions Work each problem according to the instructions given.

a. Solve: $4 + 3y < 12$

b. Solve: $4 - 3y < 12$

c. Solve for y: $4x + 3y = 12$

d. Graph: $y < -\frac{4}{3}x + 4$

26. Paying Attention to Instructions Work each problem according to the instructions given.

a. Solve: $3x + 2 \geq 6$

b. Solve: $-3x + 2 \geq 6$

c. Solve for y: $3x + 2y = 6$

d. Graph: $y \geq -\frac{3}{2}x + 3$

27. Find the equation of the line shown in part a, then use this information to find the inequalities whose solution sets are shown in parts b and c.

a.

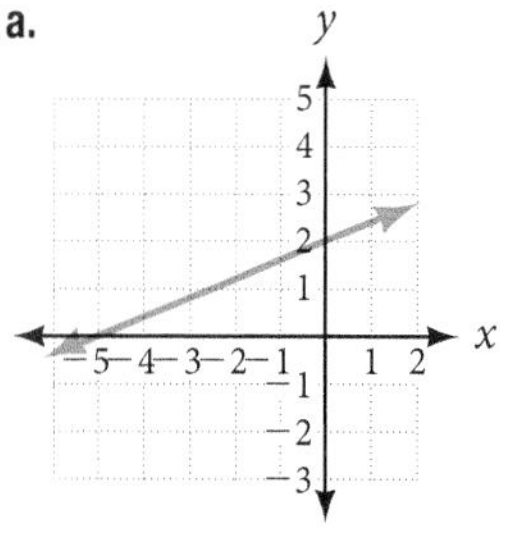

b.

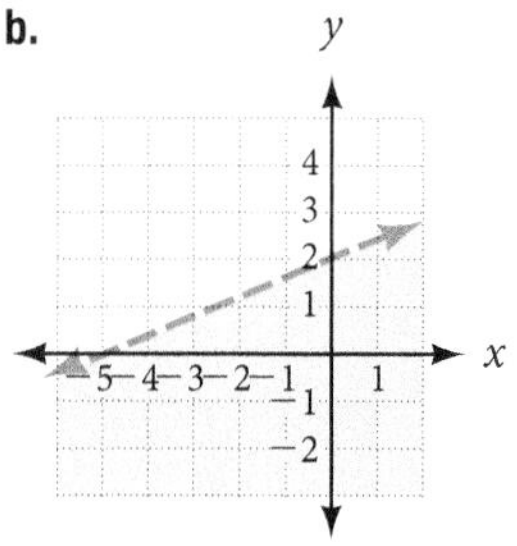

c.

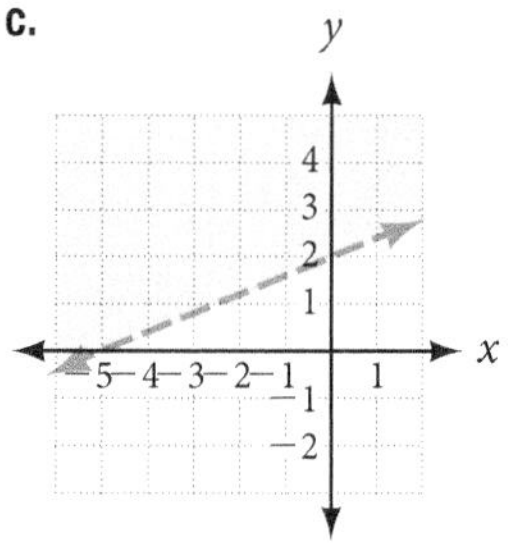

28. Find the equation of the line shown in part a, then use this information to find the inequalities whose solution sets are shown in parts b and c.

a.

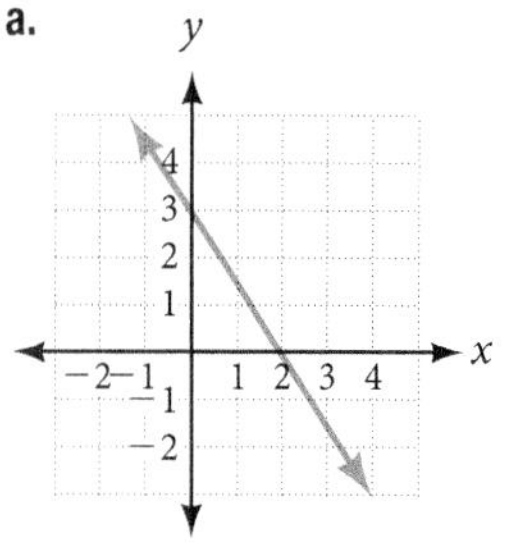

b.

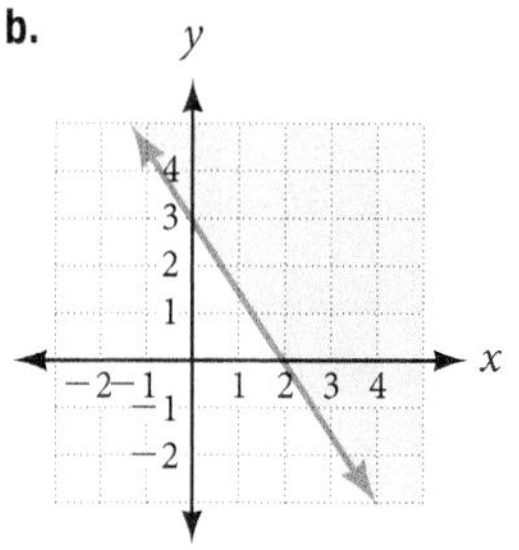

c.

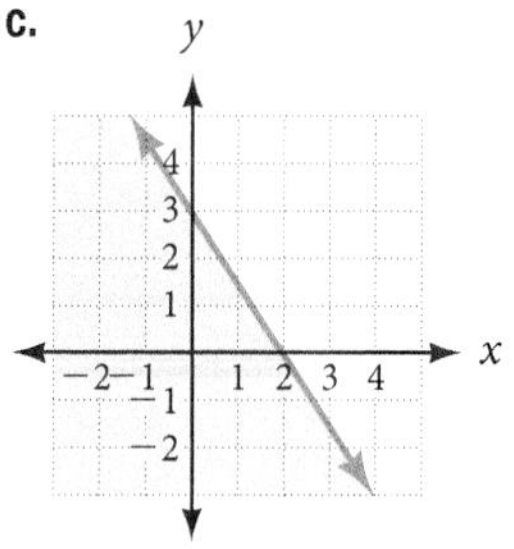

Learning Objectives Assessment

The following problems can be used to help assess if you have successfully met the learning objectives for this section.

29. The boundary for the solution set of $x - 2y > 4$ should be drawn as:

a. a solid line.

b. a dashed line.

30. The solution set for $x - 2y > 4$ is:

a. the line $x - 2y = 4$ and the region above the line.

b. the line $x - 2y = 4$ and the region below the line.

c. only the region below the line $x - 2y = 4$.

d. only the region above the line $x - 2y = 4$.

Maintaining Your Skills

31. Simplify the expression $7 - 3(2x - 4) - 8$.

32. Find the value of $x^2 - 2xy + y^2$ when $x = 3$ and $y = -4$.

Solve each equation.

33. $-\frac{3}{2}x = 12$

34. $2x - 4 = 5x + 2$

35. $8 - 2(x + 7) = 2$

36. $3(2x - 5) - (2x - 4) = 6 - (4x + 5)$

37. Solve the formula $P = 2l + 2w$ for w.

Solve each inequality, and graph the solution set.

38. $-4x < 20$

39. $3 - 2x > 5$

40. $3 - 4(x - 2) \geq -5x + 6$

41. Solve the formula $3x - 2y \leq 12$ for y.

42. What number is 12% of 2,000?

43. **Geometry** The length of a rectangle is 5 inches more than 3 times the width. If the perimeter is 26 inches, find the length and width.

Chapter 3 Summary

EXAMPLES

1. For the point $(-4, -5)$, the x-coordinate is -4 and the y-coordinate is -5. Also, the point lies in the third quadrant, QIII.

Rectangular Coordinate System [3.1]

A pair of numbers enclosed in parentheses and separated by a comma, such as $(-2, 1)$, is called an ***ordered pair*** of numbers. The first number in the pair is called the ***x-coordinate*** of the ordered pair; the second number is called the ***y-coordinate***.

Ordered pairs of numbers are important in the study of mathematics because they give us a way to visualize solutions to equations. The visual component of ordered pairs is called the rectangular coordinate system.

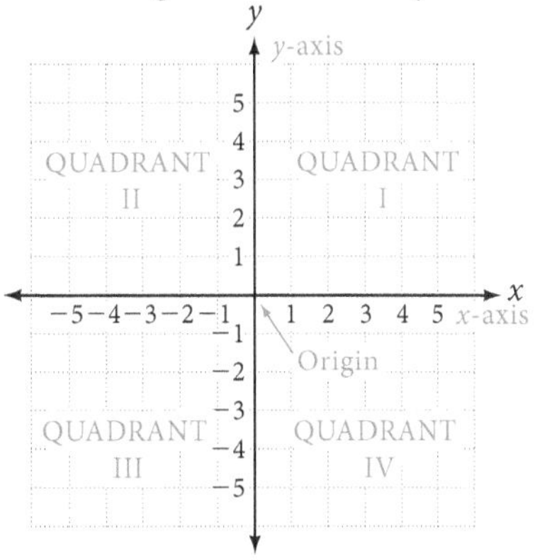

FIGURE 6

2. The equation $3x + 2y = 6$ is an example of a linear equation in two variables.

Linear Equation in Two Variables [3.2]

A linear equation in two variables is any equation that can be put in the form $ax + by = c$. The graph of every linear equation is a straight line.

3. The graph of $y = -\frac{2}{3}x - 1$ is shown below.

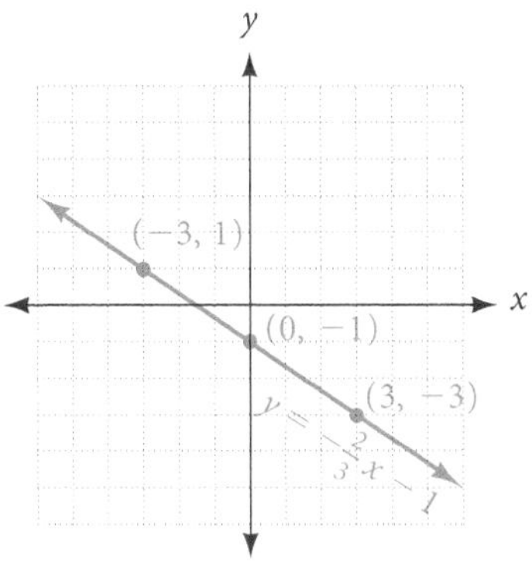

Strategy for Graphing Linear Equations in Two Variables [3.2]

Step 1: Find any three ordered pairs that satisfy the equation. This can be done by using a convenient number for one variable and solving for the other variable.

Step 2: Graph the three ordered pairs found in step 1. Actually, we need only two points to graph a line. The third point serves as a check. If all three points do not line up, there is a mistake in our work.

Step 3: Draw a line through the three points graphed in step 2.

Intercepts [3.3]

The x-intercept of a line is the x-coordinate of the point where the graph crosses the x-axis. The y-intercept is the y-coordinate of the point where the graph crosses the y-axis. We find the y-intercept by substituting $x = 0$ into the equation and solving for y. The x-intercept is found by letting $y = 0$ and solving for x.

4. To find the x-intercept for $3x + 2y = 6$, we let $y = 0$ and get

$$3x = 6$$
$$x = 2$$

In this case the x-intercept is 2, and the graph crosses the x-axis at (2, 0).

Slope of a Line [3.4]

The *slope* of the line containing the points (x_1, y_1) and (x_2, y_2) is given by

$$\text{Slope} = m = \frac{y_2 - y_1}{x_2 - x_1} = \frac{\text{rise}}{\text{run}}$$

5. The slope of the line through $(3, -5)$ and $(-2, 1)$ is

$$m = \frac{-5 - 1}{3 - (-2)} = \frac{-6}{5} = -\frac{6}{5}$$

Vertical and Horizontal Lines [3.2, 3.4]

The graph of $x = a$ is a vertical line passing through $(a, 0)$. A vertical line has no slope.

The graph of $y = b$ is a horizontal line passing through $(0, b)$. A horizontal line has slope $m = 0$.

6. The graph of $x = 3$ is a vertical line, and the graph of $y = 3$ is a horizontal line. The figure below shows both graphs.

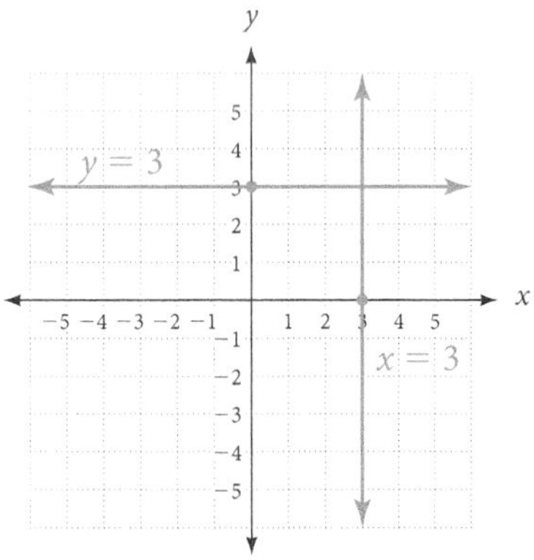

Parallel and Perpendicular Lines [3.4]

Two distinct, nonvertical lines are parallel if they have the same slope. They are perpendicular if the product of their slopes is -1.

7. A line parallel to $y = \frac{2}{3}x + 1$ would have a slope of $m = \frac{2}{3}$. A perpendicular line would have a slope of $m = -\frac{3}{2}$.

Slope-Intercept Form of a Line [3.5]

The equation of the line with a slope of m and a y-intercept of b is

$$y = mx + b$$

8. The equation of the line with a slope of 2 and a y-intercept of 5 is

$$y = 2x + 5$$

Point-Slope Form of a Line [3.6]

9. The equation of the line through (1, 2) with a slope of 3 is

$$y - 2 = 3(x - 1)$$
$$y - 2 = 3x - 3$$
$$y = 3x - 1$$

If a line has a slope of m and contains the point (x_1, y_1), the equation can be written as

$$y - y_1 = m(x - x_1)$$

To Graph a Linear Inequality in Two Variables [3.7]

9. Graph $x - y \geq 3$.

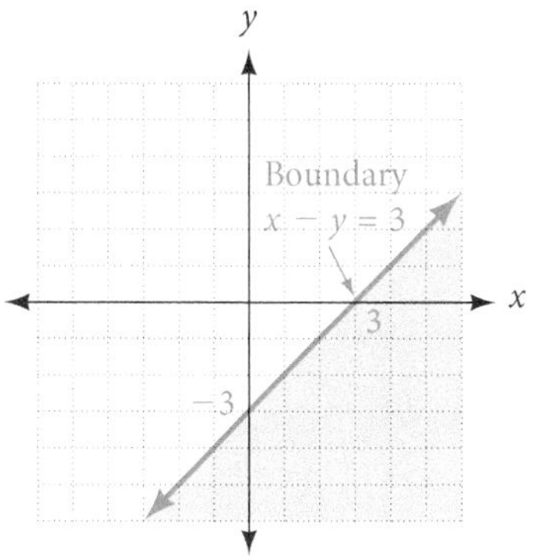

Step 1: Replace the inequality symbol with an equal sign. The resulting equation represents the boundary for the solution set.

Step 2: Graph the boundary found in step 1, using a *solid line* if the original inequality symbol was either ≤ or ≥. Use a *dashed line* otherwise.

Step 3: Choose any convenient point not on the boundary and substitute the coordinates into the *original* inequality. If the resulting statement is *true*, the graph lies on the *same* side of the boundary as the chosen point. If the resulting statement is *false*, the solution set lies on the *opposite* side of the boundary.

Step 4: Shade the side of the boundary line containing solutions.

Chapter 3 Test

Graph the ordered pairs. [3.1]

1. (2, −1) **2.** (−4, 3) **3.** (−3, −2) **4.** (0, −4)

5. Fill in the following ordered pairs for the equation $3x - 2y = 6$. [3.2]

(0,) (, 0) (4,) (, −6)

6. Which of the following ordered pairs are solutions to $y = -3x + 7$? [3.2]

(0, 7) (2, −1) (4, −5) (−5, −3)

Graph each line. [3.2]

7. $y = -\frac{1}{2}x + 4$

8. $2x - 5y = 10$

9. $x = -3$

10. $y = 2$

Find the x- and y-intercepts. [3.3]

11. $8x - 4y = 16$

12. $y = \frac{3}{2}x + 6$

13. $y = 3$

14. $x = -2$

Find the slope of the line through each pair of points. [3.4]

15. (3, 2), (−5, 6)

16. (0, 9), (7, 1)

Find the slope of each line. [3.4]

17.

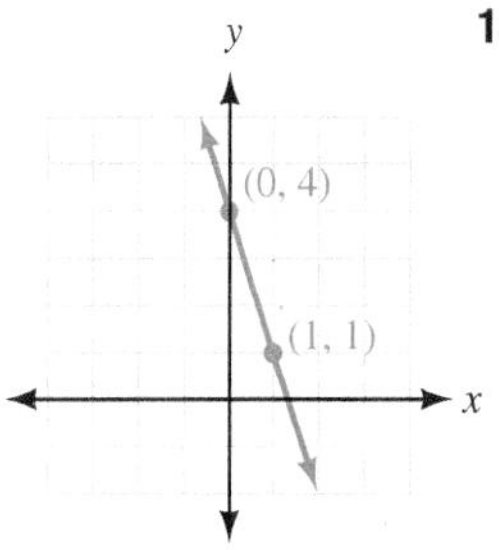

18.

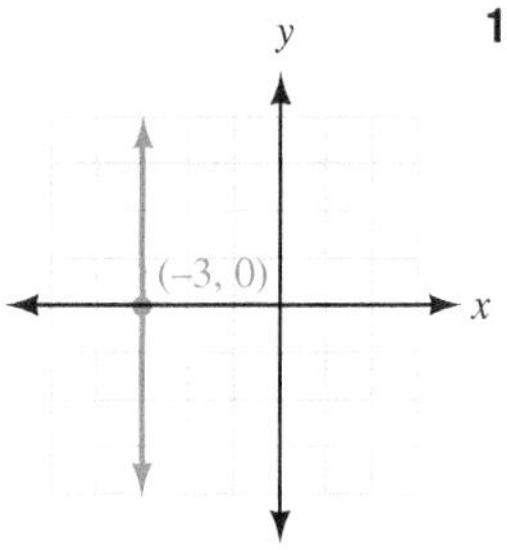

19.

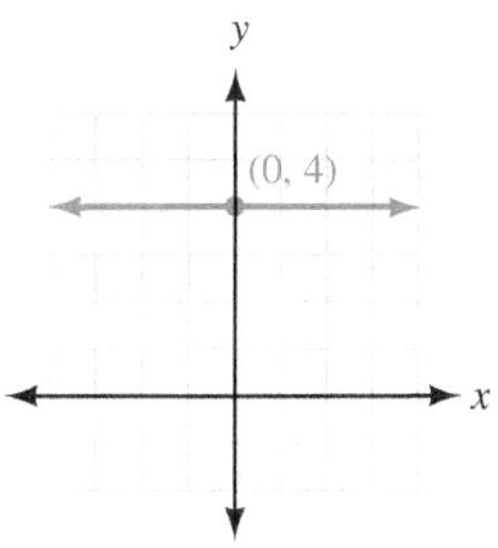

Find the slope and y-intercept for each line. [3.5]

20. $y = -\frac{1}{2}x + 6$

21. $y = 3x$

22. Find the slope of a line parallel to the line $y = \frac{3}{7}x - 5$. [3.4, 3.5]

23. Find the slope of a line perpendicular to $y = -4x + 3$. [3.4, 3.5]

24. Find the equation of the line with a slope of 3 and y-intercept −5. [3.5]

25. Find the equation of the line through (4, 1) with a slope of $-\frac{1}{2}$. [3.5, 3.6]

26. Find the equation of the line passing through the points (3, −4) and (−6, 2). [3.6]

27. A line has an x-intercept 3 and contains the point (−2, 6). Find its equation. [3.6]

28. Find the equation of the line passing through (1, −2) and parallel to the line $y = \frac{3}{2}x + 3$. Write your answer in standard form. [3.5, 3.6]

29. Find the equation of the line passing through $(-4, 3)$ and perpendicular to the line $3x + 2y = 6$. Write your answer in standard form. [3.5, 3.6]

30. Find the equation of the line with slope $m = 0$ passing through $(5, -5)$. [3.5, 3.6]

31. Find the equation of the vertical line passing through $(-2, -4)$. [3.5, 3.6]

Graph the solution set for each linear inequality in two variables. [3.7]

32. $y > x - 6$ **33.** $6x - 9y \leq 18$ **34.** $x \leq 2$

Systems of Linear Equations

4

Chapter Outline

4.1 Solving Linear Systems by Graphing

4.2 The Elimination Method

4.3 The Substitution Method

4.4 Applications

Two companies offer Internet access to their customers. Company A charges \$10 a month plus \$3 for every hour of Internet connection. Company B charges \$18 a month plus \$1 for every hour of Internet connection. To compare the monthly charges of the two companies we form what is called a system of equations. Here is that system.

$$y = 3x + 10$$
$$y = x + 18$$

The first equation gives us information on company A; the second equation gives information on company B. Tables 1 and 2 and the graphs in Figure 1 give us additional information about this system of equations.

TABLE 1

Company A

Hours	Cost
0	\$10
1	\$13
2	\$16
3	\$19
4	\$22
5	\$25
6	\$28
7	\$31
8	\$34
9	\$37
10	\$40

TABLE 2

Company B

Hours	Cost
0	\$18
1	\$19
2	\$20
3	\$21
4	\$22
5	\$23
6	\$24
7	\$25
8	\$26
9	\$27
10	\$28

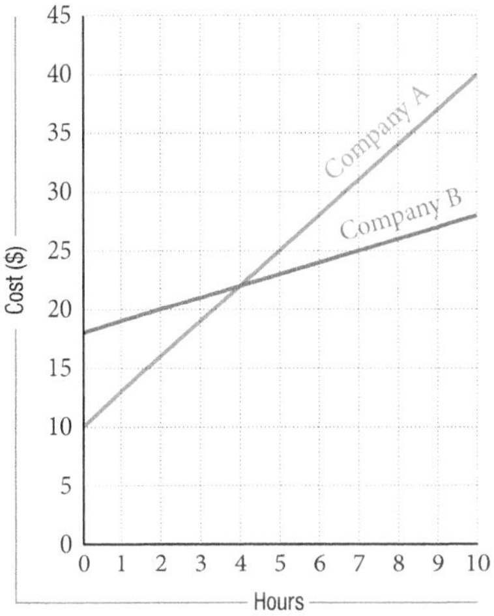

FIGURE 1

As you can see from looking at the tables and at the graphs in Figure 1, the monthly charges for the two companies will be equal if Internet use is exactly 4 hours. In this chapter we work with systems of linear equations.

Success Skills

The study skills for this chapter concern the way you approach new situations in mathematics. The first study skill applies to your natural instincts for what does and doesn't work in mathematics. The second study skill gives you a way of testing your instincts.

1. **Don't Let Your Intuition Fool You** As you become more experienced and more successful in mathematics, you will be able to trust your mathematical intuition. For now, though, it can get in the way of success. For example, if you ask a beginning algebra student to "subtract 3 from -5" many will answer -2 or 2. Both answers are incorrect, even though they may seem intuitively true.

2. **Test Properties About Which You Are Unsure** From time to time you will be in a situation in which you would like to apply a property or rule, but you are not sure if it is true. You can always test a property or statement by substituting numbers for variables. For instance, I always have students that rewrite $(x + 3)^2$ as $x^2 + 9$, thinking that the two expressions are equivalent. The fact that the two expressions are not equivalent becomes obvious when we substitute 10 for x in each one.

   When $x = 10$, the expression $(x + 3)^2$ is $(10 + 3)^2 = 13^2 = 169$

   When $x = 10$, the expression $x^2 + 9 = 10^2 + 9 = 100 + 9 = 109$

   It is not unusual, nor is it wrong, to try occasionally to apply a property that doesn't exist. If you have any doubt about generalizations you are making, test them by replacing variables with numbers and simplifying.

Solving Linear Systems by Graphing 4.1

Learning Objectives

In this section, we will learn how to:

1. Determine if an ordered pair is a solution to a system of linear equations.
2. Solve a linear system by graphing.
3. Identify an inconsistent system.
4. Identify a system with dependent equations.

Introduction

Two linear equations considered at the same time make up what is called a *system of linear equations*. Both equations have graphs that are lines. The following are systems of linear equations:

$$\begin{aligned} x + y &= 3 \\ 3x + 4y &= 2 \end{aligned} \qquad \begin{aligned} y &= 2x + 1 \\ y &= 3x + 2 \end{aligned} \qquad \begin{aligned} 2x - y &= 1 \\ 3x - 2y &= 6 \end{aligned}$$

Solutions to a Linear System

A solution to a system of linear equations in two variables is an ordered pair that satisfies every equation in the system. For systems with two equations, the ordered pair must make both equations true statements.

VIDEO EXAMPLES

SECTION 4.1

EXAMPLE 1 Determine whether each ordered pair is a solution to the following system:

$$x + y = 3$$

$$3x + 4y = 2$$

a. $(-2, 5)$ **b.** $(10, -7)$

SOLUTION We must determine if the ordered pair satisfies both equations.

a.

When	$x = -2$	When	$x = -2$
and	$y = 5$	and	$y = 5$
the equation	$x + y = 3$	the equation	$3x + 4y = 2$
becomes	$-2 + 5 \stackrel{?}{=} 3$	becomes	$3(-2) + 4(5) \stackrel{?}{=} 2$
	$3 = 3$		$-6 + 20 \stackrel{?}{=} 2$
			$14 \neq 2$

The ordered pair $(-2, 5)$ satisfies the first equation but not the second, so it is not a solution to the system.

b.

When	$x = 10$	When	$x = 10$
and	$y = -7$	and	$y = -7$
the equation	$x + y = 3$	the equation	$3x + 4y = 2$
becomes	$10 + (-7) \stackrel{?}{=} 3$	becomes	$3(10) + 4(-7) \stackrel{?}{=} 2$
	$3 = 3$		$30 + (-28) \stackrel{?}{=} 2$
			$2 = 2$

The ordered pair $(10, -7)$ makes both equations true, so it is a solution to the system of equations.

Solving Systems by Graphing

The solution set for a system of linear equations contains all ordered pairs that are solutions to both equations. Each linear equation has a graph that is a line. If the lines intersect at a point, then this point lies on both lines and the corresponding ordered pair must therefore satisfy both equations, making it a solution to the system. So we can solve a linear system graphically by locating intersection points. If we graph both equations on the same coordinate system, we can read the coordinates of the point of intersection and have the solution to our system. Here is an example.

EXAMPLE 2 Solve the following system by graphing.

$$x + y = 4$$
$$x - y = -2$$

SOLUTION On the same set of coordinate axes we graph each equation separately. Figure 1 shows both graphs, without showing the work necessary to get them. We can see from the graphs that they appear to intersect at the point (1, 3).

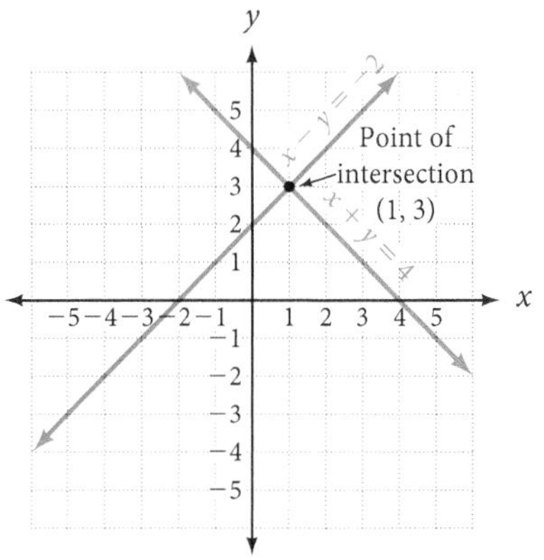

FIGURE 1

We can verify that the ordered pair (1, 3) is a solution to the system by substituting the coordinates $x = 1$, $y = 3$ into both equations to see if they work.

When	$x = 1$	When	$x = 1$
and	$y = 3$	and	$y = 3$
the equation	$x + y = 4$	the equation	$x - y = -2$
becomes	$1 + 3 \stackrel{?}{=} 4$	becomes	$1 - 3 \stackrel{?}{=} -2$
	$4 = 4$		$-2 = -2$

The point (1, 3) satisfies both equations, so it is a solution to the system. From Figure 1 we can see that the two lines intersect at a single point, so (1, 3) is the *only* solution to the system. The solution set is $\{(1, 3)\}$.

Here are some steps to follow in solving linear systems by graphing.

HOW TO *Solve a Linear System by Graphing*

Step 1: Graph the first equation by any of the methods described in Chapter 3.
Step 2: Graph the second equation on the same set of axes used for the first equation.
Step 3: Identify the coordinates of the point where the two graphs appear to intersect.
Step 4: Verify the solution in both equations.

EXAMPLE 3 Solve the following system by graphing.

$$x + 2y = 8$$
$$2x - 3y = 2$$

SOLUTION Graphing each equation on the same coordinate system, we have the lines shown in Figure 2.

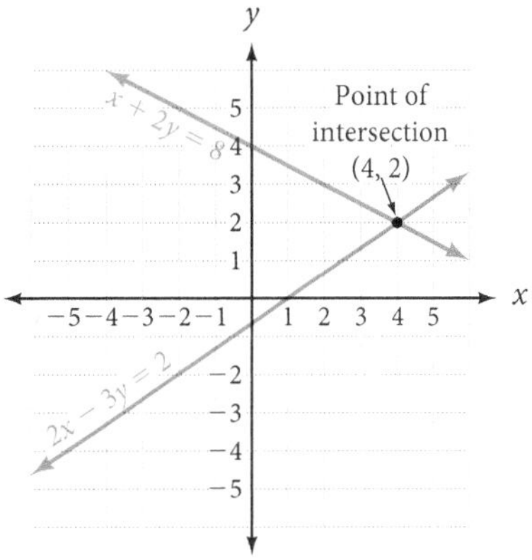

FIGURE 2

From Figure 2, we can see the solution for our system appears to be the point (4, 2). We check this solution as follows.

When	$x = 4$	When	$x = 4$
and	$y = 2$	and	$y = 2$
the equation	$x + 2y = 8$	the equation	$2x - 3y = 2$
becomes	$4 + 2(2) \stackrel{?}{=} 8$	becomes	$2(4) - 3(2) \stackrel{?}{=} 2$
	$4 + 4 \stackrel{?}{=} 8$		$8 - 6 \stackrel{?}{=} 2$
	$8 = 8$		$2 = 2$

The point (4, 2) satisfies both equations and, therefore, must be the solution to our system. The solution set is $\{(4, 2)\}$.

EXAMPLE 4 Solve the following system by graphing.

$$y = 2x - 3$$
$$x = 3$$

SOLUTION Graphing both equations on the same set of axes, we have Figure 3.

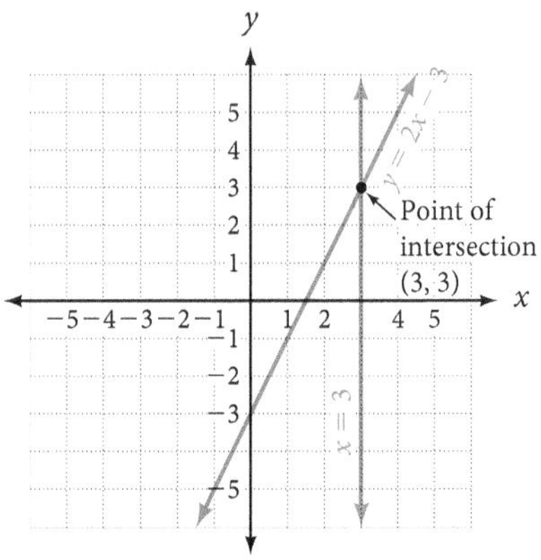

FIGURE 3

The graphs appear to intersect at the point (3, 3). Because (3, 3) satisfies both equations (check for yourself), it is the solution to the system. The solution set is $\{(3, 3)\}$.

EXAMPLE 5 Solve the following system by graphing.

$$y = x - 2$$
$$y = x + 1$$

SOLUTION Graphing both equations produces the lines shown in Figure 4. We can see in Figure 4 that the lines are parallel and therefore do not intersect. Our system has no ordered pair as a solution because there is no ordered pair that satisfies both equations. The solution set is the empty set, which we denote by the symbol $\varnothing$.

Note We know that the lines in Example 5 must be parallel because they both have slope $m = 1$.

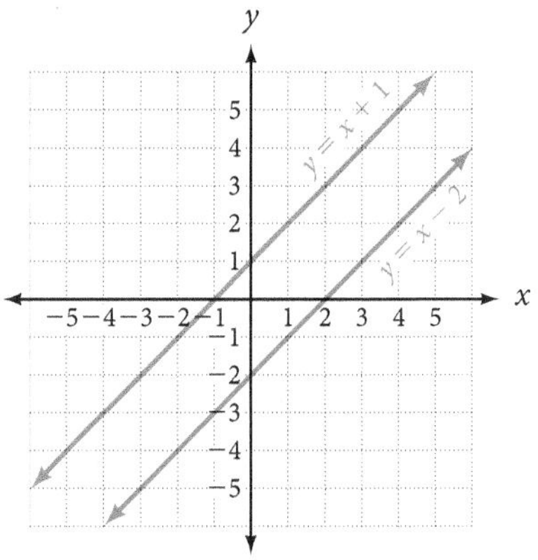

FIGURE 4

EXAMPLE 6 Graph the system.

$$2x + y = 4$$
$$4x + 2y = 8$$

SOLUTION Both graphs are shown in Figure 5. The two graphs coincide. The reason becomes apparent when we multiply both sides of the first equation by 2:

$$2x + y = 4$$
$$2(2x + y) = 2(4) \quad \text{Multiply both sides by 2}$$
$$4x + 2y = 8$$

The equations are equivalent and have the same solution set. Any point that lies on one line also lies on the other. Therefore, every point on either line is a solution to the system. Because a line consists of an infinite set of points, the system has an infinite number of solutions.

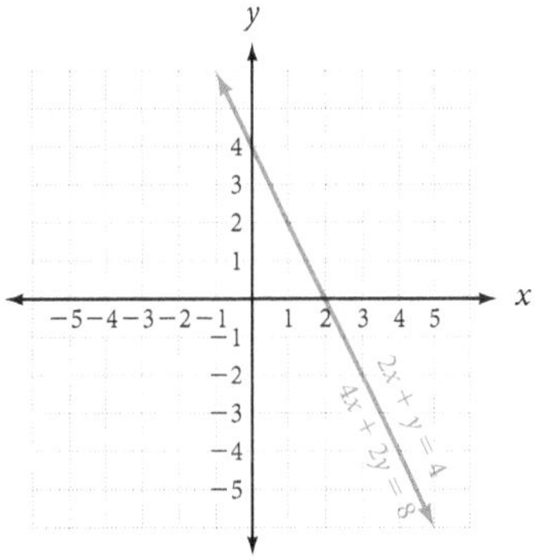

FIGURE 5

Note When the lines coincide and there are an infinite number of solutions, either of the equations can be used to write the solution set. In Example 6, we could also have written $\{(x, y) \mid 4x + 2y = 8\}$.

We can use set-builder notation to write the solution set as $\{(x, y) \mid 2x + y = 4\}$.

We sometimes use special vocabulary to describe the special cases shown in Examples 5 and 6. When a system of equations has no solution because the lines are parallel (as in Example 5), we say the system is *inconsistent*. When the lines coincide (as in Example 6), we say the equations are *dependent*. The two special cases illustrated in the previous two examples do not happen often. Usually, a system has a single ordered pair as a solution.

Here is a summary of three possible types of solutions to a system of equations in two variables.

One Solution

Lines intersect at a single point

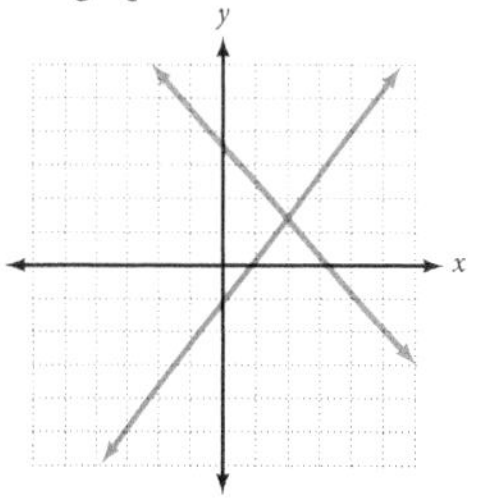

No Solution

Lines are parallel and never cross

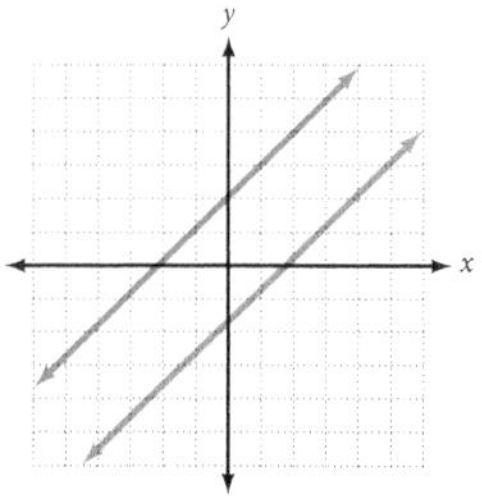

System is inconsistent

Infinite Solutions

Lines coincide

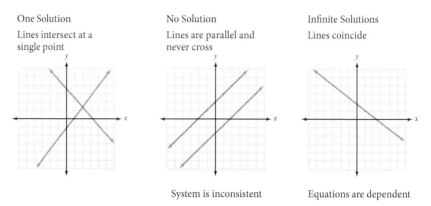

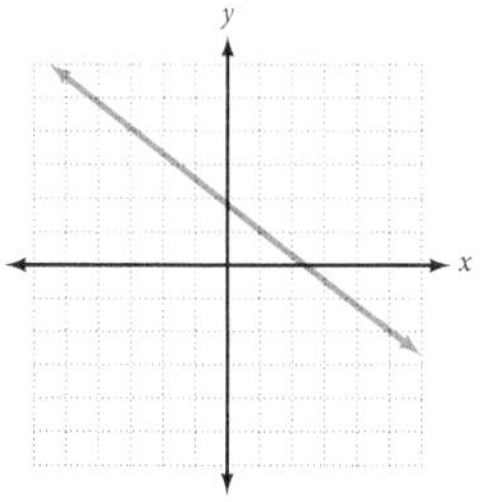

Equations are dependent

Solving a system of linear equations by graphing is useful only when the point of intersection has integers for coordinates. In the next two sections, we will introduce algebraic methods for solving linear systems that will allow us to find the solution even when the ordered pair does not have convenient values for coordinates.

Getting Ready for Class

After reading through the preceding section, respond in your own words and in complete sentences.

A. What is a system of two linear equations in two variables?

B. What is a solution to a system of linear equations?

C. How do we solve a system of linear equations by graphing?

D. Under what conditions will a system of linear equations not have a solution?

For each of the following systems of linear equations, determine if any of the given ordered pairs are solutions to the system.

1. $x - y = 1$
$x + 2y = 4$
a. $(-1, -2)$ **b.** $(2, 1)$ **c.** $(2, 2)$

2. $x + y = 2$
$x - y = 4$
a. $(0, 2)$ **b.** $(-1, -2)$ **c.** $(3, -1)$

3. $2x - 4y = 6$
$-x + 2y = -3$
a. $(5, 1)$ **b.** $(-1, -1)$ **c.** $(3, 0)$

4. $2x - y = -3$
$y = 2x + 3$
a. $(2, -1)$ **b.** $(0, 3)$ **c.** $(-1, 1)$

Solve the following systems of linear equations by graphing, then write the solution set.

5. $x + y = 3$
$x - y = 1$

6. $x + y = 2$
$x - y = 4$

7. $x + y = 1$
$-x + y = 3$

8. $x + y = 1$
$x - y = -5$

9. $x + y = 8$
$-x + y = 2$

10. $x + y = 6$
$-x + y = -2$

11. $3x - 2y = 6$
$x - y = 1$

12. $5x - 2y = 10$
$x - y = -1$

13. $6x - 2y = 12$
$3x + y = -6$

14. $4x - 2y = 8$
$2x + y = -4$

15. $4x + y = 4$
$3x - y = 3$

16. $5x - y = 10$
$2x + y = 4$

17. $x + 2y = 0$
$2x - y = 0$

18. $3x + y = 0$
$5x - y = 0$

19. $3x - 5y = 15$
$-2x + y = 4$

20. $2x - 4y = 8$
$2x - y = -1$

21. $y = 2x + 1$
$y = -2x - 3$

22. $y = 3x - 4$
$y = -2x + 1$

23. $x + 3y = 3$
$y = x + 5$

24. $2x + y = -2$
$y = x + 4$

25. $x + y = 2$
$x = -3$

26. $x + y = 6$
$y = 2$

27. $x = -4$
$y = 6$

28. $x = 5$
$y = -1$

29. $x + y = 4$
$2x + 2y = -6$

30. $x - y = 3$
$2x - 2y = 6$

31. $4x - 2y = 8$
$2x - y = 4$

32. $3x - 6y = 6$
$x - 2y = 4$

33. As you probably have guessed by now, it can be difficult to solve a system of equations by graphing if the solution to the system contains a fraction. The solution to the following system is $\left(\frac{1}{2}, 1\right)$. Solve the system by graphing.

$$y = -2x + 2$$
$$y = 4x - 1$$

34. The solution to the following system is $\left(\frac{1}{3}, -2\right)$. Solve the system by graphing.

$$y = 3x - 3$$
$$y = -3x - 1$$

35. A second difficulty can arise in solving a system of equations by graphing if one or both of the equations is difficult to graph. The solution to the following system is (2, 1). Solve the system by graphing.

$$3x - 8y = -2$$
$$x - y = 1$$

36. The solution to the following system is $(-3, 2)$. Solve the system by graphing.

$$2x + 5y = 4$$
$$x - y = -5$$

37. Find a and b so that $(2, -3)$ is a solution to the system of equations

$$x - y = a$$
$$2x + 3y = b$$

38. Find c and d so that $(-4, 1)$ is a solution to the system of equations

$$3x + y = c$$
$$x - 5y = d$$

39. Consider the following system of equations:

$$y = 3x + 1$$
$$y = ax + b$$

a. Find the values for a and b so that the system has an infinite number of solutions.

b. For what values of a and b will the system be inconsistent?

c. For what values of a will the solution set to the system be a single ordered pair?

40. Figure 6 shows the graphs of both equations in the following linear system.

$$-31x + 61y = 183$$
$$-30x + 59y = -118$$

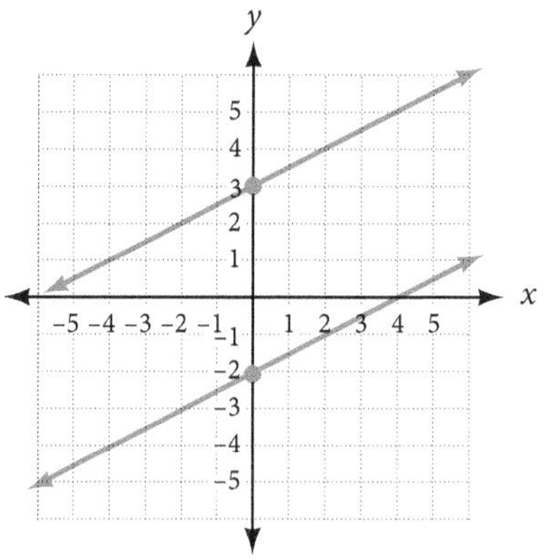

FIGURE 6

The two lines appear to be parallel. Explain how we know this system does have a solution.

Applying the Concepts

41. Job Comparison Jane is deciding between two sales positions. She can work for Marcy's and receive \$8.00 per hour, or she can work for Gigi's, where she earns \$6.00 per hour but also receives a \$50 commission per week. The two lines in the following figure represent the money Jane will make for working at each of the jobs.

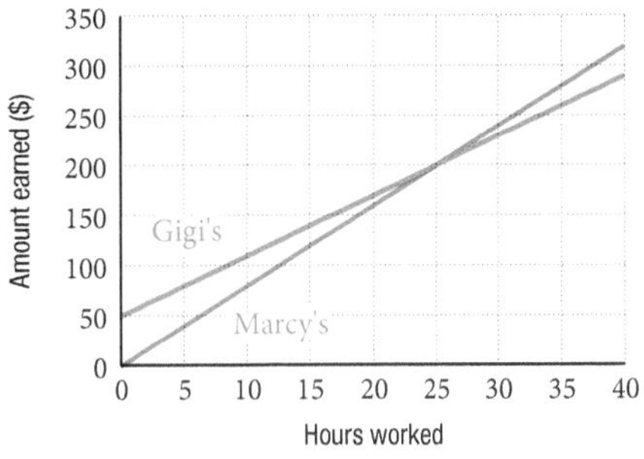

a. From the figure, how many hours would Jane have to work to earn the same amount at each of the positions?

b. If Jane expects to work less than 20 hours a week, which job should she choose?

c. If Jane expects to work more than 30 hours a week, which job should she choose?

42. Truck Rental You need to rent a moving truck for two days. Rider Moving Trucks charges \$50 per day and \$0.50 per mile. UMove Trucks charges \$45 per day and \$0.75 per mile. The following figure represents the cost of renting each of the trucks for two days.

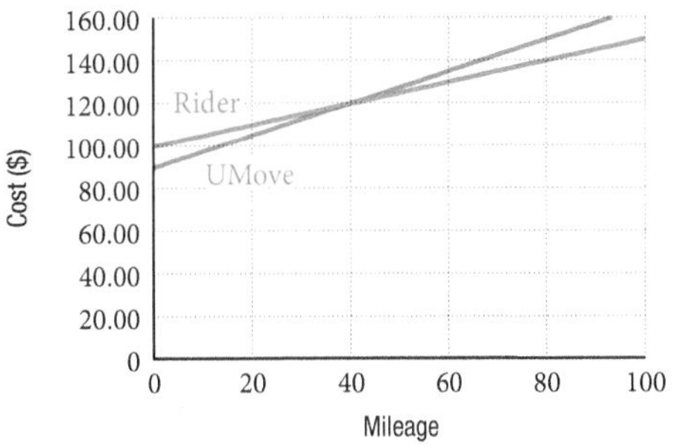

a. From the figure, after how many miles would the trucks cost the same?

b. Which company will give you a better deal if you drive less than 30 miles?

c. Which company will give you a better deal if you drive more than 60 miles?

Learning Objectives Assessment

The following problems can be used to help assess if you have successfully met the learning objectives for this section.

43. Which ordered pair is a solution to the following system?

$$x - y = 5$$

$$x + y = 3$$

a. $(4, -1)$ **b.** $(6, 1)$ **c.** $(2, -3)$ **d.** $(-4, 1)$

44. Which graph represents the solution to the following system of equations?

$$x - 2y = 4$$
$$3x + 2y = 6$$

a.

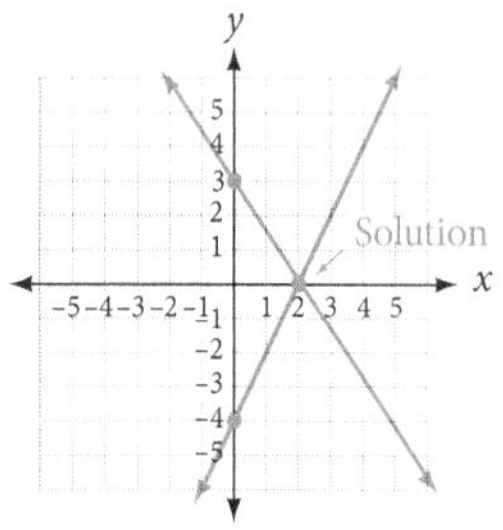

b.

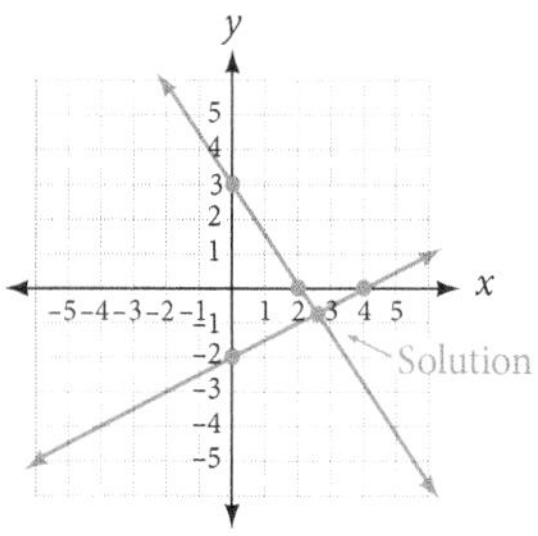

c.

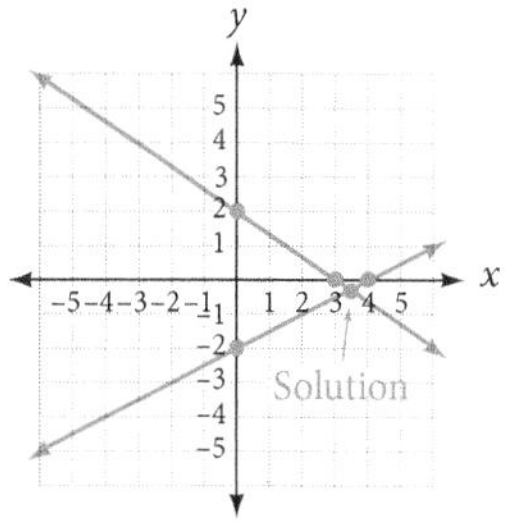

d.

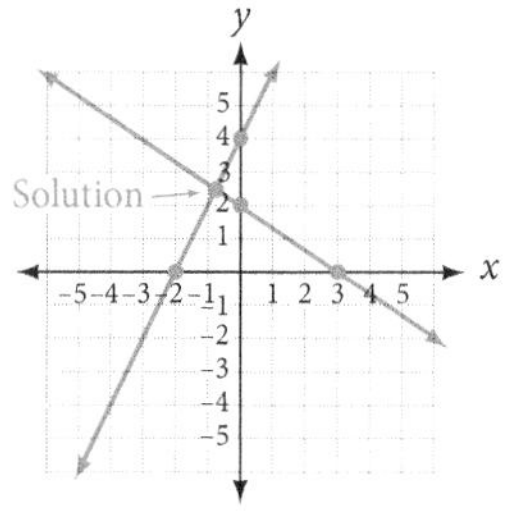

45. The graph of an inconsistent system of linear equations will be:

a. two lines that coincide

b. two distinct, parallel lines

c. two lines that intersect at a single point

d. two lines with one of the lines vertical

46. If the graph of a system of linear equations consists of two lines that coincide, then the solution set:

a. is all real numbers

b. is the empty set

c. contains an infinite number of ordered pairs

d. is undefined

Getting Ready For the Next Section

Simplify each of the following.

47. $(x + y) + (x - y)$

48. $(x + 2y) + (-x + y)$

49. $3(2x - y) + (x + 3y)$

50. $3(2x + 4y) - 2(3x + 5y)$

51. $-4(3x + 5y) + 5(5x + 4y)$

52. $(3x + 8y) - (3x - 2y)$

53. $6\left(\frac{1}{2}x - \frac{1}{3}y\right)$

54. $12\left(\frac{1}{4}x + \frac{2}{3}y\right)$

55. Let $x + y = 4$. If $x = 3$, find y.

56. Let $x + 2y = 4$. If $y = 3$, find x.

57. Let $x + 3y = 3$. If $x = 3$, find y.

58. Let $2x + 4y = -1$. If $y = \frac{1}{2}$, find x.

59. Let $3x + 5y = -7$. If $x = 6$, find y.

60. Let $3x - 2y = 12$. If $y = 6$, find x.

The Elimination Method 4.2

Learning Objectives

In this section, we will learn how to:

1. Solve a system of linear equations using the elimination method.

Introduction

The addition property of equality states that if equal quantities are added to both sides of an equation, the solution set is unchanged. In the past we have used this property to help solve equations in one variable. We will now use it to solve systems of linear equations. Here is another way to state the addition property of equality.

Let A, B, C, and D represent algebraic expressions.

$$\begin{array}{ll} \text{If} & A = B \\ \text{and} & C = D \\ \text{then} & A + C = B + D \end{array}$$

Because C and D are equal (that is, they represent the same number), what we have done is added the same amount to both sides of the equation $A = B$. Let's see how we can use this form of the addition property of equality to solve a system of linear equations.

VIDEO EXAMPLES

SECTION 4.2

EXAMPLE 1 Solve the following system.

$$\begin{aligned} x + y &= 4 \\ x - y &= 2 \end{aligned}$$

SOLUTION The system is in the form of the addition property of equality as described in this section. It looks like this:

$$\begin{aligned} A &= B \\ C &= D \end{aligned}$$

where A is $x + y$, B is 4, C is $x - y$, and D is 2.

Note When adding the left sides together and the right sides together vertically, notice that we are really just combining like terms.

We use the addition property of equality to add the left sides together and the right sides together, performing the addition in a vertical format.

$$\begin{array}{r} x + y = 4 \\ + \; x - y = 2 \\ \hline 2x + 0 = 6 \end{array}$$

We now solve the resulting equation for x.

$$\begin{aligned} 2x + 0 &= 6 \\ 2x &= 6 \\ x &= 3 \end{aligned}$$

The value we get for x is the value of the x-coordinate of the point of intersection of the two lines $x + y = 4$ and $x - y = 2$. To find the y-coordinate, we simply substitute $x = 3$ into either of the two original equations. Using the first equation, we get

$$\begin{aligned} 3 + y &= 4 \\ y &= 1 \end{aligned}$$

The solution to our system is the ordered pair (3, 1). It satisfies both equations.

When	$x = 3$	When	$x = 3$
and	$y = 1$	and	$y = 1$
the equation	$x + y = 4$	the equation	$x - y = 2$
becomes	$3 + 1 \stackrel{?}{=} 4$	becomes	$3 - 1 \stackrel{?}{=} 2$
	$4 = 4$		$2 = 2$

If we were to graph both lines, we would see that they intersect at the point (3, 1) as shown in Figure 1.

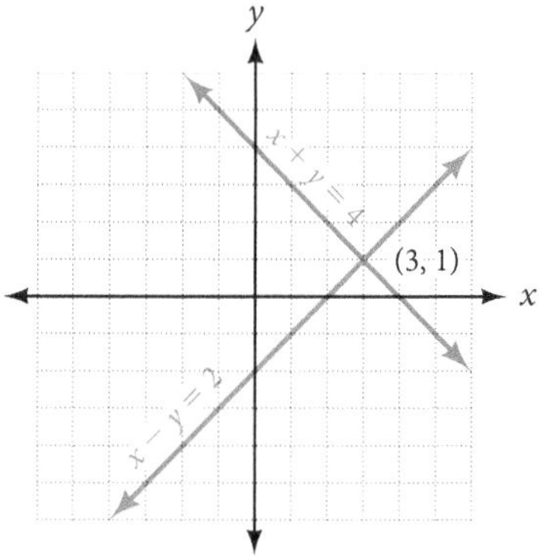

FIGURE 1

The most important part of this method of solving linear systems is eliminating one of the variables when we add the left and right sides together. In our first example, the equations were such that the y variable was eliminated when we added the left and right sides together. This is because the coefficients of the two terms containing y were opposites.

We will not always be able to eliminate a variable by adding the equations as given. We may have to multiply both sides of one (or both) equations by some value to ensure that one of the variables has coefficients that are opposites. The next examples show how this is done.

EXAMPLE 2 Solve the following system.

$$x + 2y = 4$$
$$x - y = -5$$

SOLUTION Notice that if we were to add the equations together as they are, the resulting equation would have terms in both x and y. Let's eliminate the variable x by multiplying both sides of the second equation by -1 before we add the equations together. (As you will see, we can choose to eliminate either the x or the y variable.) Multiplying both sides of the second equation by -1 will not change its solution, so we do not need to be concerned that we have altered the system.

$$x + 2y = 4 \xrightarrow{\text{No change}} x + 2y = 4$$
$$x - y = -5 \xrightarrow[\text{Multiply by } -1]{} -x + y = 5$$

$0 + 3y = 9$ Add left and right sides

$3y = 9$

$y = 3$ $\Big\{$ y-Coordinate of the point of intersection

Substituting $y = 3$ into either of the two original equations, we get $x = -2$. The solution to the system is $(-2, 3)$. It satisfies both equations.

EXAMPLE 3 Solve the following system.

$$2x - y = 6$$
$$x + 3y = 3$$

SOLUTION Let's eliminate the y variable from the two equations. We can do this by multiplying the first equation by 3 and leaving the second equation unchanged.

$$2x - y = 6 \xrightarrow{\text{3 times both sides}} 6x - 3y = 18$$
$$x + 3y = 3 \xrightarrow[\text{No change}]{} x + 3y = 3$$

The important thing about our system now is that the coefficients (the numbers in front) of the y variables are opposites. When we add the terms on each side of the equal sign, then the terms in y will add to zero and be eliminated.

$$\begin{aligned} 6x - 3y &= 18 \\ x + 3y &= 3 \\ \hline 7x \quad &= 21 \end{aligned} \qquad \text{Add corresponding terms}$$

This gives us $x = 3$. Using this value of x in the second equation of our original system, we have

$$\begin{aligned} 3 + 3y &= 3 \\ 3y &= 0 \\ y &= 0 \end{aligned}$$

The solution to our system is the ordered pair (3, 0).

EXAMPLE 4 Solve the system.

$$2x + 4y = -1$$
$$3x + 5y = -2$$

SOLUTION Let's eliminate x from the two equations. If we multiply the first equation by 3 and the second by -2, the coefficients of x will be 6 and -6, respectively. The x terms in the two equations will then add to zero.

$$2x + 4y = -1 \xrightarrow{\text{Multiply by 3}} 6x + 12y = -3$$
$$3x + 5y = -2 \xrightarrow[\text{Multiply by } -2]{} -6x - 10y = 4$$

Note If you are having trouble understanding this method of solution, it is probably because you can't see why we chose to multiply by 3 and -2 in the first step of Example 4. Look at the result of doing so: the $6x$ and $-6x$ will add to 0. We chose to multiply by 3 and -2 because they produce $6x$ and $-6x$, which will add to 0.

We now add the left and right sides of our new system together.

$$\begin{aligned} 6x + 12y &= -3 \\ -6x - 10y &= 4 \\ \hline 2y &= 1 \\ y &= \frac{1}{2} \end{aligned}$$

Substituting $y = \frac{1}{2}$ into the first equation in our original system, we have

$$\begin{aligned} 2x + 4\left(\frac{1}{2}\right) &= -1 \\ 2x + 2 &= -1 \\ 2x &= -3 \\ x &= -\frac{3}{2} \end{aligned}$$

The solution to our system is $\left(-\frac{3}{2}, \frac{1}{2}\right)$. It is the only ordered pair that satisfies both equations.

EXAMPLE 5 Solve the system.

$$\begin{aligned} 3x + 5y &= 2 \\ 5x + 4y &= 1 \end{aligned}$$

SOLUTION Let's eliminate y by multiplying the first equation by -4 and the second equation by 5.

$$\begin{array}{rcrl} 3x + 5y = 2 & \xrightarrow{\text{Multiply by } -4} & -12x - 20y &= -8 \\ 5x + 4y = 1 & \xrightarrow[\text{Multiply by } 5]{} & 25x + 20y &= 5 \\ \hline & & 13x \phantom{{}+20y} &= -3 \\ & & x &= -\frac{3}{13} \end{array}$$

If we substitute $x = -\frac{3}{13}$ into the first equation, we have

$$\begin{aligned} 3\left(-\frac{3}{13}\right) + 5y &= 2 \\ -\frac{9}{13} + 5y &= 2 \\ 5y &= 2 + \frac{9}{13} \\ 5y &= \frac{35}{13} \\ y &= \frac{7}{13} \end{aligned}$$

As an alternative to working with the fraction in the substitution step above, we can repeat the elimination process, but eliminate x instead of y. To do so, we can multiply the first equation by 5 and the second equation by -3.

$$\begin{array}{rcrl} 3x + 5y = 2 & \xrightarrow{\text{Multiply by } 5} & 15x + 25y &= 10 \\ 5x + 4y = 1 & \xrightarrow[\text{Multiply by } -3]{} & -15x - 12y &= -3 \\ \hline & & 13y &= 7 \\ & & y &= \frac{7}{13} \end{array}$$

Notice we get the same result for y. Using either approach, we obtain the solution $\left(-\frac{3}{13}, \frac{7}{13}\right)$.

EXAMPLE 6 Solve the system.

$$\frac{1}{2}x - \frac{1}{3}y = 2$$
$$\frac{1}{4}x + \frac{2}{3}y = 6$$

SOLUTION Although we could solve this system without clearing the equations of fractions, there is probably less chance for error if we have only integer coefficients to work with. So let's begin by multiplying both sides of the first equation by 6 and both sides of the second equation by 12, to clear each equation of fractions.

$$\frac{1}{2}x - \frac{1}{3}y = 2 \xrightarrow{\text{Multiply by 6}} 3x - 2y = 12$$
$$\frac{1}{4}x + \frac{2}{3}y = 6 \xrightarrow[\text{Multiply by 12}]{} 3x + 8y = 72$$

Now we can eliminate x by multiplying the top equation by -1 and leaving the bottom equation unchanged.

$$3x - 2y = 12 \xrightarrow{\text{Multiply by } -1} -3x + 2y = -12$$
$$3x + 8y = 72 \xrightarrow[\text{No change}]{} 3x + 8y = 72$$
$$10y = 60$$
$$y = 6$$

We can substitute $y = 6$ into any equation that contains both x and y. Let's use $3x - 2y = 12$.

$$3x - 2(6) = 12$$
$$3x - 12 = 12$$
$$3x = 24$$
$$x = 8$$

The solution to the system is (8, 6).

Our next two examples will show what happens when we apply the elimination method to a system of equations consisting of distinct parallel lines and to a system in which the lines coincide.

EXAMPLE 7 Solve the system.

$$2x - y = 2$$
$$4x - 2y = 12$$

SOLUTION Let us choose to eliminate y from the system. We can do this by multiplying the first equation by -2 and leaving the second equation unchanged.

$$2x - y = 2 \xrightarrow{\text{Multiply by } -2} -4x + 2y = -4$$
$$4x - 2y = 12 \xrightarrow[\text{No change}]{} 4x - 2y = 12$$

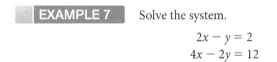

If we add both sides of the resulting system, we have

$$
\begin{aligned}
-4x + 2y &= -4 \\
4x - 2y &= 12 \\
\hline
0 + 0 &= 8 \\
0 &= 8 \quad \text{A false statement}
\end{aligned}
$$

Both variables have been eliminated and we end up with the false statement $0 = 8$. We cannot solve for x or y. The system of equations is equivalent to an equation that is always false and therefore has no solution. There are no ordered pairs that satisfy both equations, so the solution set is $\varnothing$.

The reason this happened is that we have tried to solve a system that consists of two parallel lines. Figure 2 is a visual representation of the situation and is conclusive evidence that there is no solution to our system.

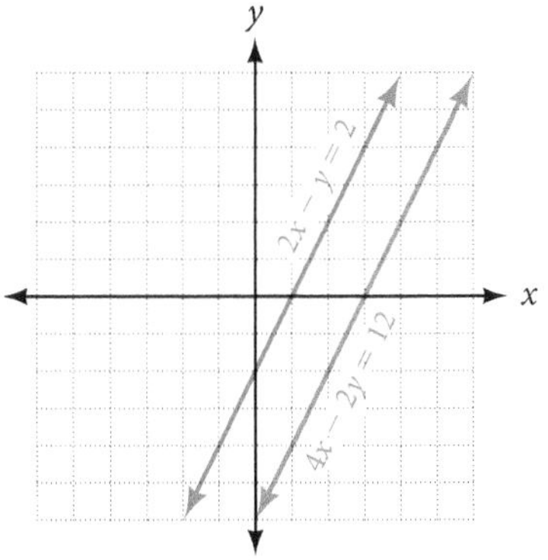

FIGURE 2

EXAMPLE 8 Solve the system.

$$
\begin{aligned}
4x - 3y &= 2 \\
8x - 6y &= 4
\end{aligned}
$$

SOLUTION Multiplying the top equation by -2 and adding, we can eliminate the variable x.

$$
\begin{aligned}
4x - 3y = 2 &\xrightarrow{\text{Multiply by } -2} & -8x + 6y &= -4 \\
8x - 6y = 4 &\xrightarrow[\text{No change}]{} & 8x - 6y &= 4 \\
& & \hline
& & 0 &= 0
\end{aligned}
$$

Both variables have been eliminated, and the resulting statement $0 = 0$ is true. The system of equations is equivalent to a statement that is always true, meaning that there is a solution corresponding to every real number. It does not imply that *any* ordered pair will satisfy both equations, but that there are an infinite number of ordered pairs that do.

In this case the lines coincide because the equations are equivalent. The solution set consists of all ordered pairs that satisfy either equation, which we can express as $\{(x, y) \mid 4x - 3y = 2\}$.

The preceding two examples illustrate the two special cases in which the graphs of the equations in the system either coincide or are parallel.

Here is a summary of our results from these two examples:

Both variables are eliminated and the resulting statement is false.	$\leftrightarrow$	The lines are parallel and there is no solution to the system.
Both variables are eliminated and the resulting statement is true.	$\leftrightarrow$	The lines coincide and there is an infinite number of solutions to the system.

The main idea in solving a system of linear equations by the elimination method is to use the multiplication property of equality on one or both of the original equations, if necessary, to make the coefficients of either variable opposites. The following box shows some steps to follow when solving a system of linear equations by the elimination method.

HOW TO *Solve a System of Linear Equations by the Elimination Method*

Step 1: Decide which variable to eliminate. (In some cases one variable will be easier to eliminate than the other. With some practice you will notice which one it is.)

Step 2: If necessary, use the multiplication property of equality on each equation separately to make the coefficients of the variable that is to be eliminated opposites.

Step 3: Add the respective left and right sides of the system together.

Step 4: Solve for the variable remaining.

Step 5: Substitute the value of the variable from step 4 into either of the original equations and solve for the other variable.

Step 6: Check your solution in both equations, if necessary.

Getting Ready for Class

After reading through the preceding section, respond in your own words and in complete sentences.

A. How is the addition property of equality used in the elimination method of solving a system of linear equations?

B. What happens when we use the elimination method to solve a system of linear equations consisting of two parallel lines?

C. What does it mean when we solve a system of linear equations by the elimination method and we end up with the statement $0 = 0$?

D. What is the first step in solving a system of linear equations that contains fractions?

Problem Set 4.2

Solve the following systems of linear equations by elimination.

1. $x + y = 3$
   $x - y = 1$
2. $x + y = -2$
   $x - y = 6$
3. $x + y = 10$
   $-x + y = 4$
4. $x - y = 1$
   $-x - y = -7$
5. $x - y = 7$
   $-x - y = 3$
6. $x - y = 4$
   $2x + y = 8$
7. $x + y = -1$
   $3x - y = -3$
8. $2x - y = -2$
   $-2x - y = 2$
9. $3x + 2y = 1$
   $-3x - 2y = -1$
10. $-2x - 4y = 1$
    $2x + 4y = -1$

Solve each of the following systems by eliminating the variable y.

11. $3x - y = 4$
    $2x + 2y = 24$
12. $2x + y = 3$
    $3x + 2y = 1$
13. $5x - 3y = -2$
    $10x - y = 1$
14. $4x - y = -1$
    $2x + 4y = 13$
15. $11x - 4y = 11$
    $5x + y = 5$
16. $3x - y = 7$
    $10x - 5y = 25$

Solve each of the following systems by eliminating the variable x.

17. $3x - 5y = 7$
    $-x + y = -1$
18. $4x + 2y = 32$
    $x + y = -2$
19. $-x - 8y = -1$
    $-2x + 4y = 13$
20. $-x + 10y = 1$
    $-5x + 15y = -9$
21. $-3x - y = 7$
    $6x + 7y = 11$
22. $-5x + 2y = -6$
    $10x + 7y = 34$

Solve each of the following systems of linear equations by the elimination method.

23. $6x - y = -8$
    $2x + y = -16$
24. $5x - 3y = -3$
    $3x + 3y = -21$
25. $x + 3y = 9$
    $2x - y = 4$
26. $x + 2y = 0$
    $2x - y = 0$
27. $x - 6y = 3$
    $4x + 3y = 21$
28. $8x + y = -1$
    $4x - 5y = 16$
29. $2x + 9y = 2$
    $5x + 3y = -8$
30. $5x + 2y = 11$
    $7x + 8y = 7$
31. $\frac{1}{3}x + \frac{1}{4}y = \frac{7}{6}$
    $\frac{3}{2}x - \frac{1}{3}y = \frac{7}{3}$
32. $\frac{7}{12}x - \frac{1}{2}y = \frac{1}{6}$
    $\frac{2}{5}x - \frac{1}{3}y = \frac{11}{15}$
33. $3x + 2y = -1$
    $6x + 4y = 0$
34. $8x - 2y = 2$
    $4x - y = 2$
35. $11x + 6y = 17$
    $5x - 4y = 1$
36. $3x - 8y = 7$
    $10x - 5y = 45$
37. $\frac{1}{2}x + \frac{1}{6}y = \frac{1}{3}$
    $-x - \frac{1}{3}y = -\frac{1}{6}$
38. $\frac{1}{3}x + \frac{1}{2}y = \frac{2}{3}$
    $-\frac{2}{3}x - y = -\frac{4}{3}$

39. Multiply both sides of the second equation in the following system by 100, and then solve as usual.

$$x + y = 22$$
$$0.05x + 0.10y = 1.70$$

40. Multiply both sides of the second equation in the following system by 100, and then solve as usual.

$$x + y = 15{,}000$$
$$0.06x + 0.07y = 980$$

Learning Objectives Assessment

The following problems can be used to help assess if you have successfully met the learning objectives for this section.

41. Solve the following system using the elimination method.

$$4x - y = -3$$
$$2x + 3y = -5$$

a. $(1, 7)$ **b.** $(0, 3)$ **c.** $(-1, -1)$ **d.** $\varnothing$

42. Solve the following system using the elimination method.

$$x - 2y = -4$$
$$-3x + 6y = 4$$

a. $\varnothing$ **b.** $(0, 2)$ **c.** $(2, 3)$ **d.** $\{(x, y) \mid x - 2y = -4\}$

Getting Ready for the Next Section

Solve. Round to the nearest hundredth when necessary.

43. $x + (2x - 1) = 2$

44. $2(2y + 7) - 3y = 12$

45. $2(3y - 1) - 3y = 4$

46. $-2x + 4(3x + 6) = 14$

47. $4x + 2(-2x + 4) = 8$

48. $1.5x + 15 = 0.75x + 24.95$

Solve each equation for the indicated variable.

49. $x - 3y = -1$ for x

50. $-3x + y = 6$ for y

51. Let $y = 2x - 1$. If $x = 1$, find y.

52. Let $y = 2x - 8$. If $x = 5$, find y.

53. Let $x = 3y - 1$. If $y = 2$, find x.

54. Let $y = 3x + 6$. If $y = -6$, find x.

Let $y = 1.5x + 15$.

55. If $x = 13$, find y.

56. If $x = 14$, find y.

Let $y = 0.75x + 24.95$.

57. If $x = 12$, find y.

58. If $x = 16$, find y.

SPOTLIGHT ON SUCCESS ***Student Instructor Penelope***

Never give up on something that you can't go a day without thinking about.
— Sir Winston Churchill

Since I was young, math has been a part of my life. Both my parents have Mathematics degrees, with one of them earning a doctorate in Math. Least to say, math was ingrained in my mind and a subject I had a knack for. As a child, I was so proud that I knew the square root of 144. Now this math is simple, but back then, I felt smart for knowing the answer. That excitement stayed alive through elementary school. Honestly, there were times when I became frustrated, such as when learning division and factoring for the first time, but I was still enthusiastic.

In middle school and high school, I lost my ability to enjoy math. With classes getting progressively more difficult and having to worry about the future, my focus veered away from wanting to learn more to forcing myself to master the material for the grades. The weight of having parents with math backgrounds became quite heavy. This created high standards and expectations from friends and family that turned into expectations that I placed on myself. I felt like I had to do well, to be more than proficient in math so that I would not let them down. It was not until I went to university that I found my motivation and inspiration again.

While working to attain a degree in Business Administration at Cal Poly, I decided to take the Calculus series. In a college setting, math was even more difficult to understand but I did not let that deter me. I persevered and found myself in the Proofs in Mathematics course, where I learned to truly appreciate math again. The class utilized a different way of thinking than I was used to. It was a challenging class; I struggled throughout most of it, but I was eager to learn the new material despite my grades not being what I desired. It was at that moment that I did not care how my grades ended up. Enjoying and understanding what I was learning was my main priority.

The Substitution Method 4.3

Learning Objectives

In this section, we will learn how to:

1. Solve a system of linear equations using the substitution method.

Introduction

There is a third method of solving systems of equations. It is the substitution method, and, like the elimination method, it can be used on any system of linear equations. Some systems, however, lend themselves more to the substitution method than others do.

VIDEO EXAMPLES

SECTION 4.3

EXAMPLE 1 Solve the following system.

$$x + y = 2$$
$$y = 2x - 1$$

SOLUTION If we were to solve this system by the methods used in the previous section, we would have to rearrange the terms of the second equation so that similar terms would be in the same column. There is no need to do this, however, because the second equation tells us that y is $2x - 1$. We can replace the variable y in the first equation with the expression $2x - 1$ from the second equation; that is, we substitute $2x - 1$ from the second equation for y in the first equation. Here is what it looks like:

$$x + (2x - 1) = 2$$

Note Sometimes this method of solving systems of equations is confusing the first time you see it. If you are confused, you may want to read through this first example more than once.

The equation we end up with contains only the variable x. The variable y has been eliminated by substitution.

Solving the resulting equation, we have

$$x + (2x - 1) = 2$$
$$3x - 1 = 2$$
$$3x = 3$$
$$x = 1$$

This is the x-coordinate of the solution to our system. To find the y-coordinate, we can substitute $x = 1$ into the second equation of our system.

$$y = 2(1) - 1$$
$$y = 2 - 1$$
$$y = 1$$

The solution to our system is the ordered pair (1, 1). It satisfies both of the original equations. Figure 1 provides visual evidence that the substitution method yields the correct solution.

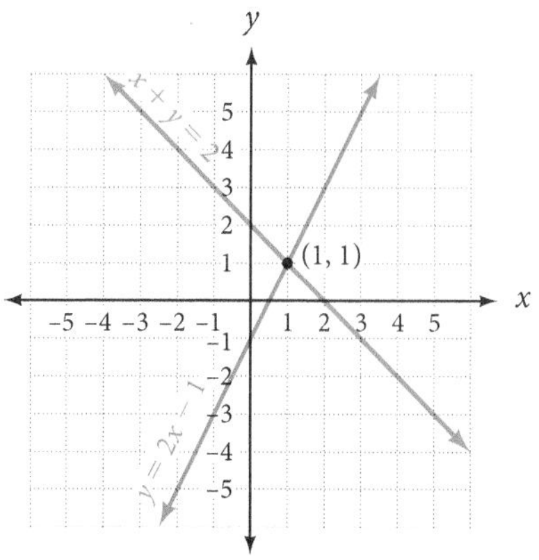

FIGURE 1

EXAMPLE 2 Solve the following system by the substitution method.

$$2x - 3y = 12$$
$$x = 2y + 7$$

SOLUTION This time, the second equation tells us x is $2y + 7$. We can replace the variable x in the first equation with $2y + 7$. Substituting $2y + 7$ from the second equation for x in the first equation, we have

$$2(2y + 7) - 3y = 12$$
$$4y + 14 - 3y = 12$$
$$y + 14 = 12$$
$$y = -2$$

To find the x-coordinate of our solution, we substitute $y = -2$ into the second equation in the original system.

When $y = -2$

the equation $x = 2y + 7$

becomes $x = 2(-2) + 7$

$x = -4 + 7 = 3$

The solution to our system is $(3, -2)$.

From the first two examples, we see that substitution works using either variable. In general, we usually choose the variable that is easiest to work with. In many cases, one of the variables will be easier to isolate than the other. The next two examples illustrate this situation.

EXAMPLE 3 Solve the following system by solving the first equation for x and then using the substitution method:

$$x - 3y = -1$$
$$2x - 3y = 4$$

SOLUTION Looking at both equations, we observe that the term containing x in the first equation is the only variable term with a coefficient of 1. Therefore, it will be easiest to isolate x in the first equation. We solve the first equation for x by adding $3y$ to both sides to get

$$x = 3y - 1$$

Substituting this expression for x in the second equation, we have

$$2(3y - 1) - 3y = 4$$
$$6y - 2 - 3y = 4$$
$$3y - 2 = 4$$
$$3y = 6$$
$$y = 2$$

Next, we find x.

When $\quad y = 2$

the equation $\quad x = 3y - 1$

becomes $\quad x = 3(2) - 1$

$$x = 6 - 1$$
$$x = 5$$

The solution to our system is (5, 2).

EXAMPLE 4 Solve by substitution.

$$-2x + 4y = 14$$
$$-3x + y = 6$$

SOLUTION We can solve either equation for either variable. If we look at the system closely, it becomes apparent that solving the second equation for y is the easiest way to go. If we add $3x$ to both sides of the second equation, we have

$$y = 3x + 6$$

Substituting the expression $3x + 6$ back into the first equation in place of y yields the following result.

$$-2x + 4(3x + 6) = 14$$
$$-2x + 12x + 24 = 14$$
$$10x + 24 = 14$$
$$10x = -10$$
$$x = -1$$

Substituting $x = -1$ into the equation $y = 3x + 6$ leaves us with

$$y = 3(-1) + 6$$
$$y = -3 + 6$$
$$y = 3$$

The solution to our system is $(-1, 3)$.

Here are the steps to use in solving a system of equations by the substitution method.

HOW TO ***Solving a System of Linear Equations by the Substitution Method***

Step 1: Solve either of the equations for x or y. (This step is not necessary if one of the equations is already in the correct form, as in Examples 1 and 2.)

Step 2: Substitute the expression for the variable obtained in step 1 into the other equation and solve it for the remaining variable.

Step 3: Substitute the value from step 2 into the equation obtained in step 1 to find the value of the other variable.

Step 4: Check your results, if necessary.

EXAMPLE 5 Solve by substitution.

$$4x + 2y = 8$$
$$y = -2x + 4$$

SOLUTION Substituting the expression $-2x + 4$ for y from the second equation into the first equation, we have

$$4x + 2(-2x + 4) = 8$$
$$4x - 4x + 8 = 8$$
$$8 = 8 \quad \text{A true statement}$$

Both variables have been eliminated, and we are left with a true statement. Recall from the last section that a true statement in this situation tells us the lines coincide; that is, the equations $4x + 2y = 8$ and $y = -2x + 4$ have exactly the same graph. Any point on that graph has coordinates that satisfy both equations and is a solution to the system. We can write the solution set as $\{(x, y) \mid 4x + 2y = 8\}$.

EXAMPLE 6 The following table shows two monthly contract rates charged by GTE Wireless for cellular phone use. At how many minutes will the two rates cost the same amount?

	Flat Rate	Plus	Per Minute Charge
Plan 1	\$15		\$1.50
Plan 2	\$24.95		\$0.75

SOLUTION If we let y = the monthly charge for x minutes of phone use, then the equations for each plan are

Plan 1: $y = 1.5x + 15$

Plan 2: $y = 0.75x + 24.95$

We can solve this system by substitution by replacing the variable y in Plan 2 with the expression $1.5x + 15$ from Plan 1. If we do so, we have

$$1.5x + 15 = 0.75x + 24.95$$
$$0.75x + 15 = 24.95$$
$$0.75x = 9.95$$
$$x = 13.27 \quad \text{to the nearest hundredth}$$

The monthly bill is based on the number of minutes you use the phone, with any fraction of a minute moving you up to the next minute. If you talk for a total of 13 minutes, you are billed for 13 minutes. If you talk for 13 minutes, 10 seconds, you are billed for 14 minutes. The number of minutes on your bill will always be a whole number. So, to calculate the cost for talking 13.27 minutes, we would replace x with 14 and find y. Let's compare the two plans at $x = 13$ minutes and at $x = 14$ minutes.

Plan 1:	$y = 1.5x + 15$	Plan 2:	$y = 0.75x + 24.95$
When	$x = 13, y = \$34.50$	When	$x = 13, y = \$34.70$
When	$x = 14, y = \$36.00$	When	$x = 14, y = \$35.45$

The two plans will never give the same cost for talking x minutes. If you talk 13 or fewer minutes, Plan 1 will cost less. If you talk for more than 13 minutes, you will be billed for 14 minutes, and Plan 2 will cost less than Plan 1.

Getting Ready for Class

After reading through the preceding section, respond in your own words and in complete sentences.

A. What is the first step in solving a system of linear equations by substitution?

B. When would substitution be more efficient than the elimination method in solving two linear equations?

C. What does it mean when we solve a system of linear equations by the substitution method and we end up with the statement $8 = 8$?

D. How would you begin solving the following system using the substitution method?

$$x + y = 2$$
$$y = 2x - 1$$

Problem Set 4.3

Solve the following systems by substitution. Substitute the expression in the second equation into the first equation and solve.

1. $x + y = 11$
$y = 2x - 1$

2. $x - y = -3$
$y = 3x + 5$

3. $x + y = 20$
$y = 5x + 2$

4. $3x - y = -1$
$x = 2y - 7$

5. $-2x + y = -1$
$y = -4x + 8$

6. $4x - y = 5$
$y = -4x + 1$

7. $3x - 2y = -2$
$x = -y + 6$

8. $2x - 3y = 17$
$x = -y + 6$

9. $5x - 4y = -16$
$y = 4$

10. $6x + 2y = 18$
$x = 3$

11. $5x + 4y = 7$
$y = -3x$

12. $10x + 2y = -6$
$y = -5x$

Solve the following systems by solving one of the equations for x or y and then using the substitution method.

13. $x + 3y = 4$
$x - 2y = -1$

14. $x - y = 5$
$x + 2y = -1$

15. $2x + y = 1$
$x - 5y = 17$

16. $2x - 2y = 2$
$x - 3y = -7$

17. $3x + 5y = -3$
$x - 5y = -5$

18. $2x - 4y = -4$
$x + 2y = 8$

19. $5x + 3y = 0$
$x - 3y = -18$

20. $x - 3y = -5$
$x - 2y = 0$

21. $-3x - 9y = 7$
$x + 3y = 12$

22. $2x + 6y = -18$
$x + 3y = -9$

Solve each system by substitution. You can eliminate the decimals or fractions if you like, but you don't have to. The solution will be the same in either case.

23. $0.05x + 0.10y = 1.70$
$x + y = 22$

24. $0.20x + 0.50y = 3.60$
$x + y = 12$

25. $\frac{1}{4}x + \frac{1}{3}y = -\frac{1}{2}$
$x - y = 2$

26. $\frac{1}{2}x + \frac{3}{5}y = \frac{7}{10}$
$x - y = 3$

Solve the following systems using either the substitution method or the elimination method. Choose the method you feel is best suited for the problem.

27. $5x - 8y = 7$
$2x - y = 5$

28. $7x - 6y = -1$
$x - 2y = -1$

29. $5x - 6y = -4$
$x = y$

30. $2x - 4y = 0$
$y = x$

31. $4x + 2y = 3$
$3x - 12y = -9$

32. $3x + 4y = 10$
$16x - 2y = 30$

33. $-3x + 2y = 6$
$-3x + y = 0$

34. $-2x - y = -3$
$3x + y = 0$

35. $y = -x + 3$
$y = 2x - 12$

36. $y = 2x + 1$
$y = -2x + 1$

37. $7x - 11y = 16$
$y = 10$

38. $9x - 7y = -14$
$x = 7$

39. $-4x + 4y = -8$
$x - y = 2$

40. $-4x + 2y = -10$
$2x - y = 5$

41. $3x + 7y = 2$
$4x + 2y = -1$

42. $2x - 5y = 3$
$3x + 4y = -2$

Applying the Concepts

43. Gas Mileage Daniel is trying to decide whether to buy a car or a truck. The truck he is considering will cost him \$250 a month in loan payments, and it gets 20 miles per gallon in gas mileage. The car will cost \$340 a month in loan payments, but it gets 35 miles per gallon in gas mileage. Daniel estimates that he will pay \$4.20 per gallon for gas. This means that the monthly cost to drive the truck x miles will be $y = \frac{4.20}{20}x + 250$. The total monthly cost to drive the car x miles will be $y = \frac{4.20}{35}x + 340$. The following figure shows the graph of each equation:

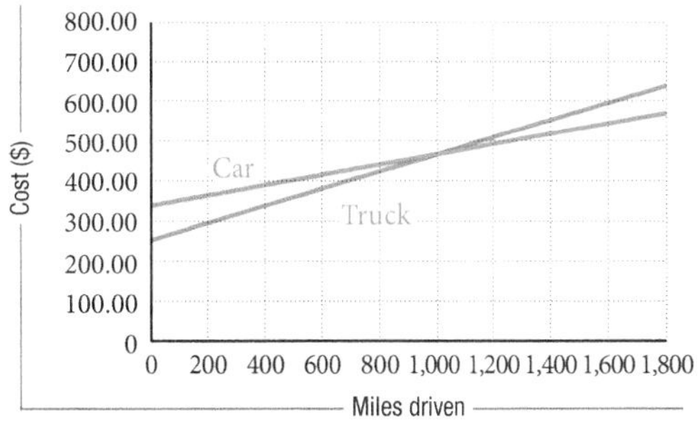

a. At how many miles do the car and the truck cost the same to operate?

b. If Daniel drives more than 1,200 miles, which will be cheaper?

c. If Daniel drives fewer than 800 miles, which will be cheaper?

d. Why do the graphs appear in the first quadrant only?

44. Video Production Pat runs a small company that duplicates videotapes. The daily cost and daily revenue for a company duplicating videos are shown in the following figure. The daily cost for duplicating x videos is $y = \frac{6}{5}x + 20$; the daily revenue (the amount of money he brings in each day) for duplicating x videos is $y = 1.7x$. The graphs of the two lines are shown in the following figure:

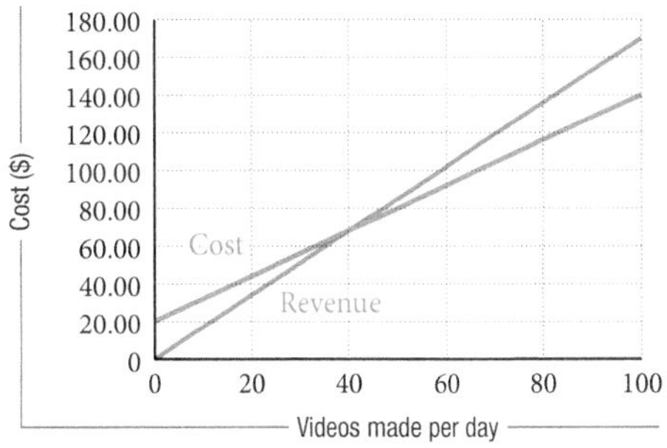

a. Pat will "break even" when his cost and his revenue are equal. How many videos does he need to duplicate to break even?

b. Pat will incur a loss when his revenue is less than his cost. If he duplicates 30 videos in one day, will he incur a loss?

c. Pat will make a profit when his revenue is larger than his costs. For what values of x will Pat make a profit?

d. Why do the graphs appear in the first quadrant only?

Learning Objectives Assessment

The following problems can be used to help assess if you have successfully met the learning objectives for this section.

45. Which substitution could be used to solve the following system?

$$2x + 5y = 4$$
$$x - 4y = -2$$

a. $x = -5y + 2$ **b.** $x = 4y - 2$ **c.** $y = x + 2$ **d.** $y = -x + \frac{1}{2}$

46. Solve the following system using the substitution method.

$$2x + y = -1$$
$$x - 2y = 7$$

a. $(-1, 1)$ **b.** $(2, -5)$ **c.** $(1, -3)$ **d.** $\varnothing$

Getting Ready for the Next Section

47. One number is eight more than five times another; their sum is 26. Find the numbers.

48. One number is three less than four times another; their sum is 27. Find the numbers.

49. The difference of two positive numbers is nine. The larger number is six less than twice the smaller number. Find the numbers.

50. The difference of two positive numbers is 17. The larger number is one more than twice the smaller number. Find the numbers.

51. The length of a rectangle is five inches more than three times the width. The perimeter is 58 inches. Find the length and width.

52. The length of a rectangle is three inches less than twice the width. The perimeter is 36 inches. Find the length and width.

53. John has \$1.70 in nickels and dimes in his pocket. He has four more nickels than he does dimes. How many of each does he have?

54. Jamie has \$2.65 in dimes and quarters in her pocket. She has two more dimes than she does quarters. How many of each does she have?

Applications 4.4

Learning Objectives

In this section, we will learn how to:

1. Solve number problems using a system of equations.
2. Solve interest problems using a system of equations.
3. Solve coin problems using a system of equations.
4. Solve mixture problems using a system of equations.

Introduction

I often have heard students remark about the word problems in beginning algebra: "What does this have to do with real life?" Most of the word problems we will encounter don't have much to do with "real life." We are actually just practicing. Ultimately, all problems requiring the use of algebra are word problems; that is, they are stated in words first, then translated to symbols. The problem then is solved by some system of mathematics, like algebra. Most real applications involve calculus or higher levels of mathematics. So, if the problems we solve are upsetting or frustrating to you, then you are probably taking them too seriously.

The word problems in this section have two unknown quantities. We will write two equations in two variables (each of which represents one of the unknown quantities), which of course is a system of equations. We then solve the system by one of the methods developed in the previous sections of this chapter. Here are the steps to follow in solving these word problems.

Note If this Blueprint for Problem Solving looks familiar to you, it should. We introduced it in Section 2.6 when we solved word problems with one variable. In that chapter, all of our problems involved one equation in one variable. Now we are solving problems using a system of equations (two equations in two variables). As you work through the examples in this section, you will see some word problems that are almost identical to ones we solved earlier. The difference is the method we use to solve them.

HOW TO *Use the Blueprint for Problem Solving For a System of Equations*

Step 1: *Read* the problem, and then mentally *list* the items that are known and the items that are unknown.

Step 2: ***Assign variables*** to each of the unknown items; that is, let x be one of the unknown items and y be the other unknown item. Then ***translate*** the other ***information*** in the problem to expressions involving the two variables.

Step 3: *Reread* the problem, and then ***write a system of equations,*** using the items and variables listed in steps 1 and 2, that describes the situation.

Step 4: ***Solve the system*** found in step 3.

Step 5: *Write* your ***answers*** using complete sentences.

Step 6: *Reread* the problem, and ***check*** your solution with the original words in the problem.

Remember, the more problems you work, the more problems you will be able to work. If you have trouble getting started on the problem set, come back to the examples and work through them yourself. The examples are similar to the problems found in the problem set.

Number Problem

VIDEO EXAMPLES

SECTION 4.4

EXAMPLE 1 One number is 2 more than 5 times another number. Their sum is 20. Find the two numbers.

SOLUTION Applying the steps in our blueprint, we have

Step 1: We know that the two numbers have a sum of 20 and that one of them is 2 more than 5 times the other. We don't know what the numbers themselves are.

Step 2: Let x represent one of the numbers and y represent the other. "One number is 2 more than 5 times another" translates to

$$y = 5x + 2$$

"Their sum is 20" translates to

$$x + y = 20$$

Step 3: The system that describes the situation must be

$$\begin{aligned} x + y &= 20 \\ y &= 5x + 2 \end{aligned}$$

Step 4: We can solve this system by substituting the expression $5x + 2$ in the second equation for y in the first equation:

$$\begin{aligned} x + (5x + 2) &= 20 \\ 6x + 2 &= 20 \\ 6x &= 18 \\ x &= 3 \end{aligned}$$

Using $x = 3$ in either of the first two equations and then solving for y, we get $y = 17$.

Step 5: So 17 and 3 are the numbers we are looking for.

Step 6: The number 17 is 2 more than 5 times 3, and the sum of 17 and 3 is 20.

Interest Problem

EXAMPLE 2 Mr. Hicks had \$15,000 to invest. He invested some at 6% and the rest at 7%. If he earns \$980 in interest, how much did he invest at each rate?

SOLUTION Remember, step 1 is done mentally.

Step 1: We do not know the specific amounts invested in the two accounts. We do know that their sum is \$15,000 and that the interest rates on the two accounts are 6% and 7%. We also know that the total interest earned is \$980.

Step 2: Let x be the amount invested at 6% and y be the amount invested at 7%. Because Mr. Hicks invested a total of \$15,000, we have

$$x + y = 15{,}000$$

The interest he earns comes from 6% of the amount invested at 6% and 7% of the amount invested at 7%. To find 6% of x, we multiply x by 0.06, which gives us $0.06x$. To find 7% of y, we multiply 0.07 times y and get $0.07y$.

$$\begin{matrix} \text{Interest} \\ \text{at 6\%} \end{matrix} + \begin{matrix} \text{interest} \\ \text{at 7\%} \end{matrix} = \begin{matrix} \text{total} \\ \text{interest} \end{matrix}$$

$$0.06x + 0.07y = 980$$

Step 3: The system is

$$\begin{aligned} x + y &= 15{,}000 \\ 0.06x + 0.07y &= 980 \end{aligned}$$

Step 4: We multiply the first equation by -6 and the second by 100 to eliminate x:

$$\begin{array}{rcr} x + \quad y = 15{,}000 & \xrightarrow{\text{Multiply by } -6} & -6x - 6y = -90{,}000 \\ 0.06x + 0.07y = 980 & \xrightarrow[\text{Multiply by } 100]{} & 6x + 7y = 98{,}000 \\ \hline & & y = 8{,}000 \end{array}$$

Substituting $y = 8{,}000$ into the first equation and solving for x, we get $x = 7{,}000$.

Step 5: He invested \$7,000 at 6% and \$8,000 at 7%.

Step 6: Checking our solutions in the original problem, we have: The sum of \$7,000 and \$8,000 is \$15,000, the total amount he invested. To complete our check, we find the total interest earned from the two accounts:

The interest on \$7,000 at 6% is $0.06(7{,}000) = \$420$
The interest on \$8,000 at 7% is $0.07(8{,}000) = \$560$
The total interest is \$980

Coin Problem

EXAMPLE 3 John has \$1.70 all in dimes and nickels. He has a total of 22 coins. How many of each kind does he have?

SOLUTION

Step 1: We know that John has 22 coins that are dimes and nickels. We know that a dime is worth 10 cents and a nickel is worth 5 cents. We do not know the specific number of dimes and nickels he has.

Step 2: Let x be the number of nickels and y be the number of dimes. The total number of coins is 22, so

$$x + y = 22$$

The total amount of money he has is \$1.70, which comes from nickels and dimes:

$$\begin{array}{ccccc} \text{Amount of money in nickels} & + & \text{amount of money in dimes} & = & \text{total amount of money} \\ 0.05x & + & 0.10y & = & 1.70 \end{array}$$

Step 3: The system that represents the situation is

$$\begin{aligned} x + y &= 22 && \text{The number of coins} \\ 0.05x + 0.10y &= 1.70 && \text{The value of the coins} \end{aligned}$$

Step 4: We multiply the first equation by -5 and the second by 100 to eliminate the variable x:

$$\begin{array}{rcr} x + \quad y = \quad 22 & \xrightarrow{\text{Multiply by } -5} & -5x - 5y = -110 \\ 0.05x + 0.10y = 1.70 & \xrightarrow[\text{Multiply by } 100]{} & 5x + 10y = 170 \\ \hline & & 5y = 60 \\ & & y = 12 \end{array}$$

Substituting $y = 12$ into our first equation, we get $x = 10$.

Step 5: John has 12 dimes and 10 nickels.
Step 6: Twelve dimes and 10 nickels total 22 coins.

$$\begin{aligned} \text{12 dimes are worth } 12(0.10) &= 1.20 \\ \underline{\text{10 nickels are worth } 10(0.05)} &= \underline{0.50} \\ \text{The total value is } & \ \ \$1.70 \end{aligned}$$

Mixture Problem

EXAMPLE 4 How much of a 20% alcohol solution and 50% alcohol solution must be mixed to get 12 gallons of 30% alcohol solution?

SOLUTION To solve this problem we must first understand that a 20% alcohol solution is 20% alcohol and 80% water.

Step 1: We know there are two solutions that together must total 12 gallons. 20% of one of the solutions is alcohol and the rest is water, whereas the other solution is 50% alcohol and 50% water. We do not know how many gallons of each individual solution we need.

Step 2: Let x be the number of gallons of 20% alcohol solution needed and y be the number of gallons of 50% alcohol solution needed. Because the total number of gallons we will end up with is 12, and this 12 gallons must come from the two solutions we are mixing, our first equation is

$$x + y = 12$$

To obtain our second equation, we look at the amount of alcohol in our two original solutions and our final solution. The amount of alcohol in the x gallons of 20% solution is $0.20x$, and the amount of alcohol in y gallons of 50% solution is $0.50y$. The amount of alcohol in the 12 gallons of 30% solution is 0.30(12). Because the amount of alcohol we start with must equal the amount of alcohol we end up with, our second equation is

$$0.20x + 0.50y = 0.30(12)$$

The information we have so far can also be summarized with a table. Sometimes by looking at a table like the one that follows it is easier to see where the equations come from.

	20% Solution	50% Solution	Final Solution
Number of Gallons	x	y	12
Gallons of Alcohol	$0.20x$	$0.50y$	0.30(12)

Step 3: Our system of equations is

$$\begin{aligned} x + y &= 12 \\ 0.20x + 0.50y &= 0.30(12) \end{aligned}$$

Step 4: We can solve this system by substitution. Solving the first equation for y and substituting the result into the second equation, we have

$$0.20x + 0.50(12 - x) = 0.30(12)$$

Multiplying each side by 10 gives us an equivalent equation that is a little easier to work with.

$$\begin{aligned} 2x + 5(12 - x) &= 3(12) \\ 2x + 60 - 5x &= 36 \\ -3x + 60 &= 36 \\ -3x &= -24 \\ x &= 8 \end{aligned}$$

If x is 8, then y must be 4 because $x + y = 12$.

Step 5: It takes 8 gallons of 20% alcohol solution and 4 gallons of 50% alcohol solution to produce 12 gallons of 30% alcohol solution.

Step 6: Try it and see.

Getting Ready for Class

After reading through the preceding section, respond in your own words and in complete sentences.

A. If you were to apply the Blueprint for Problem Solving from Section 2.6 to the examples in this section, what would be the first step?

B. If you were to apply the Blueprint for Problem Solving from Section 2.6 to the examples in this section, what would be the last step?

C. Which method of solving systems of equations do you prefer? Why?

D. Write an application problem for which the solution depends on solving a system of equations.

Problem Set 4.4

Solve the following word problems. Be sure to show the equations used.

Number Problems

1. Two numbers have a sum of 25. One number is 5 more than the other. Find the numbers.
2. The difference of two numbers is 6. Their sum is 30. Find the two numbers.
3. The sum of two numbers is 15. One number is 4 times the other. Find the numbers.
4. The difference of two positive numbers is 28. One number is 3 times the other. Find the two numbers.
5. Two positive numbers have a difference of 5. The larger number is one more than twice the smaller. Find the two numbers.
6. One number is 2 more than 3 times another. Their sum is 26. Find the two numbers.
7. One number is 5 more than 4 times another. Their sum is 35. Find the two numbers.
8. The difference of two positive numbers is 8. The larger is twice the smaller decreased by 7. Find the two numbers.

Interest Problems

9. Mr. Wilson invested money in two accounts. His total investment was \$20,000. If one account pays 6% in interest and the other pays 8% in interest, how much does he have in each account if he earned a total of \$1,380 in interest in 1 year?
10. A total of \$11,000 was invested. Part of the \$11,000 was invested at 4%, and the rest was invested at 7%. If the investments earn \$680 per year, how much was invested at each rate?
11. A woman invested 4 times as much at 5% as she did at 6%. The total amount of interest she earns in 1 year from both accounts is \$520. How much did she invest at each rate?
12. Ms. Hagan invested twice as much money in an account that pays 7% interest as she did in an account that pays 6% in interest. Her total investment pays her \$1,000 a year in interest. How much did she invest at each rate?

Coin Problems

13. Ron has 14 coins with a total value of \$2.30. The coins are nickels and quarters. How many of each coin does he have?
14. Diane has \$0.95 in dimes and nickels. She has a total of 11 coins. How many of each kind does she have?
15. Suppose Tom has 21 coins totaling \$3.45. If he has only dimes and quarters, how many of each type does he have?
16. A coin collector has 31 dimes and nickels with a total face value of \$2.40. How many of each coin does she have?

Mixture Problems

17. How many liters of 50% alcohol solution and 20% alcohol solution must be mixed to obtain 18 liters of 30% alcohol solution?

	50% Solution	20% Solution	Final Solution
Number of Liters	x	y	18
Liters of Alcohol			

18. How many liters of 10% alcohol solution and 5% alcohol solution must be mixed to obtain 40 liters of 8% alcohol solution?

	10% Solution	5% Solution	Final Solution
Number of Liters	x	y	40
Liters of Alcohol			

19. A mixture of 8% disinfectant solution is to be made from 10% and 7% disinfectant solutions. How much of each solution should be used if 30 gallons of 8% solution are needed?

20. How much 50% antifreeze solution and 40% antifreeze solution should be combined to give 50 gallons of 46% antifreeze solution?

Miscellaneous Problems

21. For a Saturday matinee, adult tickets cost \$5.50 and kids under 12 pay only \$4.00. If 70 tickets are sold for a total of \$310, how many of the tickets were adult tickets and how many were sold to kids under 12?

22. The Bishop's Peak 4-H club is having its annual fundraising dinner. Adults pay \$15 apiece and children pay \$10 apiece. If the number of adult tickets sold is twice the number of children's tickets sold, and the total income for the dinner is \$1,600, how many of each kind of ticket did the 4-H club sell?

23. A farmer has 96 feet of fence with which to make a corral. If he arranges it into a rectangle that is twice as long as it is wide, what are the dimensions?

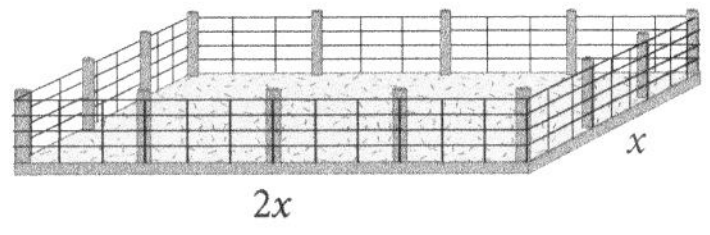

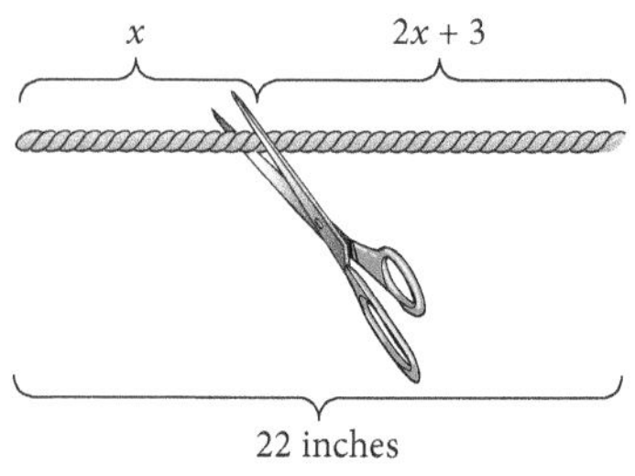

24. If a 22-inch rope is to be cut into two pieces so that one piece is 3 inches longer than twice the other, how long is each piece?

25. A gambler finishes a session of blackjack with \$5 chips and \$25 chips. If he has 45 chips in all, with a total value of \$465, how many of each kind of chip does the gambler have?

26. Tyler has been saving his winning lottery tickets. He has 23 tickets that are worth a total of \$175. If each ticket is worth either \$5 or \$10, how many of each does he have?

27. Mary Jo spends \$2,550 to buy stock in two companies. She pays \$11 a share to one of the companies and \$20 a share to the other. If she ends up with a total of 150 shares, how many shares did she buy at \$11 a share and how many did she buy at \$20 a share?

28. Kelly sells 62 shares of stock she owns for a total of \$433. If the stock was in two different companies, one selling at \$6.50 a share and the other at \$7.25 a share, how many of each did she sell?

Learning Objectives Assessment

The following problems can be used to help assess if you have successfully met the learning objectives for this section.

For Problems 49-52, determine which system of equations is an appropriate model for solving the given problem.

29. Two numbers have a sum of 49 and a difference of 7. Find the numbers.

a. $x + y = 49$
$x - y = 7$

b. $xy = 49$
$y - x = 7$

c. $x + y = 49$
$y = 7 - x$

d. $x + y = 7$
$x - y = 49$

30. A total of \$8,000 was invested in two accounts, one earning 3% interest and the other 5% interest. If the investments earn a total of \$352 in interest after one year, find the amount invested at each rate.

a. $x + y = 8{,}000$
$0.03x + 0.05y = 352$

b. $x + y = 352$
$0.03x + 0.05y = 8{,}000$

c. $x + y = 8{,}000$
$3x + 5y = 352$

d. $x + y = 352$
$3x + 5y = 8{,}000$

31. Valerie has \$2.50 in nickels and quarters. She has a total of 14 coins. How many of each coin does she have?

a. $x + y = 250$
$25x + 5y = 14$

b. $x + y = 2.50$
$0.25x + 0.05y = 14$

c. $x + y = 14$
$0.25x + 0.05y = 2.50$

d. $x + y = 14$
$25x + 5y = 2.50$

32. How many liters of 40% alcohol solution and 75% alcohol solution must be mixed to obtain 20 liters of 50% alcohol solution?

a. $x + y = 20$
$0.40x + 0.75y = 0.50$

b. $x + y = 20$
$0.40x + 0.75y = 0.50(20)$

c. $x + y = 50$
$40x + 75y = 20$

d. $x + y = 20$
$40x + 75y = 50$

Maintaining Your Skills

Simplify.

33. $6(3 + 4) + 5$

34. $[(1 + 2)(2 + 3)] + (4 \div 2)$

35. $1^2 + 2^2 + 3^2$

36. $(1 + 2 + 3)^2$

37. $5(6 + 3 \cdot 2) + 4 + 3 \cdot 2$

38. $(1 + 2)^3 + [(2 \cdot 3) + (4 \cdot 5)]$

39. $(1^3 + 2^3) + [(2 \cdot 3) + (4 \cdot 5)]$

40. $[2(3 + 4 + 5)] \div 3$

41. $(2 \cdot 3 + 4 + 5) \div 3$

42. $10^4 + 10^3 + 10^2 + 10^1$

43. $6 \cdot 10^3 + 5 \cdot 10^2 + 4 \cdot 10^1$

44. $5 \cdot 10^3 + 2 \cdot 10^2 + 8 \cdot 10^1$

45. $1 \cdot 10^3 + 7 \cdot 10^2 + 6 \cdot 10^1$

46. $4(2 - 1) + 5(3 - 2)$

47. $4 \cdot 2 - 1 + 5 \cdot 3 - 2$

48. $2^3 + 3^2 \cdot 4 - 5$

49. $(2^3 + 3^2) \cdot 4 - 5$

50. $4^2 - 2^4 + (2 \cdot 2)^2$

51. $2(2^2 + 3^2) + 3(3^2)$

52. $2 \cdot 2^2 + 3^2 + 3 \cdot 3^2$

SPOTLIGHT ON SUCCESS *Student Instructor Stephanie*

For success, attitude is equally as important as ability.
—Harry F. Banks

Math has always fascinated me. From addition to calculus, I've taken great interest in the material and great pride in my work. Whenever I struggled with concepts, I asked questions and worked problems over and over until they became second nature. I used to assume this was how everyone dealt with concepts they didn't understand. However, in high school, I noticed how easily students got discouraged with mathematics. In my senior year calculus and statistics classes, I was surrounded by bright students who simply gave up on trying to fully understand the material because it seemed confusing or difficult. Even if we shared a similar level of academic ability, the difference between these students' grades and my own reflected a difference in attitude. I noticed many students giving up without really trying to understand the concepts because they lacked confidence and didn't feel they were capable. They began coming to me for help. Though I was glad to help them with the math, I had a greater goal to help them believe they could succeed on their own. Soon the students I tutored gained more understanding and achieved success by simply paying more attention in class and working extra problems outside of class. It was amazing how much improvement I saw in both their confidence levels and their grades. It goes to show that a little extra effort and a positive attitude can truly make a difference.

Chapter 4 Summary

EXAMPLES

1. The solution to the system

$$x + 2y = 4$$
$$x - y = 1$$

is the ordered pair (2, 1). It is the only ordered pair that satisfies both equations.

Definitions [4.1]

1. A system of linear equations, as the term is used in this book, is two linear equations that each contain the same two variables.
2. The solution set for a system of equations is the set of all ordered pairs that satisfy *both* equations. The solution set to a system of linear equations will be one of the following:
   - One ordered pair when the graphs of the two equations intersect at only one point (this is the most common situation)
   - No ordered pairs when the graphs of the two equations are distinct, parallel lines
   - An infinite number of ordered pairs when the graphs of the two equations coincide (are the same line)

2. Solving the system in Example 1 by graphing looks like

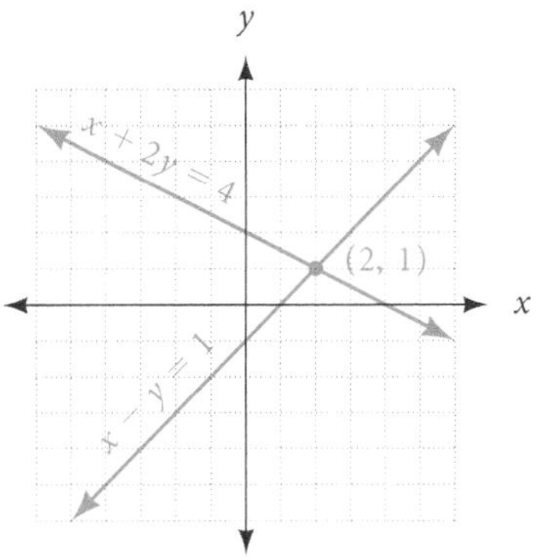

Solving a System by Graphing [4.1]

Step 1: Graph the first equation.

Step 2: Graph the second equation on the same set of axes.

Step 3: Identify the coordinates of the point where the graphs appear to cross each other (the coordinates of the point of intersection).

Step 4: Check the solution to verify that it satisfies *both* equations.

3. We can eliminate the y variable from the system in Example 1 by multiplying both sides of the second equation by 2 and adding the result to the first equation

$$\begin{array}{lcl} x + 2y = 4 & & x + 2y = 4 \\ x - y = 1 & \xrightarrow[\text{Multiply by 2}]{} & \underline{2x - 2y = 2} \\ & & 3x \quad\quad = 6 \\ & & x \quad\quad = 2 \end{array}$$

Substituting $x = 2$ into either of the original two equations gives $y = 1$. The solution is (2, 1).

Solving a System by the Elimination Method [4.2]

Step 1: Look the system over to decide which variable will be easier to eliminate.

Step 2: If necessary, use the multiplication property of equality on each equation separately to ensure that the coefficients of the variable to be eliminated are opposites.

Step 3: Add the respective left and right sides of the system together.

Step 4: Solve for the variable remaining.

Step 5: Substitute the solution from step 4 into either of the original equations, and solve for the other variable.

Step 6: Check your solution in both equations, if necessary.

Solving a System by the Substitution Method [4.3]

4. We can apply the substitution method to the system in Example 1 by first solving the second equation for x to get $x = y + 1$. Substituting this expression for x into the first equation, we have

$$\begin{aligned}(y + 1) + 2y &= 4\\ 3y + 1 &= 4\\ 3y &= 3\\ y &= 1\end{aligned}$$

Using $y = 1$ in either of the original equations gives $x = 2$.

Step 1: Solve either of the equations for one of the variables (this step is not necessary if one of the equations has the correct form already).

Step 2: Substitute the results of step 1 into the other equation, and solve for the remaining variable.

Step 3: Substitute the value from step 2 into the equation obtained in step 1 to find the value of the other variable.

Step 4: Check your solution, if necessary.

Special Cases [4.2, 4.3]

In some cases, using the elimination or substitution method eliminates both variables. The situation is interpreted as follows.

1. If the resulting statement is *false*, then the lines are parallel and there is no solution to the system.
2. If the resulting statement is *true*, then the equations represent the same line (the lines coincide). In this case any ordered pair that satisfies either equation is a solution to the system.

COMMON MISTAKE

The most common mistake encountered in solving linear systems is the failure to complete the problem. Here is an example.

$$\begin{aligned}x + y &= 8\\ x - y &= 4\\ 2x &= 12\\ x &= 6\end{aligned}$$

This is only half the solution. To find the other half, we must substitute the 6 back into one of the original equations and then solve for y.

Remember, solutions to systems of linear equations always consist of ordered pairs. We need an x-coordinate and a y-coordinate; $x = 6$ can never be a solution to a system of linear equations.

Chapter 4 Test

1. Write the solution to the system which is graphed below. [4.1]

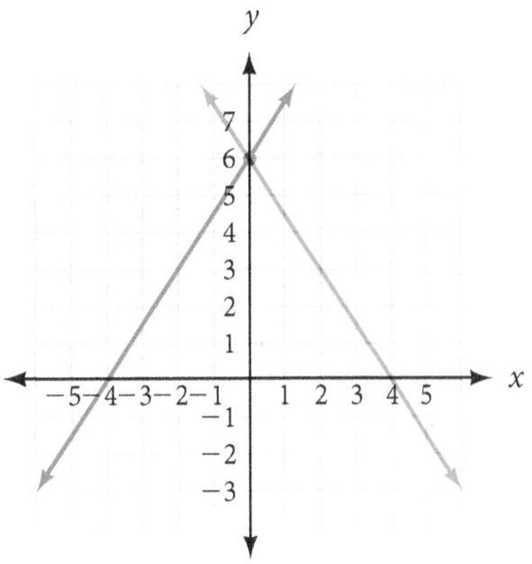

Solve each system by graphing. [4.1]

2. $4x - 2y = 8$
$y = \frac{2}{3}x$

3. $3x - 2y = 13$
$y = 4$

4. $2x - 2y = -12$
$-3x - y = 2$

Solve each system by the elimination method. [4.2]

5. $x - y = -9$
$2x + 3y = 7$

6. $3x - y = 1$
$5x - y = 3$

7. $2x + 3y = -3$
$x + 6y = 12$

8. $2x + 3y = 4$
$4x + 6y = 8$

Solve each system by the substitution method. [4.3]

9. $3x - y = 12$
$y = 2x - 8$

10. $3x - 6y = 3$
$x = 4y - 17$

11. $2x - 3y = -18$
$3x + y = -5$

12. $2x - 3y = 13$
$x - 4y = -1$

Solve the following word problems. In each case, be sure to show the system of equations that describes the situation. [4.4]

13. **Number Problem** The sum of two numbers is 18. One number is 2 more than 3 times the other. Find the two numbers.

14. **Investing** Dave has $2,000 to invest. He would like to earn $135.20 per year in interest. How much should he invest at 6% if the rest is to be invested at 7%?

15. **Coin Problem** Maria has 19 coins that total $1.35. If the coins are all nickels and dimes, how many of each type does she have?

16. **Mixture Problem** How much 40% antifreeze solution and 70% antifreeze solution should be combined to obtain 30 gallons of 50% antifreeze solution?

17. **Fencing Problem** A rancher wants to build a rectangular corral using 198 feet of fence. If the length of the corral is to be 15 feet longer than twice the width, find the dimensions of the corral.

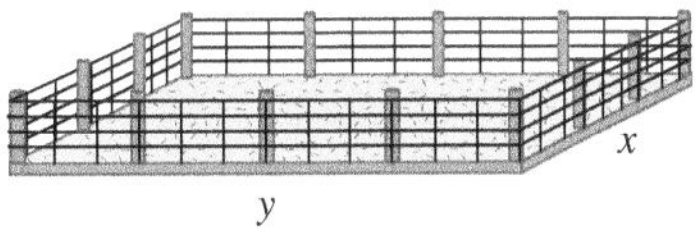

5

Exponents and Polynomials

Chapter Outline

5.1 Multiplication with Exponents and Scientific Notation

5.2 Division with Exponents

5.3 Operations with Monomials

5.4 Addition and Subtraction of Polynomials

5.5 Multiplication with Polynomials

5.6 Binomial Squares and Other Special Products

5.7 Division with Polynomials

If you were given a penny on the first day of September, and then each day after that you were given twice the amount of money you received the day before, how much money would you receive on September 30th? To begin, Table 1 and Figure 1 show the amount of money you would receive on each of the first 10 days of the month. As you can see, on the tenth day of the month you would receive \$5.12.

TABLE 1

Money That Doubles Each Day

Day	Money (in cents)
1	$1 = 2^0$
2	$2 = 2^1$
3	$4 = 2^2$
4	$8 = 2^3$
5	$16 = 2^4$
6	$32 = 2^5$
7	$64 = 2^6$
8	$128 = 2^7$
9	$256 = 2^8$
10	$512 = 2^9$

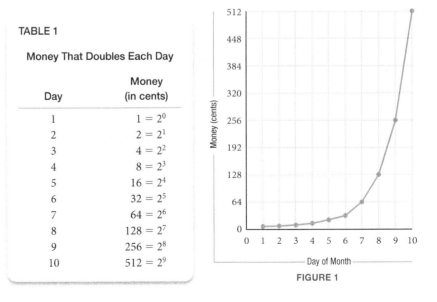

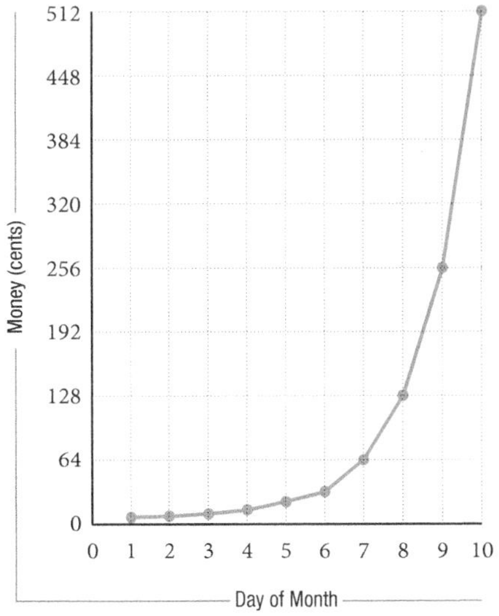

FIGURE 1

To find the amount of money on day 30, we could continue to double the amount on each of the next 20 days. Or, we could notice the pattern of exponents in the second column of the table and reason that the amount of money on day 30 would be 2^{29} cents, which is a very large number. In fact, 2^{29} cents is \$5,368,709.12—a little less than \$5.4 million. When you are finished with this chapter, you will have a good working knowledge of exponents.

The study skills for this chapter are about attitude. They are points of view that point toward success.

1. **Be Focused, Not Distracted** I have students who begin their assignments by asking themselves, "Why am I taking this class?" Or, "When am I ever going to use this stuff?" If you are asking yourself similar questions, you may be distracting yourself away from doing the things that will produce the results you want in this course. Don't dwell on questions and evaluations of the class that can be used as excuses for not doing well. If you want to succeed in this course, focus your energy and efforts toward success, rather than distracting yourself away from your goals.
2. **Be Resilient** Don't let setbacks keep you from your goals. You want to put yourself on the road to becoming a person who can succeed in this class or any class in college. Failing a test or quiz, or having a difficult time on some topics, is normal. No one goes through college without some setbacks. Don't let a temporary disappointment keep you from succeeding in this course. A low grade on a test or quiz is simply a signal that some reevaluation of your study habits needs to take place.
3. **Intend to Succeed** I always have a few students who simply go through the motions of studying without intending to master the material. It is more important to them to look like they are studying than to actually study. You need to study with the intention of being successful in the course. Intend to master the material, no matter what it takes.

Multiplication with Exponents and Scientific Notation 5.1

Learning Objectives

In this section, we will learn how to:

1. Multiply like bases by adding exponents.
2. Simplify the power of a power.
3. Simplify the power of a product.
4. Write numbers in scientific notation.

Introduction

Recall that an *exponent* is a number written just above and to the right of another number, which is called the *base*. In the expression 5^2, for example, the exponent is 2 and the base is 5. The expression 5^2 is read "5 to the second power" or "5 squared." The meaning of the expression is

$$5^2 = 5 \cdot 5 = 25$$

In the expression 5^3, the exponent is 3 and the base is 5. The expression 5^3 is read "5 to the third power" or "5 cubed." The meaning of the expression is

$$5^3 = 5 \cdot 5 \cdot 5 = 125$$

Here are some further examples.

VIDEO EXAMPLES

SECTION 5.1

EXAMPLE 1 Write each expression as a single number.

a. 4^3 **b.** -3^4 **c.** $(-2)^5$ **d.** $\left(-\frac{3}{4}\right)^2$

SOLUTION

a. $4^3 = 4 \cdot 4 \cdot 4 = 16 \cdot 4 = 64$ Exponent 3, base 4

b. $-3^4 = -(3 \cdot 3 \cdot 3 \cdot 3) = -81$ Exponent 4, base 3

c. $(-2)^5 = (-2)(-2)(-2)(-2)(-2) = -32$ Exponent 5, base -2

d. $\left(-\frac{3}{4}\right)^2 = \left(-\frac{3}{4}\right)\left(-\frac{3}{4}\right) = \frac{9}{16}$ Exponent 2, base $-\frac{3}{4}$

Note Many students have difficulty simplifying expressions like the one in Example 1b, because they tend to treat it like the expression in Example 1c. How are -3^4 and $(-3)^4$ different? In the first case the base is 3. In the second case, the base is -3. It helps if we understand that

$$-3^4 = -1(3^4)$$

Following the order of operations, we must evaluate the exponent first (giving us 81), and then take the opposite of this value to get -81.

Multiplication with Exponents

We can simplify our work with exponents by developing some properties of exponents. We want to list the things we know are true about exponents and then use these properties to simplify expressions that contain exponents.

The first property of exponents applies to products with the same base. We can use the definition of exponents, as indicating repeated multiplication, to simplify expressions like $7^4 \cdot 7^2$.

$$\begin{aligned} 7^4 \cdot 7^2 &= (7 \cdot 7 \cdot 7 \cdot 7)(7 \cdot 7) \\ &= (7 \cdot 7 \cdot 7 \cdot 7 \cdot 7 \cdot 7) \\ &= 7^6 \qquad \text{Notice: } 4 + 2 = 6 \end{aligned}$$

As you can see, multiplication with the same base resulted in addition of exponents. We can summarize this result with the following property.

PROPERTY *Property 1 for Exponents*

If a is any real number and r and s are integers, then

$$a^r \cdot a^s = a^{r+s}$$

In words: To multiply two expressions with the same base, add exponents and use the common base.

Here is an example using Property 1.

EXAMPLE 2 Use Property 1 to simplify the following expressions. Leave your answers in terms of exponents:

a. $5^3 \cdot 5^6$ **b.** $x^7 \cdot x^8$ **c.** $3^4 \cdot 3^8 \cdot 3^5$

SOLUTION

a. $5^3 \cdot 5^6 = 5^{3+6} = 5^9$

b. $x^7 \cdot x^8 = x^{7+8} = x^{15}$

c. $3^4 \cdot 3^8 \cdot 3^5 = 3^{4+8+5} = 3^{17}$

Note In Example 2, notice that in each case the base in the original problem is the same base that appears in the answer and that it is written only once in the answer. A very common mistake that people make when they first begin to use Property 1 is to write a 2 in front of the base in the answer. For example, people making this mistake would get $2x^{15}$ or $(2x)^{15}$ as the result in Example 2b. To avoid this mistake, you must be sure you understand the meaning of Property 1 exactly as it is written.

Another common type of expression involving exponents is one in which an expression containing an exponent is raised to another power. The expression $(5^3)^2$ is an example:

$$\begin{aligned}(5^3)^2 &= (5^3)(5^3) \\ &= 5^{3+3} \\ &= 5^6 \qquad \text{Notice: } 3 \cdot 2 = 6\end{aligned}$$

This result offers justification for the second property of exponents.

PROPERTY *Property 2 for Exponents*

If a is any real number and r and s are integers, then

$$(a^r)^s = a^{r \cdot s}$$

In words: A power raised to another power is the base raised to the product of the powers.

EXAMPLE 3 Simplify the following expressions:

a. $(4^5)^6$ **b.** $(x^3)^5$ **c.** $(x^4)^3(x^2)^5$

SOLUTION

a. $(4^5)^6 = 4^{5 \cdot 6} = 4^{30}$

b. $(x^3)^5 = x^{3 \cdot 5} = x^{15}$

c. $(x^4)^3(x^2)^5 = x^{12} \cdot x^{10}$ Property 2

$= x^{22}$ Property 1

The third property of exponents applies to expressions in which the product of two or more numbers or variables is raised to a power. Let's look at how the expression $(2x)^3$ can be simplified:

$$\begin{aligned}(2x)^3 &= (2x)(2x)(2x)\\ &= (2\cdot 2\cdot 2)(x\cdot x\cdot x)\\ &= 2^3\cdot x^3 \qquad \text{Notice: The exponent 3 distributes over the product } 2x\\ &= 8x^3\end{aligned}$$

We can generalize this result into a third property of exponents.

PROPERTY ***Property 3 for Exponents***

If a and b are any two real numbers and r is an integer, then

$$(ab)^r = a^r b^r$$

In words: The power of a product is the product of the powers.

Here are some examples using Property 3 to simplify expressions.

EXAMPLE 4 Simplify the following expressions:

a. $(3y)^2$ **b.** $\left(-\frac{1}{4}x^2y^3\right)^2$ **c.** $(x^2y^5)^3(x^4y)^2$

SOLUTION

a. $(3y)^2 = 3^2y^2$ Property 3

$= 9y^2$

b. $\left(-\frac{1}{4}x^2y^3\right)^2 = \left(-\frac{1}{4}\right)^2(x^2)^2(y^3)^2$ Property 3

$= \frac{1}{16}x^4y^6$ Property 2

c. $(x^2y^5)^3(x^4y)^2 = (x^2)^3(y^5)^3\cdot(x^4)^2y^2$ Property 3

$= x^6y^{15}\cdot x^8y^2$ Property 2

$= (x^6x^8)(y^{15}y^2)$ Commutative and associative properties

$= x^{14}y^{17}$ Property 1

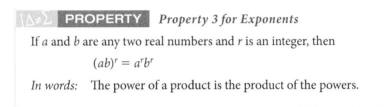

Note If we include units with the dimensions of the diagrams, then the units for the area will be square units and the units for volume will be cubic units. More specifically,

If a square has a side 5 inches long, then its area will be
$A = (5 \text{ inches})^2 = 25 \text{ inches}^2$
where the unit inches^2 stands for square inches.

If a cube has a side 5 inches long, then its volume will be
$V = (5 \text{ inches})^3 = 125 \text{ inches}^3$
where the unit inches^3 stands for cubic inches.

If a rectangular solid has a length of 5 inches, a width of 4 inches, and a height of 3 inches, then its volume is
$V = (5 \text{ in.})(4 \text{ in.})(3 \text{ in.})$
$= 60 \text{ inches}^3$

FACTS FROM GEOMETRY *Volume of a Rectangular Solid*

It is easy to see why the phrase "five squared" is associated with the expression 5^2. Simply find the area of the square shown in Figure 1 with a side of 5.

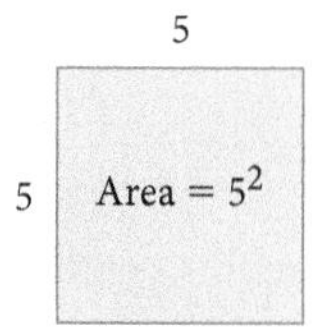

FIGURE 1

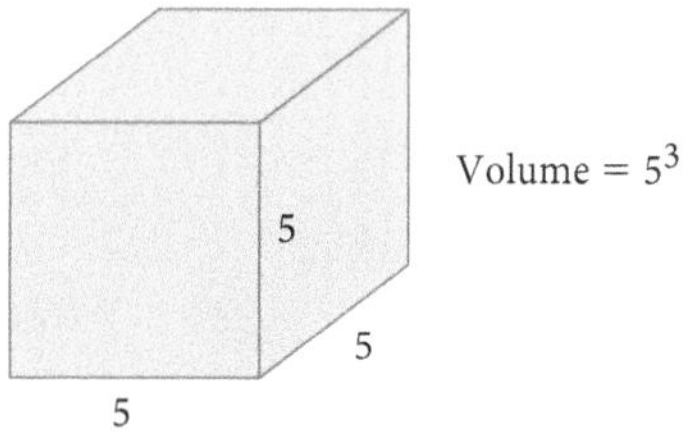

FIGURE 2

To see why the phrase "five cubed" is associated with the expression 5^3, we have to find the ***volume*** of a cube for which all three dimensions are 5 units long. The volume of a cube is a measure of the space occupied by the cube. To calculate the volume of the cube shown in Figure 2, we multiply the three dimensions together to get $5 \cdot 5 \cdot 5 = 5^3$.

The cube shown in Figure 2 is a special case of a general category of three dimensional geometric figures called ***rectangular solids***. Rectangular solids have rectangles for sides, and all connecting sides meet at right angles. The three dimensions are length, width, and height. To find the volume of a rectangular solid, we find the product of the three dimensions.

Scientific Notation

Many branches of science require working with very large numbers. In astronomy, for example, distances commonly are given in light-years. A light-year is the distance light travels in a year. It is approximately

5,880,000,000,000 miles

This number is difficult to use in calculations because of the number of zeros it contains. Scientific notation provides a way of writing very large numbers in a more manageable form.

DEFINITION *scientific notation*

A number is in ***scientific notation*** when it is written as the product of a number between 1 and 10 and an integer power of 10. A number written in scientific notation has the form

$$n \times 10^r$$

where $1 \le n < 10$ and r is an integer.

EXAMPLE 5 Write 376,000 in scientific notation.

SOLUTION We must rewrite 376,000 as the product of a number between 1 and 10 and a power of 10. To do so, we move the decimal point 5 places to the left so that it appears between the 3 and the 7. Then we multiply this number by 10^5. The number that results has the same value as our original number and is written in scientific notation:

$$376{,}000 = 3.76 \times 100{,}000 = 3.76 \times 10^5$$

Moved 5 places

Decimal point originally here

Keeps track of the 5 places we moved the decimal point

Note Moving the decimal point one position to the left in a number is equivalent to *dividing* the number by a factor of 10. To preserve the value of the original number, we must *multiply* this new decimal by a factor of 10.

In Example 5, when moving the decimal point five places to the left we are dividing the number by five factors of 10.

$$3.76 = \frac{376{,}000}{10^5}$$

Therefore, we must multiply the new decimal 3.76 by five factors of 10 so that the result is the same value of our original number.

$$3.76 \times 10^5 = \frac{376{,}000}{\cancel{10^5}} \cdot \cancel{10^5}$$
$$= 376{,}000$$

EXAMPLE 6 Write 4.52×10^3 in expanded form.

SOLUTION Since 10^3 is 1,000, we can think of this as simply a multiplication problem; that is,

$$4.52 \times 10^3 = 4.52 \times 1{,}000 = 4{,}520$$

On the other hand, we can think of the exponent 3 as indicating the number of places we need to move the decimal point to write our number in expanded form. Since our exponent is 3, we move the decimal point three places to the right:

$$4.520 \times 10^3 = 4{,}520$$

The zero we add does not change the value of the decimal.

Getting Ready for Class

After reading through the preceding section, respond in your own words and in complete sentences.

A. Explain the difference between -5^2 and $(-5)^2$.

B. How do you multiply two expressions containing exponents when they each have the same base?

C. Explain the difference between $2x^5$ and $(2x)^5$.

D. When is a number written in scientific notation?

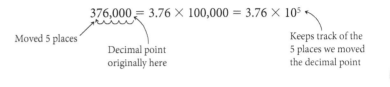

Problem Set 5.1

Name the base and exponent in each of the following expressions. Then use the definition of exponents as repeated multiplication to simplify.

1. 4^2 **2.** 6^2 **3.** $(0.3)^2$ **4.** $(0.03)^2$ **5.** 4^3 **6.** 10^3

7. $(-5)^2$ **8.** -5^2 **9.** -2^3 **10.** $(-2)^3$ **11.** 3^4 **12.** $(-3)^4$

13. $\left(\frac{2}{3}\right)^2$ **14.** $\left(\frac{2}{3}\right)^3$ **15.** $\left(\frac{1}{2}\right)^4$ **16.** $\left(\frac{4}{5}\right)^2$

17. a. Complete the following table.

Number x	1	2	3	4	5	6	7
Square x^2							

b. Using the results of part **a**, fill in the blank in the following statement: For numbers larger than 1, the square of the number is____ than the number.

18. a. Complete the following table.

Number x	$\frac{1}{2}$	$\frac{1}{3}$	$\frac{1}{4}$	$\frac{1}{5}$	$\frac{1}{6}$	$\frac{1}{7}$	$\frac{1}{8}$
Square x^2							

b. Using the results of part **a**, fill in the blank in the following statement: For numbers between 0 and 1, the square of the number is _____ than the number.

Use Property 1 to simplify the following expressions.

19. $x^4 \cdot x^5$ **20.** $x^7 \cdot x^3$ **21.** $y^{10} \cdot y^{20}$

22. $y^{30} \cdot y^{30}$ **23.** $2^5 \cdot 2^4 \cdot 2^3$ **24.** $4^2 \cdot 4^3 \cdot 4^4$

25. $x^4 \cdot x^6 \cdot x^8 \cdot x^{10}$ **26.** $x^{20} \cdot x^{18} \cdot x^{16} \cdot x^{14}$

Use Property 2 for exponents to write each of the following problems with a single exponent. (Assume all variables are positive numbers.)

27. $(x^2)^5$ **28.** $(x^5)^2$ **29.** $(5^4)^3$ **30.** $(5^3)^4$ **31.** $(y^3)^3$ **32.** $(y^2)^2$

33. $(2^5)^{10}$ **34.** $(10^5)^2$ **35.** $(a^3)^x$ **36.** $(a^5)^x$ **37.** $(b^x)^y$ **38.** $(b^r)^s$

Use Property 3 for exponents to simplify each of the following expressions.

39. $(4x)^2$ **40.** $(2x)^4$ **41.** $(2y)^5$ **42.** $(5y)^2$

43. $(-3x)^4$ **44.** $(-3x)^3$ **45.** $(0.5ab)^2$ **46.** $(0.4ab)^2$

47. $(4xyz)^3$ **48.** $(5xyz)^3$

Simplify the following expressions by using the properties of exponents.

49. $(2x^4)^3$ **50.** $(3x^5)^2$ **51.** $(4a^3)^2$ **52.** $(5a^2)^2$

53. $(x^2)^3(x^4)^2$ **54.** $(x^5)^2(x^3)^5$ **55.** $(a^3)^1(a^2)^4$ **56.** $(a^4)^1(a^1)^3$

57. $(4x^2y^3)^2$ **58.** $(9x^3y^5)^2$ **59.** $\left(\frac{2}{3}a^4b^5\right)^3$ **60.** $\left(\frac{3}{4}ab^7\right)^3$

61. Complete the following table, and then construct a line graph of the information in the table.

Number x	-3	-2	-1	0	1	2	3
Square x^2							

62. Complete the table, and then construct a line graph of the information in the table.

Number x	-3	-2	-1	0	1	2	3
Cube x^3							

63. Complete the table. When you are finished, notice how the points in this table could be used to refine the line graph you created in Problem 61.

Number x	-2.5	-1.5	-0.5	0	0.5	1.5	2.5
Square x^2							

64. Complete the following table. When you are finished, notice that this table contains exactly the same entries as the table from Problem 63. This table uses fractions, whereas the table from Problem 63 uses decimals.

Number x	$-\frac{5}{2}$	$-\frac{3}{2}$	$-\frac{1}{2}$	0	$\frac{1}{2}$	$\frac{3}{2}$	$\frac{5}{2}$
Square x^2							

Write each number in scientific notation.

65. 43,200

66. 432,000

67. -570

68. $-5,700$

69. 238,000

70. 2,380,000

Write each number in expanded form.

71. 2.49×10^3

72. 2.49×10^4

73. -3.52×10^2

74. -3.52×10^5

75. 2.8×10^4

76. 2.8×10^3

Applying the Concepts

77. **Volume of a Cube** Find the volume of a cube if each side is 3 inches long.

78. **Volume of a Cube** Find the volume of a cube if each side is 4 feet long.

79. **Volume of a Cube** A bottle of perfume is packaged in a box that is in the shape of a cube. Find the volume of the box if each side is 2.5 inches long. Round to the nearest tenth.

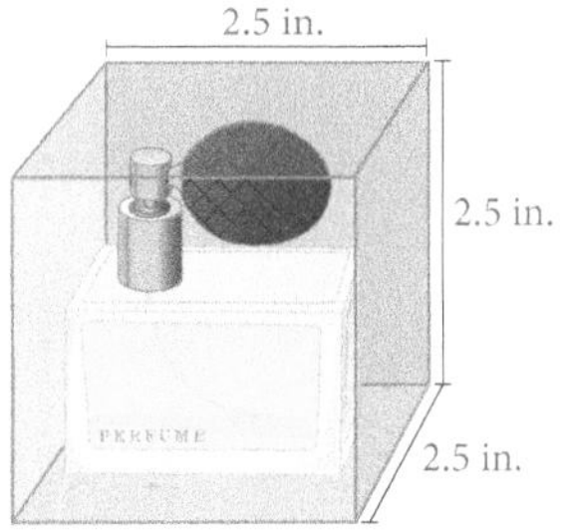

80. **Volume of a Cube** A television set is packaged in a box that is in the shape of a cube. Find the volume of the box if each side is 18 inches long.

81. **Volume of a Box** A rented videotape is in a plastic container that has the shape of a rectangular solid. Find the volume of the container if the length is 8 inches, the width is 4.5 inches, and the height is 1 inch.

82. **Volume of a Box** Your textbook is in the shape of a rectangular solid. Find the volume in cubic inches.

83. **Volume of a Box** If a box has a volume of 42 cubic feet, is it possible for you to fit inside the box? Explain your answer.

84. **Volume of a Box** A box has a volume of 45 cubic inches. Will a can of soup fit inside the box? Explain your answer.

85. **Age in seconds** If you are 21 years old, you have been alive for more than 650,000,000 seconds. Write this last number in scientific notation.

86. **Distance Around the Earth** The distance around the Earth at the equator is more than 130,000,000 feet. Write this number in scientific notation.

87. **Lifetime Earnings** If you earn at least \$12 an hour and work full-time for 30 years, you will make at least 7.4×10^5 dollars. Write this last number in expanded form.

88. **Heart Beats per Year** If your pulse is 72, then in one year your heart will beat at least 3.78×10^7 times. Write this last number in expanded form.

89. **Investing** If you put \$1,000 into a savings account every year from the time you are 25 years old until you are 55 years old, you will have more than 1.8×10^5 dollars in the account when you reach 55 years of age (assuming 10% annual interest). Write 1.8×10^5 in expanded form.

90. **Investing** If you put \$20 into a savings account every month from the time you are 20 years old until you are 30 years old, you will have more than 3.27×10^3 dollars in the account when you reach 30 years of age (assuming 6% annual interest compounded monthly). Write 3.27×10^3 in expanded form.

Displacement The displacement, in cubic inches, of a car engine is given by the formula

$$d = \pi \cdot s \cdot c \cdot \left(\frac{1}{2} \cdot b\right)^2$$

where s is the stroke and b is the bore, as shown in the figure, and c is the number of cylinders.

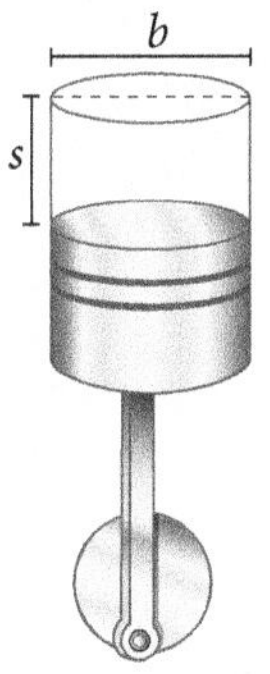

Calculate the engine displacement for each of the following cars. Use 3.14 to approximate π. Round to the nearest whole number.

91. **Ferrari Modena** 8 cylinders, 3.35 inches of bore, 3.11 inches of stroke

92. **Audi A8** 8 cylinders, 3.32 inches of bore, 3.66 inches of stroke

93. **Mitsubishi Eclipse** 6 cylinders, 3.59 inches of bore, 2.99 inches of stroke

94. **Porsche 911 GT3** 6 cylinders, 3.94 inches of bore, 3.01 inches of stroke

Learning Objectives Assessment

The following problems can be used to help assess if you have successfully met the learning objectives for this section.

95. Simplify: $x^2 \cdot x^4 \cdot x^6$.

a. x^{12} **b.** x^{48} **c.** $3x^{12}$ **d.** $3x^{48}$

96. Simplify: $(y^7)^3$.

a. y^4 **b.** y^{10} **c.** y^{21} **d.** $3y^7$

97. Simplify: $(-2xy)^4$.

a. $16x^4y^4$ **b.** $-16x^4y^4$ **c.** $-2xy^4$ **d.** $-8xy$

98. Write 340,000 in scientific notation.

a. 34×10^4 **b.** 3.4×10^5 **c.** 0.34×10^6 **d.** 340×10^3

Getting Ready for the Next Section

Subtract.

99. $4 - 7$

100. $-4 - 7$

101. $4 - (-7)$

102. $-4 - (-7)$

103. $15 - 20$

104. $15 - (-20)$

105. $-15 - (-20)$

106. $-15 - 20$

107. $2(3) - 4$

108. $5(3) - 10$

109. $4(3) - 3(2)$

110. $-8 - 2(3)$

111. $2(5 - 3)$

112. $2(3) - 4 - 3(-4)$

113. $5 + 4(-2) - 2(-3)$

114. $2(3) + 4(5) - 5(2)$

Division with Exponents 5.2

Learning Objectives

In this section, we will learn how to:

1. Simplify expressions containing negative exponents.
2. Divide like bases by subtracting exponents.
3. Simplify the power of a quotient.
4. Simplify expressions containing exponents of 1 or 0.
5. Use scientific notation with negative exponents.

Introduction

In Section 5.1 we found that multiplication with the same base results in addition of exponents; that is, $a^r \cdot a^s = a^{r+s}$. Since division is the inverse operation of multiplication, we can expect division with the same base to result in subtraction of exponents.

To develop the properties for exponents under division, we again apply the definition of exponents:

$$\begin{aligned} \frac{x^5}{x^3} &= \frac{x \cdot x \cdot x \cdot x \cdot x}{x \cdot x \cdot x} \\ &= \frac{x \cdot x \cdot x}{x \cdot x \cdot x}(x \cdot x) \\ &= 1(x \cdot x) \\ &= x^2 \qquad \text{Notice: } 5 - 3 = 2 \end{aligned}$$

$$\begin{aligned} \frac{2^4}{2^7} &= \frac{2 \cdot 2 \cdot 2 \cdot 2}{2 \cdot 2 \cdot 2 \cdot 2 \cdot 2 \cdot 2 \cdot 2} \\ &= \frac{2 \cdot 2 \cdot 2 \cdot 2}{2 \cdot 2 \cdot 2 \cdot 2} \cdot \frac{1}{2 \cdot 2 \cdot 2} \\ &= \frac{1}{2 \cdot 2 \cdot 2} \\ &= \frac{1}{2^3} \qquad \text{Notice: } 7 - 4 = 3 \end{aligned}$$

In both cases division with the same base resulted in subtraction of the smaller exponent from the larger. The problem is deciding whether the answer is a fraction. The following discussion will help us resolve this problem.

Negative Exponents

Consider the product $3^3 \cdot 3^{-1}$. We do not know yet what 3^{-1} represents. But according to Property 1 for exponents, it should be the case that

$$3^3 \cdot 3^{-1} = 3^{3+(-1)} = 3^2$$

Simplifying the positive exponents, we have

$$27 \cdot 3^{-1} = 9$$

If we solve this equation for 3^{-1}, we find

$$3^{-1} = \frac{9}{27} = \frac{1}{3}$$

It would appear that 3^{-1} represents the reciprocal of 3. This leads us to the following definition.

DEFINITION ***Negative Exponents***

If $a \neq 0$ and r is a positive integer, then

$$a^{-1} = \frac{1}{a}$$

and

$$a^{-r} = \frac{1}{a^r} = \left(\frac{1}{a}\right)^r$$

Notice that a negative exponent indicates a reciprocal of some kind. The following examples illustrate how we use this definition to simplify expressions that contain negative exponents.

VIDEO EXAMPLES

SECTION 5.2

EXAMPLE 1 Write each expression with a positive exponent and then simplify:

a. 2^{-3} **b.** $(-5)^{-2}$ **c.** -5^{-2} **d.** $3x^{-6}$ **e.** $(3x)^{-6}$

SOLUTION

a. $2^{-3} = \frac{1}{2^3} = \frac{1}{8}$ Notice: Negative exponents do not indicate negative numbers. They indicate reciprocals

b. $(-5)^{-2} = \frac{1}{(-5)^2} = \frac{1}{25}$

c. $-5^{-2} = -(5^{-2}) = -\left(\frac{1}{5^2}\right) = -\frac{1}{25}$

d. $3x^{-6} = 3 \cdot \frac{1}{x^6} = \frac{3}{x^6}$ x is the base

e. $(3x)^{-6} = \frac{1}{(3x)^6} = \frac{1}{3^6x^6} = \frac{1}{729x^6}$ $3x$ is the base

Division with Exponents

Now let us look back to the problem in the introduction to this section and try to work it again with the help of a negative exponent. We know that $\frac{2^4}{2^7} = \frac{1}{2^3}$. Let us decide now that with division of the same base, we will always subtract the exponent in the denominator from the exponent in the numerator and see if this conflicts with what we know is true.

$$\frac{2^4}{2^7} = 2^{4-7} \quad \text{Subtracting the bottom exponent from the top exponent}$$

$$= 2^{-3} \quad \text{Subtraction}$$

$$= \frac{1}{2^3} \quad \text{Definition of negative exponents}$$

Subtracting the exponent in the denominator from the exponent in the numerator and then using the definition of negative exponents gives us the same result we obtained previously. We can now continue the list of properties of exponents we started in Section 5.1.

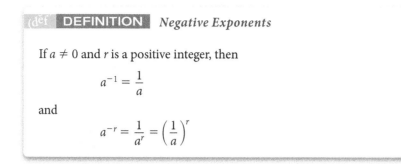

PROPERTY *Property 4 for Exponents*

If a is any real number and r and s are integers, then

$$\frac{a^r}{a^s} = a^{r-s} \qquad (a \neq 0)$$

In words: To divide with the same base, subtract the exponent in the denominator from the exponent in the numerator and raise the base to the exponent that results.

The following examples show how we use Property 4 and the definition for negative exponents to simplify expressions involving division.

EXAMPLE 2 Simplify the following expressions:

a. $\frac{x^9}{x^6}$ **b.** $\frac{x^4}{x^{10}}$ **c.** $\frac{2^{15}}{2^{20}}$

SOLUTION

a. $\frac{x^9}{x^6} = x^{9-6} = x^3$

b. $\frac{x^4}{x^{10}} = x^{4-10} = x^{-6} = \frac{1}{x^6}$

c. $\frac{2^{15}}{2^{20}} = 2^{15-20} = 2^{-5} = \frac{1}{2^5} = \frac{1}{32}$

Our final property of exponents is similar to Property 3 from Section 5.1, but it involves division instead of multiplication. After we have stated the property, we will give a proof of it. The proof shows why this property is true.

PROPERTY *Property 5 for Exponents*

If a and b are any two real numbers ($b \neq 0$) and r is an integer, then

$$\left(\frac{a}{b}\right)^r = \frac{a^r}{b^r}$$

In words: A quotient raised to a power is the quotient of the powers.

Proof

$\left(\frac{a}{b}\right)^r = \left(a \cdot \frac{1}{b}\right)^r$ By the definition of division

$= a^r \cdot \left(\frac{1}{b}\right)^r$ By Property 3

$= a^r \cdot b^{-r}$ By the definition of negative exponents

$= a^r \cdot \frac{1}{b^r}$ By the definition of negative exponents

$= \frac{a^r}{b^r}$ By the definition of division

EXAMPLE 3 Simplify the following expressions.

a. $\left(\frac{x}{2}\right)^3$ **b.** $\left(\frac{5}{y}\right)^2$ **c.** $\left(\frac{2}{3}\right)^4$

SOLUTION

a. $\left(\frac{x}{2}\right)^3 = \frac{x^3}{2^3} = \frac{x^3}{8}$

b. $\left(\frac{5}{y}\right)^2 = \frac{5^2}{y^2} = \frac{25}{y^2}$

c. $\left(\frac{2}{3}\right)^4 = \frac{2^4}{3^4} = \frac{16}{81}$

Zero and One as Exponents

We have two special exponents left to deal with before our rules for exponents are complete: 0 and 1. To obtain an expression for x^1, we will solve a problem two different ways:

$$\left.\begin{aligned}\frac{x^3}{x^2} &= \frac{x \cdot x \cdot x}{x \cdot x} = x \\ \frac{x^3}{x^2} &= x^{3-2} = x^1\end{aligned}\right\} \quad \text{Hence } x^1 = x$$

Stated generally, this rule says that $a^1 = a$. This seems reasonable and we will use it since it is consistent with our property of division using the same base.

We use the same procedure to obtain an expression for x^0:

$$\left.\begin{aligned}\frac{5^2}{5^2} &= \frac{25}{25} = 1 \\ \frac{5^2}{5^2} &= 5^{2-2} = 5^0\end{aligned}\right\} \quad \text{Hence } 5^0 = 1$$

It seems, therefore, that the best definition of x^0 is 1 for all x except $x = 0$. In the case of $x = 0$, we have 0^0, which we will not define. This definition will probably seem awkward at first. Most people would like to define x^0 as 0 when they first encounter it. Remember, the zero in this expression is an exponent, so x^0 does not mean to multiply by zero. Thus, we can make the general statement that $a^0 = 1$ for all real numbers except $a = 0$.

déf **DEFINITION** *Zero and One as Exponents*

$a^1 = a$

$a^0 = 1 \quad (a \neq 0)$

Here are some examples involving the exponents 0 and 1.

EXAMPLE 4 Simplify the following expressions:

a. 8^0 **b.** 8^1 **c.** $4^0 + 4^1$ **d.** $(2x^2y)^0$

SOLUTION

a. $8^0 = 1$

b. $8^1 = 8$

c. $4^0 + 4^1 = 1 + 4 = 5$

d. $(2x^2y)^0 = 1$

Here is a summary of the definitions and properties of exponents we have developed so far. For each definition or property in the list, a and b are real numbers, and r and s are integers.

Definitions		Properties	
$a^{-1} = \frac{1}{a}$	$a \neq 0$	**1.** $a^r \cdot a^s = a^{r+s}$	
$a^{-r} = \frac{1}{a^r} = \left(\frac{1}{a}\right)^r$	$a \neq 0$	**2.** $(a^r)^s = a^{r \cdot s}$	
$a^1 = a$		**3.** $(ab)^r = a^r b^r$	
$a^0 = 1$	$a \neq 0$	**4.** $\frac{a^r}{a^s} = a^{r-s}$	$a \neq 0$
		5. $\left(\frac{a}{b}\right)^r = \frac{a^r}{b^r}$	$b \neq 0$

Here are some additional examples. These examples use a combination of the preceding properties and definitions.

Note Because of the order of operations, we must simplify the exponent in the numerator in Example 5 and denominator in Example 6 before using Property 4 to do the division.

EXAMPLES Simplify each expression. Write all answers with positive exponents only:

5. $\frac{(5x^3)^2}{x^4} = \frac{25(x^3)^2}{x^4}$ Property 3

$= \frac{25x^6}{x^4}$ Property 2

$= 25x^2$ Property 4

6. $\frac{x^{-8}}{(x^2)^3} = \frac{x^{-8}}{x^6}$ Property 2

$= x^{-8-6}$ Property 4

$= x^{-14}$ Subtraction

$= \frac{1}{x^{14}}$ Definition of negative exponents

7. $\left(\frac{y^5}{y^3}\right)^2 = \frac{(y^5)^2}{(y^3)^2}$ Property 5

$= \frac{y^{10}}{y^6}$ Property 2

$= y^4$ Property 4

Notice in Example 7 that we could have simplified inside the parentheses first and then raised the result to the second power:

$$\left(\frac{y^5}{y^3}\right)^2 = (y^2)^2 = y^4$$

8. $(3x^5)^{-2} = \dfrac{1}{(3x^5)^2}$ Definition of negative exponents

$= \dfrac{1}{9x^{10}}$ Properties 2 and 3

9. $x^{-8} \cdot x^5 = x^{-8+5}$ Property 1

$= x^{-3}$ Addition

$= \dfrac{1}{x^3}$ Definition of negative exponents

10. $\dfrac{(a^3)^2a^{-4}}{(a^{-4})^3} = \dfrac{a^6a^{-4}}{a^{-12}}$ Property 2

$= \dfrac{a^2}{a^{-12}}$ Property 1

$= a^{14}$ Property 4

Applications

In the next two examples we use division to compare the area and volume of geometric figures.

EXAMPLE 11 Suppose you have two squares, one of which is larger than the other. If the length of a side of the larger square is 3 times as long as the length of a side of the smaller square, how many of the smaller squares will it take to cover up the larger square?

SOLUTION If we let x represent the length of a side of the smaller square, then the length of a side of the larger square is $3x$. The area of each square, along with a diagram of the situation, is given in Figure 1.

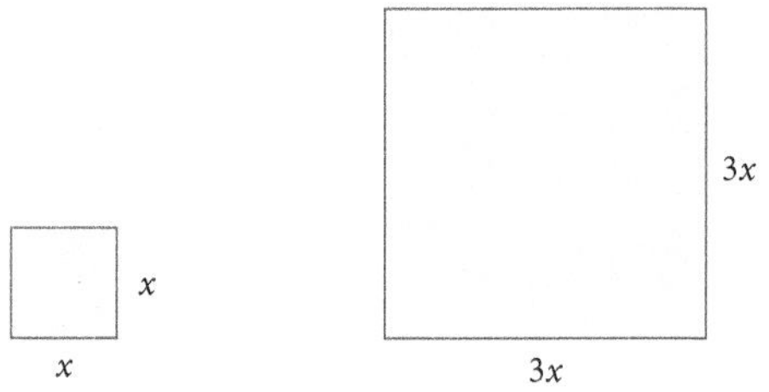

Square 1: $A = x^2$ Square 2: $A = (3x)^2 = 9x^2$

FIGURE 1

To find out how many smaller squares it will take to cover up the larger square, we divide the area of the larger square by the area of the smaller square.

$$\frac{\text{Area of square 2}}{\text{Area of square 1}} = \frac{9x^2}{x^2} = 9$$

It will take 9 of the smaller squares to cover the larger square.

EXAMPLE 12 Suppose you have two boxes, each of which is a cube. If the length of a side in the second box is 3 times as long as the length of a side of the first box, how many of the smaller boxes will fit inside the larger box?

SOLUTION If we let x represent the length of a side of the smaller box, then the length of a side of the larger box is $3x$. The volume of each box, along with a diagram of the situation, is given in Figure 2.

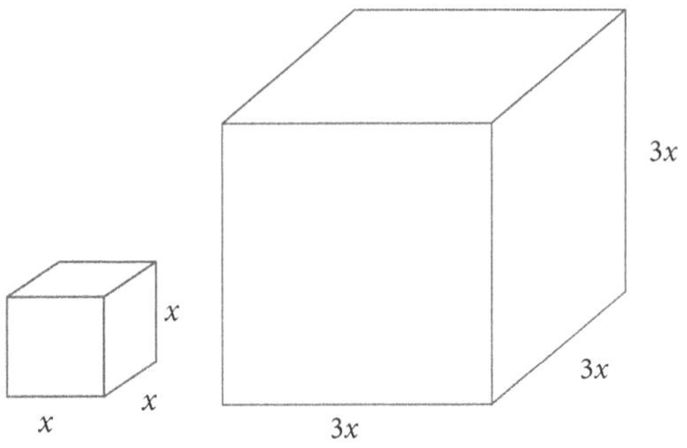

Box 1: $V = x^3$ Box 2: $V = (3x)^3 = 27x^3$

FIGURE 2

To find out how many smaller boxes will fit inside the larger box, we divide the volume of the larger box by the volume of the smaller box.

$$\frac{\text{Volume of box 2}}{\text{Volume of box 1}} = \frac{27x^3}{x^3} = 27$$

We can fit 27 of the smaller boxes inside the larger box.

More on Scientific Notation

Now that we have completed our list of definitions and properties of exponents, we can expand the work we did previously with scientific notation.

Recall that a number is in scientific notation when it is written in the form

$$n \times 10^r$$

where $1 \le n < 10$ and r is an integer.

Since negative exponents give us reciprocals, we can use negative exponents to write very small numbers in scientific notation. For example, the number 0.00057, when written in scientific notation, is equivalent to 5.7×10^{-4}. Here's why:

$$5.7 \times 10^{-4} = 5.7 \times \frac{1}{10^4} = 5.7 \times \frac{1}{10{,}000} = \frac{5.7}{10{,}000} = 0.00057$$

The table below lists some other numbers in both scientific notation and expanded form.

Number Written the Long Way		Number Written Again in Scientific Notation
376,000	=	3.76×10^5
49,500	=	4.95×10^4
3,200	=	3.2×10^3
591	=	5.91×10^2
46	=	4.6×10^1
8	=	8×10^0
0.47	=	4.7×10^{-1}
0.093	=	9.3×10^{-2}
0.00688	=	6.88×10^{-3}
0.0002	=	2×10^{-4}
0.000098	=	9.8×10^{-5}

Notice that in each case, when the number is written in scientific notation, the decimal point in the first number is placed so that the number is between 1 and 10. The exponent on 10 in the second number keeps track of the number of places we moved the decimal point in the original number to get a number between 1 and 10:

$$376{,}000 = 3.76 \times 10^5$$

Large number = positive exponent

Moved 5 places.

Decimal point was originally here.

Keeps track of the 5 places we moved the decimal point.

$$0.00688 = 6.88 \times 10^{-3}$$

Small number = negative exponent

Moved 3 places.

Keeps track of the 3 places we moved the decimal point.

EXAMPLE 13 Write each number in scientific notation.

a. 36×10^{-4} **b.** 0.36×10^5

SOLUTION

a. 36×10^{-4} is not in scientific notation because 36 is not between 1 and 10. We first write 36 in scientific notation, and then simplify the result using Property 1 for exponents.

$$\begin{aligned} 36 \times 10^{-4} &= (3.6 \times 10^1) \times 10^{-4} \\ &= 3.6 \times 10^{1+(-4)} \\ &= 3.6 \times 10^{-3} \end{aligned}$$

b. This number is not in scientific notation because 0.36 is less than 1. Following our process above, we have

$$\begin{aligned} 0.36 \times 10^5 &= (3.6 \times 10^{-1}) \times 10^5 \\ &= 3.6 \times 10^{-1+5} \\ &= 3.6 \times 10^4 \end{aligned}$$

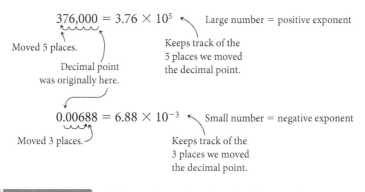

Getting Ready for Class

After reading through the preceding section, respond in your own words and in complete sentences.

A. How do you divide two expressions containing exponents when they each have the same base?

B. Explain the difference between 3^2 and 3^{-2}.

C. If a positive base is raised to a negative exponent, can the result be a negative number?

D. Explain what happens when we use zero as an exponent.

Problem Set 5.2

Simplify each expression.

1. 5^{-1} **2.** 4^{-1} **3.** x^{-1} **4.** y^{-1} **5.** 3^{-2} **6.** 3^{-3}

7. 6^{-2} **8.** 2^{-6} **9.** 8^{-2} **10.** 3^{-4} **11.** 5^{-3} **12.** 9^{-2}

13. a^{-4} **14.** b^{-5} **15.** -4^{-2} **16.** -6^{-3} **17.** $-x^{-1}$ **18.** $-y^{-7}$

19. $(-5)^{-3}$ **20.** $(-2)^{-4}$ **21.** $(-4)^{-2}$ **22.** $(-3)^{-3}$ **23.** $2x^{-3}$ **24.** $5x^{-1}$

25. $(2x)^{-3}$ **26.** $(5x)^{-1}$ **27.** $(5y)^{-2}$ **28.** $5y^{-2}$ **29.** 10^{-2} **30.** 10^{-3}

31. Complete the following table.

Number x	Square x^2	Power of 2 2^x
-3		
-2		
-1		
0		
1		
2		
3		

32. Complete the following table.

Number x	Cube x^3	Power of 3 3^x
-3		
-2		
-1		
0		
1		
2		
3		

Use Property 4 to simplify each of the following expressions. Write all answers that contain exponents with positive exponents only.

33. $\frac{5^1}{5^3}$ **34.** $\frac{7^6}{7^8}$ **35.** $\frac{x^{10}}{x^4}$ **36.** $\frac{x^4}{x^{10}}$ **37.** $\frac{4^3}{4^0}$ **38.** $\frac{4^0}{4^3}$

39. $\frac{(2x)^7}{(2x)^4}$ **40.** $\frac{(2x)^4}{(2x)^7}$ **41.** $\frac{6^{11}}{6}$ **42.** $\frac{8^7}{8}$ **43.** $\frac{6}{6^{11}}$ **44.** $\frac{8}{8^7}$

45. $\frac{2^{-5}}{2^3}$ **46.** $\frac{2^{-5}}{2^{-3}}$ **47.** $\frac{2^5}{2^{-3}}$ **48.** $\frac{2^{-3}}{2^{-5}}$ **49.** $\frac{(3x)^{-5}}{(3x)^{-8}}$ **50.** $\frac{(2x)^{-10}}{(2x)^{-15}}$

Simplify the following expressions. Any answers that contain exponents should contain positive exponents only.

51. $(3xy)^4$ **52.** $(4xy)^3$ **53.** 10^0 **54.** 10^1

55. $(2a^2b)^1$ **56.** $(2a^2b)^0$ **57.** $(7y^3)^{-2}$ **58.** $(5y^4)^{-2}$

59. $x^{-3}x^{-5}$ **60.** $x^{-6} \cdot x^8$ **61.** $y^7 \cdot y^{-10}$ **62.** $y^{-4} \cdot y^{-6}$

63. $\frac{(x^2)^3}{x^4}$ **64.** $\frac{(x^5)^3}{x^{10}}$ **65.** $\frac{(a^4)^3}{(a^3)^2}$ **66.** $\frac{(a^5)^3}{(a^5)^2}$

67. $\frac{y^7}{(y^2)^8}$ **68.** $\frac{y^2}{(y^3)^4}$ **69.** $\left(\frac{y^7}{y^2}\right)^8$ **70.** $\left(\frac{y^2}{y^3}\right)^4$

71. $\frac{(x^{-2})^3}{x^{-5}}$ **72.** $\frac{(x^2)^{-3}}{x^{-5}}$ **73.** $\left(\frac{x^{-2}}{x^{-5}}\right)^3$ **74.** $\left(\frac{x^2}{x^{-5}}\right)^{-3}$

75. $\frac{(a^3)^2(a^4)^5}{(a^5)^2}$ **76.** $\frac{(a^4)^8(a^2)^5}{(a^3)^4}$ **77.** $\frac{(a^{-2})^3(a^4)^2}{(a^{-3})^{-2}}$ **78.** $\frac{(a^{-5})^{-3}(a^7)^{-1}}{(a^{-3})^5}$

79. Complete the following table, and then construct a line graph of the information in the table.

Number x	-3	-2	-1	0	1	2	3
Power of 2 2^x							

80. Complete the following table, and then construct a line graph of the information in the table.

Number x	-3	-2	-1	0	1	2	3
Power of 3 3^x							

Write each of the following numbers in scientific notation.

81. 0.0048 **82.** 0.000048 **83.** 25 **84.** 35

85. 0.25 **86.** 0.35 **87.** 0.000009 **88.** 0.0009

89. Complete the following table.

Expanded Form	Scientific Notation $n \times 10^r$
0.000357	3.57×10^{-4}
0.00357	
0.0357	
0.357	
3.57	
35.7	
357	
3,570	
35,700	

90. Complete the following table.

Expanded Form	Scientific Notation $n \times 10^r$
0.000123	1.23×10^{-4}
	1.23×10^{-3}
	1.23×10^{-2}
	1.23×10^{-1}
	1.23×10^{0}
	1.23×10^{1}
	1.23×10^{2}
	1.23×10^{3}
	1.23×10^{4}

Write each of the following numbers in expanded form.

91. 4.23×10^{-3} **92.** 4.23×10^{3} **93.** 8×10^{-5} **94.** 8×10^{5}

95. 4.2×10^{0} **96.** 4.2×10^{1} **97.** 2.4×10^{-1} **98.** 2.4×10^{-6}

Applying the Concepts

Scientific Notation Problems

99. Some home computers can do a calculation in 2×10^{-3} seconds. Write this number in expanded form.

100. Some of the cells in the human body have a radius of 3×10^{-5} inches. Write this number in expanded form.

101. Margin of Victory Since 1993, the Nascar races with the smallest margin of victory are shown here. Write each number in scientific notation.

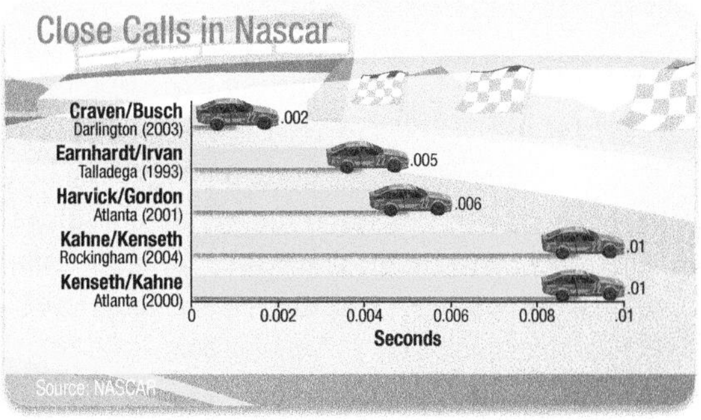

102. Some cameras used in scientific research can take one picture every 0.000000167 second. Write this number in scientific notation.

103. The number 25×10^3 is not in scientific notation because 25 is larger than 10. Write 25×10^3 in scientific notation.

104. The number 0.25×10^3 is not in scientific notation because 0.25 is less than 1. Write 0.25×10^3 in scientific notation.

105. The number 23.5×10^4 is not in scientific notation because 23.5 is not between 1 and 10. Rewrite 23.5×10^4 in scientific notation.

106. The number 375×10^3 is not in scientific notation because 375 is not between 1 and 10. Rewrite 375×10^3 in scientific notation.

107. The number 0.82×10^{-3} is not in scientific notation because 0.82 is not between 1 and 10. Rewrite 0.82×10^{-3} in scientific notation.

108. The number 0.93×10^{-2} is not in scientific notation because 0.93 is not between 1 and 10. Rewrite 0.93×10^{-2} in scientific notation.

Comparing Areas Suppose you have two squares, one of which is larger than the other. Suppose further that the side of the larger square is twice as long as the side of the smaller square.

109. If the length of the side of the smaller square is 10 inches, give the area of each square. Then find the number of smaller squares it will take to cover the larger square.

110. How many smaller squares will it take to cover the larger square if the length of the side of the smaller square is 1 foot?

111. If the length of the side of the smaller square is x, find the area of each square. Then find the number of smaller squares it will take to cover the larger square.

112. Suppose the length of the side of the larger square is 1 foot. How many smaller squares will it take to cover the larger square?

Comparing Volumes Suppose you have two boxes, each of which is a cube. Suppose further that the length of a side of the second box is twice as long as the length of a side of the first box.

113. If the length of a side of the first box is 6 inches, give the volume of each box. Then find the number of smaller boxes that will fit inside the larger box.

114. How many smaller boxes can be placed inside the larger box if the length of a side of the second box is 1 foot?

115. If the length of a side of the first box is x, find the volume of each box. Then find the number of smaller boxes that will fit inside the larger box.

116. Suppose the length of a side of the larger box is 12 inches. How many smaller boxes will fit inside the larger box?

Learning Objectives Assessment

The following problems can be used to help assess if you have successfully met the learning objectives for this section.

117. Simplify: 4^{-3}

a. -12 **b.** $\frac{1}{64}$ **c.** 64 **d.** -64

118. Simplify: $\frac{x^2}{x^{-5}}$

a. x^7 **b.** x^{-3} **c.** x^3 **d.** $\frac{1}{x^7}$

119. Simplify $\left(\frac{x}{3}\right)^4$

a. $\frac{x^4}{81}$ **b.** $\frac{x^4}{3}$ **c.** $\frac{x}{81}$ **d.** $\frac{4x}{3}$

120. Simplify: $2^0 - 2^1$

a. 0 **b.** -2 **c.** $\frac{1}{2}$ **d.** -1

121. Write 0.00123 in scientific notation.

a. 123×10^5 **b.** 123×10^{-5} **c.** 1.23×10^3 **d.** 1.23×10^{-3}

Getting Ready for the Next Section

Simplify.

122. $3(4.5)$

123. $\frac{1}{2} \cdot \frac{5}{7}$

124. $\frac{4}{5}(10)$

125. $\frac{9.6}{3}$

126. $6.8(3.9)$

127. $9 - 20$

128. $-3 + 15$

129. $2x \cdot x \cdot \frac{1}{2}x$

130. $x^5 \cdot x^3$

131. $y^2 \cdot y$

132. $\frac{x^3}{x^2}$

133. $\frac{x^2}{x}$

134. $\frac{y^3}{y^5}$

135. $\frac{x^2}{x^5}$

Write in expanded form.

136. 3.4×10^2

137. 6.0×10^{-4}

Operations with Monomials 5.3

Learning Objectives

In this section, we will learn how to:

1. State the degree of a monomial.
2. Multiply and divide monomials.
3. Multiply and divide numbers written in scientific notation.
4. Add and subtract monomials.

Introduction

We have developed all the tools necessary to perform the four basic operations on the simplest of polynomials: monomials.

DEFINITION *monomial*

A ***monomial*** is a one-term expression that is either a constant (number) or the product of a constant and one or more variables raised to whole number exponents.

The following are examples of monomials:

$$-3 \qquad 15x \qquad -23x^2y \qquad 49x^4y^2z^4 \qquad \frac{3}{4}a^2b^3$$

The numerical part of each monomial is called the *numerical coefficient*, or just *coefficient*. Monomials are also called *terms*.

DEFINITION *degree of a monomial*

The ***degree*** of a monomial in one variable is the exponent of the variable. If the monomial contains multiple variables, then the ***degree*** is the sum of the exponents on the variables. A constant term is defined to have a degree of zero.

VIDEO EXAMPLES

SECTION 5.3

EXAMPLES State the coefficient and degree of each monomial.

1. $8x^5$	Coefficient $= 8$	Degree $= 5$
2. $-\frac{1}{2}x$	Coefficient $= -\frac{1}{2}$	Degree $= 1$
3. $14x^4y^3$	Coefficient $= 14$	Degree $= 7$
4. -3	Coefficient $= -3$	Degree $= 0$

Multiplication and Division of Monomials

There are two basic steps involved in the multiplication of monomials. First, we rewrite the products using the commutative and associative properties. Then, we simplify by multiplying coefficients and adding exponents of like bases.

EXAMPLE 5 Multiply:

a. $(-3x^2)(4x^3)$ **b.** $\left(\frac{4}{5}x^5 \cdot y^2\right)(10x^3 \cdot y)$

SOLUTION

a. $(-3x^2)(4x^3) = (-3 \cdot 4)(x^2 \cdot x^3)$ Commutative and associative properties

$= -12x^5$ Multiply coefficients, add exponents

b. $\left(\frac{4}{5}x^5 \cdot y^2\right)(10x^3 \cdot y) = \left(\frac{4}{5} \cdot 10\right)(x^5 \cdot x^3)(y^2 \cdot y)$ Commutative and associative properties

$= 8x^8y^3$ Multiply coefficients, add exponents

You can see that in each case the work was the same—multiply coefficients and add exponents of the same base. We can expect division of monomials to proceed in a similar way. Since our properties are consistent, division of monomials will result in division of coefficients and subtraction of exponents of like bases.

EXAMPLE 6 Divide:

a. $\frac{15x^3}{3x^2}$ **b.** $\frac{39x^2y^3}{3xy^5}$

SOLUTION

a. $\frac{15x^3}{3x^2} = \frac{15}{3} \cdot \frac{x^3}{x^2}$ Write as separate fractions

$= 5x$ Divide coefficients, subtract exponents

b. $\frac{39x^2y^3}{3xy^5} = \frac{39}{3} \cdot \frac{x^2}{x} \cdot \frac{y^3}{y^5}$ Write as separate fractions

$= 13 \cdot x \cdot \frac{1}{y^2}$ Divide coefficients, subtract exponents

$= \frac{13x}{y^2}$ Write answer as a single fraction

Note Notice that when dividing monomials, the result may not be a monomial. In Example 6b, the expression

$$\frac{13x}{y^2} = 13xy^{-2}$$

is not a monomial because it contains a negative exponent.

In Example 6b, the expression $\frac{y^3}{y^5}$ simplifies to $\frac{1}{y^2}$ because of Property 4 for exponents and the definition of negative exponents. If we were to show all the work in this simplification process, it would look like this:

$\frac{y^3}{y^5} = y^{3-5}$ Property 4 for exponents

$= y^{-2}$ Subtraction

$= \frac{1}{y^2}$ Definition of negative exponents

The point of this explanation is this: Even though we may not show all the steps when simplifying an expression involving exponents, the result we obtain still can be justified using the properties of exponents. We have not introduced any new properties in Example 6; we have just not shown the details of each simplification.

EXAMPLE 7 Divide $25a^5b^3$ by $50a^2b^7$.

SOLUTION

$$\frac{25a^5b^3}{50a^2b^7} = \frac{25}{50} \cdot \frac{a^5}{a^2} \cdot \frac{b^3}{b^7} \quad \text{Write as separate fractions}$$

$$= \frac{1}{2} \cdot a^3 \cdot \frac{1}{b^4} \quad \text{Divide coefficients, subtract exponents}$$

$$= \frac{a^3}{2b^4} \quad \text{Write answer as a single fraction}$$

Notice in Example 7 that dividing 25 by 50 results in $\frac{1}{2}$. This is the same result we would obtain if we reduced the fraction $\frac{25}{50}$ to lowest terms, and there is no harm in thinking of it that way. Also, notice that the expression $\frac{b^3}{b^7}$ simplifies to $\frac{1}{b^4}$ by Property 4 for exponents and the definition of negative exponents, even though we have not shown the steps involved in doing so.

EXAMPLE 8 Simplify: $\dfrac{(6x^4y)(3x^7y^5)}{9x^5y^2}$.

SOLUTION We begin by multiplying the two monomials in the numerator:

$$\frac{(6x^4y)(3x^7y^5)}{9x^5y^2} = \frac{18x^{11}y^6}{9x^5y^2} \quad \text{Simplify numerator}$$

$$= 2x^6y^4 \quad \text{Divide}$$

EXAMPLE 9 Simplify each expression.

a. $(2y)^3(3y)^2$ **b.** $\dfrac{(6xy^4)^2}{(2x^3y^2)^3}$

SOLUTION In both cases, we follow the order of operations by applying properties 2 and 3 for exponents first. Then we can perform the multiplication and division.

a. $(2y)^3(3y)^2 = 8y^3 \cdot 9y^2$ Property 3 for exponents

$= 72y^5$ Multiply coefficients, add exponents

b. $\dfrac{(6xy^4)^2}{(2x^3y^2)^3} = \dfrac{36x^2y^8}{8x^9y^6}$ Properties 2 and 3

$= \dfrac{9}{2} \cdot \dfrac{1}{x^7} \cdot y^2$ Divide coefficients, subtract exponents

$= \dfrac{9y^2}{2x^7}$ Write answer as a single fraction

Multiplication and Division of Numbers Written in Scientific Notation

We multiply and divide numbers written in scientific notation using the same steps we used to multiply and divide monomials.

EXAMPLE 10 Multiply: $(4 \times 10^7)(2 \times 10^{-4})$.

SOLUTION Since multiplication is commutative and associative, we can rearrange the order of these numbers and group them as follows:

$$(4 \times 10^7)(2 \times 10^{-4}) = (4 \times 2)(10^7 \times 10^{-4})$$

$$= 8 \times 10^3$$

Notice that we add exponents, $7 + (-4) = 3$, when we multiply with the same base.

EXAMPLE 11 Divide: $\dfrac{9.6 \times 10^{12}}{3 \times 10^4}$.

SOLUTION We group the numbers between 1 and 10 separately from the powers of 10 and proceed as we did in Example 6:

$$\begin{aligned}\frac{9.6 \times 10^{12}}{3 \times 10^4} &= \frac{9.6}{3} \times \frac{10^{12}}{10^4} \\ &= 3.2 \times 10^8\end{aligned}$$

Notice that the procedure we used in both of these examples is very similar to multiplication and division of monomials, for which we multiplied or divided coefficients and added or subtracted exponents.

EXAMPLE 12 Simplify: $\dfrac{(6.8 \times 10^5)(3.9 \times 10^{-7})}{1.7 \times 10^{-4}}$.

SOLUTION We group the numbers between 1 and 10 separately from the powers of 10:

$$\begin{aligned}\frac{(6.8)(3.9)}{1.7} \times \frac{(10^5)(10^{-7})}{10^{-4}} &= 15.6 \times 10^{5+(-7)-(-4)} \\ &= 15.6 \times 10^2\end{aligned}$$

Our result is not in scientific notation because 15.6 is greater than 10. We convert our answer into scientific notation as follows.

$$\begin{aligned}15.6 \times 10^2 &= (1.56 \times 10^1) \times 10^2 \\ &= 1.56 \times 10^3\end{aligned}$$

Addition and Subtraction of Monomials

Addition and subtraction of monomials will be almost identical since subtraction is defined as addition of the opposite. With multiplication and division of monomials, the key was rearranging the numbers and variables using the commutative and associative properties. With addition, the key is application of the distributive property. We sometimes use the phrase *combine monomials* to describe addition and subtraction of monomials.

DEFINITION ***similar terms***

Two terms (monomials) with the same variable part (same variables raised to the same powers) are called ***similar*** (or like) ***terms***.

You can add only similar terms. This is because the distributive property (which is the key to addition of monomials) cannot be applied to terms that are not similar.

EXAMPLE 13 Combine the following monomials.

a. $-3x^2 + 15x^2$ **b.** $9x^2y - 20x^2y$ **c.** $5x^2 + 8y^2$

SOLUTION

a. $-3x^2 + 15x^2 = (-3 + 15)x^2$ Distributive property

$= 12x^2$ Add coefficients

b. $9x^2y - 20x^2y = (9 - 20)x^2y$ Distributive property

$= -11x^2y$ Add coefficients

c. $5x^2 + 8y^2$ In this case we cannot apply the distributive property, so we cannot add the monomials

EXAMPLE 14 Simplify: $\frac{14x^5}{2x^2} + \frac{15x^8}{3x^5}$.

SOLUTION Simplifying each expression separately and then combining similar terms gives

$$\frac{14x^5}{2x^2} + \frac{15x^8}{3x^5} = 7x^3 + 5x^3 \quad \text{Divide}$$

$$= 12x^3 \quad \text{Add}$$

Application

EXAMPLE 15 A rectangular solid is twice as long as it is wide and one-half as high as it is wide. Write an expression for the volume.

SOLUTION We begin by making a diagram of the object (Figure 1) with the dimensions labeled as given in the problem.

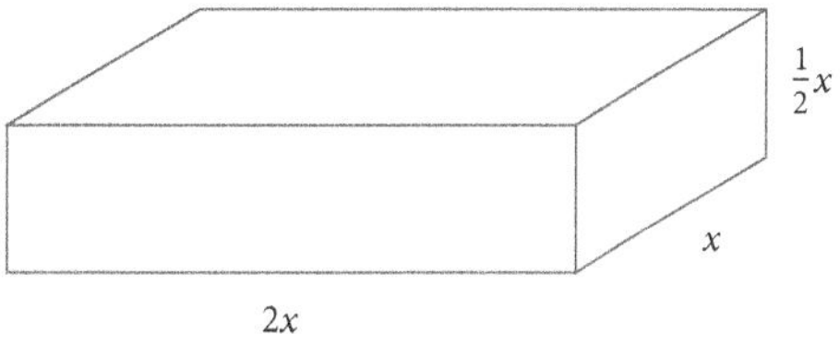

FIGURE 1

The volume is the product of the three dimensions:

$$V = 2x \cdot x \cdot \frac{1}{2}x = x^3$$

The box has the same volume as a cube with side x, as shown in Figure 2.

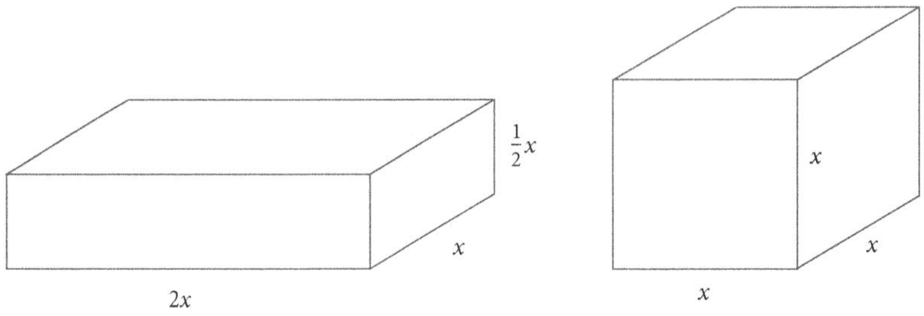

Equal Volumes

FIGURE 2

Getting Ready for Class

After reading through the preceding section, respond in your own words and in complete sentences.

A. What is a monomial?

B. Describe how you would multiply $3x^2$ and $5x^2$.

C. Describe how you would add $3x^2$ and $5x^2$.

D. Describe how you would multiply two numbers written in scientific notation.

Problem Set 5.3

State the coefficient and degree for each of the following monomials.

1. $7x^3$ **2.** $-2x^8$ **3.** $-x$ **4.** $\frac{x}{3}$ **5.** $\frac{1}{2}xy$

6. $\frac{3}{4}x^2y$ **7.** $-4a^6b^5$ **8.** $-9a^3b^{11}$ **9.** 8 **10.** -5

Multiply.

11. $(3x^4)(4x^3)$ **12.** $(6x^5)(-2x^2)$ **13.** $(-2y^4)(8y^7)$

14. $(5y^{10})(2y^5)$ **15.** $(8x)(4x)$ **16.** $(7x)(5x)$

17. $(10a^3)(10a)(2a^2)$ **18.** $(5a^4)(10a)(10a^4)$ **19.** $(6ab^2)(-4a^2b)$

20. $(-5a^3b)(4ab^4)$ **21.** $(4x^2y)(3x^3y^3)(2xy^4)$ **22.** $(5x^6)(-10xy^4)(-2x^2y^6)$

Divide. Write all answers with positive exponents only.

23. $\frac{15x^3}{5x^2}$ **24.** $\frac{25x^5}{5x^4}$ **25.** $\frac{18y^9}{3y^{12}}$ **26.** $\frac{24y^4}{8y^7}$

27. $\frac{32a^3}{64a^4}$ **28.** $\frac{25a^5}{75a^6}$ **29.** $\frac{21a^2b^3}{-7ab^5}$ **30.** $\frac{32a^5b^6}{8ab^5}$

31. $\frac{3x^3y^2z}{27xy^2z^3}$ **32.** $\frac{5x^5y^4z}{30x^3yz^2}$

33. Fill in the table.

a	b	ab	$\frac{a}{b}$	$\frac{b}{a}$
10	$5x$			
$20x^3$	$6x^2$			
$25x^5$	$5x^4$			
$3x^{-2}$	$3x^2$			
$-2y^4$	$8y^7$			

34. Fill in the table.

a	b	ab	$\frac{a}{b}$	$\frac{b}{a}$
$10y$	$2y^2$			
$10y^2$	$2y$			
$5y^3$	15			
5	$15y^3$			
$4y^{-3}$	$4y^3$			

Find each product. Write all answers in scientific notation.

35. $(3 \times 10^3)(2 \times 10^5)$ **36.** $(4 \times 10^8)(1 \times 10^6)$

37. $(3.5 \times 10^4)(5 \times 10^{-6})$ **38.** $(7.1 \times 10^5)(2 \times 10^{-8})$

39. $(5.5 \times 10^{-3})(2.2 \times 10^{-4})$ **40.** $(3.4 \times 10^{-2})(4.5 \times 10^{-6})$

Find each quotient. Write all answers in scientific notation.

41. $\dfrac{8.4 \times 10^5}{2 \times 10^2}$

42. $\dfrac{9.6 \times 10^{20}}{3 \times 10^6}$

43. $\dfrac{6 \times 10^8}{2 \times 10^{-2}}$

44. $\dfrac{8 \times 10^{12}}{4 \times 10^{-3}}$

45. $\dfrac{2.5 \times 10^{-6}}{5 \times 10^{-4}}$

46. $\dfrac{4.5 \times 10^{-8}}{9 \times 10^{-4}}$

Combine like terms by adding or subtracting as indicated.

47. $3x^2 + 5x^2$

48. $4x^3 + 8x^3$

49. $8x^5 - 19x^5$

50. $75x^6 - 50x^6$

51. $2a + a - 3a$

52. $5a + a - 6a$

53. $10x^3 - 8x^3 + 2x^3$

54. $7x^5 + 8x^5 - 12x^5$

55. $20ab^2 - 19ab^2 + 30ab^2$

56. $18a^3b^2 - 20a^3b^2 + 10a^3b^2$

57. Fill in the table.

a	b	ab	a + b
$5x$	$3x$		
$4x^2$	$2x^2$		
$3x^3$	$6x^3$		
$2x^4$	$-3x^4$		
x^5	$7x^5$		

58. Fill in the table.

a	b	ab	a − b
$2y$	$3y$		
$-2y$	$3y$		
$4y^2$	$5y^2$		
y^3	$-3y^3$		
$5y^4$	$7y^4$		

Simplify. Write all answers with positive exponents only.

59. $(2x)^3(2x)^4$

60. $(3x)^2(3x)^3$

61. $(3x^2)^3(2x)^4$

62. $(3x)^3(2x^3)^2$

63. $(4xy^3)^2(2x^5y)^3$

64. $(3x^3y^2)^3(2x^2y^3)^2$

65. $\dfrac{(2x)^5}{(2x)^3}$

66. $\dfrac{(3x)^2}{(3x)^4}$

67. $\dfrac{(3x^2)^3}{(2x)^4}$

68. $\dfrac{(3x)^3}{(2x^3)^2}$

69. $\dfrac{(2x^2y)^4}{(6xy^4)^2}$

70. $\dfrac{(4x^2y^3)^3}{(2x^3y^2)^4}$

71. $\dfrac{(3x^2)(8x^5)}{6x^4}$

72. $\dfrac{(7x^3)(6x^8)}{14x^5}$

73. $\dfrac{(9a^2b)(2a^3b^4)}{18a^5b^7}$

74. $\dfrac{(21a^5b)(2a^8b^4)}{14ab}$

75. $\dfrac{(4x^3y^2)(9x^4y^{10})}{(3x^5y)(2x^6y)}$

76. $\dfrac{(5x^4y^4)(10x^3y^3)}{(25xy^5)(2xy^7)}$

Simplify each expression, and write all answers in scientific notation.

77. $\dfrac{(6 \times 10^8)(3 \times 10^5)}{9 \times 10^7}$

78. $\dfrac{(8 \times 10^4)(5 \times 10^{10})}{2 \times 10^7}$

79. $\dfrac{(5 \times 10^3)(4 \times 10^{-5})}{2 \times 10^{-2}}$

80. $\dfrac{(7 \times 10^6)(4 \times 10^{-4})}{1.4 \times 10^{-3}}$

81. $\dfrac{(2.8 \times 10^{-7})(3.6 \times 10^4)}{2.4 \times 10^3}$

82. $\dfrac{(5.4 \times 10^2)(3.5 \times 10^{-9})}{4.5 \times 10^6}$

Simplify.

83. $\dfrac{18x^4}{3x} + \dfrac{21x^7}{7x^4}$

84. $\dfrac{24x^{10}}{6x^4} + \dfrac{32x^7}{8x}$

85. $\dfrac{45a^6}{9a^4} - \dfrac{50a^8}{2a^6}$

86. $\dfrac{16a^9}{4a} - \dfrac{28a^{12}}{4a^4}$

87. $\dfrac{6x^7y^4}{3x^2y^2} + \dfrac{8x^5y^8}{2y^6}$

88. $\dfrac{40x^{10}y^{10}}{8x^2y^5} + \dfrac{10x^8y^8}{5y^3}$

Learning Objectives Assessment

The following problems can be used to help assess if you have successfully met the learning objectives for this section.

89. State the degree of $5x^7y^2$.

a. 7 **b.** 9 **c.** 5 **d.** 14

90. Simplify: $\dfrac{(2x^3)(9x^4)}{6x^5}$.

a. $3x^2$ **b.** $3x^7$ **c.** $\dfrac{27x^2}{4}$ **d.** $\dfrac{27x^7}{4}$

91. Simplify: $(2.6 \times 10^{-5})(3.8 \times 10^3)$.

a. 6.4×10^{-2} **b.** 6.4×10^{-15} **c.** 9.88×10^{-15} **d.** 9.88×10^{-2}

92. Simplify: $15a^2b^3 - 6a^2b^3$.

a. $9a^4b^9$ **b.** $-90a^4b^9$ **c.** $9a^2b^3$ **d.** 9

Getting Ready for the Next Section

Simplify.

93. $3 - 8$

94. $-5 + 7$

95. $-1 + 7$

96. $1 - 8$

97. $3(5)^2 + 1$

98. $3(-2)^2 - 5(-2) + 4$

99. $2x^2 + 4x^2$

100. $3x^2 - x^2$

101. $-5x + 7x$

102. $x - 2x$

103. $-(2x + 9)$

104. $-(4x^2 - 2x - 6)$

105. Find the value of $2x + 3$ when $x = 4$.

106. Find the value of $(3x)^2$ when $x = 3$.

SPOTLIGHT ON SUCCESS *Student Instructor Gordon*

Math takes time. This fact holds true in the smallest of math problems as much as it does in the most math intensive careers. I see proof in each video I make. My videos get progressively better with each take, though I still make mistakes and find aspects I can improve on with each new video. In order to keep trying to improve in spite of any failures or lack of improvement, something else is needed. For me it is the sense of a specific goal in sight, to help me maintain the desire to put in continued time and effort.

When I decided on the number one university I wanted to attend, I wrote the name of that school in bold block letters on my door, written to remind myself daily of my ultimate goal. Stuck in the back of my head, this end result pushed me little by little to succeed and meet all of the requirements for the university I had in mind. And now I can say I'm at my dream school bringing with me that skill.

I recognize that others may have much more difficult circumstances than my own to endure, with the goal of improving or escaping those circumstances, and I deeply respect that. But that fact demonstrates to me how easy but effective it is, in comparison, to "stay with the problems longer" with a goal in mind of something much more easily realized, like a good grade on a test. I've learned to set goals, small or big, and to stick with them until they are realized.

Addition and Subtraction of Polynomials 5.4

Learning Objectives

In this section, we will learn how to:

1. State the degree and leading coefficient for a polynomial in one variable.
2. Evaluate a polynomial.
3. Add polynomials.
4. Subtract polynomials.

In this section we will extend what we learned in Section 5.3 to expressions called polynomials.

Polynomials

We begin this section with the definition of a polynomial.

DEFINITION *polynomial*

A ***polynomial*** is a finite sum of monomials (terms).

Here are some examples of polynomials:

$$3x^2 + 2x + 1 \qquad 15x^2y + 21xy^2 - 7 \qquad 3a - 2b + 4c - 5d$$

Polynomials can be further classified by the number of terms they contain. A polynomial with two terms is called a *binomial*. If it has three terms, it is a *trinomial*. As stated before, a *monomial* has only one term.

To write a polynomial in *standard form*, we write the terms in order of degree from highest degree to lowest degree. Once we have done so, the first term in the polynomial is called the *leading term*.

DEFINITION *degree of a polynomial*

The ***degree*** of a polynomial is the degree of the leading term once the polynomial is written in standard form.

For a polynomial in one variable, the degree of the polynomial is the highest power to which the variable is raised.

Various degrees of polynomials:

$3x^5 + 2x^3 + 1$	A trinomial of degree 5
$2x + 1$	A binomial of degree 1
$3x^2 + 2x + 1$	A trinomial of degree 2
$3x^5$	A monomial of degree 5
-9	A monomial of degree 0

VIDEO EXAMPLES

SECTION 5.4

EXAMPLE 1 Write $2x - 5 + 3x^4 - x^2$ in standard form. Then state the degree of the polynomial and the leading coefficient.

SOLUTION For standard form, we write the terms in order from highest degree to lowest.

$$3x^4 - x^2 + 2x - 5 \qquad \text{Standard form}$$

The leading term is $3x^4$, so the degree is 4 and the leading coefficient is 3.

Evaluating Polynomials

Evaluating a polynomial means finding the value of a polynomial for a given value of the variable. To find the value of the polynomial $3x^2 + 1$ when x is 5, we replace x with 5 and simplify the result:

When	$x = 5$
the polynomial	$3x^2 + 1$
becomes	$3(5)^2 + 1 = 3(25) + 1$
	$= 75 + 1$
	$= 76$

EXAMPLE 2 Find the value of $3x^2 - 5x + 4$ when $x = -2$.

SOLUTION

When	$x = -2$
the polynomial	$3x^2 - 5x + 4$
becomes	$3(-2)^2 - 5(-2) + 4 = 3(4) + 10 + 4$
	$= 12 + 10 + 4$
	$= 26$

Addition

There are no new rules for adding one or more polynomials. We rely only on our previous knowledge. Here are some examples.

EXAMPLE 3 Add: $(2x^2 - 5x + 3) + (4x^2 + 7x - 8)$.

SOLUTION We use the commutative and associative properties to group similar terms together and then apply the distributive property to add.

$$(2x^2 - 5x + 3) + (4x^2 + 7x - 8)$$

$$= (2x^2 + 4x^2) + (-5x + 7x) + (3 - 8) \qquad \text{Commutative and associative properties}$$

$$= (2 + 4)x^2 + (-5 + 7)x + (3 - 8) \qquad \text{Distributive property}$$

$$= 6x^2 + 2x - 5 \qquad \text{Addition}$$

The results here indicate that to add two polynomials, we add coefficients of similar terms.

EXAMPLE 4 Add $x^2 + 3x$ and $2x + 6$.

SOLUTION The only similar terms here are the two middle terms. We combine them as usual to get

$$x^2 + 3x + 2x + 6 = x^2 + 5x + 6$$

Subtraction

You will recall from Chapter 1 the definition of subtraction: $a - b = a + (-b)$. To subtract one expression from another, we simply add its opposite. The letters a and b in the definition can each represent polynomials.

Before we can subtract, we need to be able to find the *opposite* of a polynomial. The opposite of a polynomial is the opposite of each of its terms. This is equivalent to multiplying the polynomial by a factor of -1 and then distributing.

EXAMPLE 5 Find the opposite of $3x^2 - 5x + 8$.

SOLUTION To find the opposite, we take the opposite of each term.

$$\begin{aligned} -(3x^2 - 5x + 8) &= -1(3x^2 - 5x + 8) \\ &= -3x^2 + 5x - 8 \end{aligned}$$

Now we are ready to subtract polynomials. When you subtract one polynomial from another you add the opposite of each of its terms.

EXAMPLE 6 Subtract: $(3x^2 + x + 4) - (x^2 + 2x + 3)$.

SOLUTION To subtract $x^2 + 2x + 3$, we change the sign of each of its terms and add.

$$(3x^2 + x + 4) - (x^2 + 2x + 3)$$

$$= 3x^2 + x + 4 - x^2 - 2x - 3 \quad \text{Take the opposite of each term in the second polynomial}$$

$$= (3x^2 - x^2) + (x - 2x) + (4 - 3)$$

$$= 2x^2 - x + 1$$

EXAMPLE 7 Subtract $-4x^2 + 5x - 7$ from $x^2 - x - 1$.

SOLUTION The polynomial $x^2 - x - 1$ comes first, then the subtraction sign, and finally the polynomial $-4x^2 + 5x - 7$ in parentheses.

$$(x^2 - x - 1) - (-4x^2 + 5x - 7)$$

$$= x^2 - x - 1 + 4x^2 - 5x + 7 \quad \text{Take the opposite of each term in the second polynomial}$$

$$= (x^2 + 4x^2) + (-x - 5x) + (-1 + 7)$$

$$= 5x^2 - 6x + 6$$

Getting Ready for Class

After reading through the preceding section, respond in your own words and in complete sentences.

A. What are similar terms?

B. What is the degree of a polynomial?

C. How you would find the value of $3x^2 - 5x + 4$ when x is -2?

D. Describe how you would subtract one polynomial from another.

Identify each of the following polynomials as a trinomial, binomial, or monomial, and give the degree in each case.

1. $2x^3 - 3x^2 + 1$
2. $4x^2 - 4x + 1$
3. $5 + 8a - 9a^3$
4. $6 + 12x^3 + x^4$
5. $2x - 1$
6. $4 + 7x$
7. $45x^2 - 1$
8. $3a^3 + 8$
9. $7a^2$
10. $90x$
11. -4
12. 56

Write each polynomial in standard form. Then identify the degree and leading coefficient.

13. $2 + 5x^2$
14. $5 - 4x^3$
15. $3x^2 - x^3 - 6x$
16. $2x^3 - 6 + x^4 - x$
17. $6x^2 - 1 + x$
18. $4x^2 - 9$

Evaluate each polynomial for the given value of the variable.

19. $4x + 2, x = 3$
20. $2x - 4, x = 5$
21. $x^2 - 9, x = -1$
22. $16 - y^2, y = -2$
23. $3a^2 - 2a + 4, a = 2$
24. $4a^2 + 5a - 9, a = 0$
25. $x^2 - 2x + 1, x = 3$
26. $(x - 1)^2, x = 3$

Perform the following additions.

27. $(2x^2 + 3x + 4) + (3x^2 + 2x + 5)$
28. $(x^2 + 5x + 6) + (x^2 + 3x + 4)$
29. $(3a^2 - 4a + 1) + (2a^2 - 5a + 6)$
30. $(5a^2 - 2a + 7) + (4a^2 - 3a + 2)$
31. $(x^2 + 4x) + (2x + 8)$
32. $(y^2 - 18y) + (2y - 12)$
33. $(x^2 - 3x) + (3x - 9)$
34. $(x^2 - 5x) + (5x - 25)$

Find the opposite of each polynomial.

35. $10x - 5$
36. $2x + 6$
37. $5x^2 + x - 2$
38. $3x^2 - 4x + 1$
39. $3 + 2x - x^2$
40. $12 - x - x^2$

Perform the indicated operation.

41. $(6x^3 - 4x^2 + 2x) - (9x^2 - 6x + 3)$
42. $(5x^3 + 2x^2 + 3x) - (2x^2 + 5x + 1)$
43. $(a^2 - a - 1) - (-a^2 + a + 1)$
44. $(5a^2 - a - 6) - (-3a^2 - 2a + 4)$
45. $(6x^2 - 3x) - (10x - 5)$
46. $(x^2 + 5x) - (3x + 15)$
47. $(3y^2 - 5y) - (6y - 10)$
48. $(10x^2 + 30x) - (2x + 6)$

49. $\left(\frac{2}{3}x^2 - \frac{1}{5}x - \frac{3}{4}\right) + \left(\frac{4}{3}x^2 - \frac{4}{5}x + \frac{7}{4}\right)$

50. $\left(\frac{3}{8}x^3 - \frac{5}{7}x^2 - \frac{2}{5}\right) + \left(\frac{5}{8}x^3 - \frac{2}{7}x^2 + \frac{7}{5}\right)$

51. $\left(\frac{5}{9}x^3 + \frac{1}{3}x^2 - 2x + 1\right) - \left(\frac{2}{3}x^3 + x^2 + \frac{1}{2}x - \frac{3}{4}\right)$

52. $\left(4x^3 - \frac{2}{5}x^2 + \frac{3}{8}x - 1\right) - \left(\frac{9}{2}x^3 + \frac{1}{4}x^2 - x + \frac{5}{6}\right)$

53. $(4y^2 - 3y + 2) + (5y^2 + 12y - 4) - (13y^2 - 6y + 20)$

54. $(2y^2 - 7y - 8) - (6y^2 + 6y - 8) + (4y^2 - 2y + 3)$

55. Subtract $10x^2 + 23x - 50$ from $11x^2 - 10x + 13$.

56. Subtract $2x^2 - 3x + 5$ from $4x^2 - 5x + 10$.

57. Subtract $3y^2 + 7y - 15$ from $11y^2 + 11y + 11$.

58. Subtract $15y^2 - 8y - 2$ from $3y^2 - 3y + 2$.

59. Add $50x^2 - 100x - 150$ to $25x^2 - 50x + 75$.

60. Add $7x^2 - 8x + 10$ to $-8x^2 + 2x - 12$.

61. Subtract $2x + 1$ from the sum of $3x - 2$ and $11x + 5$.

62. Subtract $3x - 5$ from the sum of $5x + 2$ and $9x - 1$.

Applying the Concepts

63. Packaging A crystal ball with a diameter of 6 inches is being packaged for shipment. If the crystal ball is placed inside a circular cylinder with radius 3 inches and height 6 inches, how much volume will need to be filled with padding? (The volume of a sphere with radius r is $\frac{4}{3}\pi r^3$, and the volume of a right circular cylinder with radius r and height h is $\pi r^2 h$.) Use 3.14 to approximate π.

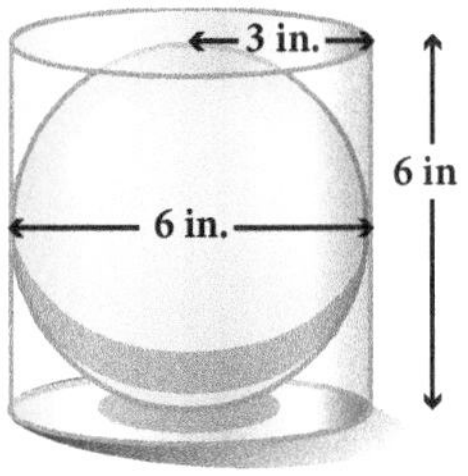

64. Packaging Suppose the circular cylinder of Problem 63 has a radius of 4 inches and a height of 7 inches. How much volume will need to be filled with padding?

Learning Objectives Assessment

The following problems can be used to help assess if you have successfully met the learning objectives for this section.

65. State the degree of the polynomial $2x^3 + x + 3x^4 - 4$.

a. 3 **b.** 2 **c.** -4 **d.** 4

66. Evaluate $2x^2 - 5x - 3$ if $x = -2$.

a. 15 **b.** -1 **c.** -5 **d.** 3

67. Add: $(2x^3 - x^2 + 4) + (3x^2 + x - 2)$.

a. $5x^5 - x^3 + 2$ **b.** $2x^3 + 2x^2 + x + 2$

c. $3x^3 + 2x + 2$ **d.** $2x^3 - 4x^2 - x + 6$

68. Subtract: $(2x^3 - x^2 + 4) - (3x^2 + x - 2)$.

a. $2x^3 - 4x^2 - x + 6$ **b.** $2x^3 + 2x^2 + x + 2$

c. $-x^5 - x^3 + 6$ **d.** $2x^3 - 4x^2 + x + 2$

Getting Ready for the Next Section

Simplify.

69. $(-5)(-1)$

70. $3(-4)$

71. $(-1)(6)$

72. $(-7) \cdot 8$

73. $(5x)(-4x)$

74. $(3x)(2x)$

75. $3x(-7)$

76. $3x(-1)$

77. $5x + (-3x)$

78. $-3x - 10x$

79. $3(2x - 6)$

80. $-4x(x + 5)$

SPOTLIGHT ON SUCCESS *Student Instructor Breylor*

There are three ingredients in the good life: learning, earning and yearning.

—Christopher Morley

It can be hard to improve yourself in life, no matter what you are doing. To succeed and prosper, it is helpful to think about Christopher Morley's quote above. I love to learn new things and value what I earn from learning, but sometimes life can get busy and I think, "I know enough. I can slow down." The real key to improvement is perseverance and yearning for more. Training in martial arts is a passion of mine, and it continues to enthuse me to this day. However, obstacles often pop up that can distract me from my training. In the moment, I find it easy to think, "I'll just skip today." Then I ask myself, "Where would that thinking take me?" I strive for improvement, I yearn for it, and skipping a day of learning will not help me reach my goals and earn the success I seek. This thinking relates to all aspects of life, math included. Improvement only happens if someone has a desire to get better, or a yearning for the knowledge to come.

Multiplication with Polynomials 5.5

Learning Objectives

In this section, we will learn how to:

1. Multiply a polynomial by a monomial.
2. Multiply two binomials using the FOIL method.
3. Multiply polynomials.
4. Solve application problems involving multiplication with polynomials.

Introduction

We begin our discussion of multiplication of polynomials by finding the product of a monomial and a trinomial.

VIDEO EXAMPLES

SECTION 5.5

EXAMPLE 1 Multiply: $3x^2(2x^2 + 4x + 5)$.

SOLUTION Applying the distributive property gives us

$$3x^2(2x^2 + 4x + 5) = 3x^2(2x^2) + 3x^2(4x) + 3x^2(5) \quad \text{Distributive property}$$
$$= 6x^4 + 12x^3 + 15x^2 \quad \text{Multiplication}$$

Multiplying Binomials

The distributive property is the key to multiplication of polynomials. We can use it to find the product of any two polynomials. There are some shortcuts we can use in certain situations, however. Let's look at an example that involves the product of two binomials.

EXAMPLE 2 Multiply: $(3x - 5)(2x - 1)$.

SOLUTION
$$(3x - 5)(2x - 1) = 3x(2x - 1) - 5(2x - 1)$$
$$= 3x(2x) + 3x(-1) + (-5)(2x) + (-5)(-1)$$
$$= 6x^2 - 3x - 10x + 5$$
$$= 6x^2 - 13x + 5$$

If we look closely at the second and third lines of work in this example, we can see that the terms in the answer come from all possible products of terms in the first binomial with terms in the second binomial. This result is generalized as follows.

RULE

To multiply any two polynomials, multiply each term in the first with each term in the second.

There are several ways we can put this rule to work.

FOIL Method

If we look at the original problem in Example 2 and then to the answer, we see that the first term in the answer came from multiplying the first terms in each binomial:

$$3x \cdot 2x = 6x^2 \qquad \text{First}$$

The middle term in the answer came from adding the products of the two outside terms with the two inside terms in each binomial:

$$\begin{aligned} 3x(-1) &= -3x \qquad \text{Outside} \\ -5(2x) &= -10x \qquad \text{Inside} \\ \hline &= -13x \end{aligned}$$

The last term in the answer came from multiplying the two last terms:

$$-5(-1) = 5 \qquad \text{Last}$$

The word FOIL is a mnemonic to help us remember how we obtain the four products that result when we multiply two binomials:

F	O	I	L
First	Outside	Inside	Last

To summarize the FOIL method, we will multiply another two binomials.

EXAMPLE 3 Multiply: $(2x + 3)(5x - 4)$.

SOLUTION $(2x + 3)(5x - 4) = \underbrace{2x(5x)}_{\text{First}} + \underbrace{2x(-4)}_{\text{Outside}} + \underbrace{3(5x)}_{\text{Inside}} + \underbrace{3(-4)}_{\text{Last}}$

$$= 10x^2 - 8x + 15x - 12$$

$$= 10x^2 + 7x - 12$$

With practice $-8x + 15x = 7x$ can be done mentally.

EXAMPLE 4 Multiply:

a. $4a^2(2a^2 - 3a + 5) = 4a^2(2a^2) + 4a^2(-3a) + 4a^2(5)$

$$= 8a^4 - 12a^3 + 20a^2$$

b. $(x - 2)(y + 3) = \underset{F}{x(y)} + \underset{O}{x(3)} + \underset{I}{(-2)(y)} + \underset{L}{(-2)(3)}$

$$= xy + 3x - 2y - 6$$

c. $(x + y)(a - b) = \underset{F}{x(a)} + \underset{O}{x(-b)} + \underset{I}{y(a)} + \underset{L}{y(-b)}$

$$= xa - xb + ya - yb$$

d. $(5x - 1)(2x + 6) = \underset{F}{5x(2x)} + \underset{O}{5x(6)} + \underset{I}{(-1)(2x)} + \underset{L}{(-1)(6)}$

$$= 10x^2 + 30x + (-2x) + (-6)$$

$$= 10x^2 + 28x - 6$$

Multiplying Polynomials

The FOIL method can be applied only when multiplying two binomials. To find products of polynomials with more than two terms, we once again apply the distributive property.

EXAMPLE 5 Multiply: $(x^2 - 4x + 2)(3x^2 + x - 5)$.

SOLUTION First, we use the distributive property to multiply each term in the first polynomial by the second polynomial.

$$(x^2 - 4x + 2)(3x^2 + x - 5) =$$

$$= x^2(3x^2 + x - 5) - 4x(3x^2 + x - 5) + 2(3x^2 + x - 5)$$

Now we use the distributive property three more times to perform the multiplication by each monomial.

$$= 3x^4 + x^3 - 5x^2 + (-12x^3 - 4x^2 + 20x) + (6x^2 + 2x - 10)$$

$$= 3x^4 + (x^3 - 12x^3) + (-5x^2 - 4x^2 + 6x^2) + (20x + 2x) - 10$$

$$= 3x^4 - 11x^3 - 3x^2 + 22x - 10$$

The column method of multiplying two polynomials is very similar to long multiplication with whole numbers. It is just another way of finding all possible products of terms in one polynomial with terms in another polynomial.

EXAMPLE 6 Multiply: $(2x + 3)(3x^2 - 2x + 1)$.

SOLUTION

$$\begin{array}{rl}
3x^2 - 2x + 1 & \\
\times \qquad\qquad 2x + 3 & \\
\hline
6x^3 - 4x^2 + 2x \qquad & \leftarrow 2x(3x^2 - 2x + 1) \\
9x^2 - 6x + 3 & \leftarrow 3(3x^2 - 2x + 1) \\
\hline
6x^3 + 5x^2 - 4x + 3 & \leftarrow \text{Add similar terms}
\end{array}$$

A third way to multiply polynomials is the rectangle method. We create a rectangle with each term in the first polynomial at the start of a row and each term in the second polynomial at the top of a column. We then fill in each box in the rectangle by multiplying the corresponding monomials. Our next example illustrates this process.

EXAMPLE 7 Multiply: $(3x - 2)(4x^2 + x + 3)$.

SOLUTION We create a rectangle with two rows and three columns, placing the terms of $3x - 2$ along one side and the terms of $4x^2 + x + 3$ along the other as follows.

	$4x^2$	x	3
$3x$			
-2			

The entry in each box is found by multiplying the monomial at the left with the monomial above.

	$4x^2$	x	3
$3x$	$12x^3$	$3x^2$	$9x$
-2	$-8x^2$	$-2x$	-6

Notice that our rectangle contains six terms:

(binomial $\times$ trinomial $=$ 2 terms $\times$ 3 terms $=$ 6 terms)

Combining like terms gives us the final result.

$$= 12x^3 + (3x^2 - 8x^2) + (9x - 2x) - 6$$

$$= 12x^3 - 5x^2 + 7x - 6$$

It will be to your advantage to become very fast and accurate at multiplying polynomials. Use the method you are most comfortable with and make the fewest mistakes with.

Applications

EXAMPLE 8 The length of a rectangle is 3 more than twice the width. Write an expression for the area of the rectangle.

SOLUTION We begin by drawing a rectangle and labeling the width with x. Since the length is 3 more than twice the width, we label the length with $2x + 3$.

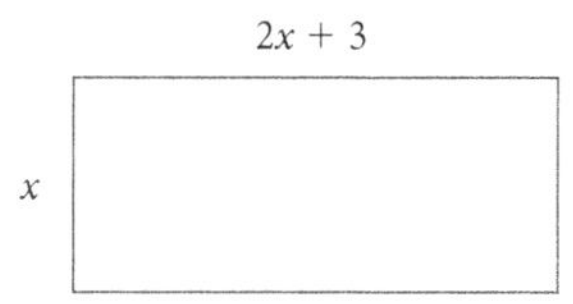

The area A of a rectangle is the product of the length and width, so we write our formula for the area of this rectangle as

$$A = x(2x + 3)$$

$$A = 2x^2 + 3x \quad \text{Multiply}$$

Revenue

Suppose that a store sells x items at p dollars per item. The total amount of money obtained by selling the items is called the *revenue*. It can be found by multiplying the number of items sold, x, by the price per item, p. For example, if 100 items are sold for \$6 each, the revenue is $100(6) = \$600$. Similarly, if 500 items are sold for \$8 each, the total revenue is $500(8) = \$4{,}000$. If we denote the revenue with the letter R, then the formula that relates R, x, and p is

$$R = xp$$

In words: Revenue = (number of items sold)(price of each item)

EXAMPLE 9 A store selling cases for cell phones knows from past experience that it can sell x cases each day at a price of p dollars per case, according to the equation $x = 800 - 100p$. Write a formula for the daily revenue that involves only the variables R and p.

SOLUTION From our previous discussion we know that the revenue R is given by the formula

$$R = xp$$

But, since $x = 800 - 100p$, we can substitute $800 - 100p$ for x in the revenue equation to obtain

$$R = (800 - 100p)p$$

$$R = 800p - 100p^2$$

This last formula gives the revenue, R, in terms of the price, p.

Getting Ready for Class

After reading through the preceding section, respond in your own words and in complete sentences.

A. How do we multiply two polynomials?

B. Describe how the distributive property is used to multiply a monomial and a polynomial.

C. Describe how you would use the foil method to multiply two binomials.

D. Explain how you would multiply two trinomials.

Problem Set 5.5

Multiply the following by applying the distributive property.

1. $2x(3x + 1)$
2. $4x(2x - 3)$
3. $2x^2(3x^2 - 2x + 1)$
4. $5x(4x^3 - 5x^2 + x)$
5. $2ab(a^2 - ab + 1)$
6. $3a^2b(a^3 + a^2b^2 + b^3)$
7. $y^2(3y^2 + 9y + 12)$
8. $5y(2y^2 - 3y + 5)$
9. $4x^2y(2x^3y + 3x^2y^2 + 8y^3)$
10. $6xy^3(2x^2 + 5xy + 12y^2)$

Multiply the following binomials using the FOIL method.

11. $(x + 3)(x + 4)$
12. $(x + 2)(x + 5)$
13. $(x + 6)(x + 1)$
14. $(x + 1)(x + 4)$
15. $\left(x + \frac{1}{2}\right)\left(x + \frac{3}{2}\right)$
16. $\left(x + \frac{3}{5}\right)\left(x + \frac{2}{5}\right)$
17. $(a + 5)(a - 3)$
18. $(a - 8)(a + 2)$
19. $(x - a)(y + b)$
20. $(x + a)(y - b)$
21. $(x + 6)(x - 6)$
22. $(x + 3)(x - 3)$
23. $\left(y + \frac{5}{6}\right)\left(y - \frac{5}{6}\right)$
24. $\left(y - \frac{4}{7}\right)\left(y + \frac{4}{7}\right)$
25. $(2x - 3)(x - 4)$
26. $(3x - 5)(x - 2)$
27. $(a + 2)(2a - 1)$
28. $(a - 6)(3a + 2)$
29. $(2x - 5)(3x - 2)$
30. $(3x + 6)(2x - 1)$
31. $(2x + 3)(a + 4)$
32. $(2x - 3)(a - 4)$
33. $(5x - 4)(5x + 4)$
34. $(6x + 5)(6x - 5)$
35. $\left(2x - \frac{1}{2}\right)\left(x + \frac{3}{2}\right)$
36. $\left(4x - \frac{3}{2}\right)\left(x + \frac{1}{2}\right)$
37. $(1 - 2a)(3 - 4a)$
38. $(1 - 3a)(3 + 2a)$

Multiply the following polynomials using one of the three methods described in this section.

39. $(a - 3)(a^2 - 3a + 2)$
40. $(a + 5)(a^2 + 2a + 3)$
41. $(x + 2)(x^2 - 2x + 4)$
42. $(x + 3)(x^2 - 3x + 9)$
43. $(2x + 1)(x^2 + 8x + 9)$
44. $(3x - 2)(x^2 - 7x + 8)$
45. $(5x^2 + 2x + 1)(x^2 - 3x + 5)$
46. $(2x^2 + x + 1)(x^2 - 4x + 3)$
47. $(3x^2 - 5x - 2)(2x^2 + x - 1)$
48. $(4x^2 + 2x - 3)(3x^2 - 2x + 3)$
49. $(a^3 + a + 2)(a^2 - 3a + 4)$
50. $(a^3 - 2a^2 - 3)(2a^2 + a + 1)$
51. $(x^2 + 3)(2x^2 - 5)$
52. $(4x^3 - 8)(5x^3 + 4)$
53. $(3a^4 + 2)(2a^2 + 5)$
54. $(7a^4 - 8)(4a^3 - 6)$
55. $(x + 3)(x + 4)(x + 5)$
56. $(x - 3)(x - 4)(x - 5)$

Simplify each expression.

57. $(x - 3)(x - 2) + 2$
58. $(2x - 5)(3x + 2) - 4$
59. $(2x - 3)(4x + 3) + 4$
60. $(3x + 8)(5x - 7) + 52$
61. $(x + 4)(x - 5) + (-5)(2)$
62. $(x + 3)(x - 4) + (-4)(2)$
63. $2(x - 3) + x(x + 2)$
64. $5(x + 3) + 1(x + 4)$
65. $3x(x + 1) - 2x(x - 5)$
66. $4x(x - 2) - 3x(x - 4)$
67. $x(x + 2) - 3$
68. $2x(x - 4) + 6$
69. $a(a - 3) + 6$
70. $a(a - 4) + 8$

For each of the following problems, fill in the area of each small rectangle and square, and then add the results together to find the indicated product.

71. $(x + 2)(x + 3)$

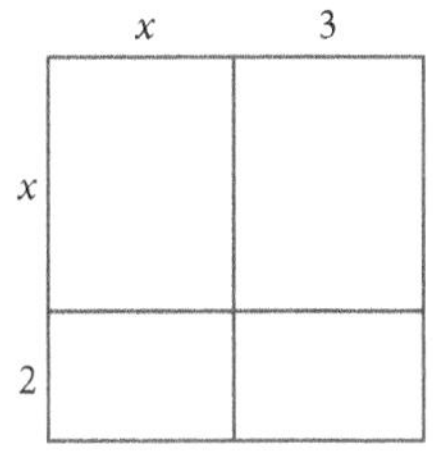

72. $(x + 4)(x + 5)$

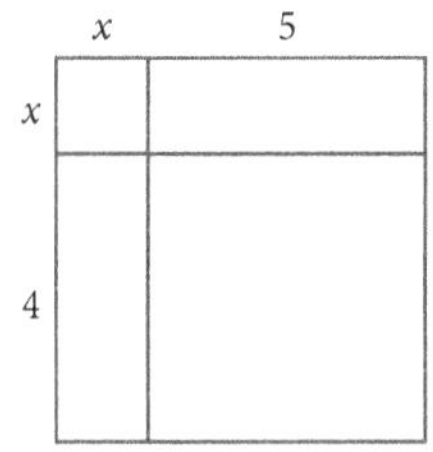

73. $(x + 1)(2x + 2)$

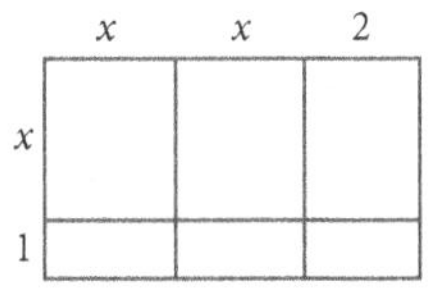

74. $(2x + 1)(2x + 2)$

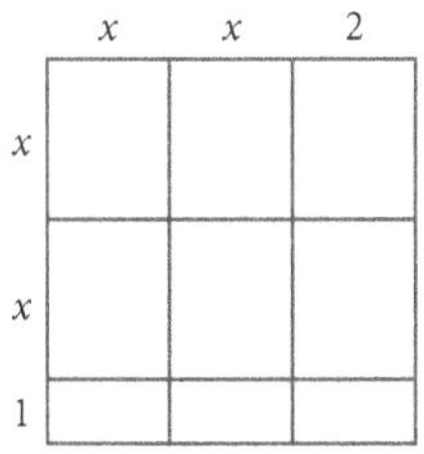

75. Paying Attention to Instructions Work each problem according to the instructions given.

a. Add $2x + 5$ and $3x - 4$.

b. Subtract $2x + 5$ from $3x - 4$.

c. Solve $2x + 5 = 3x - 4$.

d. Multiply $2x + 5$ by $3x - 4$.

76. Paying Attention to Instructions Work each problem according to the instructions given.

a. Add $2x - 1$ and $x^2 + 3x + 4$.

b. Solve $2x - 1 = 0$

c. Multiply $2x - 1$ by $x^2 + 3x + 4$.

d. Subtract $2x - 1$ from $x^2 + 3x + 4$.

Applying the Concepts

77. Area The length of a rectangle is 5 units more than twice the width. Write an expression for the area of the rectangle.

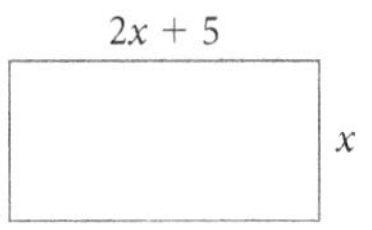

78. Area The length of a rectangle is 2 more than three times the width. Write an expression for the area of the rectangle.

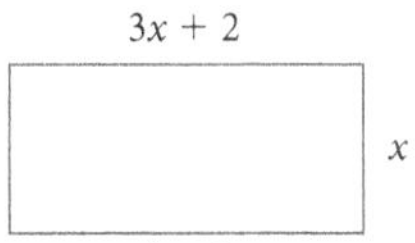

79. Area The width and length of a rectangle are given by two consecutive integers. Write an expression for the area of the rectangle.

80. Area The width and length of a rectangle are given by two consecutive even integers. Write an expression for the area of the rectangle.

81. Revenue A stationery store can sell x binders each day at a price of p dollars per binder, according to the equation $x = 100 - 10p$. Write a formula for the daily revenue that involves only the variables R and p.

82. Revenue A surf shop can sell x packs of board wax each day at a price of p dollars per pack, according to the equation $x = 40 - 5p$. Write a formula for the daily revenue that involves only the variables R and p.

Learning Objectives Assessment

The following problems can be used to help assess if you have successfully met the learning objectives for this section.

83. Multiply: $3x^4(2x^2 - 5x + 4)$.

a. $5x^6 - 2x^5 + 7x^4$ **b.** $3x^{15}$

c. $6x^8 - 3x^4$ **d.** $6x^6 - 15x^5 + 12x^4$

84. Multiply $(2x + 7)(x - 4)$ using the FOIL method.

a. $2x^2 - 28$ **b.** $2x^2 + 5x + 3$

c. $2x^2 - 15x - 28$ **d.** $2x^2 - x - 28$

85. Multiply: $(x + 4)(2x^2 - x - 3)$.

a. $2x^3 + 8x^2 - x - 3$ **b.** $2x^3 - 9x^2 + x - 12$

c. $2x^3 + 7x^2 - 7x - 12$ **d.** $2x^2 + 1$

86. Area The length of a rectangle is 4 units less than twice the width. Write an expression for the area of the rectangle if x represents the width.

a. $2x^2 - 4x$ **b.** $2x^2 - 8x$ **c.** $4x - 2x^2$ **d.** $x^3 - 4x$

Getting Ready for the Next Section

Simplify.

87. $13 \cdot 13$

88. $3x \cdot 3x$

89. $2(x)(-5)$

90. $2(2x)(-3)$

91. $6x + (-6x)$

92. $3x + (-3x)$

93. $(2x)(-3) + (2x)(3)$

94. $(2x)(-5y) + (2x)(5y)$

Multiply.

95. $-4(3x - 4)$

96. $-2x(2x + 7)$

97. $(x - 1)(x + 2)$

98. $(x + 5)(x - 6)$

99. $(x + 3)(x + 3)$

100. $(3x - 2)(3x - 2)$

Binomial Squares and Other Special Products 5.6

Learning Objectives

In this section, we will learn how to:

1. Find the square of a binomial.
2. Multiply binomials that differ only in the sign between their terms.

Introduction

In this section we will combine the results of the last section with our definition of exponents to find some special products.

VIDEO EXAMPLES

SECTION 5.6

EXAMPLE 1 Find the square of $(3x - 2)$.

SOLUTION To square $(3x - 2)$, we multiply it by itself:

$(3x - 2)^2 = (3x - 2)(3x - 2)$	Definition of exponents
$= 9x^2 - 6x - 6x + 4$	FOIL method
$= 9x^2 - 12x + 4$	Combine similar terms

Notice that the first and last terms in the answer are the square of the first and last terms in the original problem and that the middle term is twice the product of the two terms in the original binomial.

EXAMPLE 2 Expand and multiply each expression.

a. $(a + b)^2$ **b.** $(a - b)^2$

SOLUTION

a. $(a + b)^2 = a^2 + ab + ab + b^2$

$= a^2 + 2ab + b^2$

b. $(a - b)^2 = a^2 - ab - ab + b^2$

$= a^2 - 2ab + b^2$

Note A very common mistake when squaring binomials is to write

$$(a + b)^2 = a^2 + b^2$$

which just isn't true. The mistake becomes obvious when we substitute 2 for a and 3 for b:

$$(2 + 3)^2 \neq 2^2 + 3^2$$

$$25 \neq 13$$

Exponents do not distribute over addition or subtraction.

Binomial Squares

Binomial squares having the form of Example 2 occur very frequently in algebra. It will be to your advantage to memorize the following rule for squaring a binomial.

RULE

The square of a binomial is a trinomial containing the square of the first term, the square of the last term, and twice the product of the two original terms. In symbols this rule is written as follows:

$(a + b)^2$	$=$	a^2	$+$	$2ab$	$+$	b^2
		Square of first term		Twice product of the two terms		Square of last term
$(a - b)^2$	$=$	a^2	$-$	$2ab$	$+$	b^2

EXAMPLES Multiply using the preceding rule:

		First term squared		*Twice their product*		*Last term squared*		*Answer*
3. $(x-5)^2$	=	x^2	−	$2(x)(5)$	+	5^2	=	$x^2 - 10x + 25$
4. $(x+2)^2$	=	x^2	+	$2(x)(2)$	+	2^2	=	$x^2 + 4x + 4$
5. $(2x-3y)^2$	=	$(2x)^2$	−	$2(2x)(3y)$	+	$(3y)^2$	=	$4x^2 - 12xy + 9y^2$
6. $(5x^2-4)^2$	=	$(5x^2)^2$	−	$2(5x^2)(4)$	+	4^2	=	$25x^4 - 40x^2 + 16$

Difference of Squares

Another special product that occurs frequently is $(a + b)(a - b)$. The only difference in the two binomials is the sign between the two terms. Here are some examples.

EXAMPLES Multiply using the FOIL method:

7. $(2x - 3)(2x + 3) = 4x^2 + 6x - 6x - 9$ FOIL method
$= 4x^2 - 9$

8. $(x - 5)(x + 5) = x^2 + 5x - 5x - 25$ FOIL method
$= x^2 - 25$

9. $(3x - 1)(3x + 1) = 9x^2 + 3x - 3x - 1$ FOIL method
$= 9x^2 - 1$

Notice that in each case the middle term is zero and therefore we get a binomial instead of a trinomial. The answers all turn out to be the difference of two squares. Here is a rule to help you memorize the result.

RULE

When multiplying two binomials that differ only in the sign between their terms, subtract the square of the last term from the square of the first term.

$$(a - b)(a + b) = a^2 - b^2$$

Here are some problems that result in the difference of two squares.

EXAMPLES Multiply using the preceding rule:

		First term squared		*Last term squared*		*Answer*
10. $(x+3)(x-3)$	=	x^2	−	3^2	=	$x^2 - 9$
11. $(a+2)(a-2)$	=	a^2	−	2^2	=	$a^2 - 4$
12. $(1+9a)(1-9a)$	=	1^2	−	$(9a)^2$	=	$1 - 81a^2$
13. $(2x-5y)(2x+5y)$	=	$(2x)^2$	−	$(5y)^2$	=	$4x^2 - 25y^2$
14. $(3a^2-7b^2)(3a^2+7b^2)$	=	$(3a^2)^2$	−	$(7b^2)^2$	=	$9a^4 - 49b^4$

Although all the problems in this section can be worked correctly using the methods in the previous section, they can be done much faster if the two rules are *memorized.* Here is a summary of the two rules:

$$(a + b)^2 = (a + b)(a + b) = a^2 + 2ab + b^2$$
$$(a - b)^2 = (a - b)(a - b) = a^2 - 2ab + b^2$$
$$(a - b)(a + b) = a^2 - b^2$$

EXAMPLE 15 Write an expression in symbols for the sum of the squares of three consecutive even integers. Then, simplify that expression.

SOLUTION If we let $x =$ the first of the even integers, then $x + 2$ is the next consecutive even integer, and $x + 4$ is the one after that. An expression for the sum of their squares is

$x^2 + (x + 2)^2 + (x + 4)^2$	Sum of squares
$= x^2 + (x^2 + 4x + 4) + (x^2 + 8x + 16)$	Expand squares
$= 3x^2 + 12x + 20$	Add similar terms

Getting Ready for Class

After reading through the preceding section, respond in your own words and in complete sentences.

A. Explain why $(x + 3)^2 \neq x^2 + 9$.

B. What kind of products result in the difference of two squares?

C. When multiplied out, how will $(x + 3)^2$ and $(x - 3)^2$ differ?

D. Explain how to use the rule for binomial squares to find $(2x - 3)^2$.

Problem Set 5.6

Perform the indicated operations.

1. $(x-2)^2$
2. $(x+2)^2$
3. $(a+3)^2$
4. $(a-3)^2$
5. $(x-5)^2$
6. $(x-4)^2$
7. $\left(a-\frac{1}{2}\right)^2$
8. $\left(a+\frac{1}{2}\right)^2$
9. $(x+10)^2$
10. $(x-10)^2$
11. $(a+0.8)^2$
12. $(a-0.4)^2$
13. $(2x-1)^2$
14. $(3x+2)^2$
15. $(4a+5)^2$
16. $(4a-5)^2$
17. $(3x-2)^2$
18. $(2x-3)^2$
19. $(3a+5b)^2$
20. $(5a-3b)^2$
21. $(4x-5y)^2$
22. $(5x+4y)^2$
23. $(x^2+5)^2$
24. $(x^2+3)^2$
25. $(a^3+1)^2$
26. $(a^3-2)^2$
27. $(7m^2+2n)^2$
28. $(2m^2-7n)^2$
29. $(6x^2-10y^2)^2$
30. $(10x^2+6y^2)^2$

Comparing Expressions Fill in each table.

31.

x	$(x+3)^2$	x^2+9	x^2+6x+9
1			
2			
3			
4			

32.

x	$(x-5)^2$	x^2+25	$x^2-10x+25$
1			
2			
3			
4			

33.

a	1	3	3	4
b	1	5	4	5
$(a+b)^2$				
a^2+b^2				
a^2+ab+b^2				
$a^2+2ab+b^2$				

34.

a	2	5	2	4
b	1	2	5	3
$(a-b)^2$				
a^2-b^2				
$a^2-2ab+b^2$				

Multiply.

35. $(a + 5)(a - 5)$
36. $(a - 6)(a + 6)$
37. $(y - 1)(y + 1)$
38. $(y - 2)(y + 2)$
39. $(9 + x)(9 - x)$
40. $(10 - x)(10 + x)$
41. $(2x + 5)(2x - 5)$
42. $(3x + 5)(3x - 5)$
43. $\left(4x + \frac{1}{3}\right)\left(4x - \frac{1}{3}\right)$
44. $\left(6x + \frac{1}{4}\right)\left(6x - \frac{1}{4}\right)$
45. $(2a + 7b)(2a - 7b)$
46. $(3a + 10b)(3a - 10b)$
47. $(6 - 7x)(6 + 7x)$
48. $(7 - 6x)(7 + 6x)$
49. $(x^2 + 3)(x^2 - 3)$
50. $(x^2 + 2)(x^2 - 2)$
51. $(a^2 + 4b^2)(a^2 - 4b^2)$
52. $(a^2 + 9b^2)(a^2 - 9b^2)$
53. $(5y^4 - 8)(5y^4 + 8)$
54. $(7y^5 + 6)(7y^5 - 6)$

Multiply and simplify.

55. $(x + 3)(x - 3) + (x - 5)(x + 5)$
56. $(x - 7)(x + 7) + (x - 4)(x + 4)$
57. $(2x + 3)^2 - (4x - 1)^2$
58. $(3x - 5)^2 - (2x + 3)^2$
59. $(a + 1)^2 - (a + 2)^2 + (a + 3)^2$
60. $(a - 1)^2 + (a - 2)^2 - (a - 3)^2$
61. $(2x + 3)^3$
62. $(3x - 2)^3$

Applying the Concepts

63. Shortcut The formula for the difference of two squares can be used as a shortcut to multiplying certain whole numbers if they have the correct form. Use the difference of two squares formula to multiply 49(51) by first writing 49 as $(50 - 1)$ and 51 as $(50 + 1)$.

64. Shortcut Use the difference of two squares formula to multiply 101(99) by first writing 101 as $(100 + 1)$ and 99 as $(100 - 1)$.

65. Comparing Expressions Evaluate the expression $(x + 3)^2$ and the expression $x^2 + 6x + 9$ for $x = 2$.

66. Comparing Expressions Evaluate the expression $x^2 - 25$ and the expression $(x - 5)(x + 5)$ for $x = 6$.

67. Number Problem Write an expression for the sum of the squares of two consecutive integers. Then, simplify that expression.

68. Number Problem Write an expression for the sum of the squares of two consecutive odd integers. Then, simplify that expression.

69. Number Problem Write an expression for the sum of the squares of three consecutive integers. Then, simplify that expression.

70. Number Problem Write an expression for the sum of the squares of three consecutive odd integers. Then, simplify that expression.

For each problem, fill in the area of each small rectangle and square, and then add the results together to find the indicated product.

71. Area We can use the concept of area to further justify our rule for squaring a binomial. The length of each side of the square shown in the figure is $a + b$. (The longer line segment has length a and the shorter line segment has length b.) The area of the whole square is $(a + b)^2$. However, the whole area is the sum of the areas of the two smaller squares and the two smaller rectangles that make it up. Write the area of the two smaller squares and the two smaller rectangles and then add them together to verify the formula $(a + b)^2 = a^2 + 2ab + b^2$.

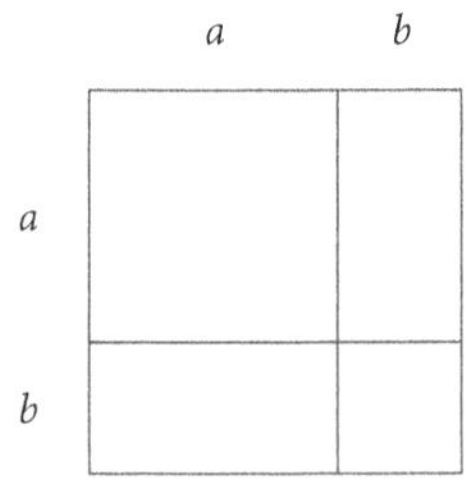

72. Area The length of each side of the large square shown in the figure is $x + 5$. Therefore, its area is $(x + 5)^2$. Find the area of the two smaller squares and the two smaller rectangles that make up the large square, then add them together to verify the formula $(x + 5)^2 = x^2 + 10x + 25$.

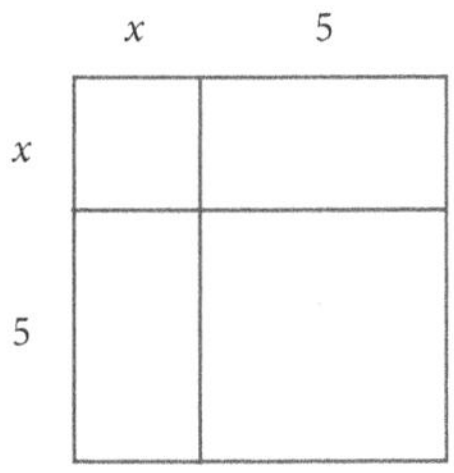

Learning Objectives Assessment

The following problems can be used to help assess if you have successfully met the learning objectives for this section.

73. Simplify: $(3x - 4y)^2$.

a. $9x^2 + 16y^2$ **b.** $9x^2 - 24xy + 16y^2$

c. $9x^2 - 16y^2$ **d.** $9x^2 - 12xy + 16y^2$

74. Multiply: $(5x + 2)(5x - 2)$.

a. $25x^2 + 20x - 4$ **b.** $25x^2 - 20x - 4$

c. $25x^2 + 4$ **d.** $25x^2 - 4$

Getting Ready for the Next Section

Simplify.

75. $\frac{10x^3}{5x}$ **76.** $\frac{15x^2}{5x}$ **77.** $\frac{3x^2}{3}$ **78.** $\frac{4x^2}{2}$

79. $\frac{9x^2}{3x}$ **80.** $\frac{3x^4}{3x^2}$ **81.** $\frac{24x^3y^2}{8x^2y}$ **82.** $\frac{4xy^3}{8x^2y}$

Divide.

83. $27\overline{)3{,}962}$

84. $13\overline{)18{,}780}$

Multiply.

85. $(x - 3)x$

86. $(x - 3)(-2)$

87. $2x^2(x - 5)$

88. $10x(x - 5)$

Subtract.

89. $(x^2 - 5x) - (x^2 - 3x)$

90. $(2x^3 + 0x^2) - (2x^3 - 10x^2)$

91. $(-2x + 8) - (-2x + 6)$

92. $(4x - 14) - (4x - 10)$

Division with Polynomials 5.7

Learning Objectives

In this section, we will learn how to:

1. Divide a polynomial by a monomial.
2. Divide a polynomial by a polynomial.

Dividing by a Monomial

To divide a polynomial by a monomial, we will use the definition of division and apply the distributive property. Follow the steps in this example closely.

VIDEO EXAMPLES

SECTION 5.7

EXAMPLE 1 Divide $10x^3 - 15x^2$ by $5x$.

SOLUTION

$\dfrac{10x^3 - 15x^2}{5x} = (10x^3 - 15x^2)\dfrac{1}{5x}$	Division by $5x$ is the same as multiplication by $\frac{1}{5x}$
$= 10x^3\left(\dfrac{1}{5x}\right) - 15x^2\left(\dfrac{1}{5x}\right)$	Distribute $\frac{1}{5x}$ to both terms
$= \dfrac{10x^3}{5x} - \dfrac{15x^2}{5x}$	Multiplication by $\frac{1}{5x}$ is the same as division by $5x$
$= 2x^2 - 3x$	Division of monomials as done in Section 5.3

If we were to leave out the first steps, the problem would look like this:

$$\frac{10x^3 - 15x^2}{5x} = \frac{10x^3}{5x} - \frac{15x^2}{5x}$$
$$= 2x^2 - 3x$$

The problem is much shorter and clearer this way. You may leave out the first two steps from Example 1 when working problems in this section. They are part of Example 1 only to help show you why the following rule is true.

RULE

To divide a polynomial by a monomial, simply divide each term in the polynomial by the monomial.

Here are some further examples using our rule for division of a polynomial by a monomial.

EXAMPLE 2 Divide: $\dfrac{3x^2 - 6}{3}$.

SOLUTION We begin by writing the 3 in the denominator under each term in the numerator. Then we simplify the result:

$\dfrac{3x^2 - 6}{3} = \dfrac{3x^2}{3} - \dfrac{6}{3}$	Divide each term in the numerator by 3
$= x^2 - 2$	Simplify

EXAMPLE 3 Find the quotient of $27x^3 - 9x^2$ and $3x$.

SOLUTION We again are asked to divide the first polynomial by the second one:

$$\frac{27x^3 - 9x^2}{3x} = \frac{27x^3}{3x} - \frac{9x^2}{3x} \qquad \text{Divide each term by } 3x$$

$$= 9x^2 - 3x \qquad \text{Simplify}$$

EXAMPLE 4 Divide: $(15x^2y - 21xy^2) \div (-3xy)$.

SOLUTION This is the same type of problem we have shown in the first three examples; it is just worded a little differently. Note that when we divide each term in the first polynomial by $-3xy$, the negative sign must be taken into account:

$$\frac{15x^2y - 21xy^2}{-3xy} = \frac{15x^2y}{-3xy} - \frac{21xy^2}{-3xy} \qquad \text{Divide each term by } -3xy$$

$$= -5x - (-7y) \qquad \text{Simplify}$$

$$= -5x + 7y \qquad \text{Simplify}$$

EXAMPLE 5 Divide: $\dfrac{24x^3y^2 + 16x^2y^2 - 4xy^3}{8x^2y}$.

SOLUTION Writing $8x^2y$ under each term in the numerator and then simplifying, we have

$$\frac{24x^3y^2 + 16x^2y^2 - 4xy^3}{8x^2y} = \frac{24x^3y^2}{8x^2y} + \frac{16x^2y^2}{8x^2y} - \frac{4xy^3}{8x^2y}$$

$$= 3xy + 2y - \frac{y^2}{2x}$$

From the first five examples, it is clear that to divide a polynomial by a monomial, we must divide each term in the polynomial by the monomial. Often, students taking algebra for the first time will make the following mistake:

$$\frac{x+2}{2} = x + 1 \qquad \text{Mistake}$$

The mistake here is in not dividing both terms in the numerator by 2. The correct way to divide $x + 2$ by 2 looks like this:

$$\frac{x+2}{2} = \frac{x}{2} + \frac{2}{2} = \frac{x}{2} + 1 \qquad \text{Correct}$$

Dividing a Polynomial by a Polynomial

Since long division for polynomials is very similar to long division with whole numbers, we will begin by reviewing a division problem with whole numbers. You may realize when looking at Example 6 that you don't have a very good idea why you proceed as you do with long division. What you do know is that the process always works. We are going to approach the explanations for the following examples in much the same manner; that is, we won't always be sure why the steps we will use are important, only that they always produce the correct result.

EXAMPLE 6 Divide: $27\overline{)3{,}962}$.

SOLUTION The divisor is 27 and the dividend is 3,962.

$$\begin{array}{r l}
1 & \leftarrow \text{Estimate 27 into 39} \\
27\overline{)3{,}962} & \\
2\,7 & \leftarrow \text{Multiply } 1 \times 27 = 27 \\
\hline
1\,2 & \leftarrow \text{Subtract } 39 - 27 = 12
\end{array}$$

$$\begin{array}{r l}
1 & \\
27\overline{)3{,}962} & \\
2\,7\downarrow & \\
\hline
1\,26 & \leftarrow \text{Bring down the 6}
\end{array}$$

These are the four basic steps in long division. Estimate, multiply, subtract, and bring down the next term. To finish the problem, we simply perform the same four steps again:

$$\begin{array}{r l}
14 & \leftarrow \text{4 is the estimate} \\
27\overline{)3{,}962} & \\
2\,7\downarrow & \\
\hline
1\,26 & \\
1\,08\downarrow & \leftarrow \text{Multiply to get 108} \\
\hline
182 & \leftarrow \text{Subtract to get 18, then bring down the 2}
\end{array}$$

One more time.

$$\begin{array}{r l}
146 & \leftarrow \text{6 is the estimate} \\
27\overline{)3{,}962} & \\
2\,7 & \\
\hline
1\,26 & \\
1\,08 & \\
\hline
182 & \\
162 & \leftarrow \text{Multiply to get 162} \\
\hline
20 & \leftarrow \text{Subtract to get 20}
\end{array}$$

Since there is nothing left to bring down, we have our answer. The quotient is 146 and the remainder is 20. We write the result in the form

$$\frac{\text{Dividend}}{\text{Divisor}} = \text{Quotient} + \frac{\text{Remainder}}{\text{Divisor}}$$

In this case, we have

$$\frac{3{,}962}{27} = 146 + \frac{20}{27} \quad \text{or} \quad 146\frac{20}{27}$$

Here is how it works with polynomials.

EXAMPLE 7 Divide: $\dfrac{x^2 - 5x + 8}{x - 3}$.

SOLUTION The dividend is $x^2 - 5x + 8$ and the divisor is $x - 3$.

$$\begin{array}{rl} x & \leftarrow \text{Estimate } x^2 \div x = x \\ x - 3\overline{)\ x^2 - 5x + 8} & \\ \overset{-}{\cancel{+}}\, x^2 \overset{+}{\cancel{-}}\, 3x & \leftarrow \text{Multiply } x(x - 3) = x^2 - 3x \\ \hline -\ 2x & \leftarrow \text{Subtract } (x^2 - 5x) - (x^2 - 3x) = -2x \end{array}$$

$$\begin{array}{rl} x & \\ x - 3\overline{)\ x^2 - 5x + 8} & \\ \overset{-}{\cancel{+}}\, x^2 \overset{+}{\cancel{-}}\, 3x \quad \downarrow & \\ \hline -\ 2x + 8 & \leftarrow \text{Bring down the 8} \end{array}$$

Notice that to subtract one polynomial from another, we add its opposite. That is why we change the signs on $x^2 - 3x$ and add what we get to $x^2 - 5x$. (To subtract the second polynomial, simply change the signs and add.)

We perform the same four steps again:

$$\begin{array}{rl} x - 2 & \leftarrow -2 \text{ is the estimate } (-2x \div x = -2) \\ x - 3\overline{)\ x^2 - 5x + 8} & \\ \overset{-}{\cancel{+}}\, x^2 \overset{+}{\cancel{-}}\, 3x \quad \downarrow & \\ \hline -\ 2x + 8 & \\ \overset{+}{\cancel{-}}\, 2x \overset{-}{\cancel{+}}\, 6 & \leftarrow \text{Multiply } -2(x - 3) = -2x + 6. \\ \hline 2 & \leftarrow \text{Subtract } (-2x + 8) - (-2x + 6) = 2 \end{array}$$

Since there is nothing left to bring down, we have our answer: The quotient is $x - 2$ and the remainder is 2. We write

$$\frac{x^2 - 5x + 8}{x - 3} = x - 2 + \frac{2}{x - 3}$$

To check our answer, we multiply $(x - 3)(x - 2)$ to get $x^2 - 5x + 6$. Then, adding on the remainder, 2, we have $x^2 - 5x + 8$.

EXAMPLE 8 Divide: $\dfrac{6x^3 - 11x^2 - 14x + 3}{2x^2 - 5x - 1}$.

SOLUTION

$$\begin{array}{r} 3x + 2 \\ 2x^2 - 5x - 1\overline{)6x^3 - 11x^2 - 14x + 3} \\ \overset{-}{\cancel{+}}\, 6x^3 \overset{+}{\cancel{-}}\, 15x^2 \overset{+}{\cancel{-}}\, 3x \quad \downarrow \\ \hline +\ 4x^2 - 11x + 3 \\ \overset{-}{\cancel{+}}\, 4x^2 \overset{+}{\cancel{-}}\, 10x \overset{+}{\cancel{-}}\, 2 \\ \hline -x + 5 \end{array}$$

$$\frac{6x^3 - 11x^2 - 14x + 3}{2x^2 - 5x - 1} = 3x + 2 + \frac{-x + 5}{2x^2 - 5x - 1}$$

Polynomial Division with Missing Powers

One last step is sometimes necessary. The two polynomials in a division problem must both be in descending powers of the variable and cannot skip any powers from the highest power down to the constant term. If any powers of the variable or the constant term is missing from either the dividend or the quotient, we can insert zero terms to act as placeholders. This will ensure that like terms are lined up vertically in the long division process.

EXAMPLE 9 Divide: $\dfrac{2x^3 - 3x + 2}{x - 5}$.

SOLUTION Notice that the dividend (the numerator) is missing an x^2 term. We write $2x^3 - 3x + 2$ as $2x^3 + 0x^2 - 3x + 2$. Adding $0x^2$ does not change our original problem, and it will make sure we are always combining like terms.

$$
\begin{array}{r l}
 & 2x^2 \qquad\qquad\qquad \leftarrow \text{Estimate } 2x^3 \div x = 2x^2 \\
x - 5 \,\big)\!\! & \overline{2x^3 + 0x^2 - 3x + 2} \\
 & \underline{\not{-}\,2x^3 \not{+}\,10x^2} \quad \downarrow \qquad \leftarrow \text{Multiply } 2x^2(x - 5) = 2x^3 - 10x^2 \\
 & \qquad + 10x^2 - 3x \qquad \leftarrow \text{Subtract: } (2x^3 + 0x^2) - (2x^3 - 10x^2) = 10x^2 \\
 & \qquad\qquad\quad \uparrow \\
 & \qquad\qquad \text{Bring down the next term}
\end{array}
$$

Adding the term $0x^2$ gives us a column in which to write $10x^2$. (Remember, you can add and subtract only similar terms.)

Here is the completed problem:

$$
\begin{array}{r l}
 & 2x^2 + 10x + 47 \\
x - 5 \,\big)\!\! & \overline{2x^3 + 0x^2 - 3x + 2} \\
 & \underline{\not{-}\,2x^3 \not{+}\,10x^2} \\
 & \qquad + 10x^2 - 3x \\
 & \qquad \underline{\not{-}\,10x^2 \not{+}\,50x} \\
 & \qquad\qquad + 47x + 2 \\
 & \qquad\qquad \underline{\not{-}\,47x \not{+}\,235} \\
 & \qquad\qquad\qquad 237
\end{array}
$$

Our answer is
$$\frac{2x^3 - 3x + 2}{x - 5} = 2x^2 + 10x + 47 + \frac{237}{x - 5}$$

EXAMPLE 10 Divide: $\dfrac{6x^3 + 2x^2 - x}{2x^2 - 3}$.

SOLUTION The dividend is missing the constant term and the divisor is missing an x term. We write the dividend as $6x^3 + 2x^2 - x + 0$ and the divisor as $2x^2 + 0x - 3$.

$$
\begin{array}{r l}
 & 3x + 1 \\
2x^2 + 0x - 3 \,\big)\!\! & \overline{6x^3 + 2x^2 - x + 0} \\
 & \underline{\not{-}\,6x^3 \not{-}\,0x^2 \not{+}\,9x} \quad \downarrow \\
 & \quad + 2x^2 + 8x + 0 \\
 & \quad \underline{\not{-}\,2x^2 \not{-}\,0x \not{+}\,3} \\
 & \qquad\qquad 8x + 3
\end{array}
$$

The result is
$$\frac{6x^3 + 2x^2 - x}{2x^2 - 3} = 3x + 1 + \frac{8x + 3}{2x^2 - 3}$$

As you can see, long division with polynomials is a mechanical process. Once you have done it correctly a couple of times, it becomes very easy to produce the correct answer.

Getting Ready for Class

After reading through the preceding section, respond in your own words and in complete sentences.

A. What property of real numbers is key to dividing a polynomial by a monomial?

B. Why is our answer to Example 5 not a polynomial?

C. How is division of two polynomials similar to long division with whole numbers?

D. How do we use 0 when dividing the polynomial $2x^3 - 3x + 2$ by $x - 5$?

Divide the following polynomials by $5x$.

1. $5x^2 - 10x$
2. $10x^3 - 15x$
3. $25x^2y - 10xy$
4. $15xy^2 + 20x^2y$
5. $35x^5 - 30x^4 + 25x^3$
6. $75x^6 + 50x^3 - 25x$

Divide the following by $-2a$.

7. $8a^2 - 4a$
8. $a^3 - 6a^2$
9. $12a^3b - 6a^2b^2 + 14ab^3$
10. $4ab^3 - 16a^2b^2 - 22a^3b$
11. $a^2 + 2ab + b^2$
12. $a^2b - 2ab^2 + b^3$

Perform the following monomial divisions (find the following quotients).

13. $\dfrac{6x + 8y}{2}$
14. $\dfrac{9x - 3y}{3}$
15. $\dfrac{7y - 21}{-7}$
16. $\dfrac{14y - 12}{2}$
17. $\dfrac{10xy - 8x}{2x}$
18. $\dfrac{12xy^2 - 18x}{-6x}$
19. $\dfrac{x^2y - x^3y^2}{x}$
20. $\dfrac{x^2y - x^3y^2}{x^2}$
21. $\dfrac{a^2b^2 - ab^2}{-ab^2}$
22. $\dfrac{a^2b^2c - ab^2c^2}{abc}$
23. $\dfrac{x^3 - 3x^2y + xy^2}{x}$
24. $\dfrac{x^2 - 3xy^2 + xy^3}{x}$
25. $\dfrac{10a^2 - 15a^2b + 25a^2b^2}{5a^2}$
26. $\dfrac{6x^2a + 12x^2b - 6x^2c}{36x^2}$
27. $\dfrac{26x^2y^2 - 13xy}{-13xy}$
28. $\dfrac{6x^2y^2 - 3xy}{6xy}$
29. $\dfrac{5a^2x - 10ax^2 + 15a^2x^2}{20a^2x^2}$
30. $\dfrac{12ax - 9bx + 18cx}{6x^2}$
31. $\dfrac{16x^5 + 8x^2 + 12x}{12x^3}$
32. $\dfrac{27x^2 - 9x^3 - 18x^4}{-18x^3}$

Divide. Assume all variables represent positive numbers.

33. $\dfrac{9a^{5m} - 27a^{3m}}{3a^{2m}}$
34. $\dfrac{26a^{3m} - 39a^{5m}}{13a^{3m}}$
35. $\dfrac{10x^{5m} - 25x^{3m} + 35x^{m}}{5x^{m}}$
36. $\dfrac{18x^{2m} + 24x^{4m} - 30x^{6m}}{6x^{2m}}$

Simplify each numerator, and then divide.

37. $\dfrac{2x^3(3x + 2) - 3x^2(2x - 4)}{2x^2}$
38. $\dfrac{5x^2(6x - 3) + 6x^3(3x - 1)}{3x}$
39. $\dfrac{(x + 2)^2 - (x - 2)^2}{2x}$
40. $\dfrac{(x - 3)^2 - (x + 3)^2}{3x}$
41. $\dfrac{(x + 5)^2 + (x + 5)(x - 5)}{2x}$
42. $\dfrac{(x - 4)^2 + (x + 4)(x - 4)}{2x}$

Use long division to perform each division.

43. $\dfrac{x^2 - 5x + 6}{x - 3}$

44. $\dfrac{x^2 - 5x + 6}{x - 2}$

45. $\dfrac{a^2 + 9a + 20}{a + 5}$

46. $\dfrac{a^2 + 9a + 20}{a + 4}$

47. $\dfrac{2x^2 + 5x - 3}{2x - 1}$

48. $\dfrac{4x^2 + 4x - 3}{2x - 1}$

49. $\dfrac{x^3 - 9x^2 + 27x - 27}{x^2 - 6x + 9}$

50. $\dfrac{x^3 + 15x^2 + 75x + 125}{x^2 + 10x + 25}$

51. $\dfrac{x^4 - x^3 - x^2 - 2x - 6}{x^2 + 2}$

52. $\dfrac{x^4 + 3x^3 + 2x^2 - 6x - 8}{x^2 - 2}$

53. $\dfrac{x + 3}{x - 2}$

54. $\dfrac{x - 3}{x + 1}$

55. $\dfrac{3x + 4}{x + 2}$

56. $\dfrac{4x - 1}{x - 3}$

57. $\dfrac{x^2 + 5x + 8}{x + 3}$

58. $\dfrac{x^2 + 5x + 4}{x + 3}$

59. $\dfrac{x^2 + 2x + 1}{x - 2}$

60. $\dfrac{x^2 + 6x + 9}{x - 3}$

61. $\dfrac{x^2 + 5x - 6}{x + 1}$

62. $\dfrac{x^2 - x - 6}{x + 1}$

63. $\dfrac{2x^2 - 2x + 5}{2x + 4}$

64. $\dfrac{15x^2 + 19x - 4}{3x + 8}$

65. $\dfrac{6a^2 + 5a + 1}{2a + 3}$

66. $\dfrac{4a^2 + 4a + 3}{2a + 1}$

67. $\dfrac{6a^3 - 13a^2 - 4a + 15}{3a - 5}$

68. $\dfrac{2a^3 - a^2 + 3a + 2}{2a + 1}$

69. $\dfrac{x^2 - 6x + 9}{x^2 - 3x - 2}$

70. $\dfrac{x^2 + 10x + 25}{x^2 + 5x + 4}$

71. $\dfrac{2a^3 - 9a^2 - 5a + 4}{2a^2 + a + 3}$

72. $\dfrac{4a^3 - 8a^2 - 5a + 1}{2a^2 + a - 2}$

Use long division to find the quotients, filling in missing terms with zero place holders.

73. $\dfrac{x^3 + 4x + 5}{x + 1}$

74. $\dfrac{x^3 + 4x^2 - 8}{x + 2}$

75. $\dfrac{x^3 - 1}{x - 1}$

76. $\dfrac{x^3 + 1}{x + 1}$

77. $\dfrac{x^3 - 8}{x - 2}$

78. $\dfrac{x^3 + 27}{x + 3}$

79. $\dfrac{a^2 + 3a + 2}{a^2 + 1}$

80. $\dfrac{a^2 + 4a + 3}{a^2 - 5}$

81. $\dfrac{a^3 + 3a^2 + 1}{a^2 - 2}$

82. $\dfrac{a^3 - a^2 + 3}{a^2 + 1}$

83. $\dfrac{4a^4 + 4a^2 - 2}{2a^2 - 1}$

84. $\dfrac{6a^4 - 2a^2 + 9}{3a^2 + 2}$

85. Comparing Expressions Evaluate the expression $\frac{10x + 15}{5}$ and the expression $2x + 3$ when $x = 2$.

86. Comparing Expressions Evaluate the expression $\frac{6x^2 + 4x}{2x}$ and the expression $3x + 2$ when $x = 5$.

87. Comparing Expressions Show that the expression $\frac{3x + 8}{2}$ is not the same as the expression $3x + 4$ by replacing x with 10 in both expressions and simplifying the results.

88. Comparing Expressions Show that the expression $\frac{x + 10}{x}$ is not equal to 10 by replacing x with 5 and simplifying.

Long Division Use the information in the table to find the monthly payment for auto insurance for the cities below. Round to the nearest cent.

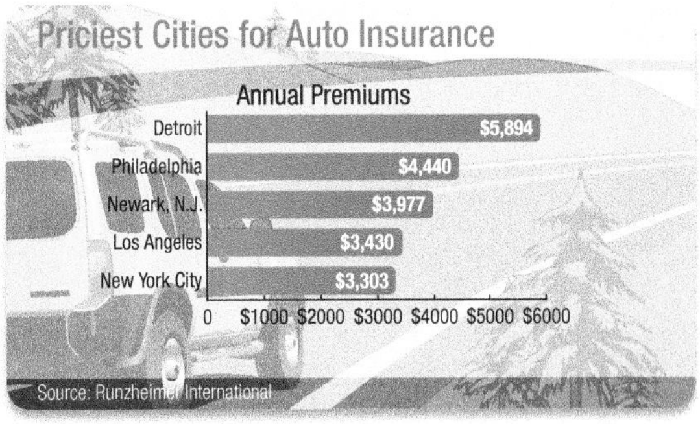

89. Detroit **90.** Philadelphia **91.** Newark, N.J. **92.** Los Angeles

Learning Objectives Assessment

The following problems can be used to help assess if you have successfully met the learning objectives for this section.

93. Divide: $\frac{12x^3y^2 - 3xy^3}{6x^2y^2}$.

a. $2x - 3xy^3$ **b.** $\frac{3}{2}x^2y^3$ **c.** $2x - 2xy$ **d.** $2x - \frac{y}{2x}$

94. Divide: $\frac{5x^2 - 3x + 2}{x + 4}$.

a. $5x - 23 + \frac{94}{x + 4}$ **b.** $5x + 17 + \frac{70}{x + 4}$

c. $5x - 3 + \frac{2}{x + 4}$ **d.** $2x + \frac{1}{2}$

Maintaining Your Skills

Simplify each expression. (Write all answers with positive exponents only.)

95. $(5x^3)^2(2x^6)^3$

96. 2^{-3}

97. $\frac{x^4}{x^{-3}}$

98. $\frac{(20x^2y^3)(5x^4y)}{(2xy^5)(10x^2y^3)}$

99. $(2 \times 10^{-4})(4 \times 10^5)$

100. $\frac{9 \times 10^{-3}}{3 \times 10^{-2}}$

101. $20ab^2 - 16ab^2 + 6ab^2$

102. Subtract $6x^2 - 5x - 7$ from $9x^2 + 3x - 2$.

Multiply.

103. $2x^2(3x^2 + 3x - 1)$

104. $(2x + 3)(5x - 2)$

105. $(3y - 5)^2$

106. $(a - 4)(a^2 + 4a + 16)$

107. $(2a^2 + 7)(2a^2 - 7)$

108. Divide: $15x^{10} - 10x^8 + 25x^6$ by $5x^6$.

Chapter 5 Summary

Exponents: Definition and Properties [5.1, 5.2]

Integer exponents indicate repeated multiplications.

$a^r \cdot a^s = a^{r+s}$ — To multiply with the same base, you add exponents

$\dfrac{a^r}{a^s} = a^{r-s}$ — To divide with the same base, you subtract exponents

$(ab)^r = a^r \cdot b^r$ — Exponents distribute over multiplication

$\left(\dfrac{a}{b}\right)^r = \dfrac{a^r}{b^r}$ — Exponents distribute over division

$(a^r)^s = a^{r \cdot s}$ — A power of a power is the product of the powers

$a^{-r} = \dfrac{1}{a^r}$ — Negative exponents imply reciprocals

EXAMPLES

1. a. $2^3 = 2 \cdot 2 \cdot 2 = 8$

b. $x^5 \cdot x^3 = x^{5+3} = x^8$

c. $\frac{x^5}{x^3} = x^{5-3} = x^2$

d. $(3x)^2 = 3^2 \cdot x^2 = 9x^2$

e. $\left(\frac{2}{3}\right)^3 = \frac{2^3}{3^3} = \frac{8}{27}$

f. $(x^5)^3 = x^{5 \cdot 3} = x^{15}$

g. $3^{-2} = \frac{1}{3^2} = \frac{1}{9}$

Scientific Notation [5.1, 5.2]

A number is in scientific notation when it is written as the product of a number between 1 and 10 and an integer power of 10.

2. $768{,}000 = 7.68 \times 10^5$
$0.00039 = 3.9 \times 10^{-4}$

Multiplication of Monomials [5.3]

To multiply two monomials, multiply coefficients and add exponents.

3. $(5x^2)(3x^4) = 15x^6$

Division of Monomials [5.3]

To divide two monomials, divide coefficients and subtract exponents.

4. $\frac{12x^9}{4x^5} = 3x^4$

Addition of Polynomials [5.4]

To add two polynomials, add coefficients of similar terms.

5. $(3x^2 - 2x + 1) + (2x^2 + 7x - 3)$
$= 5x^2 + 5x - 2$

Subtraction of Polynomials [5.4]

To subtract one polynomial from another, add the opposite of the second to the first.

6. $(3x + 5) - (4x - 3)$
$= 3x + 5 - 4x + 3$
$= -x + 8$

Multiplication of Polynomials [5.5]

To multiply a polynomial by a monomial, we apply the distributive property. To multiply two binomials we use the FOIL method. To multiply any two polynomials, we multiply each term in the first polynomial by each term in the second polynomial.

7. a. $2a^2(5a^2 + 3a - 2)$
$= 10a^4 + 6a^3 - 4a^2$

b. $(x + 2)(3x - 1)$
$= 3x^2 - x + 6x - 2$
$= 3x^2 + 5x - 2$

c.
$$\begin{array}{r} 2x^2 - 3x + 4 \phantom{{}- 8} \\ 3x - 2 \phantom{{}- 8} \\ \hline 6x^3 - 9x^2 + 12x \phantom{{}- 8} \\ - 4x^2 + 6x - 8 \\ \hline 6x^3 - 13x^2 + 18x - 8 \end{array}$$

Special Products [5.6]

$(a + b)^2 = a^2 + 2ab + b^2$ }
$(a - b)^2 = a^2 - 2ab + b^2$ } Binomial squares
$(a + b)(a - b) = a^2 - b^2$ Difference of two squares

8. $(x + 3)^2 = x^2 + 6x + 9$
$(x - 3)^2 = x^2 - 6x + 9$
$(x + 3)(x - 3) = x^2 - 9$

Dividing a Polynomial by a Monomial [5.7]

To divide a polynomial by a monomial, divide each term in the polynomial by the monomial.

9. $\dfrac{12x^3 - 18x^2}{6x} = \dfrac{12x^3}{6x} - \dfrac{18x^2}{6x}$
$= 2x^2 - 3x$

Long Division with Polynomials [5.7]

Division with polynomials is similar to long division with whole numbers. The steps in the process are estimate, multiply, subtract, and bring down the next term.

10.
$$\begin{array}{r} x - 2 \\ x - 3 \overline{) \; x^2 - 5x + 8} \\ \underline{\not{-}\, x^2 \not{+}\, 3x \quad \downarrow} \\ -2x + 8 \\ \underline{\not{+}\, 2x \not{-}\, 6} \\ 2 \end{array}$$

$\dfrac{x^2 - 5x + 8}{x} - 3$
$= x - 2 + \dfrac{2}{x - 3}$

Chapter 5 Test

Simplify each of the following expressions. [5.1]

1. $(-2)^5$
2. -4^2
3. $x^9 \cdot x^{14}$
4. $(4x^2y^3)^2$

Simplify each expression. Write all answers with positive exponents only. [5.2]

5. 4^{-2}
6. $(4a^5b^3)^0$
7. $\dfrac{x^{-4}}{x^{-7}}$
8. $\left(\dfrac{x}{3}\right)^3$
9. $\dfrac{(x^{-3})^2(x^{-5})^{-3}}{(x^{-3})^{-4}}$
10. Write 0.04307 in scientific notation. [5.2]
11. Write 7.63×10^6 in expanded form. [5.1]

Simplify. Write all answers with positive exponents only. [5.3]

12. $(6a^2b)(-4ab^3)$
13. $\dfrac{17x^2y^5z^3}{51x^4y^2z}$
14. $\dfrac{(3a^3b)(4a^2b^5)}{24a^2b^4}$
15. $\dfrac{28x^4}{4x} + \dfrac{30x^7}{6x^4}$
16. $\dfrac{(1.1 \times 10^5)(3 \times 10^{-2})}{4.4 \times 10^{-5}}$

Add or subtract as indicated. [5.4]

17. $(9x^2 - 2x) + (7x + 4)$
18. $(4x^2 + 5x - 6) - (2x^2 - x - 4)$
19. Subtract $2x + 7$ from $7x + 3$. [5.4]
20. Find the value of $3a^2 + 4a + 6$ when a is -3. [5.4]

Multiply. [5.5]

21. $3x^2(5x^2 - 2x + 4)$
22. $\left(x + \dfrac{1}{4}\right)\left(x - \dfrac{1}{3}\right)$
23. $(2x - 3)(5x + 6)$
24. $(x + 4)(x^2 - 4x + 16)$

Multiply. [5.6]

25. $(x - 6)^2$
26. $(2a + 4b)^2$
27. $(3x - 6)(3x + 6)$
28. $(x^2 - 4)(x^2 + 4)$
29. Divide $18x^3 - 36x^2 + 6x$ by $6x$. [5.7]

Divide. [5.7]

30. $\dfrac{9x^2 - 6x - 4}{3x - 1}$
31. $\dfrac{4x^3 + 3x^2 + 1}{x^2 + 2}$
32. **Volume** Find the volume of a cube if the length of a side is 3.2 inches. Round to the nearest hundredth. [5.1]
33. **Volume** Find the volume of a rectangular solid if the length is three times the width, and the height is one third the width. [5.3]

6

Factoring

Chapter Outline

6.1 The Greatest Common Factor and Factoring by Grouping

6.2 Factoring Trinomials

6.3 More on Factoring Trinomials

6.4 Special Factoring Patterns

6.5 Factoring: A General Review

6.6 Solving Equations by Factoring

6.7 Applications

If you watch professional football on television, you will hear the announcers refer to "hang time" when the punter punts the ball. Hang time is the amount of time the ball is in the air, and it depends on only one thing—the initial vertical velocity imparted to the ball by the kicker's foot. We can find the hang time of a football by solving equations. Table 1 shows the equations to solve for hang time, given various initial vertical velocities. Figure 1 is a visual representation of the equations in Table 1. In Figure 1, you can find hang time on the horizontal axis.

TABLE 1

Hang Time for a Football

Initial Vertical Velocity	Equation in Factored Form	Hang Time
16	$16t(1 - t) = 0$	1
32	$16t(2 - t) = 0$	2
48	$16t(3 - t) = 0$	3
64	$16t(4 - t) = 0$	4
80	$16t(5 - t) = 0$	5

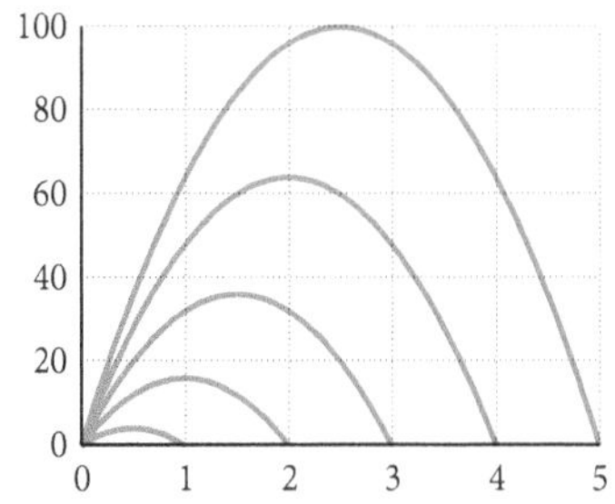

FIGURE 1

The equations in the second column of the table are in what is called "factored form." Once the equation is in factored form, hang time can be read from the second factor. In this chapter, we develop techniques that allow us to factor a variety of polynomials. Factoring is the key to solving equations like the ones in Table 1.

Success Skills

This is the last chapter in which we will mention study skills. You know by now what works best for you and what you have to do to achieve your goals for this course. From now on, it is simply a matter of sticking with the things that work for you and avoiding the things that do not. It seems simple, but as with anything that takes effort, it is up to you to see that you maintain the skills that get you where you want to be in the course.

If you intend to take more classes in mathematics and want to ensure your success in those classes, then you can work toward this goal: ***Become the type of student who can learn mathematics on his or her own.*** Most people who have degrees in mathematics were students who could learn mathematics on their own. This doesn't mean that you must learn it all on your own, or that you study alone, or that you don't ask questions. It means that you know your resources, both internal and external, and you can count on those resources when you need them. Attaining this goal gives you independence and puts you in control of your success in any math class you take.

The Greatest Common Factor and Factoring by Grouping 6.1

Learning Objectives

In this section, we will learn how to:

1. Find the greatest common factor for a polynomial.
2. Factor the greatest common factor from a polynomial.
3. Factor a polynomial using grouping.

Introduction

In Chapter 1, we used the following diagram to illustrate the relationship between multiplication and factoring.

Multiplication

$$\text{Factors} \longrightarrow 3 \cdot 5 = 15 \longleftarrow \text{Product}$$

Factoring

A similar relationship holds for multiplication of polynomials. Reading the following diagram from left to right, we say the product of the binomials $x + 2$ and $x + 3$ is the trinomial $x^2 + 5x + 6$. However, if we read in the other direction, we can say that $x^2 + 5x + 6$ factors into the product of $x + 2$ and $x + 3$.

Multiplication

$$\text{Factors} \longrightarrow (x + 2)(x + 3) = x^2 + 5x + 6 \longleftarrow \text{Product}$$

Factoring

In this chapter we develop a systematic method of factoring polynomials. When factoring polynomials, our objective is to change the polynomial (a sum of terms) into a product of two or more factors (quantities connected by multiplication).

The Greatest Common Factor

In this section we will apply the distributive property to polynomials to factor from them what is called the greatest common factor.

(def **DEFINITION** ***greatest common factor***

The ***greatest common factor*** (GCF) for a polynomial is the largest monomial that divides (is a factor of) each term of the polynomial.

We use the term *largest monomial* to mean the monomial with the greatest coefficient and highest power of the variable that will divide evenly into each term.

VIDEO EXAMPLES

SECTION 6.1

EXAMPLE 1 Find the greatest common factor for the polynomial:

$$3x^5 + 12x^2$$

SOLUTION The terms of the polynomial are $3x^5$ and $12x^2$. The largest number that divides the coefficients is 3, and the highest power of x that is a factor of x^5 and x^2 is x^2. Therefore, the greatest common factor for $3x^5 + 12x^2$ is $3x^2$; that is, $3x^2$ is the largest monomial that divides each term of $3x^5 + 12x^2$.

Notice in Example 1 that the variable part of the GCF, x^2, contains the smaller of the two exponents appearing on the variable in the polynomial. In general, this will always be true. To find the variable part of the GCF, choose the smallest exponent of each variable that is common to every term in the polynomial.

EXAMPLE 2 Find the greatest common factor for:

$$8a^3b^2 + 16a^2b^3 + 20a^3b^3$$

SOLUTION The largest number that divides each of the coefficients is 4. Because both variables a and b appear in every term, the GCF will contain the smallest exponent appearing on these variables, which is a^2 and b^2. The greatest common factor for $8a^3b^2 + 16a^2b^3 + 20a^3b^3$ is $4a^2b^2$. It is the largest monomial that is a factor of each term.

EXAMPLE 3 Find the greatest common factor for $2x + 9$.

SOLUTION The largest number that divides the coefficients is 1. Because the variable x is not common to both terms, no variable will appear in the GCF. Therefore, the greatest common factor for $2x + 9$ is 1.

Factoring the GCF from a Polynomial

Once we have recognized the greatest common factor of a polynomial, we can apply the distributive property and factor it out of each term. We rewrite the polynomial as the product of its greatest common factor with the polynomial that remains after the greatest common factor has been divided from each term in the original polynomial.

EXAMPLE 4 Factor the greatest common factor from $3x - 15$.

SOLUTION The greatest common factor for the terms $3x$ and 15 is 3. We can rewrite both $3x$ and 15 so that the greatest common factor 3 is showing in each term. It is important to realize that $3x$ means $3 \cdot x$. The 3 and the x are not "stuck" together:

$$3x - 15 = 3 \cdot x - 3 \cdot 5$$

Now, applying the distributive property, we have:

$$3 \cdot x - 3 \cdot 5 = 3(x - 5)$$

To check a factoring problem like this, we can multiply 3 and $x - 5$ to get $3x - 15$, which is what we started with. Factoring is simply a procedure by which we change sums and differences into products. In this case we changed the difference $3x - 15$ into the product $3(x - 5)$. Note, however, that we have not changed the meaning or value of the expression. The expression we end up with is equivalent to the expression we started with.

EXAMPLE 5 Factor the greatest common factor from:

$$5x^3 - 15x^2$$

Note An alternative approach to Example 5 is to view it from the perspective of division:

$$5x^3 - 15x^2 = 5x^2\left(\frac{5x^3}{5x^2} - \frac{15x^2}{5x^2}\right)$$
$$= 5x^2(x - 3)$$

SOLUTION The greatest common factor is $5x^2$. We rewrite the polynomial as:

$$5x^3 - 15x^2 = 5x^2 \cdot x - 5x^2 \cdot 3$$

Then we apply the distributive property to get:

$$5x^2 \cdot x - 5x^2 \cdot 3 = 5x^2(x - 3)$$

To check our work, we simply multiply $5x^2$ and $(x - 3)$ to get $5x^3 - 15x^2$, which is our original polynomial.

EXAMPLE 6 Factor the greatest common factor from:

$$16x^5 - 20x^4 + 8x^3$$

SOLUTION The greatest common factor is $4x^3$. We rewrite the polynomial so we can see the greatest common factor $4x^3$ in each term; then we apply the distributive property to factor it out.

$$16x^5 - 20x^4 + 8x^3 = 4x^3 \cdot 4x^2 - 4x^3 \cdot 5x + 4x^3 \cdot 2$$
$$= 4x^3(4x^2 - 5x + 2)$$

Notice that the terms in the grouping symbol are what remains when the GCF of $4x^3$ is divided out of each term in the trinomial.

EXAMPLE 7 Factor the greatest common factor from:

$$-6x^3y - 18x^2y^2 + 12xy^3$$

SOLUTION Notice that the leading term is negative. When this is the case, we will include a negative sign with our GCF. As a result, the greatest common factor is $-6xy$. We rewrite the polynomial in terms of $-6xy$ and then apply the distributive property as follows.

$$-6x^3y - 18x^2y^2 + 12xy^3 = -6xy(x^2) - 6xy(3xy) - 6xy(-2y^2)$$
$$= -6xy(x^2 + 3xy - 2y^2)$$

Again, notice the terms in the grouping symbols are the quotients that result when $-6xy$ is divided out of each term in the original trinomial:

$$-6xy(x^2 + 3xy - 2y^2)$$
$$\nearrow \qquad \uparrow \qquad \nwarrow$$
$$\frac{-6x^3y}{-6xy} \qquad \frac{-18x^2y^2}{-6xy} \qquad \frac{12xy^3}{-6xy}$$

To check our answer, we can multiply $x^2 + 3xy - 2y^2$ by $-6xy$ and verify the resulting product is equal to the original polynomial.

EXAMPLE 8 Factor the greatest common factor from:

$$3a^2b - 6a^3b^2 + 9a^3b^3$$

SOLUTION The greatest common factor is $3a^2b$:

$$3a^2b - 6a^3b^2 + 9a^3b^3 = 3a^2b(1) - 3a^2b(2ab) + 3a^2b(3ab^2)$$
$$= 3a^2b(1 - 2ab + 3ab^2)$$

Factoring by Grouping

To develop our next method of factoring, called *factoring by grouping*, we start by examining the polynomial $xc + yc$. The greatest common factor for the two terms is c. Factoring c from each term we have:

$$xc + yc = c(x + y)$$

But suppose that c itself was a more complicated expression, such as $a + b$, so that the expression we were trying to factor was $x(a + b) + y(a + b)$, instead of $xc + yc$. The greatest common factor for $x(a + b) + y(a + b)$ is the binomial $(a + b)$. Factoring this common factor from each term looks like this:

$$x(a + b) + y(a + b) = (a + b)(x + y)$$

To see how all of this applies to factoring polynomials, consider the polynomial

$$xy + 3x + 2y + 6$$

There is no greatest common factor other than the number 1. However, if we group the terms together two at a time, we can factor an x from the first two terms and a 2 from the last two terms:

$$xy + 3x + 2y + 6 = x(y + 3) + 2(y + 3)$$

The expression on the right can be thought of as having two terms: $x(y + 3)$ and $2(y + 3)$. Each of these expressions contains the common factor $y + 3$, which can be factored out using the distributive property:

$$x(y + 3) + 2(y + 3) = (y + 3)(x + 2)$$

This last expression is in factored form. The process we used to obtain it is called factoring by grouping. Here are some additional examples.

EXAMPLE 9 Factor: $ax + bx + ay + by$.

SOLUTION We begin by factoring x from the first two terms and y from the last two terms:

$$ax + bx + ay + by = x(a + b) + y(a + b)$$

We can see that the binomial $a + b$ is now a common factor, which we factor out to obtain

$$x(a + b) + y(a + b) = (a + b)(x + y)$$

To convince yourself that this is factored correctly, multiply the two factors $(a + b)$ and $(x + y)$.

Note In Example 10, if we factored out a y instead of a $-y$, we would have

$$2x^2 + 5ax - 2xy - 5ay = x(2x + 5a) + y(-2x - 5a)$$

Grouping would not work because the resulting binomials, $2x + 5a$ and $-2x - 5a$, are not the same. These binomials must be identical so that we have a common factor for the second step.

EXAMPLE 10 Factor: $2x^2 + 5ax - 2xy - 5ay$.

SOLUTION From the first two terms we factor x. From the second two terms we factor $-y$ because the leading term in that pair is negative. Remember, we are actually dividing each term in the second pair by $-y$. This means that the resulting terms in the grouping symbol will both be positive.

$$\begin{aligned} 2x^2 + 5ax - 2xy - 5ay &= x(2x + 5a) - y(2x + 5a) \\ &= (2x + 5a)(x - y) \end{aligned}$$

Another way to accomplish the same result is to use the commutative property to interchange the middle two terms, and then factor by grouping:

$$\begin{aligned} 2x^2 + 5ax - 2xy - 5ay &= 2x^2 - 2xy + 5ax - 5ay && \text{Commutative property} \\ &= 2x(x - y) + 5a(x - y) \\ &= (x - y)(2x + 5a) \end{aligned}$$

This is the same result we obtained previously.

In some cases, grouping may not work with the terms in the original order. If so, we can try rearranging the terms and try again. The last example illustrates this situation.

EXAMPLE 11 Factor: $3ax - 10 + 15x - 2a$.

SOLUTION The first pair of terms, $3ax - 10$, has a GCF of 1. The same is true of the last pair of terms. This is not helpful if we are to factor the expression. However, if we rearrange terms we can make common factors (other than 1) appear.

$$\begin{aligned} 3ax - 10 + 15x - 2a &= 3ax + 15x - 10 - 2a && \text{Commutative property} \\ &= 3x(a + 5) - 2(5 + a) && \text{Grouping} \\ &= 3x(a + 5) - 2(a + 5) && 5 + a = a + 5 \\ &= (3x - 2)(a + 5) \end{aligned}$$

Getting Ready for Class

After reading through the preceding section, respond in your own words and in complete sentences.

A. What is the greatest common factor for a polynomial?

B. After factoring a polynomial, how can you check your result?

C. When would you try to factor by grouping?

D. What is the relationship between multiplication and factoring?

Problem Set 6.1

Find the greatest common factor for each expression.

1. $3x - 12$

2. $6x + 14$

3. $2x^3 - 18x$

4. $12y^3 - 3y$

5. $8a^3b^4 - 12a^4b + 20a^2b^3$

6. $6a^3b^3 + 12a^2b^4 - 8ab^5$

7. $3x(2x + 1) - 4(2x + 1)$

8. $4x^2(x - 3) - 9(x - 3)$

Factor the following by taking out the greatest common factor.

9. $15x + 25$

10. $14x + 21$

11. $6a + 9$

12. $8a + 10$

13. $4x - 8y$

14. $9x - 12y$

15. $3x^2 - 6x - 9$

16. $2x^2 + 6x + 4$

17. $3a^2 - 3a - 60$

18. $2a^2 - 18a + 28$

19. $24y^2 - 52y + 24$

20. $18y^2 + 48y + 32$

21. $9x^2 - 8x^3$

22. $7x^3 - 4x^2$

23. $13a^2 - 26a^3$

24. $5a^2 - 10a^3$

25. $21x^2y - 28xy^2$

26. $30xy^2 - 25x^2y$

27. $22a^2b^2 - 11ab^2$

28. $15x^3 - 25x^2 + 30x$

29. $7x^3 + 21x^2 - 28x$

30. $16x^4 - 20x^2 - 16x$

31. $121y^4 - 11x^4$

32. $25a^4 - 5b^4$

33. $100x^4 - 50x^3 + 25x^2$

34. $36x^5 + 72x^3 - 81x^2$

35. $8a^2 + 16b^2 + 32c^2$

36. $9a^2 - 18b^2 - 27c^2$

37. $4a^2b - 16ab^2 + 32a^2b^2$

38. $5ab^2 + 10a^2b^2 + 15a^2b$

39. $121a^3b^2 - 22a^2b^3 + 33a^3b^3$

40. $20a^4b^3 - 18a^3b^4 + 22a^4b^4$

41. $12x^2y^3 - 72x^5y^3 - 36x^4y^4$

42. $49xy - 21x^2y^2 + 35x^3y^3$

Factor by grouping.

43. $xy + 5x + 3y + 15$

44. $xy + 2x + 4y + 8$

45. $xy + 6x + 2y + 12$

46. $xy + 2y + 6x + 12$

47. $ab + 7a - 3b - 21$

48. $ab + 3b - 7a - 21$

49. $ax - bx + ay - by$

50. $ax - ay + bx - by$

51. $2ax + 5a - 2x - 5$

52. $3ax + 21x - a - 7$

53. $27by - 6y + 9b - 2$

54. $4by - 24y + b - 6$

55. $3xb - 4b - 6x + 8$

56. $3xb - 4b - 15x + 20$

57. $x^2 + 2a + 2x + ax$

58. $x^2 + 3a + ax + 3x$

59. $x^2 + ab - ax - bx$

60. $x^2 - ab + ax - bx$

Factor by grouping. You can group the terms together two at a time or three at a time. Either way will produce the same result.

61. $ax + ay + bx + by + cx + cy$

62. $ax + bx + cx + ay + by + cy$

Factor the following polynomials by grouping the terms together two at a time.

63. $6x^2 + 9x + 4x + 6$

64. $6x^2 - 9x - 4x + 6$

65. $20x^2 - 2x + 50x - 5$

66. $20x^2 + 25x + 4x + 5$

67. $20x^2 + 4x + 25x + 5$

68. $20x^2 + 4x - 25x - 5$

69. $x^3 + 2x^2 + 3x + 6$

70. $x^3 - 5x^2 - 4x + 20$

71. $6x^3 - 4x^2 + 15x - 10$

72. $8x^3 - 12x^2 + 14x - 21$

73. The greatest common factor of the binomial $3x + 6$ is 3. The greatest common factor of the binomial $2x + 4$ is 2. What is the greatest common factor of their product $(3x + 6)(2x + 4)$ when it has been multiplied out?

74. The greatest common factors of the binomials $4x + 2$ and $5x + 10$ are 2 and 5, respectively. What is the greatest common factor of their product $(4x + 2)(5x + 10)$ when it has been multiplied out?

75. The following factorization is incorrect. Find the mistake, and correct the right-hand side:

$$12x^2 + 6x + 3 = 3(4x^2 + 2x)$$

76. Find the mistake in the following factorization, and then rewrite the right-hand side correctly:

$$10x^2 + 2x + 6 = 2(5x^2 + 3)$$

Applying the Concepts

77. Investing If you invest \$1,000 in an account with an annual interest rate of r compounded annually, the amount of money you have in the account after one year is:

$$A = 1{,}000 + 1{,}000r$$

Write this formula again with the right side in factored form. Then, find the amount of money in this account at the end of one year if the interest rate is 12%.

78. Investing If you invest P dollars in an account with an annual interest rate of 8% compounded annually, then the amount of money in that account after one year is given by the formula:

$$A = P + 0.08P$$

Rewrite this formula with the right side in factored form, and then find the amount of money in the account at the end of one year if \$500 was the initial investment.

79. Biological Growth If 1,000,000 bacteria are placed in a petri dish and the bacteria have a growth rate of r (a percent expressed as a decimal) per hour, then 1 hour later the amount of bacteria will be $A = 1{,}000{,}000 + 1{,}000{,}000r$ bacteria.

a. Factor the right side of the equation.

b. If $r = 30\%$, find the number of bacteria present after one hour.

80. Biological Growth If there are B E. coli bacteria present initially in a petri dish and their growth rate is r (a percent expressed as a decimal) per hour, then after one hour there will be $A = B + Br$ bacteria present.

a. Factor the right side of this equation.

b. The following bar graph shows the number of E. coli bacteria present initially and the number of bacteria present hours later. Use the bar chart to find B and A in the preceding equation.

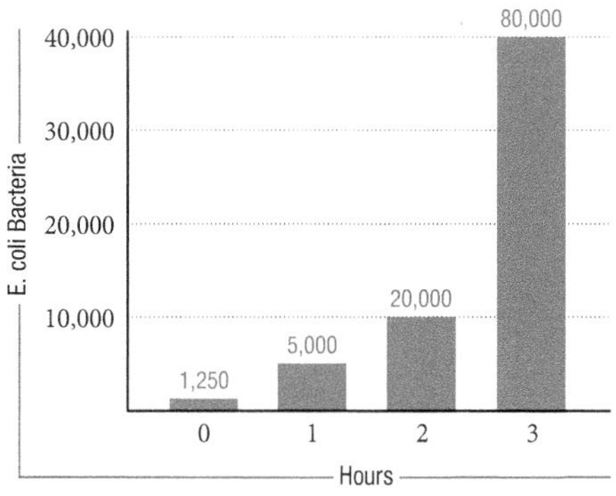

Learning Objectives Assessment

The following problems can be used to help assess if you have successfully met the learning objectives for this section.

81. Find the greatest common factor for $9a^2b^4 - 15a^3b$.

a. $3a^2b$ **b.** $3a^3b^4$ **c.** ab **d.** $45a^5b^5$

82. Factor: $10x^2y - 25xy^3$

a. $5xy(2x - 5y^2)$ **b.** $5x^2y^3(2y^2 - 5x)$

c. $2xy(5x - 12y^2)$ **d.** $-15xy^2$

83. Factor by grouping: $2ax + 6x - 5a - 15$

a. $(2x - 5)(a + 3)(a - 3)$ **b.** $(2x - 5)(a + 3)^2$

c. $(2x - 5)(a + 3)$ **d.** $(2x + 5)(a + 3)$

Getting Ready for the Next Section

Multiply.

84. $(x - 7)(x + 2)$

85. $(x - 7)(x - 2)$

86. $(x - 3)(x + 2)$

87. $(x + 3)(x - 2)$

88. $(x + 3)(x^2 - 3x + 9)$

89. $(x - 2)(x^2 + 2x + 4)$

90. $(2x + 1)(x^2 + 4x - 3)$

91. $(3x + 2)(x^2 - 2x - 4)$

92. $3x^4(6x^3 - 4x^2 + 2x)$

93. $2x^4(5x^3 + 4x^2 - 3x)$

94. $\left(x + \frac{1}{3}\right)\left(x + \frac{2}{3}\right)$

95. $\left(x + \frac{1}{4}\right)\left(x + \frac{3}{4}\right)$

96. $(6x + 4y)(2x - 3y)$

97. $(8a - 3b)(4a - 5b)$

98. $(9a + 1)(9a - 1)$

99. $(7b + 1)(7b + 1)$

100. $(x - 9)(x - 9)$

101. $(x - 8)(x - 8)$

102. $(x + 2)(x^2 - 2x + 4)$

103. $(x - 3)(x^2 + 3x + 9)$

SPOTLIGHT ON SUCCESS ***Student Instructor Stefanie***

Never confuse a single defeat with a final defeat.
—F. Scott Fitzgerald

The idea that has worked best for my success in college, and more specifically in my math courses, is to stay positive and be resilient. I have learned that a 'bad' grade doesn't make me a failure; if anything it makes me strive to do better. That is why I never let a bad grade on a test or even in a class get in the way of my overall success.

By sticking with this positive attitude, I have been able to achieve my goals. My grades have never represented how well I know the material. This is because I have struggled with test anxiety and it has consistently lowered my test scores in a number of courses. However, I have not let it defeat me. When I applied to graduate school, I did not meet the grade requirements for my top two schools, but that did not stop me from applying.

One school asked that I convince them that my knowledge of mathematics was more than my grades indicated. If I had let my grades stand in the way of my goals, I wouldn't have been accepted to both of my top two schools, and wouldn't be attending one of them in the Fall, on my way to becoming a mathematics teacher.

Factoring Trinomials 6.2

Learning Objectives

In this section, we will learn how to:

1. Factor trinomials whose leading coefficient is 1.

Introduction

Note As you will see as we progress through the book, factoring is a tool that is used in solving a number of problems. Before seeing how it is used, however, we first must learn how to do it. So, in this section and the two sections that follow, we will be developing our factoring skills.

In this section, we will factor trinomials in which the coefficient of the squared term is 1. The more familiar we are with multiplication of binomials the easier factoring trinomials will be.

Recall multiplication of binomials using the FOIL method from Chapter 4:

$$\begin{aligned} & \quad\ \text{F} \quad\ \ \text{O} \quad\ \ \text{I} \quad\ \ \text{L} \\ (x+3)(x+4) &= x^2 + \underbrace{4x + 3x} + 12 \\ &= x^2 + \quad 7x \quad + 12 \end{aligned}$$

$$\begin{aligned} & \quad\ \text{F} \quad\ \ \text{O} \quad\ \ \text{I} \quad\ \ \text{L} \\ (x-5)(x+2) &= x^2 + \underbrace{2x - 5x} - 10 \\ &= x^2 - \quad 3x \quad - 10 \end{aligned}$$

The first term in the answer is the product of the first terms in each binomial. The last term in the answer is the product of the last terms in each binomial. The middle term in the answer comes from adding the product of the outside terms to the product of the inside terms.

Let's have a and b represent real numbers and look at the product of $(x + a)$ and $(x + b)$:

$$\begin{aligned} (x+a)(x+b) &= x^2 + ax + bx + ab \\ &= x^2 + (a+b)x + ab \end{aligned}$$

The coefficient of the middle term is the sum of a and b. The last term is the product of a and b. Writing this as a factoring problem, we have:

$$x^2 + \underset{\text{Sum}}{(a+b)}x + \underset{\text{Product}}{ab} = (x+a)(x+b)$$

VIDEO EXAMPLES

SECTION 6.2

To factor a trinomial in which the coefficient of x^2 is 1, we need only find the numbers a and b whose sum is the coefficient of the middle term and whose product is the constant term (last term).

EXAMPLE 1 Factor: $x^2 + 8x + 12$.

SOLUTION The coefficient of x^2 is 1. We need two numbers whose sum is 8 and whose product is 12. Because the product is positive and the sum is also positive, both numbers must be positive. The numbers are 6 and 2:

$$x^2 + 8x + 12 = (x + 6)(x + 2)$$

We can easily check our work by multiplying $(x + 6)$ and $(x + 2)$.

$$\begin{aligned} \text{Check:} \quad (x+6)(x+2) &= x^2 + 2x + 6x + 12 \\ &= x^2 + 8x + 12 \end{aligned}$$

Note Because multiplication is commutative, it does not matter which order we write the factors. In Example 1, we could have said $x^2 + 8x + 12 = (x + 2)(x + 6)$ and still be correct.

EXAMPLE 2 Factor: $x^2 - 7x + 12$.

Note Again, we can check our results by multiplying our factors to see if their product is the original polynomial.

SOLUTION Observe the leading coefficient is 1. We need two numbers whose sum is -7 and whose product is 12. This time, because the product is positive but the sum is negative, both numbers must be negative. The numbers that work are -3 and -4:

$$x^2 - 7x + 12 = (x - 3)(x - 4)$$

EXAMPLE 3 Factor: $x^2 - 2x - 15$.

SOLUTION The coefficient of x^2 is again 1. We need to find a pair of numbers whose sum is -2 and whose product is -15. Because the product is negative, the two numbers must have opposite signs. Here are all the possibilities for products that are -15.

Products	Sums
$-1(15) = -15$	$-1 + 15 = 14$
$1(-15) = -15$	$1 + (-15) = -14$
$-5(3) = -15$	$-5 + 3 = -2$
$5(-3) = -15$	$5 + (-3) = 2$

The third line gives us what we want. Notice that, for the sum to be negative, the number with the larger absolute value is negative. The factors of $x^2 - 2x - 15$ are $(x - 5)$ and $(x + 3)$:

$$x^2 - 2x - 15 = (x - 5)(x + 3)$$

EXAMPLE 4 Factor $x^2 + 3x - 18$.

SOLUTION We must find a pair of numbers whose sum is 3 and whose product is -18. Here are the possibilities.

Products	Sums
$-1(18) = -18$	$-1 + 18 = 17$
$1(-18) = -18$	$1 + (-18) = -17$
$-2(9) = -18$	$-2 + 9 = 7$
$2(-9) = -18$	$2 + (-9) = -7$
$-3(6) = -18$	$-3 + 6 = 3$
$3(-6) = -18$	$3 + (-6) = -3$

From the fifth line we see that the numbers are -3 and 6. Notice that the number with the larger absolute value is positive. The factors of $x^2 + 3x - 18$ are $(x - 3)$ and $(x + 6)$.

$$x^2 + 3x - 18 = (x - 3)(x + 6)$$

Based upon our work from the first four examples, we can make the following observations about factoring trinomials whose leading coefficient is 1.

RULE ***Factoring trinomials with leading coefficient of 1***

To factor $x^2 + bx + c$:

1. If c is positive, then the numbers in the factors will have the same sign as b.
2. If c is negative, then the numbers in the factors will have opposite signs, and the number with the larger absolute value will have the same sign as b.

EXAMPLE 5 Factor: $x^2 + 8xy + 12y^2$.

SOLUTION We need two numbers whose product is 12 and whose sum is 8 (see Example 1 in this section). However, because the third term contains y^2, the last terms in the two binomials must each include a factor of y:

$$x^2 + 8xy + 12y^2 = (x + 6y)(x + 2y)$$

You should convince yourself that these factors are correct by finding their product. When you do, notice that the product of the outer terms and the product of the inner terms both contain the expression xy.

Note Trinomials in which the coefficient of the second-degree term is 1 are the easiest to factor. Success in factoring any type of polynomial is directly related to the amount of time spent working the problems. The more we practice, the more accomplished we become at factoring.

To conclude this section, we look at some trinomials that include a greatest common factor. We factor out the GCF first, and then (if possible) factor the remaining trinomial.

EXAMPLE 6 Factor: $2x^2 + 10x - 28$.

SOLUTION We begin by factoring out the greatest common factor, which is 2:

$$2x^2 + 10x - 28 = 2(x^2 + 5x - 14)$$

Now, we factor the remaining trinomial by finding a pair of numbers whose sum is 5 and whose product is -14. The numbers must have opposite signs, with the larger of the two positive. The only choices are -1 and 14, or -2 and 7. Since $-2 + 7 = 5$, we see that the factors of $x^2 + 5x - 14$ are $(x + 7)$ and $(x - 2)$. Here is the complete problem:

$$\begin{aligned} 2x^2 + 10x - 28 &= 2(x^2 + 5x - 14) \\ &= 2(x + 7)(x - 2) \end{aligned}$$

Note In Example 6 we began by factoring out the greatest common factor. The first step in factoring any trinomial is to look for the greatest common factor. If the trinomial in question has a greatest common factor other than 1, we factor it out first and then try to factor the trinomial that remains.

EXAMPLE 7 Factor: $3x^3 - 3x^2 - 18x$.

SOLUTION We begin by factoring out the greatest common factor, which is $3x$. Then we factor the remaining trinomial. Without showing the table of products and sums as we did in Examples 3 and 4, here is the complete problem:

$$\begin{aligned} 3x^3 - 3x^2 - 18x &= 3x(x^2 - x - 6) \\ &= 3x(x - 3)(x + 2) \end{aligned}$$

EXAMPLE 8 Factor: $-y^2 + 11y - 24$.

SOLUTION The leading coefficient is -1, not 1, so we begin by factoring out a GCF of -1 so that the remaining trinomial has a positive leading term.

$$-y^2 + 11y - 24 = -1(y^2 - 11y + 24)$$

To factor the remaining trinomial, we need two negative numbers whose product is 24 and whose sum is -11. The numbers are -3 and -8. Here is the complete problem:

$$\begin{aligned} -y^2 + 11y - 24 &= -1(y^2 - 11y + 24) \\ &= -1(y - 3)(y - 8) \end{aligned}$$

Getting Ready for Class

After reading through the preceding section, respond in your own words and in complete sentences.

A. When the leading coefficient of a trinomial is 1, what is the relationship between the other two coefficients and the factors of the trinomial?

B. How can you check to see that you have factored a trinomial correctly?

C. Describe how you would find the factors of $x^2 + 8x + 12$.

D. If the third term of the trinomial is negative, how do we decide which of the two factors has a plus sign and which has a minus sign?

Factor the following trinomials.

1. $x^2 + 7x + 12$
2. $x^2 + 7x + 10$
3. $x^2 + 3x + 2$
4. $x^2 + 7x + 6$
5. $a^2 + 10a + 21$
6. $a^2 - 7a + 12$
7. $x^2 - 7x + 10$
8. $x^2 - 3x + 2$
9. $y^2 - 10y + 21$
10. $y^2 - 7y + 6$
11. $x^2 - x - 12$
12. $x^2 - 4x - 5$
13. $y^2 + y - 12$
14. $y^2 + 3y - 18$
15. $x^2 + 5x - 14$
16. $x^2 - 5x - 24$
17. $r^2 - 8r - 9$
18. $r^2 - r - 2$
19. $x^2 - x - 30$
20. $x^2 + 8x + 12$
21. $a^2 + 15a + 56$
22. $a^2 - 9a + 20$
23. $y^2 - y - 42$
24. $y^2 + y - 42$
25. $x^2 + 13x + 42$
26. $x^2 - 13x + 42$
27. $x^2 + 5xy + 6y^2$
28. $x^2 - 5xy + 6y^2$
29. $x^2 - 9xy + 20y^2$
30. $x^2 + 9xy + 20y^2$
31. $a^2 + 2ab - 8b^2$
32. $a^2 - 2ab - 8b^2$
33. $a^2 - 10ab + 25b^2$
34. $a^2 + 6ab + 9b^2$
35. $a^2 + 10ab + 25b^2$
36. $a^2 - 6ab + 9b^2$
37. $x^2 + 2xa - 48a^2$
38. $x^2 - 3xa - 10a^2$
39. $x^2 - 5xb - 36b^2$
40. $x^2 - 13xb + 36b^2$

Factor the following problems completely. First, factor out the greatest common factor, and then factor the remaining trinomial.

41. $2x^2 + 6x + 4$
42. $3x^2 - 6x - 9$
43. $3a^2 - 3a - 60$
44. $2a^2 - 18a + 28$
45. $100x^2 - 500x + 600$
46. $100x^2 - 900x + 2{,}000$
47. $100p^2 - 1{,}300p + 4{,}000$
48. $100p^2 - 1{,}200p + 3{,}200$
49. $x^4 - x^3 - 12x^2$
50. $x^4 - 11x^3 + 24x^2$
51. $2r^3 + 4r^2 - 30r$
52. $5r^3 + 45r^2 + 100r$
53. $2y^4 - 6y^3 - 8y^2$
54. $3r^3 - 3r^2 - 6r$
55. $x^5 + 4x^4 + 4x^3$
56. $x^5 + 13x^4 + 42x^3$
57. $3y^4 - 12y^3 - 15y^2$
58. $5y^4 - 10y^3 + 5y^2$
59. $4x^4 - 52x^3 + 144x^2$
60. $3x^3 - 3x^2 - 18x$
61. $-a^2 - 11a - 30$
62. $-a^2 + 11a - 18$
63. $56 - x - x^2$
64. $44 + 7x - x^2$

Factor completely.

65. $x^4 - 5x^2 + 6$

66. $x^6 - 2x^3 - 15$

67. $x^2 - 80x - 2{,}000$

68. $x^2 - 190x - 2{,}000$

69. $x^2 - x + \frac{1}{4}$

70. $x^2 - \frac{2}{3}x + \frac{1}{9}$

71. $x^2 + 0.6x + 0.08$

72. $x^2 + 0.8x + 0.15$

73. If one of the factors of $x^2 + 24x + 128$ is $x + 8$, what is the other factor?

74. If one factor of $x^2 + 260x + 2{,}500$ is $x + 10$, what is the other factor?

75. What polynomial, when factored, gives $(4x + 3)(x - 1)$?

76. What polynomial factors to $(4x - 3)(x + 1)$?

Learning Objectives Assessment

The following problems can be used to help assess if you have successfully met the learning objectives for this section.

77. Factor: $x^2 + 8x + 12$.

a. $(x + 4)(x + 3)$ **b.** $(x + 2)(x + 6)$

c. $(x - 2)(x - 6)$ **d.** $(x + 10)(x - 2)$

78. Factor: $x^2 - 9x + 8$.

a. $(x + 1)(x + 8)$ **b.** $(x - 2)(x - 4)$

c. $(x + 2)(x + 4)$ **d.** $(x - 1)(x - 8)$

79. Factor: $a^2 - 5a - 6$.

a. $(a + 6)(a - 1)$ **b.** $(a - 2)(a - 3)$

c. $(a - 6)(a + 1)$ **d.** $(a + 2)(a - 3)$

80. Factor: $a^2 + 10a - 24$.

a. $(a + 4)(a + 6)$ **b.** $(a - 4)(a - 6)$

c. $(a - 12)(a + 2)$ **d.** $(a + 12)(a - 2)$

Getting Ready for the Next Section

Multiply using the FOIL method.

81. $(6a + 1)(a + 2)$

82. $(6a - 1)(a - 2)$

83. $(3a + 2)(2a + 1)$

84. $(3a - 2)(2a - 1)$

85. $(6a + 2)(a + 1)$

86. $(3a + 1)(2a + 2)$

More on Factoring Trinomials 6.3

Learning Objectives

In this section, we will learn how to:

1. Factor trinomials by trial and error.
2. Factor trinomials by grouping.
3. Solve applications that involve factoring trinomials.

Introduction

We will now consider trinomials whose greatest common factor is 1 and whose leading coefficient (the coefficient of the squared term) is a number other than 1.

Trial and Error

Suppose we want to factor the trinomial $2x^2 - 5x - 3$. We know the factors (if they exist) will be a pair of binomials. The product of their first terms is $2x^2$, so the first terms must be $2x$ and x. The product of their last terms is -3, so the last terms must be 3 and 1, with the binomials having opposite signs. Let us list all the possible factors along with the trinomial that would result if we were to multiply them together. Remember, the middle term comes from the product of the inside terms plus the product of the outside terms.

Binomial Factors	First Term	Middle Term	Last Term
$(2x - 3)(x + 1)$	$2x^2$	$-x$	-3
$(2x + 3)(x - 1)$	$2x^2$	$+x$	-3
$(2x - 1)(x + 3)$	$2x^2$	$+5x$	-3
$(2x + 1)(x - 3)$	$2x^2$	$-5x$	-3

We can see from the last line that the factors of $2x^2 - 5x - 3$ are $(2x + 1)$ and $(x - 3)$. There is no straightforward way, as there was in the previous section, to find the factors, other than by trial and error or by simply listing all the possibilities. We look for possible factors that, when multiplied, will give the correct first and last terms, and then we see if we can adjust them to give the correct middle term.

VIDEO EXAMPLES

SECTION 6.3

EXAMPLE 1 Factor: $6a^2 + 7a + 2$.

SOLUTION We list all the possible pairs of factors that, when multiplied together, give a trinomial whose first term is $6a^2$ and whose last term is $+2$. The first terms can either be $6a$ and $1a$, or $2a$ and $3a$. The last terms must be 1 and 2, with both binomials positive.

Binomial Factors	First Term	Middle Term	Last Term
$(6a + 1)(a + 2)$	$6a^2$	$+13a$	$+2$
$(6a + 2)(a + 1)$	$6a^2$	$+8a$	$+2$
$(3a + 1)(2a + 2)$	$6a^2$	$+8a$	$+2$
$(3a + 2)(2a + 1)$	$6a^2$	$+7a$	$+2$

We can rule out the second line because the first factor has a 2 common to each term and so could be factored again, giving $2(3a + 1)(a + 1)$. Since our original trinomial, $6a^2 + 7a + 2$, did *not* have a greatest common factor of 2, neither of its factors will. Likewise, we can rule out the third line because $2a + 2$ has a greatest common factor of 2 also. The fourth line gives the correct middle term, so the factors of $6a^2 + 7a + 2$ are $(3a + 2)$ and $(2a + 1)$.

Check: $(3a + 2)(2a + 1) = 6a^2 + 7a + 2$

Note Remember, we can always check our results by multiplying the factors we have and comparing that product with our original polynomial.

EXAMPLE 2 Factor: $4x^2 - x - 3$.

SOLUTION The choices for the first terms are $4x$ and x, or $2x$ and $2x$. The last terms must be 1 and 3 with the binomials having opposite signs. We list all the possible factors that, when multiplied, give a trinomial whose first term is $4x^2$ and whose last term is -3.

Binomial Factors	First Term	Middle Term	Last Term
$(4x + 1)(x - 3)$	$4x^2$	$-11x$	-3
$(4x - 1)(x + 3)$	$4x^2$	$+11x$	-3
$(4x + 3)(x - 1)$	$4x^2$	$-x$	-3
$(4x - 3)(x + 1)$	$4x^2$	$+x$	-3
$(2x + 1)(2x - 3)$	$4x^2$	$-4x$	-3
$(2x - 1)(2x + 3)$	$4x^2$	$+4x$	-3

The third line shows that the factors are $(4x + 3)$ and $(x - 1)$.

Check: $(4x + 3)(x - 1) = 4x^2 - x - 3$

You will find that the more practice you have at factoring this type of trinomial, the faster you will get the correct factors. You will pick up some shortcuts along the way, or you may come across a system of eliminating some factors as possibilities. Whatever works best for you is the method you should use. Factoring is a very important tool, and you must be good at it.

EXAMPLE 3 Factor: $12y^3 + 10y^2 - 12y$.

SOLUTION We begin by factoring out the greatest common factor, $2y$:

$$12y^3 + 10y^2 - 12y = 2y(6y^2 + 5y - 6)$$

Note Once again, the first step in any factoring problem is to factor out the greatest common factor if it is other than 1.

The choices for the first terms are $3y$ and $2y$, or $6y$ and y. The choices for the last terms are 2 and 3, or 1 and 6. The binomials must have opposite signs. We now list all possible factors of a trinomial with the first term $6y^2$ and last term -6, along with the associated middle terms.

Possible Factors	Middle Term When Multiplied
$(3y + 2)(2y - 3)$	$-5y$
$(3y - 2)(2y + 3)$	$+5y$
$(6y + 1)(y - 6)$	$-35y$
$(6y - 1)(y + 6)$	$+35y$

Notice that we did not list any possibilities where either factor would have a greatest common factor, such as $(6y + 2)(y - 3)$. The second line gives the correct factors. The complete problem is:

$$\begin{aligned} 12y^3 + 10y^2 - 12y &= 2y(6y^2 + 5y - 6) \\ &= 2y(3y - 2)(2y + 3) \end{aligned}$$

Factoring Trinomials by Grouping

Factoring trinomials by trial and error works well as long as there are not too many possibilities to test for the factors. When the coefficients of the first and third terms have many possible factors, it may take a considerable time to find the correct combination, and it can be easy to overlook some possibilities. In this case, we can use a second method to factor the trinomial that involves grouping (sometimes this process is referred to as the AC method or splitting the middle term).

The main idea behind this second method is to write the middle term as a sum of two new terms. This changes the trinomial into a polynomial with 4 terms, allowing grouping to be used.

Let's revisit the trinomial from the beginning of this section. To factor $2x^2 - 5x - 3$, we begin by multiplying the coefficients of the first and third terms:

$$\text{(first coefficient)(third coefficient)} = 2(-3) = -6$$

Then we list all possible factors of -6 and their corresponding sums.

Factors	Sum
$1(-6)$	$1 + (-6) = -5$
$-1(6)$	$-1 + (6) = 5$
$2(-3)$	$2 + (-3) = -1$
$-2(3)$	$-2 + 3 = 1$

We want the factors whose sum gives us the coefficient of the middle term. The first line shows that the correct factors are 1 and -6. Writing $-5x$ as $x + (-6x)$, we have

$$2x^2 - 5x - 3 = 2x^2 + x - 6x - 3$$

It does not matter in which order we write the new terms. It would work just as well if we wrote $2x^2 - 6x + x - 3$.

We can now factor the resulting expression using grouping. Here is the complete process:

$$\begin{aligned} 2x^2 - 5x - 3 &= 2x^2 + x - 6x - 3 \\ &= x(2x + 1) - 3(2x + 1) \\ &= (2x + 1)(x - 3) \end{aligned}$$

Notice we obtain the same result as we did using trial and error.

EXAMPLE 4 Factor: $6x^2 - 17x + 12$.

SOLUTION The product of the first and third coefficients is $6(12) = 72$. We need factors of 72 whose sum is -17. Because the product is positive and the middle term is negative, both factors must be negative. Here are the possible factors:

Factors	Sum
$-1(-72)$	$-1 + (-72) = -73$
$-2(-36)$	$-2 + (-36) = -38$
$-3(-24)$	$-3 + (-24) = -27$
$-4(-18)$	$-4 + (-18) = -22$
$-6(-12)$	$-6 + (-12) = -18$
$-8(-9)$	$-8 + (-9) = -17$

The last line shows that the correct factors are -8 and -9. Writing $-17x$ as $-8x + (-9x)$, we have:

$$\begin{aligned} 6x^2 - 17x + 12 &= 6x^2 - 8x - 9x + 12 \\ &= 2x(3x - 4) - 3(3x - 4) \\ &= (2x - 3)(3x - 4) \end{aligned}$$

EXAMPLE 5 Factor: $8x^2 + 26x - 15$.

SOLUTION We need two numbers whose product is $8(-15) = -120$ and whose sum is 26. The numbers must have opposite signs, and the number with the larger absolute value must be positive for the sum to be positive. We begin listing possible factors and their sums.

Factors	Sum
$-1(120)$	$-1 + (120) = 119$
$-2(60)$	$-2 + (60) = 58$
$-3(40)$	$-3 + (40) = 37$
$-4(30)$	$-4 + (30) = 26$

The fourth line shows the correct numbers. Writing $26x$ as $-4x + 30x$, we have:

$$\begin{aligned} 8x^2 + 26x - 15 &= 8x^2 - 4x + 30x - 15 \\ &= 4x(2x - 1) + 15(2x - 1) \\ &= (2x - 1)(4x + 15) \end{aligned}$$

EXAMPLE 6 Factor: $30x^2y - 5xy^2 - 10y^3$.

SOLUTION The greatest common factor is $5y$:

$$30x^2y - 5xy^2 - 10y^3 = 5y(6x^2 - xy - 2y^2)$$

Writing $-xy$ as $3xy - 4xy$, we have

$$\begin{aligned} &= 5y(6x^2 + 3xy - 4xy - 2y^2) \\ &= 5y[3x(2x + y) - 2y(2x + y)] \\ &= 5y[(2x + y)(3x - 2y)] \\ &= 5y(2x + y)(3x - 2y) \end{aligned}$$

Note: In Example 6, we came up with $3xy - 4xy$ by mentally reviewing the possible factors and their sums.

Application

EXAMPLE 7 A ball is tossed into the air with an upward velocity of 16 feet per second from the top of a building 32 feet high. The equation that gives the height h of the ball above the ground at any time t is

$$h = 32 + 16t - 16t^2$$

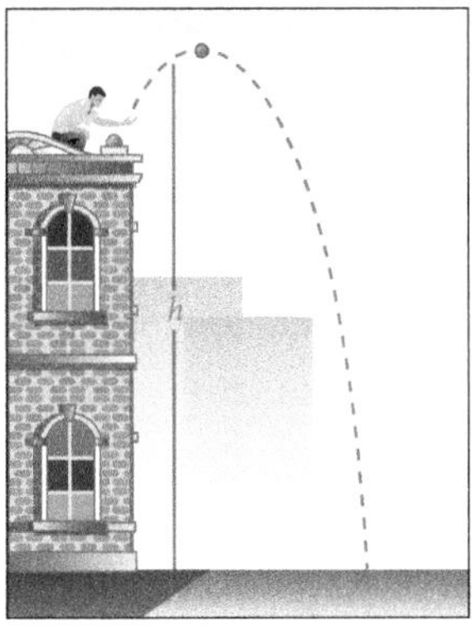

Factor the right side of this equation and then find h when t is 2.

SOLUTION We begin by factoring out the greatest common factor, -16. Then, we factor the trinomial that remains:

$$h = -16t^2 + 16t + 32$$

$$h = -16(t^2 - t - 2)$$

$$h = -16(t - 2)(t + 1)$$ Letting $t = 2$ in the equation, we have

$$h = -16(0)(3) = 0$$

When t is 2, h is 0.

Getting Ready for Class

After reading through the preceding section, respond in your own words and in complete sentences.

A. Describe the criteria you would use to set up a table of possible factors of a trinomial using trial and error.

B. Describe the criteria you would use to set up a table of possible factors for a trinomial using the grouping method.

C. In factoring $2a^2 + a - 6$ by grouping, explain why the correct pair of numbers is 4 and -3, not 2 and 3.

D. What does it mean if you factor a trinomial and one of your factors has a greatest common factor of 3?

Problem Set 6.3

Factor the following trinomials using either method.

1. $2x^2 + 7x + 3$
2. $2x^2 + 5x + 3$
3. $2a^2 - a - 3$
4. $2a^2 + a - 3$
5. $3x^2 + 2x - 5$
6. $3x^2 - 2x - 5$
7. $3y^2 + 2y - 5$
8. $3y^2 + 14y - 5$
9. $6x^2 + 13x + 6$
10. $6x^2 - 13x + 6$
11. $4x^2 - 12xy + 9y^2$
12. $4x^2 + 12xy + 9y^2$
13. $4y^2 - 11y - 3$
14. $4y^2 + y - 3$
15. $20x^2 - 41x + 20$
16. $20x^2 + 9x - 20$
17. $20a^2 + 48ab - 5b^2$
18. $20a^2 + 29ab + 5b^2$
19. $20x^2 - 21x - 5$
20. $20x^2 - 48x - 5$
21. $12m^2 + 16m - 3$
22. $12m^2 + 20m + 3$
23. $20x^2 + 37x + 15$
24. $20x^2 + 13x - 15$
25. $12a^2 - 25ab + 12b^2$
26. $12a^2 + 7ab - 12b^2$
27. $3x^2 - xy - 14y^2$
28. $3x^2 + 19xy - 14y^2$
29. $14x^2 + 29x - 15$
30. $14x^2 + 11x - 15$
31. $6x^2 - 43x + 55$
32. $6x^2 - 7x - 55$
33. $15t^2 - 67t + 38$
34. $15t^2 - 79t - 34$

Factor each of the following completely. Look first for the greatest common factor.

35. $4x^2 + 2x - 6$
36. $6x^2 - 51x + 63$
37. $24a^2 - 50a + 24$
38. $18a^2 + 48a + 32$
39. $10 + 13x - 3x^2$
40. $10 - x - 3x^2$
41. $-12x^2 + 10x + 8$
42. $-18x^2 + 51x - 15$
43. $10x^3 - 23x^2 + 12x$
44. $10x^4 + 7x^3 - 12x^2$
45. $6x^4 - 11x^3 - 10x^2$
46. $6x^3 + 19x^2 + 10x$
47. $10a^3 - 6a^2 - 4a$
48. $6a^3 + 15a^2 + 9a$
49. $15x^3 - 102x^2 - 21x$
50. $2x^4 - 24x^3 + 64x^2$
51. $35y^3 - 60y^2 - 20y$
52. $14y^4 - 32y^3 + 8y^2$
53. $15a^4 - 2a^3 - a^2$
54. $10a^5 - 17a^4 + 3a^3$
55. $24x^2y - 6xy - 45y$
56. $8x^2y^2 + 26xy^2 + 15y^2$
57. $12x^2y - 34xy^2 + 14y^3$
58. $12x^2y - 46xy^2 + 14y^3$
59. Evaluate the expression $2x^2 + 7x + 3$ and the expression $(2x + 1)(x + 3)$ for $x = 2$.
60. Evaluate the expression $2a^2 - a - 3$ and the expression $(2a - 3)(a + 1)$ for $a = 5$.
61. What polynomial factors to $(2x + 3)(2x - 3)$?
62. What polynomial factors to $(5x + 4)(5x - 4)$?
63. What polynomial factors to $(x + 3)(x - 3)(x^2 + 9)$?
64. What polynomial factors to $(x + 2)(x - 2)(x^2 + 4)$?

Applying the Concepts

65. Archery Margaret shoots an arrow into the air. The equation for the height (in feet) of the tip of the arrow is:

$$h = 8 + 62t - 16t^2$$

Factor the right side of this equation. Then fill in the table for various heights of the arrow, using the factored form of the equation.

Time t (seconds)	0	1	2	3	4
Height h (feet)					

66. Coin Toss At the beginning of every football game, the referee flips a coin to see who will kick off. The equation that gives the height (in feet) of the coin tossed in the air is:

$$h = 6 + 29t - 16t^2$$

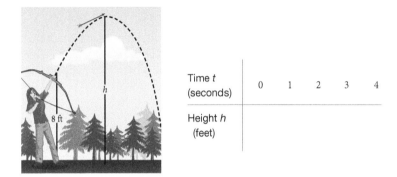

a. Factor this equation.

b. Use the factored form of the equation to find the height of the coin after 0 seconds, 1 second, and 2 seconds.

67. Constructing a Box Yesterday I was experimenting with how to cut and fold a certain piece of cardboard to make a box with different volumes. Unfortunately, today I have lost both the cardboard and most of my notes. I remember that I made the box by cutting equal squares from the corners then folding up the side flaps.

I don't remember how big the cardboard was, and I can only find the last page of notes, which says that if x is the length of a side of a small square (in inches), then the volume is $V = 99x - 40x^2 + 4x^3$.

a. Factor the right side of this expression completely.

b. What were the dimensions of the original piece of cardboard?

68. Constructing a Box Repeat Problem 67 if the remaining formula is $V = 15x - 16x^2 + 4x^3$.

Learning Objectives Assessment

The following problems can be used to help assess if you have successfully met the learning objectives for this section.

69. Factor $4x^2 - 5x - 6$ by trial and error. Which factor appears in the answer?

a. $2x + 3$ **b.** $2x - 1$ **c.** $x - 3$ **d.** $x - 2$

70. In factoring $6x^2 + 11x - 7$ by grouping, what is the correct way to rewrite the middle term $11x$?

a. $-3x + 14x$ **b.** $9x + 2x$ **c.** $12x - x$ **d.** $18x - 7x$

71. A water balloon is tossed in the air. The equation $h = 8 + 28t - 16t^2$ gives the height of the water balloon after t seconds. If the equation is factored completely, which factor appears in the answer?

a. $t + 2$ **b.** $t - 2$ **c.** $16t + 1$ **d.** $8t - 1$

Getting Ready for the Next Section

Multiply each of the following.

72. $(x + 3)(x - 3)$

73. $(x - 4)(x + 4)$

74. $(2x - 3y)(2x + 3y)$

75. $(5x - 6y)(5x + 6y)$

76. $(x^2 + 4)(x + 2)(x - 2)$

77. $(x^2 + 9)(x + 3)(x - 3)$

78. $(x + 3)^2$

79. $(x - 4)^2$

80. $(2x + 3)^2$

81. $(3x - y)^2$

82. $(4x - 2y)^2$

83. $(5x - 6y)^2$

84. **a.** 1^3 **b.** 2^3 **c.** 3^3 **d.** 4^3 **e.** 5^3

85. **a.** $(-1)^3$ **b.** $(-2)^3$ **c.** $(-3)^3$ **d.** $(-4)^3$ **e.** $(-5)^3$

86. **a.** $x(x^2 - 2x + 4)$ **b.** $2(x^2 - 2x + 4)$ **c.** $(x + 2)(x^2 - 2x + 4)$

87. **a.** $x(x^2 + 2x + 4)$ **b.** $-2(x^2 + 2x + 4)$ **c.** $(x - 2)(x^2 + 2x + 4)$

88. **a.** $x(x^2 - 3x + 9)$ **b.** $3(x^2 - 3x + 9)$ **c.** $(x + 3)(x^2 - 3x + 9)$

89. **a.** $x(x^2 + 3x + 9)$ **b.** $-3(x^2 + 3x + 9)$ **c.** $(x - 3)(x^2 + 3x + 9)$

Special Factoring Patterns 6.4

Learning Objectives

In this section, we will learn how to:

1. Factor a perfect square trinomial.
2. Factor a difference of two squares.
3. Factor a sum or difference of two cubes.

Introduction

In Chapter 5, we listed the following three special products:

$$(a + b)^2 = (a + b)(a + b) = a^2 + 2ab + b^2$$
$$(a - b)^2 = (a - b)(a - b) = a^2 - 2ab + b^2$$
$$(a + b)(a - b) = a^2 - b^2$$

Since factoring is the reverse of multiplication, we can also consider the three special products as three special factoring patterns:

$$a^2 + 2ab + b^2 = (a + b)^2$$
$$a^2 - 2ab + b^2 = (a - b)^2$$
$$a^2 - b^2 = (a + b)(a - b)$$

We begin by considering the first two of these formulas.

Perfect Square Trinomials

Any trinomial of the form $a^2 + 2ab + b^2$ or $a^2 - 2ab + b^2$ is called a *perfect square trinomial* because it can be factored as the square of a binomial. For convenience, we repeat both formulas as follows.

RULE *Perfect Square Trinomials*

$$a^2 + 2ab + b^2 = (a + b)^2$$
$$a^2 - 2ab + b^2 = (a - b)^2$$

Perfect square trinomials can be factored using the methods of the two previous sections, but if we recognize that a trinomial is a perfect square trinomial, it can often be factored more quickly using the special pattern.

VIDEO EXAMPLES

SECTION 6.4

EXAMPLE 1 Factor: $25x^2 - 60x + 36$.

SOLUTION We notice that the first and last terms are the perfect squares $(5x)^2$ and $(6)^2$, and that the middle term involves the product $2(5x)(6) = 60x$. This means we have a perfect square trinomial with $a = 5x$ and $b = 6$. Thus, we have

$$25x^2 - 60x + 36 = (5x)^2 - 2(5x)(6) + (6)^2$$
$$= (5x - 6)^2$$

The trinomial $25x^2 - 60x + 36$ factors to $(5x - 6)(5x - 6) = (5x - 6)^2$.

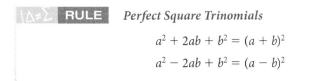

EXAMPLE 2 Factor: $5x^2 + 30x + 45$.

SOLUTION We begin by factoring out the greatest common factor, which is 5. Then we notice that the trinomial that remains is a perfect square trinomial with $a = x$ and $b = 3$:

$$5x^2 + 30x + 45 = 5(x^2 + 6x + 9)$$
$$= 5(x^2 + 2(x)(3) + 3^2)$$
$$= 5(x + 3)^2$$

Note As we have indicated before, perfect square trinomials like the ones in Examples 1 and 2 can be factored by the methods developed in previous sections. Recognizing that they factor to binomial squares simply saves time in factoring.

Factoring Binomials

Our third special product gives us one way to factor certain binomials whose terms are both perfect squares.

RULE *Difference of Squares*

$$a^2 - b^2 = (a + b)(a - b)$$

Notice that this pattern requires the binomial to have a subtraction. Other than a possible greatest common factor, the expression $a^2 + b^2$ is prime, meaning there is no special pattern by which it can be factored.

EXAMPLE 3 Factor: $16x^2 - 25$.

SOLUTION We can see that both terms are perfect squares. This fact becomes even more obvious if we rewrite the problem as:

$$16x^2 - 25 = (4x)^2 - (5)^2$$

The first term is the square of the quantity $4x$, and the last term is the square of 5. The completed problem looks like this:

$$16x^2 - 25 = (4x)^2 - (5)^2$$
$$= (4x + 5)(4x - 5)$$

To check our results, we multiply:

$$(4x + 5)(4x - 5) = 16x^2 - 20x + 20x - 25$$
$$= 16x^2 - 25$$

EXAMPLE 4 Factor: $1 - 36a^2$.

SOLUTION We rewrite the two terms to show they are perfect squares and then factor. Remember, 1 is its own square, $1^2 = 1$.

$$1 - 36a^2 = (1)^2 - (6a)^2$$
$$= (1 + 6a)(1 - 6a)$$

To check our results, we multiply:

$$(1 + 6a)(1 - 6a) = 1 - 6a + 6a - 36a^2$$
$$= 1 - 36a^2$$

EXAMPLE 5 Factor: $4x^2 + 16y^2$.

SOLUTION We begin by factoring out the greatest common factor of 4:

$$4x^2 + 16y^2 = 4(x^2 + 4y^2)$$

Both terms of the resulting binomial, $x^2 + 4y^2$, are perfect squares. However, because the binomial is a *sum* of squares, it cannot be factored any further.

EXAMPLE 6 Factor: $x^4 - y^4$.

SOLUTION x^4 is the perfect square $(x^2)^2$, and y^4 is $(y^2)^2$:

$$\begin{aligned} x^4 - y^4 &= (x^2)^2 - (y^2)^2 \\ &= (x^2 + y^2)(x^2 - y^2) \end{aligned}$$

The factor $(x^2 - y^2)$ is itself the difference of two squares and therefore can be factored again. The factor $(x^2 + y^2)$ is the *sum* of two squares and cannot be factored again. The complete problem is this:

$$\begin{aligned} x^4 - y^4 &= (x^2)^2 - (y^2)^2 \\ &= (x^2 + y^2)(x^2 - y^2) \\ &= (x^2 + y^2)(x + y)(x - y) \end{aligned}$$

Note If you think the sum of two squares $x^2 + y^2$ factors, you should try it. Write down the factors you think it has, and then multiply them using the FOIL method. You won't get $x^2 + y^2$.

EXAMPLE 7 Factor: $(x - 3)^2 - 25$.

SOLUTION This example has the form $a^2 - b^2$, where a is $x - 3$ and b is 5. We factor it according to the formula for the difference of two squares:

$$\begin{aligned} (x - 3)^2 - 25 &= (x - 3)^2 - 5^2 && \text{Write 25 as } 5^2 \\ &= [(x - 3) + 5][(x - 3) - 5] && \text{Factor} \\ &= (x + 2)(x - 8) && \text{Simplify} \end{aligned}$$

Notice in this example we could have expanded $(x - 3)^2$, subtracted 25, and then factored to obtain the same result:

$$\begin{aligned} (x - 3)^2 - 25 &= x^2 - 6x + 9 - 25 && \text{Expand } (x - 3)^2 \\ &= x^2 - 6x - 16 && \text{Simplify} \\ &= (x - 8)(x + 2) && \text{Factor} \end{aligned}$$

Just as we can factor certain binomials whose terms are both perfect squares, we can also factor binomials whose terms are both perfect cubes. The formulas that allow us to factor the sum of two cubes and the difference of two cubes are not as simple as the formula for factoring the difference of two squares.

RULE *Sum or Difference of Cubes*

$$a^3 + b^3 = (a + b)(a^2 - ab + b^2)$$
$$a^3 - b^3 = (a - b)(a^2 + ab + b^2)$$

Let's begin our work with these two formulas by showing that they are true. To do so, we multiply out the right side of each formula.

EXAMPLE 8 Verify the two formulas.

SOLUTION We verify the formulas by multiplying the right sides and comparing the results with the left sides:

$$\begin{array}{rrrrl} & a^2 & -\,ab & +\,b^2 & \\ \times & & a & +\,b & \\ \hline & a^3 & -\,a^2b & +\,ab^2 & \\ + & & a^2b & -\,ab^2 & +\,b^3 \\ \hline & a^3 & & & +\,b^3 \end{array} \qquad \begin{array}{rrrrl} & a^2 & +\,ab & +\,b^2 & \\ \times & & a & -\,b & \\ \hline & a^3 & +\,a^2b & +\,ab^2 & \\ + & & -\,a^2b & -\,ab^2 & -\,b^3 \\ \hline & a^3 & & & -\,b^3 \end{array}$$

The first formula is correct. The second formula is correct.

Here are some examples that use the formulas for factoring the sum and difference of two cubes.

EXAMPLE 9 Factor: $x^3 - 8$.

SOLUTION Since both terms are perfect cubes, we write them as such and apply the formula:

$$\begin{aligned} x^3 - 8 &= x^3 - 2^3 \\ &= (x - 2)(x^2 + 2x + 2^2) && \text{Formula using } a = x,\ b = 2 \\ &= (x - 2)(x^2 + 2x + 4) \end{aligned}$$

EXAMPLE 10 Factor: $y^3 + 27$.

SOLUTION Proceeding as we did in Example 9, we first write 27 as 3^3. Then, we apply the formula for factoring the sum of two cubes, which is $a^3 + b^3 = (a + b)(a^2 - ab + b^2)$:

$$\begin{aligned} y^3 + 27 &= y^3 + 3^3 \\ &= (y + 3)(y^2 - 3y + 3^2) && \text{Formula using } a = y,\ b = 3 \\ &= (y + 3)(y^2 - 3y + 9) \end{aligned}$$

Here are some additional examples using the formulas for factoring the sum and difference of two cubes.

EXAMPLE 11 Factor: $64 + t^3$.

SOLUTION The first term is the cube of 4 and the second term is the cube of t. Therefore,

$$\begin{aligned} 64 + t^3 &= 4^3 + t^3 \\ &= (4 + t)(4^2 - 4t + t^2) \\ &= (4 + t)(16 - 4t + t^2) \end{aligned}$$

EXAMPLE 12 Factor: $27x^3 + 125y^3$.

SOLUTION Writing both terms as perfect cubes, we have

$$\begin{aligned} 27x^3 + 125y^3 &= (3x)^3 + (5y)^3 \\ &= (3x + 5y)((3x)^2 - 3x(5y) + (5y)^2) \\ &= (3x + 5y)(9x^2 - 15xy + 25y^2) \end{aligned}$$

EXAMPLE 13 Factor: $a^3 - \frac{1}{8}$.

SOLUTION The first term is the cube of a, whereas the second term is the cube of $\frac{1}{2}$:

$$\begin{aligned} a^3 - \frac{1}{8} &= a^3 - \left(\frac{1}{2}\right)^3 \\ &= \left(a - \frac{1}{2}\right)\left(a^2 + \frac{1}{2}a + \left(\frac{1}{2}\right)^2\right) \\ &= \left(a - \frac{1}{2}\right)\left(a^2 + \frac{1}{2}a + \frac{1}{4}\right) \end{aligned}$$

EXAMPLE 14 Factor: $x^6 - y^6$.

SOLUTION We have a choice of how we want to write the two terms to begin. We can write the expression as the difference of two squares, $(x^3)^2 - (y^3)^2$, or as the difference of two cubes, $(x^2)^3 - (y^2)^3$. It is better to use the difference of two squares if we have a choice. Then we factor again, using the formulas for the difference and sum of two cubes.

$$\begin{aligned} x^6 - y^6 &= (x^3)^2 - (y^3)^2 \\ &= (x^3 - y^3)(x^3 + y^3) \\ &= (x - y)(x^2 + xy + y^2)(x + y)(x^2 - xy + y^2) \end{aligned}$$

Getting Ready for Class

After reading through the preceding section, respond in your own words and in complete sentences.

A. What is a perfect square trinomial?

B. What are the different ways that a binomial can be factored?

C. Describe how you factor the difference of two squares.

D. How are you going to remember that the sum of two cubes factors, while the sum of two squares is prime?

Problem Set 6.4

Factor each trinomial.

1. $x^2 - 2x + 1$
2. $x^2 - 6x + 9$
3. $x^2 + 2x + 1$
4. $x^2 + 6x + 9$
5. $a^2 - 10a + 25$
6. $a^2 + 10a + 25$
7. $y^2 + 4y + 4$
8. $y^2 - 8y + 16$
9. $x^2 - 4x + 4$
10. $x^2 + 8x + 16$
11. $m^2 - 12m + 36$
12. $m^2 + 12m + 36$
13. $4a^2 + 12a + 9$
14. $9a^2 - 12a + 4$
15. $49x^2 - 14x + 1$
16. $64x^2 - 16x + 1$
17. $9y^2 - 30y + 25$
18. $25y^2 + 30y + 9$
19. $x^2 + 10xy + 25y^2$
20. $25x^2 + 10xy + y^2$
21. $9a^2 + 6ab + b^2$
22. $9a^2 - 6ab + b^2$

Factor the following by first factoring out the greatest common factor.

23. $3a^2 + 18a + 27$
24. $4a^2 - 16a + 16$
25. $2x^2 + 20xy + 50y^2$
26. $3x^2 + 30xy + 75y^2$
27. $5x^3 + 30x^2y + 45xy^2$
28. $12x^2y - 36xy^2 + 27y^3$

Factor each binomial.

29. $x^2 - 9$
30. $x^2 - 25$
31. $a^2 - 36$
32. $a^2 - 64$
33. $x^2 - 49$
34. $x^2 - 121$
35. $4a^2 - 16$
36. $4a^2 + 16$
37. $9x^2 + 25$
38. $16x^2 - 36$
39. $25x^2 - 169$
40. $x^2 - y^2$
41. $9a^2 - 16b^2$
42. $49a^2 - 25b^2$
43. $9 - m^2$
44. $16 - m^2$
45. $25 - 4x^2$
46. $36 - 49y^2$
47. $2x^2 - 18$
48. $3x^2 - 27$
49. $x^3 - y^3$
50. $x^3 + y^3$
51. $a^3 + 8$
52. $a^3 - 8$
53. $27 + x^3$
54. $27 - x^3$
55. $y^3 - 1$
56. $y^3 + 1$
57. $64 - y^3$
58. $64 + y^3$
59. $125h^3 - t^3$
60. $t^3 + 125h^3$
61. $x^3 - 216$
62. $216 + x^3$
63. $2y^3 - 54$
64. $81 + 3y^3$
65. $64 + 27a^3$
66. $27 - 64a^3$
67. $8x^3 - 27y^3$
68. $27x^3 - 8y^3$
69. $32a^2 - 128$
70. $3a^3 - 48a$
71. $8x^2y - 18y$
72. $50a^2b - 72b$
73. $2a^3 - 128b^3$
74. $128a^3 + 2b^3$
75. $2x^3 + 432y^3$
76. $432x^3 - 2y^3$
77. $10a^3 - 640b^3$
78. $640a^3 + 10b^3$
79. $10r^3 - 1{,}250$
80. $10r^3 + 1{,}250$
81. $t^3 + \frac{1}{27}$
82. $t^3 - \frac{1}{27}$
83. $27x^3 - \frac{1}{27}$
84. $8x^3 + \frac{1}{8}$
85. $64a^3 + 125b^3$
86. $125a^3 - 27b^3$
87. $\frac{1}{8}x^3 - \frac{1}{27}y^3$
88. $\frac{1}{27}x^3 + \frac{1}{8}y^3$
89. $a^4 - b^4$
90. $a^4 - 16$
91. $16m^4 - 81$
92. $81 - m^4$
93. $3x^3y - 75xy^3$
94. $2xy^3 - 8x^3y$
95. $a^6 - b^6$
96. $x^6 - 64y^6$
97. $64x^6 - y^6$
98. $x^6 - (3y)^6$
99. $x^6 - (5y)^6$
100. $(4x)^6 - (7y)^6$

Factor by grouping the first three terms together.

101. $x^2 + 6x + 9 - y^2$

102. $x^2 + 10x + 25 - y^2$

103. $x^2 + 2xy + y^2 - 9$

104. $a^2 + 2ab + b^2 - 25$

105. Find a value for b so that the polynomial $x^2 + bx + 49$ factors to $(x + 7)^2$.

106. Find a value of b so that the polynomial $x^2 + bx + 81$ factors to $(x + 9)^2$.

107. Find the value of c for which the polynomial $x^2 + 10x + c$ factors to $(x + 5)^2$.

108. Find the value of a for which the polynomial $ax^2 + 12x + 9$ factors to $(2x + 3)^2$.

Applying the Concepts

109. Area

a. What is the area of the following figure?

b. Factor the answer from part **a**.

c. Find a way to cut the figure into two pieces and put them back together to show that the factorization in part **b**. is correct.

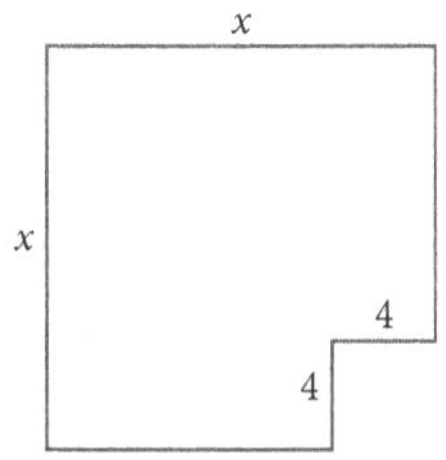

110. Area

a. What is the area of the following figure?

b. Factor the expression from part **a**.

c. Cut and rearrange the figure to show that the factorization is correct.

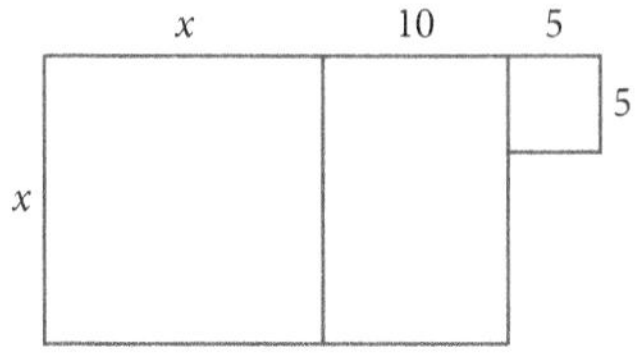

Find the area for the shaded regions; then write your result in factored form.

111.

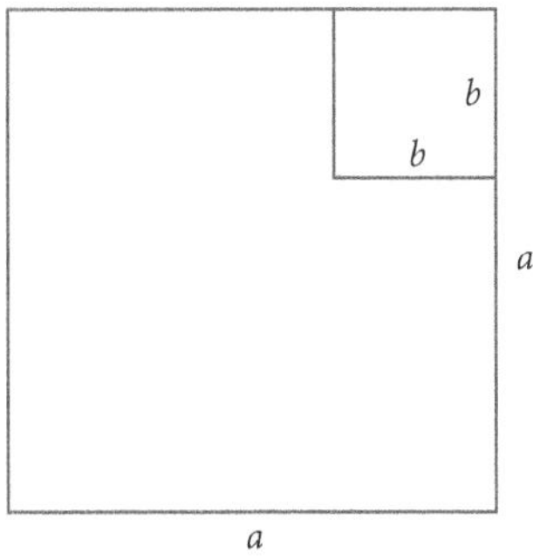

112.

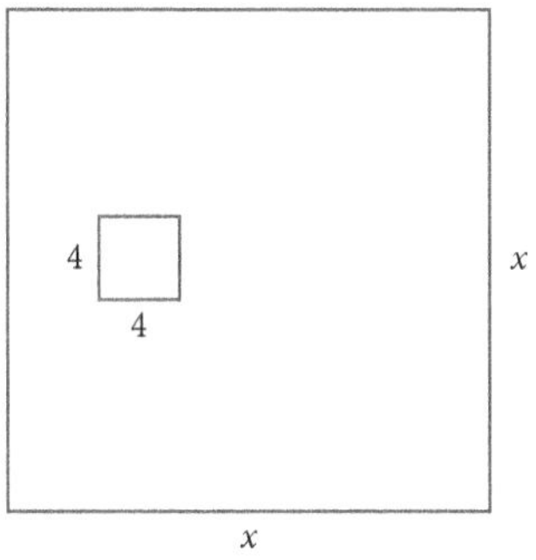

Learning Objectives Assessment

The following problems can be used to help assess if you have successfully met the learning objectives for this section.

113. Factor: $9x^2 - 48x + 64$.

a. $(3x + 8)(3x - 8)$ **b.** $(3x + 8)^2$
c. $(3x - 8)^2$ **d.** $(9x - 64)(x + 1)$

114. Factor: $4 - 9a^2$.

a. Prime **b.** $(2 + 3a)^2$
c. $(2 - 3a)^2$ **d.** $(2 + 3a)(2 - 3a)$

115. Factor: $4y^2 + 25$.

a. $(2y + 5)(2y - 5)$ **b.** Prime
c. $(2y + 5)^2$ **d.** $(2y - 5)$

116. Factor: $8x^3 + 1$.

a. Prime **b.** $(2x + 1)(2x^2 - 4x + 1)$
c. $(2x + 1)(4x^2 + 2x + 1)$ **d.** $(2x + 1)(4x^2 - 2x + 1)$

Getting Ready for the Next Section

Multiply each of the following.

117. $2x^3(x + 2)(x - 2)$

118. $3x^2(x + 3)(x - 3)$

119. $3x^2(x - 3)^2$

120. $2x^2(x + 5)^2$

121. $y(y^2 + 25)$

122. $y^3(y^2 + 36)$

123. $(5a - 2)(3a + 1)$

124. $(3a - 4)(2a - 1)$

125. $4x^2(x - 5)(x + 2)$

126. $6x(x - 4)(x + 2)$

127. $2ab^3(b^2 - 4b + 1)$

128. $2a^3b(a^2 + 3a + 1)$

Factoring: A General Review 6.5

Learning Objectives

In this section, we will learn how to:

1. Use a factoring strategy to factor a variety of polynomials.

Introduction

In this section we will review the different methods of factoring that we presented in the previous sections of the chapter. This section is important because it will give you an opportunity to factor a variety of polynomials. Prior to this section, the polynomials you worked with were grouped together according to the method used to factor them. What usually happens in a situation like this is that you become proficient at factoring the kind of polynomial you are working with at the time but have trouble when given a variety of polynomials to factor.

We begin this section with a strategy that can be used in factoring polynomials of any type. When you have finished this section and the problem set that follows, you want to be proficient enough at factoring that the checklist is second nature to you.

HOW TO *Factor a polynomial*

Step 1: If the polynomial has a greatest common factor other than 1, then factor out the greatest common factor.

Step 2: If the polynomial has two terms (it is a binomial), then see if it is the difference of squares or a sum or difference of cubes.

Step 3: If the polynomial has three terms (a trinomial), then either it is a perfect square trinomial, which will factor into the square of a binomial, or it is not a perfect square trinomial, in which case you use the methods developed in Sections 6.2 and 6.3.

Step 4: If the polynomial has more than three terms, try to factor it by grouping.

Step 5: As a final check, see if any of the factors you have written can be factored further. If you have overlooked a common factor, you can catch it here.

Here are some examples illustrating how we use the checklist.

VIDEO EXAMPLES

SECTION 6.5

EXAMPLE 1 Factor: $2x^5 - 8x^3$.

SOLUTION First, we check to see if the greatest common factor is other than 1. Since the greatest common factor is $2x^3$, we begin by factoring it out. Once we have done so, we notice that the binomial that remains is the difference of two squares:

$$2x^5 - 8x^3 = 2x^3(x^2 - 4) \qquad \text{Factor out the greatest common factor, } 2x^3$$

$$= 2x^3(x + 2)(x - 2) \qquad \text{Factor the difference of two squares}$$

Note that the greatest common factor $2x^3$ that we factored from each term in the first step of Example 1 remains as part of the answer to the problem. This is because it is one of the factors of the original binomial. Remember, the expression we end up with when factoring must be equal to the expression we start with. We can't just drop a factor and expect the resulting expression to equal the original expression.

EXAMPLE 2 Factor: $3x^4 - 18x^3 + 27x^2$.

SOLUTION Step 1 is to factor out the greatest common factor, $3x^2$. After we have done so, we notice that the trinomial that remains is a perfect square trinomial, which will factor as the square of a binomial:

$$3x^4 - 18x^3 + 27x^2 = 3x^2(x^2 - 6x + 9) \quad \text{Factor out } 3x^2$$
$$= 3x^2(x - 3)^2 \quad x^2 - 6x + 9 \text{ is the square of } x - 3$$

EXAMPLE 3 Factor: $y^4 + 27y$.

SOLUTION We begin by factoring out the y that is common to both terms. The binomial that remains after we have done so is the sum of two cubes:

$$y^4 + 27y = y(y^3 + 27) \quad \text{Factor out the greatest common factor of } y$$
$$= y(y + 3)(y^2 - 3y + 9) \quad \text{Sum of cubes pattern}$$

EXAMPLE 4 Factor: $6a^2 - 11a + 4$.

SOLUTION Here we have a trinomial that does not have a greatest common factor other than 1. Since it is not a perfect square trinomial, we factor it by trial and error or by grouping. Without showing all the different possibilities, here is the answer by grouping:

$$6a^2 - 11a + 4 = 6a^2 - 3a - 8a + 4$$
$$= 3a(2a - 1) - 4(2a - 1)$$
$$= (3a - 4)(2a - 1)$$

EXAMPLE 5 Factor: $6x^3 - 12x^2 - 48x$.

SOLUTION This trinomial has a greatest common factor of $6x$. The trinomial that remains after the $6x$ has been factored from each term must be factored using the rules from Section 6.2.

$$6x^3 - 12x^2 - 48x = 6x(x^2 - 2x - 8)$$
$$= 6x(x - 4)(x + 2)$$

EXAMPLE 6 Factor: $2ab^5 + 8ab^4 + 2ab^3$.

SOLUTION The greatest common factor is $2ab^3$. We begin by factoring it from each term. After that we find the trinomial that remains cannot be factored further:

$$2ab^5 + 8ab^4 + 2ab^3 = 2ab^3(b^2 + 4b + 1)$$

EXAMPLE 7 Factor: $xy + 8x + 3y + 24$.

SOLUTION Since our polynomial has four terms, we try factoring by grouping:

$$\begin{aligned} xy + 8x + 3y + 24 &= x(y + 8) + 3(y + 8) \\ &= (y + 8)(x + 3) \end{aligned}$$

Getting Ready for Class

After reading through the preceding section, respond in your own words and in complete sentences.

A. What is the first step in factoring any polynomial?

B. If a polynomial has four terms, what method of factoring should you try?

C. If a polynomial has two terms, what method of factoring should you try?

D. What is the last step in factoring any polynomial?

Problem Set 6.5

Factor each of the following polynomials completely; that is, once you are finished factoring, none of the factors you obtain should be factorable any further. Also, note that the even-numbered problems are not necessarily similar to the odd-numbered problems that precede them in this problem set.

1. $x^2 - 81$
2. $x^2 - 18x + 81$
3. $x^2 + 2x - 15$
4. $15x^2 + 11x - 6$
5. $x^2 + 6x + 9$
6. $12x^2 - 11x + 2$
7. $y^2 - 10y + 25$
8. $21y^2 - 25y - 4$
9. $2a^3b + 6a^2b + 2ab$
10. $6a^2 - ab - 15b^2$
11. $x^2 + x + 1$
12. $2x^2 - 4x + 2$
13. $12a^2 - 75$
14. $16a^3 - 250$
15. $9x^2 - 12xy + 4y^2$
16. $x^3 - x^2$
17. $4x^3 + 16xy^2$
18. $16x^2 + 49y^2$
19. $2y^3 + 20y^2 + 50y$
20. $3y^2 - 9y - 30$
21. $a^6 + 4a^4b^2$
22. $5a^2 - 45b^2$
23. $xy + 3x + 4y + 12$
24. $xy + 7x + 6y + 42$
25. $x^4 - 16$
26. $x^4 - 81$
27. $xy - 5x + 2y - 10$
28. $xy - 7x + 3y - 21$
29. $5a^2 + 10ab + 5b^2$
30. $3a^3b^2 + 15a^2b^2 + 3ab^2$
31. $64 + x^3$
32. $49 + x^2$
33. $3x^2 + 15xy + 18y^2$
34. $3x^2 + 27xy + 54y^2$
35. $2x^2 + 15x - 38$
36. $2x^2 + 7x - 85$
37. $100x^2 - 300x + 200$
38. $100x^2 - 400x + 300$
39. $x^2 - 64$
40. $x^3 - 64$
41. $x^2 + 3x + ax + 3a$
42. $x^2 + 4x + bx + 4b$
43. $49a^7 - 9a^5$
44. $a^4 - 1$
45. $49x^2 + 9y^2$
46. $12x^4 - 62x^3 + 70x^2$
47. $25a^3 + 20a^2 + 3a$
48. $36a^4 - 100a^2$
49. $xa - xb + ay - by$
50. $xy - bx + ay - ab$
51. $48a^4b - 3a^2b$
52. $18a^4b^2 - 12a^3b^3 + 8a^2b^4$
53. $5x^5 - 40x^2$
54. $16x^3 + 16x^2 + 3x$
55. $3x^2 + 35xy - 82y^2$
56. $3x^2 + 37xy - 86y^2$
57. $16x^5 - 44x^4 + 30x^3$
58. $16x^2 + 16x - 1$
59. $2x^2 + 2ax + 3x + 3a$
60. $2x^2 + 2ax + 5x + 5a$
61. $y^4 - 1$
62. $25y^7 - 16y^5$
63. $12x^4y^2 + 36x^3y^3 + 27x^2y^4$
64. $16x^3y^2 - 4xy^2$

Learning Objectives Assessment

The following problems can be used to help assess if you have successfully met the learning objectives for this section.

For problems 65-68, match each polynomial with the factoring method by which it could be factored.

65. $6x^2 + 7x - 10$

66. $x^2 - 16y^2$

67. $4x^2 + 64y^2$

68. $x^2 - 6x + xy - by$

a. Greatest common factor

b. Grouping

c. Trial and error

d. Difference of squares

Getting Ready for the Next Section

Solve each equation.

69. $3x - 6 = 9$

70. $5x - 1 = 14$

71. $2x + 3 = 0$

72. $4x - 5 = 0$

73. $4x + 3 = 0$

74. $3x - 1 = 0$

SPOTLIGHT ON SUCCESS *Student Instructor Ryan*

You do not determine a man's greatness by his talent or wealth, as the world does, but rather by what it takes to discourage him.
— Dr. Jerry Falwell

From very early in school, I seemed to have a knack for math, but I also had a knack for laziness and procrastination. In rare times, I would really focus in class and nail all the material, but more often I would spend my time goofing off with friends and coasting on what came easily to me. My parents tried their hardest to motivate me, to do anything they could to help me succeed, but when it came down to it, it was on me to control the outcome.

At one point, my dad voiced his frustrations with having to pay for such a good education when I wasn't taking advantage of it. He told me that if I didn't get above a 3.8 GPA, that I would be attending a different school the following fall. It was the motivation I needed. Finally, I understood the importance of trying my best in school. That semester I achieved the goal my dad had given to me.

But it wasn't that easy. I encountered numerous things in high school that could have derailed my academic journey. I felt abandoned by people I considered family, I suffered multiple injuries requiring surgery, and I experienced the death of a classmate and teammate. There was so much going on that I could have just shut down and gone back to coasting by, but luckily, I had already learned my lesson. I managed to stay positive and work hard through all the challenges. Everything paid off senior year when I was accepted to California Polytechnic State University and managed to pass every AP test I took. This journey has taught me that one of the most important things we can do is to work hard no matter what the circumstances; if we refuse to be discouraged, then we can achieve greatness.

Solving Equations by Factoring 6.6

Learning Objectives

In this section, we will learn how to:

1. Use the Zero-Factor Property to solve a quadratic equation.
2. Solve a quadratic equation by factoring.

Introduction

In this section we will use the methods of factoring developed in previous sections, along with a special property of 0, to solve quadratic equations.

DEFINITION *quadratic equation*

Any equation that can be put in the form $ax^2 + bx + c = 0$, where a, b, and c are real numbers ($a \neq 0$), is called a ***quadratic equation.*** The equation $ax^2 + bx + c = 0$ is called ***standard form*** for a quadratic equation.

Notice that a quadratic equation in standard form is simply a polynomial of degree 2 that has been set equal to zero.

The number 0 has a special property. If we multiply two numbers and the product is 0, then one or both of the original two numbers must be 0. In symbols, this property looks like this.

PROPERTY *Zero-Factor Property*

Let a and b represent real numbers. If $a \cdot b = 0$, then $a = 0$ or $b = 0$.

Note The zero-factor property allows us to solve a quadratic equation by replacing it with two linear equations in one variable.

Suppose we want to solve the quadratic equation $x^2 + 5x + 6 = 0$. We can factor the left side into $(x + 2)(x + 3)$. Then we have:

$$x^2 + 5x + 6 = 0$$
$$(x + 2)(x + 3) = 0$$

Now, $(x + 2)$ and $(x + 3)$ both represent real numbers. Their product is 0; therefore, either $(x + 3)$ is 0 or $(x + 2)$ is 0. Either way we have a solution to our equation. We use the property of 0 stated above to finish the problem:

$$x^2 + 5x + 6 = 0$$
$$(x + 2)(x + 3) = 0$$
$$x + 2 = 0 \quad \text{or} \quad x + 3 = 0$$
$$x = -2 \quad \text{or} \quad x = -3$$

Our solution set is $\{-2, -3\}$. Our equation has two solutions. To check our solutions we have to check each one separately to see that they both produce a true statement when used in place of the variable:

When	$x = -3$
the equation	$x^2 + 5x + 6 = 0$
becomes	$(-3)^2 + 5(-3) + 6 \stackrel{?}{=} 0$
	$9 + (-15) + 6 \stackrel{?}{=} 0$
	$0 = 0$
When	$x = -2$
the equation	$x^2 + 5x + 6 = 0$
becomes	$(-2)^2 + 5(-2) + 6 \stackrel{?}{=} 0$
	$4 + (-10) + 6 \stackrel{?}{=} 0$
	$0 = 0$

Since substituting $x = 3$ and $x = -2$ into the original equation produces true statements, both solutions check. We have solved a quadratic equation by replacing it with two linear equations in one variable.

HOW TO *Strategy for solving a quadratic equation by factoring*

Step 1: Put the equation in standard form; that is, 0 on one side and decreasing powers of the variable on the other.
Step 2: Factor completely.
Step 3: Use the zero-factor property to set each variable factor from step 2 equal to 0.
Step 4: Solve each equation produced in step 3.
Step 5: Check each solution, if necessary.

VIDEO EXAMPLES

SECTION 6.6

EXAMPLE 1 Solve: $(2x - 5)(x + 4) = 0$.

SOLUTION The first two steps have already been done. That is, the equation has 0 on one side and the other side is factored. We can proceed directly with the zero-factor property in Step 3.

Step 3: Set each factor equal to 0:

$$2x - 5 = 0 \quad \text{or} \quad x + 4 = 0$$

Step 4: Solve each of the equations from Step 3:

$$2x - 5 = 0 \quad \text{or} \quad x + 4 = 0$$

$$2x = 5 \quad \text{or} \quad x = -4$$

$$x = \frac{5}{2}$$

Step 5: Check solutions:

Check: $x = \frac{5}{2}$

$$\left(2 \cdot \frac{5}{2} - 5\right)\left(\frac{5}{2} + 4\right) \stackrel{?}{=} 0$$
$$(5 - 5)\left(\frac{5}{2} + \frac{8}{2}\right) \stackrel{?}{=} 0$$
$$0\left(\frac{13}{2}\right) \stackrel{?}{=} 0$$
$$0 = 0$$

Check: $x = -4$

$$(2(-4) - 5)(-4 + 4) \stackrel{?}{=} 0$$
$$(-8 - 5)(0) \stackrel{?}{=} 0$$
$$-13(0) \stackrel{?}{=} 0$$
$$0 = 0$$

EXAMPLE 2 Solve the equation $2x^2 - 5x = 12$.

SOLUTION

Step 1: Begin by adding -12 to both sides, so the equation is in standard form:

$$2x^2 - 5x = 12$$
$$2x^2 - 5x - 12 = 0$$

Step 2: Factor the left side completely:

$$(2x + 3)(x - 4) = 0$$

Step 3: Set each factor to 0:

$$2x + 3 = 0 \quad \text{or} \quad x - 4 = 0$$

Step 4: Solve each of the equations from Step 3:

$$2x + 3 = 0 \qquad x - 4 = 0$$
$$2x = -3 \qquad x = 4$$
$$x = -\frac{3}{2}$$

Step 5: Substitute each solution into $2x^2 - 5x = 12$ to check:

Check: $x = -\frac{3}{2}$

$$2\left(-\frac{3}{2}\right)^2 - 5\left(-\frac{3}{2}\right) \stackrel{?}{=} 12$$
$$2\left(\frac{9}{4}\right) + 5\left(\frac{3}{2}\right) \stackrel{?}{=} 12$$
$$\frac{9}{2} + \frac{15}{2} \stackrel{?}{=} 12$$
$$\frac{24}{2} \stackrel{?}{=} 12$$
$$12 = 12$$

Check: $x = 4$

$$2(4)^2 - 5(4) \stackrel{?}{=} 12$$
$$2(16) - 20 \stackrel{?}{=} 12$$
$$32 - 20 \stackrel{?}{=} 12$$
$$12 = 12$$

EXAMPLE 3 Solve for a: $16a^2 - 25 = 0$.

SOLUTION The equation is already in standard form:

$$16a^2 - 25 = 0$$

$$(4a - 5)(4a + 5) = 0 \quad \text{Factor the left side}$$

$$4a - 5 = 0 \quad \text{or} \quad 4a + 5 = 0 \quad \text{Set each factor equal to 0}$$

$$4a = 5 \qquad 4a = -5 \quad \text{Solve the resulting equations}$$

$$a = \frac{5}{4} \qquad a = -\frac{5}{4}$$

The solutions are $\frac{5}{4}$ and $-\frac{5}{4}$.

EXAMPLE 4 Solve: $4x^2 = 8x$.

SOLUTION We begin by adding $-8x$ to each side of the equation to put it in standard form. Then we factor the left side of the equation by factoring out the greatest common factor.

$$4x^2 = 8x$$

$$4x^2 - 8x = 0 \quad \text{Add } -8x \text{ to each side}$$

$$4x(x - 2) = 0 \quad \text{Factor the left side}$$

$$4x = 0 \quad \text{or} \quad x - 2 = 0 \quad \text{Set each factor equal to 0}$$

$$x = 0 \quad \text{or} \quad x = 2 \quad \text{Solve the resulting equations}$$

The solutions are 0 and 2.

EXAMPLE 5 Solve: $x(2x + 3) = 44$.

SOLUTION We must multiply out the left side first and then put the equation in standard form:

$$x(2x + 3) = 44$$

$$2x^2 + 3x = 44 \quad \text{Multiply out the left side}$$

$$2x^2 + 3x - 44 = 0 \quad \text{Add } -44 \text{ to each side}$$

$$(2x + 11)(x - 4) = 0 \quad \text{Factor the left side}$$

$$2x + 11 = 0 \quad \text{or} \quad x - 4 = 0 \quad \text{Set each factor equal to 0}$$

$$2x = -11 \quad \text{or} \quad x = 4 \quad \text{Solve the resulting equations}$$

$$x = -\frac{11}{2}$$

The two solutions are $-\frac{11}{2}$ and 4.

Note In Example 5, even though the left side of the equation was factored originally, we cannot say:

$$x = 44 \text{ or } 2x + 3 = 44$$

It simply is not true. The fact that the two numbers have a product of 44 does not mean one of the numbers must equal 44. For example $2 \cdot 22 = 44$. The zero-factor property is only valid when the product is equal to 0.

EXAMPLE 6 Solve for x: $5^2 = x^2 + (x+1)^2$.

SOLUTION Before we can put this equation in standard form we must square the binomial. Remember, to square a binomial, we use the formula $(a+b)^2 = a^2 + 2ab + b^2$:

$$
\begin{aligned}
5^2 &= x^2 + (x+1)^2 \\
25 &= x^2 + x^2 + 2x + 1 && \text{Expand } 5^2 \text{ and } (x+1)^2 \\
25 &= 2x^2 + 2x + 1 && \text{Simplify the right side} \\
0 &= 2x^2 + 2x - 24 && \text{Add } -25 \text{ to each side} \\
0 &= x^2 + x - 12 && \text{Divide both sides by 2} \\
0 &= (x+4)(x-3) && \text{Factor completely} \\
x + 4 = 0 \quad &\text{or} \quad x - 3 = 0 && \text{Set each factor equal to 0} \\
x = -4 \quad &\text{or} \quad x = 3
\end{aligned}
$$

Notice that it makes no difference which side of the equation is 0 when we write the equation in standard form.

Although the equation in the next example is not a quadratic equation, it can be solved by the method shown in the first six examples.

EXAMPLE 7 Solve $24x^3 = -10x^2 + 6x$ for x.

SOLUTION First, we write the equation in standard form:

$$
\begin{aligned}
24x^3 + 10x^2 - 6x &= 0 && \text{Standard form} \\
2x(12x^2 + 5x - 3) &= 0 && \text{Factor out } 2x \\
2x(3x-1)(4x+3) &= 0 && \text{Factor remaining trinomial} \\
2x = 0 \quad \text{or} \quad 3x - 1 = 0 \quad &\text{or} \quad 4x + 3 = 0 && \text{Set factors equal to 0} \\
x = 0 \quad \text{or} \quad x = \frac{1}{3} \quad &\text{or} \quad x = -\frac{3}{4} && \text{Solutions}
\end{aligned}
$$

Getting Ready for Class

After reading through the preceding section, respond in your own words and in complete sentences.

A. When is an equation in standard form?

B. What is the first step in solving an equation by factoring?

C. Describe the zero-factor property in your own words.

D. Describe how you would solve the equation $2x^2 - 5x = 12$.

Problem Set 6.6

The following equations are already in factored form. Use the zero-factor property to solve.

1. $(x + 2)(x - 1) = 0$

2. $(x + 3)(x + 2) = 0$

3. $(a - 4)(a - 5) = 0$

4. $(a + 6)(a - 1) = 0$

5. $x(x + 1)(x - 3) = 0$

6. $x(2x + 1)(x - 5) = 0$

7. $(3x + 2)(2x + 3) = 0$

8. $(4x - 5)(x - 6) = 0$

9. $m(3m + 4)(3m - 4) = 0$

10. $m(2m - 5)(3m - 1) = 0$

11. $2y(3y + 1)(5y + 3) = 0$

12. $3y(2y - 3)(3y - 4) = 0$

Solve the following equations.

13. $x^2 + 3x + 2 = 0$

14. $x^2 - x - 6 = 0$

15. $x^2 - 9x + 20 = 0$

16. $x^2 + 2x - 3 = 0$

17. $a^2 - 2a - 24 = 0$

18. $a^2 - 11a + 30 = 0$

19. $100x^2 - 500x + 600 = 0$

20. $100x^2 - 300x + 200 = 0$

21. $x^2 = -6x - 9$

22. $x^2 = 10x - 25$

23. $a^2 - 16 = 0$

24. $a^2 - 36 = 0$

25. $2x^2 + 5x - 12 = 0$

26. $3x^2 + 14x - 5 = 0$

27. $9x^2 + 12x + 4 = 0$

28. $12x^2 - 24x + 9 = 0$

29. $a^2 + 25 = 10a$

30. $a^2 + 16 = 8a$

31. $0 = 20 + 3x - 2x^2$

32. $0 = 2 + x - 6x^2$

33. $3m^2 = 20 - 7m$

34. $2m^2 = -18 + 15m$

35. $4x^2 - 49 = 0$

36. $16x^2 - 25 = 0$

37. $x^2 + 6x = 0$

38. $x^2 + 5x = 0$

39. $3x - x^2 = 0$

40. $8x - x^2 = 0$

41. $2x^2 = 8x$

42. $2x^2 = 10x$

43. $3x^2 = 15x$

44. $5x^2 = 15x$

45. $1{,}400 = 400 + 700x - 100x^2$

46. $2{,}700 = 700 + 900x - 100x^2$

47. $6x^2 = -5x + 4$

48. $9x^2 = 12x - 4$

49. $x(2x - 3) = 20$

50. $x(3x - 5) = 12$

51. $t(t + 2) = 80$

52. $t(t + 2) = 99$

53. $4{,}000 = (1{,}300 - 100p)p$

54. $3{,}200 = (1{,}200 - 100p)p$

55. $x(14 - x) = 48$

56. $x(12 - x) = 32$

57. $(x + 5)^2 = 2x + 9$

58. $(x + 7)^2 = 2x + 13$

59. $(y - 6)^2 = y - 4$

60. $(y + 4)^2 = y + 6$

61. $10^2 = (x + 2)^2 + x^2$

62. $15^2 = (x + 3)^2 + x^2$

63. $2x^3 + 11x^2 + 12x = 0$

64. $3x^3 + 17x^2 + 10x = 0$

65. $4y^3 - 2y^2 - 30y = 0$

66. $9y^3 + 6y^2 - 24y = 0$

67. $8x^3 + 16x^2 = 10x$

68. $24x^3 - 22x^2 = -4x$

69. $20a^3 = -18a^2 + 18a$

70. $12a^3 = -2a^2 + 10a$

71. $16t^2 - 32t + 12 = 0$

72. $16t^2 - 64t + 48 = 0$

Simplify each side as much as possible, then solve the equation.

73. $(a - 5)(a + 4) = -2a$

74. $(a + 2)(a - 3) = -2a$

75. $3x(x + 1) - 2x(x - 5) = -42$

76. $4x(x - 2) - 3x(x - 4) = -3$

77. $2x(x + 3) = x(x + 2) - 3$

78. $3x(x - 3) = 2x(x - 4) + 6$

79. $a(a - 3) + 6 = 2a$

80. $a(a - 4) + 8 = 2a$

81. $15(x + 20) + 15x = 2x(x + 20)$

82. $15(x + 8) + 15x = 2x(x + 8)$

83. $15 = a(a + 2)$

84. $6 = a(a - 5)$

Use factoring by grouping to solve the following equations.

85. $x^3 + 3x^2 - 4x - 12 = 0$

86. $x^3 + 5x^2 - 9x - 45 = 0$

87. $x^3 + x^2 - 16x - 16 = 0$

88. $4x^3 + 12x^2 - 9x - 27 = 0$

89. **Paying Attention to Instructions** Work each problem according to the instructions given.

a. Solve: $2x^2 + 7x - 4 = 0$.

b. Factor: $2x^2 + 7x - 4$.

c. Solve: $2x^2 + 7x - 4 = -7$.

d. Solve: $2x + 7x - 4 = 0$.

90. **Paying Attention to Instructions** Work each problem according to the instructions given.

a. Multiply: $(3x + 2)(x - 4)$.

b. Solve: $(3x + 2)(x - 4) = 0$.

c. Solve: $(3x + 2)(x - 4) = 17$.

d. Evaluate $(3x + 2)(x - 4)$ if $x = 2$.

Learning Objectives Assessment

The following problems can be used to help assess if you have successfully met the learning objectives for this section.

91. Solve: $(5x + 2)(x - 3) = 0$

a. $-3, \frac{2}{5}$ **b.** $-\frac{2}{5}, 3$ **c.** $\frac{3}{5}, 2$ **d.** $5x^2 - 13x - 6$

92. Solve: $(x - 4)(x - 2) = 3$

a. $2, 4$ **b.** $5, 7$ **c.** $1, 5$ **d.** $-5, -1$

Getting Ready for the Next Section

Write each sentence as an algebraic equation.

93. The product of two consecutive integers is 72.
94. The product of two consecutive even integers is 80.
95. The product of two consecutive odd integers is 99.
96. The product of two consecutive odd integers is 63.
97. The product of two consecutive even integers is 10 less than 5 times their sum.
98. The product of two consecutive odd integers is 1 less than 4 times their sum.

The following word problems are taken from the book *Academic Algebra*, written by William J. Milne and published by the American Book Company in 1901. Solve each problem.

99. **Cost of a Bicycle and a Suit** A bicycle and a suit cost $90. How much did each cost, if the bicycle cost 5 times as much as the suit?

100. **Cost of a Cow and a Calf** A man bought a cow and a calf for $36, paying 8 times as much for the cow as for the calf. What was the cost of each?

101. **Cost of a House and a Lot** A house and a lot cost $3,000. If the house cost 4 times as much as the lot, what was the cost of each?

102. **Daily Wages** A plumber and two helpers together earned $7.50 per day. How much did each earn per day, if the plumber earned 4 times as much as each helper?

Applications 6.7

Learning Objectives

In this section, we will learn how to:

1. Solve number problems.
2. Solve area problems.
3. Solve problems involving the Pythagorean theorem.
4. Solve cost and revenue problems.

In this section we will look at some application problems, the solutions to which require solving a quadratic equation. We will also introduce the Pythagorean theorem, one of the oldest theorems in the history of mathematics. The person whose name we associate with the theorem, Pythagoras (of Samos), was a Greek philosopher and mathematician who lived from about 560 B.C. to 480 B.C. According to the British philosopher Bertrand Russell, Pythagoras was "intellectually one of the most important men that ever lived."

Also in this section, the solutions to the examples show only the essential steps from our Blueprint for Problem Solving that we introduced in Section 2.6. Recall that Step 1 is done mentally; we read the problem and mentally list the items that are known and the items that are unknown. This is an essential part of problem solving. However, now that you have had experience with application problems, you are doing Step 1 automatically.

Number Problems

VIDEO EXAMPLES

SECTION 6.7

EXAMPLE 1 The product of two consecutive odd integers is 63. Find the integers.

SOLUTION Let x be the first odd integer; then $x + 2$ is the second odd integer. An equation that describes the situation is:

$$x(x + 2) = 63 \qquad \text{Their product is 63}$$

We solve the equation:

$$x(x + 2) = 63$$
$$x^2 + 2x = 63$$
$$x^2 + 2x - 63 = 0$$
$$(x - 7)(x + 9) = 0$$
$$x - 7 = 0 \quad \text{or} \quad x + 9 = 0$$
$$x = 7 \quad \text{or} \quad x = -9$$

If the first odd integer is 7, the next odd integer is $7 + 2 = 9$. If the first odd integer is -9, the next consecutive odd integer is $-9 + 2 = -7$. We have two pairs of consecutive odd integers that are solutions. They are 7, 9 and $-9, -7$.

We check to see that their products are 63:

$$7(9) = 63$$
$$-7(-9) = 63$$

Suppose we know that the sum of two numbers is 50. We want to find a way to represent each number using only one variable. If we let x represent one of the two numbers, how can we represent the other? Let's suppose for a moment that x turns out to be 30. Then the other number will be 20, because their sum is 50; that is, if two numbers add up to 50 and one of them is 30, then the other must be $50 - 30 = 20$. Generalizing this to any number x, we see that if two numbers have a sum of 50 and one of the numbers is x, then the other must be $50 - x$. The table that follows shows some additional examples.

If two numbers have a sum of	and one of them is	then the other must be
50	x	$50 - x$
100	x	$100 - x$
10	y	$10 - y$
12	n	$12 - n$

Now, let's look at an example that uses this idea.

EXAMPLE 2 The sum of two numbers is 13. Their product is 40. Find the numbers.

SOLUTION If we let x represent one of the numbers, then $13 - x$ must be the other number because their sum is 13. Since their product is 40, we can write:

$$x(13 - x) = 40 \quad \text{The product of the two numbers is 40}$$

$$13x - x^2 = 40 \quad \text{Multiply the left side}$$

$$x^2 - 13x = -40 \quad \text{Multiply both sides by } -1 \text{ and reverse the order of the terms on the left side}$$

$$x^2 - 13x + 40 = 0 \quad \text{Add 40 to each side}$$

$$(x - 8)(x - 5) = 0 \quad \text{Factor the left side}$$

$$x - 8 = 0 \quad \text{or} \quad x - 5 = 0$$

$$x = 8 \quad \text{or} \quad x = 5$$

The two solutions are 8 and 5. If x is 8, then the other number is

$$13 - x = 13 - 8 = 5$$

Likewise, if x is 5, the other number is $13 - x = 13 - 5 = 8$. Therefore, the two numbers we are looking for are 8 and 5. Their sum is 13 and their product is 40.

Area Problems

Many word problems dealing with area can best be described algebraically by quadratic equations.

EXAMPLE 3 The length of a rectangle is 3 more than twice the width. The area is 44 square inches. Find the dimensions (find the length and width).

SOLUTION As shown in Figure 1, let x be the width of the rectangle. Then $2x + 3$ is the length of the rectangle because the length is three more than twice the width.

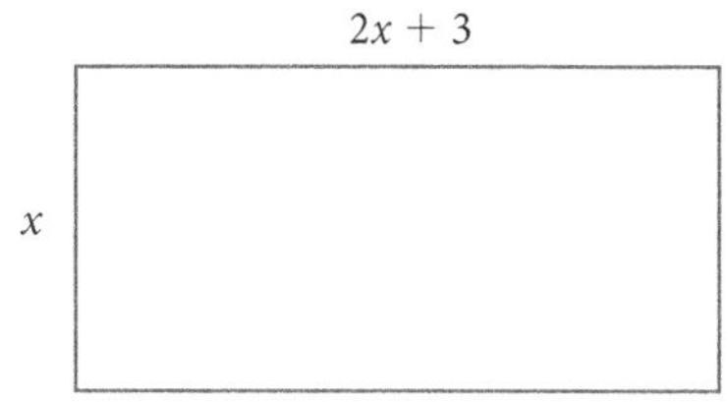

FIGURE 1

Since the area is 44 square inches, an equation that describes the situation is

$$x(2x + 3) = 44 \qquad \text{Width} \cdot \text{length} = \text{area}$$

We now solve the equation:

$$\begin{aligned} x(2x + 3) &= 44 \\ 2x^2 + 3x &= 44 \\ 2x^2 + 3x - 44 &= 0 \\ (2x + 11)(x - 4) &= 0 \\ 2x + 11 = 0 \quad &\text{or} \quad x - 4 = 0 \\ x = -\frac{11}{2} \quad &\text{or} \quad x = 4 \end{aligned}$$

The solution $x = -\frac{11}{2}$ cannot be used since length and width are always given in positive units. The width is 4. The length is 3 more than twice the width or $2(4) + 3 = 11$.

$$\text{Width} = 4 \text{ inches}$$
$$\text{Length} = 11 \text{ inches}$$

The solutions check in the original problem since $4(11) = 44$.

EXAMPLE 4 The numerical value of the area of a square is twice its perimeter. What is the length of its side?

SOLUTION As shown in Figure 2, let x be the length of its side. Then x^2 is the area of the square and $4x$ is the perimeter of the square:

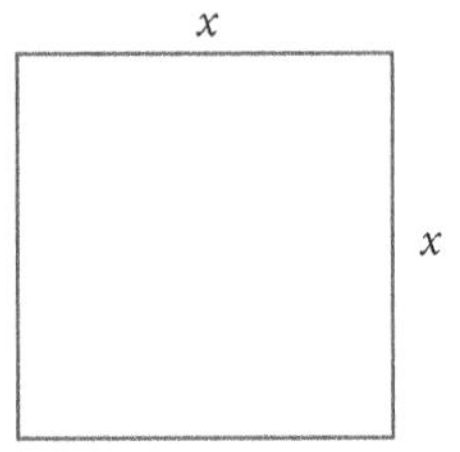

FIGURE 2

An equation that describes the situation is

$$x^2 = 2(4x) \quad \text{The area is 2 times the perimeter}$$
$$x^2 = 8x$$
$$x^2 - 8x = 0$$
$$x(x - 8) = 0$$
$$x = 0 \quad \text{or} \quad x = 8$$

Since $x = 0$ does not make sense in our original problem, we use $x = 8$. If the side has length 8, then the perimeter is $4(8) = 32$ and the area is $8^2 = 64$. Since 64 is twice 32, our solution is correct.

The Pythagorean Theorem

FACTS FROM GEOMETRY *The Pythagorean Theorem*

Next, we will work some problems involving the Pythagorean theorem, which we mentioned in the introduction to this section. It may interest you to know that Pythagoras formed a secret society around the year 540 B.C. Known as the Pythagoreans, members kept no written record of their work; everything was handed down by spoken word. They influenced not only mathematics, but religion, science, medicine, and music as well. Among other things, they discovered the correlation between musical notes and the reciprocals of counting numbers, $\frac{1}{2}$, $\frac{1}{3}$, $\frac{1}{4}$, and so on. In their daily lives, they followed strict dietary and moral rules to achieve a higher rank in future lives.

PROPERTY *Pythagorean Theorem*

In any right triangle (Figure 3), the square of the longest side (called the hypotenuse) is equal to the sum of the squares of the other two sides (called legs).

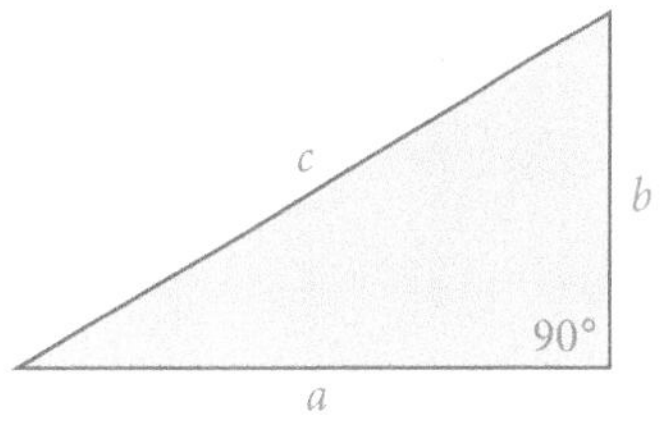

$$c^2 = a^2 + b^2$$

FIGURE 3

EXAMPLE 5 The three sides of a right triangle are three consecutive integers. Find the lengths of the three sides.

SOLUTION Let x be the first integer (shortest side)

then $x + 1 =$ the next consecutive integer

and $x + 2 =$ the last consecutive integer (longest side)

A diagram of the triangle is shown in Figure 4.

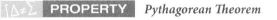

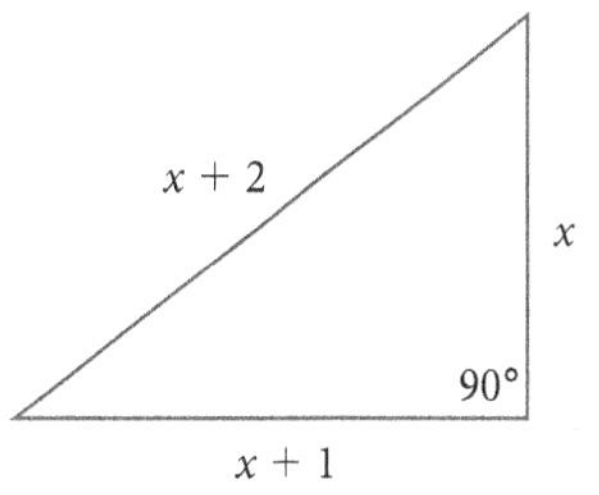

FIGURE 4

The Pythagorean theorem tells us that the square of the longest side $(x + 2)^2$ is equal to the sum of the squares of the two shorter sides, $(x + 1)^2 + x^2$. Here is the equation:

$$(x + 2)^2 = (x + 1)^2 + x^2$$

$$x^2 + 4x + 4 = x^2 + 2x + 1 + x^2 \quad \text{Expand squares}$$

$$x^2 - 2x - 3 = 0 \quad \text{Standard form}$$

$$(x - 3)(x + 1) = 0 \quad \text{Factor}$$

$$x - 3 = 0 \quad \text{or} \quad x + 1 = 0 \quad \text{Set factors to 0}$$

$$x = 3 \quad \text{or} \quad x = -1$$

Since a triangle cannot have a side with a negative number for its length, we must not use -1 for a solution to our original problem; therefore, the shortest side is 3. The other two sides are the next two consecutive integers, 4 and 5.

Note Many students make the mistake of assuming if

$$(x + 2)^2 = (x + 1)^2 + x^2$$

then

$$x + 2 = (x + 1) + x$$

Unfortunately, this is simply not true. We do not get an equivalent equation by "dropping" the squares. For instance,

$$5^2 = 4^2 + 3^2$$

is a true statement, but $5 \neq 4 + 3$.

EXAMPLE 6 The hypotenuse of a right triangle is 13 inches, and one leg is 7 inches longer than the other leg. Find the lengths of the two legs.

SOLUTION If we let x be the length of the shortest leg, then the longer leg must be $x + 7$. A diagram of the triangle is shown in Figure 5.

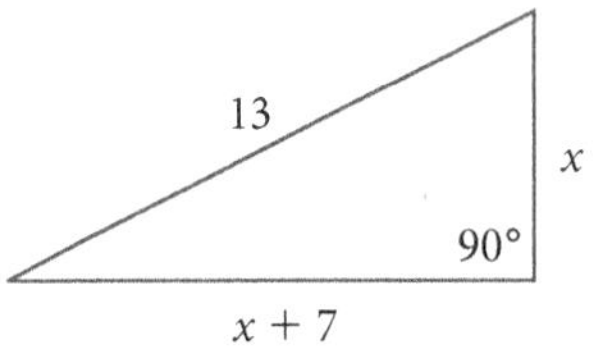

FIGURE 5

The Pythagorean theorem tells us that the square of the longest side, 13^2, is equal to the sum of the squares of the two shorter sides, $x^2 + (x + 7)^2$. Here is the equation:

$$13^2 = x^2 + (x + 7)^2 \quad \text{Pythagorean theorem}$$
$$169 = x^2 + x^2 + 14x + 49 \quad \text{Expand } 13^2 \text{ and } (x + 7)^2$$
$$169 = 2x^2 + 14x + 49 \quad \text{Simplify the right side}$$
$$0 = 2x^2 + 14x - 120 \quad \text{Add } -169 \text{ to each side}$$
$$0 = x^2 + 7x - 60 \quad \text{Divide both sides by 2}$$
$$0 = (x + 12)(x - 5) \quad \text{Factor completely}$$
$$x + 12 = 0 \quad \text{or} \quad x - 5 = 0 \quad \text{Set factors equal to 0}$$
$$x = -12 \quad \text{or} \quad x = 5$$

Since a triangle cannot have a side with a negative number for its length, we cannot use -12; therefore, the shortest leg must be 5 inches. The other leg is $x + 7 = 5 + 7 = 12$ inches.

Cost and Revenue

EXAMPLE 7 A company can manufacture x hundred items for a total cost of $C = 300 + 500x - 100x^2$. How many items were manufactured if the total cost is \$900?

SOLUTION We are looking for x when C is 900. We begin by substituting 900 for C in the cost equation. Then we solve for x:

$$\text{When} \quad C = 900$$
$$\text{the equation} \quad C = 300 + 500x - 100x^2$$
$$\text{becomes} \quad 900 = 300 + 500x - 100x^2$$

We can write this equation in standard form by adding -300, $-500x$, and $100x^2$ to each side. The result looks like this:

$$100x^2 - 500x + 600 = 0$$
$$x^2 - 5x + 6 = 0 \quad \text{Divide both sides by 100}$$
$$(x - 2)(x - 3) = 0 \quad \text{Factor completely}$$
$$x - 2 = 0 \quad \text{or} \quad x - 3 = 0 \quad \text{Set variable factors to 0}$$
$$x = 2 \quad \text{or} \quad x = 3$$

Our solutions are 2 and 3, which means that the company can manufacture 2 hundred items or 3 hundred items for a total cost of \$900.

EXAMPLE 8 A manufacturer of small portable radios knows that the number of radios she can sell each week is related to the price of the radios by the equation $x = 1{,}300 - 100p$ (x is the number of radios and p is the price per radio). What price should she charge for the radios to have a weekly revenue of \$4,000?

SOLUTION First, we must find the revenue equation. The equation for total revenue is $R = xp$, where x is the number of units sold and p is the price per unit. Since we want R in terms of p, we substitute $1{,}300 - 100p$ for x in the equation $R = xp$:

$$\text{If} \quad R = xp$$
$$\text{and} \quad x = 1{,}300 - 100p$$
$$\text{then} \quad R = (1{,}300 - 100p)p$$

We want to find p when R is 4,000. Substituting 4,000 for R in the equation gives us:

$$4{,}000 = (1{,}300 - 100p)p$$

If we multiply out the right side, we have:

$$4{,}000 = 1{,}300p - 100p^2$$

To write this equation in standard form, we add $100p^2$ and $-1{,}300p$ to each side:

$$100p^2 - 1{,}300p + 4{,}000 = 0 \quad \text{Add } 100p^2 \text{ and } -1{,}300p$$
$$p^2 - 13p + 40 = 0 \quad \text{Divide both sides by 100}$$
$$(p - 5)(p - 8) = 0 \quad \text{Factor completely}$$
$$p - 5 = 0 \quad \text{or} \quad p - 8 = 0 \quad \text{Set factors equal to 0}$$
$$p = 5 \quad \text{or} \quad p = 8$$

If she sells the radios for \$5 each or for \$8 each, she will have a weekly revenue of \$4,000.

Getting Ready for Class

After reading through the preceding section, respond in your own words and in complete sentences.

A. What are consecutive integers?

B. Explain the Pythagorean theorem in words.

C. If the sum of two numbers is 20 and x is one of the numbers, what is the other number?

D. If $x^2 + (x + 2)^2 = (x + 4)^2$, we cannot conclude that $x + (x + 2) = (x + 4)$. Show that the expressions are not equivalent by substituting 6 for x in each one.

Problem Set 6.7

Solve the following word problems. Be sure to show the equation used.

Number Problems

1. The product of two consecutive even integers is 80. Find the two integers.
2. The product of two consecutive integers is 72. Find the two integers.
3. The product of two consecutive odd integers is 99. Find the two integers.
4. The product of two consecutive integers is 132. Find the two integers.
5. The product of two consecutive even integers is 10 less than 5 times their sum. Find the two integers.
6. The product of two consecutive odd integers is 1 less than 4 times their sum. Find the two integers.
7. The sum of two numbers is 14. Their product is 48. Find the numbers.
8. The sum of two numbers is 12. Their product is 32. Find the numbers.
9. One number is 2 more than 5 times another. Their product is 24. Find the numbers.
10. One number is 1 more than twice another. Their product is 55. Find the numbers.
11. One number is 4 times another. Their product is 4 times their sum. Find the numbers.
12. One number is 2 more than twice another. Their product is 2 more than twice their sum. Find the numbers.

Geometry Problems

13. The length of a rectangle is 1 more than the width. The area is 12 square inches. Find the dimensions.
14. The length of a rectangle is 3 more than twice the width. The area is 44 square inches. Find the dimensions.
15. The height of a triangle is twice the base. The area is 9 square inches. Find the base.
16. The height of a triangle is 2 more than twice the base. The area is 20 square feet. Find the base.
17. The hypotenuse of a right triangle is 10 inches. The lengths of the two legs are given by two consecutive even integers. Find the lengths of the two legs.
18. The hypotenuse of a right triangle is 15 inches. One of the legs is 3 inches more than the other. Find the lengths of the two legs.
19. The shorter leg of a right triangle is 5 meters. The hypotenuse is 1 meter longer than the longer leg. Find the length of the longer leg.
20. The shorter leg of a right triangle is 12 yards. If the hypotenuse is 20 yards, how long is the other leg?

Business Problems

21. A company can manufacture x hundred items for a total cost of $C = 400 + 700x - 100x^2$. Find x if the total cost is \$1,400.

22. If the total cost C of manufacturing x hundred items is given by the equation $C = 700 + 900x - 100x^2$, find x when C is \$2,700.

23. The relationship between the number of calculators a company sells per week, x, and the price p of each calculator is given by the equation $x = 1{,}700 - 100p$. At what price should the calculators be sold if the weekly revenue is to be \$7,000?

24. The relationship between the number of pencil sharpeners a company can sell each week, x, and the price p of each sharpener is given by the equation $x = 1{,}800 - 100p$. At what price should the sharpeners be sold if the weekly revenue is to be \$7,200?

Other Applications

25. Pythagorean Theorem A 13-foot ladder is placed so that it reaches to a point on the wall that is 2 feet higher than twice the distance from the base of the wall to the base of the ladder.

a. How far from the wall is the base of the ladder?

b. How high does the ladder reach?

26. Constructing a Box I have a piece of cardboard that is twice as long as it is wide. If I cut a 2-inch by 2-inch square from each corner and fold up the resulting flaps, I get a box with a volume of 32 cubic inches. What are the dimensions of the cardboard?

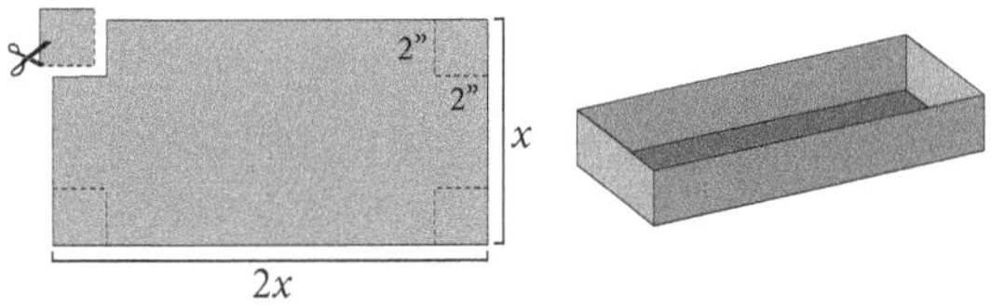

27. Projectile Motion A gun fires a bullet almost straight up from the edge of a 100-foot cliff. If the bullet leaves the gun with a speed of 396 feet per second, its height at time t is given by $h(t) = -16t^2 + 396t + 100$, measured from the ground below the cliff.

a. When will the bullet land on the ground below the cliff? (*Hint:* What is its height when it lands? Remember that we are measuring from the ground below, not from the cliff.)

b. Make a table showing the bullet's height every five seconds, from the time it is fired ($t = 0$) to the time it lands. (*Note:* It is faster to substitute into the factored form.)

28. **Height of a Projectile** If a rocket is fired vertically into the air with a speed of 240 feet per second, its height at time t seconds is given by $h(t) = -16t^2 + 240t$. At what time(s) will the rocket be the following number of feet above the ground?
 a. 704 feet
 b. 896 feet
 c. Why do parts **a.** and **b.** each have two answers?
 d. How long will the rocket be in the air? (*Hint:* How high is it when it hits the ground?)
 e. When the equation for part **d.** is solved, one of the answers is $t = 0$ seconds. What does this represent?

Learning Objectives Assessment

The following problems can be used to help assess if you have successfully met the learning objectives for this section.

29. One number is 1 less than twice another. Their product is 120. Find the numbers. One of the numbers is:
 a. 8 b. 12 c. 6 d. 5
30. The length of a rectangle is 5 feet more than the width. The area of the rectangle is 36 square feet. Find the width.
 a. 2 b. 3 c. 4 d. 6
31. The hypotenuse of a right triangle is 17 meters. One leg is 7 meters longer than the other. Which of the following equations can be used as a model to solve this problem?
 a. $x(x + 7) = 17$ b. $x + (x + 7) = 17$
 c. $x^2 + (x - 7)^2 = 17$ d. $x^2 + (x + 7)^2 = 289$
32. The relationship between the number of lattes a café can sell each week, x, and the price, p, of each latte is given by the equation $x = 700 - 100p$. At what price should the lattes be sold if the weekly revenue is to be \$1,200?
 a. \$3 or \$4 b. \$3 or \$5 c. \$4 or \$5 d. \$2 or \$5

Maintaining Your Skills

33. Fill in each ordered pair so that it is a solution to $y = \frac{1}{2}x + 3$.

$$(-2, \ ), (0, \ ), (2, \ )$$

34. Graph the line $y = \frac{1}{2}x + 3$.
35. Graph the line $x = -2$.
36. Graph $3x - 2y = 6$.
37. Find the slope of the line through (2, 5) and (0, 1).
38. Find the slope and y-intercept for the line $2x - 5y = 10$.
39. Find the equation of the line through $(-2, 1)$ with slope $\frac{1}{2}$.
40. Write the equation of the line with slope -2 and y-intercept $\frac{3}{2}$.
41. Find the equation of the line through (2, 5) and (0, 1).
42. Graph the solution set for $2x - y < 4$.

Chapter 6 Summary

EXAMPLES

Greatest Common Factor [6.1]

1. $8x^4 - 10x^3 + 6x^2$
$= 2x^2 \cdot 4x^2 - 2x^2 \cdot 5x + 2x^2 \cdot 3$
$= 2x^2(4x^2 - 5x + 3)$

The largest monomial that divides each term of a polynomial is called the greatest common factor for that polynomial. We begin all factoring by factoring out the greatest common factor.

Factoring by Grouping [6.1]

2. $2x + ax + 2y + ay$
$= x(2 + a) + y(2 + a)$
$= (2 + a)(x + y)$

Try factoring the greatest common factor from pairs of terms to create a common binomial factor that can be factored out.

Factoring Trinomials [6.2, 6.3]

3. $x^2 + 5x + 6 = (x + 2)(x + 3)$
$x^2 - 5x + 6 = (x - 2)(x - 3)$
$6x^2 - x - 2 = (2x + 1)(3x - 2)$
$6x^2 + 7x + 2 = (2x + 1)(3x + 2)$

One method of factoring a trinomial is to list all pairs of binomials whose product of the first terms gives the first term of the trinomial and whose product of the last terms gives the last term of the trinomial. We then choose the pair that gives the correct middle term for the original trinomial.

Perfect Square Trinomials [6.4]

4. $x^2 + 10x + 25 = (x + 5)^2$
$x^2 - 10x + 25 = (x - 5)^2$

$$a^2 + 2ab + b^2 = (a + b)^2$$
$$a^2 - 2ab + b^2 = (a - b)^2$$

Factoring Binomials [6.4]

5. $x^2 - 25 = (x + 5)(x - 5)$
$x^3 - 27 = (x - 3)(x^2 + 3x + 9)$
$x^3 + 27 = (x + 3)(x^2 - 3x + 9)$

$a^2 - b^2 = (a + b)(a - b)$ Difference of two squares

$a^3 - b^3 = (a - b)(a^2 + ab + b^2)$ Difference of two cubes

$a^3 + b^3 = (a + b)(a^2 - ab + b^2)$ Sum of two cubes

Strategy for Factoring a Polynomial [6.5]

6. a. $2x^5 - 8x^3 = 2x^3(x^2 - 4)$
$= 2x^3(x + 2)(x - 2)$

b. $3x^4 - 18x^3 + 27x^2$
$= 3x^2(x^2 - 6x + 9)$
$= 3x^2(x - 3)^2$

c. $6x^3 - 12x^2 - 48x$
$= 6x(x^2 - 2x - 8)$
$= 6x(x - 4)(x + 2)$

d. $x^2 + ax + bx + ab$
$= x(x + a) + b(x + a)$
$= (x + a)(x + b)$

Step 1: If the polynomial has a greatest common factor other than 1, then factor out the greatest common factor.

Step 2: If the polynomial has two terms (it is a binomial), then see if it is the difference of squares or the sum or difference of cubes, and then factor accordingly. Remember, if it is the sum of squares, it will not factor.

Step 3: If the polynomial has three terms (a trinomial), then it is either a perfect square trinomial that will factor into the square of a binomial, or it is not a perfect square trinomial, in which case you use the methods developed in Sections 6.2 and 6.3.

Step 4: If the polynomial has more than three terms, then try to factor it by grouping.

Step 5: As a final check, see if any of the factors you have written can be factored further. If you have overlooked a common factor, you can catch it here.

Strategy for Solving a Quadratic Equation [6.6]

7. Solve $x^2 - 6x = -8$.

$$x^2 - 6x + 8 = 0$$
$$(x - 4)(x - 2) = 0$$
$$x - 4 = 0 \quad \text{or} \quad x - 2 = 0$$
$$x = 4 \quad \text{or} \quad x = 2$$

Both solutions check.

Step 1: Write the equation in standard form $ax^2 + bx + c = 0$.

Step 2: Factor completely.

Step 3: Use the zero-factor property and set each variable factor equal to 0.

Step 4: Solve the equations found in step 3.

Step 5: Check solutions, if necessary.

The Pythagorean Theorem [6.7]

8. The hypotenuse of a right triangle is 5 inches, and the lengths of the two legs (the other two sides) are given by two consecutive integers. Find the lengths of the two legs.

If we let x = the length of the shortest side, then the other side must be $x + 1$. The Pythagorean theorem tells us that

$$5^2 = x^2 + (x + 1)^2$$
$$25 = x^2 + x^2 + 2x + 1$$
$$25 = 2x^2 + 2x + 1$$
$$0 = 2x^2 + 2x - 24$$
$$0 = 2(x^2 + x - 12)$$
$$0 = 2(x + 4)(x - 3)$$
$$x + 4 = 0 \quad \text{or} \quad x - 3 = 0$$
$$x = -4 \quad \text{or} \quad x = 3$$

Since a triangle cannot have a side with a negative number for its length, we cannot use -4. One leg is $x = 3$ and the other leg is $x + 1 = 3 + 1 = 4$.

In any right triangle, the square of the longest side (called the hypotenuse) is equal to the sum of the squares of the other two sides (called legs).

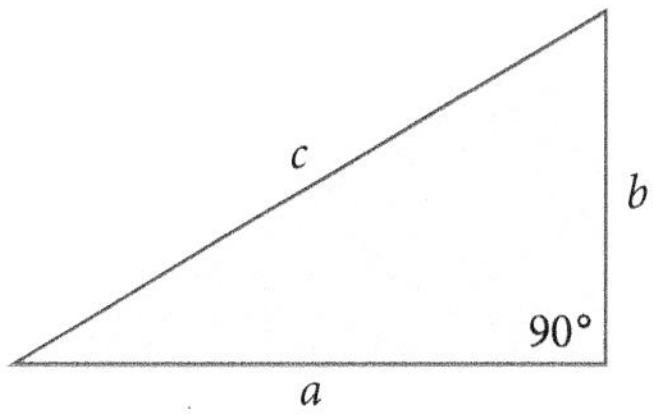

$$c^2 = a^2 + b^2$$

COMMON MISTAKE

It is a mistake to apply the zero-factor property to numbers other than zero. For example, consider the equation $(x - 3)(x + 4) = 18$. A fairly common mistake is to attempt to solve it with the following steps:

$$(x - 3)(x + 4) = 18$$

$$x - 3 = 18 \quad \text{or} \quad x + 4 = 18 \leftarrow \text{Mistake}$$

$$x = 21 \quad \text{or} \quad x = 14$$

These are obviously not solutions, as a quick check will verify:

Check: $x = 21$		Check: $x = 14$
$(21 - 3)(21 + 4) \stackrel{?}{=} 18$		$(14 - 3)(14 + 4) \stackrel{?}{=} 18$
$18 \cdot 25 \stackrel{?}{=} 18$		$11 \cdot 18 \stackrel{?}{=} 18$
$450 = 18$	← False statements →	$198 = 18$

The mistake is in setting each factor equal to 18. It is not necessarily true that when the product of two numbers is 18, either one of them is itself 18. The correct solution looks like this:

$$(x - 3)(x + 4) = 18$$

$$x^2 + x - 12 = 18$$

$$x^2 + x - 30 = 0$$

$$(x + 6)(x - 5) = 0$$

$$x + 6 = 0 \quad \text{or} \quad x - 5 = 0$$

$$x = -6 \quad \text{or} \quad x = 5$$

To avoid this mistake, remember that before you factor a quadratic equation, you must write it in standard form. It is in standard form only when 0 is on one side and decreasing powers of the variable are on the other.

Chapter 6 Test

Factor out the greatest common factor. [6.1]

1. $6x + 18$

2. $12a^2b - 24ab + 8ab^2$

Factor by grouping. [6.1]

3. $x^2 + 3ax - 2bx - 6ab$

4. $15y - 5xy - 12 + 4x$

Factor the following completely. [6.2–6.5]

5. $x^2 + x - 12$

6. $x^2 - 4x - 21$

7. $x^2 - 25$

8. $x^4 - 16$

9. $x^2 + 36$

10. $18x^2 - 32y^2$

11. $x^3 + 4x^2 - 3x - 12$

12. $x^2 + bx - 3x - 3b$

13. $4x^2 - 6x - 10$

14. $4n^2 + 13n - 12$

15. $12c^2 + c - 6$

16. $12x^3 + 12x^2 - 9x$

17. $x^3 + 125y^3$

18. $54b^3 - 128$

Solve the following equations. [6.6]

19. $x^2 - 2x - 15 = 0$

20. $x^2 - 7x + 12 = 0$

21. $x^2 - 25 = 0$

22. $x^2 = 5x + 14$

23. $x^2 + x = 30$

24. $y^3 = 9y$

25. $2x^2 = -5x + 12$

26. $15x^3 - 65x^2 - 150x = 0$

Solve the following word problems. Be sure to show the system of equations used. [6.7]

27. Number Problem Two numbers have a sum of 18. Their product is 72. Find the numbers.

28. Consecutive Integers The product of two consecutive even integers is 14 more than their sum. Find the integers.

29. Geometry The length of a rectangle is 1 foot more than 3 times the width. The area is 52 square feet. Find the dimensions.

30. Geometry One leg of a right triangle is 2 feet more than the other. The hypotenuse is 10 feet. Find the lengths of the two legs.

31. Production Cost A company can manufacture x hundred items for a total cost C, given the equation $C = 100 + 500x - 100x^2$. How many items can be manufactured if the total cost is to be $700?

32. Price and Revenue A manufacturer knows that the number of items he can sell each week, x, is related to the price p of each item by the equation $x = 800 - 100p$. What price should he charge for each item to have a weekly revenue of $1,500? (*Remember:* $R = xp$.)

7

Rational Expressions

Chapter Outline

7.1 Reducing Rational Expressions to Lowest Terms

7.2 Multiplication and Division of Rational Expressions

7.3 Addition and Subtraction of Rational Expressions

7.4 Complex Fractions

7.5 Equations Involving Rational Expressions

7.6 Proportions

7.7 Applications

7.8 Variation

First Bank of San Luis Obispo charges \$2.00 per month and \$0.15 per check for a regular checking account. If we write x checks in one month, the total monthly cost of the checking account will be $C = 2.00 + 0.15x$. From this formula we see that the more checks we write in a month, the more we pay for the account. But, it is also true that the more checks we write in a month, the lower the cost per check. To find the average cost per check, we divide the total cost by the number of checks written:

$$\text{Average cost} = A = \frac{C}{x} = \frac{2.00 + 0.15x}{x}$$

We can use this formula to create Table 1 and Figure 1, giving us a visual interpretation of the relationship between the number of checks written and the average cost per check.

TABLE 1

Number of Checks	Average Cost Per Check
1	2.15
2	1.15
5	0.55
10	0.35
15	0.28
20	0.25

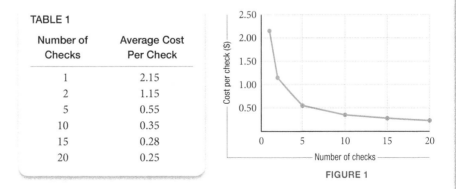

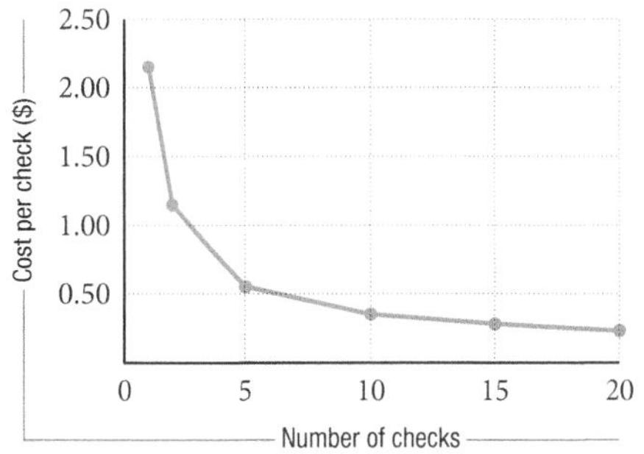

FIGURE 1

As you can see, if we write one check per month, the cost per check is relatively high, \$2.15. However, if we write 20 checks per month, each check costs us only \$0.25. Using average cost per check is a good way to compare different checking accounts. The expression $\frac{2.00 + 0.15x}{x}$ in the average cost formula is a rational expression. When you have finished this chapter you will have a good working knowledge of rational expressions.

Success Skills

iStockphoto.com © IPGGutenbergUKLtd

If you have made it this far, then you have the study skills necessary to be successful in this course. Success skills are more general in nature and will help you with all your classes and ensure your success in college as well.

Let's start with a question:

Question: What quality is most important for success in any college course?

Answer: Independence. You want to become an independent learner.

We all know people like this. They are generally happy. They don't worry about getting the right instructor, or whether or not things work out every time. They have a confidence that comes from knowing that they are responsible for their success or failure in the goals they set for themselves.

Here are some of the qualities of an independent learner:

- Intends to succeed.
- Doesn't let setbacks deter them.
- Knows their resources.
  - Instructor's office hours
  - Math lab
  - Student Solutions Manual
  - Group study
  - Internet
- Doesn't mistake activity for achievement.
- Has a positive attitude.

There are other traits as well. The first step in becoming an independent learner is doing a little self-evaluation and then making of list of traits that you would like to acquire. What skills do you have that align with those of an independent learner? What attributes do you have that keep you from being an independent learner? What qualities would you like to obtain that you don't have now?

Reducing Rational Expressions to Lowest Terms 7.1

Learning Objectives

In this section, we will learn how to:

1. Evaluate a rational expression.
2. Determine the values for which a rational expression is undefined.
3. Reduce rational expressions.
4. Solve problems involving ratios.

Introduction

In Chapter 1 we defined the set of rational numbers to be the set of all numbers that could be put in the form $\frac{a}{b}$, where a and b are integers ($b \neq 0$):

$$\text{Rational numbers} = \left\{ \frac{a}{b} \,\middle|\, a \text{ and } b \text{ are integers, } b \neq 0 \right\}$$

We will now extend this idea to fractions involving polynomials.

Rational Expressions

DEFINITION *Rational Expression*

A ***rational expression*** is any expression that can be put in the form $\frac{P}{Q}$, where P and Q are polynomials and $Q \neq 0$:

$$\text{Rational expressions} = \left\{ \frac{P}{Q} \,\middle|\, P \text{ and } Q \text{ are polynomials, } Q \neq 0 \right\}$$

Each of the following is an example of a rational expression:

$$\frac{2x+3}{x} \qquad \frac{x^2-6x+9}{x^2-4} \qquad \frac{5}{x^2+6} \qquad \frac{2x^2+3x+4}{2}$$

For the rational expression

$$\frac{x^2-6x+9}{x^2-4}$$

the polynomial on top, $x^2 - 6x + 9$, is called the numerator, and the polynomial on the bottom, $x^2 - 4$, is called the denominator. The same is true of the other rational expressions.

Evaluating Rational Expressions

We can find the value of a rational expression for a given value of the variable by substituting the given value in place of the variable and simplifying the result.

VIDEO EXAMPLES

SECTION 7.1

EXAMPLE 1 Evaluate $\dfrac{x^2 - 6x + 9}{x^2 - 4}$ for $x = 3$ and $x = -2$.

SOLUTION We replace x with each given value and simplify.

When	$x = 3$	When	$x = -2$
the expression	$\dfrac{x^2 - 6x + 9}{x^2 - 4}$	the expression	$\dfrac{x^2 - 6x + 9}{x^2 - 4}$
becomes	$\dfrac{(3)^2 - 6(3) + 9}{(3)^2 - 4}$	becomes	$\dfrac{(-2)^2 - 6(-2) + 9}{(-2)^2 - 4}$
	$= \dfrac{9 - 18 + 9}{9 - 4}$		$= \dfrac{4 + 12 + 9}{4 - 4}$
	$= \dfrac{0}{5}$		$= \dfrac{25}{0}$
	$= 0$		which is undefined

The value of the rational expression is 0 when $x = 3$, and when $x = -2$ the rational expression is undefined.

EXAMPLE 2 In the introduction to this chapter, we saw that the average cost per check of writing x checks in one month at a certain bank is given by the rational expression

$$A = \frac{2.00 + 0.15x}{x}$$

Find the average cost per check if 8 checks are written in a month.

SOLUTION We substitute $x = 8$ into the expression and simplify.

$$\begin{aligned} A &= \frac{2.00 + 0.15(8)}{8} \\ &= \frac{2.00 + 1.20}{8} \\ &= \frac{3.20}{8} \\ &= 0.40 \end{aligned}$$

The average cost is \$0.40, or 40¢, per check.

Restricting the Variable

As we saw in Example 1, some values of the variable may cause a rational expression to be undefined. We must be careful that we do not use a value of the variable that will give us a denominator of zero. Remember, division by zero is not defined.

EXAMPLE 3 State the restrictions on the variable:

$$\frac{x + 2}{x - 3}$$

SOLUTION The variable x can be any real number except $x = 3$ since, when $x = 3$, the denominator is $3 - 3 = 0$. We state this restriction by writing $x \neq 3$.

EXAMPLE 4 Determine any values of the variable for which the rational expression $\frac{5}{x^2 - x - 6}$ will be undefined.

SOLUTION We must find any values of x that will make the denominator, $x^2 - x - 6$, equal to zero. That is, we need to solve the equation

$$x^2 - x - 6 = 0$$

This is a quadratic equation, which we can solve by factoring using the zero-factor property from the previous chapter.

$$x^2 - x - 6 = 0$$

$$(x - 3)(x + 2) = 0 \qquad \text{Factor}$$

$$x - 3 = 0 \quad \text{or} \quad x + 2 = 0 \qquad \text{Zero-factor property}$$

$$x = 3 \quad \text{or} \quad x = -2$$

Our restrictions are $x \neq 3$ and $x \neq -2$ since either one makes $x^2 - x - 6 = 0$.

We will not always list each restriction on a rational expression, but we should be aware of them and keep in mind that no rational expression can have a denominator of zero.

Reducing Rational Expressions

The two fundamental properties of rational expressions are listed next. We will use these two properties many times in this chapter.

PROPERTY *Properties of Rational Expressions*

Property 1
Multiplying the numerator and denominator of a rational expression by the same nonzero quantity will not change the value of the rational expression.

Property 2
Dividing the numerator and denominator of a rational expression by the same nonzero quantity will not change the value of the rational expression.

We can use Property 2 to reduce rational expressions to lowest terms. Since this process is almost identical to the process of reducing fractions to lowest terms, let's recall how the fraction $\frac{6}{15}$ is reduced to lowest terms:

$$\frac{6}{15} = \frac{2 \cdot 3}{5 \cdot 3} \qquad \text{Factor numerator and denominator}$$

$$= \frac{2 \cdot \cancel{3}}{5 \cdot \cancel{3}} \qquad \text{Divide out the common factor, 3}$$

$$= \frac{2}{5} \qquad \text{Reduce to lowest terms}$$

The same procedure applies to reducing rational expressions to lowest terms. The process is summarized in the following rule.

RULE

To reduce a rational expression to lowest terms, first factor the numerator and denominator completely and then divide both the numerator and denominator by any factors they have in common.

EXAMPLE 5 Reduce $\frac{x^2 - 9}{x^2 + 5x + 6}$ to lowest terms.

SOLUTION We begin by factoring:

$$\frac{x^2 - 9}{x^2 + 5x + 6} = \frac{(x - 3)(x + 3)}{(x + 2)(x + 3)}$$

Note It is convenient to draw a line through the factors as we divide them out. It is especially helpful when the problems become longer.

Note Students sometimes make the mistake of dividing out common terms:

$$\frac{x^2 - 9}{x^2 + 5x + 6} \neq \frac{-9}{5x + 6}$$

This does not give us an equivalent expression. We can only divide out common factors, which usually requires that we factor the rational expression first.

Notice that both polynomials contain the factor $(x + 3)$. If we divide the numerator by $(x + 3)$, we are left with $(x - 3)$. If we divide the denominator by $(x + 3)$, we are left with $(x + 2)$. The complete problem looks like this:

$$\frac{x^2 - 9}{x^2 + 5x + 6} = \frac{(x - 3)(x + 3)}{(x + 2)(x + 3)}$$ Factor the numerator and denominator completely

$$= \frac{(x - 3)(x + 3)}{(x + 2)(x + 3)}$$ Divide out the common factor, $x + 3$

$$= \frac{x - 3}{x + 2}$$

EXAMPLE 6 Reduce to lowest terms: $\frac{10a + 20}{20 - 5a^2}$.

SOLUTION We begin by factoring out the greatest common factor from the numerator and denominator:

$$\frac{10a + 20}{20 - 5a^2} = \frac{10(a + 2)}{-5(a^2 - 4)}$$ Factor the greatest common factor from the numerator and denominator

$$= \frac{10(a + 2)}{-5(a + 2)(a - 2)}$$ Factor the denominator as the difference of two squares

$$= -\frac{2}{a - 2}$$ Divide out the common factors 5 and $a + 2$

EXAMPLE 7 Reduce $\frac{2x^3 + 2x^2 - 24x}{x^3 + 2x^2 - 8x}$ to lowest terms.

SOLUTION We begin by factoring the numerator and denominator completely. Then we divide out all factors common to the numerator and denominator. Here is what it looks like:

$$\frac{2x^3 + 2x^2 - 24x}{x^3 + 2x^2 - 8x} = \frac{2x(x^2 + x - 12)}{x(x^2 + 2x - 8)}$$ Factor out the greatest common factor first

$$= \frac{2x(x - 3)(x + 4)}{x(x - 2)(x + 4)}$$ Factor the remaining trinomials

$$= \frac{2(x - 3)}{x - 2}$$ Divide out the factors common to the numerator and denominator

EXAMPLE 8 Reduce $\frac{x-5}{x^2-25}$ to lowest terms. Also, state any restrictions on the variable.

Note Even though the rational expression in Example 8 can be reduced, the original expression is undefined for both $x = -5$ and $x = 5$. When determining any restrictions on the variable, we must work with the original denominator prior to reducing the expression.

SOLUTION First, we reduce the expression by dividing out common factors.

$$\frac{x-5}{x^2-25} = \frac{x-5}{(x-5)(x+5)} \qquad \text{Factor numerator and denominator completely}$$

$$= \frac{1}{x+5} \qquad \text{Divide out the common factor, } x-5$$

To find any restrictions on the variable, we must find any values of x that make the *original* expression undefined. This will be the case if $x^2 - 25 = 0$.

$$x^2 - 25 = 0$$

$$(x+5)(x-5) = 0 \qquad \text{Factor}$$

$$x + 5 = 0 \quad \text{or} \quad x - 5 = 0 \qquad \text{Zero-factor property}$$

$$x = -5 \quad \text{or} \quad x = 5$$

Our restrictions are $x \neq -5$ and $x \neq 5$.

Ratios

For the rest of this section we will concern ourselves with *ratios*, a topic closely related to reducing fractions and rational expressions to lowest terms. Let's start with a definition.

DEFINITION *Ratio*

If a and b are any two numbers, $b \neq 0$, then the ***ratio*** of a and b is

$$\frac{a}{b}$$

As you can see, ratios are another name for fractions or rational numbers. They are a way of comparing quantities. Since we also can think of $\frac{a}{b}$ as the quotient of a and b, ratios are also quotients. The following table gives some ratios in words and as fractions.

Ratio	As a Fraction	In Lowest Terms	
25 to 75	$\frac{25}{75}$	$\frac{1}{3}$	
8 to 2	$\frac{8}{2}$	$\frac{4}{1}$	With ratios it is common to leave the 1 in the denominator.
20 to 16	$\frac{20}{16}$	$\frac{5}{4}$	

EXAMPLE 9 A solution of hydrochloric acid (HCl) and water contains 49 milliliters of water and 21 milliliters of HCl. Find the ratio of HCl to water and of HCl to the total volume of the solution.

SOLUTION The ratio of HCl to water is 21 to 49, or

$$\frac{21}{49} = \frac{3}{7}$$

The amount of total solution volume is $49 + 21 = 70$ milliliters. Therefore, the ratio of HCl to total solution is 21 to 70, or

$$\frac{21}{70} = \frac{3}{10}$$

Rate Equation

Many of the problems in this chapter will use what is called the *rate equation*. You use this equation on an intuitive level when you are estimating how long it will take you to drive long distances. For example, if you drive at a steady speed of 50 miles per hour for 2 hours, you will travel 100 miles. Here is the rate equation:

$$\text{Distance} = \text{rate} \cdot \text{time}$$

$$d = r \cdot t$$

The rate equation has two equivalent forms, the most common of which is obtained by solving for r. Here it is:

$$r = \frac{d}{t}$$

The rate r in the rate equation is the ratio of distance to time and also is referred to as *average speed*. The units for rate are miles per hour, feet per second, kilometers per hour, and so on.

EXAMPLE 10 The Comstock Express chair lift at the Northstar California ski resort in Lake Tahoe is 5,649 feet long. If a ride on this chair lift takes 6 minutes, what is the average speed of the lift in feet per minute?

SOLUTION To find the speed of the lift, we find the ratio of distance covered to time. (Our answer is rounded to the nearest whole number.)

$$\text{Rate} = \frac{\text{distance}}{\text{time}} = \frac{5{,}649 \text{ feet}}{6 \text{ minutes}} = \frac{5{,}649}{6} \text{ feet/minute} \approx 942 \text{ feet/minute}$$

Note how we separate the numerical part of the problem from the units. In the next section, we will convert this rate to miles per hour.

Getting Ready for Class

After reading through the preceding section, respond in your own words and in complete sentences.

A. For what values of the variable is a rational expression undefined?

B. What are the properties we use to manipulate rational expressions?

C. How do you reduce a rational expression to lowest terms?

D. What is a ratio?

Evaluate each rational expression for the given values of the variable.

1. $\frac{x-3}{x+4}$ for $x = 2$, $x = 3$, and $x = -4$

2. $\frac{x+2}{x-5}$ for $x = -2$, $x = 4$, and $x = 5$

3. $\frac{2x+1}{x^2+x-2}$ for $x = -3$, $x = 0$, and $x = 1$

4. $\frac{x^2-4}{x^2+4}$ for $x = 0$, $x = 2$, and $x = -1$

Determine any values of the variable for which the rational expression is undefined.

5. $\frac{x-2}{x}$

6. $\frac{3x}{x-6}$

7. $\frac{x+1}{(x+2)(x-3)}$

8. $\frac{x-1}{x(x-4)}$

9. $\frac{2x+5}{3x^2-2x-1}$

10. $\frac{x^2-9}{x^2+9}$

11. $\frac{x+4}{x^2+4}$

12. $\frac{6x+10}{4x^2-25}$

Reduce the following rational expressions to lowest terms, if possible. Also, specify any restrictions on the variable.

13. $\frac{5}{5x-10}$

14. $\frac{-4}{2x-8}$

15. $\frac{a-3}{a^2-9}$

16. $\frac{a+4}{a^2-16}$

17. $\frac{x+5}{x^2-25}$

18. $\frac{x-2}{x^2-4}$

19. $\frac{2x^2-8}{4}$

20. $\frac{5x-10}{x-2}$

21. $\frac{2x-10}{3x-6}$

22. $\frac{4x-8}{x-2}$

23. $\frac{10a+20}{5a+10}$

24. $\frac{11a+33}{6a+18}$

Reduce each rational expression to lowest terms.

25. $\frac{5x^2-5}{4x+4}$

26. $\frac{7x^2-28}{2x+4}$

27. $\frac{x-3}{x^2-6x+9}$

28. $\frac{x^2-10x+25}{x-5}$

29. $\frac{3x+15}{3x^2+24x+45}$

30. $\frac{5x+15}{5x^2+40x+75}$

31. $\frac{a^2-3a}{a^3-8a^2+15a}$

32. $\frac{a^2-3a}{a^3+2a^2-15a}$

33. $\frac{3x-2}{9x^2-4}$

34. $\frac{2x-3}{4x^2-9}$

35. $\frac{x^2+8x+15}{x^2+5x+6}$

36. $\frac{x^2-8x+15}{x^2-x-6}$

37. $\frac{2m^3-2m^2-12m}{m^2-5m+6}$

38. $\frac{2m^3+4m^2-6m}{m^2-m-12}$

39. $\frac{x^3+3x^2-4x}{x^3-16x}$

40. $\frac{3a^2-8a+4}{9a^3-4a}$

41. $\frac{4x^3-10x^2+6x}{2x^3+x^2-3x}$

42. $\frac{3a^3-8a^2+5a}{4a^3-5a^2+1a}$

43. $\dfrac{4x^2 - 12x + 9}{4x^2 - 9}$ **44.** $\dfrac{5x^2 + 18x - 8}{5x^2 + 13x - 6}$ **45.** $\dfrac{x + 3}{x^4 - 81}$

46. $\dfrac{x^2 + 9}{x^4 - 81}$ **47.** $\dfrac{3x^2 + x - 10}{x^4 - 16}$ **48.** $\dfrac{5x^2 - 16x + 12}{x^4 - 16}$

49. $\dfrac{42x^3 - 20x^2 - 48x}{6x^2 - 5x - 4}$ **50.** $\dfrac{36x^3 + 132x^2 - 135x}{6x^2 + 25x - 9}$ **51.** $\dfrac{x^3 - y^3}{x^2 - y^2}$

52. $\dfrac{x^3 + y^3}{x^2 - y^2}$ **53.** $\dfrac{x^3 + 8}{x^2 - 4}$ **54.** $\dfrac{x^3 - 125}{x^2 - 25}$

55. $\dfrac{x^3 + 8}{x^2 + x - 2}$ **56.** $\dfrac{x^2 - 2x - 3}{x^3 - 27}$

To reduce each of the following rational expressions to lowest terms, you will have to use factoring by grouping. Be sure to factor each numerator and denominator completely before dividing out any common factors. (Remember, factoring by grouping takes two steps.)

57. $\dfrac{xy + 3x + 2y + 6}{xy + 3x + 5y + 15}$ **58.** $\dfrac{xy + 7x + 4y + 28}{xy + 3x + 4y + 12}$

59. $\dfrac{x^2 - 3x + ax - 3a}{x^2 - 3x + bx - 3b}$ **60.** $\dfrac{x^2 - 6x + ax - 6a}{x^2 - 7x + ax - 7a}$

The next two problems are intended to give you practice reading, and paying attention to, the instructions that accompany the problems you're working. Working these problems is an excellent way to get ready for a test or quiz.

61. Paying Attention to Instructions Work each problem according to the instructions given:

a. Add: $(x^2 - 4x) + (4x - 16)$ **b.** Subtract: $(x^2 - 4x) - (4x - 16)$

c. Multiply: $(x^2 - 4x)(4x - 16)$ **d.** Reduce: $\dfrac{x^2 - 4x}{4x - 16}$

62. Paying Attention to Instructions Work each problem according to the instructions given:

a. Add: $(9x^2 - 3x) + (6x - 2)$ **b.** Subtract: $(9x^2 - 3x) - (6x - 2)$

c. Multiply: $(9x^2 - 3x)(6x - 2)$ **d.** Reduce: $\dfrac{9x^2 - 3x}{6x - 2}$

Write each ratio as a fraction in lowest terms.

63. 8 to 6 **64.** 6 to 8 **65.** 200 to 250 **66.** 250 to 200

67. 32 to 4 **68.** 4 to 32

Applying the Concepts

69. **Cost and Average Cost** As we mentioned in the introduction to this chapter, if a bank charges \$2.00 per month and \$0.15 per check for one of its checking accounts, then the total monthly cost to write x checks is $C = 2.00 + 0.15x$, and the average cost of each of the x checks written is $A = \frac{2.00 + 0.15x}{x}$. Compare these two formulas by filling in the following table. Round to the nearest cent.

Checks Written x	Total Cost $2.00 + 0.15x$	Cost per Check $\frac{2.00 + 0.15x}{x}$
0		
5		
10		
15		
20		

70. **Cost and Average Cost** A rewritable Blu-ray drive for a computer costs \$60. An individual BD-RE disc for the drive costs \$4.00 and can store 50 gigabytes of information. The total cost of filling x discs with information is $C = 60 + 4x$ dollars. The average cost per gigabyte of information is given by $A = \frac{4x + 60}{50x}$. Compare the total cost and average cost per gigabyte of storage by completing the following table. Round all answers to the nearest cent.

Discs Purchased x	Total Cost $60 + 4x$	Cost per Gigabyte $\frac{4x + 60}{50x}$
0		
5		
10		
15		
20		

71. **Speed of a Car** A car travels 122 miles in 3 hours. Find the average speed of the car in miles per hour. Round to the nearest tenth.

72. **Speed of a Bullet** A bullet fired from a gun travels a distance of 4,500 feet in 3 seconds. Find the average speed of the bullet in feet per second.

73. Baseball For the four pitchers mentioned in the chart, calculate the number of strikeouts per inning. Round to the nearest hundredth.

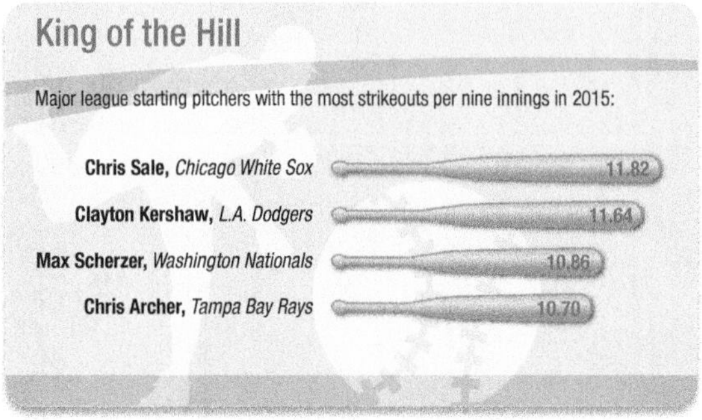

74. Ferris Wheel A person riding a Ferris Wheel travels once around the wheel, a distance of 188 feet, in 30 seconds. What is the average speed of the rider in feet per second? Round to the nearest tenth.

75. Jogging A jogger covers a distance of 3 miles in 24 minutes. Find the average speed of the jogger in miles per minute.

76. Fuel Consumption An economy car travels 168 miles on 3.5 gallons of gas. Give the average fuel consumption of the car in miles per gallon.

Learning Objectives Assessment

The following problems can be used to help assess if you have successfully met the learning objectives for this section.

77. Evaluate $\frac{2x-3}{x^2-4}$ for $x = -1$.

a. 1 **b.** Undefined **c.** $-\frac{3}{5}$ **d.** $\frac{5}{3}$

78. For what values of x is $\frac{x-5}{x+6}$ undefined?

a. 5 **b.** -6 **c.** -6 and 5 **d.** none

79. Reduce to lowest terms: $\frac{x^2-4x}{x^2-16}$.

a. $\frac{x}{4}$ **b.** $\frac{x}{x+4}$

c. $\frac{x}{x-4}$ **d.** $\frac{1}{4}$

80. In 2002, the Hubbard Glacier surged forward 170 meters in six weeks (42 days). Find the approximate average speed by which the glacier advanced during this period.

a. 4 meters per day **b.** 0.25 meters per day

c. 17 meters per day **d.** 7,140 meters per day

Getting Ready for the Next Section

Perform the indicated operation.

81. $\frac{3}{4} \cdot \frac{10}{21}$

82. $\frac{2}{9} \cdot \frac{15}{22}$

83. $\frac{4}{5} \div \frac{8}{9}$

84. $\frac{3}{5} \div \frac{15}{7}$

Factor completely.

85. $x^2 - 9$

86. $x^2 - 25$

87. $3x - 9$

88. $2x - 4$

89. $x^2 - x - 20$

90. $x^2 + 7x + 12$

91. $a^2 + 5a$

92. $a^2 - 4a$

Reduce to lowest terms.

93. $\frac{a(a+5)(a-5)(a+4)}{a^2 + 5a}$

94. $\frac{a(a+2)(a-4)(a+5)}{a^2 - 4a}$

Multiply. Give the answers as decimals rounded to the nearest tenth.

95. $\frac{5603}{11} \cdot \frac{1}{2580} \cdot \frac{60}{1}$

96. $\frac{772}{2.2} \cdot \frac{1}{2580} \cdot \frac{60}{1}$

SPOTLIGHT ON SUCCESS *Student Instructor Aaron*

Sometimes you have to take a step back
in order to get a running start forward.
—Anonymous

As a high school senior I was encouraged to go to college immediately after graduating. I earned good grades in high school and I knew that I would have a pretty good group of schools to pick from. Even though I felt like "more school" was not quite what I wanted, the counselors had so much faith and had done this process so many times that it was almost too easy to get the applications out. I sent out applications to schools I knew I could get into and a "dream school."

One night in my email inbox there was a letter of acceptance from my dream school. There was just one problem with getting into this school. It was going to be difficult and I still had senioritis. Going into my first quarter of college was as exciting and difficult as I knew it would be. But after my first quarter I could see that this was not the time for me to be here. I was interested in the subject matter but I could not find my motivating purpose like I had in high school. Instead of dropping out completely, I decided a community college would be a good way for me to stay on track. Without necessarily knowing my direction, I could take the general education classes and get those out of the way while figuring out exactly what and where I felt a good place for me to be.

Now I know what I want to go to school for and the next time I walk onto a four year campus it will be on my terms with my reasons for being there driving me to succeed. I encourage everyone to continue school after high school, even if you have no clue as to what you want to study. There are always stepping stones, like community colleges, that can help you get a clearer picture of what you want to strive for.

Multiplication and Division of Rational Expressions 7.2

Learning Objectives

In this section, we will learn how to:

1. Multiply rational expressions.
2. Divide rational expressions.
3. Use a conversion factor to convert units.

Introduction

Recall that to multiply two fractions we simply multiply numerators and multiply denominators and then reduce to lowest terms, if possible:

$$\frac{3}{4} \cdot \frac{10}{21} = \frac{30}{84} \quad \begin{array}{l} \leftarrow \text{Multiply numerators} \\ \leftarrow \text{Multiply denominators} \end{array}$$

$$= \frac{5}{14} \quad \leftarrow \text{Reduce to lowest terms}$$

Recall also that the same result can be achieved by factoring numerators and denominators first and then dividing out the factors they have in common:

$$\frac{3}{4} \cdot \frac{10}{21} = \frac{3}{2 \cdot 2} \cdot \frac{2 \cdot 5}{3 \cdot 7} \qquad \text{Factor}$$

$$= \frac{3 \cdot 2 \cdot 5}{2 \cdot 2 \cdot 3 \cdot 7} \qquad \begin{array}{l} \text{Multiply numerators} \\ \text{Multiply denominators} \end{array}$$

$$= \frac{5}{14} \qquad \text{Divide out common factors}$$

We can apply the second process to the product of two rational expressions, as the following example illustrates.

VIDEO EXAMPLES

SECTION 7.2

EXAMPLE 1 Multiply: $\frac{x-2}{x+3} \cdot \frac{x^2-9}{2x-4}$.

SOLUTION We begin by factoring numerators and denominators as much as possible. Then we multiply the numerators and denominators. The last step consists of dividing out all factors common to the numerator and denominator:

$$\frac{x-2}{x+3} \cdot \frac{x^2-9}{2x-4} = \frac{x-2}{x+3} \cdot \frac{(x-3)(x+3)}{2(x-2)} \qquad \text{Factor completely}$$

$$= \frac{(x-2)(x-3)(x+3)}{(x+3)(2)(x-2)} \qquad \text{Multiply numerators and denominators}$$

$$= \frac{x-3}{2} \qquad \text{Divide out common factors}$$

EXAMPLE 2 Multiply $\frac{3a+6}{a^2} \cdot \frac{a}{2a+4}$.

SOLUTION

$$\frac{3a+6}{a^2} \cdot \frac{a}{2a+4}$$

$$= \frac{3(a+2)}{a^2} \cdot \frac{a}{2(a+2)} \quad \text{Factor completely}$$

$$= \frac{3(a+2)a}{a^2(2)(a+2)} \quad \text{Multiply}$$

$$= \frac{3}{2a} \quad \text{Divide numerator and denominator by common factors } a(a+2)$$

In Chapter 1 we defined division as the equivalent of multiplication by the reciprocal. This is how it looks with fractions:

$$\frac{4}{5} \div \frac{8}{9} = \frac{4}{5} \cdot \frac{9}{8} \quad \text{Multiply by the reciprocal of the divisor}$$

$$= \frac{2 \cdot 2 \cdot 3 \cdot 3}{5 \cdot 2 \cdot 2 \cdot 2}$$

$$= \frac{9}{10} \quad \text{Factor and divide out common factors}$$

The same idea holds for division with rational expressions. The rational expression that follows the division symbol is called the *divisor;* to divide, we multiply by the reciprocal of the divisor.

EXAMPLE 3 Divide: $\frac{x^2+7x+12}{x^2-16} \div \frac{x^2+6x+9}{2x-8}$.

SOLUTION We begin by taking the reciprocal of the divisor and writing the problem again in terms of multiplication. We then factor, multiply, and, finally, divide out all factors common to the numerator and denominator of the resulting expression. The complete problem looks like this:

$$\frac{x^2+7x+12}{x^2-16} \div \frac{x^2+6x+9}{2x-8}$$

$$= \frac{x^2+7x+12}{x^2-16} \cdot \frac{2x-8}{x^2+6x+9} \quad \text{Multiply by the reciprocal of the divisor}$$

$$= \frac{(x+3)(x+4)}{(x-4)(x+4)} \cdot \frac{2(x-4)}{(x+3)(x+3)} \quad \text{Factor}$$

$$= \frac{(x+3)(x+4)(2)(x-4)}{(x-4)(x+4)(x+3)(x+3)} \quad \text{Multiply}$$

$$= \frac{2}{x+3} \quad \text{Divide out common factors}$$

As you can see, factoring is the single most important tool we use in working with rational expressions. Most of the work we have done or will do with rational expressions is accomplished most easily if the rational expressions are in factored form. Here are some more examples of multiplication and division with rational expressions.

EXAMPLE 4 Divide: $\dfrac{3x-9}{x^2-x-20} \div \dfrac{15-2x-x^2}{x^2-25}$.

SOLUTION

$$\frac{3x-9}{x^2-x-20} \div \frac{15-2x-x^2}{x^2-25}$$

$$= \frac{3x-9}{x^2-x-20} \cdot \frac{x^2-25}{15-2x-x^2}$$ Multiply by the reciprocal of the divisor

$$= \frac{3x-9}{x^2-x-20} \cdot \frac{x^2-25}{-1(x^2+2x-15)}$$ Factor a -1 from the second denominator

$$= \frac{3(x-3)(x-5)(x+5)}{(x+4)(x-5)(-1)(x+5)(x-3)}$$ Factor and multiply

$$= -\frac{3}{x+4}$$ Divide out common factors

In Example 4 we factored and multiplied the two expressions in a single step. This saves writing the problem one extra time.

EXAMPLE 5 Multiply: $(49-x^2)\left(\dfrac{x+4}{x-7}\right)$.

SOLUTION We can think of the polynomial $49-x^2$ as having a denominator of 1. Thinking of $49-x^2$ in this way allows us to proceed as we did in previous examples:

$$(49-x^2)\left(\frac{x+4}{x-7}\right)$$

$$= \frac{49-x^2}{1} \cdot \frac{x+4}{x-7}$$ Write x^2-49 with denominator 1

$$= \frac{-1(x^2-49)}{1} \cdot \frac{x+4}{x-7}$$ Factor and multiply

$$= \frac{-1(x+7)(x-7)(x+4)}{x-7}$$

$$= -1(x+7)(x+4)$$ Divide out common factors

In this section let's agree to leave our answers in factored form.

EXAMPLE 6 Multiply: $a(a+5)(a-5)\left(\dfrac{a+4}{a^2+5a}\right)$.

SOLUTION We can think of the expression $a(a+5)(a-5)$ as having a denominator of 1:

$$a(a+5)(a-5)\left(\frac{a+4}{a^2+5a}\right)$$

$$= \frac{a(a+5)(a-5)}{1} \cdot \frac{a+4}{a^2+5a}$$

$$= \frac{a(a+5)(a-5)(a+4)}{a(a+5)}$$ Factor and multiply

$$= (a-5)(a+4)$$ Divide out common factors

Unit Analysis

Unit analysis is a method of converting between units of measure by multiplying by the number 1. Here is our first illustration: Suppose you are flying in a commercial airliner and the pilot tells you the plane has reached its cruising altitude of 35,000 feet. How many miles is the plane above the ground?

If you know that 1 mile is 5,280 feet, then it is simply a matter of deciding what to do with the two numbers, 5,280 and 35,000. By using unit analysis, this decision is unnecessary:

$$35{,}000 \text{ feet} = \frac{35{,}000 \text{ feet}}{1} \cdot \frac{1 \text{ mile}}{5{,}280 \text{ feet}}$$

We treat the units common to the numerator and denominator in the same way we treat factors common to the numerator and denominator; common units can be divided out, just as common factors are. In the previous expression, we have feet common to the numerator and denominator. Dividing them out leaves us with miles only. Here is the complete problem:

$$\begin{aligned} 35{,}000 \text{ feet} &= \frac{35{,}000 \text{ feet}}{1} \cdot \frac{1 \text{ mile}}{5{,}280 \text{ feet}} \\ &= \frac{35{,}000}{5{,}280} \text{ miles} \\ &= 6.6 \text{ miles to the nearest tenth of a mile} \end{aligned}$$

The expression $\frac{1 \text{ mile}}{5{,}280 \text{ feet}}$ is called a *conversion factor*. It is simply the number 1 written in a convenient form. Because it is the number 1, we can multiply any other number by it and always be sure we have not changed that number. The key to unit analysis is choosing the right conversion factors.

Note Realizing that conversion factors, such as $\frac{1 \text{ mile}}{5{,}280}$ feet, are equivalent to the number 1 may take some getting used to. For example, you know that $\frac{2}{2} = 1$ since the numerator and denominator are equal. The same goes for conversion factors; since 1 mile = 5,280 feet, the numerator and denominator are equal, and the fraction is equivalent to 1.

EXAMPLE 7 The Mall of America in the Twin Cities covers 78 acres of land. If 1 square mile = 640 acres, how many square miles does the Mall of America cover? Round your answer to the nearest hundredth of a square mile.

SOLUTION We are starting with acres and want to end up with square miles. We need to multiply by a conversion factor that will allow acres to divide out and leave us with square miles:

$$\begin{aligned} 78 \text{ acres} &= \frac{78 \text{ acres}}{1} \cdot \frac{1 \text{ square mile}}{640 \text{ acres}} \\ &= \frac{78}{640} \text{ square miles} \\ &= 0.12 \text{ square miles to the nearest hundredth} \end{aligned}$$

The next example is a continuation of Example 10 from Section 7.1.

EXAMPLE 8 The Comstock Express chair lift at the Northstar California ski resort in Lake Tahoe is 5,649 feet long. If a ride on this chair lift takes 6 minutes, what is the average speed of the lift in miles per hour?

SOLUTION First, we find the speed of the lift in feet per second, as we did in Example 10 of Section 7.1, by taking the ratio of distance to time.

$$\text{Rate} = \frac{\text{distance}}{\text{time}} = \frac{5{,}649 \text{ feet}}{6 \text{ minutes}} = \frac{5{,}649}{6} \text{ feet per minute}$$

$$\approx 942 \text{ feet per minute}$$

Next, we convert feet per minute to miles per hour. To do this, we need to know that

$$1 \text{ mile} = 5{,}280 \text{ feet}$$

$$1 \text{ hour} = 60 \text{ minutes}$$

$$\text{Average speed} = 942 \text{ feet per minute} = \frac{942 \text{ feet}}{1 \text{ minute}} \cdot \frac{1 \text{ mile}}{5{,}280 \text{ feet}} \cdot \frac{60 \text{ minutes}}{1 \text{ hour}}$$

$$= \frac{942 \cdot 60}{5{,}280} \text{ miles per hour}$$

$$= 10.7 \text{ mph to the nearest tenth}$$

Getting Ready for Class

After reading through the preceding section, respond in your own words and in complete sentences.

A. How do we multiply rational expressions?

B. Explain the steps used to divide rational expressions.

C. What part does factoring play in multiplying and dividing rational expressions?

D. Why are all conversion factors the same as the number 1?

Problem Set 7.2

Multiply or divide as indicated. Be sure to reduce all answers to lowest terms. (The numerator and denominator of the answer should not have any factors in common.)

1. $\frac{x+y}{3} \cdot \frac{6}{x+y}$

2. $\frac{x-1}{x+1} \cdot \frac{5}{x-1}$

3. $\frac{2x+10}{x^2} \cdot \frac{x^3}{4x+20}$

4. $\frac{3x^4}{3x-6} \cdot \frac{x-2}{x^2}$

5. $\frac{9}{2a-8} \div \frac{3}{a-4}$

6. $\frac{8}{a^2-25} \div \frac{16}{a+5}$

7. $\frac{x+1}{x^2-9} \div \frac{2x+2}{x+3}$

8. $\frac{-11}{x-2} \div \frac{22}{2x^2-8}$

9. $\frac{a^2+5a}{7a} \cdot \frac{4a^2}{a^2+4a}$

10. $\frac{4a^2+4a}{a^2-25} \cdot \frac{a^2-5a}{8a}$

11. $\frac{y^2-5y+6}{2y+4} \div \frac{2y-6}{y+2}$

12. $\frac{y^2-7y}{3y^2-48} \div \frac{y^2-9}{y^2-7y+12}$

13. $\frac{2x-8}{x^2-4} \cdot \frac{x^2+6x+8}{x-4}$

14. $\frac{x^2+5x+1}{7x-7} \cdot \frac{x-1}{x^2+5x+1}$

15. $\frac{x-1}{6+x-x^2} \cdot \frac{x^2+5x+6}{x^2-1}$

16. $\frac{x^2-3x-10}{x^2-4x+3} \cdot \frac{x^2-5x+6}{10+3x-x^2}$

17. $\frac{a^2+10a+25}{a+5} \div \frac{a^2-25}{a-5}$

18. $\frac{a^2+a-2}{a^2+5a+6} \div \frac{a-1}{a}$

19. $\frac{y^3-5y^2}{y^4+3y^3+2y^2} \div \frac{y^2-5y+6}{y^2-2y-3}$

20. $\frac{y^2-5y}{y^2+7y+12} \div \frac{y^3-7y^2+10y}{y^2+9y+18}$

21. $\frac{2x^2+17x+21}{x^2+2x-35} \cdot \frac{25-x^2}{2x^2-7x-15}$

22. $\frac{x^2-13x+42}{4x^2-31x+21} \cdot \frac{4x^2+5x-6}{4-x^2}$

23. $\frac{2x^2+10x+12}{4x^2+24x+32} \cdot \frac{2x^2+18x+40}{x^2+8x+15}$

24. $\frac{3x^2-3}{6x^2+18x+12} \cdot \frac{2x^2-8}{x^2-3x+2}$

25. $\frac{2a^2+7a+3}{a^2-16} \div \frac{4a^2+8a+3}{2a^2-5a-12}$

26. $\frac{3a^2+7a-20}{a^2+3a-4} \div \frac{3a^2-2a-5}{a^2-2a+1}$

27. $\frac{4y^2-12y+9}{36-y^2} \div \frac{2y^2-5y+3}{y^2+5y-6}$

28. $\frac{5y^2-6y+1}{1-y^2} \div \frac{16y^2-9}{4y^2+7y+3}$

29. $\frac{x^2-1}{6x^2+18x+12} \cdot \frac{7x^2+17x+6}{x+1} \cdot \frac{6x+30}{7x^2-11x-6}$

30. $\frac{4x^2-1}{3x-15} \cdot \frac{4x^2-17x-15}{4x^2-9x-9} \cdot \frac{3x-3}{x^2-9}$

31. $\frac{18x^3+21x^2-60x}{21x^2-25x-4} \cdot \frac{28x^2-17x-3}{16x^3+28x^2-30x}$

32. $\frac{56x^3+54x^2-20x}{8x^2-2x-15} \cdot \frac{6x^2+5x-21}{63x^3+129x^2-42x}$

The next two problems are intended to give you practice reading, and paying attention to, the instructions that accompany the problems you are working. Working these problems is an excellent way to get ready for a test or quiz.

33. Paying Attention to Instructions Work each problem according to the instructions given:

a. Simplify: $\frac{9-1}{27-1}$

b. Reduce: $\frac{x^2-1}{x^3-1}$

c. Multiply: $\frac{x^2-1}{x^3-1} \cdot \frac{x-2}{x+1}$

d. Divide: $\frac{x^2-1}{x^3-1} \div \frac{x-1}{x^2+x+1}$

34. Paying Attention to Instructions Work each problem according to the instructions given:

a. Simplify: $\frac{16 - 9}{16 + 24 + 9}$

b. Reduce: $\frac{4x^2 - 9}{4x^2 + 12x + 9}$

c. Multiply: $\frac{4x^2 - 9}{4x^2 + 12x + 9} \cdot \frac{2x + 3}{2x - 3}$

d. Divide: $\frac{4x^2 - 9}{4x^2 + 12x + 9} \div \frac{2x + 3}{2x - 3}$

Multiply the following expressions using the method shown in Examples 5 and 6 in this section.

35. $(x^2 - 9)\left(\frac{2}{x + 3}\right)$

36. $(x^2 - 9)\left(\frac{-3}{x - 3}\right)$

37. $(x^2 - x - 6)\left(\frac{x + 1}{x - 3}\right)$

38. $(x^2 - 2x - 8)\left(\frac{x + 3}{x - 4}\right)$

39. $(x^2 - 4x - 5)\left(\frac{-2x}{x + 1}\right)$

40. $(x^2 - 6x + 8)\left(\frac{4x}{x - 2}\right)$

Each of the following problems involves some factoring by grouping. Remember, before you can divide out factors common to the numerators and denominators of a product, you must factor completely.

41. $\frac{x^2 - 9}{x^2 - 3x} \cdot \frac{2x + 10}{xy + 5x + 3y + 15}$

42. $\frac{x^2 - 16}{x^2 - 4x} \cdot \frac{3x + 18}{xy + 6x + 4y + 24}$

43. $\frac{2x^2 + 4x}{x^2 - y^2} \cdot \frac{x^2 + 3x + xy + 3y}{x^2 + 5x + 6}$

44. $\frac{x^2 - 25}{3x^2 + 3xy} \cdot \frac{x^2 + 4x + xy + 4y}{x^2 + 9x + 20}$

45. $\frac{x^3 - 3x^2 + 4x - 12}{x^4 - 16} \cdot \frac{3x^2 + 5x - 2}{3x^2 - 10x + 3}$

46. $\frac{x^3 - 5x^2 + 9x - 45}{x^4 - 81} \cdot \frac{5x^2 + 18x + 9}{5x^2 - 22x - 15}$

Simplify each expression. Work inside parentheses first, and then divide out common factors.

47. $\left(1 - \frac{1}{2}\right)\left(1 - \frac{1}{3}\right)\left(1 - \frac{1}{4}\right)\left(1 - \frac{1}{5}\right)$

48. $\left(1 + \frac{1}{2}\right)\left(1 + \frac{1}{3}\right)\left(1 + \frac{1}{4}\right)\left(1 + \frac{1}{5}\right)$

The dots in the following problems represent factors not written that are in the same pattern as the surrounding factors. Simplify.

49. $\left(1 - \frac{1}{2}\right)\left(1 - \frac{1}{3}\right)\left(1 - \frac{1}{4}\right)\ldots\left(1 - \frac{1}{99}\right)\left(1 - \frac{1}{100}\right)$

50. $\left(1 - \frac{1}{3}\right)\left(1 - \frac{1}{4}\right)\left(1 - \frac{1}{5}\right)\ldots\left(1 - \frac{1}{98}\right)\left(1 - \frac{1}{99}\right)$

Applying the Concepts

51. Mount Whitney The top of Mount Whitney, the highest point in California, is 14,494 feet above sea level. Give this height in miles to the nearest tenth of a mile.

52. Motor Displacement The relationship between liters and cubic inches, both of which are measures of volume, is 0.0164 liters = 1 cubic inch. If a Ford Mustang has a motor with a displacement of 4.9 liters, what is the displacement in cubic inches? Round your answer to the nearest cubic inch.

53. Speed of Sound The speed of sound is 1,088 feet per second. Convert the speed of sound to miles per hour. Round your answer to the nearest whole number.

54. Average Speed A car travels 122 miles in 3 hours. Find the average speed of the car in feet per second. Round to the nearest whole number.

55. Ferris Wheel The first Ferris wheel was built in 1893. It was a large wheel with a circumference of 785 feet. If one trip around the circumference of the wheel took 20 minutes, find the average speed of a rider in miles per hour. Round to the nearest hundredth.

56. Unit Analysis If we assume light travels 186,000 miles in 1 second, we can find the number of miles in 1 light-year by converting 186,000 miles/second to miles/year. Find the number of miles in 1 light-year. Write your answer in expanded form and in scientific notation.

57. Ferris Wheel A Ferris wheel called Colossus has a circumference of 518 feet. If a trip around the circumference of Colossus takes 40 seconds, find the average speed of a rider in miles per hour. Round to the nearest tenth.

58. Fitness Walking The guidelines for fitness now indicate that a person who walks 10,000 steps daily is physically fit. According to *The Walking Site* on the Internet, "The average person's stride length is approximately 2.5 feet long. That means it takes just over 2,000 steps to walk one mile, and 10,000 steps is close to 5 miles." Use your knowledge of unit analysis to determine if these facts are correct.

Learning Objectives Assessment

The following problems can be used to help assess if you have successfully met the learning objectives for this section.

59. Multiply: $\dfrac{x^2 - x - 6}{2x - 10} \cdot \dfrac{6x - 30}{x^2 + 3x + 2}$.

a. $\dfrac{6x^3 - 36x^2 + 6x + 180}{2x^3 - 4x^2 + 26x - 20}$ **b.** $\dfrac{3(x - 3)}{x + 1}$

c. $\dfrac{3(x - 6)}{x + 2}$ **d.** -13

60. Divide: $\dfrac{x^2 - 3x - 4}{6} \div \dfrac{2x - 8}{x + 1}$.

a. $\dfrac{(x - 4)^2}{3}$ **b.** $\dfrac{x(x - 1)}{3}$ **c.** $\dfrac{(x + 1)^2}{12}$ **d.** $\dfrac{1}{3}$

61. Convert 10 miles per hour to feet per minute.

a. 0.00003 ft/min **b.** 880 ft/min
c. 3,168,000 ft/min **d.** 0.11 ft/min

Getting Ready for the Next Section

Perform the indicated operation.

62. $\frac{1}{5} + \frac{3}{5}$ **63.** $\frac{1}{7} + \frac{5}{7}$ **64.** $\frac{1}{10} + \frac{3}{14}$ **65.** $\frac{1}{21} + \frac{4}{15}$

66. $\frac{1}{10} - \frac{3}{14}$ **67.** $\frac{1}{21} - \frac{4}{15}$

Multiply.

68. $2(x - 3)$ **69.** $x(x + 2)$ **70.** $(x + 4)(x - 5)$ **71.** $(x + 3)(x - 4)$

Reduce to lowest terms.

72. $\frac{x + 3}{x^2 - 9}$ **73.** $\frac{x + 7}{x^2 - 49}$

74. $\frac{x^2 - x - 30}{50 - 2x^2}$ **75.** $\frac{x^2 - x - 20}{32 - 2x^2}$

Simplify.

76. $(x + 4)(x - 5) - 10$ **77.** $(x + 3)(x - 4) - 8$

Addition and Subtraction of Rational Expressions 7.3

Learning Objectives

In this section, we will learn how to:

1. Add and subtract rational expressions with a common denominator.
2. Identify the least common denominator for a set of rational expressions.
3. Add and subtract rational expressions that do not have a common denominator.

Introduction

In Chapter 1, we combined fractions having the same denominator by combining their numerators and putting the result over the common denominator. We use the same process to add two rational expressions with the same denominator.

Addition and Subtraction with Common Denominators

VIDEO EXAMPLES

SECTION 7.3

EXAMPLE 1 Add: $\frac{5}{x} + \frac{3}{x}$.

SOLUTION The two rational expressions have a common denominator of x. Adding numerators, we have:

$$\frac{5}{x} + \frac{3}{x} = \frac{8}{x}$$

EXAMPLE 2 Add: $\frac{x}{x^2 - 9} + \frac{3}{x^2 - 9}$.

SOLUTION Since both expressions have the same denominator, we add numerators and reduce to lowest terms:

$$\frac{x}{x^2 - 9} + \frac{3}{x^2 - 9} = \frac{x + 3}{x^2 - 9}$$

$$= \frac{x + 3}{(x + 3)(x - 3)}$$

$$= \frac{1}{x - 3}$$

Reduce to lowest terms by factoring the denominator and then dividing out the common factor $x + 3$

EXAMPLE 3 Subtract: $\frac{6x - 13}{x^2 - 2x - 3} - \frac{2x - 1}{x^2 - 2x - 3}$.

SOLUTION The rational expressions have a common denominator, so we can subtract numerators and then reduce:

$$\frac{6x - 13}{x^2 - 2x - 3} - \frac{2x - 1}{x^2 - 2x - 3} = \frac{6x - 13 - (2x - 1)}{x^2 - 2x - 3}$$ Subtract numerators

$$= \frac{6x - 13 - 2x + 1}{x^2 - 2x - 3}$$ Distribute the negative sign

$$= \frac{4x - 12}{x^2 - 2x - 3}$$ Combine like terms

$$= \frac{4(x - 3)}{(x + 1)(x - 3)}$$ Factor

$$= \frac{4}{x + 1}$$ Divide out common factor of $x - 3$

Least Common Denominator

Remember, it is the distributive property that allows us to add rational expressions by simply adding numerators. Because of this, we must begin all addition problems involving rational expressions by first making sure all the expressions have the same denominator.

DEFINITION *least common denominator*

The ***least common denominator*** (LCD) for a set of denominators is the simplest quantity that is evenly divisible by all the denominators.

If all of the denominators have been factored, then the LCD will be the product of each factor raised to the highest exponent it appears within any of the denominators.

Before we attempt to add or subtract rational expressions with different denominators, we will practice finding the least common denominator.

EXAMPLE 4 Find the least common denominator for the rational expressions

$$\frac{2}{x} \quad \text{and} \quad \frac{4}{x-3}$$

SOLUTION There are only two factors, x and $x - 3$. Since they both appear to the first power, the LCD is $x(x - 3)$. This is the simplest quantity divisible by both x and $x - 3$.

Note Because of the subtraction sign, the x appearing in the denominator $x - 3$ is a term, not a factor. That is why the LCD in Example 4 requires an additional x.

EXAMPLE 5 Find the least common denominator for the rational expressions

$$\frac{x-5}{2x^2+4x+2} \quad \text{and} \quad \frac{3x}{2x^2-2}$$

SOLUTION Factoring both denominators, we have

$$2x^2 + 4x + 2 = 2(x^2 + 2x + 1) = 2(x + 1)^2 \quad \text{and} \quad 2x^2 - 2 = 2(x^2 - 1) = 2(x + 1)(x - 1)$$

The factors appearing in the two denominators are 2, $x + 1$, and $x - 1$. The factors 2 and $x - 1$ only appear to the first power. The highest exponent of the factor $x + 1$ is 2. So we have:

$$\text{LCD} = 2(x - 1)(x + 1)^2$$

Notice that, even though the factor 2 appears in both denominators, it does not show up twice in the LCD. Likewise, the LCD only contains $(x + 1)^2$, not $(x + 1)^3$.

Addition and Subtraction with Different Denominators

Now that we are able to identify the least common denominator for a set of rational expressions, we are ready to add and subtract when the denominators are not identical. We will begin with an example that does not involve any variables.

EXAMPLE 6 Add: $\frac{1}{10} + \frac{3}{14}$.

SOLUTION

Step 1: Find the LCD for 10 and 14. To do so, we factor each denominator and build the LCD from the factors:

$$\left.\begin{aligned} 10 &= 2 \cdot 5 \\ 14 &= 2 \cdot 7 \end{aligned}\right\} \quad \text{LCD} = 2 \cdot 5 \cdot 7 = 70$$

We know the LCD is divisible by 10 because it contains the factors 2 and 5. It is also divisible by 14 because it contains the factors 2 and 7.

Step 2: Change to equivalent fractions that each have a denominator of 70. To accomplish this task, we multiply the numerator and denominator of each fraction by any factors of the LCD that are not also factors of its denominator:

Original Fractions		*Denominators in Factored Form*		*Multiply by Factor Needed to Obtain LCD*		*These Have the Same Value as the Original Fractions*
$\frac{1}{10}$	=	$\frac{1}{2 \cdot 5}$	=	$\frac{1}{2 \cdot 5} \cdot \frac{7}{7}$	=	$\frac{7}{70}$
$\frac{3}{14}$	=	$\frac{3}{2 \cdot 7}$	=	$\frac{3}{2 \cdot 7} \cdot \frac{5}{5}$	=	$\frac{15}{70}$

The fraction $\frac{7}{70}$ has the same value as the fraction $\frac{1}{10}$. Likewise, the fractions $\frac{15}{70}$ and $\frac{3}{14}$ are equivalent; they have the same value.

Step 3: Add numerators and put the result over the LCD:

$$\frac{7}{70} + \frac{15}{70} = \frac{7 + 15}{70} = \frac{22}{70}$$

Step 4: Reduce to lowest terms:

$$\frac{22}{70} = \frac{11}{35} \quad \text{Divide numerator and denominator by 2}$$

> *Note* If you have had difficulty in the past with addition and subtraction of fractions with different denominators, this is the time to get it straightened out. Go over Example 6 as many times as is necessary for you to understand the process.

The main idea in adding fractions is to write each fraction again with the LCD for a denominator. Once we have done that, we simply add numerators. The same process can be used to add rational expressions, as the next example illustrates.

EXAMPLE 7 Subtract: $\frac{3}{x} - \frac{1}{2}$.

SOLUTION

Step 1: The LCD for x and 2 is $2x$. It is the smallest expression divisible by x and by 2.

Step 2: To change to equivalent expressions with the denominator $2x$, we multiply the first fraction by $\frac{2}{2}$ and the second by $\frac{x}{x}$:

$$\frac{3}{x} \cdot \frac{2}{2} = \frac{6}{2x}$$

$$\frac{1}{2} \cdot \frac{x}{x} = \frac{x}{2x}$$

Step 3: Subtracting numerators of the rational expressions in step 2, we have

$$\frac{6}{2x} - \frac{x}{2x} = \frac{6 - x}{2x}$$

Step 4: Since $6 - x$ and $2x$ do not have any factors in common, we cannot reduce any further. Here is the complete problem:

$$\begin{aligned}\frac{3}{x} - \frac{1}{2} &= \frac{3}{x} \cdot \frac{2}{2} - \frac{1}{2} \cdot \frac{x}{x} \\ &= \frac{6}{2x} - \frac{x}{2x} \\ &= \frac{6 - x}{2x}\end{aligned}$$

EXAMPLE 8 Add: $\dfrac{5}{2x - 6} + \dfrac{x}{x - 3}$.

SOLUTION If we factor $2x - 6$, we have $2x - 6 = 2(x - 3)$. The LCD is $2(x - 3)$. We need only multiply the second rational expression in our problem by $\frac{2}{2}$ to have two expressions with the same denominator:

$$\begin{aligned}\frac{5}{2x - 6} + \frac{x}{x - 3} &= \frac{5}{2(x - 3)} + \frac{x}{x - 3} \\ &= \frac{5}{2(x - 3)} + \frac{2}{2}\left(\frac{x}{x - 3}\right) \\ &= \frac{5}{2(x - 3)} + \frac{2x}{2(x - 3)} \\ &= \frac{2x + 5}{2(x - 3)}\end{aligned}$$

EXAMPLE 9 Add: $\dfrac{1}{x + 4} + \dfrac{8}{x^2 - 16}$.

SOLUTION After writing each denominator in factored form, we find that the least common denominator is $(x + 4)(x - 4)$. To change the first rational expression to an equivalent rational expression with the common denominator, we multiply its numerator and denominator by $x - 4$:

$$\frac{1}{x + 4} + \frac{8}{x^2 - 16}$$

$$= \frac{1}{x + 4} + \frac{8}{(x + 4)(x - 4)} \qquad \text{Factor each denominator}$$

$$= \frac{1}{x + 4} \cdot \frac{x - 4}{x - 4} + \frac{8}{(x + 4)(x - 4)} \qquad \text{Change to equivalent rational expressions}$$

$$= \frac{x - 4}{(x + 4)(x - 4)} + \frac{8}{(x + 4)(x - 4)} \qquad \text{Simplify}$$

$$= \frac{x + 4}{(x + 4)(x - 4)} \qquad \text{Add numerators}$$

$$= \frac{1}{x - 4} \qquad \text{Divide out common factor } x + 4$$

EXAMPLE 10 Subtract: $\dfrac{2x-1}{x^2+5x+6} - \dfrac{x+1}{x^2-9}$.

SOLUTION

Step 1: We factor each denominator and build the LCD from the factors:

$$\left.\begin{aligned} x^2+5x+6 &= (x+2)(x+3) \\ x^2-9 &= (x+3)(x-3) \end{aligned}\right\} \quad \text{LCD} = (x+2)(x+3)(x-3)$$

Step 2: Change to equivalent rational expressions:

$$\frac{2x-1}{x^2+5x+6} = \frac{2x-1}{(x+2)(x+3)} \cdot \frac{(x-3)}{(x-3)} = \frac{2x^2-7x+3}{(x+2)(x+3)(x-3)}$$

$$\frac{x+1}{x^2-9} = \frac{x+1}{(x+3)(x-3)} \cdot \frac{(x+2)}{(x+2)} = \frac{x^2+3x+2}{(x+2)(x+3)(x-3)}$$

Note In Step 2 of Example 10, we are using the FOIL method to find the product in each numerator.

Step 3: Subtract numerators of the rational expressions produced in step 2:

$$\frac{2x^2-7x+3}{(x+2)(x+3)(x-3)} - \frac{x^2+3x+2}{(x+2)(x+3)(x-3)}$$

$$= \frac{2x^2-7x+3-(x^2+3x+2)}{(x+2)(x+3)(x-3)}$$

$$= \frac{2x^2-7x+3-x^2-3x-2}{(x+2)(x+3)(x-3)} \qquad \text{Distribute the negative sign}$$

$$= \frac{x^2-10x+1}{(x+2)(x+3)(x-3)} \qquad \text{Combine like terms}$$

The numerator and denominator do not have any factors in common, so the expression cannot be simplified any further.

EXAMPLE 11 Subtract: $\dfrac{x+4}{2x+10} - \dfrac{5}{x^2-25}$.

SOLUTION We begin by factoring each denominator:

$$\frac{x+4}{2x+10} - \frac{5}{x^2-25} = \frac{x+4}{2(x+5)} - \frac{5}{(x+5)(x-5)}$$

The LCD is $2(x+5)(x-5)$. Completing the problem, we have:

$$= \frac{x+4}{2(x+5)} \cdot \frac{(x-5)}{(x-5)} + \frac{-5}{(x+5)(x-5)} \cdot \frac{2}{2}$$

$$= \frac{x^2-x-20}{2(x+5)(x-5)} + \frac{-10}{2(x+5)(x-5)}$$

$$= \frac{x^2-x-30}{2(x+5)(x-5)}$$

To see if this expression will reduce, we factor the numerator into $(x-6)(x+5)$:

$$= \frac{(x-6)(x+5)}{2(x+5)(x-5)}$$

$$= \frac{x-6}{2(x-5)}$$

Note In the last step we reduced the rational expression to lowest terms by dividing out the common factor of $x+5$.

EXAMPLE 12 Write an expression for the sum of a number and its reciprocal, and then simplify that expression.

SOLUTION If we let x be the number, then its reciprocal is $\frac{1}{x}$. To find the sum of the number and its reciprocal, we add them:

$$x + \frac{1}{x}$$

The first term x can be thought of as having a denominator of 1. Since the denominators are 1 and x, the least common denominator is x.

$$x + \frac{1}{x} = \frac{x}{1} + \frac{1}{x} \quad \text{Write } x \text{ as } \frac{x}{1}$$

$$= \frac{x}{1} \cdot \frac{x}{x} + \frac{1}{x} \quad \text{The LCD is } x$$

$$= \frac{x^2}{x} + \frac{1}{x}$$

$$= \frac{x^2 + 1}{x} \quad \text{Add numerators}$$

Getting Ready for Class

After reading through the preceding section, respond in your own words and in complete sentences.

A. How do we add two rational expressions that have the same denominator?

B. What is the least common denominator for two fractions?

C. What role does factoring play in finding a least common denominator?

D. Explain how to find a common denominator for two rational expressions.

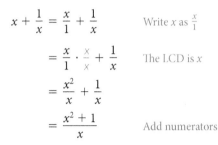

Problem Set 7.3

Find the following sums and differences.

1. $\frac{3}{x} + \frac{4}{x}$
2. $\frac{5}{x} + \frac{3}{x}$
3. $\frac{9}{a} - \frac{5}{a}$
4. $\frac{8}{a} - \frac{7}{a}$
5. $\frac{1}{x+1} + \frac{x}{x+1}$
6. $\frac{x}{x-3} - \frac{3}{x-3}$
7. $\frac{y^2}{y-1} - \frac{1}{y-1}$
8. $\frac{y^2}{y+3} - \frac{9}{y+3}$
9. $\frac{x^2}{x+2} + \frac{4x+4}{x+2}$
10. $\frac{x^2-6x}{x-3} + \frac{9}{x-3}$
11. $\frac{x^2}{x-2} - \frac{4x-4}{x-2}$
12. $\frac{x^2}{x-5} - \frac{10x-25}{x-5}$
13. $\frac{x+2}{x+6} - \frac{x-4}{x+6}$
14. $\frac{x+5}{x+2} - \frac{x+3}{x+2}$

Find the least common denominator for each pair of rational expresssions.

15. $\frac{2}{x}, \frac{x}{5}$
16. $\frac{x}{2}, \frac{3}{x+1}$
17. $\frac{4}{x}, \frac{x+5}{x-3}$
18. $\frac{4x}{x+2}, \frac{1}{x-2}$
19. $\frac{3}{2y^2}, \frac{y-1}{y(y+4)}$
20. $\frac{9}{(y+3)^2}, \frac{y-6}{4(y+3)}$
21. $\frac{a+5}{a^2-6a+8}, \frac{2a+1}{a^2+a-6}$
22. $\frac{a-2}{a^2-1}, \frac{4a}{a^2-6a+5}$
23. $\frac{4x-3}{x^2+6x+9}, \frac{x-1}{x^2+5x+6}$
24. $\frac{x+4}{3x^3-12x^2}, \frac{5x+2}{x^2-8x+16}$

Find the following sums and differences.

25. $\frac{y}{2} - \frac{2}{y}$
26. $\frac{3}{y} + \frac{y}{3}$
27. $\frac{a}{3} + \frac{1}{2}$
28. $\frac{2a}{5} + \frac{2}{3}$
29. $\frac{x}{x+1} + \frac{3}{4}$
30. $\frac{x}{x-3} + \frac{1}{3}$
31. $\frac{x+1}{x-2} - \frac{4x+7}{5x-10}$
32. $\frac{3x+1}{2x-6} - \frac{x+2}{x-3}$
33. $\frac{4x-2}{3x+12} - \frac{x-2}{x+4}$
34. $\frac{6x+5}{5x-25} - \frac{x+2}{x-5}$
35. $\frac{6}{x(x-2)} + \frac{3}{x}$
36. $\frac{10}{x(x+5)} - \frac{2}{x}$
37. $\frac{4}{a} - \frac{12}{a^2+3a}$
38. $\frac{5}{a} + \frac{20}{a^2-4a}$
39. $\frac{2}{x+5} - \frac{10}{x^2-25}$
40. $\frac{6}{x^2-1} + \frac{3}{x+1}$
41. $\frac{x-4}{x-3} + \frac{6}{x^2-9}$
42. $\frac{x+1}{x-1} - \frac{4}{x^2-1}$
43. $\frac{a-4}{a-3} + \frac{5}{a^2-a-6}$
44. $\frac{a+2}{a+1} + \frac{7}{a^2-5a-6}$
45. $\frac{8}{x^2-16} - \frac{7}{x^2-x-12}$
46. $\frac{6}{x^2-9} - \frac{5}{x^2-x-6}$
47. $\frac{4y}{y^2+6y+5} - \frac{3y}{y^2+5y+4}$
48. $\frac{3y}{y^2+7y+10} - \frac{2y}{y^2+6y+8}$
49. $\frac{4x+1}{x^2+5x+4} - \frac{x+3}{x^2+4x+3}$
50. $\frac{2x-1}{x^2+x-6} - \frac{x+2}{x^2+5x+6}$
51. $\frac{x-3}{x^2+4x+4} - \frac{x+1}{x^2-4}$
52. $\frac{2x-1}{x^2-6x+9} - \frac{x+3}{x^2+x-12}$
53. $\frac{1}{x} + \frac{x}{3x+9} - \frac{3}{x^2+3x}$
54. $\frac{1}{x} + \frac{x}{2x+4} - \frac{2}{x^2+2x}$

55. Paying Attention to Instructions Work each problem according to the instructions given.

a. Multiply: $\frac{4}{9} \cdot \frac{1}{6}$

b. Divide: $\frac{4}{9} \div \frac{1}{6}$

c. Add: $\frac{4}{9} + \frac{1}{6}$

d. Multiply: $\frac{x+2}{x-2} \cdot \frac{3x+10}{x^2-4}$

e. Divide: $\frac{x+2}{x-2} \div \frac{3x+10}{x^2-4}$

f. Subtract: $\frac{x+2}{x-2} - \frac{3x+10}{x^2-4}$

56. Paying Attention to Instructions Work each problem according to the instructions given.

a. Multiply: $\frac{9}{25} \cdot \frac{1}{15}$

b. Divide: $\frac{9}{25} \div \frac{1}{15}$

c. Subtract: $\frac{9}{25} - \frac{1}{15}$

d. Multiply: $\frac{3x-2}{3x+2} \cdot \frac{15x+6}{9x^2-4}$

e. Divide: $\frac{3x-2}{3x+2} \div \frac{15x+6}{9x^2-4}$

f. Subtract: $\frac{3x+2}{3x-2} - \frac{15x+6}{9x^2-4}$

Complete the following tables.

57.

Number x	Reciprocal $\frac{1}{x}$	Sum $1 + \frac{1}{x}$	Quotient $\frac{x+1}{x}$
1			
2			
3			
4			

58.

Number x	Reciprocal $\frac{1}{x}$	Difference $1 - \frac{1}{x}$	Quotient $\frac{x-1}{x}$
1			
2			
3			
4			

Add and subtract as indicated.

59. $1 + \frac{1}{x+2}$ **60.** $1 - \frac{1}{x+2}$ **61.** $1 - \frac{1}{x+3}$ **62.** $1 + \frac{1}{x+3}$

The following problems involve more than one operation. Simplify inside the parentheses first, then multiply.

63. $\left(1 - \frac{1}{x}\right)\left(1 - \frac{1}{x+1}\right)\left(1 - \frac{1}{x+2}\right)$ **64.** $\left(1 + \frac{1}{x}\right)\left(1 + \frac{1}{x+1}\right)\left(1 + \frac{1}{x+2}\right)$

65. $\left(1 + \frac{1}{x+3}\right)\left(1 + \frac{1}{x+2}\right)\left(1 + \frac{1}{x+1}\right)$

66. $\left(1 - \frac{1}{x+3}\right)\left(1 - \frac{1}{x+2}\right)\left(1 - \frac{1}{x+1}\right)$

Applying the Concepts

67. Number Problem Write an expression for the sum of a number and twice its reciprocal. Then, simplify that expression. (If the reciprocal of a number is $\frac{1}{x}$, then twice that is $\frac{2}{x}$, not $\frac{1}{2x}$.)

68. Number Problem Write an expression for the sum of a number and 3 times its reciprocal. Then, simplify that expression.

69. Number Problem One number is twice another. Write an expression for the sum of their reciprocals. Then, simplify that expression. Hint: The numbers are x and $2x$. Their reciprocals are, respectively, $\frac{1}{x}$ and $\frac{1}{2x}$.

70. Number Problem One number is three times another. Write an expression for the sum of their reciprocals. Then, simplify that expression.

Learning Objectives Assessment

The following problems can be used to help assess if you have successfully met the learning objectives for this section.

71. Add: $\frac{x^2 - 4x}{x - 5} + \frac{x - 10}{x - 5}$.

a. $\frac{x^2 - 3x - 10}{(x-5)^2}$ **b.** $\frac{x^2 - 5x + 10}{x - 5}$ **c.** $x + 2$ **d.** $\frac{x+2}{2}$

72. Subtract: $\frac{3x + 2}{x + 4} - \frac{2x - 3}{x + 4}$.

a. $\frac{x+5}{x+4}$ **b.** $\frac{x-1}{x+4}$ **c.** $x + 5$ **d.** $x - 1$

73. Find the least common denominator for $\frac{x+2}{x^3 + 4x^2}$ and $\frac{x-3}{x^2 - x}$.

a. $(x^3 + 4x^2)(x^2 - x)$ **b.** $x^3(x + 4)(x - 1)$

c. $x^2(x + 4)(x - 1)$ **d.** $x(x + 4)(x - 1)$

74. Add: $\frac{5x}{x^2 + 7x + 6} + \frac{3}{x^2 + 4x - 12}$.

a. $\frac{5x + 3}{2x + 11x - 6}$ **b.** $5x^2 - 7x + 3$

c. $\frac{5x^3 + 23x - 39x + 18}{(x+1)(x-2)(x+6)}$ **d.** $\frac{5x^2 - 7x + 3}{(x+1)(x+6)(x-2)}$

Getting Ready for the Next Section

Simplify.

75. $\frac{1}{2} \div \frac{2}{3}$

76. $\frac{1}{3} \div \frac{3}{4}$

77. $1 + \frac{1}{2}$

78. $1 + \frac{2}{3}$

79. $y^5 \cdot \frac{2x^3}{y^2}$

80. $y^7 \cdot \frac{3x^5}{y^4}$

81. $\frac{2x^3}{y^2} \cdot \frac{y^5}{4x}$

82. $\frac{3x^5}{y^4} \cdot \frac{y^7}{6x^2}$

Factor.

83. $x^2y + x$

84. $xy^2 + y$

Reduce.

85. $\frac{2x^3y^2}{4x}$

86. $\frac{3x^5y^3}{6x^2}$

87. $\frac{x^2 - 4}{x^2 - x - 6}$

88. $\frac{x^2 - 9}{x^2 - 5x + 6}$

Complex Fractions 7.4

Learning Objectives

In this section, we will learn how to:

1. Simplify complex fractions using the LCD method.
2. Simplify complex fractions using the division method.

Introduction

A complex fraction is a fraction or rational expression that contains other fractions in its numerator or denominator. Each of the following is a *complex fraction*:

$$\frac{\frac{1}{2}}{\frac{2}{3}} \qquad \frac{x + \frac{1}{y}}{y + \frac{1}{x}} \qquad \frac{\frac{a+1}{a^2-9}}{\frac{2}{a+3}}$$

We will begin this section by simplifying the first of these complex fractions.

VIDEO EXAMPLES

SECTION 7.4

EXAMPLE 1 Simplify $\frac{\frac{1}{2}}{\frac{2}{3}}$.

SOLUTION There are two methods we can use to solve this problem.

LCD Method

We can multiply the numerator and denominator of this complex fraction by the LCD for both fractions. In this case the LCD is 6:

$$\frac{\frac{1}{2}}{\frac{2}{3}} = \frac{6 \cdot \frac{1}{2}}{6 \cdot \frac{2}{3}} = \frac{3}{4}$$

Division Method

We can treat this as a division problem. Instead of dividing by $\frac{2}{3}$, we can multiply by its reciprocal $\frac{3}{2}$:

$$\frac{\frac{1}{2}}{\frac{2}{3}} = \frac{1}{2} \div \frac{2}{3} = \frac{1}{2} \cdot \frac{3}{2} = \frac{3}{4}$$

Using either method, we obtain the same result.

EXAMPLE 2 Simplify: $\dfrac{\dfrac{2x^3}{y^2}}{\dfrac{4x}{y^5}}$.

SOLUTION

LCD Method

The LCD for each rational expression is y^5. Multiplying the numerator and denominator of the complex fraction by y^5, we have:

$$\frac{\frac{2x^3}{y^2}}{\frac{4x}{y^5}} = \frac{y^5 \cdot \frac{2x^3}{y^2}}{y^5 \cdot \frac{4x}{y^5}} = \frac{2x^3y^3}{4x} = \frac{x^2y^3}{2}$$

Division Method

Instead of dividing by $\frac{4x}{y^5}$ we can multiply by its reciprocal, $\frac{y^5}{4x}$:

$$\frac{\frac{2x^3}{y^2}}{\frac{4x}{y^5}} = \frac{2x^3}{y^2} \div \frac{4x}{y^5} = \frac{2x^3}{y^2} \cdot \frac{y^5}{4x} = \frac{x^2y^3}{2}$$

Again the result is the same using either method.

EXAMPLE 3 Simplify: $\dfrac{x + \dfrac{1}{y}}{y + \dfrac{1}{x}}$

SOLUTION

Division Method

To apply the division method as we did in the first two examples, we first have to simplify the numerator and denominator separately to obtain a single rational expression for both before we can multiply by the reciprocal. Here are the steps:

$$\frac{x + \frac{1}{y}}{y + \frac{1}{x}} = \left(x + \frac{1}{y}\right) \div \left(y + \frac{1}{x}\right)$$

$$= \left(\frac{xy}{y} + \frac{1}{y}\right) \div \left(\frac{xy}{x} + \frac{1}{x}\right)$$

$$= \frac{xy + 1}{y} \div \frac{xy + 1}{x}$$

$$= \frac{xy + 1}{y} \cdot \frac{x}{xy + 1}$$

$$= \frac{x}{y}$$

LCD Method

It is much easier, in this case, to multiply the numerator and denominator by the LCD, which is xy:

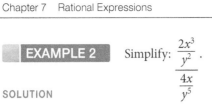

$$\frac{x + \frac{1}{y}}{y + \frac{1}{x}} = \frac{xy\left(x + \frac{1}{y}\right)}{xy\left(y + \frac{1}{x}\right)}$$ Multiply numerator and denominator by xy

$$= \frac{xy \cdot x + xy \cdot \frac{1}{y}}{xy \cdot y + xy \cdot \frac{1}{x}}$$ Distributive property

$$= \frac{x^2y + x}{xy^2 + y}$$ Simplify

We can factor x from $x^2y + x$ and y from $xy^2 + y$ and then reduce to lowest terms:

$$= \frac{x(xy + 1)}{y(xy + 1)}$$

$$= \frac{x}{y}$$

Both the division method and the LCD method can be used to simplify any complex fraction. In choosing which method to use, you may find these guidelines helpful:

> If the numerator and denominator of the complex fraction are single fractions, the division method is usually faster.
>
> If the numerator or denominator contains more than one fraction (two or more terms), the LCD method is typically faster.

EXAMPLE 4 Simplify: $\dfrac{1 - \frac{4}{x^2}}{1 - \frac{1}{x} - \frac{6}{x^2}}$

SOLUTION The easiest way to simplify this complex fraction is to multiply the numerator and denominator by the LCD, x^2:

$$\frac{1 - \frac{4}{x^2}}{1 - \frac{1}{x} - \frac{6}{x^2}} = \frac{x^2\left(1 - \frac{4}{x^2}\right)}{x^2\left(1 - \frac{1}{x} - \frac{6}{x^2}\right)}$$ Multiply numerator and denominator by x^2

$$= \frac{x^2 \cdot 1 - x^2 \cdot \frac{4}{x^2}}{x^2 \cdot 1 - x^2 \cdot \frac{1}{x} - x^2 \cdot \frac{6}{x^2}}$$ Distributive property

$$= \frac{x^2 - 4}{x^2 - x - 6}$$ Simplify

$$= \frac{(x - 2)(x + 2)}{(x - 3)(x + 2)}$$ Factor

$$= \frac{x - 2}{x - 3}$$ Reduce

In the introduction to Chapter 1, we mentioned a sequence of numbers that was related closely to the numbers in the Fibonacci sequence. In our next example, we find the relationship between that sequence and the numbers in the Fibonacci sequence.

EXAMPLE 5 Simplify each term in the following sequence, and then explain how this sequence is related to the Fibonacci sequence:

$$1+\frac{1}{1+1},\quad 1+\frac{1}{1+\frac{1}{1+1}},\quad 1+\frac{1}{1+\frac{1}{1+\frac{1}{1+1}}},\ldots$$

SOLUTION We can simplify our work somewhat if we notice that the first term $1+\frac{1}{1+1}$ is the denominator of the complex fraction in the second term and that the second term is the denominator of the complex fraction in the third term:

$$\text{First term: } 1+\frac{1}{1+1}=1+\frac{1}{2}=\frac{2}{2}+\frac{1}{2}=\frac{3}{2}$$

$$\text{Second term: } 1+\frac{1}{1+\frac{1}{1+1}}=1+\frac{1}{\frac{3}{2}}=1+\frac{2}{3}=\frac{3}{3}+\frac{2}{3}=\frac{5}{3}$$

$$\text{Third term: } 1+\frac{1}{1+\frac{1}{1+\frac{1}{1+1}}}=1+\frac{1}{\frac{5}{3}}=1+\frac{3}{5}=\frac{5}{5}+\frac{3}{5}=\frac{8}{5}$$

Here are the simplified numbers for the first three terms in our sequence:

$$\frac{3}{2},\frac{5}{3},\frac{8}{5},\ldots$$

Recall the Fibonacci sequence:

$$1, 1, 2, 3, 5, 8, 13, 21, \ldots$$

As you can see, each term in the sequence we have simplified is the ratio of two consecutive numbers in the Fibonacci sequence. If the pattern continues in this manner, the next number in our sequence will be $\frac{13}{8}$.

Getting Ready for Class

After reading through the preceding section, respond in your own words and in complete sentences.

A. What is a complex fraction?

B. Explain the division method of simplifying complex fractions.

C. How is a least common denominator used to simplify a complex fraction?

D. What types of complex fractions can be rewritten as division problems?

Problem Set 7.4

Simplify each complex fraction using either method.

1. $\dfrac{\frac{3}{4}}{\frac{1}{8}}$

2. $\dfrac{\frac{1}{3}}{\frac{5}{6}}$

3. $\dfrac{\frac{2}{3}}{4}$

4. $\dfrac{5}{\frac{1}{2}}$

5. $\dfrac{\frac{x^2}{y}}{\frac{x}{y^3}}$

6. $\dfrac{\frac{x^5}{y^3}}{\frac{x^2}{y^8}}$

7. $\dfrac{\frac{4x^3}{y^6}}{\frac{8x^2}{y^7}}$

8. $\dfrac{\frac{6x^4}{y}}{\frac{2x}{y^5}}$

9. $\dfrac{y + \frac{1}{x}}{x + \frac{1}{y}}$

10. $\dfrac{y - \frac{1}{x}}{x - \frac{1}{y}}$

11. $\dfrac{1 + \frac{1}{a}}{1 - \frac{1}{a}}$

12. $\dfrac{\frac{1}{a} - 1}{\frac{1}{a} + 1}$

13. $\dfrac{\frac{x + 1}{x^2 - 9}}{\frac{2}{x + 3}}$

14. $\dfrac{\frac{3}{x - 5}}{\frac{x + 1}{x^2 - 25}}$

15. $\dfrac{\frac{1}{a + 2}}{\frac{1}{a^2 - a - 6}}$

16. $\dfrac{\frac{1}{a^2 + 5a + 6}}{\frac{1}{a + 3}}$

17. $\dfrac{1 - \frac{9}{y^2}}{1 - \frac{1}{y} - \frac{6}{y^2}}$

18. $\dfrac{1 - \frac{4}{y^2}}{1 - \frac{2}{y} - \frac{8}{y^2}}$

19. $\dfrac{\frac{1}{y} + \frac{1}{x}}{\frac{1}{xy}}$

20. $\dfrac{\frac{1}{xy}}{\frac{1}{y} - \frac{1}{x}}$

21. $\dfrac{1 - \frac{1}{a^2}}{1 - \frac{1}{a}}$

22. $\dfrac{1 + \frac{1}{a}}{1 - \frac{1}{a^2}}$

23. $\dfrac{\frac{1}{10x} - \frac{y}{10x^2}}{\frac{1}{10} - \frac{y}{10x}}$

24. $\dfrac{\frac{1}{2x} + \frac{y}{2x^2}}{\frac{1}{4} + \frac{y}{4x}}$

25. $\dfrac{\frac{1}{a + 1} + 2}{\frac{1}{a + 1} + 3}$

26. $\dfrac{\frac{2}{a + 1} + 3}{\frac{3}{a + 1} + 4}$

27. Simplify each term in the following sequence.

$$2 + \frac{1}{2 + 1},\ 2 + \frac{1}{2 + \frac{1}{2 + 1}},\ 2 + \frac{1}{2 + \frac{1}{2 + \frac{1}{2 + 1}}},\ \ldots$$

28. Simplify each term in the following sequence.

$$2 + \frac{3}{2 + 3},\ 2 + \frac{3}{2 + \frac{3}{2 + 3}},\ 2 + \frac{3}{2 + \frac{3}{2 + \frac{3}{2 + 3}}},\ \ldots$$

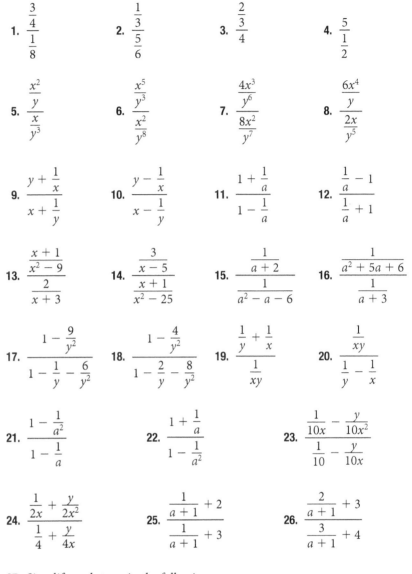

Complete the following tables.

29.

Number x	Reciprocal $\frac{1}{x}$	Quotient $\frac{x}{\frac{1}{x}}$	Square x^2
1			
2			
3			
4			

30.

Number x	Reciprocal $\frac{1}{x}$	Quotient $\frac{\frac{1}{x}}{x}$	Square x^2
1			
2			
3			
4			

31.

Number x	Reciprocal $\frac{1}{x}$	Sum $1 + \frac{1}{x}$	Quotient $\frac{1 + \frac{1}{x}}{\frac{1}{x}}$
1			
2			
3			
4			

32.

Number x	Reciprocal $\frac{1}{x}$	Difference $1 - \frac{1}{x}$	Quotient $\frac{1 - \frac{1}{x}}{\frac{1}{x}}$
1			
2			
3			
4			

Learning Objectives Assessment

The following problems can be used to help assess if you have successfully met the learning objectives for this section.

33. Simplify using the division method: $\dfrac{\dfrac{5x^2}{x^2 - 16}}{\dfrac{2x}{x + 4}}$.

a. $\dfrac{10x^3}{(x + 4)^2(x - 4)}$

b. $\dfrac{2(x - 4)}{5x}$

c. $\dfrac{5x}{2}$

d. $\dfrac{5x}{2(x - 4)}$

34. Simplify using the LCD method: $\dfrac{\dfrac{2}{3} + \dfrac{4}{x}}{\dfrac{3}{x} + \dfrac{1}{2}}$.

a. 2

b. 1

c. $\dfrac{4}{3}$

d. $\dfrac{3(x + 2)}{2(x + 3)}$

Getting Ready for the Next Section

Simplify.

35. $6\left(\dfrac{1}{2}\right)$

36. $10\left(\dfrac{1}{5}\right)$

37. $\dfrac{0}{5}$

38. $\dfrac{0}{2}$

39. $\dfrac{5}{0}$

40. $\dfrac{2}{0}$

41. $1 - \dfrac{5}{2}$

42. $1 - \dfrac{5}{3}$

Use the distributive property to simplify.

43. $6\left(\dfrac{x}{3} + \dfrac{5}{2}\right)$

44. $10\left(\dfrac{x}{2} + \dfrac{3}{5}\right)$

45. $x^2\left(1 - \dfrac{5}{x}\right)$

46. $x^2\left(1 - \dfrac{3}{x}\right)$

Solve.

47. $2x + 15 = 3$

48. $15 = 3x - 3$

49. $-2x - 9 = x - 3$

50. $a^2 - a - 22 = -2$

SPOTLIGHT ON SUCCESS *Instructor Edwin*

You never fail until you stop trying.
—Albert Einstein

Coming to the United States at the age of 10 and not knowing how to speak English was a very difficult hurdle to overcome. However, with hard work and dedication I was able to rise above those obstacles. When I came to the U.S. our school did not have a strong English development program as it was known at that time, English as a Second Language (ESL). The approach back then was "sink or swim." When my self-esteem was low, my mom and my three older sisters were always there for me and they would always encourage me to do well. My mom was a single parent, and her number one priority was that we would receive a good education. My mother's perseverance is what has made me the person I am today. At a young age I was able to see that she had overcome more than what my situation was, and I would always tell myself, "if Mom can do it, I could also do it." Not only did she not have an education, but she also saved us from a civil war that was happening in my home country of El Salvador.

When things in school got hard, I would always reflect on all the hard work, sacrifice and effort of my mother. I would just tell myself that I should not have any excuses and that I needed to keep going. If my mother, who worked as a housekeeper, could send all four of her kids to college doesn't motivate you, I don't know what does. It definitely motivated me. The day everything began to change for me was when I was in eighth grade. I was sitting in my biology class not paying attention to the teacher because I was really focusing on a piece of paper on the wall. It said, "You never fail until you stop trying." I read it over and over, trying to digest what the quote meant. With my limited English I was doing my best to translate what it meant in my native language. It finally clicked! I was able to figure out what those seven words meant. I memorized the quote and began to apply it to my academics and to real-life situations. I began to really focus in my studies. I wanted to do well in school, and most important I wanted to improve my English. To this day I always reflect to that quote when I feel I can't do something.

I was able to finish junior high successfully. Going to high school was a lot easier and I ended up with very good grades and eventually I was accepted to an excellent college. I was never the smartest student on campus, but I always did well because I never quit. I earned my college degree and now I teach at a dual immersion elementary school. I have that same quote in my classroom and I constantly remind my students to never stop trying.

Equations Involving Rational Expressions 7.5

Learning Objectives

In this section, we will learn how to:

1. Solve an equation containing rational expressions using the LCD.
2. Identify extraneous solutions to rational equations.

Introduction

The first step in solving an equation that contains one or more rational expressions is to find the LCD for all denominators in the equation. Once the LCD has been found, we multiply both sides of the equation by the LCD in order to clear fractions. The resulting equation should be equivalent to the original one (unless we inadvertently multiplied by zero) and free from any denominators except the number 1.

VIDEO EXAMPLES

SECTION 7.5

EXAMPLE 1 Solve $\frac{x}{3} + \frac{5}{2} = \frac{1}{2}$ for x.

SOLUTION The LCD for 3 and 2 is 6. If we multiply both sides by 6, we have:

$$6\left(\frac{x}{3} + \frac{5}{2}\right) = 6\left(\frac{1}{2}\right) \qquad \text{Multiply both sides by 6}$$

$$6\left(\frac{x}{3}\right) + 6\left(\frac{5}{2}\right) = 6\left(\frac{1}{2}\right) \qquad \text{Distributive property}$$

$$2x + 15 = 3$$

$$2x = -12$$

$$x = -6$$

We can check our solution by replacing x with -6 in the original equation:

$$-\frac{6}{3} + \frac{5}{2} \stackrel{?}{=} \frac{1}{2}$$

$$\frac{1}{2} = \frac{1}{2}$$

Multiplying both sides of an equation containing fractions by the LCD clears the equation of all denominators, because the LCD has the property that all denominators will divide it evenly.

EXAMPLE 2 Solve for x: $\frac{3}{x-1} = \frac{3}{5}$.

SOLUTION The LCD for $(x - 1)$ and 5 is $5(x - 1)$. Multiplying both sides by $5(x - 1)$, we have:

$$5(x-1) \cdot \frac{3}{x-1} = 5(x-1) \cdot \frac{3}{5}$$

$$5 \cdot 3 = (x-1) \cdot 3$$

$$15 = 3x - 3$$

$$18 = 3x$$

$$6 = x$$

If we substitute $x = 6$ into the original equation, we have:

$$\frac{3}{6-1} \stackrel{?}{=} \frac{3}{5}$$

$$\frac{3}{5} = \frac{3}{5}$$

The solution set is $\{6\}$.

EXAMPLE 3 Solve: $1 + \dfrac{6}{x^2} = \dfrac{5}{x}$.

SOLUTION The LCD is x^2. Multiplying both sides by x^2, we have

$$x^2\left(1 + \frac{6}{x^2}\right) = x^2\left(\frac{5}{x}\right) \qquad \text{Multiply both sides by } x^2$$

$$x^2(1) + x^2\left(\frac{6}{x^2}\right) = x^2\left(\frac{5}{x}\right) \qquad \text{Apply distributive property to the left side}$$

$$x^2 + 6 = 5x \qquad \text{Simplify each side}$$

We have a quadratic equation, which we write in standard form, factor, and solve as we did in Chapter 6.

$$x^2 - 5x + 6 = 0 \qquad \text{Standard form}$$

$$(x-2)(x-3) = 0 \qquad \text{Factor}$$

$$x - 2 = 0 \quad \text{or} \quad x - 3 = 0 \qquad \text{Set factors equal to 0}$$

$$x = 2 \quad \text{or} \quad x = 3$$

The two possible solutions are 2 and 3. Checking each in the original equation, we find they both give true statements. They are both solutions to the original equation:

Check $x = 2$

$$1 + \frac{6}{4} \stackrel{?}{=} \frac{5}{2}$$

$$1 + \frac{3}{2} \stackrel{?}{=} \frac{5}{2}$$

$$\frac{2}{2} + \frac{3}{2} \stackrel{?}{=} \frac{5}{2}$$

$$\frac{5}{2} = \frac{5}{2}$$

Check $x = 3$

$$1 + \frac{6}{9} \stackrel{?}{=} \frac{5}{3}$$

$$1 + \frac{2}{3} \stackrel{?}{=} \frac{5}{3}$$

$$\frac{3}{3} + \frac{2}{3} \stackrel{?}{=} \frac{5}{3}$$

$$\frac{5}{3} = \frac{5}{3}$$

EXAMPLE 4 Solve: $\frac{x}{x^2 - 9} - \frac{3}{x - 3} = \frac{1}{x + 3}$.

SOLUTION The factors of $x^2 - 9$ are $(x + 3)(x - 3)$, so the LCD is $(x + 3)(x - 3)$:

$$(x + 3)(x - 3) \cdot \frac{x}{(x + 3)(x - 3)} - (x + 3)(x - 3) \cdot \frac{3}{x - 3}$$

$$= (x + 3)(x - 3) \cdot \frac{1}{x + 3}$$

$$x - (x + 3)(3) = (x - 3)1$$

$$x - 3x - 9 = x - 3$$

$$-2x - 9 = x - 3$$

$$-3x = 6$$

$$x = -2$$

The solution is $x = -2$. It checks when replaced for x in the original equation.

EXAMPLE 5 Solve: $\frac{x}{x - 3} + \frac{3}{2} = \frac{3}{x - 3}$.

SOLUTION We begin by multiplying each term on both sides of the equation by $2(x - 3)$:

$$2(x - 3) \cdot \frac{x}{x - 3} + 2(x - 3) \cdot \frac{3}{2} = 2(x - 3) \cdot \frac{3}{x - 3}$$

$$2x + (x - 3) \cdot 3 = 2 \cdot 3$$

$$2x + 3x - 9 = 6$$

$$5x - 9 = 6$$

$$5x = 15$$

$$x = 3$$

Our only possible solution is $x = 3$. If we substitute $x = 3$ into the original equation, we get:

$$\frac{3}{3 - 3} + \frac{3}{2} \stackrel{?}{=} \frac{3}{3 - 3}$$

$$\frac{3}{0} + \frac{3}{2} = \frac{3}{0}$$

Two of the terms are undefined, so the equation is meaningless. Because $x = 3$ does not satisfy the equation, it is not a solution. Therefore, the equation has no solution, and the solution set is $\varnothing$.

The value of $x = 3$ in Example 5 is called an *extraneous solution*. When we multiply both sides of an equation by an LCD that contains a variable, we could be multiplying both sides by zero. The equation produced by doing this is not equivalent to the original equation, and it may result in extraneous solutions.

We always must check our solution when we multiply both sides of an equation by an expression containing the variable to make sure we have not multiplied both sides by zero.

EXAMPLE 6 Solve $\dfrac{a+4}{a^2+5a} = \dfrac{-2}{a^2-25}$ for a.

SOLUTION Factoring each denominator, we have:

$$a^2 + 5a = a(a+5)$$
$$a^2 - 25 = (a+5)(a-5)$$

The LCD is $a(a+5)(a-5)$. Multiplying both sides of the equation by the LCD gives us:

$$a(a+5)(a-5)\cdot\frac{a+4}{a(a+5)} = a(a+5)(a-5)\cdot\frac{-2}{(a+5)(a-5)}$$
$$(a-5)(a+4) = -2a$$
$$a^2 - a - 20 = -2a$$

The result is a quadratic equation, which we write in standard form, factor, and solve:

$$a^2 + a - 20 = 0 \qquad \text{Add } 2a \text{ to both sides}$$
$$(a+5)(a-4) = 0 \qquad \text{Factor}$$
$$a + 5 = 0 \quad \text{or} \quad a - 4 = 0 \qquad \text{Set factors equal to 0}$$
$$a = -5 \quad \text{or} \quad a = 4$$

The two possible solutions are -5 and 4. There is no problem with $a = 4$. It checks when substituted for a in the original equation. However, -5 is not a solution. Substituting -5 into the original equation gives:

$$\frac{-5+4}{(-5)^2+5(-5)} \stackrel{?}{=} \frac{-2}{(-5)^2-25}$$
$$\frac{-1}{0} = \frac{-2}{0}$$

Because $a = -5$ does not satisfy the original equation, it is an extraneous solution. The only valid solution to the equation is $a = 4$.

Getting Ready for Class

After reading through the preceding section, respond in your own words and in complete sentences.

A. What is the first step in solving an equation that contains rational expressions?

B. Explain how the LCD is used to clear an equation of fractions.

C. Why do we sometimes get extraneous solutions when solving rational equations?

D. How do we check for extraneous solutions to an equation containing rational expressions?

Solve the following equations. Be sure to check for extraneous solutions if you multiply both sides by an expression that contains the variable.

1. $\frac{x}{3} + \frac{1}{2} = -\frac{1}{2}$
2. $\frac{x}{2} + \frac{4}{3} = -\frac{2}{3}$
3. $\frac{4}{a} = \frac{1}{5}$

4. $\frac{2}{3} = \frac{6}{a}$
5. $\frac{3}{x} + 1 = \frac{2}{x}$
6. $\frac{4}{x} + 3 = \frac{1}{x}$

7. $\frac{3}{a} - \frac{2}{a} = \frac{1}{5}$
8. $\frac{7}{a} + \frac{1}{a} = 2$
9. $\frac{3}{x} + 2 = \frac{1}{2}$

10. $\frac{5}{x} + 3 = \frac{4}{3}$
11. $\frac{1}{y} - \frac{1}{2} = -\frac{1}{4}$
12. $\frac{3}{y} - \frac{4}{5} = -\frac{1}{5}$

13. $1 - \frac{8}{x} = \frac{-15}{x^2}$
14. $1 - \frac{3}{x} = \frac{-2}{x^2}$
15. $\frac{x}{2} - \frac{4}{x} = -\frac{7}{2}$

16. $\frac{x}{2} - \frac{5}{x} = -\frac{3}{2}$
17. $\frac{x-3}{2} + \frac{2x}{3} = \frac{5}{6}$
18. $\frac{x-2}{3} + \frac{5x}{2} = 5$

19. $\frac{x+1}{3} + \frac{x-3}{4} = \frac{1}{6}$
20. $\frac{x+2}{3} + \frac{x-1}{5} = -\frac{3}{5}$
21. $\frac{6}{x+2} = \frac{3}{5}$

22. $\frac{4}{x+3} = \frac{1}{2}$
23. $\frac{3}{y-2} = \frac{2}{y-3}$
24. $\frac{5}{y+1} = \frac{4}{y+2}$

25. $\frac{x}{x-2} + \frac{2}{3} = \frac{2}{x-2}$
26. $\frac{x}{x-5} + \frac{1}{5} = \frac{5}{x-5}$

27. $\frac{x}{x-2} + \frac{3}{2} = \frac{9}{2(x-2)}$
28. $\frac{x}{x+1} + \frac{4}{5} = \frac{-14}{5(x+1)}$

29. $\frac{5}{x+2} + \frac{1}{x+3} = \frac{-1}{x^2+5x+6}$
30. $\frac{3}{x-1} + \frac{2}{x+3} = \frac{-3}{x^2+2x-3}$

31. $\frac{8}{x^2-4} + \frac{3}{x+2} = \frac{1}{x-2}$
32. $\frac{10}{x^2-25} - \frac{1}{x-5} = \frac{3}{x+5}$

33. $\frac{a}{2} + \frac{3}{a-3} = \frac{a}{a-3}$
34. $\frac{a}{2} + \frac{4}{a-4} = \frac{a}{a-4}$

35. $\frac{6}{y^2-4} = \frac{4}{y^2+2y}$
36. $\frac{2}{y^2-9} = \frac{5}{y^2+3y}$

37. $\frac{2}{a^2-9} = \frac{3}{a^2+a-12}$
38. $\frac{2}{a^2-1} = \frac{6}{a^2-2a-3}$

39. $\frac{3x}{x-5} - \frac{2x}{x+1} = \frac{-42}{x^2-4x-5}$
40. $\frac{4x}{x-4} - \frac{3x}{x-2} = \frac{-3}{x^2-6x+8}$

41. $\frac{2x}{x+2} = \frac{x}{x+3} - \frac{3}{x^2+5x+6}$
42. $\frac{3x}{x-4} = \frac{2x}{x-3} + \frac{6}{x^2-7x+12}$

43. Solve each equation.

a. $5x - 1 = 0$
b. $\frac{5}{x} - 1 = 0$
c. $\frac{x}{5} - 1 = \frac{2}{3}$
d. $\frac{5}{x} - 1 = \frac{2}{3}$
e. $\frac{5}{x^2} + 5 = \frac{26}{x}$

44. Solve each equation.

a. $2x - 3 = 0$
b. $2 - \frac{3}{x} = 0$
c. $\frac{x}{3} - 2 = \frac{1}{2}$
d. $\frac{3}{x} - 2 = \frac{1}{2}$
e. $\frac{1}{x} + \frac{3}{x^2} = 2$

45. Paying Attention to Instructions Work each problem according to the instructions given.

a. Divide: $\frac{7}{a^2 - 5a - 6} \div \frac{a + 2}{a + 1}$

b. Add: $\frac{7}{a^2 - 5a - 6} + \frac{a + 2}{a + 1}$

c. Solve: $\frac{7}{a^2 - 5a - 6} + \frac{a + 2}{a + 1} = 2$

46. Paying Attention to Instructions Work each problem according to the instructions given.

a. Divide: $\frac{6}{x^2 - 9} \div \frac{x - 4}{x - 3}$

b. Add: $\frac{6}{x^2 - 9} + \frac{x - 4}{x - 3}$

c. Solve: $\frac{6}{x^2 - 9} + \frac{x - 4}{x - 3} = \frac{3}{4}$

Learning Objectives Assessment

The following problems can be used to help assess if you have successfully met the learning objectives for this section.

47. Solve: $\frac{2x}{x + 3} - \frac{1}{2} = \frac{6}{x + 3}$.

a. 5 **b.** $\frac{3(x - 5)}{2(x + 3)}$ **c.** 3 **d.** $\varnothing$

48. Which value is an extraneous solution to the following equation?

$$\frac{x}{x + 4} + \frac{1}{x - 4} = \frac{8}{(x + 4)(x - 4)}$$

a. -1 **b.** -4 **c.** 4 **d.** 0

Getting Ready for the Next Section

Solve.

49. $21 = 6x$ **50.** $72 = 2x$ **51.** $x^2 + x = 6$ **52.** $x^2 + 2x = 8$

Proportions 7.6

Learning Objectives

In this section, we will learn how to:

1. Solve a proportion.
2. Solve applied problems involving proportions.

Introduction

A proportion is two equal ratios; that is, if $\frac{a}{b}$ and $\frac{c}{d}$ are ratios, then:

$$\frac{a}{b} = \frac{c}{d}$$

is a proportion.

Each of the four numbers in a proportion is called a *term* of the proportion. We number the terms as follows:

$$\begin{array}{r} \text{First term} \rightarrow a \\ \text{Second term} \rightarrow b \end{array} = \begin{array}{l} c \leftarrow \text{Third term} \\ d \leftarrow \text{Fourth term} \end{array}$$

The first and fourth terms are called the *extremes*, and the second and third terms are called the *means:*

$$\text{Means} \quad \frac{a}{b} = \frac{c}{d} \quad \text{Extremes}$$

For example, in the proportion:

$$\frac{3}{8} = \frac{12}{32}$$

the extremes are 3 and 32, and the means are 8 and 12.

PROPERTY ***Means-Extremes Property***

Let a, b, c, and d be real numbers with $b \neq 0$ and $d \neq 0$.

If $\quad \frac{a}{b} = \frac{c}{d}$

then $\quad ad = bc$

In words: In any proportion, the product of the extremes is equal to the product of the means.

This property of proportions comes from the multiplication property of equality. We can use it to solve for a missing term in a proportion.

VIDEO EXAMPLES

SECTION 7.6

EXAMPLE 1 Solve the proportion $\frac{3}{x} = \frac{6}{7}$ for x.

SOLUTION We could solve for x by using the method developed in Section 7.5; that is, multiplying both sides by the LCD of $7x$. Instead, let's use our new means-extremes property:

$$\frac{3}{x} = \frac{6}{7} \qquad \text{Extremes are 3 and 7; means are } x \text{ and 6}$$

$$21 = 6x \qquad \text{Product of extremes = product of means}$$

$$\frac{21}{6} = x \qquad \text{Divide both sides by 6}$$

$$x = \frac{7}{2} \qquad \text{Reduce to lowest terms}$$

If x is replaced with $\frac{7}{2}$ in the proportion, the result is a true statement. Therefore, the solution set is $\left\{\frac{7}{2}\right\}$.

EXAMPLE 2 Solve for x: $\frac{x+1}{2} = \frac{3}{x}$.

SOLUTION Again, we want to point out that we could solve for x by using the method we used in Section 7.5. Using the means-extremes property is simply an alternative to the method developed in Section 7.5:

$$\frac{x+1}{2} = \frac{3}{x} \qquad \text{Extremes are } x + 1 \text{ and } x\text{; means are 2 and 3}$$

$$x^2 + x = 6 \qquad \text{Product of extremes = product of means}$$

$$x^2 + x - 6 = 0 \qquad \text{Standard form for a quadratic equation}$$

$$(x + 3)(x - 2) = 0 \qquad \text{Factor}$$

$$x + 3 = 0 \quad \text{or} \quad x - 2 = 0 \qquad \text{Set factors equal to 0}$$

$$x = -3 \quad \text{or} \quad x = 2$$

Because both values satisfy the proportion, we have two solutions: -3 and 2.

Applications

EXAMPLE 3 A manufacturer knows that during a production run, 8 out of every 100 parts produced by a certain machine will be defective. If the machine produces 1,450 parts, how many can be expected to be defective?

SOLUTION The ratio of defective parts to total parts produced is $\frac{8}{100}$. If we let x represent the number of defective parts out of the total of 1,450 parts, then we can write this ratio again as $\frac{x}{1{,}450}$. This gives us a proportion to solve:

$$\text{Defective parts in numerator / Total parts in denominator} \qquad \frac{x}{1{,}450} = \frac{8}{100} \qquad \text{Extremes are } x \text{ and 100; means are 1,450 and 8}$$

$$100x = 11{,}600 \qquad \text{Product of extremes = product of means}$$

$$x = 116$$

The manufacturer can expect 116 defective parts out of the total of 1,450 parts if the machine usually produces 8 defective parts for every 100 parts it produces.

EXAMPLE 4 A woman drives her car 270 miles in 6 hours. If she continues at the same rate, how far will she travel in 10 hours?

SOLUTION We let x represent the distance traveled in 10 hours. Using x, we translate the problem into the following proportion:

$$\begin{matrix}\text{Miles} \longrightarrow \\ \text{Hours} \longrightarrow\end{matrix} \frac{x}{10} = \frac{270}{6} \begin{matrix}\longleftarrow \text{Miles} \\ \longleftarrow \text{Hours}\end{matrix}$$

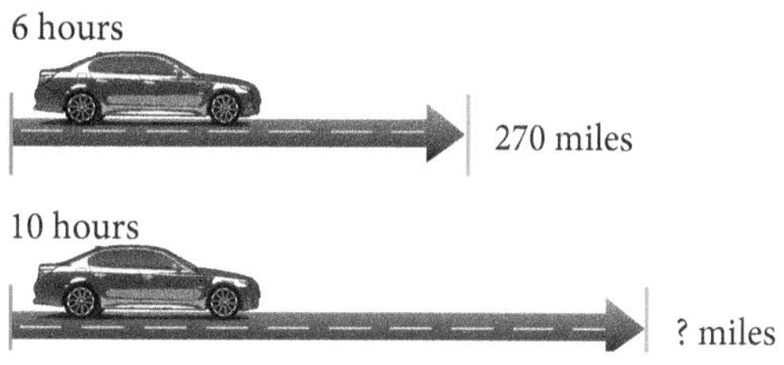

Notice that the two ratios in the proportion compare the same quantities. That is, both ratios compare miles to hours. In words this proportion says:

$$\underbrace{x \text{ miles is to 10 hours}}_{\frac{x}{10}} \text{ as } \underbrace{\text{270 miles is to 6 hours}}_{\frac{270}{6}}$$

$$\frac{x}{10} = \frac{270}{6}$$

Next, we solve the proportion.

$$6x = 2700$$

$$x = 450 \text{ miles}$$

If the woman continues at the same rate, she will travel 450 miles in 10 hours.

EXAMPLE 5 A baseball player gets 8 hits in the first 18 games of the season. If he continues at the same rate, how many hits will he get in 45 games?

SOLUTION We let x represent the number of hits he will get in 45 games. Then

$$\underbrace{x \text{ is to } 45}_{} \text{ as } \underbrace{8 \text{ is to } 18}_{}$$

$$\begin{matrix}\text{Hits} \rightarrow \\ \text{Games} \rightarrow\end{matrix} \frac{x}{45} = \frac{8}{18} \begin{matrix}\leftarrow \text{Hits} \\ \leftarrow \text{Games}\end{matrix}$$

Notice again that the two ratios are comparing the same quantities, hits to games. We solve the proportion as follows:

$$18x = 360$$

$$x = 20$$

If he continues to hit at the rate of 8 hits in 18 games, he will get 20 hits in 45 games.

Getting Ready for Class

After reading through the preceding section, respond in your own words and in complete sentences.

A. What is a proportion?

B. What are the means and extremes of a proportion?

C. What is the relationship between the means and the extremes in a proportion? (It is called the means-extremes property of proportions.)

D. Explain in your own words how to solve a proportion.

Problem Set 7.6

Solve each of the following proportions.

1. $\frac{x}{2} = \frac{6}{12}$
2. $\frac{x}{4} = \frac{6}{8}$
3. $\frac{2}{5} = \frac{4}{x}$
4. $\frac{3}{8} = \frac{9}{x}$
5. $\frac{10}{20} = \frac{20}{x}$
6. $\frac{15}{60} = \frac{60}{x}$
7. $\frac{a}{3} = \frac{5}{12}$
8. $\frac{a}{2} = \frac{7}{20}$
9. $\frac{2}{x} = \frac{6}{7}$
10. $\frac{4}{x} = \frac{6}{7}$
11. $\frac{x+1}{3} = \frac{4}{x}$
12. $\frac{x+1}{6} = \frac{7}{x}$
13. $\frac{x}{2} = \frac{8}{x}$
14. $\frac{x}{9} = \frac{4}{x}$
15. $\frac{4}{a+2} = \frac{a}{2}$
16. $\frac{3}{a+2} = \frac{a}{5}$
17. $\frac{1}{x} = \frac{x-5}{6}$
18. $\frac{1}{x} = \frac{x-6}{7}$
19. $\frac{26}{x-3} = \frac{38}{x+3}$
20. $\frac{9}{x-2} = \frac{11}{x+2}$

Applying the Concepts

21. **Baseball** A baseball player gets 6 hits in the first 18 games of the season. If he continues hitting at the same rate, how many hits will he get in the first 45 games?
22. **Basketball** A basketball player makes 8 of 12 free throws in the first game of the season. If she shoots with the same accuracy in the second game, how many of the 15 free throws she attempts will she make?
23. **Mixture Problem** A solution contains 12 milliliters of alcohol and 16 milliliters of water. If another solution is to have the same concentration of alcohol in water but is to contain 28 milliliters of water, how much alcohol must it contain?
24. **Mixture Problem** A solution contains 15 milliliters of HCl and 42 milliliters of water. If another solution is to have the same concentration of HCl in water but is to contain 140 milliliters of water, how much HCl must it contain?
25. **Nutrition** If 100 grams of ice cream contains 13 grams of fat, how much fat is in 350 grams of ice cream?
26. **Nutrition** A 6-ounce serving of grapefruit juice contains 159 grams of water. How many grams of water are in 20 ounces of grapefruit juice?
27. **Map Reading** A map is drawn so that every 3.5 inches on the map corresponds to an actual distance of 100 miles. If the actual distance between the two cities is 420 miles, how far apart are they on the map?

28. Map Reading The scale on a map indicates that 1 inch on the map corresponds to an actual distance of 105 miles. Two cities are 4.5 inches apart on the map. What is the actual distance between the two cities?

29. Distance A man drives his car 245 miles in 5 hours. At this rate, how far will he travel in 7 hours?

30. Distance An airplane flies 1,380 miles in 3 hours. How far will it fly in 5 hours?

Learning Objectives Assessment

The following problems can be used to help assess if you have successfully met the learning objectives for this section.

31. Solve: $\frac{4}{x+3} = \frac{x}{10}$.

a. 8, −5 **b.** 5 **c.** 8 **d.** −8, 5

32. Nutrition If 12 ounces of a popular soda contain 45 milligrams of sodium, how many milligrams of sodium will 34 ounces of the soda contain?

a. 96 **b.** 67 **c.** 127.5 **d.** 9.1

Getting Ready for the Next Section

Solve.

33. $\frac{1}{x} + \frac{1}{2x} = \frac{9}{2}$

34. $\frac{50}{x+5} = \frac{30}{x-5}$

35. $\frac{1}{10} - \frac{1}{15} = \frac{1}{x}$

36. $\frac{15}{x} + \frac{15}{x+20} = 2$

Find the value of $y = \frac{-6}{x}$ for the given value of x.

37. $x = -6$

38. $x = -3$

39. $x = 2$

40. $x = 1$

Applications 7.7

Learning Objectives

In this section, we will learn how to:

1. Solve number problems involving rational expressions.
2. Solve motion problems involving rational expressions.
3. Solve work problems involving rational expressions.

In this section we will solve some word problems whose equations involve rational expressions. Like the other word problems we have encountered, the more you work with them, the easier they become.

Number Problems

VIDEO EXAMPLES

SECTION 7.7

EXAMPLE 1 One number is twice another. The sum of their reciprocals is $\frac{9}{2}$. Find the two numbers.

SOLUTION Let x represent the smaller number. The larger then must be $2x$. Their reciprocals are $\frac{1}{x}$ and $\frac{1}{2x}$, respectively. An equation that describes the situation is:

$$\frac{1}{x} + \frac{1}{2x} = \frac{9}{2}$$

We can multiply both sides by the LCD of $2x$ and then solve the resulting equation:

$$2x\left(\frac{1}{x}\right) + 2x\left(\frac{1}{2x}\right) = 2x\left(\frac{9}{2}\right)$$

$$2 + 1 = 9x$$

$$3 = 9x$$

$$x = \frac{3}{9} = \frac{1}{3}$$

The smaller number is $\frac{1}{3}$. The other number is twice as large, or $\frac{2}{3}$. If we add their reciprocals, we have:

$$\frac{3}{1} + \frac{3}{2} = \frac{6}{2} + \frac{3}{2} = \frac{9}{2}$$

The solutions check with the original problem.

Motion Problems

Recall from Section 7.1 that if an object travels at a constant rate r for a specified time t, then the distance traveled is given by the rate equation

$$\text{Distance} = \text{Rate} \cdot \text{Time}$$

$$d = rt$$

If we know the distance traveled and the rate, then we can find the time by dividing the distance by the rate:

$$\text{Time} = \frac{\text{Distance}}{\text{Rate}}$$

$$t = \frac{d}{r}$$

The next two examples use this version of the rate equation to solve problems involving motion.

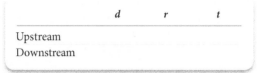

EXAMPLE 2 A boat travels 30 miles up a river in the same amount of time it takes to travel 50 miles down the same river. If the current is 5 miles per hour, what is the speed of the boat in still water?

SOLUTION The easiest way to work a problem like this is with a table. The top row of the table is labeled with d for distance, r for rate, and t for time. The left column of the table is labeled with the two trips: upstream and downstream. Here is what the table looks like:

	d	r	t
Upstream			
Downstream			

The next step is to read the problem over again and fill in as much of the table as we can with the information in the problem. The distance the boat travels upstream is 30 miles and the distance downstream is 50 miles. Since we are asked for the speed of the boat in still water, we will let that be x. If the speed of the boat in still water is x, then its speed upstream (against the current) must be $x - 5$, and its speed downstream (with the current) must be $x + 5$. Putting these four quantities into the appropriate positions in the table, we have

	d	r	t
Upstream	30	$x - 5$	
Downstream	50	$x + 5$	

Note There are two things to note about this problem. The first is that to solve the equation $d = r \cdot t$ for t, we divide each side by r, like this:

$$\frac{d}{r} = \frac{r \cdot t}{r}$$
$$\frac{d}{r} = t$$

Secondly, the speed of the boat in still water is the rate at which it would be traveling if there were no current; that is, it is the speed of the boat through the water. Since the water itself is moving at 5 miles per hour, the boat is going 5 miles per hour slower when it travels against the current and 5 miles per hour faster when it travels with the current.

The last positions in the table are filled in by using the equation $t = \dfrac{d}{r}$.

	d	r	t
Upstream	30	$x - 5$	$\frac{30}{x-5}$
Downstream	50	$x + 5$	$\frac{50}{x+5}$

Reading the problem again, we find that the time for the trip upstream is equal to the time for the trip downstream. Setting these two quantities equal to each other, we have our equation:

$$\text{time (downstream)} = \text{time (upstream)}$$
$$\frac{50}{x+5} = \frac{30}{x-5}$$

Notice the resulting equation is a proportion. We can solve for x using the means-extremes property.

$$50(x - 5) = 30(x + 5)$$

$$50x - 250 = 30x + 150$$

$$20x = 400$$

$$x = 20$$

The speed of the boat in still water is 20 miles per hour.

EXAMPLE 3 Tina is training for a biathlon. To train for the bicycle portion, she rides her bike 15 miles up a hill and then 15 miles back down the same hill. The complete trip takes her 2 hours. If her downhill speed is 20 miles per hour faster than her uphill speed, how fast does she ride uphill?

SOLUTION Again, we make a table. As in the previous example, we label the top row with distance, rate, and time. We label the left column with the two trips, uphill and downhill.

	d	r	t
Uphill			
Downhill			

Next, we fill in the table with as much information as we can from the problem. We know the distance traveled is 15 miles uphill and 15 miles downhill, which allows us to fill in the distance column. To fill in the rate column, we first note that she rides 20 miles per hour faster downhill than uphill. Therefore, if we let x be her rate uphill, then her rate downhill is $x + 20$. Filling in the table with this information gives us

	d	r	t
Uphill	15	x	
Downhill	15	$x + 20$	

Since time is distance divided by rate, $t = \frac{d}{r}$, we can fill in the last column in the table.

	d	r	t
Uphill	15	x	$\frac{15}{x}$
Downhill	15	$x + 20$	$\frac{15}{x + 20}$

Rereading the problem, we find that the total time (the time riding uphill plus the time riding downhill) is two hours. We write our equation as follows:

$$\text{time (uphill)} + \text{time (downhill)} = 2$$

$$\frac{15}{x} \quad + \quad \frac{15}{x + 20} \quad = 2$$

We solve this equation for x by first finding the LCD and then multiplying each term in the equation by it to clear the equation of all denominators. The LCD is $x(x + 20)$. Here is our solution:

$$x(x + 20)\frac{15}{x} + x(x + 20)\frac{15}{x + 20} = 2 \cdot [x(x + 20)]$$

$$x(x + 20)\frac{15}{x} + x(x + 20)\frac{15}{x + 20} = 2 \cdot [x(x + 20)]$$

$$15(x + 20) + 15x = 2x(x + 20)$$

$$15x + 300 + 15x = 2x^2 + 40x$$

$$0 = 2x^2 + 10x - 300$$

$$0 = x^2 + 5x - 150 \qquad \text{Divide both sides by 2}$$

$$0 = (x + 15)(x - 10)$$

$$x + 15 = 0 \quad \text{or} \quad x - 10 = 0$$

$$x = -15 \qquad\qquad x = 10$$

Since we cannot have a negative speed, the only valid solution is $x = 10$. Tina rides her bike at a rate of 10 miles per hour when going uphill. (Her downhill speed is $x + 20 = 30$ miles per hour.)

Work Problems

Work problems involve two or more people or objects attempting to complete some task. If a person or object works at a constant rate, and the time it takes them to complete the task is given by t, then the fraction of the task completed in one unit of time will be $\frac{1}{t}$. That is, the rate at which they work is found by taking the reciprocal of the time it takes to do the job.

EXAMPLE 4 Allison can clean the house in 5 hours. Working together, she and Kaitlin can clean the house in 3 hours. How long would it take Kaitlin, working alone, to clean the house?

SOLUTION Let x be the amount of time it takes Kaitlin to clean the house. We now consider the fraction of the task completed in 1 hour for each girl working alone, and for the girls working together.

1. If Allison can clean the house in 5 hours, then in 1 hour she cleans $\frac{1}{5}$ of the house.
2. If Kaitlin can clean the house in x hours, then in 1 hour she cleans $\frac{1}{x}$ of the house.
3. If it takes 3 hours for both girls to clean the house working together, then in 1 hour they clean $\frac{1}{3}$ of the house.

Therefore, in 1 hour we have

$$\frac{1}{5} + \frac{1}{x} = \frac{1}{3}$$

Amount of the house cleaned by Allison ($\frac{1}{5}$); Amount of the house cleaned by Kaitlin ($\frac{1}{x}$); Amount of the house cleaned by both girls ($\frac{1}{3}$)

Multiplying both sides of the equation by the LCD of $15x$, we have:

$$15x\left(\frac{1}{5}\right) + 15x\left(\frac{1}{x}\right) = 15x\left(\frac{1}{3}\right)$$

$$3x + 15 = 5x$$

$$15 = 2x$$

$$7.5 = x$$

It would take Kaitlin 7.5 hours to clean the house by herself.

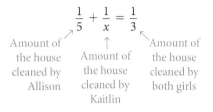

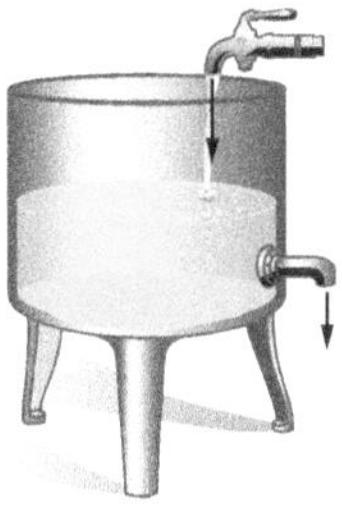

EXAMPLE 5 An inlet pipe can fill a water tank in 10 hours, while an outlet pipe can empty the same tank in 15 hours. By mistake, both pipes are left open. How long will it take to fill the water tank with both pipes open?

SOLUTION Let x be the amount of time to fill the tank with both pipes open.

One method of solving this type of problem is to think in terms of how much of the job is done by a pipe in 1 hour.

1. If the inlet pipe fills the tank in 10 hours, then in 1 hour the inlet pipe fills $\frac{1}{10}$ of the tank.
2. If the outlet pipe empties the tank in 15 hours, then in 1 hour the outlet pipe empties $\frac{1}{15}$ of the tank.
3. If it takes x hours to fill the tank with both pipes open, then in 1 hour the tank is $\frac{1}{x}$ full.

Here is how we set up the equation. In 1 hour,

$$\frac{1}{10} - \frac{1}{15} = \frac{1}{x}$$

Amount of water let in by inlet pipe ($\frac{1}{10}$); Amount of water let out by outlet pipe ($\frac{1}{15}$); Total amount of water in tank ($\frac{1}{x}$)

The LCD for our equation is $30x$. We multiply both sides by the LCD and solve:

$$30x\left(\frac{1}{10}\right) - 30x\left(\frac{1}{15}\right) = 30x\left(\frac{1}{x}\right)$$

$$3x - 2x = 30$$

$$x = 30$$

It takes 30 hours with both pipes open to fill the tank.

Note In solving a problem of this type, we have to assume that the thing doing the work (whether it is a pipe, a person, or a machine) is working at a constant rate; that is, as much work gets done in the first hour as is done in the last hour and any other hour in between.

Getting Ready for Class

After reading through the preceding section, respond in your own words and in complete sentences.

A. If we know the distance an object travels at a constant rate, how can we find the time it traveled?

B. How does the current of a river affect the speed of a motor boat traveling against the current?

C. How does the current of a river affect the speed of a motor boat traveling in the same direction as the current?

D. What is the relationship between the total number of minutes it takes for a drain to empty a sink and the amount of water that drains out of the sink in 1 minute?

Problem Set 7.7

Number Problems

1. One number is 3 times as large as another. The sum of their reciprocals is $\frac{16}{3}$. Find the two numbers.
2. If $\frac{3}{5}$ is added to twice the reciprocal of a number, the result is 1. Find the number.
3. The sum of a number and its reciprocal is $\frac{13}{6}$. Find the number.
4. The sum of a number and 10 times its reciprocal is 7. Find the number.
5. If a certain number is added to both the numerator and denominator of the fraction $\frac{7}{9}$, the result is $\frac{5}{7}$. Find the number.
6. The numerator of a certain fraction is 2 more than the denominator. If $\frac{1}{3}$ is added to the fraction, the result is 2. Find the fraction.
7. The sum of the reciprocals of two consecutive even integers is $\frac{5}{12}$. Find the integers.
8. The sum of the reciprocals of two consecutive integers is $\frac{7}{12}$. Find the two integers.

Motion Problems

9. A boat travels 26 miles up the river in the same amount of time it takes to travel 38 miles down the same river. If the current is 3 miles per hour, what is the speed of the boat in still water?

	d	*r*	*t*
Upstream			
Downstream			

10. A boat can travel 9 miles up a river in the same amount of time it takes to travel 11 miles down the same river. If the current is 2 miles per hour, what is the speed of the boat in still water?

	d	*r*	*t*
Upstream			
Downstream			

11. An airplane flying against the wind travels 140 miles in the same amount of time it would take the same plane to travel 160 miles with the wind. If the wind speed is a constant 20 miles per hour, how fast would the plane travel in still air?
12. An airplane flying against the wind travels 500 miles in the same amount of time that it would take to travel 600 miles with the wind. If the speed of the wind is 50 miles per hour, what is the speed of the plane in still air?

13. One plane can travel 20 miles per hour faster than another. One of them goes 285 miles in the same time it takes the other to go 255 miles. What are their speeds?
14. One car travels 300 miles in the same amount of time it takes a second car, traveling 5 miles per hour slower than the first, to go 275 miles. What are the speeds of the cars?
15. Tina, whom we mentioned in Example 3 of this section, is training for a biathlon. To train for the running portion of the race, she runs 8 miles each day, over the same course. The first 2 miles of the course are on level ground, while the last 6 miles are downhill. She runs 3 miles per hour slower on level ground than she runs downhill. If the complete course takes 1 hour, how fast does she run on the downhill part of the course?
16. Jerri is training for the same biathlon as Tina (Example 3 and Problem 15). To train for the bicycle portion of the race, she rides 24 miles out a straight road, then turns around and rides 24 miles back. The trip out is against the wind, whereas the trip back is with the wind. If she rides 10 miles per hour faster with the wind than she does against the wind, and the complete trip out and back takes 2 hours, how fast does she ride when she rides against the wind?
17. To train for the running of a triathlon, Jerri jogs 1 hour each day over the same 9-mile course. Five miles of the course is downhill, whereas the other 4 miles is on level ground. Jerri figures that she runs 2 miles per hour faster downhill than she runs on level ground. Find the rate at which Jerri runs on level ground.
18. Travis paddles his kayak in the harbor at Morro Bay, California, where the incoming tide has caused a current in the water. From the point where he enters the water, he paddles 1 mile against the current, then turns around and paddles 1 mile back to where he started. His average speed when paddling with the current is 4 miles per hour faster than his speed against the current. If the complete trip (out and back) takes him 1.2 hours, find his average speed when he paddles against the current.

Work Problems

19. Jason can wax the family car in 90 minutes, while his brother Kevin can wax the car in 60 minutes. How long would it take Jason and Kevin to wax the car working together?
20. Logan can tile a bathroom in 4 hours. His partner Lance can do the job in 6 hours. How long would it take Logan and Lance to tile a bathroom working together?
21. Valerie can install a car stereo in 45 minutes. Working together with her trainee, Peggy, the installation only takes 30 minutes. How long does it take Peggy to do the installation on her own?
22. It takes Marie 15 hours to install wood flooring in a small house. If she works together with Curtis, they can do the installation in 9 hours. How long would it take Curtis to do the job by himself?
23. An inlet pipe can fill a pool in 12 hours, while an outlet pipe can empty it in 15 hours. If both pipes are left open, how long will it take to fill the pool?

24. A water tank can be filled in 20 hours by an inlet pipe and emptied in 25 hours by an outlet pipe. How long will it take to fill the tank if both pipes are left open?

25. A bathtub can be filled by the cold water faucet in 10 minutes and by the hot water faucet in 12 minutes. How long does it take to fill the tub if both faucets are open?

26. A water faucet can fill a sink in 12 minutes, whereas the drain can empty it in 6 minutes. If the sink is full, how long will it take to empty if both the faucet and the drain are open?

27. A sink can be filled by the cold water faucet in 3 minutes. The drain can empty a full sink in 4 minutes. If the sink is empty and both the cold water faucet and the drain are open, how long will it take the sink to fill?

28. A bathtub can be filled by the cold water faucet in 9 minutes and by the hot water faucet in 10 minutes. The drain can empty the tub in 5 minutes. Can the tub be filled if both faucets and the drain are open?

Learning Objectives Assessment

The following problems can be used to help assess if you have successfully met the learning objectives for this section.

29. The sum of a number and twice its reciprocal is $\frac{41}{12}$. Which equation can be used as a model to find the number?

a. $2\left(x + \frac{1}{x}\right) = \frac{41}{12}$

b. $x + \frac{1}{2x} = \frac{41}{12}$

c. $x + \frac{2}{x} = \frac{41}{12}$

d. $x + \frac{1}{x + 2} = \frac{41}{12}$

30. A kayaker can paddle 6 miles against a current in the same time it takes to paddle 10 miles with the current. If the current is 1 mile per hour, which equation can be used as a model to find the rate of the kayaker in still water?

a. $\frac{6}{x - 1} = \frac{10}{x + 1}$

b. $\frac{6}{x + 1} = \frac{10}{x - 1}$

c. $\frac{6}{1 - x} = \frac{10}{1 + x}$

d. $\frac{6}{1 + x} = \frac{10}{1 - x}$

31. A hot tub can be filled in 10 hours by the inlet pipe. If a hose is used, the hot tub can be filled in 8 hours. Which equation could be used to find the time required to fill the hot tub using both the inlet pipe and hose?

a. $\frac{1}{10} + \frac{1}{x} = \frac{1}{8}$

b. $10 + 8 = \frac{1}{x}$

c. $\frac{1}{10} - \frac{1}{8} = \frac{1}{x}$

d. $\frac{1}{10} + \frac{1}{8} = \frac{1}{x}$

Getting Ready for the Next Section

Use the formula $y = 5x$ to find y when

32. $x = 4$

33. $x = 3$

Use the formula $y = \frac{20}{x}$ to find y when

34. $x = 10$

35. $x = 5$

Use the formula $y = 2x^2$ to find x when

36. $y = 50$

37. $y = 72$

Use the formula $y = Kx$ to find K when

38. $y = 15$ and $x = 3$

39. $y = 72$ and $x = 4$

Use the formula $y = Kx^2$ to find K when

40. $y = 32$ and $x = 4$

41. $y = 45$ and $x = 3$

Variation 7.8

Learning Objectives

In this section, we will learn how to:

1. Express a direct variation in symbols.
2. Express an inverse variation in symbols.
3. Solve variation problems.
4. Solve application problems involving direct or inverse variation.

Direct Variation

Two variables are said to *vary directly* if one is a constant multiple of the other. For instance, y varies directly as x if $y = Kx$, where K is a constant. The constant K is called the *constant of variation*. The following examples give the relationship between direct variation statements and their equivalent algebraic equations.

EXAMPLES

Statement	Equation
1. y varies directly as x	$y = Kx$
2. y varies directly as the square of x	$y = Kx^2$
3. s varies directly as the square root of t	$s = K\sqrt{t}$
4. r varies directly as the cube of s	$r = Ks^3$

Any time we run across a statement similar to those in the previous examples, we immediately can write an equivalent expression involving variables and a constant of variation K.

VIDEO EXAMPLES

SECTION 7.8

EXAMPLE 5 Suppose y varies directly as x. When y is 15, x is 3. Find y when x is 4.

SOLUTION From the first sentence we can write the relationship between x and y as

$$y = Kx$$

We now use the second sentence to find the value of K. Since y is 15 when x is 3, we have

$$15 = K(3) \qquad \text{or} \qquad K = 5$$

Now we can rewrite the relationship between x and y more specifically as

$$y = 5x$$

To find the value of y when x is 4 we simply substitute $x = 4$ into our last equation.

Substituting $x = 4$
into $y = 5x$
we have $y = 5(4)$
$y = 20$

EXAMPLE 6 Suppose y varies directly as the square of x. When x is 4, y is 32. Find x when y is 50.

SOLUTION The first sentence gives us

$$y = Kx^2$$

Since y is 32 when x is 4, we have

$$32 = K(4)^2$$

$$32 = 16K$$

$$K = 2$$

The equation now becomes

$$y = 2x^2$$

When y is 50, we have

$$50 = 2x^2$$

$$25 = x^2$$

$$x = \pm 5$$

There are two possible solutions, $x = 5$ or $x = -5$.

Inverse Variation

Two variables are said to *vary inversely* if one is a constant multiple of the reciprocal of the other. For example, y varies inversely as x if $y = \frac{K}{x}$, where K is a real number constant. Again, K is called the constant of variation. The examples that follow give some inverse variation statements and their associated algebraic equations.

EXAMPLES

Statement	Equation
7. y varies inversely as x	$y = \frac{K}{x}$
8. y varies inversely as the square of x	$y = \frac{K}{x^2}$
9. F varies inversely as the square root of t	$F = \frac{K}{\sqrt{t}}$
10. r varies inversely as the cube of s	$r = \frac{K}{s^3}$

EXAMPLE 11 Suppose y varies inversely as x. When y is 4, x is 5. Find y when x is 10.

SOLUTION The first sentence gives us the relationship between x and y:

$$y = \frac{K}{x}$$

We use the second sentence to find the value of the constant K:

$$4 = \frac{K}{5} \quad \text{or} \quad K = 20$$

We can now write the relationship between x and y more specifically as:

$$y = \frac{20}{x}$$

We use this equation to find the value of y when x is 10.

$$\begin{aligned} \text{Substituting} \quad & x = 10 \\ \text{into} \quad & y = \frac{20}{x} \\ \text{we have} \quad & y = \frac{20}{10} \\ & y = 2 \end{aligned}$$

Applications

EXAMPLE 12 The cost of a certain kind of candy varies directly with the weight of the candy. If 12 ounces of the candy cost \$1.68, how much will 16 ounces cost?

SOLUTION Let x = the number of ounces of candy and y = the cost of the candy. Then $y = Kx$. Since y is 1.68 when x is 12, we have

$$\begin{aligned} 1.68 &= K \cdot 12 \\ K &= \frac{1.68}{12} \\ &= 0.14 \end{aligned}$$

The equation must be

$$y = 0.14x$$

When x is 16, we have

$$\begin{aligned} y &= 0.14(16) \\ &= 2.24 \end{aligned}$$

The cost of 16 ounces of candy is \$2.24.

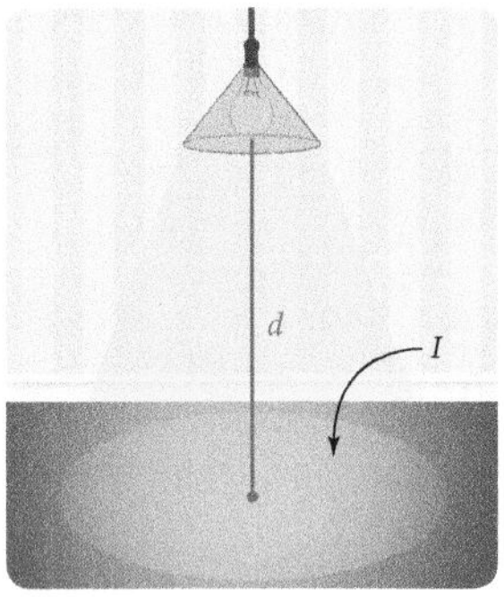

EXAMPLE 13 The intensity (I) of light from a source varies inversely as the square of the distance (d) from the source. Ten feet away from the source the intensity is 200 footcandles. What is the intensity 5 feet from the source?

SOLUTION

$$I = \frac{K}{d^2}$$

Since $I = 200$ when $d = 10$, we have

$$200 = \frac{K}{10^2}$$

$$200 = \frac{K}{100}$$

$$K = 20{,}000$$

The equation becomes

$$I = \frac{20{,}000}{d^2}$$

When $d = 5$, we have

$$I = \frac{20{,}000}{5^2}$$

$$= \frac{20{,}000}{25}$$

$$= 800 \text{ footcandles}$$

Getting Ready for Class

After reading through the preceding section, respond in your own words and in complete sentences.

A. What does it mean when we say "y varies directly with x"?

B. Give an example of a sentence that is a direct variation statement.

C. Translate the equation $y = \frac{K}{x}$ into words.

D. Give an example of an everyday situation where one quantity varies inversely with another.

Determine whether each of the following equations represents a direct variation or inverse variation. Also, specify the constant of variation.

1. $y = 10x$ **2.** $E = 14c^2$ **3.** $t = \frac{40}{r}$ **4.** $R = \frac{1}{T}$

5. $V = \frac{4}{3}\pi r^3$ **6.** $b = \frac{1}{2\sqrt{n}}$ **7.** $I = \frac{1}{9d^2}$ **8.** $S = \frac{P}{4}$

Express each sentence algebraically as an equation. Use K for the constant of variation.

9. C varies directly as the square of r.
10. F varies inversely as the square of d.
11. P varies inversely as V.
12. y varies directly as the cube root of x.
13. R varies directly as the square root of n.
14. f varies inversely as the sum of a and b.

For each of the following problems, y varies directly as x.

15. If $y = 10$ when $x = 5$, find y when x is 4.
16. If $y = -18$ when $x = 6$, find y when x is 3.
17. If $y = 30$ when $x = -15$, find x when y is 8.
18. If $y = 30$ when $x = 4$, find y when x is 7.

For each of the following problems, y varies inversely as x.

19. If $y = 5$ when $x = 2$, find y when x is 5.
20. If $y = 2$ when $x = 1$, find y when x is 4.
21. If $y = 5$ when $x = 3$, find x when y is 15.
22. If $y = 15$ when $x = 2$, find x when y is 6.

Solve each of the following variation problems by first expressing the relationship algebraically as an equation.

23. If y varies directly as the square of x, and $y = 75$ when $x = 5$, find y when x is 1.

24. If m varies directly as the square of n, and $m = -72$ when $n = 6$, find m when n is 3.

25. If z varies inversely as the square of w, and $z = 5$ when $w = 2$, find z when w is 6.

26. If y varies inversely as the square of t, and $y = 4$ when $t = 3$, find y when t is 2.

27. If F varies directly as the square root of h, and $F = 24$ when $h = 4$, find F when h is 25.

28. If H varies inversely as the square root of d, and $H = 2$ when $d = 9$, find H when d is 4.

Applying the Concepts

29. **Tension in a Spring** The tension t in a spring varies directly with the distance d the spring is stretched. If the tension is 42 pounds when the spring is stretched 2 inches, find the tension when the spring is stretched twice as far.
30. **Fill Time** The time t it takes to fill a bucket varies directly with the volume g of the bucket. If it takes 1 minute to fill a 4-gallon bucket, how long will it take to fill a 6-gallon bucket?
31. **Electricity** The power P in an electric circuit varies directly with the square of the current I. If $P = 30$ when $I = 2$, find P when $I = 7$.
32. **Electricity** The resistance R in an electric circuit varies directly with the voltage V. If $R = 20$ when $V = 120$, find R when $V = 240$.
33. **Wages** The amount of money M a woman makes per week varies directly with the number of hours h she works per week. If she works 20 hours and earns \$185, how much does she make if she works 30 hours?
34. **Volume** The volume V of a gas varies directly as the temperature T. If $V = 3$ when $T = 150$, find V when T is 200.
35. **Weight** The weight F of a body varies inversely with the square of the distance d between the body and the center of the Earth. If a man weighs 150 pounds 4,000 miles from the center of the Earth, how much will he weigh at a distance of 5,000 miles from the center of the Earth?
36. **Light Intensity** The intensity I of a light source varies inversely with the square of the distance d from the source. Four feet from the source, the intensity is 9 footcandles. What is the intensity 3 feet from the source?
37. **Electricity** The current I in an electric circuit varies inversely with the resistance R. If a current of 30 amperes is produced by a resistance of 2 ohms, what current will be produced by a resistance of 5 ohms?
38. **Pressure** The pressure exerted by a gas on the container in which it is held varies inversely with the volume of the container. A pressure of 40 pounds per square inch is exerted on a container of volume 2 cubic feet. What is the pressure on a container whose volume is 8 cubic feet?

Learning Objectives Assessment

The following problems can be used to help assess if you have successfully met the learning objectives for this section.

39. Express as an equation: "T varies directly as r."

    a. $T = \frac{1}{r}$ **b.** $T = r$ **c.** $T = Kr$ **d.** $T = \frac{K}{r}$

40. Express as an equation: "f varies inversely as the square of h."

    a. $f = Kh^2$ **b.** $f = \frac{K}{h^2}$ **c.** $f = \frac{1}{\sqrt{h}}$ **d.** $f = \sqrt{h}$

41. If y varies directly as the square of x, and $y = 27$ when $x = 3$, find y when x is 5.

    a. 75 **b.** 45 **c.** $\frac{81}{5}$ **d.** $\frac{243}{25}$

42. The current I in an electric circuit varies inversely with the resistance R. If a current of 36 amperes is produced by a resistance of 2 ohms, what current will be produced by a resistance of 9 ohms?

a. 8 amperes **b.** 2 amperes **c.** 16 amperes **d.** 162 amperes

Maintaining Your Skills

43. Reduce to lowest terms: $\frac{x^2 - x - 6}{x^2 - 9}$.

44. Divide using long division: $\frac{x^2 - 2x + 6}{x - 4}$.

Perform the indicated operations.

45. $\frac{x^2 - 25}{x + 4} \cdot \frac{2x + 8}{x^2 - 9x + 20}$

46. $\frac{3x + 6}{x^2 - 4x + 3} \div \frac{x^2 + x - 2}{x^2 + 2x - 3}$

47. $\frac{x}{x^2 - 16} + \frac{4}{x^2 - 16}$

48. $\frac{2}{x^2 - 1} - \frac{5}{x^2 + 3x - 4}$

49. $\dfrac{1 - \frac{25}{x^2}}{1 - \frac{8}{x} + \frac{15}{x^2}}$

Solve each equation.

50. $\frac{x}{2} - \frac{5}{x} = -\frac{3}{2}$

51. $\frac{x}{x^2 - 9} - \frac{3}{x - 3} = \frac{1}{x + 3}$

52. Speed of a Boat A boat travels 30 miles up a river in the same amount of time it takes to travel 50 miles down the same river. If the current is 5 miles per hour, what is the speed of the boat in still water?

53. Filling a Pool A pool can be filled by an inlet pipe in 8 hours. The drain will empty the pool in 12 hours. How long will it take to fill the pool if both the inlet pipe and the drain are open?

54. Mixture Problem If 30 liters of a certain solution contains 2 liters of alcohol, how much alcohol is in 45 liters of the same solution?

55. y varies directly with x. If $y = 8$ when x is 12, find y when x is 36.

Chapter 7 Summary

EXAMPLES

1. We reduce rational expressions to lowest terms by factoring the numerator and denominator and then dividing out any factors they have in common:

$$\frac{x-3}{x^2-9} = \frac{x-3}{(x-3)(x+3)} = \frac{1}{x+3}$$

Rational Expressions [7.1]

Any expression of the form $\frac{P}{Q}$, where P and Q are polynomials ($Q \neq 0$), is a rational expression.

Multiplying or dividing the numerator and denominator of a rational expression by the same nonzero quantity always produces a rational expression equivalent to the original one.

2. The rational expression

$$\frac{x-3}{x^2-9}$$

is undefined if

$$x^2 - 9 = 0$$

Solving for x, we have

$$(x+3)(x-3) = 0$$

$$x + 3 = 0 \quad \text{or} \quad x - 3 = 0$$

$$x = -3 \qquad\qquad x = 3$$

The restrictions are $x \neq -3$, $x \neq 3$.

Restricting the Variable [7.1]

A rational expression is undefined only for values of the variable that make the denominator equal to zero. To find the restriction on the variable, we set the denominator equal to 0 and solve the resulting equation.

3. $$\frac{x-1}{x^2+2x-3} \cdot \frac{x^2-9}{x-2}$$

$$= \frac{x-1}{(x+3)(x-1)} \cdot \frac{(x-3)(x+3)}{x-2}$$

$$= \frac{x-3}{x-2}$$

Multiplication [7.2]

To multiply two rational expressions, multiply numerators, multiply denominators, and divide out any factors common to the numerator and denominator:

$$\text{For rational expressions } \frac{P}{Q} \text{ and } \frac{R}{S}, \quad \frac{P}{Q} \cdot \frac{R}{S} = \frac{PR}{QS}$$

4. $$\frac{2x}{x^2-25} \div \frac{4}{x-5}$$

$$= \frac{2x}{(x-5)(x+5)} \cdot \frac{(x-5)}{(2 \cdot 2)}$$

$$= \frac{x}{2(x+5)}$$

Division [7.2]

To divide by a rational expression, simply multiply by its reciprocal:

$$\text{For rational expressions } \frac{P}{Q} \text{ and } \frac{R}{S}, \quad \frac{P}{Q} \div \frac{R}{S} = \frac{P}{Q} \cdot \frac{S}{R} = \frac{PS}{QR}$$

5. The least common denominator for

$$\frac{3}{(x+1)(x-1)} \text{ and } \frac{4}{(x+1)^2}$$

is $(x+1)^2(x-1)$.

Least Common Denominator [7.3]

The least common denominator (LCD) for a set of denominators is the simplest expression that is evenly divisible by all the denominators. If the denominators are factored, then the LCD will be the product of each factor raised to the highest exponent it appears within any of the denominators.

Addition and Subtraction [7.3]

6. $\dfrac{4}{x-1} - \dfrac{x}{2}$

$= \dfrac{4}{x-1} \cdot \dfrac{2}{2} - \dfrac{x}{2} \cdot \dfrac{x-1}{x-1}$

$= \dfrac{8}{2(x-1)} - \dfrac{x^2-x}{2(x-1)}$

$= \dfrac{8-x^2+x}{2(x-1)}$

$= -\dfrac{x^2-x-8}{2(x-1)}$

To add or subtract two rational expressions, find a common denominator, change each expression to an equivalent expression having the common denominator, and then add/subtract numerators and reduce if possible:

For rational expressions $\dfrac{P}{S}$ and $\dfrac{Q}{S}$,

$$\frac{P}{S} + \frac{Q}{S} = \frac{P+Q}{S} \quad \text{and} \quad \frac{P}{S} - \frac{Q}{S} = \frac{P-Q}{S}$$

When subtracting, be sure to distribute the negative sign on the second numerator.

Complex Fractions [7.4]

7. $\dfrac{1-\dfrac{4}{x}}{x-\dfrac{16}{x}} = \dfrac{x\left(1-\dfrac{4}{x}\right)}{x\left(x-\dfrac{16}{x}\right)}$

$= \dfrac{x-4}{x^2-16}$

$= \dfrac{x-4}{(x-4)(x+4)}$

$= \dfrac{1}{x+4}$

A rational expression that contains a fraction in its numerator or denominator is called a complex fraction. The most common method of simplifying a complex fraction is to multiply the top and bottom by the LCD for all denominators.

Equations [7.5]

8. Solve $\dfrac{1}{2} + \dfrac{3}{x} = 5$.

$2x\left(\dfrac{1}{2}\right) + 2x\left(\dfrac{3}{x}\right) = 2x(5)$

$x + 6 = 10x$

$6 = 9x$

$x = \dfrac{2}{3}$

To solve equations involving rational expressions, first find the least common denominator (LCD) for all denominators. Then multiply both sides by the LCD and solve as usual. Check all solutions in the original equation to be sure there are no extraneous solutions.

Ratio and Proportion [7.1, 7.6]

9. Solve for x: $\dfrac{3}{x} = \dfrac{5}{20}$.

$3 \cdot 20 = 5 \cdot x$

$60 = 5x$

$x = 12$

The ratio of a to b is:

$$\frac{a}{b}$$

Two equal ratios form a proportion. In the proportion

$$\frac{a}{b} = \frac{c}{d}$$

a and d are the *extremes*, and b and c are the *means*. In any proportion the product of the extremes is equal to the product of the means.

If $\dfrac{a}{b} = \dfrac{c}{d}$, then $ad = bc$.

Direct Variation [7.8]

10. If y varies directly as the square of x, then

$$y = Kx^2$$

The variable y is said to vary directly as the variable x if $y = Kx$, where K is a real number. The constant K is called the constant of variation.

Inverse Variation [7.8]

11. If y varies inversely as the cube of x, then

$$y = \frac{K}{x^3}$$

The variable y is said to vary inversely as the variable x if $y = \frac{K}{x}$, where K is a real number.

Chapter 7 Test

Evaluate each rational expression when $x = -2$. [7.1]

1. $\dfrac{2x + 5}{3x^2 - 2x - 1}$

2. $\dfrac{x^2 - 4}{x^2 + 4}$

Determine any values of the variable for which the given rational expression is undefined. [7.1]

3. $\dfrac{2x}{x - 5}$

4. $\dfrac{x + 1}{x^2 + x - 12}$

Reduce to lowest terms. [7.1]

5. $\dfrac{x^2 - 9}{x^2 - 6x + 9}$

6. $\dfrac{15a + 30}{5a^2 - 10a - 40}$

Multiply or divide as indicated. [7.2]

7. $\dfrac{2x - 6}{3} \cdot \dfrac{9}{4x - 12}$

8. $\dfrac{x^2 - 9}{x - 4} \div \dfrac{x + 3}{x^2 - 16}$

9. $\dfrac{x^2 + x - 6}{x^2 + 4x + 3} \div \dfrac{x^2 + 2x - 8}{2x^2 - x - 3}$

10. $(x^2 - 16)\left(\dfrac{x - 1}{x - 4}\right)$

Add or subtract as indicated. [7.3]

11. $\dfrac{7}{x - 8} - \dfrac{9}{x - 8}$

12. $\dfrac{x}{x^2 - 16} + \dfrac{3}{3x - 12}$

13. $\dfrac{3}{(x - 3)(x + 3)} - \dfrac{1}{(x - 3)(x - 1)}$

Simplify each complex fraction. [7.4]

14. $\dfrac{1 + \dfrac{2}{x}}{1 - \dfrac{2}{x}}$

15. $\dfrac{1 - \dfrac{9}{x^2}}{1 - \dfrac{1}{x} - \dfrac{6}{x^2}}$

Solve the following equations. [7.5]

16. $\dfrac{3}{5} = \dfrac{x + 3}{7}$

17. $\dfrac{25}{x - 3} = \dfrac{7}{x}$

18. $\dfrac{6}{x - 3} - \dfrac{5}{x + 1} = \dfrac{7}{x^2 - 2x - 3}$

19. **Mixture** A solution of alcohol and water contains 29 milliliters of alcohol and 87 milliliters of water. What is the ratio of alcohol to water and the ratio of alcohol to total volume? [7.6]

20. **Ratio** A manufacturer knows that during a production run 6 out of every 150 parts produced by a certain machine will be defective. If the machine produces 2,550 parts, how many can be expected to be defective? [7.6]

21. **Speed of a Boat** It takes a boat 3 hours to travel upstream. It takes the same boat 2 hours to travel the same distance downstream. If the current is 3 miles per hour, what is the speed of the boat in still water? [7.7]

22. **Emptying a Pool** An inlet pipe can fill a pool in 12 hours, whereas an outlet pipe can empty it in 8 hours. If the pool is full and both pipes are open, how long will it take to empty?[7.7]

23. **Direct Variation** Suppose y varies directly as the cube of x. If y is 16 when x is 2, find y when x is 3. [7.8]

24. **Inverse Variation** If y varies inversely as the square of x, and y is 8 when x is 3, find y when x is 6. [7.8]

8

Roots and Radical Expressions

Chapter Outline

8.1 Definitions and Common Roots
8.2 Simplified Form and Properties of Radicals
8.3 Addition and Subtraction of Radical Expressions
8.4 Multiplication of Radicals
8.5 Division of Radicals
8.6 Equations Involving Radicals

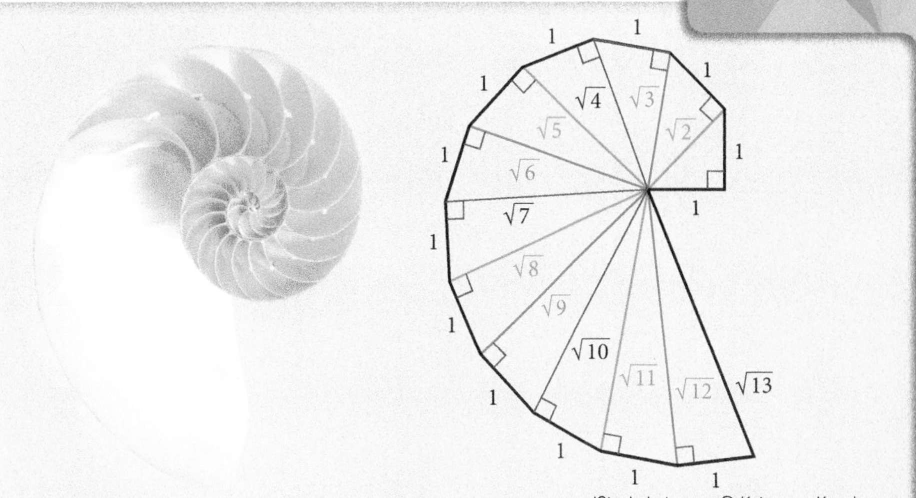

The diagram above is called the *spiral of roots*. The spiral of roots mimics the shell of the chambered nautilus, an animal that has survived largely unchanged for millions of years. The mathematical diagram is constructed using the Pythagorean theorem, which we introduced in Chapter 6. The spiral of roots gives us a way to visualize positive square roots, one of the topics we will cover in this chapter. Table 1 gives the lengths of the diagonals in the spiral of roots, accurate to the nearest hundredth. If we take each of the diagonals in the spiral of roots and place it above the corresponding whole number on the x-axis and then connect the tops of all these segments with a smooth curve, we have the graph shown in Figure 1. This curve is also the graph of the equation $y = \sqrt{x}$.

TABLE 1

Approximate Length of Diagonals

Number	Positive Square Root
1	1
2	1.41
3	1.73
4	2
5	2.24
6	2.45
7	2.65
8	2.83
9	3
10	3.16

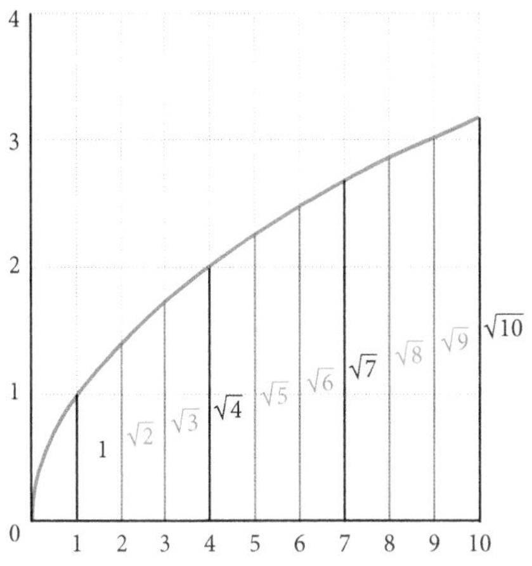

FIGURE 1

Never mistake activity for achievement.

— John Wooden, legendary UCLA basketball coach

You may think that the John Wooden quote above has to do with being productive and efficient, or using your time wisely, but it is really about being honest with yourself. I have had students come to me after failing a test saying, "I can't understand why I got such a low grade after I put so much time in studying." One student even had help from a tutor and felt she understood everything that we covered. After asking her a few questions, it became clear that she spent all her time studying with a tutor and the tutor was doing most of the work. The tutor can work all the homework problems, but the student cannot. She has mistaken activity for achievement.

Can you think of situations in your life when you are mistaking activity for achievement?

How would you describe someone who is mistaking activity for achievement in the way they study for their math class?

Which of the following best describes the idea behind the John Wooden quote:

- Always be efficient.
- Don't kid yourself.
- Take responsibility for your own success.
- Study with purpose.

Definitions and Common Roots 8.1

Learning Objectives

In this section, we will learn how to:

1. Find the positive and negative square roots of a number.
2. Find cube roots and higher roots of numbers.
3. Simplify radicals containing variables.
4. Simplify expressions containing radicals.
5. Find the hypotenuse of a triangle using the Pythagorean theorem.

Introduction

Figure 1 shows a square in which each of the four sides is 1 inch long. To find the square of the length of the diagonal c, we apply the Pythagorean theorem:

Note Although $\sqrt{2}$ is a real number, it cannot be expressed as a quotient of two integers. This discovery was a shock to the Pythagoreans, and they went to great lengths to protect this secret because it threatened the basis of their philosophical beliefs. In fact, there is a legend that Hippasus, one of the Pythagoreans, was murdered for divulging this secret.

$$c^2 = 1^2 + 1^2$$
$$c^2 = 2$$

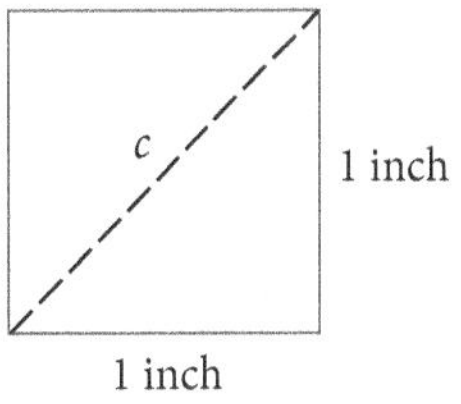

FIGURE 1

Because we know that c is positive and that its square is 2, we call c the *positive square root* of 2, and we write $c = \sqrt{2}$. In this section, we will explore the concept of square roots and other types of roots.

Square Roots

In Chapter 5, we developed notation (exponents) that would take us from a number to its square. If we wanted the square of 5, we wrote $5^2 = 25$. In this section, we will use another type of notation that will take us in the reverse direction—from the square of a number back to the number itself.

In general, we are interested in going from a number, say, 25, back to the number we squared to get 25. Since the square of 5 is 25, we say 5 is a square root of 25. The notation we use looks like this:

$$\sqrt{25} = 5$$

Notation In the expression $\sqrt{25}$, 25 is called the *radicand*; $\sqrt{}$ is the *radical sign*; and the complete expression, $\sqrt{25}$, is called the *radical*.

DEFINITION ***positive square root, negative square root***

If x represents any positive real number, then the expression $\sqrt{x}$ is the ***positive square root*** of x (*or the* ***principal square root*** of x). It is the *positive* number we square to get x.

The expression $-\sqrt{x}$ is the ***negative square root*** of x. It is the negative number we square to get x.

We can generalize this in symbols by writing.

$$\text{If } \sqrt{x} \text{ is the number we square to get } x, \text{ then } (\sqrt{x})^2 = x.$$

Square Roots of Positive Numbers

Every positive number has two square roots, one positive and the other negative. Some books refer to the positive square root of a number as the principal root.

The positive square root of 25 is 5 and can be written $\sqrt{25} = 5$. The negative square root of 25 is -5 and can be written $-\sqrt{25} = -5$.

If we want to consider the negative square root of a number, we must put a negative sign in front of the radical. It is a common mistake to think of $\sqrt{25}$ as meaning either 5 or -5. The expression $\sqrt{25}$ means the positive square root of 25, which is 5. If we want the negative square root, we write $-\sqrt{25}$ to begin with.

Note We can also interpret $-\sqrt{25}$ as the opposite of $\sqrt{25}$.

$$\begin{aligned} -\sqrt{25} &= \text{the opposite of } \sqrt{25} \\ &= \text{the opposite of } 5 \\ &= -5 \end{aligned}$$

VIDEO EXAMPLES

SECTION 8.1

EXAMPLE 1 Find the following roots.

a. $\sqrt{36}$ **b.** $-\sqrt{36}$ **c.** $\sqrt{\frac{1}{36}}$

SOLUTION

a. $\sqrt{36} = 6$ *6 is the positive number we square to get 36*

b. $-\sqrt{36} = -6$ *−6 is the negative number we square to get 36*

c. $\sqrt{\frac{1}{36}} = \frac{1}{6}$ *$\frac{1}{6}$ is the positive number we square to get $\frac{1}{36}$*

Square Roots That Are Irrational Numbers

The positive square root of 17 is written $\sqrt{17}$. The negative square root of 17 is written $-\sqrt{17}$. We have no other exact representation for these two roots. Since 17 itself is not a perfect square (the square of an integer), its two square roots, $\sqrt{17}$ and $-\sqrt{17}$, are irrational numbers. They have a place on the real number line but cannot be written as the ratio of two integers. The square roots of any number that is not itself a perfect square are irrational numbers. Here are some additional examples:

Number	Positive Square Root	Negative Square Root	Roots Are
9	3	-3	Rational numbers
49	7	-7	Rational numbers
5	$\sqrt{5}$	$-\sqrt{5}$	Irrational numbers
23	$\sqrt{23}$	$-\sqrt{23}$	Irrational numbers
$\frac{1}{16}$	$\frac{1}{4}$	$-\frac{1}{4}$	Rational numbers

Square Root of Zero

The number 0 is the only real number with one square root. It is also its own square root:

$$\sqrt{0} = 0$$

Square Roots of Negative Numbers

Negative numbers have square roots, but their square roots are not real numbers. They do not have a place on the real number line.

The expression $\sqrt{-4}$ does not represent a real number since there is no real number we can square and end up with -4. The same is true of square roots of any negative number.

Other Roots

There are many other roots of numbers besides square roots, although square roots are the most commonly used. The cube root of a number is the number we cube (raise to the third power) to get the original number. The cube root of 8 is 2 since $2^3 = 8$. The cube root of -27 is -3 since $(-3)^3 = -27$. We can go as high as we want with roots. The fourth root of 16 is 2 because $2^4 = 16$. Here is a general definition for all types of roots.

DEFINITION ***n*th root, index**

The ***nth root*** of x, written $\sqrt[n]{x}$, is the real number we raise to the nth power to get x. That is, $a = \sqrt[n]{x}$ if $a^n = x$.

The number n is called the ***index*** of the radical. If n is an even number then

$\sqrt[n]{x}$ is the positive nth root of x

$-\sqrt[n]{x}$ is the negative nth root of x

Note With square roots, the index ($n = 2$) is not written. Anytime we see a radical with no index indicated, it is assumed to be a square root.

It is important to understand that even and odd roots behave differently. If n is even, then $x > 0$ and $\sqrt[n]{x}$ is always positive. If n is odd, then x can be any real number (positive or negative), and $\sqrt[n]{x}$ will be the same sign as x.

Based on our previous discussion, we can now write

$$\sqrt[3]{8} = 2$$

$$\sqrt[3]{-27} = -3$$

$$\sqrt[4]{16} = 2$$

Here is a list of the most common roots. They are the roots that will come up most often in the remainder of the book, and they should be memorized.

Square Roots		Cube Roots	Fourth Roots
$\sqrt{1} = 1$	$\sqrt{49} = 7$	$\sqrt[3]{1} = 1$	$\sqrt[4]{1} = 1$
$\sqrt{4} = 2$	$\sqrt{64} = 8$	$\sqrt[3]{8} = 2$	$\sqrt[4]{16} = 2$
$\sqrt{9} = 3$	$\sqrt{81} = 9$	$\sqrt[3]{27} = 3$	$\sqrt[4]{81} = 3$
$\sqrt{16} = 4$	$\sqrt{100} = 10$	$\sqrt[3]{64} = 4$	$\sqrt[4]{256} = 4$
$\sqrt{25} = 5$	$\sqrt{121} = 11$	$\sqrt[3]{125} = 5$	$\sqrt[4]{625} = 5$
$\sqrt{36} = 6$	$\sqrt{144} = 12$		

Remember, with even roots—square roots, fourth roots, sixth roots, and so on—we cannot have negative numbers *under* the radical sign. With odd roots, negative numbers under the radical sign are allowed.

Note At first it may be difficult to see the difference in some of these examples. Generally, we have to be careful with even roots of negative numbers; that is, if the index on the radical is an even number, then we cannot have a negative number under the radical sign. That is why $\sqrt{-4}$ and $\sqrt[4]{-16}$ are not real numbers.

EXAMPLE 2 Find the following roots, if possible.

a. $\sqrt[3]{-8}$ **b.** $\sqrt[3]{-64}$ **c.** $\sqrt{-4}$

d. $-\sqrt{4}$ **e.** $\sqrt[4]{-16}$ **f.** $-\sqrt[4]{16}$

SOLUTION

a. $\sqrt[3]{-8} = -2$ Because $(-2)^3 = -8$

b. $\sqrt[3]{-64} = -4$ Because $(-4)^3 = -64$

c. $\sqrt{-4}$ Not a real number since there is no real number whose square is -4

d. $-\sqrt{4} = -2$ The opposite of $\sqrt{4}$ is $-(\sqrt{4}) = -2$.

e. $\sqrt[4]{-16}$ Not a real number since there is no real number that can be raised to the fourth power to obtain -16

f. $-\sqrt[4]{16} = -2$ The opposite of $\sqrt[4]{16}$ is $-(\sqrt[4]{16}) = -2$.

Square Root of a Perfect Square

Consider the following two statements:

$$\sqrt{3^2} = \sqrt{9} = 3 \quad \text{and} \quad \sqrt{(-3)^2} = \sqrt{9} = 3$$

Whether we operate on 3 or -3, the result is the same; both expressions simplify to 3. The other operation we have worked with in the past that produces the same result is absolute value. That is,

$$|3| = 3 \quad \text{and} \quad |-3| = 3$$

This leads us to a property of radicals.

Note We are not going to do much with this property. We are discussing it here just to be complete. In the remaining examples we assume that our variables under radicals are not negative numbers.

PROPERTY ***Square Root of a Square***

If a is a real number, then $\sqrt{a^2} = |a|$.

The result of this discussion is simply this:

If we know a is positive, then $\sqrt{a^2} = a$.

If we know a is negative, then $\sqrt{a^2} = |a|$.

If we don't know if a is positive or negative, then $\sqrt{a^2} = |a|$.

Variables Under the Radical Sign

To simplify our work in this chapter, unless we say otherwise, we will assume that all variables that appear under a radical sign represent nonnegative numbers. That way we can simplify expressions involving radicals that contain variables without having to worry about absolute value symbols. Here are some examples.

EXAMPLE 3 Simplify: $\sqrt{25a^2}$, $(a \geq 0)$.

SOLUTION We are looking for the expression we square to get $25a^2$. Since the square of 5 is 25 and the square of a is a^2, we can square $5a$ and get $25a^2$:

$$\sqrt{25a^2} = 5a \qquad \text{Because } (5a)^2 = 25a^2$$

EXAMPLE 4 Simplify: $\sqrt{16a^2b^2}$, $(a, b \geq 0)$.

SOLUTION We want an expression whose square is $16a^2b^2$. That expression is $4ab$:

$$\sqrt{16a^2b^2} = 4ab \qquad \text{Because } (4ab)^2 = 16a^2b^2$$

EXAMPLE 5 Simplify: $\sqrt[3]{125a^3}$.

SOLUTION We are looking for the expression we cube to get $125a^3$. That expression is $5a$:

$$\sqrt[3]{125a^3} = 5a \qquad \text{Because } (5a)^3 = 125a^3$$

EXAMPLE 6 Simplify: $\sqrt{x^6}$, $(x \geq 0)$.

SOLUTION The number we square to obtain x^6 is x^3.

$$\sqrt{x^6} = x^3 \qquad \text{Because } (x^3)^2 = x^6$$

As we progress through this chapter you will see more and more expressions that involve the product of a number and a radical. Here are some examples:

$$3\sqrt{2} \qquad \frac{1}{2}\sqrt{5} \qquad 5\sqrt{7} \qquad 3x\sqrt{2x} \qquad 2ab\sqrt{5a}$$

All of these are products. The first expression $3\sqrt{2}$ is the product of 3 and $\sqrt{2}$. That is,

$$3\sqrt{2} = 3 \cdot \sqrt{2}$$

The 3 and the $\sqrt{2}$ are not stuck together in some mysterious way. The expression $3\sqrt{2}$ is simply the product of two numbers, one of which is rational, and the other is irrational. If we use a calculator, we find a decimal approximation of $\sqrt{2}$ is 1.414. Therefore,

$$3\sqrt{2} \approx 3(1.414) = 4.242$$

The symbol $\approx$ tells us we are using an approximation. We say $3\sqrt{2}$ is approximately 4.242.

EXAMPLE 7 Use a calculator to find decimal approximations of each of the following, rounded to the nearest thousandth:

a. $\sqrt{12}$ **b.** $2\sqrt{3}$ **c.** $\sqrt{45}$ **d.** $3\sqrt{5}$

SOLUTION Using a calculator and rounding to the nearest thousandth, we have

a. $\sqrt{12} \approx 3.464$

b. $2\sqrt{3} \approx 2(1.732) = 3.464$

c. $\sqrt{45} \approx 6.708$

d. $3\sqrt{5} \approx 3(2.236) = 6.708$

It is no coincidence that the approximations for $\sqrt{12}$ and $2\sqrt{3}$ are the same. These two expressions are, in fact, equal. We will see how to change one of them into the other as we progress through the chapter. The same is true of the expressions $\sqrt{45}$ and $3\sqrt{5}$; they are also equal.

Simplifying Radical Expressions

Next, consider the expression $4 + 6\sqrt{3}$. This expression is the sum of two terms. The first term is 4 and the second term is $6\sqrt{3}$ (the product of 6 and $\sqrt{3}$). These two terms have a factor of 2 in common, which we could factor out if we wanted to.

$$4 + 6\sqrt{3} = 2 \cdot 2 + 2 \cdot 3\sqrt{3} = 2(2 + 3\sqrt{3})$$

Note that this is exactly the same process we would use to factor the binomial $4 + 6x$:

$$4 + 6x = 2 \cdot 2 + 2 \cdot 3x = 2(2 + 3x)$$

As long as you treat expressions such as $6\sqrt{3}$ as the product of two numbers, you can apply many of the techniques you have learned already to a number of new expressions, as the next example demonstrates.

EXAMPLE 8 Simplify each expression.

a. $\sqrt{36} + \sqrt{64}$ **b.** $\sqrt{36 + 64}$ **c.** $\dfrac{3\sqrt{5}}{6}$ **d.** $\dfrac{4 + 6\sqrt{3}}{4}$

SOLUTION

a.
$$\begin{aligned} \sqrt{36} + \sqrt{64} &= 6 + 8 && \text{Evaluate the square roots} \\ &= 14 && \text{Add} \end{aligned}$$

b.
$$\begin{aligned} \sqrt{36 + 64} &= \sqrt{100} && \text{Add} \\ &= 10 && \text{Evaluate the root} \end{aligned}$$

Observe in part **b** that the radical sign acts as a grouping symbol. We must simplify the radicand first, and then we can find the square root. Also notice that $\sqrt{36 + 64} \neq \sqrt{36} + \sqrt{64}$. In general, the square root of a sum does not equal the sum of the square roots. For the last two expressions, we can simplify by dividing out factors that are common to the numerator and denominator.

c.
$$\begin{aligned} \frac{3\sqrt{5}}{6} &= \frac{3\sqrt{5}}{3 \cdot 2} && \text{Factor denominator} \\ &= \frac{\sqrt{5}}{2} && \text{Divide out common factor} \end{aligned}$$

d.
$$\begin{aligned} \frac{4 + 6\sqrt{3}}{4} &= \frac{2(2 + 3\sqrt{3})}{2 \cdot 2} && \text{Factor numerator and denominator} \\ &= \frac{2 + 3\sqrt{3}}{2} && \text{Divide out common factor} \end{aligned}$$

Notice that it would be a mistake to reduce any further. Although 2 is a factor of the denominator, it is not a factor of the numerator. The 2 in the numerator is a term, not a factor.

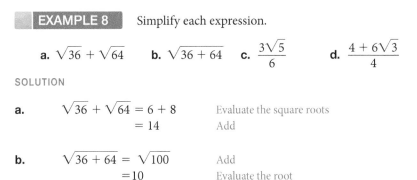

The Pythagorean Theorem Revisited

Now that we have some experience working with square roots, we can rewrite the Pythagorean theorem using a square root. In Figure 2, if triangle ABC is a right triangle with angle $C = 90°$, then the length of the longest side (the hypotenuse) is the *square root* of the sum of the squares of the other two sides.

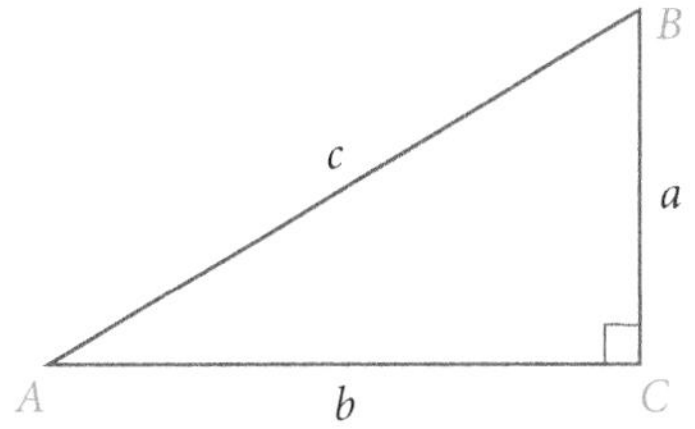

$$c = \sqrt{a^2 + b^2}$$

FIGURE 2

EXAMPLE 9 A tent pole is 8 feet in length and makes an angle of 90° with the ground. One end of a rope is attached to the top of the pole, and the other end of the rope is anchored to the ground 6 feet from the bottom of the pole. Find the length of the rope.

SOLUTION The diagram in Figure 3 is a visual representation of the situation. To find the length of the rope, we apply the Pythagorean theorem.

$$x = \sqrt{6^2 + 8^2}$$
$$x = \sqrt{36 + 64}$$
$$x = \sqrt{100}$$
$$x = 10 \text{ feet}$$

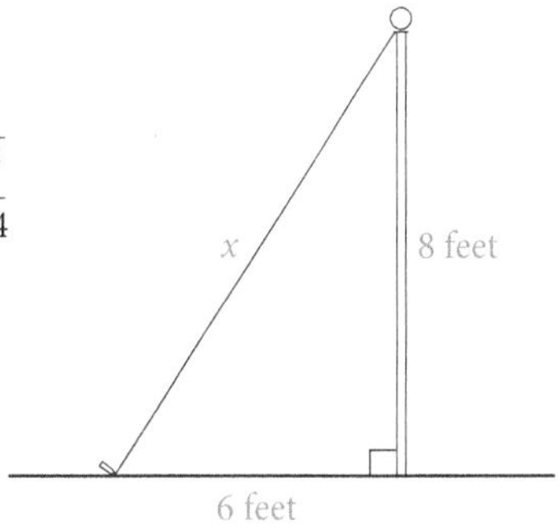

FIGURE 3

The Spiral of Roots

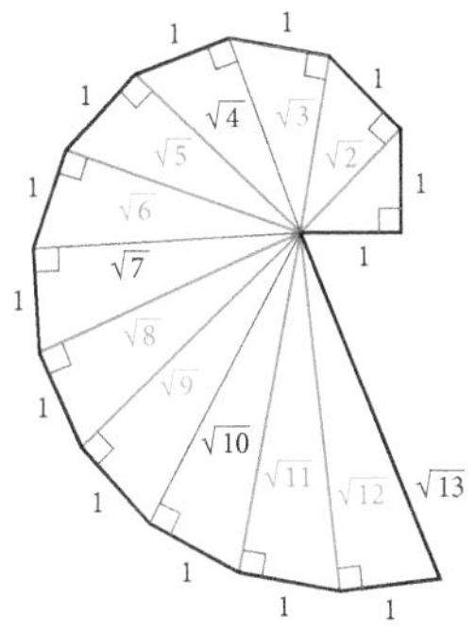

Associating numbers, such as $\sqrt{2}$, with the diagonal of a square or rectangle allows us to analyze some interesting items from geometry. To visualize the square roots of the positive integers, we can construct the spiral of roots, which we mentioned in the introduction to this chapter. To begin, we draw two line segments, each of length 1, at right angles to each other. Then we use the Pythagorean theorem to find the length of the diagonal (or hypotenuse). Figure 4 illustrates.

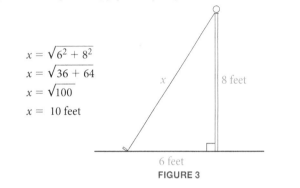

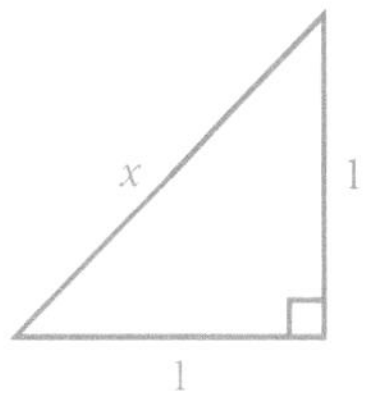

$$x = \sqrt{1^2 + 1^2}$$
$$= \sqrt{2}$$

FIGURE 4

Next, we construct a second triangle by connecting a line segment of length 1 to the end of the first diagonal so that the angle formed is a right angle. We find the length of the second diagonal using the Pythagorean theorem. Figure 5 illustrates this procedure. Continuing to draw new triangles by connecting line segments of length 1 to the end of each new diagonal, so that the angle formed is a right angle, the spiral of roots begins to appear (see Figure 6).

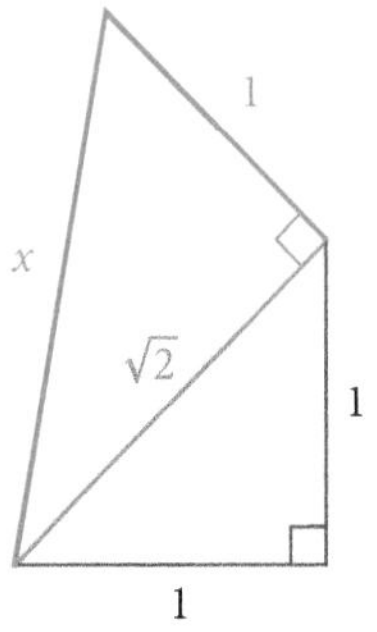

$$
\begin{aligned}
x &= \sqrt{(\sqrt{2})^2 + 1^2} \\
&= \sqrt{2 + 1} \\
&= \sqrt{3}
\end{aligned}
$$

FIGURE 5

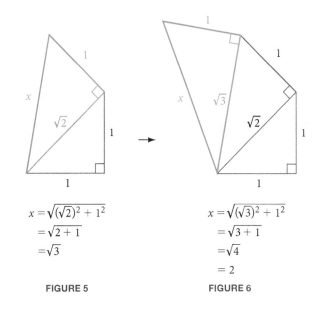

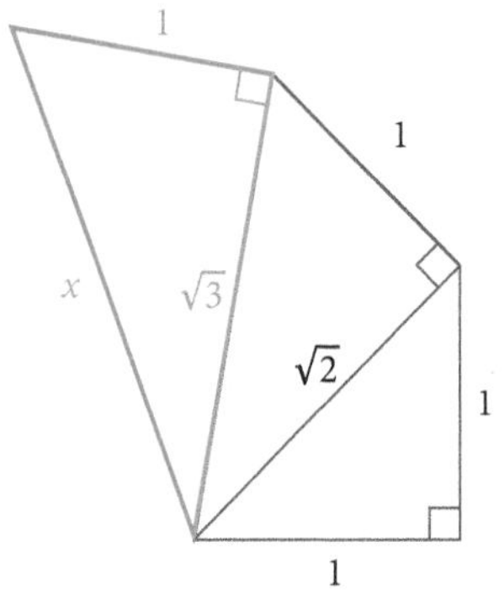

$$
\begin{aligned}
x &= \sqrt{(\sqrt{3})^2 + 1^2} \\
&= \sqrt{3 + 1} \\
&= \sqrt{4} \\
&= 2
\end{aligned}
$$

FIGURE 6

Getting Ready for Class

After reading through the preceding section, respond in your own words and in complete sentences.

A. Every real number has two square roots. Explain the notation we use to tell them apart. Use the square roots of 3 for examples.

B. Explain why a square root of -4 is not a real number.

C. Why is the statement $4 + 6\sqrt{3} = 10\sqrt{3}$ not correct?

D. What kinds of square roots represent irrational numbers?

Problem Set 8.1

Find the following roots. If the root does not exist as a real number, write "not a real number."

1. $\sqrt{4}$ **2.** $\sqrt{49}$ **3.** $-\sqrt{4}$ **4.** $-\sqrt{49}$

5. $\sqrt{-16}$ **6.** $\sqrt{-36}$ **7.** $-\sqrt{144}$ **8.** $\sqrt{256}$

9. $\sqrt{625}$ **10.** $-\sqrt{625}$ **11.** $\sqrt{-25}$ **12.** $\sqrt{-144}$

13. $-\sqrt{64}$ **14.** $-\sqrt{25}$ **15.** $-\sqrt{100}$ **16.** $\sqrt{121}$

17. $\sqrt{\frac{1}{4}}$ **18.** $\sqrt{\frac{1}{25}}$ **19.** $-\sqrt{\frac{1}{100}}$ **20.** $\sqrt{-\frac{1}{64}}$

21. $\sqrt{\frac{9}{49}}$ **22.** $\sqrt{\frac{25}{81}}$ **23.** $-\sqrt{\frac{4}{121}}$ **24.** $-\sqrt{\frac{9}{16}}$

25. $-\sqrt{1{,}225}$ **26.** $-\sqrt{1{,}681}$ **27.** $\sqrt[3]{1}$ **28.** $-\sqrt[4]{81}$

29. $\sqrt[3]{-8}$ **30.** $\sqrt[3]{125}$ **31.** $-\sqrt[3]{125}$ **32.** $-\sqrt[3]{-8}$

33. $\sqrt[3]{-1}$ **34.** $-\sqrt[3]{-1}$ **35.** $\sqrt[3]{-27}$ **36.** $-\sqrt[3]{27}$

37. $\sqrt[3]{\frac{1}{64}}$ **38.** $\sqrt[3]{-\frac{1}{64}}$ **39.** $-\sqrt[4]{16}$ **40.** $\sqrt[4]{-16}$

41. $\sqrt[4]{81}$ **42.** $-\sqrt[4]{625}$ **43.** $\sqrt[5]{32}$ **44.** $\sqrt[6]{64}$

Assume all variables represent nonnegative numbers, and find the following roots.

45. $\sqrt{y^2}$ **46.** $\sqrt{x^2}$ **47.** $\sqrt{25x^2}$ **48.** $\sqrt{81x^2}$

49. $\sqrt{a^2b^2}$ **50.** $\sqrt{x^2y^2}$ **51.** $\sqrt{(a+b)^2}$ **52.** $\sqrt{(x+y)^2}$

53. $\sqrt{81x^2y^2}$ **54.** $\sqrt{36x^2y^2}$ **55.** $\sqrt[3]{x^3}$ **56.** $\sqrt[3]{a^3}$

57. $\sqrt[3]{8x^3}$ **58.** $\sqrt[3]{27x^3}$ **59.** $\sqrt{x^4}$ **60.** $\sqrt{x^6}$

61. $\sqrt{36a^6}$ **62.** $\sqrt{64a^4}$ **63.** $\sqrt{25a^8b^4}$ **64.** $\sqrt{16a^4b^8}$

Use a calculator to find decimal approximations of each of the following. Round your answers to the nearest thousandth.

65. $\sqrt{2}$ **66.** $\sqrt{3}$ **67.** $-\sqrt{17}$ **68.** $-\sqrt{31}$

69. a. $\sqrt{18}$ **b.** $3\sqrt{2}$ **70. a.** $\sqrt{20}$ **b.** $2\sqrt{5}$

71. a. $\sqrt{50}$ **b.** $5\sqrt{2}$ **72. a.** $\sqrt{27}$ **b.** $3\sqrt{3}$

Factor the greatest common factor from each expression.

73. a. $8 + 6x$ **b.** $8 + 6\sqrt{3}$ **74. a.** $10 + 5y$ **b.** $10 + 5\sqrt{6}$

Simplify each expression.

75. $\sqrt{9} + \sqrt{16}$ **76.** $\sqrt{64} + \sqrt{36}$ **77.** $\sqrt{9 + 16}$

78. $\sqrt{64 + 36}$ **79.** $\sqrt{144} + \sqrt{25}$ **80.** $\sqrt{25} - \sqrt{16}$

81. $\sqrt{144 + 25}$ **82.** $\sqrt{25 - 16}$

83. Use the approximation $\sqrt{5} \approx 2.236$ and find approximations for each of the following expressions.

a. $\frac{1+\sqrt{5}}{2}$ **b.** $\frac{1-\sqrt{5}}{2}$ **c.** $\frac{1+\sqrt{5}}{2} + \frac{1-\sqrt{5}}{2}$

84. Use the approximation $\sqrt{3} \approx 1.732$ and find approximations for each of the following expressions.

a. $\dfrac{1+\sqrt{3}}{2}$ **b.** $\dfrac{1-\sqrt{3}}{2}$ **c.** $\dfrac{1+\sqrt{3}}{2}+\dfrac{1-\sqrt{3}}{2}$

85. Evaluate each root.

a. $\sqrt{9}$ **b.** $\sqrt{900}$ **c.** $\sqrt{0.09}$

86. Evaluate each root.

a. $\sqrt[3]{27}$ **b.** $\sqrt[3]{0.027}$ **c.** $\sqrt[3]{27{,}000}$

Simplify each of the following pairs of expressions.

87. $\dfrac{5+\sqrt{49}}{2}$ and $\dfrac{5-\sqrt{49}}{2}$

88. $\dfrac{3+\sqrt{25}}{4}$ and $\dfrac{3-\sqrt{25}}{4}$

Simplify each expression.

89. a. $\dfrac{2\sqrt{3}}{10}$ **b.** $\dfrac{5\sqrt{2}}{15}$ **c.** $\dfrac{15+6\sqrt{3}}{12}$ **d.** $\dfrac{5+10\sqrt{6}}{5}$

Simplify each expression.

90. a. $\dfrac{5\sqrt{7}}{30}$ **b.** $\dfrac{7\sqrt{6}}{35}$ **c.** $\dfrac{8+4\sqrt{5}}{6}$ **d.** $\dfrac{14+7\sqrt{3}}{21}$

Applying the Concepts

Find x in each of the following right triangles. Round to the nearest tenth if necessary.

91.

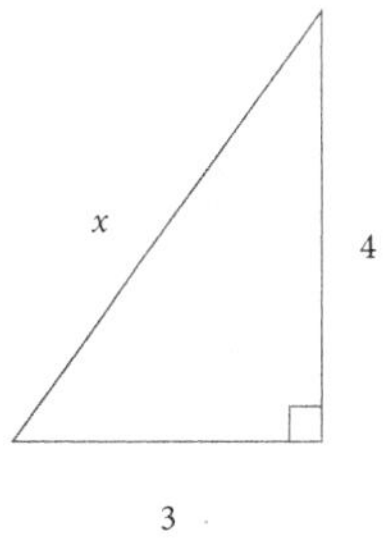

92.

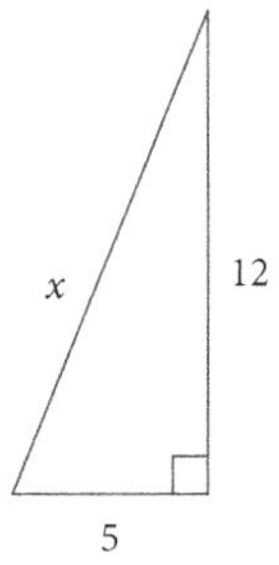

93.

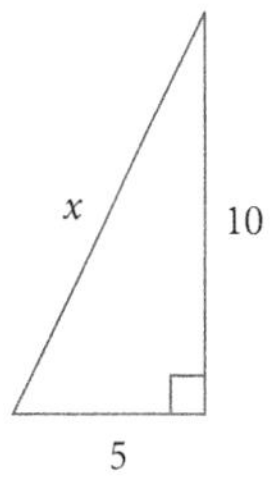

94.

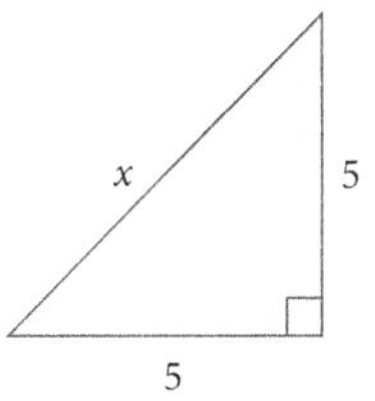

95. Geometry One end of a wire is attached to the top of a 24-foot pole; the other end of the wire is anchored to the ground 18 feet from the bottom of the pole. If the pole makes an angle of 90° with the ground, find the length of the wire.

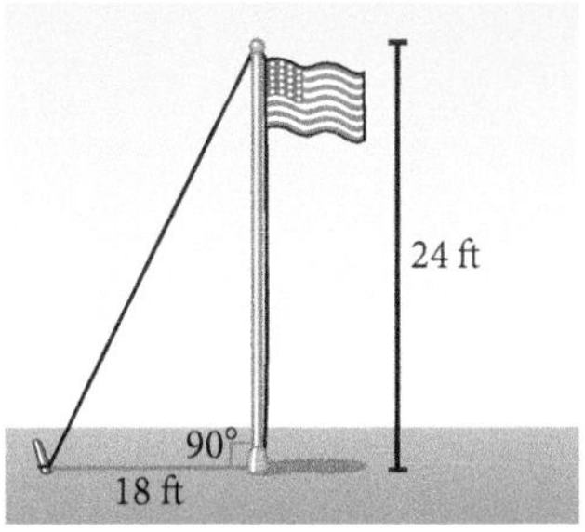

96. Geometry Two children are trying to cross a stream. They want to use a log that goes from one bank to the other. If the right bank is 5 feet higher than the left bank and the stream is 12 feet wide, how long must a log be to just barely reach?

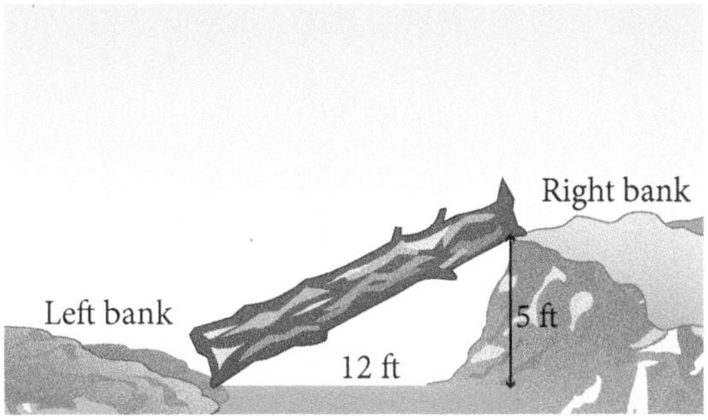

97. Spiral of Roots Construct your own spiral of roots by using a ruler. Draw the first triangle by using two 1-inch lines. The first diagonal will have a length of $\sqrt{2}$ inches. Each new triangle will be formed by drawing a 1-inch line segment at the end of the previous diagonal so that the angle formed is 90°.

98. Spiral of Roots Construct a spiral of roots by using line segments of length 2 inches.

99. Number Sequence Simplify the terms in the following sequence. The result will be a sequence that gives the lengths of the diagonals in the spiral of roots you constructed in Problem 97.

$$\sqrt{1^2+1},\ \sqrt{(\sqrt{2})^2+1},\ \sqrt{(\sqrt{3})^2+1},\ldots$$

100. Number Sequence Simplify the terms in the following sequence. The result will be a sequence that gives the lengths of the diagonals in the spiral of roots you constructed in Problem 100.

$$\sqrt{2^2+4},\ \sqrt{(2\sqrt{2})^2+4},\ \sqrt{(2\sqrt{3})^2+4},\ldots$$

Learning Objectives Assessment

The following problems can be used to help assess if you have successfully met the learning objectives for this section.

101. Find: $\sqrt{64}$.

a. 4 **b.** 8 **c.** 2 **d.** 32

102. Find: $\sqrt[3]{-64}$.

a. -8 **b.** 4 **c.** -4 **d.** Not a real number

103. Simplify: $\sqrt{16x^2}$. Assume x represents a nonnegative number.

a. $4x$ **b.** $-4x$ **c.** $4x^2$ **d.** $16x$

104. Simplify: $\sqrt{25 - 9}$.

a. 2 **b.** 8 **c.** 4 **d.** 16

105. The legs of a right triangle have lengths 1 meter and 3 meters. Find the length of the hypotenuse.

a. 4 meters **b.** 2 meters **c.** 10 meters **d.** $\sqrt{10}$ meters

Getting Ready for the Next Section

Simplify.

106. $3 \cdot \sqrt{16}$

107. $6 \cdot \sqrt{4}$

Factor each of the following numbers into the product of two numbers, one of which is a perfect square. Remember from Chapter 1, a perfect square is 1, 4, 9, 16, 25, 36, ...)

108. 75

109. 12

110. 50

111. 20

112. 40

113. 18

114. x^2

115. x^3

116. $12x^2$

117. $20x^3$

118. Factor $25x^2y^2$ from $50x^3y^2$.

119. Factor $25x^2y^2$ from $75x^2y^3$.

Simplified Form and Properties of Radicals 8.2

Learning Objectives

In this section, we will learn how to:

1. Use the product property of radicals to remove perfect squares from a square root.
2. Use the product property of radicals to remove *n*th powers from the radicand of an *n*th root.
3. Use the quotient property of radicals to eliminate fractions from under the radical sign.
4. Simplify fractions containing radical expressions.
5. Show that the ratio of length to width in a rectangle is the golden ratio.

Introduction

A *radical expression* is any expression containing a radical, whether it is a square root, a cube root, or a higher root. The simplified form for a radical expression is the form that is easiest to work with. A radical expression is in *simplified form* if it has three special characteristics.

def DEFINITION *simplified form*

A radical expression is in ***simplified form*** if

1. There are no perfect squares that are factors of the quantity under the square root sign, no perfect cubes that are factors of the quantity under the cube root sign, and so on. We want as little as possible under the radical sign.
2. There are no fractions under the radical sign.
3. There are no radicals in the denominator.

As we will see, simplified form is not always the least complicated expression. In many cases, the simplified expression looks more complicated than the original expression. The important thing about simplified form for radicals is that simplified expressions are easier to work with.

Note Simplified form for radicals is the form that we work toward when simplifying radicals. The properties of radicals are the tools we use to get us to simplified form.

The Product Property

The first step in putting a radical expression in simplified form is to take as much out from under the radical sign as possible. To do this, we use the first of two properties of radicals.

Consider the following two problems:

$$\sqrt{9 \cdot 16} = \sqrt{144} = 12$$
$$\sqrt{9} \cdot \sqrt{16} = 3 \cdot 4 = 12$$

Since the answers to both are equal, the original problems also must be equal; that is, $\sqrt{9 \cdot 16} = \sqrt{9} \cdot \sqrt{16}$. We can generalize this property as follows.

PROPERTY *Product Property for Square Roots*

If x and y represent nonnegative real numbers, then it is always true that

$$\sqrt{ab} = \sqrt{a}\sqrt{b}$$

In words: The square root of a product is the product of the square roots.

VIDEO EXAMPLES

SECTION 8.2

Note Working a problem like the one in Example 1 depends on recognizing the largest perfect square that divides (is a factor of) the number under the radical sign. The set of perfect squares is the set

$$\{1, 4, 9, 16, 25, 36, \ldots\}$$

To simplify an expression like $\sqrt{20}$, we first must find the largest number in this set that is a factor of the number under the radical sign.

We can use this property to simplify a radical expression whose radicand contains a perfect square.

EXAMPLE 1 Simplify: $\sqrt{20}$.

SOLUTION To simplify $\sqrt{20}$, we want to take as much out from under the radical sign as possible. We begin by looking for the largest perfect square that is a factor of 20. The largest perfect square that divides 20 is 4, so we write 20 as $4 \cdot 5$:

$$\sqrt{20} = \sqrt{4 \cdot 5}$$

Next, we apply the product property of radicals and write

$$\sqrt{4 \cdot 5} = \sqrt{4}\sqrt{5}$$

And since $\sqrt{4} = 2$, we have

$$\sqrt{4}\sqrt{5} = 2\sqrt{5}$$

The expression $2\sqrt{5}$ is the simplified form of $\sqrt{20}$ since we have taken as much out from under the radical sign as possible.

EXAMPLE 2 Simplify: $\sqrt{75}$.

SOLUTION Since 25 is the largest perfect square that divides 75, we have

$\sqrt{75} = \sqrt{25 \cdot 3}$	Factor 75 into $25 \cdot 3$
$= \sqrt{25}\sqrt{3}$	Product property xfor square roots
$= 5\sqrt{3}$	$\sqrt{25} = 5$

The expression $5\sqrt{3}$ is the simplified form for $\sqrt{75}$ since we have taken as much out from under the radical sign as possible.

The next two examples involve square roots of expressions that contain variables. Remember, we are assuming that all variables that appear under a radical sign represent nonnegative numbers.

EXAMPLE 3 Simplify: $\sqrt{25x^3}$.

SOLUTION The largest perfect square that is a factor of $25x^3$ is $25x^2$. We write $25x^3$ as $25x^2 \cdot x$ and apply the product property of square roots.

$\sqrt{25x^3} = \sqrt{25x^2 \cdot x}$	Factor $25x^3$ into $25x^2 \cdot x$
$= \sqrt{25x^2}\sqrt{x}$	Product property for square roots
$= 5x\sqrt{x}$	$\sqrt{25x^2} = 5x$

EXAMPLE 4 Simplify: $\sqrt{18y^4}$.

SOLUTION The largest perfect square that is a factor of $18y^4$ is $9y^4$. We write $18y^4$ as $9y^4 \cdot 2$ and apply the product property of square roots:

$$\begin{aligned} \sqrt{18y^4} &= \sqrt{9y^4 \cdot 2} && \text{Factor } 18y^4 \text{ into } 9y^4 \cdot 2 \\ &= \sqrt{9y^4}\,\sqrt{2} && \text{Product property for square roots} \\ &= 3y^2\sqrt{2} && \sqrt{9y^4} = 3y^2 \end{aligned}$$

EXAMPLE 5 Simplify: $3\sqrt{32}$.

SOLUTION We want to get as much out from under $\sqrt{32}$ as possible. Since 16 is the largest perfect square that divides 32, we have:

$$\begin{aligned} 3\sqrt{32} &= 3\sqrt{16 \cdot 2} && \text{Factor 32 into } 16 \cdot 2 \\ &= 3\sqrt{16}\sqrt{2} && \text{Product property for square roots} \\ &= 3 \cdot 4\sqrt{2} && \sqrt{16} = 4 \\ &= 12\sqrt{2} && 3 \cdot 4 = 12 \end{aligned}$$

Although we have stated the product property for radicals in terms of square roots only, it holds for higher roots as well. If we were to state this property again for cube roots, it would look like this:

$$\sqrt[3]{ab} = \sqrt[3]{a}\,\sqrt[3]{b}$$

Here is the general form of the product property for radicals.

PROPERTY ***Product Property for Radicals***

If a and b represent nonnegative real numbers, then

$$\sqrt[n]{a \cdot b} = \sqrt[n]{a} \cdot \sqrt[n]{b}$$

In words: The nth root of a product equals the product of the nth roots.

EXAMPLE 6 Simplify: $\sqrt[3]{24x^3}$.

SOLUTION Since we are simplifying a cube root, we look for the largest perfect cube that is a factor of $24x^3$. Since 8 is a perfect cube, the largest perfect cube that is a factor of $24x^3$ is $8x^3$.

$$\begin{aligned} \sqrt[3]{24x^3} &= \sqrt[3]{8x^3 \cdot 3} && \text{Factor } 24x^3 \text{ into } 8x^3 \cdot 3 \\ &= \sqrt[3]{8x^3}\,\sqrt[3]{3} && \text{Product property for radicals} \\ &= 2x\sqrt[3]{3} && \sqrt[3]{8x^3} = 2x \end{aligned}$$

The Quotient Property

The second property of radicals has to do with division. The property becomes apparent when we consider the following two problems:

$$\sqrt{\frac{64}{16}} = \sqrt{4} = 2$$

$$\frac{\sqrt{64}}{\sqrt{16}} = \frac{8}{4} = 2$$

Since the answers in each case are equal, the original problems also must be equal:

$$\sqrt{\frac{64}{16}} = \frac{\sqrt{64}}{\sqrt{16}}$$

Here is the property in general.

PROPERTY ***Quotient Property for Radicals***

If a and b both represent nonnegative real numbers and $b \neq 0$, then it is always true that

$$\sqrt[n]{\frac{a}{b}} = \frac{\sqrt[n]{a}}{\sqrt[n]{b}}$$

In words: The nth root of a quotient equals the quotient of the nth roots.

We can use the quotient property for radicals to simplify radical expressions that contain fractions. This will allow us to meet the second condition of simplified form for radicals.

EXAMPLE 7 Simplify: $\sqrt{\frac{49}{81}}$.

SOLUTION We begin by applying the quotient property for radicals to separate the fraction into two separate radicals. Then we simplify each radical separately:

$$\sqrt{\frac{49}{81}} = \frac{\sqrt{49}}{\sqrt{81}} \quad \text{Quotient property for radicals}$$

$$= \frac{7}{9} \quad \sqrt{49} = 7 \text{ and } \sqrt{81} = 9$$

EXAMPLE 8 Simplify: $\sqrt[4]{\frac{81}{16}}$.

SOLUTION Proceeding as we did in Example 7, we have:

$$\sqrt[4]{\frac{81}{16}} = \frac{\sqrt[4]{81}}{\sqrt[4]{16}} \quad \text{Quotient property}$$

$$= \frac{3}{2} \quad \sqrt[4]{81} = 3 \text{ and } \sqrt[4]{16} = 2$$

EXAMPLE 9 Simplify: $\sqrt{\frac{50}{49}}$.

SOLUTION We can simplify this expression by applying both properties and working with fractions.

$$\begin{aligned}\sqrt{\frac{50}{49}} &= \frac{\sqrt{50}}{\sqrt{49}} && \text{Quotient property for radicals}\\ &= \frac{\sqrt{25\cdot 2}}{7} && \text{Factor 50 into } 25\cdot 2,\ \sqrt{49}=7\\ &= \frac{\sqrt{25}\cdot\sqrt{2}}{7} && \text{Product property for radicals}\\ &= \frac{5\sqrt{2}}{7} && \sqrt{25}=5\end{aligned}$$

EXAMPLE 10 Simplify: $\sqrt{\frac{12x^2}{25}}$.

SOLUTION Proceeding as we have in the previous three examples, we use the quotient property for radicals to separate the numerator and denominator into two separate radicals. Then, we simplify each radical separately:

$$\begin{aligned}\sqrt{\frac{12x^2}{25}} &= \frac{\sqrt{12x^2}}{\sqrt{25}} && \text{Quotient property for radicals}\\ &= \frac{\sqrt{4x^2\cdot 3}}{5} && \text{Factor } 12x^2 \text{ into } 4x^2\cdot 3,\ \sqrt{25}=5\\ &= \frac{\sqrt{4x^2}\cdot\sqrt{3}}{5} && \text{Product property for radicals}\\ &= \frac{2x\sqrt{3}}{5} && \sqrt{4x^2}=2x\end{aligned}$$

EXAMPLE 11 Simplify each expression.

a. $\frac{\sqrt{12}}{6}$ **b.** $\frac{5\sqrt{18}}{15}$ **c.** $\frac{6+\sqrt{8}}{2}$ **d.** $\frac{4+\sqrt{36}}{2}$ **e.** $\frac{-2+\sqrt{48}}{4}$

SOLUTION In each case, we simplify the radical first, then we factor and reduce to lowest terms.

a.
$$\begin{aligned}\frac{\sqrt{12}}{6} &= \frac{2\sqrt{3}}{6} && \text{Simplify the radical } \sqrt{12}=\sqrt{4\cdot 3}=\sqrt{4}\sqrt{3}=2\sqrt{3}\\ &= \frac{2\sqrt{3}}{2\cdot 3} && \text{Factor denominator}\\ &= \frac{\sqrt{3}}{3} && \text{Divide out common factor}\end{aligned}$$

b.
$$\begin{aligned}\frac{5\sqrt{18}}{15} &= \frac{5\cdot 3\sqrt{2}}{15} && \sqrt{18}=\sqrt{9\cdot 2}=\sqrt{9}\sqrt{2}=3\sqrt{2}\\ &= \frac{5\cdot 3\sqrt{2}}{3\cdot 5} && \text{Factor denominator}\\ &= \sqrt{2} && \text{Divide out common factors}\end{aligned}$$

c. $\dfrac{6+\sqrt{8}}{2} = \dfrac{6+2\sqrt{2}}{2}$ $\qquad \sqrt{8} = \sqrt{4 \cdot 2} = \sqrt{4}\sqrt{2} = 2\sqrt{2}$

$= \dfrac{2(3+\sqrt{2})}{2}$ $\qquad$ Factor numerator

$= 3 + \sqrt{2}$ $\qquad$ Divide out common factor

d. $\dfrac{4+\sqrt{36}}{2} = \dfrac{4+6}{2}$ $\qquad \sqrt{36} = 6$

$= \dfrac{10}{2}$

$= 5$

e. $\dfrac{-2+\sqrt{48}}{4} = \dfrac{-2+4\sqrt{3}}{4}$ $\qquad \sqrt{48} = \sqrt{16 \cdot 3} = \sqrt{16}\sqrt{3} = 4\sqrt{3}$

$= \dfrac{2(-1+2\sqrt{3})}{2 \cdot 2}$ $\qquad$ Factor numerator and denominator

$= \dfrac{-1+2\sqrt{3}}{2}$ $\qquad$ Divide out common factor

EXAMPLE 12 Simplify each expression.

a. $\dfrac{-4-\sqrt{40}}{4}$ **b.** $\dfrac{-6+\sqrt{48}}{2}$ **c.** $\dfrac{6-\sqrt{108}}{18}$

SOLUTION In each case the first step is to simplify the radical. Then we factor, if possible, and simplify by dividing out common factors.

a. $\dfrac{-4-\sqrt{40}}{4} = \dfrac{-4-2\sqrt{10}}{4} = \dfrac{2(-2-\sqrt{10})}{2 \cdot 2} = \dfrac{-2-\sqrt{10}}{2}$

b. $\dfrac{-6+\sqrt{48}}{2} = \dfrac{-6+4\sqrt{3}}{2} = \dfrac{2(-3+2\sqrt{3})}{2} = -3+2\sqrt{3}$

c. $\dfrac{6-\sqrt{108}}{18} = \dfrac{6-6\sqrt{3}}{18} = \dfrac{6(1-\sqrt{3})}{6 \cdot 3} = \dfrac{1-\sqrt{3}}{3}$

Golden Rectangle and the Golden Ratio

The golden rectangle's origins can be traced back over 2,000 years to the Greek civilization that produced Pythagoras, Socrates, Plato, Aristotle, and Euclid. The most important mathematical work to come from that Greek civilization was Euclid's *Elements,* an elegantly written summary of all that was known about geometry at that time in history. Euclid's *Elements,* according to Howard Eves, an authority on the history of mathematics, exercised a greater influence on scientific thinking than any other work. Here is how we construct a golden rectangle from a square of side 2, using the same method that Euclid used in his *Elements.*

Step 1: Draw a square with a side of length 2. Connect the midpoint of side *CD* to corner *B* as shown in Figure 1. (Note that we have labeled the midpoint of segment *CD* with the letter *O*.)

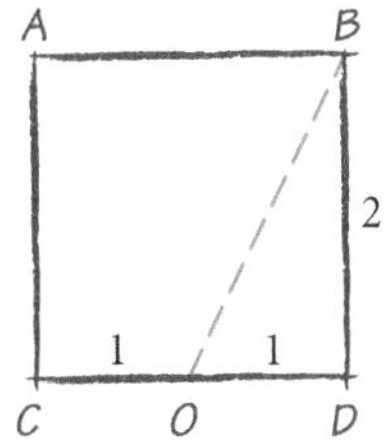

FIGURE 1

Step 2: Drop the diagonal from step 1 down so it aligns with side *CD* (see Figure 2).

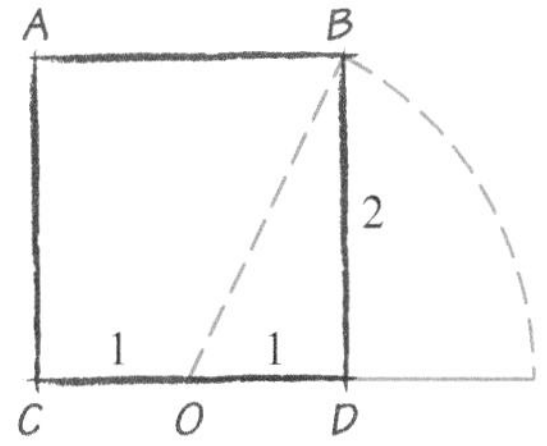

FIGURE 2

Step 3: Form rectangle *ACEF* as shown in Figure 3. This is a golden rectangle.

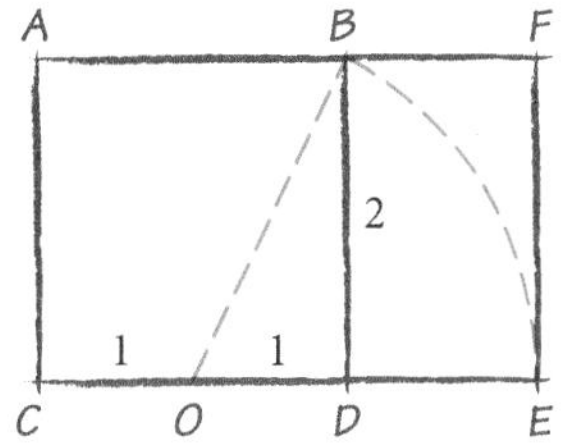

FIGURE 3

All golden rectangles are constructed from squares. Every golden rectangle, no matter how large or small it is, will have the same shape. To associate a number with the shape of the golden rectangle, we use the ratio of its length to its width. This ratio is called the *golden ratio*. To calculate the golden ratio, we must first find the length of the diagonal we used to construct the golden rectangle. Figure 4 shows the golden rectangle we constructed from a square of side 2. The length of the diagonal *OB* is found by applying the Pythagorean theorem to triangle *OBD*.

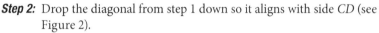

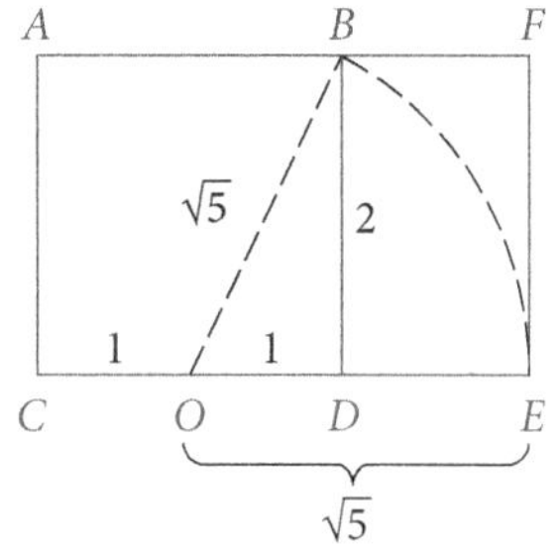

FIGURE 4

The length of segment OE is equal to the length of diagonal OB; both are $\sqrt{5}$. Because the distance from C to O is 1, the length CE of the golden rectangle is $1 + \sqrt{5}$. Now we can find the golden ratio:

$$\text{Golden ratio} = \frac{\text{length}}{\text{width}} = \frac{CE}{EF} = \frac{1 + \sqrt{5}}{2}$$

EXAMPLE 13 Construct a golden rectangle from a square of side 4. Then show that the ratio of the length to the width is the golden ratio $\frac{1+\sqrt{5}}{2}$.

SOLUTION Figure 5 shows the golden rectangle constructed from a square of side 4. The length of the diagonal OB is found from the Pythagorean theorem.

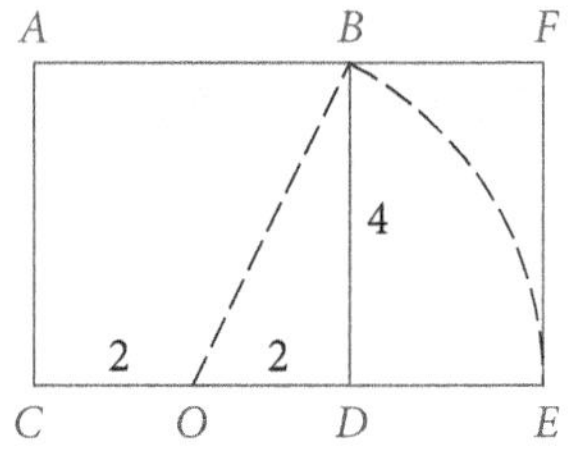

FIGURE 5

$$OB = \sqrt{2^2 + 4^2} = \sqrt{4 + 16} = \sqrt{20} = 2\sqrt{5}$$

The ratio of the length to the width for the rectangle is the golden ratio.

$$\text{Golden ratio} = \frac{CE}{EF} = \frac{2 + 2\sqrt{5}}{4} = \frac{2(1 + \sqrt{5})}{2 \cdot 2} = \frac{1 + \sqrt{5}}{2}$$

As you can see, showing that the ratio of length to width in this rectangle is the golden ratio depends on our ability to write $\sqrt{20}$ as $2\sqrt{5}$ and our ability to reduce to lowest terms by factoring and then dividing out the common factor 2 from the numerator and denominator.

Getting Ready for Class

After reading through the preceding section, respond in your own words and in complete sentences.

A. What is the product property for radicals?

B. Describe how you would apply the product property for radicals to $\sqrt{20}$.

C. What is the quotient property for radicals?

D. What is a golden rectangle?

Use the product property for radicals to simplify the following radical expressions as much as possible. Assume all variables represent nonnegative numbers.

1. $\sqrt{8}$ **2.** $\sqrt{18}$ **3.** $\sqrt{12}$ **4.** $\sqrt{27}$

5. $\sqrt[3]{24}$ **6.** $\sqrt[3]{54}$ **7.** $\sqrt{50x^2}$ **8.** $\sqrt{32x^2}$

9. $\sqrt{45a^2b^2}$ **10.** $\sqrt{128a^2b^2}$ **11.** $\sqrt[3]{54x^3}$ **12.** $\sqrt[3]{128x^3}$

13. $\sqrt{32x^4}$ **14.** $\sqrt{48x^4}$ **15.** $5\sqrt{80}$ **16.** $3\sqrt{125}$

17. $\frac{1}{2}\sqrt{28x^3}$ **18.** $\frac{2}{3}\sqrt{54x^3}$ **19.** $x\sqrt[3]{2x^4}$ **20.** $x\sqrt[3]{8x^5}$

21. $2a\sqrt[3]{27a^5}$ **22.** $3a\sqrt[3]{27a^4}$ **23.** $\frac{4}{3}\sqrt{45a^3}$ **24.** $\frac{3}{5}\sqrt{300a^3}$

25. $3\sqrt{50xy^2}$ **26.** $4\sqrt{18xy^2}$ **27.** $7\sqrt{12x^2y}$ **28.** $6\sqrt{20x^2y}$

Use the quotient property for radicals to simplify each of the following. Assume all variables represent nonnegative numbers.

29. $\sqrt{\frac{16}{25}}$ **30.** $\sqrt{\frac{81}{64}}$ **31.** $\sqrt{\frac{4}{9}}$ **32.** $\sqrt{\frac{49}{16}}$

33. $\sqrt[3]{\frac{8}{27}}$ **34.** $\sqrt[3]{\frac{64}{27}}$ **35.** $\sqrt[4]{\frac{16}{81}}$ **36.** $\sqrt[4]{\frac{81}{16}}$

37. $\sqrt{\frac{100x^2}{25}}$ **38.** $\sqrt{\frac{100x^2}{4}}$ **39.** $\sqrt{\frac{81a^2b^2}{9}}$ **40.** $\sqrt{\frac{64a^2b^2}{16}}$

41. $\sqrt[3]{\frac{27x^3}{8y^3}}$ **42.** $\sqrt[4]{\frac{625x^4}{256y^4}}$

Use combinations of the product and quotient properties for radicals to simplify the following problems as much as possible. Assume all variables represent nonnegative numbers.

43. $\sqrt{\frac{50}{9}}$ **44.** $\sqrt{\frac{32}{49}}$ **45.** $\sqrt{\frac{75}{25}}$ **46.** $\sqrt{\frac{300}{4}}$

47. $\sqrt{\frac{128}{49}}$ **48.** $\sqrt{\frac{32}{64}}$ **49.** $\sqrt{\frac{288x}{25}}$ **50.** $\sqrt{\frac{28y}{81}}$

51. $\sqrt{\frac{54a^2}{25}}$ **52.** $\sqrt{\frac{243a^2}{49}}$ **53.** $\frac{3\sqrt{50}}{2}$ **54.** $\frac{5\sqrt{48}}{3}$

55. $\frac{7\sqrt{28y^2}}{3}$ **56.** $\frac{9\sqrt{243x^2}}{2}$ **57.** $\frac{2\sqrt{20x^2y^3}}{3}$ **58.** $\frac{5\sqrt{27x^3y^2}}{2}$

59. $\frac{6\sqrt{54a^2b^3}}{5}$ **60.** $\frac{7\sqrt{75a^3b^2}}{6}$ **61.** $\frac{5\sqrt{72a^2b^2}}{\sqrt{36}}$ **62.** $\frac{2\sqrt{27a^2b^2}}{\sqrt{9}}$

Simplify each expression.

63. $\frac{\sqrt{20}}{4}$ **64.** $\frac{3\sqrt{20}}{15}$ **65.** $\frac{4+\sqrt{12}}{2}$ **66.** $\frac{2+\sqrt{9}}{5}$

67. $\frac{\sqrt{12}}{4}$ **68.** $\frac{2\sqrt{32}}{8}$ **69.** $\frac{9+\sqrt{27}}{3}$ **70.** $\frac{-6-\sqrt{64}}{2}$

71. $\frac{8+\sqrt{28}}{6}$ **72.** $\frac{-3-\sqrt{45}}{6}$ **73.** $\frac{10+\sqrt{75}}{5}$ **74.** $\frac{-6+\sqrt{45}}{3}$

75. $\frac{-2-\sqrt{27}}{6}$ **76.** $\frac{12-\sqrt{12}}{6}$ **77.** $\frac{-4-\sqrt{8}}{2}$ **78.** $\frac{6-\sqrt{48}}{8}$

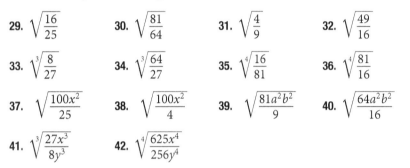

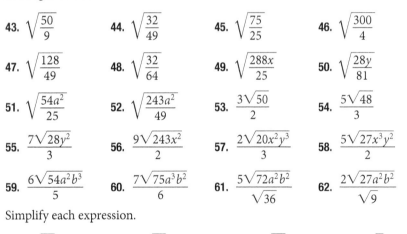

79. Simplify $\sqrt{b^2 - 4ac}$ if:

a. $a = 2, b = 4, c = -3$ **b.** $a = 1, b = 1, c = -6$

c. $a = 1, b = 1, c = -11$ **d.** $a = 3, b = 6, c = 2$

80. Simplify $\sqrt{b^2 - 4ac}$ if:

a. $a = -3, b = -4, c = 2$ **b.** $a = 1, b = -3, c = 2$

c. $a = 4, b = 8, c = 1$ **d.** $a = 1, b = -4, c = 1$

Simplify.

81. **a.** $\sqrt{32x^{10}y^5}$ **b.** $\sqrt[3]{32x^{10}y^5}$ **c.** $\sqrt[4]{32x^{10}y^5}$ **d.** $\sqrt[5]{32x^{10}y^5}$

82. **a.** $\sqrt{16x^8y^4}$ **b.** $\sqrt{16x^4y^8}$ **c.** $\sqrt[3]{16x^8y^4}$ **d.** $\sqrt[4]{16x^8y^4}$

83. **a.** $\sqrt{4}$ **b.** $\sqrt{0.04}$ **c.** $\sqrt{400}$ **d.** $\sqrt{0.0004}$

84. **a.** $\sqrt[3]{8}$ **b.** $\sqrt[3]{0.008}$ **c.** $\sqrt[3]{80}$ **d.** $\sqrt[3]{8{,}000}$

Use a calculator to help complete the following tables. If an answer needs rounding, round to the nearest thousandth.

85.

x	$\sqrt{x}$	$2\sqrt{x}$	$\sqrt{4x}$
1			
2			
3			
4			

86.

x	$\sqrt{x}$	$2\sqrt{x}$	$\sqrt{4x}$
1			
4			
9			
16			

87.

x	$\sqrt{x}$	$3\sqrt{x}$	$\sqrt{9x}$
1			
2			
3			
4			

88.

x	$\sqrt{x}$	$3\sqrt{x}$	$\sqrt{9x}$
1			
4			
9			
16			

Applications

89. Falling Time The formula that gives the number of seconds t it takes for an object to reach the ground when dropped from the top of a building h feet high is

$$t = \sqrt{\frac{h}{16}}$$

If a rock is dropped from the top of a building 25 feet high, how long will it take for the rock to hit the ground?

90. Falling Time Using the formula given in Problem 89, how long will it take for an object dropped from a building 100 feet high to reach the ground?

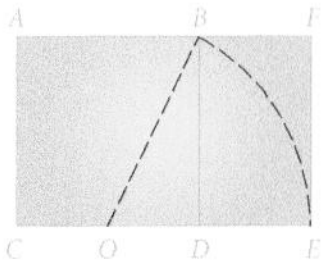

91. Golden Ratio The golden ratio is the ratio of the length to the width in any golden rectangle. The exact value of this number is $\frac{1+\sqrt{5}}{2}$. Use a calculator to find a decimal approximation of this number and round it to the nearest thousandth.

92. Golden Ratio The reciprocal of the golden ratio is $\frac{2}{1+\sqrt{5}}$. Find a decimal approximation of this number that is accurate to the nearest thousandth.

Learning Objectives Assessment

The following problems can be used to help assess if you have successfully met the learning objectives for this section.

93. Simplify: $\sqrt{80}$.

a. $2\sqrt{20}$ **b.** $16\sqrt{5}$ **c.** $5\sqrt{4}$ **d.** $4\sqrt{5}$

94. Simplify: $2\sqrt[3]{54x^3}$.

a. $6x\sqrt[3]{6}$ **b.** $54x\sqrt[3]{2}$ **c.** $6x\sqrt[3]{2}$ **d.** $6x\sqrt[3]{6x}$

95. Simplify: $\sqrt{\frac{60}{49}}$.

a. $\frac{4\sqrt{15}}{7}$ **b.** $\frac{2\sqrt{15}}{7}$ **c.** $\frac{5\sqrt{12}}{7}$ **d.** $\frac{6\sqrt{10}}{49}$

96. Simplify: $\frac{12+\sqrt{18}}{6}$.

a. $\frac{4+\sqrt{2}}{2}$ **b.** $\frac{12+\sqrt{2}}{2}$ **c.** $2+3\sqrt{2}$ **d.** $6+\sqrt{2}$

Getting Ready for the Next Section

Simplify.

97. $\sqrt{4x^3y^2}$ **98.** $\sqrt{9x^2y^3}$ **99.** $\sqrt{20}$ **100.** $\sqrt{75}$

101. $5\sqrt{12}$ **102.** $3\sqrt{45}$ **103.** $\frac{3}{2}\sqrt{24}$ **104.** $\frac{2}{3}\sqrt{54}$

105. $5x - 3x$ **106.** $4y + y$ **107.** $2x + 3y - 8x$ **108.** $9a + 11a + 6b$

Addition and Subtraction of Radical Expressions 8.3

Learning Objectives

In this section, we will learn how to:

1. Identify like radicals.
2. Add and subtract radical expressions.

Introduction

Now we will see how to use the skills we learned in the previous section to add or subtract radical expressions. To add two or more radical expressions, we apply the distributive property. Adding radical expressions is similar to adding similar terms of polynomials.

VIDEO EXAMPLES

SECTION 8.3

EXAMPLE 1 Combine terms in the expression $3\sqrt{5} - 7\sqrt{5}$.

SOLUTION The two terms $3\sqrt{5}$ and $7\sqrt{5}$ each have $\sqrt{5}$ in common. Since $3\sqrt{5}$ means 3 times $\sqrt{5}$, or $3 \cdot \sqrt{5}$, and $7\sqrt{5}$ means $7 \cdot \sqrt{5}$ we apply the distributive property:

$$3\sqrt{5} - 7\sqrt{5} = (3 - 7)\sqrt{5} \quad \text{Distributive property}$$

$$= -4\sqrt{5} \quad 3 - 7 = -4$$

When using the distributive property to combine terms containing radicals, the radical expression must be a common factor. This means that each expression must contain exactly the same radical.

Like Radical Terms

We extend our definition of similar terms from Section 2.1 to include radical expressions as follows:

DEFINITION *Like Terms Containing Radicals*

With terms containing radical expressions, ***like terms***, or ***similar terms***, must have the same variable factor and identical radical expressions (the same index and the same radicand).

Here are some examples of like terms and unlike terms containing radical expressions.

Like Terms	Unlike Terms
$\sqrt{2}, -5\sqrt{2}$	$\sqrt{3}, -5\sqrt{2}$
$\frac{1}{2}\sqrt[3]{3}, 6\sqrt[3]{3}$	$4\sqrt{2}, 6\sqrt[3]{2}$
$\sqrt{5x}, 4\sqrt{5x}$	$\sqrt{5x}, \sqrt{5y}$
$-9x\sqrt{6}, 7x\sqrt{6}$	$-9x\sqrt{6}, 7\sqrt{6x}$
$x^2\sqrt[3]{4}, 2x^2\sqrt[3]{4}$	$x^2\sqrt[3]{4}, 2x\sqrt[3]{4}$

Adding and Subtracting Radical Expressions

We can add or subtract radical expressions by using the distributive property to combine like terms.

EXAMPLE 2 Combine terms in the expression $7\sqrt{2} - 3\sqrt{2} + 6\sqrt{2}$.

SOLUTION All three terms are like terms, so we have

$$7\sqrt{2} - 3\sqrt{2} + 6\sqrt{2} = (7 - 3 + 6)\sqrt{2} \quad \text{Distributive property}$$
$$= 10\sqrt{2} \quad \text{Addition}$$

EXAMPLE 3 Simplify: $6\sqrt{5} + 5\sqrt{2} - 3\sqrt{5}$.

SOLUTION We combine the two terms that are like terms:

$$6\sqrt{5} + 5\sqrt{2} - 3\sqrt{5} = (6\sqrt{5} - 3\sqrt{5}) + 5\sqrt{2} \quad \text{Commutative and associative properties}$$
$$= 3\sqrt{5} + 5\sqrt{2} \quad \text{Combine like terms}$$

In the first three examples, each term was a radical expression in simplified form. If one or more terms are not in simplified form, we must put them into simplified form and then combine terms, if possible.

RULE

To combine two or more radical expressions, put each expression in simplified form, and then apply the distributive property, if possible.

EXAMPLE 4 Combine terms in the expression $3\sqrt{50} + 2\sqrt{32}$.

SOLUTION We begin by putting each term into simplified form:

$$3\sqrt{50} + 2\sqrt{32} = 3\sqrt{25}\sqrt{2} + 2\sqrt{16}\sqrt{2} \quad \text{Product property for radicals}$$
$$= 3 \cdot 5\sqrt{2} + 2 \cdot 4\sqrt{2} \quad \sqrt{25} = 5 \text{ and } \sqrt{16} = 4$$
$$= 15\sqrt{2} + 8\sqrt{2} \quad \text{Multiplication}$$

Applying the distributive property to the last line, we have

$$15\sqrt{2} + 8\sqrt{2} = (15 + 8)\sqrt{2} \quad \text{Distributive property}$$
$$= 23\sqrt{2} \quad 15 + 8 = 23$$

EXAMPLE 5 Combine terms in the expression: $\sqrt[3]{375} + \sqrt[3]{81} - 6\sqrt[3]{3}$.

SOLUTION

$\sqrt[3]{375} + \sqrt[3]{81} - 6\sqrt[3]{3}$	$= \sqrt[3]{125}\sqrt[3]{3} + \sqrt[3]{27}\sqrt[3]{3} - 6\sqrt[3]{3}$	Product property for radicals
	$= 5\sqrt[3]{3} + 3\sqrt[3]{3} - 6\sqrt[3]{3}$	$\sqrt[3]{125} = 5$ and $\sqrt[3]{27} = 3$
	$= (5 + 3 - 6)\sqrt[3]{3}$	Distributive property
	$= 2\sqrt[3]{3}$	Addition

The most time-consuming part of combining most radical expressions is simplifying each term in the expression. Once this has been done, applying the distributive property is simple and fast.

EXAMPLE 6 Simplify: $a\sqrt{12} + 5\sqrt{3a^2}$ (assume $a > 0$).

SOLUTION We must assume that a represents a positive number. Then we simplify each term in the expression by putting it in simplified form for radicals:

$a\sqrt{12} + 5\sqrt{3a^2}$	$= a\sqrt{4}\sqrt{3} + 5\sqrt{a^2}\sqrt{3}$	Product property for radicals
	$= a \cdot 2\sqrt{3} + 5 \cdot a\sqrt{3}$	$\sqrt{4} = 2$ and $\sqrt{a^2} = a$
	$= 2a\sqrt{3} + 5a\sqrt{3}$	Commutative property
	$= (2a + 5a)\sqrt{3}$	Distributive property
	$= 7a\sqrt{3}$	Addition

EXAMPLE 7 Simplify: $\sqrt{20x^3} - 3x\sqrt{45x} + 10\sqrt{25x^2}$ (assume $x \geq 0$).

SOLUTION

$$\sqrt{20x^3} - 3x\sqrt{45x} + 10\sqrt{25x^2}$$
$$= \sqrt{4x^2}\sqrt{5x} - 3x\sqrt{9}\sqrt{5x} + 10\sqrt{25x^2}$$
$$= 2x\sqrt{5x} - 3x \cdot 3\sqrt{5x} + 10 \cdot 5x$$
$$= 2x\sqrt{5x} - 9x\sqrt{5x} + 50x$$

Each term is now in simplified form. The best we can do next is to combine the first two terms. The last term does not have the common radical $\sqrt{5x}$.

$$2x\sqrt{5x} - 9x\sqrt{5x} + 50x = (2x - 9x)\sqrt{5x} + 50x$$
$$= -7x\sqrt{5x} + 50x$$

We have, in any case, succeeded in reducing the number of terms in our original problem.

Our last example involves a radical expression similar to those we looked at in Example 11 of the previous section. Expressions of this type often arise when solving quadratic equations, so it is important to be able to simplify them correctly.

EXAMPLE 8 Simplify: $\frac{6 + \sqrt{12}}{4}$.

SOLUTION We begin by writing $\sqrt{12}$ as $2\sqrt{3}$:

$$\frac{6 + \sqrt{12}}{4} = \frac{6 + 2\sqrt{3}}{4} \qquad \sqrt{12} = \sqrt{4 \cdot 3} = \sqrt{4}\sqrt{3} = 2\sqrt{3}$$

It is a common mistake for students to add the 6 and 2 in the numerator getting $8\sqrt{3}$. But 6 and $2\sqrt{3}$ are not like terms, so we cannot combine them. Instead, we factor 2 from the numerator and denominator and then reduce to lowest terms.

$$\frac{6 + 2\sqrt{3}}{4} = \frac{2(3 + \sqrt{3})}{2 \cdot 2}$$

$$= \frac{3 + \sqrt{3}}{2}$$

Getting Ready for Class

After reading through the preceding section, respond in your own words and in complete sentences.

A. What are similar radicals?

B. When can we add two radical expressions?

C. What is the first step when adding or subtracting expressions containing radicals?

D. It is not possible to add $\sqrt{2}$ and $\sqrt{3}$. Explain why.

In each of the following problems, simplify each term, if necessary, and then use the distributive property to combine terms, if possible.

1. $3\sqrt{2} + 4\sqrt{2}$
2. $7\sqrt{3} + 2\sqrt{3}$
3. $9\sqrt{5} - 7\sqrt{5}$
4. $6\sqrt{7} - 10\sqrt{7}$
5. $\sqrt{3} + 6\sqrt{3}$
6. $\sqrt{2} + 10\sqrt{2}$
7. $\frac{5}{8}\sqrt{5} - \frac{3}{7}\sqrt{5}$
8. $\frac{5}{6}\sqrt{11} - \frac{7}{9}\sqrt{11}$
9. $14\sqrt[3]{13} - \sqrt[3]{13}$
10. $-2\sqrt[3]{6} - 9\sqrt[3]{6}$
11. $-3\sqrt[3]{10} + 9\sqrt[3]{10}$
12. $11\sqrt[3]{11} + \sqrt[3]{11}$
13. $5\sqrt{5} + \sqrt{5}$
14. $\sqrt{6} - 10\sqrt{6}$
15. $\sqrt{8} + 2\sqrt{2}$
16. $\sqrt{20} + 3\sqrt{5}$
17. $3\sqrt{3} - \sqrt{27}$
18. $4\sqrt{5} - \sqrt{80}$
19. $5\sqrt{12} - 10\sqrt{48}$
20. $3\sqrt{300} - 5\sqrt{27}$
21. $-\sqrt[3]{54} - \sqrt[3]{2}$
22. $5\sqrt[3]{320} + 9\sqrt[3]{40}$
23. $\frac{1}{5}\sqrt{75} - \frac{1}{2}\sqrt{12}$
24. $\frac{1}{2}\sqrt{24} + \frac{1}{5}\sqrt{150}$
25. $\frac{3}{4}\sqrt{8} + \frac{3}{10}\sqrt{75}$
26. $\frac{5}{6}\sqrt{54} - \frac{3}{4}\sqrt{24}$
27. $\sqrt{27} - 2\sqrt{12} + \sqrt{3}$
28. $\sqrt{20} + 3\sqrt{45} - \sqrt{5}$
29. $\frac{5}{6}\sqrt{72} - \frac{3}{8}\sqrt{8} + \frac{3}{10}\sqrt{50}$
30. $\frac{3}{4}\sqrt{24} - \frac{5}{6}\sqrt{54} - \frac{7}{10}\sqrt{150}$
31. $5\sqrt{7} + 2\sqrt{28} - 4\sqrt{63}$
32. $3\sqrt{3} - 5\sqrt{27} + 8\sqrt{75}$
33. $5\sqrt[3]{16} - 3\sqrt[3]{128} + \sqrt[3]{432}$
34. $6\sqrt[3]{81} + \sqrt[3]{375} - 4\sqrt[3]{24}$
35. $6\sqrt{24} - 2\sqrt{12} + 5\sqrt{27}$
36. $5\sqrt{50} + 8\sqrt{12} - \sqrt{32}$
37. $6\sqrt{48} - \sqrt{72} - 3\sqrt{300}$
38. $7\sqrt{44} - 8\sqrt{99} + \sqrt{176}$

All variables in the following problems represent nonnegative real numbers. Simplify each term, and combine, if possible.

39. $\sqrt{x^3} + x\sqrt{x}$
40. $2\sqrt{x} - 2\sqrt{4x}$
41. $5\sqrt{3a^2} - a\sqrt{3}$
42. $6a\sqrt{a} + 7\sqrt{a^3}$
43. $5\sqrt{8x^3} + x\sqrt{50x}$
44. $2\sqrt{27x^2} - x\sqrt{48}$
45. $3\sqrt{75x^3y} - 2x\sqrt{3xy}$
46. $9\sqrt{24x^3y^2} - 5x\sqrt{54xy^2}$
47. $\sqrt{20ab^2} - b\sqrt{45a}$
48. $4\sqrt{a^3b^2} - 5a\sqrt{ab^2}$
49. $9\sqrt{18x^3} - 2x\sqrt{48x}$
50. $8\sqrt{72x^2} - x\sqrt{8}$
51. $2\sqrt[3]{3x^3} + x\sqrt[3]{24}$
52. $5x\sqrt[3]{x} - \sqrt[3]{27x^4}$
53. $6\sqrt[3]{64a^5b^3} - 5a\sqrt[3]{8a^2b^3}$
54. $2\sqrt[3]{250a^4b^7} + 9b^2\sqrt[3]{2a^4b}$
55. $7\sqrt{50x^2y} + 8x\sqrt{8y} - 7\sqrt{32x^2y}$
56. $6\sqrt{44x^3y^3} - 8x\sqrt{99xy^2} - 6y\sqrt{176x^3y}$

Simplify each expression.

57. $\dfrac{8 - \sqrt{24}}{6}$
58. $\dfrac{8 + \sqrt{48}}{8}$
59. $\dfrac{6 + \sqrt{8}}{2}$
60. $\dfrac{4 - \sqrt{12}}{2}$
61. $\dfrac{-10 + \sqrt{50}}{10}$
62. $\dfrac{-12 + \sqrt{20}}{6}$

63. a. $3x + 4x$ b. $3y + 4y$ c. $3\sqrt{5} + 4\sqrt{5}$
64. a. $7x + 2x$ b. $7t + 2t$ c. $7\sqrt{x} + 2\sqrt{x}$
65. a. $x + 6x$ b. $t + 6t$ c. $\sqrt{x} + 6\sqrt{x}$
66. a. $x + 10x$ b. $y + 10y$ c. $\sqrt{7} + 10\sqrt{7}$

Use a calculator to help complete the following tables. If an answer needs rounding, round to the nearest thousandth.

67.

x	$\sqrt{x^2+9}$	$x+3$
1		
2		
3		
4		
5		
6		

68.

x	$\sqrt{x^2+16}$	$x+4$
1		
2		
3		
4		
5		
6		

69.

x	$\sqrt{x+3}$	$\sqrt{x}+\sqrt{9}$
1		
2		
3		
4		
5		
6		

70.

x	$\sqrt{x+4}$	$\sqrt{x}+\sqrt{2}$
1		
2		
3		
4		
5		
6		

71. Comparing Expressions The following statement is false. Correct the right side to make the statement true.

$$4\sqrt{3} + 5\sqrt{3} = 9\sqrt{6}$$

72. Comparing Expressions The following statement is false. Correct the right side to make the statement true.

$$7\sqrt{5} - 3\sqrt{5} = 4\sqrt{25}$$

Learning Objectives Assessment

The following problems can be used to help assess if you have successfully met the learning objectives for this section.

73. Which of the following are like terms?

a. $2x\sqrt{3}, -5\sqrt{3x}$ **b.** $2x\sqrt{3}, -5x\sqrt[3]{3}$

c. $2x\sqrt{3}, -5x\sqrt{3}$ **d.** $2x^2\sqrt{3}, -5x\sqrt{3}$

74. Simplify: $5\sqrt{20} + 8\sqrt{45}$.

a. $13\sqrt{65}$ **b.** $34\sqrt{5}$ **c.** $92\sqrt{5}$ **d.** Not possible

Getting Ready for the Next Section

Simplify.

75. $\sqrt{12}$ **76.** $\sqrt{18}$ **77.** $\sqrt{20}$ **78.** $\sqrt{50}$

79. $\sqrt[3]{16}$ **80.** $\sqrt[3]{54}$

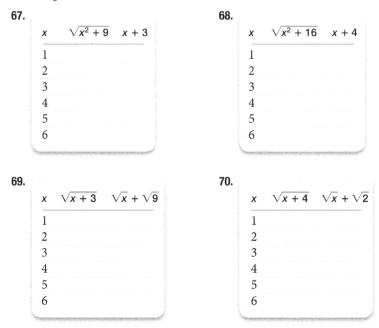

Multiply.

81. $x(2x + 5)$ **82.** $3x(x + 6)$ **83.** $(x + 1)^2$ **84.** $(x - 4)^2$

85. $(x + 8)(x - 2)$ **86.** $(x + 4)(x - 5)$ **87.** $(x + 6)(x - 6)$ **88.** $(x - 3)(x + 3)$

89. $(2x + 3)(3x - 4)$ **90.** $(3x + 1)(4x + 3)$

Combine like terms.

91. $5 + 7\sqrt{5} + 2\sqrt{5} + 14$ **92.** $3 + 5\sqrt{3} + 2\sqrt{3} + 10$

93. $x - 7\sqrt{x} + 3\sqrt{x} - 21$ **94.** $x - 6\sqrt{x} + 8\sqrt{x} - 48$

Multiplication of Radicals 8.4

Learning Objectives

In this section, we will learn how to:

1. Multiply expressions containing radicals.
2. Find nth powers of nth roots.
3. Multiply radical expressions that are conjugates.

Introduction

In this section we will look at multiplication of expressions that contain radicals. As you will see, multiplication of expressions that contain radicals is very similar to multiplication of polynomials. We will also continue to make sure that our final answers are in simplified form (which we introduced in Section 8.2).

When multiplying radicals, we use the product property, which we repeat here for convenience.

PROPERTY *Product Property for Radicals*

If a and b represent nonnegative real numbers, then

$$\sqrt[n]{a} \cdot \sqrt[n]{b} = \sqrt[n]{ab}$$

In words: The product of the nth roots is equal to the nth root of the product.

Notice that this property allows us to combine the product of two radicals into a single root as long as both radicals have the same index. Unlike adding and subtracting, when multiplying radicals the radicands do *not* need to be identical.

VIDEO EXAMPLES

SECTION 8.4

EXAMPLE 1 Multiply $\sqrt{2} \cdot \sqrt{14}$.

SOLUTION Using the product property of radicals, we have

$$\begin{aligned} \sqrt{2} \cdot \sqrt{14} &= \sqrt{2 \cdot 14} && \text{Product property for radicals} \\ &= \sqrt{28} && \text{Multiplication} \\ &= 2\sqrt{7} && \sqrt{28} = \sqrt{4 \cdot 7} = \sqrt{4}\sqrt{7} = 2\sqrt{7} \end{aligned}$$

Notice that we had to simplify $\sqrt{28}$ because 28 contains a perfect square of 4.

EXAMPLE 2 Multiply: $(3\sqrt{5})(2\sqrt{7})$.

SOLUTION We can rearrange the order and grouping of the numbers in this product by applying the commutative and associative properties. Following that, we apply the product property for radicals and multiply:

$$\begin{aligned} (3\sqrt{5})(2\sqrt{7}) &= (3 \cdot 2)(\sqrt{5}\sqrt{7}) && \text{Commutative and associative properties} \\ &= (3 \cdot 2)(\sqrt{5 \cdot 7}) && \text{Product property for radicals} \\ &= 6\sqrt{35} && \text{Multiplication} \end{aligned}$$

In actual practice, it is not necessary to show either of the first two steps, although you may want to show them on the first few problems you work, just to be sure you understand them.

EXAMPLE 3 Multiply: $\sqrt{5}(\sqrt{2} + \sqrt{5})$.

SOLUTION

$$\sqrt{5}(\sqrt{2} + \sqrt{5}) = \sqrt{5} \cdot \sqrt{2} + \sqrt{5} \cdot \sqrt{5} \qquad \text{Distributive property}$$
$$= \sqrt{10} + \sqrt{25} \qquad \text{Product property}$$
$$= \sqrt{10} + 5 \qquad \sqrt{25} = 5$$

In Example 3, notice that

$$\sqrt{5} \cdot \sqrt{5} = \sqrt{25} = 5$$

which we could also express as

$$\sqrt{5} \cdot \sqrt{5} = (\sqrt{5})^2 = 5$$

When we square a square root, the result is simply the radicand of the root. We generalize this fact with the following property.

PROPERTY ***Power Property for Radicals***

1. $(\sqrt{a})^2 = a$ for $a \geq 0$.
2. $(\sqrt[n]{a})^n = a$ where $a \geq 0$ when n is even.

EXAMPLES

4. $(\sqrt{11})^2 = 11$
5. $(\sqrt[3]{-6})^3 = -6$
6. $(\sqrt[4]{2})^4 = 2$
7. $(\sqrt[5]{3})^5 = 3$

EXAMPLE 8 Multiply: $3\sqrt{2}(2\sqrt{5} + 5\sqrt{6})$.

SOLUTION $3\sqrt{2}(2\sqrt{5} + 5\sqrt{6})$

$$= 3\sqrt{2} \cdot 2\sqrt{5} + 3\sqrt{2} \cdot 5\sqrt{6} \qquad \text{Distributive property}$$
$$= 3 \cdot 2 \cdot \sqrt{2}\sqrt{5} + 3 \cdot 5\sqrt{2}\sqrt{6} \qquad \text{Commutative property}$$
$$= 6\sqrt{10} + 15\sqrt{12} \qquad \text{Multiplication}$$
$$= 6\sqrt{10} + 15(2\sqrt{3}) \qquad \sqrt{12} = \sqrt{4 \cdot 3} = \sqrt{4}\sqrt{3} = 2\sqrt{3}$$
$$= 6\sqrt{10} + 30\sqrt{3} \qquad \text{Multiply}$$

Each item in the last line is in simplified form, so the problem is complete.

EXAMPLE 9 Multiply: $(\sqrt{5} + 2)(\sqrt{5} + 7)$.

SOLUTION We multiply using the FOIL method that we used to multiply binomials:

$$(\sqrt{5} + 2)(\sqrt{5} + 7) = \underset{\text{F}}{(\sqrt{5})^2} + \underset{\text{O}}{7\sqrt{5}} + \underset{\text{I}}{2\sqrt{5}} + \underset{\text{L}}{14}$$
$$= 5 + 9\sqrt{5} + 14$$
$$= 19 + 9\sqrt{5}$$

We must be careful not to try to simplify further by adding 19 and 9. We can add only radical expressions that have a common radical part; 19 and $9\sqrt{5}$ are not similar.

EXAMPLE 10 Multiply: $(\sqrt{x} + 3)(\sqrt{x} - 7)$.

SOLUTION Remember, we are assuming that any variables that appear under a radical represent nonnegative numbers.

$$\begin{aligned}(\sqrt{x} + 3)(\sqrt{x} - 7) &= \underset{\text{F}}{(\sqrt{x})^2} - \underset{\text{O}}{7\sqrt{x}} + \underset{\text{I}}{3\sqrt{x}} - \underset{\text{L}}{21} \\ &= x - 4\sqrt{x} - 21\end{aligned}$$

EXAMPLE 11 Expand and simplify: $(\sqrt{3} - 2)^2$.

SOLUTION Multiplying $\sqrt{3} - 2$ times itself, we have

$$\begin{aligned}(\sqrt{3} - 2)^2 &= (\sqrt{3} - 2)(\sqrt{3} - 2) \\ &= (\sqrt{3})^2 - 2\sqrt{3} - 2\sqrt{3} + 4 \\ &= 3 - 4\sqrt{3} + 4 \\ &= 7 - 4\sqrt{3}\end{aligned}$$

EXAMPLE 12 Multiply: $(\sqrt{6} + \sqrt{3})^2$.

SOLUTION We could use the FOIL method as we did in Example 11, or we could use the formula $(a + b)^2 = a^2 + 2ab + b^2$. Here is how the problem looks using the formula:

$$\begin{aligned}(\sqrt{6} + \sqrt{3})^2 &= (\sqrt{6})^2 + 2(\sqrt{6})(\sqrt{3}) + (\sqrt{3})^2 \\ &= 6 + 2\sqrt{18} + 3 \\ &= 9 + 2(3\sqrt{2}) \\ &= 9 + 6\sqrt{2}\end{aligned}$$

EXAMPLE 13 Multiply: $(\sqrt{5} + \sqrt{2})(\sqrt{5} - \sqrt{2})$.

SOLUTION We can apply the formula $(x + y)(x - y) = x^2 - y^2$ to obtain

$$\begin{aligned}(\sqrt{5} + \sqrt{2})(\sqrt{5} - \sqrt{2}) &= (\sqrt{5})^2 - (\sqrt{2})^2 \\ &= 5 - 2 \\ &= 3\end{aligned}$$

The expressions $\sqrt{5} + \sqrt{2}$ and $\sqrt{5} - \sqrt{2}$ are called *conjugates* of each other. In general, the conjugate of $a + b$ is $a - b$. The product of conjugates always results in a difference of squares:

$$(a + b)(a - b) = a^2 - b^2$$

Notice in Example 13 that with radical expressions, the product of conjugates results in a rational number. That is, the answer does not contain a radical. We will see how this is useful in the next section when we look at dividing radical expressions.

EXAMPLE 14 Multiply: $(\sqrt{a} + \sqrt{b})(\sqrt{a} - \sqrt{b})$.

SOLUTION These expressions are conjugates. The product is a difference of squares.

$$(\sqrt{a} + \sqrt{b})(\sqrt{a} - \sqrt{b}) = (\sqrt{a})^2 - (\sqrt{b})^2$$
$$= a - b$$

EXAMPLE 15 Multiply: $\sqrt[3]{9} \cdot \sqrt[3]{3}$.

SOLUTION Using the product property for radicals, we have

$$\sqrt[3]{9} \cdot \sqrt[3]{3} = \sqrt[3]{9 \cdot 3}$$
$$= \sqrt[3]{27}$$
$$= 3$$

EXAMPLE 16 Multiply: $(\sqrt[3]{4} - 5)^2$.

SOLUTION Multiplying $\sqrt[3]{4} - 5$ by itself and then using the FOIL method, we have

$$(\sqrt[3]{4} - 5)^2 = (\sqrt[3]{4} - 5)(\sqrt[3]{4} - 5)$$
$$= \sqrt[3]{4} \cdot \sqrt[3]{4} - 5\sqrt[3]{4} - 5\sqrt[3]{4} + 25$$
$$= \sqrt[3]{16} - 10\sqrt[3]{4} + 25$$
$$= 2\sqrt[3]{2} - 10\sqrt[3]{4} + 25 \qquad \sqrt[3]{16} = \sqrt[3]{8 \cdot 2} = 2\sqrt[3]{2}$$

Note In Example 16, we could also have written

$$\sqrt[3]{4} \cdot \sqrt[3]{4} = (\sqrt[3]{4})^2$$
$$= \sqrt[3]{4^2}$$
$$= \sqrt[3]{16}$$

Getting Ready for Class

After reading through the preceding section, respond in your own words and in complete sentences.

A. Describe how you would use the commutative and associative properties to multiply $3\sqrt{5}$ and $2\sqrt{7}$.

B. In order to multiply two radicals, what must they have in common?

C. Explain why $(\sqrt{5} + 7)^2$ is not the same as $5 + 49$.

D. What are conjugates?

Simplify each of the following expressions. Assume all variables represent nonnegative values.

1. $(\sqrt{2})^2$
2. $(\sqrt{6})^2$
3. $(-\sqrt{7})^2$
4. $(-\sqrt{13})^2$
5. $(\sqrt{x})^2$
6. $(\sqrt{y})^2$
7. $(\sqrt[3]{4})^3$
8. $(\sqrt[3]{9})^3$
9. $(\sqrt[3]{-10})^3$
10. $(\sqrt[3]{-2})^3$
11. $(\sqrt[3]{a})^3$
12. $(\sqrt[4]{b})^4$

Perform the following multiplications. Assume all variables represent nonnegative values. All answers should be in simplified form for radical expressions.

13. $\sqrt{3}\sqrt{2}$
14. $\sqrt{5}\sqrt{6}$
15. $\sqrt{6}\sqrt{2}$
16. $\sqrt{6}\sqrt{3}$
17. $(2\sqrt{3})(5\sqrt{7})$
18. $(3\sqrt{2})(4\sqrt{5})$
19. $(4\sqrt{3})(2\sqrt{6})$
20. $(7\sqrt{6})(3\sqrt{2})$
21. $\sqrt[3]{3}\sqrt[3]{4}$
22. $\sqrt[3]{2}\sqrt[3]{10}$
23. $\sqrt[3]{6}\sqrt[3]{9}$
24. $\sqrt[3]{4}\sqrt[3]{14}$
25. $(9\sqrt[3]{12})(2\sqrt[3]{2})$
26. $(12\sqrt[3]{9})(3\sqrt[3]{6})$
27. $\sqrt{2}(\sqrt{3} - 1)$
28. $\sqrt{3}(\sqrt{5} + 2)$
29. $\sqrt{2}(\sqrt{3} + \sqrt{2})$
30. $\sqrt{5}(\sqrt{7} - \sqrt{5})$
31. $\sqrt{3}(2\sqrt{2} + \sqrt{3})$
32. $\sqrt{11}(3\sqrt{2} - \sqrt{11})$
33. $\sqrt{3}(2\sqrt{3} - \sqrt{5})$
34. $\sqrt{7}(\sqrt{14} - \sqrt{7})$
35. $2\sqrt{3}(\sqrt{2} + \sqrt{5})$
36. $3\sqrt{2}(\sqrt{3} + \sqrt{2})$
37. $5\sqrt[3]{4}(\sqrt[3]{2} + 3\sqrt[3]{5})$
38. $4\sqrt[3]{2}(2\sqrt[3]{6} - \sqrt[3]{4})$
39. $(\sqrt{2} + 1)^2$
40. $(\sqrt{5} - 4)^2$
41. $(\sqrt{x} + 3)^2$
42. $(\sqrt{x} - 4)^2$
43. $(5 - \sqrt{2})^2$
44. $(2 + \sqrt{5})^2$
45. $\left(\sqrt{a} - \frac{1}{2}\right)^2$
46. $\left(\sqrt{a} + \frac{1}{2}\right)^2$
47. $(\sqrt{3} + \sqrt{7})^2$
48. $(\sqrt{11} - \sqrt{2})^2$
49. $(\sqrt{2} - \sqrt{10})^2$
50. $(\sqrt{3} + \sqrt{6})^2$
51. $(\sqrt{5} + 3)(\sqrt{5} + 2)$
52. $(\sqrt{7} + 4)(\sqrt{7} - 5)$
53. $(\sqrt{2} - 5)(\sqrt{2} + 6)$
54. $(\sqrt{3} + 8)(\sqrt{3} - 2)$
55. $(2\sqrt{7} + 3)(3\sqrt{7} - 4)$
56. $(3\sqrt{5} + 1)(4\sqrt{5} + 3)$
57. $(2\sqrt{x} + 4)(3\sqrt{x} + 2)$
58. $(3\sqrt{x} + 5)(4\sqrt{x} + 2)$
59. $\left(\sqrt{3} + \frac{1}{2}\right)\left(\sqrt{2} + \frac{1}{3}\right)$
60. $\left(\sqrt{5} - \frac{1}{4}\right)\left(\sqrt{3} + \frac{1}{5}\right)$
61. $\left(\sqrt{a} + \frac{1}{3}\right)\left(\sqrt{a} + \frac{2}{3}\right)$
62. $\left(\sqrt{a} + \frac{1}{4}\right)\left(\sqrt{a} + \frac{3}{4}\right)$
63. $(\sqrt{5} - 2)(\sqrt{5} + 2)$
64. $(\sqrt{6} - 3)(\sqrt{6} + 3)$
65. $(5 + \sqrt{2})(5 - \sqrt{2})$
66. $(3 - \sqrt{6})(3 + \sqrt{6})$
67. $(\sqrt{7} - \sqrt{3})(\sqrt{7} + \sqrt{3})$
68. $(\sqrt{5} + \sqrt{2})(\sqrt{5} - \sqrt{2})$
69. $(\sqrt{x} + 6)(\sqrt{x} - 6)$
70. $(\sqrt{x} + 7)(\sqrt{x} - 7)$
71. $(7\sqrt{a} + 2\sqrt{b})(7\sqrt{a} - 2\sqrt{b})$
72. $(3\sqrt{a} - 2\sqrt{b})(3\sqrt{a} + 2\sqrt{b})$

73. Paying Attention to Instructions Work each problem according to the instructions given.

a. Subtract: $(\sqrt{7} + \sqrt{3}) - (\sqrt{7} - \sqrt{3})$

b. Multiply: $(\sqrt{7} + \sqrt{3})(\sqrt{7} - \sqrt{3})$

c. Square: $(\sqrt{7} - \sqrt{3})^2$

d. Simplify: $(\sqrt{7})^2 + (\sqrt{3})^2$

74. Paying Attention to Instructions Work each problem according to the instructions given.

a. Subtract: $(\sqrt{11} - \sqrt{6}) - (\sqrt{11} + \sqrt{6})$

b. Multiply: $(\sqrt{11} - \sqrt{6})(\sqrt{11} + \sqrt{6})$

c. Square: $(\sqrt{11} - \sqrt{6})^2$

d. Simplify: $(\sqrt{11})^2 + (\sqrt{6})^2$

75. Paying Attention to Instructions Work each problem according to the instructions given.

a. Add: $(\sqrt{x} + 2) + (\sqrt{x} - 2)$

b. Multiply: $(\sqrt{x} + 2)(\sqrt{x} - 2)$

c. Square: $(\sqrt{x} + 2)^2$

d. Simplify: $(\sqrt{x})^2 + 2^2$

76. Paying Attention to Instructions Work each problem according to the instructions given.

a. Add: $(\sqrt{x} - 3) + (\sqrt{x} + 3)$

b. Multiply: $(\sqrt{x} - 3)(\sqrt{x} + 3)$

c. Square: $(\sqrt{x} + 3)^2$

d. Simplify: $(\sqrt{x})^2 + 3^2$

Applying the Concepts

77. Comparing Expressions The following statement is false. Correct the right side to make the statement true.

$$2(3\sqrt{5}) = 6\sqrt{15}$$

78. Comparing Expressions The following statement is false. Correct the right side to make the statement true.

$$5(2\sqrt{6}) = 10\sqrt{30}$$

79. Comparing Expressions The following statement is false. Correct the right side to make the statement true.

$$(\sqrt{3} + 7)^2 = 3 + 49$$

80. Comparing Expressions The following statement is false. Correct the right side to make the statement true.

$$(\sqrt{5} + \sqrt{2})^2 = 5 + 2$$

Learning Objectives Assessment

The following problems can be used to help assess if you have successfully met the learning objectives for this section.

81. Multiply: $(2\sqrt{6})(5\sqrt{10})$.

a. $40\sqrt{15}$ **b.** $20\sqrt{15}$ **c.** $7\sqrt{60}$ **d.** 28

82. Simplify: $(\sqrt{11})^2$.

a. 11 **b.** 121 **c.** $\sqrt{22}$ **d.** ± 11

83. Multiply: $(4 + \sqrt{3})(4 - \sqrt{3})$.

a. 7 **b.** 19 **c.** 13 **d.** $13 + 8\sqrt{3}$

Getting Ready for the Next Section

Simplify. Assume all variables represent nonnegative numbers.

84. $\sqrt{12}$ **85.** $\sqrt{48}$ **86.** $\sqrt{4x^2y^2}$

87. $\sqrt{25a^4}$ **88.** $\sqrt{20x^2y^3}$ **89.** $\sqrt{27x^3y^2}$

90. $(\sqrt{6})^2$ **91.** $(\sqrt{2})^2$

Perform each multiplication.

92. $\sqrt{5}\sqrt{3}$ **93.** $\sqrt{3}\sqrt{7}$ **94.** $\sqrt[3]{4}\sqrt[3]{2}$

95. $\sqrt[3]{3}\sqrt[3]{9}$ **96.** $\sqrt[3]{5xy^2}\sqrt[3]{25x^2y}$ **97.** $\sqrt[3]{6a^2b}\sqrt[3]{36ab^2}$

98. $(3 - \sqrt{5})(3 + \sqrt{5})$ **99.** $(\sqrt{3} + 2)(\sqrt{3} - 2)$

100. $(\sqrt{7} - \sqrt{2})(\sqrt{7} + \sqrt{2})$ **101.** $(\sqrt{5} + \sqrt{3})(\sqrt{5} - \sqrt{3})$

Division of Radicals

8.5

Learning Objectives

In this section, we will learn how to:

1. Divide expressions containing radicals.
2. Rationalize the denominator.

Introduction

Now we will consider expressions that involve radicals and division. To simplify radicals containing fractions, or fractions containing radicals, we rely on the quotient property of radicals, which we restate here.

PROPERTY *Quotient Property for Radicals*

If a and b represent nonnegative real numbers and $b \neq 0$, then

$$\frac{\sqrt[n]{a}}{\sqrt[n]{b}} = \sqrt[n]{\frac{a}{b}}$$

In words: The quotient of the nth roots is equal to the nth root of the quotient.

Dividing Radical Expressions

VIDEO EXAMPLES

SECTION 8.5

EXAMPLE 1 Divide: $\frac{\sqrt{60}}{\sqrt{3}}$.

SOLUTION We notice that 60 is divisible by 3. Using the quotient property for radicals, we can combine the individual square roots into a single square root. Then we can perform the division.

$$\frac{\sqrt{60}}{\sqrt{3}} = \sqrt{\frac{60}{3}} \quad \text{Quotient property}$$

$$= \sqrt{20} \quad \text{Division}$$

$$= 2\sqrt{5} \quad \sqrt{20} = \sqrt{4 \cdot 5} = \sqrt{4}\sqrt{5} = 2\sqrt{5}$$

The last step was necessary in order to express our answer as a radical in simplified form.

EXAMPLE 2 Divide: $\dfrac{6\sqrt{20}}{2\sqrt{5}}$.

SOLUTION Although there are many ways to begin this problem, we notice that 20 is divisible by 5. Using the quotient property for radicals as the first step allows us to perform this division.

$$\begin{aligned} \frac{6\sqrt{20}}{2\sqrt{5}} &= \frac{6}{2} \cdot \frac{\sqrt{20}}{\sqrt{5}} && \text{Write as separate fractions} \\ &= \frac{6}{2} \cdot \sqrt{\frac{20}{5}} && \text{Quotient property for radicals} \\ &= 3\sqrt{4} && \text{Divide} \\ &= 3 \cdot 2 && \sqrt{4} = 2 \\ &= 6 \end{aligned}$$

Note In Example 2, we are making use of a property of fractions:

$$\frac{a \cdot b}{c \cdot d} = \frac{a}{c} \cdot \frac{b}{d}$$

Notice from Example 2 that dividing radical expressions is similar to dividing monomials by monomials, where we divide the coefficients and divide the variables by subtracting exponents. With radical expressions, we divide coefficients but use the quotient property to divide the radicals.

In the following examples, we will learn how to put radical expressions involving fractions into simplified form. Specifically, we will concentrate on the second and third conditions for simplified form for radicals (no fractions under the radical sign, and no radicals in the denominator).

EXAMPLE 3 Divide: $\dfrac{5\sqrt[3]{48}}{4\sqrt[3]{2}}$.

SOLUTION We begin by dividing the coefficients and using the quotient property of radicals to divide the two cube roots.

$$\begin{aligned} \frac{5\sqrt[3]{48}}{4\sqrt[3]{2}} &= \frac{5}{4} \cdot \frac{\sqrt[3]{48}}{\sqrt[3]{2}} && \text{Write as separate fractions} \\ &= \frac{5}{4} \cdot \sqrt[3]{\frac{48}{2}} && \text{Quotient property} \\ &= \frac{5}{4} \cdot \sqrt[3]{24} && \text{Divide} \\ &= \frac{5}{4} \cdot 2\sqrt[3]{3} && \sqrt[3]{24} = \sqrt[3]{8 \cdot 3} = \sqrt[3]{8} \cdot \sqrt[3]{3} = 2\sqrt[3]{3} \\ &= \frac{5 \cdot 2\sqrt[3]{3}}{2 \cdot 2} && \text{Multiply} \\ &= \frac{5\sqrt[3]{3}}{2} && \text{Reduce} \end{aligned}$$

Rationalizing the Denominator

EXAMPLE 4 Simplify: $\sqrt{\frac{1}{2}}$.

SOLUTION The expression $\sqrt{\frac{1}{2}}$ is not in simplified form because there is a fraction under the radical sign. We can change this by applying the quotient property for radicals:

$$\sqrt{\frac{1}{2}} = \frac{\sqrt{1}}{\sqrt{2}} \qquad \text{Quotient property for radicals}$$

$$= \frac{1}{\sqrt{2}} \qquad \sqrt{1} = 1$$

The expression $\frac{1}{\sqrt{2}}$ is still not in simplified form because there is a radical sign in the denominator. If we multiply the numerator and denominator of $\frac{1}{\sqrt{2}}$ by $\sqrt{2}$, the denominator becomes $\sqrt{2} \cdot \sqrt{2} = \sqrt{4} = 2$:

$$\frac{1}{\sqrt{2}} = \frac{1}{\sqrt{2}} \cdot \frac{\sqrt{2}}{\sqrt{2}} \qquad \text{Multiply numerator and denominator by } \sqrt{2}$$

$$= \frac{\sqrt{2}}{2} \qquad 1 \cdot \sqrt{2} = \sqrt{2}$$

$$\sqrt{2} \cdot \sqrt{2} = \sqrt{4} = 2$$

If we check the expression $\frac{\sqrt{2}}{2}$ against our definition of simplified form for radicals, we find that all three rules hold. There are no perfect squares that are factors of 2. There are no fractions under the radical sign. No radicals appear in the denominator. The expression $\frac{\sqrt{2}}{2}$, therefore, must be in simplified form.

In Example 4, the denominator contained $\sqrt{2}$, which is an irrational number. By multiplying the denominator by another $\sqrt{2}$, the denominator became 2, which is a rational number. Because we changed the denominator from an irrational number to a rational one, we call this process *rationalizing the denominator*.

EXAMPLE 5 Write $\sqrt{\frac{2}{3}}$ in simplified form.

SOLUTION We proceed as we did in Example 4:

$$\sqrt{\frac{2}{3}} = \frac{\sqrt{2}}{\sqrt{3}} \qquad \text{Use the quotient property to separate radicals}$$

$$= \frac{\sqrt{2}}{\sqrt{3}} \cdot \frac{\sqrt{3}}{\sqrt{3}} \qquad \text{Rationalize the denominator}$$

$$= \frac{\sqrt{6}}{3} \qquad \sqrt{2} \cdot \sqrt{3} = \sqrt{6}$$

$$\sqrt{3} \cdot \sqrt{3} = \sqrt{9} = 3$$

EXAMPLE 6 Simplify $\sqrt{\dfrac{4x^3y^2}{3z}}$.

SOLUTION We begin by separating the numerator and denominator and then taking the perfect squares out of the numerator:

> *Note* When working with square roots of variable quantities, we will always assume that the variables represent nonnegative numbers. That way we can say that $\sqrt{x^2} = x$.

$$\sqrt{\frac{4x^3y^2}{3z}} = \frac{\sqrt{4x^3y^2}}{\sqrt{3z}} \quad \text{Quotient property for radicals}$$

$$= \frac{\sqrt{4x^2y^2} \cdot \sqrt{x}}{\sqrt{3z}} \quad \text{Product property for radicals}$$

$$= \frac{2xy\sqrt{x}}{\sqrt{3z}} \quad \sqrt{4x^2y^2} = 2xy$$

The only thing keeping our expression from being in simplified form is the $\sqrt{3z}$ in the denominator. We can take care of this by multiplying the numerator and denominator by $\sqrt{3z}$:

$$\frac{2xy\sqrt{x}}{\sqrt{3z}} = \frac{2xy\sqrt{x}}{\sqrt{3z}} \cdot \frac{\sqrt{3z}}{\sqrt{3z}} \quad \text{Rationalize the denominator}$$

$$= \frac{2xy\sqrt{3xz}}{3z} \quad \sqrt{3z} \cdot \sqrt{3z} = \sqrt{9z^2} = 3z$$

EXAMPLE 7 Rationalize the denominator in the expression

$$\frac{\sqrt{3}}{\sqrt{3} - \sqrt{2}}$$

SOLUTION To remove the two radicals in the denominator, we must multiply both the numerator and denominator by $\sqrt{3} + \sqrt{2}$. That way, when we multiply $\sqrt{3} - \sqrt{2}$ and $\sqrt{3} + \sqrt{2}$, we will obtain the difference of two squares in the denominator:

> *Note* When the denominator contains two terms, we rationalize the denominator by multiplying the numerator and denominator by the conjugate of the denominator.

$$\frac{\sqrt{3}}{\sqrt{3} - \sqrt{2}} = \frac{\sqrt{3}}{(\sqrt{3} - \sqrt{2})} \cdot \frac{(\sqrt{3} + \sqrt{2})}{(\sqrt{3} + \sqrt{2})}$$

$$= \frac{\sqrt{3}\sqrt{3} + \sqrt{3}\sqrt{2}}{(\sqrt{3})^2 - (\sqrt{2})^2}$$

$$= \frac{3 + \sqrt{6}}{3 - 2}$$

$$= \frac{3 + \sqrt{6}}{1}$$

$$= 3 + \sqrt{6}$$

EXAMPLE 8 Rationalize the denominator in the expression $\dfrac{2}{5 - \sqrt{3}}$.

SOLUTION We use the same procedure as in Example 7. Multiply the numerator and denominator by the conjugate of the denominator, which is $5 + \sqrt{3}$:

$$\left(\frac{2}{5 - \sqrt{3}}\right)\left(\frac{5 + \sqrt{3}}{5 + \sqrt{3}}\right) = \frac{10 + 2\sqrt{3}}{5^2 - (\sqrt{3})^2}$$

$$= \frac{10 + 2\sqrt{3}}{25 - 3}$$

$$= \frac{10 + 2\sqrt{3}}{22}$$

The numerator and denominator of this last expression have a factor of 2 in common. We can reduce to lowest terms by dividing out the common factor 2. Continuing, we have

$$= \frac{2(5 + \sqrt{3})}{2 \cdot 11}$$

$$= \frac{5 + \sqrt{3}}{11}$$

The final expression is in simplified form.

EXAMPLE 9 Rationalize the denominator in the expression

$$\frac{\sqrt{a} + \sqrt{b}}{\sqrt{a} - \sqrt{b}}$$

SOLUTION We remove the two radicals in the denominator by multiplying both the numerator and denominator by the conjugate of $\sqrt{a} - \sqrt{b}$, which is $\sqrt{a} + \sqrt{b}$:

$$\frac{\sqrt{a} + \sqrt{b}}{\sqrt{a} - \sqrt{b}} = \frac{\sqrt{a} + \sqrt{b}}{\sqrt{a} - \sqrt{b}}\left(\frac{\sqrt{a} + \sqrt{b}}{\sqrt{a} + \sqrt{b}}\right)$$

$$= \frac{(\sqrt{a})^2 + \sqrt{a}\sqrt{b} + \sqrt{b}\sqrt{a} + (\sqrt{b})^2}{(\sqrt{a})^2 - (\sqrt{b})^2}$$

$$= \frac{a + \sqrt{ab} + \sqrt{ab} + b}{a - b}$$

$$= \frac{a + b + 2\sqrt{ab}}{a - b}$$

EXAMPLE 10 Simplify each expression.

a. $\sqrt[3]{\dfrac{2}{3}}$ **b.** $\sqrt[3]{\dfrac{3x^4}{4y^2z}}$

SOLUTION In both cases, we begin by using the quotient property for radicals to eliminate the fraction under the radical sign.

a. $\sqrt[3]{\dfrac{2}{3}} = \dfrac{\sqrt[3]{2}}{\sqrt[3]{3}}$ Quotient property for radicals

Neither radical can be simplified. Because we do not want to leave a radical in the denominator of a fraction, we rationalize the denominator in order to change the denominator into a perfect cube.

$$\frac{\sqrt[3]{2}}{\sqrt[3]{3}} = \frac{\sqrt[3]{2}}{\sqrt[3]{3}} \cdot \frac{\sqrt[3]{9}}{\sqrt[3]{9}} \qquad \text{Rationalize the denominator}$$

$$= \frac{\sqrt[3]{18}}{\sqrt[3]{3^3}} \qquad \sqrt[3]{3} \cdot \sqrt[3]{9} = \sqrt[3]{27} = \sqrt[3]{3^3}$$

$$= \frac{\sqrt[3]{18}}{3}$$

b. Using the quotient property for radicals, we have

$$\sqrt[3]{\frac{3x^4}{4y^2z}} = \frac{\sqrt[3]{3x^4}}{\sqrt[3]{4y^2z}} \qquad \text{Quotient property for radicals}$$

$$= \frac{x\sqrt[3]{3x}}{\sqrt[3]{4y^2z}} \qquad \text{Simplify the numerator}$$

To rationalize the denominator, we must change $4y^2z$ into a perfect cube. Since $4 = 2^2$, we must multiply by one more factor of 2 and y, but two more factors of z.

$$\frac{x\sqrt[3]{3x}}{\sqrt[3]{4y^2z}} = \frac{x\sqrt[3]{3x}}{\sqrt[3]{4y^2z}} \cdot \frac{\sqrt[3]{2yz^2}}{\sqrt[3]{2yz^2}} \qquad \text{Rationalize the denominator}$$

$$= \frac{x\sqrt[3]{6xyz^2}}{\sqrt[3]{8y^3z^3}} \qquad \text{Product property for radicals}$$

$$= \frac{x\sqrt[3]{6xyz^2}}{2yz} \qquad \sqrt[3]{8y^3z^3} = \sqrt[3]{(2yz)^3} = 2yz$$

Getting Ready for Class

After reading through the preceding section, respond in your own words and in complete sentences.

A. How do we divide radical expressions?

B. What does it mean to rationalize the denominator in an expression?

C. Describe how you would put $\sqrt{\frac{1}{2}}$ in simplified form.

D. In Example 7, why can't we rationalize the denominator by multiplying the numerator and denominator by $\sqrt{3}$?

Perform the following divisions. All answers should be in simplified form for radical expressions.

1. $\frac{\sqrt{21}}{\sqrt{3}}$
2. $\frac{\sqrt{21}}{\sqrt{7}}$
3. $\frac{\sqrt{35}}{\sqrt{7}}$
4. $\frac{\sqrt{35}}{\sqrt{5}}$
5. $\frac{10\sqrt{15}}{5\sqrt{3}}$
6. $\frac{4\sqrt{12}}{8\sqrt{3}}$
7. $\frac{6\sqrt{21}}{3\sqrt{7}}$
8. $\frac{8\sqrt{50}}{16\sqrt{2}}$
9. $\frac{6\sqrt{35}}{12\sqrt{5}}$
10. $\frac{8\sqrt{35}}{16\sqrt{7}}$
11. $\frac{\sqrt[3]{36}}{\sqrt[3]{3}}$
12. $\frac{\sqrt[3]{22}}{\sqrt[3]{2}}$
13. $\frac{\sqrt[3]{72}}{6\sqrt[3]{3}}$
14. $\frac{5\sqrt[3]{108}}{9\sqrt[3]{2}}$

Simplify each of the following radical expressions. Assume all variables represent positive numbers.

15. $\sqrt{\frac{20}{5}}$
16. $\sqrt{\frac{12}{3}}$
17. $\sqrt{\frac{20}{2}}$
18. $\sqrt{\frac{48}{7}}$
19. $\sqrt{\frac{1}{2}}$
20. $\sqrt{\frac{1}{5}}$
21. $\sqrt{\frac{1}{3}}$
22. $\sqrt{\frac{1}{6}}$
23. $\sqrt{\frac{2}{5}}$
24. $\sqrt{\frac{3}{7}}$
25. $\sqrt{\frac{3}{2}}$
26. $\sqrt{\frac{5}{3}}$
27. $\sqrt{\frac{45}{6}}$
28. $\sqrt{\frac{32}{5}}$
29. $\sqrt[3]{\frac{3}{5}}$
30. $\sqrt[3]{\frac{5}{2}}$
31. $\sqrt[3]{\frac{15}{4}}$
32. $\sqrt[3]{\frac{2}{9}}$
33. $\sqrt[3]{\frac{21}{6}}$
34. $\sqrt[3]{\frac{45}{10}}$
35. $\sqrt{\frac{4x^2y^2}{2}}$
36. $\sqrt{\frac{9x^2y^2}{3}}$
37. $\sqrt{\frac{5x^2y}{3}}$
38. $\sqrt{\frac{7x^2y}{5}}$
39. $\sqrt{\frac{16a^4}{5}}$
40. $\sqrt{\frac{25a^4}{7}}$
41. $\sqrt{\frac{72a^5}{5b}}$
42. $\sqrt{\frac{12a^5}{5b}}$
43. $\sqrt{\frac{20x^2y^3}{3z}}$
44. $\sqrt{\frac{27x^2y^3}{2z}}$
45. $\sqrt[3]{\frac{12x^3y^4}{3}}$
46. $\sqrt[3]{\frac{30x^5y^6}{2}}$
47. $\sqrt[3]{\frac{3y^3}{4x}}$
48. $\sqrt[3]{\frac{5x^7}{3y^2}}$

Rationalize the denominator. All answers should be expressed in simplified form.

49. $\frac{8}{3-\sqrt{5}}$
50. $\frac{10}{5+\sqrt{5}}$
51. $\frac{\sqrt{3}}{\sqrt{2}+4}$
52. $\frac{\sqrt{2}}{\sqrt{3}-1}$
53. $\frac{\sqrt{x}+2}{\sqrt{x}-2}$
54. $\frac{\sqrt{x}-3}{\sqrt{x}+3}$
55. $\frac{\sqrt{3}}{\sqrt{5}-\sqrt{2}}$
56. $\frac{\sqrt{2}}{\sqrt{6}-\sqrt{3}}$
57. $\frac{\sqrt{5}}{\sqrt{5}+\sqrt{2}}$
58. $\frac{\sqrt{7}}{\sqrt{7}-\sqrt{2}}$
59. $\frac{\sqrt{3}+\sqrt{2}}{\sqrt{3}-\sqrt{2}}$
60. $\frac{\sqrt{5}-\sqrt{2}}{\sqrt{5}+\sqrt{2}}$
61. $\frac{\sqrt{7}-\sqrt{3}}{\sqrt{7}+\sqrt{3}}$
62. $\frac{\sqrt{11}+\sqrt{6}}{\sqrt{11}-\sqrt{6}}$
63. $\frac{\sqrt{5}-\sqrt{2}}{\sqrt{5}+\sqrt{3}}$
64. $\frac{\sqrt{7}-\sqrt{3}}{\sqrt{5}+\sqrt{2}}$

65. Use a calculator to help complete the following table. If an answer needs rounding, round to the nearest thousandth.

x	$\sqrt{x}$	$\frac{1}{\sqrt{x}}$	$\frac{\sqrt{x}}{x}$
1			
2			
3			
4			
5			
6			

66. Complete the following table. Write answers as whole numbers or fractions in lowest terms.

x	$\sqrt{x}$	$\frac{1}{\sqrt{x}}$	$\frac{\sqrt{x}}{x}$
1			
4			
9			
16			
25			
36			

Use a calculator to help complete the following tables. If an answer needs rounding, round to the nearest thousandth.

67.

x	$\sqrt{x^2}$	$\sqrt{x^3}$	$x\sqrt{x}$
1			
2			
3			
4			
5			
6			

68.

x	$\sqrt{x^2}$	$\sqrt{x^3}$	$x\sqrt{x}$
1			
4			
9			
16			
25			
36			

69. Paying Attention to Instructions Work each problem according to the instructions given.

a. Add: $(5 + \sqrt{2}) + (5 - \sqrt{2})$ **b.** Multiply: $(5 + \sqrt{2})(5 - \sqrt{2})$

c. Square: $(5 + \sqrt{2})^2$ **d.** Divide: $\dfrac{5 + \sqrt{2}}{5 - \sqrt{2}}$

70. Paying Attention to Instructions Work each problem according to the instructions given.

a. Add: $(2 + \sqrt{3}) + (2 - \sqrt{3})$ **b.** Multiply: $(2 + \sqrt{3})(2 - \sqrt{3})$

c. Square: $(2 + \sqrt{3})^2$ **d.** Divide: $\dfrac{2 + \sqrt{3}}{2 - \sqrt{3}}$

Applying The Concepts

71. How Far Can You See? The higher you are above the surface of the Earth, the farther you can see. The formula that gives the approximate distance in miles, d, that you can see from a height of h feet above the surface of the Earth is

$$d = \sqrt{\frac{3h}{2}}$$

How far can you see from a window that is 24 feet above the ground? (This is assuming that your view is unobstructed.)

72. How Far Can You See? Using the formula given in Problem 71, how far can you see from the window of an airplane flying at 6,000 feet?

Learning Objectives Assessment

The following problems can be used to help assess if you have successfully met the learning objectives for this section.

73. Divide: $\dfrac{12\sqrt{10}}{3\sqrt{2}}$.

a. $4\sqrt{5}$ **b.** $2\sqrt{10}$ **c.** $2\sqrt{5}$ **d.** 20

74. Simplify: $\sqrt{\dfrac{7}{3}}$.

a. $\dfrac{\sqrt{7}}{3}$ **b.** $\dfrac{7}{3}$ **c.** $\dfrac{7}{\sqrt{21}}$ **d.** $\dfrac{\sqrt{21}}{3}$

75. Simplify: $\sqrt[3]{\dfrac{2x}{9y}}$.

a. $\dfrac{\sqrt[3]{6xy^2}}{3y}$ **b.** $\dfrac{\sqrt[3]{2xy}}{3y}$ **c.** $\dfrac{\sqrt[3]{18xy}}{9y}$ **d.** $\dfrac{\sqrt[3]{2xy^2}}{3y}$

76. Rationalize the denominator: $\dfrac{2}{\sqrt{5} + 1}$.

a. $\dfrac{\sqrt{5}}{3}$ **b.** $\dfrac{\sqrt{5} + 1}{3}$ **c.** $\dfrac{\sqrt{5} - 1}{12}$ **d.** $\dfrac{\sqrt{5} - 1}{2}$

Getting Ready for the Next Section

Simplify.

77. 7^2

78. 5^2

79. $(-9)^2$

80. $(-4)^2$

81. $(\sqrt{x+1})^2$

82. $(\sqrt{x+2})^2$

83. $(\sqrt{2x-3})^2$

84. $(\sqrt{3x-1})^2$

85. $(x+3)^2$

86. $(x+2)^2$

Solve.

87. $3a - 2 = 4$

88. $2a - 3 = 25$

89. $x + 15 = x^2 + 6x + 9$

90. $x + 8 = x^2 + 4x + 4$

Determine whether the given numbers are solutions to the equation.

91. $\sqrt{x+15} = x + 3$

a. $x = -6$ **b.** $x = 1$

92. $\sqrt{x+8} = x + 2$

a. $x = -4$ **b.** $x = 1$

93. Evaluate $y = 3\sqrt{x}$ for $x = -4, -1, 0, 1, 4, 9, 16$.

94. Evaluate $y = \sqrt[3]{x}$ for $x = -27, -8, -1, 0, 1, 8, 27$.

Equations Involving Radicals 8.6

Learning Objectives

In this section, we will learn how to:

1. Solve equations containing square roots.
2. Solve equations containing cube roots.

Introduction

To solve equations that contain one or more radical expressions, we need an additional property. From our work with exponents we know that if two quantities are equal, then so are the squares of those quantities; that is, for real numbers a and b

$$\text{if} \qquad a = b$$
$$\text{then} \qquad a^2 = b^2$$

The only problem with squaring both sides of an equation is that occasionally we will change a false statement into a true statement. Let's take the false statement $3 = -3$ as an example.

$$3 = -3 \qquad \text{A false statement}$$
$$(3)^2 = (-3)^2 \qquad \text{Square both sides}$$
$$9 = 9 \qquad \text{A true statement}$$

As a result of squaring both sides, we may not have an equivalent equation. The solution set for the equation after squaring each side may be larger than the solution set for the original equation. This means there may be values that satisfy the equation we obtain after squaring that do not make the original equation true. We call these values *extraneous solutions*.

We can avoid this problem by always checking our solutions if, at any time during the process of solving an equation, we have squared both sides of the equation. Here is how the property is stated:

PROPERTY ***Squaring Property of Equality***

If $a = b$, then $a^2 = b^2$.

In words: We can square both sides of an equation any time it is convenient to do so, as long as we check all solutions in the original equation.

VIDEO EXAMPLES

SECTION 8.6

EXAMPLE 1 Solve for x: $\sqrt{x+1} = 7$.

SOLUTION To solve this equation by our usual methods, we must first eliminate the radical sign. We can accomplish this by squaring both sides of the equation using the squaring property of equality:

$$\sqrt{x+1} = 7$$
$$(\sqrt{x+1})^2 = 7^2 \qquad \text{Square both sides}$$
$$x + 1 = 49$$
$$x = 48$$

To check our solution, we substitute $x = 48$ into the original equation:

$$\sqrt{48 + 1} \stackrel{?}{=} 7$$
$$\sqrt{49} \stackrel{?}{=} 7$$
$$7 = 7 \qquad \text{A true statement}$$

The solution checks. The solution set is $\{48\}$.

EXAMPLE 2 Solve for x: $\sqrt{2x - 3} = -9$.

SOLUTION We square both sides and proceed as in Example 1:

$$\sqrt{2x - 3} = -9$$
$$(\sqrt{2x - 3})^2 = (-9)^2 \qquad \text{Square both sides}$$
$$2x - 3 = 81$$
$$2x = 84$$
$$x = 42$$

Checking our solution in the original equation, we have

$$\sqrt{2(42) - 3} \stackrel{?}{=} -9$$
$$\sqrt{84 - 3} \stackrel{?}{=} -9$$
$$\sqrt{81} \stackrel{?}{=} -9$$
$$9 = -9 \qquad \text{A false statement}$$

Our solution does not check because we end up with a false statement. The value $x = 42$ is an extraneous solution. This means the equation has no solution. We say that the solution set is $\varnothing$.

Note As you can see, when we check $x = 42$ in the original equation, we find that it is not a solution to the equation. Actually, it was apparent from the beginning that the equation had no solution; that is, no matter what x is, the equation

$$\sqrt{2x - 3} = -9$$

can never be true because the left side is a positive number (or zero) for any value of x, and the right side is always negative.

Squaring both sides of the equation in Example 2 produced an *extraneous solution*. This happens occasionally when we use the squaring property of equality. We can always eliminate extraneous solutions by checking each solution in the original equation.

EXAMPLE 3 Solve for a: $\sqrt{3a - 2} + 3 = 5$.

SOLUTION Before we can square both sides to eliminate the radical, we must isolate the radical on the left side of the equation. To do so, we add -3 to both sides:

$$\sqrt{3a - 2} + 3 = 5$$
$$\sqrt{3a - 2} = 2 \qquad \text{Add } -3 \text{ to both sides}$$
$$(\sqrt{3a - 2})^2 = 2^2 \qquad \text{Square both sides}$$
$$3a - 2 = 4$$
$$3a = 6$$
$$a = 2$$

Checking $a = 2$ in the original equation, we have

$$\sqrt{3 \cdot 2 - 2} + 3 \stackrel{?}{=} 5$$
$$\sqrt{4} + 3 \stackrel{?}{=} 5$$
$$5 = 5 \qquad \text{A true statement}$$

The solution set to the equation is $\{2\}$.

In Example 3, if we had not isolated the radical in the first step and instead squared both sides of the equation, we would have

$$(\sqrt{3a-2}+3)^2 = (5)^2$$

It is important to understand that squaring the left side is not equivalent to squaring each term individually. That is,

$$(\sqrt{3a-2}+3)^2 \neq (\sqrt{3a-2})^2 + 3^2$$

We cannot distribute an exponent across a sum or difference. To square the left side would require using the FOIL method:

$$\begin{aligned}(\sqrt{3a-2}+3)^2 &= (\sqrt{3a-2}+3)(\sqrt{3a-2}+3)\\ &= (\sqrt{3a-2})^2 + 3\sqrt{3a-2} + 3\sqrt{3a-2} + 9\\ &= 3a-2+6\sqrt{3a-2}+9\\ &= 3a+7+6\sqrt{3a-2}\end{aligned}$$

Notice that this would not eliminate the radical. That is why it is very important to isolate the radical before squaring both sides.

Here is a summary of the process for solving an equation containing a square root.

HOW TO *Solve Equations Containing Square Roots*

Step 1: Isolate the radical expression on one side of the equation.

Step 2: Square both sides of the equation to eliminate the radical sign.

Step 3: Solve the resulting equation for the variable.

Step 4: Check any answers for extraneous solutions.

EXAMPLE 4 Solve for x: $\sqrt{x+15} = x+3$.

SOLUTION Because the radical is already isolated, we can begin by squaring both sides:

$$\begin{aligned}(\sqrt{x+15})^2 &= (x+3)^2 && \text{Square both sides}\\ x+15 &= x^2+6x+9\end{aligned}$$

We have a quadratic equation. We put it into standard form by adding $-x$ and -15 to both sides. Then we factor and solve as usual:

$$\begin{aligned}0 &= x^2+5x-6 && \text{Standard form}\\ 0 &= (x+6)(x-1) && \text{Factor}\\ x+6=0 \quad &\text{or} \quad x-1=0 && \text{Set factors equal to 0}\\ x=-6 \quad &\text{or} \quad x=1\end{aligned}$$

We check each solution in the original equation:

Check -6	*Check 1*
$\sqrt{-6+15} \stackrel{?}{=} -6+3$	$\sqrt{1+15} \stackrel{?}{=} 1+3$
$\sqrt{9} \stackrel{?}{=} -3$	$\sqrt{16} \stackrel{?}{=} 4$
$3 = -3$	$4 = 4$
A false statement	A true statement

Since $x = -6$ does not check in the original equation, it is an extraneous solution. The only valid solution is $x = 1$. The solution set is $\{1\}$.

In our next example, we will consider an equation that contains two square roots. As we will see, this may require that we square both sides of the equation on two different occasions.

EXAMPLE 5 Solve: $\sqrt{2x+1} = \sqrt{x-4} + 3$.

SOLUTION This equation contains two radicals. Because the square root on the left side of the equation is already isolated, we can eliminate the left radical sign by squaring both sides.

$(\sqrt{2x+1})^2 = (\sqrt{x-4} + 3)^2$ Square both sides

$2x + 1 = (\sqrt{x-4} + 3)(\sqrt{x-4} + 3)$

$2x + 1 = (\sqrt{x-4})^2 + 3\sqrt{x-4} + 3\sqrt{x-4} + 9$ FOIL method

$2x + 1 = x - 4 + 6\sqrt{x-4} + 9$ Combine like radicals

$2x + 1 = x + 5 + 6\sqrt{x-4}$ Combine like terms

$x - 4 = 6\sqrt{x-4}$ Add $-x$ and -5 to both sides

Notice that, after applying the FOIL method, we still have a radical in the equation. We can now square both sides of the equation a second time to eliminate the radical sign on the right.

$(x-4)^2 = (6\sqrt{x-4})^2$ Square both sides

$x^2 - 8x + 16 = 6^2(\sqrt{x-4})^2$ Property of exponents

$x^2 - 8x + 16 = 36(x-4)$

$x^2 - 8x + 16 = 36x - 144$

$x^2 - 44x + 160 = 0$ Quadratic equation

$(x-40)(x-4) = 0$ Factor

$x - 40 = 0 \quad \text{or} \quad x - 4 = 0$ Set factors equal to 0

$x = 40 \quad \text{or} \quad x = 4$

Finally, we check both answers in the original equation:

Check: $x = 40$	Check: $x = 4$
$\sqrt{2(40)+1} \stackrel{?}{=} \sqrt{40-4} + 3$	$\sqrt{2(4)+1} \stackrel{?}{=} \sqrt{4-4} + 3$
$\sqrt{81} \stackrel{?}{=} \sqrt{36} + 3$	$\sqrt{9} \stackrel{?}{=} \sqrt{0} + 3$
$9 \stackrel{?}{=} 6 + 3$	$3 \stackrel{?}{=} 0 + 3$
$9 = 9$	$3 = 3$
A true statement	A true statement

Since both solutions check, the solution set is $\{4, 40\}$.

In the previous examples, the equations have all contained square roots. We can use a similar process for solving equations containing higher roots. The squaring property of equality can be generalized for any natural number power as follows.

Note When we raise both sides of an equation to an *odd* power, we don't get extraneous solutions. However, it is always a good idea to check your answers in the original equation.

PROPERTY ***n*th Power Property of Equality**

For any natural number n, the following is true:

$$\text{If } a = b, \text{ then } a^n = b^n.$$

In words: We can raise both sides of an equation to the nth power as long as we check all solutions in the original equation.

In our last example, we will see how to solve an equation containing a cube root by cubing both sides of the equation (raising both sides to the third power).

EXAMPLE 6 Solve: $\sqrt[3]{3 - 4x} + 2 = 5$.

SOLUTION First, we must isolate the radical. Then we can cube both sides of the equation to eliminate the cube root. Here are the steps:

$$\sqrt[3]{3 - 4x} = 3 \quad \text{Isolate the radical}$$

$$(\sqrt[3]{3 - 4x})^3 = (3)^3 \quad \text{Cube both sides}$$

$$3 - 4x = 27 \quad \text{Simplify}$$

$$-4x = 24$$

$$x = -6$$

Now we check our answer in the original equation.

$$\sqrt[3]{3 - 4(-6)} + 2 \stackrel{?}{=} 5$$

$$\sqrt[3]{3 + 24} + 2 \stackrel{?}{=} 5$$

$$\sqrt[3]{27} + 2 \stackrel{?}{=} 5$$

$$3 + 2 \stackrel{?}{=} 5$$

$$5 = 5 \quad \text{A true statement}$$

The solution set is $\{-6\}$.

Getting Ready for Class

After reading through the preceding section, respond in your own words and in complete sentences.

A. What is the squaring property of equality?

B. Why do we check our solutions to equations when we have used the squaring property of equality?

C. Why are there no solutions to the equation $\sqrt{2x - 9} = -9$?

D. Explain how to solve an equation containing a cube root.

Problem Set 8.6

Solve each equation by applying the squaring property of equality. Be sure to check all answers in the original equation for extraneous solutions.

1. $\sqrt{x+1}=2$ **2.** $\sqrt{x-3}=4$ **3.** $\sqrt{x+5}=7$

4. $\sqrt{x+8}=5$ **5.** $\sqrt{x-9}=-6$ **6.** $\sqrt{x+10}=-3$

7. $\sqrt{x-5}=-4$ **8.** $\sqrt{x+7}=-5$ **9.** $\sqrt{x-8}=0$

10. $\sqrt{x-9}=0$ **11.** $\sqrt{2x+1}=3$ **12.** $\sqrt{2x-5}=7$

13. $\sqrt{2x-3}=-5$ **14.** $\sqrt{3x-8}=-4$ **15.** $\sqrt{3x+6}=2$

16. $\sqrt{5x-1}=5$ **17.** $2\sqrt{x}=10$ **18.** $3\sqrt{x}=9$

19. $3\sqrt{a}=6$ **20.** $2\sqrt{a}=12$ **21.** $\sqrt{3x+4}-3=2$

22. $\sqrt{2x-1}+2=5$ **23.** $\sqrt{5y-4}-2=4$ **24.** $\sqrt{3y+1}+7=2$

25. $\sqrt{2x+1}+5=2$ **26.** $\sqrt{6x-8}-1=3$ **27.** $\sqrt{x+3}=x-3$

28. $\sqrt{x-3}=x-3$ **29.** $\sqrt{a+2}=a+2$ **30.** $\sqrt{a+10}=a-2$

31. $\sqrt{2x+9}=x+5$ **32.** $\sqrt{x+6}=x+4$ **33.** $\sqrt{y-4}=y-6$

34. $\sqrt{2y+13}=y+7$ **35.** $\sqrt{3x-5}=\sqrt{x+3}$ **36.** $\sqrt{4x-7}=\sqrt{2x+5}$

37. $\sqrt{a+3}=4-\sqrt{a-1}$ **38.** $\sqrt{a+6}=2+\sqrt{a-2}$

39. $\sqrt{3y-2}=\sqrt{y+7}-1$ **40.** $\sqrt{2y-3}=3-\sqrt{y+2}$

41. $\sqrt[3]{x+4}=-2$ **42.** $\sqrt[3]{x-5}=3$ **43.** $\sqrt[3]{x-7}-4=0$

44. $\sqrt[3]{x+3}+1=0$ **45.** $\sqrt[3]{2x-3}+7=2$ **46.** $\sqrt[3]{3x+2}+6=8$

Applying the Concepts

47. Solve each equation.

a. $\sqrt{y}-4=6$ **b.** $\sqrt{y-4}=6$ **c.** $\sqrt{y-4}=-6$

d. $\sqrt{y-4}=y-6$

48. Solve each equation.

a. $\sqrt{2y}+15=7$ **b.** $\sqrt{2y+15}=7$ **c.** $\sqrt{2y+15}=y$

d. $\sqrt{2y+15}=y+6$

49. Solve each equation.

a. $x-3=0$ **b.** $\sqrt{x}-3=0$ **c.** $\sqrt{x-3}=0$

d. $\sqrt{x}+3=0$ **e.** $\sqrt{x}+3=5$ **f.** $\sqrt{x}+3=-5$

g. $x-3=\sqrt{5-x}$

50. Solve each equation.

a. $x-2=0$ **b.** $\sqrt{x}-2=0$ **c.** $\sqrt{x}+2=0$

d. $\sqrt{x+2}=0$ **e.** $\sqrt{x}+2=7$ **f.** $x-2=\sqrt{2x-1}$

51. Pendulum Problem The time (in seconds) it takes for the pendulum on a clock to swing through one complete cycle is given by the formula

$$T = \frac{11}{7}\sqrt{\frac{L}{2}}$$

where L is the length of the pendulum, in feet. The following table was constructed using this formula. Draw a line graph of the information in the table.

Length L (feet)	1	2	3	4	5	6
Time T (seconds)						

52. Lighthouse Problem The higher you are above the ground, the farther you can see. If your view is unobstructed, then the distance in miles that you can see from h feet above the ground is given by the formula

$$d = \sqrt{\frac{3h}{2}}$$

The following table was constructed using this formula. Draw a line graph of the information in the table.

Height h (feet)	10	50	90	130	170	190
Distance d (miles)						

53. Number Problem The sum of a number and 2 is equal to the positive square root of 8 times the number. Find the number.

54. Number Problem The sum of twice a number and 1 is equal to 3 times the positive square root of the number. Find the number.

55. Number Problem The difference of a number and 3 is equal to twice the positive square root of the number. Find the number.

56. Number Problem The difference of a number and 2 is equal to the positive square root of the number. Find the number.

57. Pendulum Problem The number of seconds T it takes the pendulum of a grandfather clock to swing through one complete cycle is given by the formula

$$T = \frac{11}{7}\sqrt{\frac{L}{2}}$$

where L is the length, in feet, of the pendulum. Find how long the pendulum must be for one complete cycle to take 2 seconds by substituting 2 for T in the formula and then solving for L.

58. Pendulum Problem How long must the pendulum on a grandfather clock be if one complete cycle is to take 1 second?

Learning Objectives Assessment

The following problems can be used to help assess if you have successfully met the learning objectives for this section.

59. Solve: $\sqrt{2x+5} - 1 = 6$.

a. 22 **b.** 15 **c.** 10 **d.** $\varnothing$

60. Solve: $\sqrt{x+4} + 2 = 2x$.

a. $\frac{1}{4}$ **b.** $0, \frac{9}{4}$ **c.** $\frac{9}{4}$ **d.** $\varnothing$

61. Solve: $\sqrt[3]{x-3} = 2$.

a. $\varnothing$ **b.** 7 **c.** 5 **d.** 11

Maintaining Your Skills

62. Reduce to lowest terms: $\frac{x^2 - x - 6}{x^2 - 9}$.

63. Divide using long division: $\frac{x^2 - 2x + 6}{x - 4}$.

Perform the indicated operations.

64. $\frac{x^2 - 25}{x + 4} \cdot \frac{2x + 8}{x^2 - 9x + 20}$

65. $\frac{3x + 6}{x^2 - 4x + 3} \div \frac{x^2 + x - 2}{x^2 + 2x - 3}$

66. $\frac{x}{x^2 - 16} + \frac{4}{x^2 - 16}$

67. $\frac{2}{x^2 - 1} - \frac{5}{x^2 + 3x - 4}$

68. $\dfrac{1 - \frac{25}{x^2}}{1 - \frac{8}{x} + \frac{15}{x^2}}$

Solve each equation.

69. $\frac{x}{2} - \frac{5}{x} = -\frac{3}{2}$

70. $\frac{x}{x^2 - 9} - \frac{3}{x - 3} = \frac{1}{x + 3}$

71. Speed of a Boat A boat travels 30 miles up a river in the same amount of time it takes to travel 50 miles down the same river. If the current is 5 miles per hour, what is the speed of the boat in still water?

72. Filling a Pool A pool can be filled by an inlet pipe in 8 hours. The drain will empty the pool in 12 hours. How long will it take the fill the pool if both the inlet pipe and the drain are open?

73. Mixture Problem If 30 liters of a certain solution contains 2 liters of alcohol, how much alcohol is in 45 liters of the same solution?

74. y varies directly with x. If $y = 8$ when x is 12, find y when x is 36.

Chapter 8 Summary

The numbers in brackets refer to the section(s) in which the topic can be found.

EXAMPLES

Square Roots [8.1]

1. The two square roots of 25 are 5 and -5:

$$\sqrt{25} = 5 \quad \text{and} \quad -\sqrt{25} = -5$$

Every positive real number has two square roots, one positive and one negative. The positive square root of x is written $\sqrt{x}$. The negative square root of x is written $-\sqrt{x}$. In both cases the square root of x is a number we square to get x. The cube root of x is written $\sqrt[3]{x}$ and is the number we cube to get x.

*n*th Roots [8.1]

2. $\sqrt[3]{-27} = -3$

because $(-3)^3 = -27$.

The nth root of x, written $\sqrt[n]{x}$, is the real number we raise to the nth power to get x.

Notation [8.1]

3.

In the expression $\sqrt[3]{27}$, 27 is called the *radicand*, 3 is the *index*, $\sqrt{}$ is called the *radical sign*, and the whole expression $\sqrt[3]{27}$ is called the *radical*.

Properties of Radicals [8.2, 8.4]

4. a. $\sqrt{5} \cdot \sqrt{2} = \sqrt{5 \cdot 2} = \sqrt{10}$

b. $\dfrac{\sqrt{48}}{\sqrt{3}} = \sqrt{\dfrac{48}{3}} = \sqrt{16} = 4$

c. $\sqrt{7} \cdot \sqrt{7} = (\sqrt{7})^2 = 7$

d. $(\sqrt[3]{2})^3 = 2$

If a and b represent nonnegative real numbers, then

1. $\sqrt[n]{a}\,\sqrt[n]{b} = \sqrt[n]{ab}$ — The product of the nth roots is the nth root of the product

2. $\dfrac{\sqrt[n]{a}}{\sqrt[n]{b}} = \sqrt[n]{\dfrac{a}{b}} \quad (b \neq 0)$ — The quotient of the nth roots is the nth root of the quotient

3. $\sqrt{a} \cdot \sqrt{a} = (\sqrt{a})^2 = a$ — This property shows that squaring and square roots are inverse operations

4. $(\sqrt[n]{a})^n = a$ — This property shows that nth powers and roots are inverse operations

Simplified Form for Radicals [8.2]

5. Simplify $\sqrt{12}$ and $\sqrt{\dfrac{25}{9}}$.

$$\sqrt{12} = \sqrt{4 \cdot 3} = \sqrt{4}\,\sqrt{3} = 2\sqrt{3}$$

$$\sqrt{\dfrac{25}{9}} = \dfrac{\sqrt{25}}{\sqrt{9}} = \dfrac{5}{3}$$

A radical expression is in simplified form if:

1. There are no perfect squares that are factors of the quantity under the square root sign, no perfect cubes that are factors of the quantity under the cube root sign, and so on. We want as little as possible under the radical sign.

2. There are no fractions under the radical sign.

3. There are no radicals in the denominator.

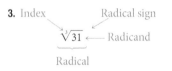

Like Terms Containing Radicals [8.3]

6. $4\sqrt{3} - \sqrt{3} + 7\sqrt{2} + 2\sqrt{2}$
$= (4 - 1)\sqrt{3} + (7 + 2)\sqrt{2}$
$= 3\sqrt{3} + 9\sqrt{2}$

With terms containing radical expressions, like terms, or similar terms, must have the same variable factor and identical radical expressions (the same index and the same radicand). We can combine like radical terms using the distributive property.

Multiplication of Radical Expressions [8.4]

7. $2\sqrt{3}(5\sqrt{6}) = (2 \cdot 5)(\sqrt{3}\sqrt{6})$
$= 10\sqrt{18}$
$= 10(3\sqrt{2})$
$= 30\sqrt{2}$

We use the product property of radicals to multiply two radical expressions.

Division of Radical Expressions [8.5]

8. $\dfrac{4\sqrt{10}}{2\sqrt{5}} = \dfrac{4}{2} \cdot \dfrac{\sqrt{10}}{\sqrt{5}}$
$= 2 \cdot \sqrt{\dfrac{10}{5}}$
$= 2\sqrt{2}$

We use the quotient property of radicals to divide two radical expressions.

Rationalizing the Denominator [8.5]

9. a. $\dfrac{3}{\sqrt{5}} = \dfrac{3}{\sqrt{5}} \cdot \dfrac{\sqrt{5}}{\sqrt{5}} = \dfrac{3\sqrt{5}}{5}$

b. $\dfrac{3}{4 + \sqrt{5}} = \dfrac{3}{4 + \sqrt{5}} \cdot \left(\dfrac{4 - \sqrt{5}}{4 - \sqrt{5}}\right)$
$= \dfrac{12 - 3\sqrt{5}}{(4)^2 - (\sqrt{5})^2}$
$= \dfrac{12 - 3\sqrt{5}}{16 - 5}$
$= \dfrac{12 - 3\sqrt{5}}{11}$

If a fraction contains a radical in the denominator, we can rationalize the denominator to eliminate the radical. If the denominator is a single term, we multiply numerator and denominator of the fraction by the radical appearing in the denominator. If the denominator contains two terms, we multiply numerator and denominator by the conjugate of the denominator.

Solving Radical Equations [8.6]

10. Solve: $\sqrt{x + 5} = x - 1$.
$(\sqrt{x + 5})^2 = (x - 1)^2$
$x + 5 = x^2 - 2x + 1$
$0 = x^2 - 3x - 4$
$0 = (x + 1)(x - 4)$
$x + 1 = 0$ or $x - 4 = 0$
$x = -1$ or $x = 4$
Check -1: $\sqrt{-1 + 5} \stackrel{?}{=} -1 - 1$
$\sqrt{4} \neq -2$
Check 4: $\sqrt{4 + 5} \stackrel{?}{=} 4 - 1$
$\sqrt{9} = 3$
The only solution is $x = 4$ ($x = -1$ is an extraneous solution.

To solve an equation combining a square root, we use the squaring property of equality to eliminate the radical:

$$\text{If } a = b, \text{ then } a^2 = b^2.$$

Here are the steps:

HOW TO ***Solve Equations Containing Square Roots***

Step 1: Isolate the radical expression on one side of the equation.

Step 2: Square both sides of the equation to eliminate the radical sign.

Step 3: Solve the resulting equation for the variable.

Step 4: Check any answers for extraneous solutions.

COMMON MISTAKE

1. A very common mistake with radicals is to think of $\sqrt{25}$ as representing both the positive and negative square roots of 25. The notation $\sqrt{25}$ stands for the *positive* square root of 25. If we want the negative square root of 25, we write $-\sqrt{25}$.
2. The most common mistake when working with radicals is to try to apply a property similar to our Product Property for radicals involving addition instead of multiplication. Here is an example:

$$\sqrt{16+9} = \sqrt{16} + \sqrt{9} \qquad \text{Mistake}$$

Although this example looks like it may be true, it isn't. If we carry it out further, the mistake becomes obvious:

$$\begin{aligned} \sqrt{16+9} &\stackrel{?}{=} \sqrt{16} + \sqrt{9} \\ \sqrt{25} &\stackrel{?}{=} 4 + 3 \\ 5 &= 7 \qquad \text{False} \end{aligned}$$

3. It is a mistake to try to simplify expressions like $2 + 3\sqrt{7}$. The 2 and 3 cannot be combined because the terms they appear in are not similar. Therefore, $2 + 3\sqrt{7} \neq 5\sqrt{7}$. The expression $2 + 3\sqrt{7}$ cannot be simplified further.

Chapter 8 Test

Find the following roots, if possible. [8.1]

1. $\sqrt{36}$ **2.** $-\sqrt{81}$ **3.** $\sqrt{-121}$

4. $\sqrt[3]{64}$ **5.** $\sqrt[3]{-27}$ **6.** $-\sqrt[4]{625}$

Put the following expressions into simplified form. Assume all variables represent nonnegative numbers. [8.2]

7. $\sqrt{72}$ **8.** $\sqrt{96}$ **9.** $4\sqrt{12y^2}$

10. $\sqrt{\dfrac{18x^3y^5}{3}}$ **11.** $\sqrt{\dfrac{128}{25}}$ **12.** $\sqrt[3]{40x^7y^8}$

Simplify each expression. [8.1, 8.2]

13. $\dfrac{8 + 2\sqrt{5}}{4}$ **14.** $\dfrac{-2 + \sqrt{24}}{6}$

Combine. [8.3]

15. $3\sqrt{20} - \sqrt{45}$ **16.** $3m\sqrt{12} + 2\sqrt{3m^2}$ **17.** $\sqrt[3]{24a^3b^3} - 5a\sqrt[3]{3b^3}$

Multiply. [8.4]

18. $\sqrt{2}(\sqrt{3} - 3)$ **19.** $(\sqrt{3} - 5)(\sqrt{3} + 2)$

20. $(\sqrt{x} - 9)(\sqrt{x} + 9)$ **21.** $(\sqrt{7} - \sqrt{2})^2$

Divide. [8.5]

22. $\dfrac{\sqrt{42}}{\sqrt{7}}$ **23.** $\dfrac{10\sqrt[3]{48}}{5\sqrt[3]{3}}$

24. $\sqrt{\dfrac{2}{7}}$ **25.** $\sqrt[3]{\dfrac{3}{2x^2}}$

Rationalize the denominator. [8.5]

26. $\dfrac{5}{\sqrt{3} - 1}$ **27.** $\dfrac{\sqrt{5} - \sqrt{7}}{\sqrt{5} + \sqrt{7}}$ **28.** $\dfrac{\sqrt{x}}{\sqrt{x} - 3}$

Solve the following equations. [8.6]

29. $\sqrt{4x - 3} - 3 = 4$ **30.** $\sqrt{2x + 8} = x + 4$

31. $\sqrt{x + 3} = \sqrt{x + 4} - 1$ **32.** $\sqrt[3]{2x + 7} = -1$

33. The difference of a number and 3 is equal to twice the positive square root of the number. Find the number. [8.6]

34. Find x in the following right triangle. [8.1]

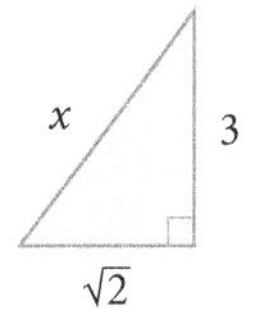

9

Quadratic Equations

Chapter Outline

9.1 Square Root Property
9.2 Completing the Square
9.3 The Quadratic Formula
9.4 Complex Numbers
9.5 Complex Solutions to Quadratic Equations
9.6 Graphing Parabolas

The city of Khiva, in Uzbekistan, is the birthplace of the mathematician al-Khwarizmi. Although he was born in Khiva, he spent most of his life in Baghdad, Iraq. The Islamic mathematician wrote on Hindu-Arabic numerals and is credited with being the first to use zero as a place holder. His book *Hisab al-jabr w'al-muqabala* is considered the first book to be written on algebra. Below is a postage stamp issued by the USSR in 1983, to commemorate the 1200th anniversary of al-Khwarizmi's birth.

"That fondness for science, ... that affability and condescension which God shows to the learned, that promptitude with which he protects and supports them in the elucidation of obscurities and in the removal of difficulties, has encouraged me to compose a short work on calculating by *al-jabr and al-muqabala,* confining it to what is easiest and most useful in arithmetic."

–al-Khwarizmi

In his book, al-Khwarizmi solves quadratic equations by a method called completing the square, one of the topics we will study in this chapter.

iStockphoto.com © Rawpixel Ltd

Think about the most successful people you have met or heard about. What are the qualities they tend to have in common? One of these qualities usually involves making a resolute commitment. If you are not firmly committed to something, then you will tend to give less than your full effort. Consider this quote from Faust by Johann Wolfgang Von Goethe:

> *Until one is committed, there is hesitancy, the chance to draw back, always ineffectiveness. Concerning all acts of initiative and creation, there is one elementary truth the ignorance of which kills countless ideas and splendid plans: that the moment one definitely commits oneself, then providence moves too. All sorts of things occur to help one that would never otherwise have occurred. A whole stream of events issues from the decision, raising in one's favor all manner of unforeseen incidents, meetings and material assistance which no man could have dreamed would have come his way. Whatever you can do or dream you can, begin it. Boldness has genius, power and magic in it.*

Successful people do not give up easily. They forge ahead, even when confronted by difficulties or when the odds seem stacked against them.

Take a moment to reflect on your own life experiences. When have you been the most successful? Can you think of a time when providence has moved in your favor, perhaps unexpectedly?

Square Root Property 9.1

Learning Objectives

In this section, we will learn how to:

1. Solve quadratic equations using the square root property.
2. Solve applied problems using the Pythagorean theorem.

Introduction

Table 1 is taken from the trail map given to skiers at the Northstar California Ski Resort in Lake Tahoe, California. The table gives the length of each chair lift at Northstar, along with the change in elevation from the beginning of the lift to the end of the lift.

Right triangles are good mathematical models for chair lifts. They are not exact models because the cables on a chair lift form curves, not straight lines. But considering the overall size of the chair lift, using a straight line to represent the cables is a good way to model the chair lift.

In this section, we will use our knowledge of right triangles, along with the new material we will develop, to solve problems involving chair lifts and a variety of other examples.

TABLE 1 From the Trail Map for the Northstar California Ski Resort

Lift Information		
Lift	Vertical Rise (feet)	Length (feet)
Big Springs Gondola	472	4,097
Comstock Express	1,253	5,649
Rendezvous Triple	646	2,920
Lookout Mountain Express	1,254	3,175
Backside Express	1,846	6,034

Review of Solving Equations by Factoring

Before we start our new work with quadratic equations, let's review how we solved quadratic equations by factoring.

VIDEO EXAMPLES

SECTION 9.1

EXAMPLE 1 Solve: $x^2 - 16 = 0$.

SOLUTION We factor the left side of the equation, set each factor equal to 0, and then solve each of the resulting equations.

$$x^2 - 16 = 0$$

$$(x - 4)(x + 4) = 0 \qquad \text{Factor left side}$$

$$x - 4 = 0 \quad \text{or} \quad x + 4 = 0 \qquad \text{Set each factor equal to 0}$$

$$x = 4 \quad \text{or} \quad x = -4 \qquad \text{Solve each equation}$$

We have two solutions, 4 and -4.

EXAMPLE 2 Solve: $2x^2 - 5x = 12$.

SOLUTION Begin by adding -12 to both sides, so the equation is in standard form:

$$2x^2 - 5x = 12$$

$$2x^2 - 5x - 12 = 0 \qquad \text{Standard form}$$

$$(2x + 3)(x - 4) = 0 \qquad \text{Factor left side}$$

$$2x + 3 = 0 \quad \text{or} \quad x - 4 = 0 \qquad \text{Set each factor equal to 0}$$

$$2x = -3 \qquad x = 4 \qquad \text{Solve the resulting equations}$$

$$x = -\frac{3}{2}$$

The solutions are 4 and $-\frac{3}{2}$.

EXAMPLE 3 Solve: $16a^2 - 25 = 0$.

SOLUTION The equation is already in standard form:

$$16a^2 - 25 = 0$$

$$(4a - 5)(4a + 5) = 0 \qquad \text{Factor the left side}$$

$$4a - 5 = 0 \quad \text{or} \quad 4a + 5 = 0 \qquad \text{Set each factor equal to 0}$$

$$4a = 5 \qquad 4a = -5 \qquad \text{Solve the resulting equations}$$

$$a = \frac{5}{4} \qquad a = -\frac{5}{4}$$

The solutions are $\frac{5}{4}$ and $-\frac{5}{4}$.

Square Root Property

Consider the equation

$$x^2 = 16$$

We could solve it by writing it in standard form, factoring the left side, and proceeding as we did with the previous three examples. We can shorten our work considerably, however, if we simply notice that x must be either the positive square root of 16 or the negative square root of 16. That is,

$$\begin{array}{llll} \text{if} & x^2 = 16 & & \\ \text{then} & x = \sqrt{16} & \text{or} & x = -\sqrt{16} \\ & x = 4 & \text{or} & x = -4 \end{array}$$

Notice that this method of solving a quadratic equation gives us the same result we obtained in Example 1 with our old method.

We generalize this result as follows:

PROPERTY *Square Root Property for Equations*

For all positive real numbers b,

$$\text{If } a^2 = b\text{, then } a = \sqrt{b} \text{ or } a = -\sqrt{b}$$

Notation A shorthand notation for

$$a = \sqrt{b} \text{ or } a = -\sqrt{b}$$

$$\text{is:} \quad a = \pm\sqrt{b}$$

which is read "a is plus or minus the square root of b."

We can use the square root property any time our quadratic equation has a squared term containing a variable and a constant term. We must make sure, however, that we include both the positive and the negative square roots when we use the square root property.

Note This method of solving quadratic equations sometimes is called *extraction of roots*.

EXAMPLE 4 Solve for x: $x^2 = 5$.

SOLUTION

$$\begin{aligned} x^2 &= 5 \\ x &= \pm\sqrt{5} && \text{Square root property} \end{aligned}$$

The two solutions are $\sqrt{5}$ and $-\sqrt{5}$.

EXAMPLE 5 Solve for a: $5a^2 = 60$.

SOLUTION We begin by dividing both sides by 5 (which is the same as multiplying both sides by $\frac{1}{5}$) because we want to isolate the squared quantity before using the square root property.

$$\begin{aligned} 5a^2 &= 60 \\ a^2 &= 12 && \text{Divide each side by 5} \\ a &= \pm\sqrt{12} && \text{Square root property} \\ a &= \pm2\sqrt{3} && \sqrt{12} = \sqrt{4 \cdot 3} = \sqrt{4}\,\sqrt{3} = 2\sqrt{3} \end{aligned}$$

We have $2\sqrt{3}$ and $-2\sqrt{3}$ as our two solutions.

EXAMPLE 6 Solve for y: $(y + 4)^2 = 25$.

SOLUTION In this case, the quantity that is squared is the binomial $y + 4$. Because the squared quantity is already isolated, we can proceed directly with the square root property. The square root of $(y + 4)^2$ is $(y + 4)$:

$$(y + 4)^2 = 25 \qquad \text{Square root property}$$
$$y + 4 = \pm 5$$

At this point we add -4 to both sides to get

$$y = -4 \pm 5$$

which we can write as

$$y = -4 + 5 \quad \text{or} \quad y = -4 - 5$$
$$y = 1 \quad \text{or} \quad y = -9$$

Our solutions are 1 and -9.

Note that we could have solved this equation by factoring:

$$(y + 4)^2 = 25$$
$$y^2 + 8y + 16 = 25 \qquad \text{Expand left side}$$
$$y^2 + 8y - 9 = 0 \qquad \text{Add } -25 \text{ to each side}$$
$$(y - 1)(y + 9) = 0 \qquad \text{Factor left side}$$
$$y - 1 = 0 \quad \text{or} \quad y + 9 = 0 \qquad \text{Set factors equal to 0}$$
$$y = 1 \quad \text{or} \quad y = -9 \qquad \text{Solve each equation}$$

Pointing out that we have more than one way to solve an equation is not meant to confuse you. Instead, we are pointing it out so you can see that the number of tools you have for solving equations is expanding.

EXAMPLE 7 Solve $(2x - 3)^2 = 25$.

SOLUTION $(2x - 3)^2 = 25$

$$2x - 3 = \pm 5 \qquad \text{Square root property}$$

Adding 3 to both sides, we have

$$2x = 3 \pm 5$$

Dividing both sides by 2 gives us

$$x = \frac{3 \pm 5}{2}$$

We separate the preceding equation into two separate statements:

$$x = \frac{3 + 5}{2} \quad \text{or} \quad x = \frac{3 - 5}{2}$$
$$x = \frac{8}{2} = 4 \quad \text{or} \quad x = \frac{-2}{2} = -1$$

We can check our answers to Example 7:

If $(2x - 3)^2 = 25$ and $x = 4$ then $(2 \cdot 4 - 3)^2 \stackrel{?}{=} 25$

If $(2x - 3)^2 = 25$ and $x = -1$ then $[2(-1) - 3]^2 \stackrel{?}{=} 25$

Note We can solve the equation in Example 7 by factoring if we first expand $(2x - 3)^2$.

$$(2x - 3)^2 = 25$$
$$4x^2 - 12x + 9 = 25$$
$$4x^2 - 12x - 16 = 0$$
$$4(x^2 - 3x - 4) = 0$$
$$4(x - 4)(x + 1) = 0$$
$$x = 4 \quad \text{or} \quad x = -1$$

$$(8-3)^2 \stackrel{?}{=} 25 \qquad (-2-3)^2 \stackrel{?}{=} 25$$
$$5^2 \stackrel{?}{=} 25 \qquad (-5)^2 \stackrel{?}{=} 25$$
$$25 = 25 \quad \text{True} \qquad 25 = 25 \quad \text{True}$$

EXAMPLE 8 Solve for y: $(4y-5)^2 = 6$.

Note In Example 8, we cannot simplify the numerator $5 \pm \sqrt{6}$ any further because 5 and $\sqrt{6}$ are not similar terms.

SOLUTION $(4y-5)^2 = 6$

$$4y - 5 = \pm\sqrt{6} \quad \text{Square root property}$$
$$4y = 5 \pm \sqrt{6} \quad \text{Add 5 to both sides}$$
$$y = \frac{5 \pm \sqrt{6}}{4} \quad \text{Divide both sides by 4}$$

Since $\sqrt{6}$ is irrational, we cannot simplify the expression further. The solution set is $\left\{\frac{5+\sqrt{6}}{4}, \frac{5-\sqrt{6}}{4}\right\}$. We can use a calculator to find approximations for each solution.

A calculator gives the decimal approximation of $\sqrt{6} \approx 2.45$. Therefore,

$$\frac{5+\sqrt{6}}{4} \approx \frac{5+2.45}{4} = \frac{7.45}{4} \approx 1.86$$
$$\frac{5-\sqrt{6}}{4} \approx \frac{5-2.45}{4} = \frac{2.55}{4} \approx 0.64$$

EXAMPLE 9 Solve for x: $(2x+6)^2 = 8$.

SOLUTION $(2x+6)^2 = 8$

$$2x + 6 = \pm\sqrt{8} \quad \text{Square root property}$$
$$2x + 6 = \pm 2\sqrt{2} \quad \sqrt{8} = \sqrt{4 \cdot 2} = 2\sqrt{2}$$
$$2x = -6 \pm 2\sqrt{2} \quad \text{Add } -6 \text{ to both sides}$$
$$x = \frac{-6 \pm 2\sqrt{2}}{2} \quad \text{Divide each side by 2}$$

We can reduce the previous expression to lowest terms by factoring a 2 from each term in the numerator and then dividing that 2 by the 2 in the denominator. This is equivalent to dividing each term in the numerator by the 2 in the denominator. Here is what it looks like:

$$x = \frac{2(-3 \pm \sqrt{2})}{2} \quad \text{Factor a 2 from each term in numerator}$$
$$x = -3 \pm \sqrt{2} \quad \text{Divide numerator and denominator by 2}$$

The two solutions are $-3 + \sqrt{2}$ and $-3 - \sqrt{2}$.

We can check our two solutions in the original equation, but the algebra in the check is more difficult than the algebra in the original solution. However, we can use a decimal approximation of $\sqrt{2}$ to see if our first solution is correct.

Note We are showing the check here so you can see that the irrational number $-3 + \sqrt{2}$ is a solution to $(2x+6)^2 = 8$. Some people don't believe it at first.

If $\sqrt{2} \approx 1.414$, then $x = -3 + \sqrt{2} \approx -3 + 1.414 = -1.586$. Substituting this number into the original equation we have:

$$[2(-1.586) + 6]^2 \stackrel{?}{=} 8$$
$$2.828^2 \stackrel{?}{=} 8$$
$$7.998 \approx 8$$

It is probably safe to say we have a correct solution.

Applications

We conclude this section by considering some applications from geometry that involve quadratic equations.

FACTS FROM GEOMETRY *More Special Triangles*

The triangles shown in Figures 1 and 2 occur frequently in mathematics.

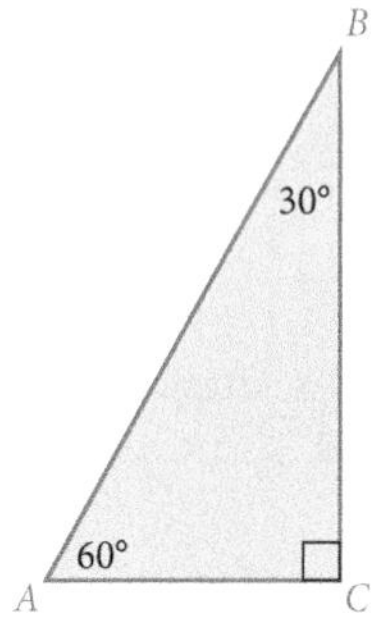

FIGURE 1

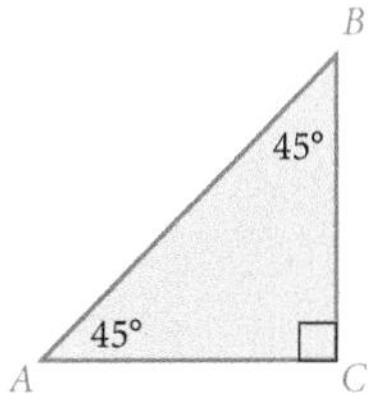

FIGURE 2

Note that both of the triangles are right triangles. We refer to the triangle in Figure 1 as a $30°–60°–90°$ triangle, and the triangle in Figure 2 as a $45°–45°–90°$ triangle.

EXAMPLE 10 If the shortest side in a $30°–60°–90°$ triangle is 1 inch, find the lengths of the other two sides.

SOLUTION In Figure 3, triangle ABC is a $30° - 60° - 90°$ triangle in which the shortest side AC is 1 inch long. Triangle DBC is also a $30° - 60° - 90°$ triangle in which the shortest side DC is 1 inch long.

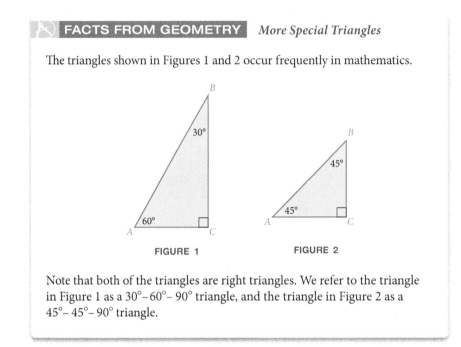

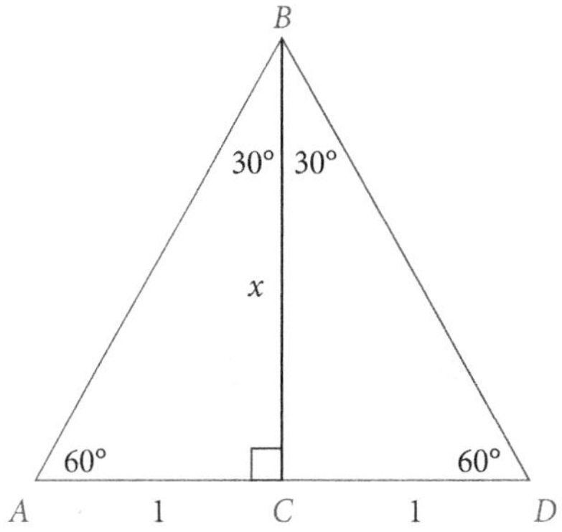

FIGURE 3

Because triangle ABD is an equilateral triangle, each side of triangle ABD is 2 inches long. Side AB in triangle ABC is therefore 2 inches. To find the length of side BC, we use the Pythagorean theorem.

$$BC^2 + AC^2 = AB^2$$
$$x^2 + 1^2 = 2^2$$
$$x^2 + 1 = 4$$
$$x^2 = 3$$
$$x = \sqrt{3} \text{ inches}$$

Note that we write only the positive square root because x is the length of a side in a triangle and is therefore a positive number.

EXAMPLE 11 Table 1 in the introduction to this section gives the vertical rise of the Comstock Express chair lift as 1,253 feet and the length of the chair lift as 5,649 feet. To the nearest foot, find the horizontal distance covered by a person riding this lift.

SOLUTION Figure 4 is a model of the Comstock Express chair lift. A rider gets on the lift at point A and exits at point B. The length of the lift is AB.

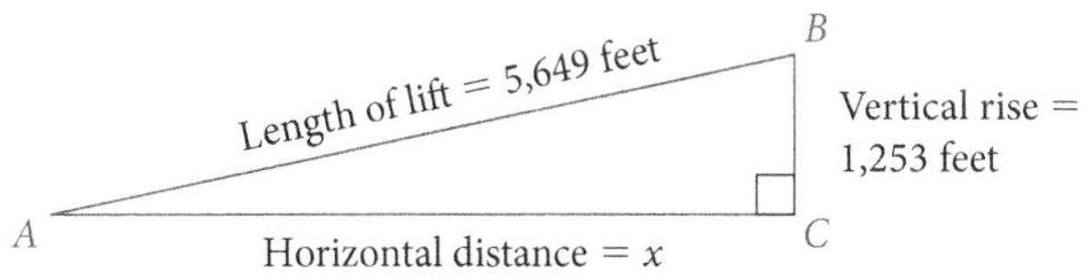

FIGURE 4

To find the horizontal distance covered by a person riding the chair lift, we use the Pythagorean theorem.

$5{,}649^2 = x^2 + 1{,}253^2$	Pythagorean theorem
$31{,}911{,}201 = x^2 + 1{,}570{,}009$	Simplify squares
$x^2 = 31{,}911{,}201 - 1{,}570{,}009$	Solve for x^2
$x^2 = 30{,}341{,}192$	Simplify the right side
$x = \sqrt{30{,}341{,}192}$	Square Root Property for Equations
$\approx 5{,}508$ feet	Round to the nearest foot

A rider getting on the lift at point A and riding to point B will cover a horizontal distance of approximately 5,508 feet.

Getting Ready for Class

After reading through the preceding section, respond in your own words and in complete sentences.

A. What is the square root property for equations?

B. Describe how you would solve the equation $x^2 = 7$.

C. What does the symbol $\pm$ stand for?

D. If you know the length of the shortest side in a 30°–60°–90° triangle, how can you go about finding the other two sides?

Problem Set 9.1

Solve each equation by factoring.

1. $x^2 - 25 = 0$
2. $x^2 - 49 = 0$
3. $4x^2 - 9 = 0$
4. $9x^2 - 4 = 0$
5. $2x^2 + 3x - 5 = 0$
6. $3x^2 + x - 2 = 0$
7. $a^2 - a - 6 = 0$
8. $a^2 + a - 2 = 0$

Solve each equation by using the Square Root Property for Equations.

9. $x^2 = 25$
10. $x^2 = 49$
11. $4x^2 = 9$
12. $9x^2 = 4$
13. $a^2 = 8$
14. $a^2 = 12$
15. $2x^2 = 24$
16. $3x^2 = 24$

Solve each of the following equations using the methods learned in this section.

17. $(x + 2)^2 = 4$
18. $(x - 3)^2 = 9$
19. $(x + 1)^2 = 16$
20. $(x + 3)^2 = 64$
21. $(a - 5)^2 = 75$
22. $(a - 4)^2 = 32$
23. $(y + 1)^2 = 12$
24. $(y - 5)^2 = 50$
25. $(2x + 1)^2 = 25$
26. $(3x - 2)^2 = 16$
27. $(4a - 5)^2 = 36$
28. $(2a + 6)^2 = 36$
29. $(3y - 1)^2 = 20$
30. $(5y - 4)^2 = 12$
31. $(3x + 6)^2 = 27$
32. $(8x - 4)^2 = 20$
33. $(3x - 9)^2 = 27$
34. $(2x + 8)^2 = 32$
35. $\left(x - \frac{1}{2}\right)^2 = \frac{7}{4}$
36. $\left(x - \frac{1}{3}\right)^2 = \frac{5}{9}$
37. $\left(a + \frac{4}{5}\right)^2 = \frac{12}{25}$
38. $\left(a + \frac{3}{7}\right)^2 = \frac{18}{49}$

Since $a^2 + 2ab + b^2$ can be written as $(a + b)^2$, each of the following equations can be solved using the square root method. The first step is to write the trinomial on the left side of the equal sign as the square of a binomial. Solve each equation.

39. $x^2 + 10x + 25 = 7$
40. $x^2 + 6x + 9 = 11$
41. $x^2 - 2x + 1 = 9$
42. $x^2 + 8x + 16 = 25$
43. $x^2 + 12x + 36 = 8$
44. $x^2 - 4x + 4 = 12$
45. Solve each equation.
    a. $2x - 1 = 0$
    b. $2x - 1 = 4$
    c. $(2x - 1)^2 = 4$
    d. $2(x - 1) = 4$
46. Solve each equation.
    a. $x + 5 = 0$
    b. $x + 5 = 8$
    c. $(x + 5)^2 = 8$
    d. $2(x + 5) = 8$

Applying the Concepts

47. Checking Solutions Use $\sqrt{2} \approx 1.414$, to check the solution

$$x = -1 + 5\sqrt{2}$$

in the equation $(x + 1)^2 = 50$.

48. Checking Solutions Use $\sqrt{6} \approx 2.449$ to check the solution

$$x = -8 + 2\sqrt{6}$$

in the equation $(x + 8)^2 = 24$.

49. Number Problem The square of the sum of a number and 3 is 16. Find the number. (There are two solutions.)

50. Number Problem The square of the sum of twice a number and 3 is 25. Find the number. (There are two solutions.)

51. Chair Lift Use Table 1 from the introduction to this section to find the horizontal distance covered by a person riding the Big Springs Gondola chair lift. Round your answer to the nearest foot.

52. Geometry The front of a tent forms an equilateral triangle with sides of 6 feet. Can a person 5 feet 8 inches tall stand up inside the tent?

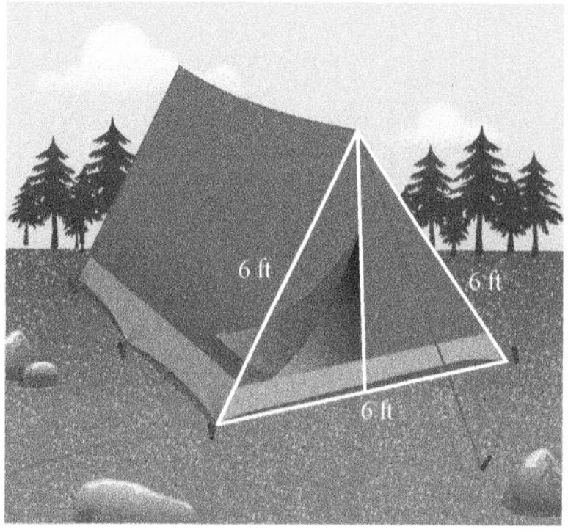

53. Geometry The base of a triangle is 8 feet, and the length of each of the sides is 5 feet. Find the height of the triangle. (The height will bisect the base.)

54. Geometry A triangle has a base 8 feet long. If the length of each of the other two sides is 6 feet, find the height of the triangle. (The height will bisect the base.)

Learning Objectives Assessment

The following problems can be used to help assess if you have successfully met the learning objectives for this section.

55. Solve $(x + 2)^2 = 18$ using the square root property.

a. $-2 + 3\sqrt{2}$ **b.** $-\sqrt{2}$ **c.** $-2 \pm 3\sqrt{2}$ **d.** $2 \pm 3\sqrt{2}$

56. The Rendezvous Triple chair lift at the Northstar California ski resort is 2,920 feet long and has a vertical rise of 646 feet. Find the horizontal distance covered by a person riding this lift to the nearest foot.

a. 2,848 feet **b.** 2,991 feet **c.** 2,274 feet **d.** 3,566 feet

Getting Ready for the Next Section

Simplify.

57. $\left(\frac{1}{2} \cdot 18\right)^2$ **58.** $\left(\frac{1}{2} \cdot 8\right)^2$ **59.** $\left[\frac{1}{2} \cdot (-2)\right]^2$ **60.** $\left[\frac{1}{2} \cdot (-10)\right]^2$

61. $\left(\frac{1}{2} \cdot 3\right)^2$ **62.** $\left(\frac{1}{2} \cdot 5\right)^2$ **63.** $\frac{2x^2 + 16}{2}$ **64.** $\frac{3x^2 - 3x}{3}$

Factor.

65. $x^2 + 6x + 9$ **66.** $x^2 + 12x + 36$ **67.** $y^2 - 3y + \frac{9}{4}$ **68.** $y^2 - 5y + \frac{25}{4}$

Completing the Square 9.2

Learning Objectives

In this section, we will learn how to:

1. Complete the square to make a perfect square trinomial.
2. Solve a quadratic equation by completing the square.

Introduction

In this section we will develop a method of solving any quadratic equation that works whether or not the equation can be factored. Since we will be working with the individual terms of trinomials, we need some new definitions so that we can keep our vocabulary straight.

DEFINITION *individual terms of trinomials*

In the trinomial $2x^2 + 3x + 4$, the first term, $2x^2$, is called the ***quadratic term***; the middle term, $3x$, is called the ***linear term***; and the last term, 4, is called the ***constant term***.

Now consider the following list of perfect square trinomials and their corresponding binomial squares:

$$x^2 + 4x + 4 = (x + 2)^2$$
$$x^2 + 6x + 9 = (x + 3)^2$$
$$x^2 - 8x + 16 = (x - 4)^2$$
$$x^2 - 10x + 25 = (x - 5)^2$$
$$x^2 + 12x + 36 = (x + 6)^2$$

In each case the coefficient of x^2 is 1. A more important observation, however, stems from the relationship between the linear terms (middle terms), the constant terms (last terms), and the second terms in the binomials. Notice that the second term in the binomial is half the coefficient of x in the linear term, and the constant term in each case is the square of half the coefficient of x in the linear term.

In summary, for every perfect square trinomial in which the coefficient of x^2 is 1, the final term is always the square of half the coefficient of the linear term. We can use this fact to build our own perfect square trinomials.

VIDEO EXAMPLES

SECTION 9.2

EXAMPLES Write the correct final term to each of the following expressions so each becomes a perfect square trinomial.

1. $x^2 - 14x$

SOLUTION The coefficient of the linear term is -14. If we take the square of half of -14, we get $(-7)^2 = 49$. Adding the 49 as the final term, we have the perfect square trinomial:

$$x^2 - 14x + 49 = (x - 7)^2$$

Notice that once we factor the trinomial, the second term in the binomial is half of -14, which is -7.

2. $x^2 + 20x$

SOLUTION Half of 20 is 10, the square of which is 100. If we add 100 at the end, we have:

$$x^2 + 20x + 100 = (x + 10)^2$$

3. $x^2 + 5x$

SOLUTION Half of 5 is $\frac{5}{2}$, the square of which is $\frac{25}{4}$:

$$x^2 + 5x + \frac{25}{4} = \left(x + \frac{5}{2}\right)^2$$

The resulting trinomial may be difficult to factor, but we can avoid factoring if we simply remember that the second term in the binomial is always half the coefficient of x in the linear term.

Let's see how we can use this process to help solve a quadratic equation.

EXAMPLE 4 Solve $x^2 - 4x + 3 = 0$ by completing the square.

SOLUTION The trinomial on the left side is not a perfect square trinomial, so we begin by adding -3 to both sides of the equation. We want just $x^2 - 4x$ on the left side so that we can add on our own final term to get a perfect square trinomial:

$$x^2 - 4x + 3 = 0$$

$$x^2 - 4x = -3 \quad \text{Add } -3 \text{ to both sides}$$

Now we can add 4 to both sides and the left side will be a perfect square:

$$x^2 - 4x + 4 = -3 + 4 \quad \text{Add 4 to both sides}$$

$$(x - 2)^2 = 1 \quad \text{Factor the left side}$$

The final line is in the form of the equations we solved in Section 9.1, so we complete the solution process using the square root property for equations.

$$x - 2 = \pm 1 \quad \text{Square root property}$$

$$x = 2 \pm 1 \quad \text{Add 2 to both sides}$$

$$x = 2 + 1 \quad \text{or} \quad x = 2 - 1$$

$$x = 3 \quad \text{or} \quad x = 1$$

The two solutions are 3 and 1.

Note The equation in Example 4 can be solved quickly by factoring:

$$x^2 - 4x + 3 = 0$$
$$(x - 3)(x - 1) = 0$$
$$x - 3 = 0 \quad \text{or} \quad x - 1 = 0$$
$$x = 3 \quad \text{or} \quad x = 1$$

We are showing you both methods here so you can see that the results are the same. This should give you confidence that our new method is working correctly.

The preceding solution method is called *completing the square.*

EXAMPLE 5 Solve $3x^2 + 24x - 27 = 0$ by completing the square.

SOLUTION We begin by moving the constant term to the other side because the trinomial is not yet a perfect square trinomial.

$$3x^2 + 24x - 27 = 0$$

$$3x^2 + 24x = 27 \quad \text{Add 27 to both sides}$$

To complete the square, we must be sure the coefficient of x^2 is 1. To accomplish this, we divide both sides by 3:

$$\frac{3x^2}{3} + \frac{24x}{3} = \frac{27}{3}$$

$$x^2 + 8x = 9$$

We now complete the square by adding the square of half the coefficient of the linear term to both sides:

$$x^2 + 8x + 16 = 9 + 16 \quad \text{Add 16 to both sides}$$
$$(x + 4)^2 = 25 \quad \text{Factor the left side}$$
$$x + 4 = \pm 5 \quad \text{Square root property}$$
$$x = -4 \pm 5 \quad \text{Add } -4 \text{ to both sides}$$
$$x = -4 + 5 \quad \text{or} \quad x = -4 - 5$$
$$x = 1 \quad \text{or} \quad x = -9$$

The solution set arrived at by completing the square is $\{1, -9\}$.

We will now summarize the preceding examples by listing the steps involved in solving quadratic equations by completing the square.

Note Step 3 is the step at which we actually complete the square.

HOW TO *Strategy for Solving a Quadratic Equation by Completing the Square*

Step 1: Put the equation in the form $ax^2 + bx = c$. This usually involves moving only the constant term to the opposite side.

Step 2: If the coefficient of the quadratic term, a, is not 1, then divide both sides by a.

Step 3: Add the square of half the coefficient of the linear term to both sides of the equation.

Step 4: Write the left hand side of the equation as a binomial square, apply the square root property, and solve.

EXAMPLE 6

Solve $y^2 - 4y = 0$:

a. by factoring **b.** by completing the square

SOLUTION Here are the two methods side-by-side. Note that the results are the same in both cases.

a.

$$y^2 - 4y = 0$$
$$y(y - 4) = 0$$
$$y = 0 \quad \text{or} \quad y - 4 = 0$$
$$y = 4$$

Solutions: 0, 4

b.

$$y^2 - 4y = 0$$
$$y^2 - 4y + 4 = 4$$
$$(y - 2)^2 = 4$$
$$y - 2 = \pm 2$$
$$y = 2 \pm 2$$
$$y = 2 + 2 = 4 \quad \text{or} \quad y = 2 - 2 = 0$$

Solutions: 0, 4

Here is one final example.

EXAMPLE 7 Solve by completing the square: $x^2 + 5x - 2 = 0$.

SOLUTION We must begin by adding 2 to both sides.

$$x^2 + 5x = 2 \qquad \text{Add 2 to each side}$$

We complete the square by adding the square of half the coefficient of the linear term to both sides:

$$x^2 + 5x + \frac{25}{4} = 2 + \frac{25}{4} \qquad \text{Half of 5 is } \frac{5}{2}\text{, the square of which is } \frac{25}{4}$$

$$\left(x + \frac{5}{2}\right)^2 = \frac{33}{4} \qquad 2 + \frac{25}{4} = \frac{8}{4} + \frac{25}{4} = \frac{33}{4}$$

$$x + \frac{5}{2} = \pm\sqrt{\frac{33}{4}} \qquad \text{Square root property}$$

$$x + \frac{5}{2} = \pm\frac{\sqrt{33}}{2} \qquad \text{Simplify the radical}$$

$$x = -\frac{5}{2} \pm \frac{\sqrt{33}}{2} \qquad \text{Add } -\frac{5}{2} \text{ to both sides}$$

$$x = \frac{-5 \pm \sqrt{33}}{2}$$

The solution set is $\left\{\frac{-5 + \sqrt{33}}{2}, \frac{-5 - \sqrt{33}}{2}\right\}$.

Note We can use a calculator to get decimal approximations to these solutions. If $\sqrt{33} \approx 5.74$, then

$$\frac{-5 + 5.74}{2} = 0.37$$

$$\frac{-5 - 5.74}{2} = -5.37$$

Getting Ready for Class

After reading through the preceding section, respond in your own words and in complete sentences.

A. What kind of quadratic equations can we solve using the method of completing the square?

B. What is the linear term in a trinomial?

C. Explain in words how you would complete the square on $x^2 - 6x = 5$.

D. How do we complete the square if the coefficient of the squared term is not equal to 1?

Copy each of the following, and fill in each blank so the left side of each is a perfect square trinomial. That is, complete the square.

1. $x^2 + 12x + \_\_\_ = (x + \_\_\_)^2$
2. $x^2 + 6x + \_\_\_ = (x + \_\_\_)^2$
3. $x^2 - 12x + \_\_\_ = (x - \_\_\_)^2$
4. $x^2 - 6x + \_\_\_ = (x - \_\_\_)^2$
5. $x^2 + 4x + \_\_\_ = (x + \_\_\_)^2$
6. $x^2 + 2x + \_\_\_ = (x + \_\_\_)^2$
7. $x^2 - 4x + \_\_\_ = (x - \_\_\_)^2$
8. $x^2 - 2x + \_\_\_ = (x - \_\_\_)^2$
9. $a^2 - 10a + \_\_\_ = (a - \_\_\_)^2$
10. $a^2 - 8a + \_\_\_ = (a - \_\_\_)^2$
11. $x^2 + 5x + \_\_\_ = (x + \_\_\_)^2$
12. $x^2 + 3x + \_\_\_ = (x + \_\_\_)^2$
13. $y^2 + y + \_\_\_ = (y + \_\_\_)^2$
14. $y^2 - y + \_\_\_ = (y - \_\_\_)^2$
15. $x^2 + \frac{1}{2}x + \_\_\_ = (x + \_\_\_)^2$
16. $x^2 - \frac{1}{3}x + \_\_\_ = (x - \_\_\_)^2$
17. $x^2 + \frac{2}{3}x + \_\_\_ = (x + \_\_\_)^2$
18. $x^2 - \frac{4}{5}x + \_\_\_ = (x - \_\_\_)^2$

Solve each of the following quadratic equations by completing the square.

19. $x^2 + 4x = 12$
20. $x^2 - 2x = 8$
21. $x^2 + 12x = -27$
22. $x^2 - 6x = 16$
23. $a^2 - 2a - 5 = 0$
24. $a^2 + 10a + 22 = 0$
25. $y^2 - 8y + 1 = 0$
26. $y^2 + 6y - 1 = 0$
27. $x^2 - 5x - 3 = 0$
28. $x^2 - 5x - 2 = 0$
29. $2x^2 - 4x - 8 = 0$
30. $3x^2 - 9x - 12 = 0$
31. $x^2 - 10x = 0$
32. $x^2 + 4x = 0$
33. $y^2 + 2y - 15 = 0$
34. $y^2 - 10y = 11$
35. $x^2 + 6x = -5$
36. $x^2 - 4x = 4$
37. $x^2 - 3x = -2$
38. $a^2 = 3a + 1$
39. $4x^2 + 8x - 4 = 0$
40. $3x^2 + 12x + 6 = 0$
41. $2x^2 + 2x - 4 = 0$
42. $4x^2 + 4x - 3 = 0$
43. Solve the equation $x^2 - 6x = 0$
    a. by factoring
    b. by completing the square
44. Solve the equation $x^2 + ax = 0$
    a. by factoring
    b. by completing the square
45. Solve the equation $x^2 + 2x = 35$
    a. by factoring
    b. by completing the square
46. Solve the equation $8x^2 - 10x - 25 = 0$
    a. by factoring
    b. by completing the square

Applying the Concepts

47. **Computer Screen** An advertisement for a portable computer indicates it has a 14-inch viewing screen. This means that the diagonal of the screen measures 14 inches. If the ratio of the height to the width of the screen is 3 to 4, we can represent the height with $3x$ and the width with $4x$, and then use the Pythagorean theorem to solve for x. Once we have x, the height will be $3x$ and the width will be $4x$. Find the height and width of this computer screen to the nearest tenth of an inch.

48. Television Screen A 25-inch television has a rectangular screen on which the diagonal measures 25 inches. The ratio of the height to the width of the screen is 3 to 4. Let the height equal $3x$ and the width equal $4x$, and then use the Pythagorean theorem to solve for x. Then find the height and width of this television screen.

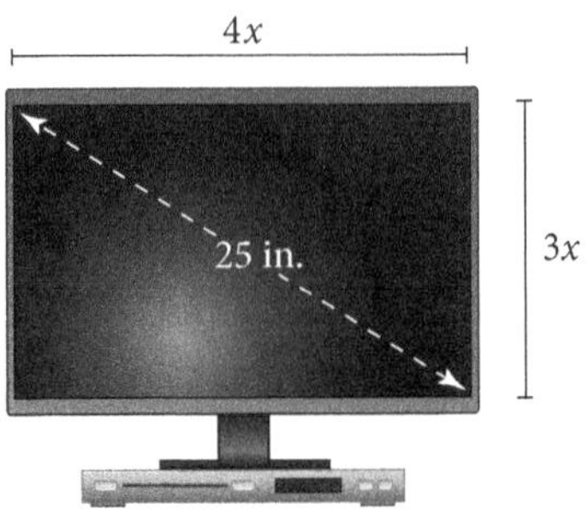

49. Finding Intercepts The graph of $y = x^2 - 2x - 1$ is shown in the following figure. To find the intercepts for this graph we let $y = 0$, and then we solve the resulting equation, which is

$$0 = x^2 - 2x - 1$$

To the nearest tenth, where does this graph cross the x-axis?

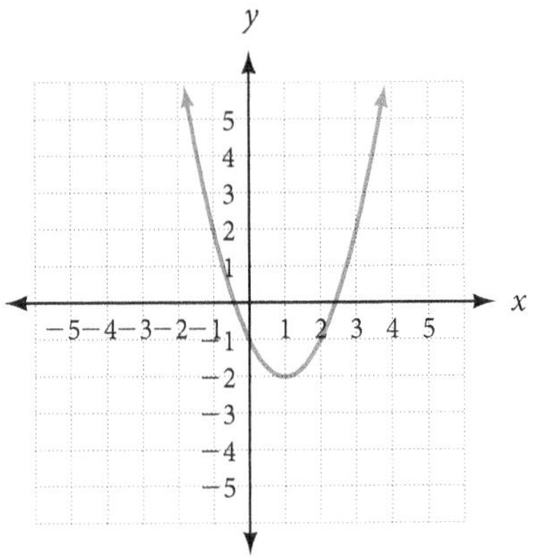

50. Finding Intercepts The graph of $y = x^2 - 4x + 2$ is shown in the following figure. To find the intercepts for this graph we let $y = 0$, and then we solve the resulting equation, which is

$$0 = x^2 - 4x + 2$$

To the nearest tenth, where does this graph cross the x-axis?

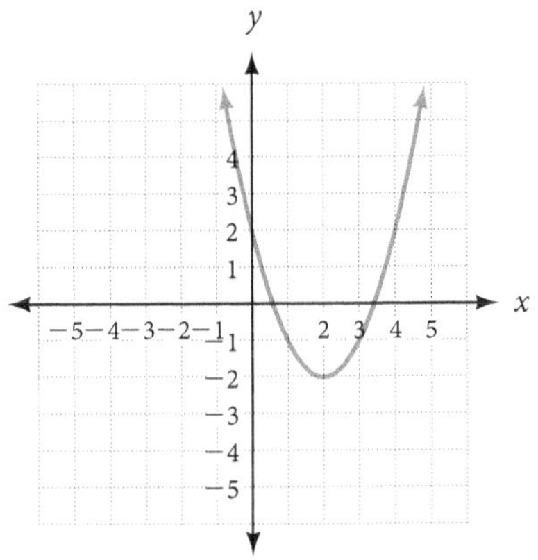

51. **Sum and Product of Solutions** The equation $x^2 + 4x - 3 = 0$ has two solutions, $-2 + \sqrt{7}$ and $-2 - \sqrt{7}$.
   a. Find the sum of these two solutions.
   b. Find the product of these two solutions.

52. **Sum and Product of Solutions** The equation $x^2 - 6x + 4 = 0$ has two solutions, $3 + \sqrt{5}$ and $3 - \sqrt{5}$.
   a. Find the sum of these two solutions.
   b. Find the product of these two solutions.

Completing the Square Visually

The following diagram can be used to show why we add 16 to $x^2 + 8x$ to complete the square:

$$x^2 + 8x = x^2 + 4x + 4x$$
$$x^2 + 8x + 16 = x^2 + 4x + 4x + 16$$
$$= (x + 4)^2$$

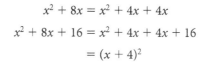

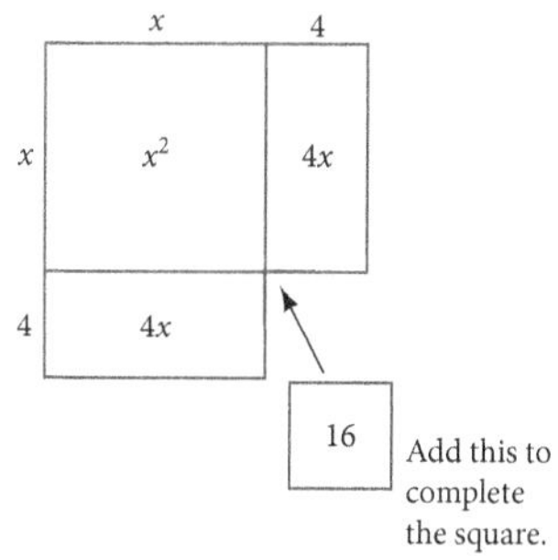

Draw a similar diagram to visualize completing the square on each of the following expressions.

53. $x^2 + 6x$
54. $x^2 + 10x$
55. $x^2 + 2x$
56. $x^2 + 4x$

Learning Objectives Assessment

The following problems can be used to help assess if you have successfully met the learning objectives for this section.

57. In order to complete the square, which value should be added to $x^2 + 8x$?
   a. 8
   b. 4
   c. 16
   d. 64

58. Which of the following equations is a correct step when solving $2x^2 + 12x - 5 = 0$?
   a. $(x + 3)^2 = \frac{5}{2}$
   b. $(x + 3)^2 = \frac{23}{2}$
   c. $(x + 6)^2 = 41$
   d. $(2x + 6)^2 = 41$

Getting Ready for the Next Section

Write the quadratic equation in standard form $(ax^2 + bx + c = 0)$.

59. $2x^2 = -4x + 3$
60. $3x^2 = -4x + 2$
61. $(x - 2)(x + 3) = 5$
62. $(x - 1)(x + 2) = 4$

Identify the coefficient of x^2, the coefficient of x, and the constant term.

63. $x^2 - 5x - 6$ **64.** $x^2 - 6x + 7$ **65.** $2x^2 + 4x - 3$ **66.** $3x^2 + 4x - 2$

Find the value of $b^2 - 4ac$ for the given values of a, b, and c.

67. $a = 1, b = -5, c = -6$

68. $a = 1, b = -6, c = 7$

69. $a = 2, b = 4, c = -3$

70. $a = 3, b = 4, c = -2$

Simplify.

71. $\dfrac{5 + \sqrt{49}}{2}$ **72.** $\dfrac{5 - \sqrt{49}}{2}$ **73.** $\dfrac{-4 - \sqrt{40}}{4}$ **74.** $\dfrac{-4 + \sqrt{40}}{4}$

The Quadratic Formula 9.3

Learning Objectives

In this section, we will learn how to:

1. Simplify the quadratic formula after substituting in values.
2. Use the quadratic formula to solve a quadratic equation.

Introduction

In this section, we will derive the quadratic formula. It is one formula that you will use in almost all types of mathematics. We will first state the formula as a theorem and then prove it. The proof is based on the method of completing the square developed in the preceding section.

PROPERTY *Quadratic Formula*

For any quadratic equation in the form $ax^2 + bx + c = 0$, where a, b, and c are real numbers and $a \neq 0$, the two solutions are:

$$x = \frac{-b + \sqrt{b^2 - 4ac}}{2a} \quad \text{and} \quad x = \frac{-b - \sqrt{b^2 - 4ac}}{2a}$$

Note This is one of the few times in the course where we actually get to show a proof. The proof shown here is the reason the quadratic formula looks the way it does. We will use the quadratic formula in every example we do in this section. As you read through the examples, you may find yourself wondering why some parts of the formula are the way they are. If that happens, come back to this proof and see for yourself.

Proof

We will prove the theorem by completing the square on:

$$ax^2 + bx + c = 0$$

Adding $-c$ to both sides, we have:

$$ax^2 + bx = -c$$

To make the coefficient of x^2 one, we divide both sides by a:

$$\frac{ax^2}{a} + \frac{bx}{a} = -\frac{c}{a}$$

$$x^2 + \frac{b}{a}x = -\frac{c}{a}$$

Now, to complete the square, we add the square of half of $\frac{b}{a}$ to both sides:

$$x^2 + \frac{b}{a}x + \left(\frac{b}{2a}\right)^2 = -\frac{c}{a} + \left(\frac{b}{2a}\right)^2 \qquad \tfrac{1}{2} \text{ of } \tfrac{b}{a} \text{ is } \tfrac{b}{2a}$$

Let's simplify the right side separately:

$$-\frac{c}{a} + \left(\frac{b}{2a}\right)^2 = -\frac{c}{a} + \frac{b^2}{4a^2}$$

The least common denominator is $4a^2$. We multiply the numerator and denominator of $-\frac{c}{a}$ by $4a$ to give it the common denominator. Then we combine numerators:

$$\frac{4a}{4a}\left(-\frac{c}{a}\right) + \frac{b^2}{4a^2} = -\frac{4ac}{4a^2} + \frac{b^2}{4a^2}$$

$$= \frac{-4ac + b^2}{4a^2}$$

$$= \frac{b^2 - 4ac}{4a^2}$$

Now, back to the equation. We use our simplified expression for the right side:

$$x^2 + \frac{b}{a}x + \left(\frac{b}{2a}\right)^2 = \frac{b^2 - 4ac}{4a^2}$$

$$\left(x + \frac{b}{2a}\right)^2 = \frac{b^2 - 4ac}{4a^2}$$

Applying the square root property, we have:

$$x + \frac{b}{2a} = \pm\frac{\sqrt{b^2 - 4ac}}{2a}$$

$$x = \frac{-b}{2a} \pm \frac{\sqrt{b^2 - 4ac}}{2a} \quad \text{Add } \frac{-b}{2a} \text{ to both sides}$$

$$x = \frac{-b \pm \sqrt{b^2 - 4ac}}{2a}$$

Note This formula is called the quadratic formula. You will see it many times if you continue taking math classes. By the time you are finished with this section and the problems in the problem set, you should have it memorized.

Our proof is now complete. What we have is this: If our equation is in the form $ax^2 + bx + c = 0$ (standard form), then the solution can always be found by using the quadratic formula:

$$x = \frac{-b \pm \sqrt{b^2 - 4ac}}{2a}$$

VIDEO EXAMPLES

SECTION 9.3

EXAMPLE 1 Simplify the quadratic formula if $a = 4$, $b = -4$, and $c = -15$.

SOLUTION Substituting in values and simplifying, we have:

$$x = \frac{-b \pm \sqrt{b^2 - 4ac}}{2a} = \frac{-(-4) \pm \sqrt{(-4)^2 - 4(4)(-15)}}{2(4)}$$

$$= \frac{4 \pm \sqrt{16 + 240}}{8}$$

$$= \frac{4 \pm \sqrt{256}}{8}$$

$$= \frac{4 \pm 16}{8}$$

So $x = \frac{4 + 16}{8}$ or $x = \frac{4 - 16}{8}$

$x = \frac{20}{8}$ $\qquad$ $x = -\frac{12}{8}$

$x = \frac{5}{2}$ $\qquad$ $x = -\frac{3}{2}$

EXAMPLE 2 Simplify the quadratic formula if $a = 3$, $b = 6$, and $c = 1$.

SOLUTION Substituting in values and simplifying, we have:

$$x = \frac{-b \pm \sqrt{b^2 - 4ac}}{2a} = \frac{-6 \pm \sqrt{6^2 - 4(3)(1)}}{2(3)}$$

$$= \frac{-6 \pm \sqrt{36 - 12}}{6}$$

$$= \frac{-6 \pm \sqrt{24}}{6}$$

$$= \frac{-6 \pm 2\sqrt{6}}{6} \qquad \sqrt{24} = \sqrt{4 \cdot 6} = 2\sqrt{6}$$

$$= \frac{2(-3 \pm \sqrt{6})}{2 \cdot 3} \qquad \text{Factor the numerator and denominator}$$

$$= \frac{-3 \pm \sqrt{6}}{3} \qquad \text{Reduce}$$

EXAMPLE 3 Solve $x^2 - 4x - 5 = 0$ by using the quadratic formula.

SOLUTION To use the quadratic formula, we must make sure the equation is in standard form; identify a, b, and c; substitute them into the formula; and work out the arithmetic.

For the equation $x^2 - 4x - 5 = 0$, $a = 1$, $b = -4$, and $c = -5$:

$$x = \frac{-b \pm \sqrt{b^2 - 4ac}}{2a}$$

$$= \frac{-(-4) \pm \sqrt{(-4)^2 - 4(1)(-5)}}{2(1)}$$

$$= \frac{4 \pm \sqrt{36}}{2}$$

$$= \frac{4 \pm 6}{2}$$

$$x = \frac{4 + 6}{2} \quad \text{or} \quad x = \frac{4 - 6}{2}$$

$$x = \frac{10}{2} \qquad x = -\frac{2}{2}$$

$$x = 5 \qquad x = -1$$

The two solutions are 5 and -1.

Note Whenever the solutions to our quadratic equations turn out to be rational numbers, as in Example 3, it means the original equation could have been solved by factoring. (We didn't solve the equation in Example 3 by factoring because we were trying to get some practice with the quadratic formula.)

EXAMPLE 4 Solve: $x^2 - 9 = 0$.

SOLUTION We have already solved this equation by factoring and by applying the square root method. This will be a third method of solving this equation. Let's rewrite the equation so we can see the coefficients more clearly.

$$x^2 - 9 = 0 \text{ is equivalent to } x^2 + 0x - 9 = 0$$

Now we see that $a = 1$, $b = 0$, and $c = -9$. Substituting these numbers into the quadratic formula, we have

$$x = \frac{-0 \pm \sqrt{0^2 - 4(1)(-9)}}{2(1)}$$

$$= \frac{0 \pm \sqrt{36}}{2}$$

$$= \pm \frac{\sqrt{36}}{2}$$

$$= \pm \frac{6}{2}$$

$$= \pm 3$$

Our two solutions are 3 and -3, which match the solutions we obtained when we solved this equation previously.

EXAMPLE 5 Solve for x: $x^2 + 2x = \frac{3}{2}$.

SOLUTION Let's make things easy on ourselves by multiplying each side by 2. That way we will not be substituting a fraction into the quadratic formula. Multiplying each side by two gives us

$$2x^2 + 4x = 3$$

$$2x^2 + 4x - 3 = 0 \qquad \text{Add } -3 \text{ to each side}$$

Now that the equation is in standard form, we see that $a = 2$, $b = 4$, and $c = -3$. Using the quadratic formula we have:

$$x = \frac{-b \pm \sqrt{b^2 - 4ac}}{2a}$$

$$= \frac{-4 \pm \sqrt{4^2 - 4(2)(-3)}}{2(2)}$$

$$= \frac{-4 \pm \sqrt{40}}{4}$$

$$= \frac{-4 \pm 2\sqrt{10}}{4}$$

We can reduce this final expression to lowest terms by factoring 2 from the numerator and denominator and then dividing it out:

$$x = \frac{\cancel{2}(-2 \pm \sqrt{10})}{\cancel{2} \cdot 2}$$

$$= \frac{-2 \pm \sqrt{10}}{2}$$

Our two solutions are $\frac{-2 + \sqrt{10}}{2}$ and $\frac{-2 - \sqrt{10}}{2}$.

EXAMPLE 6 Solve for x: $(x + 2)(x - 1) = 9$.

SOLUTION We must put the equation into standard form before we can use the formula:

$$(x + 2)(x - 1) = 9$$

$$x^2 + x - 2 = 9 \qquad \text{Multiply out the left side}$$

$$x^2 + x - 11 = 0 \qquad \text{Add } -9 \text{ to each side}$$

Now, $a = 1$, $b = 1$, and $c = -11$; therefore:

$$x = \frac{-1 \pm \sqrt{1^2 - 4(1)(-11)}}{2(1)}$$

$$= \frac{-1 \pm \sqrt{45}}{2}$$

$$= \frac{-1 \pm 3\sqrt{5}}{2}$$

The solution set is $\left\{\frac{-1 + 3\sqrt{5}}{2}, \frac{-1 - 3\sqrt{5}}{2}\right\}$.

Getting Ready for Class

After reading through the preceding section, respond in your own words and in complete sentences.

A. What is the quadratic formula?

B. Under what circumstances should the quadratic formula be applied?

C. What is standard form for a quadratic equation?

D. Generally, what is the first step in solving a quadratic equation using the quadratic formula?

Problem Set 9.3

Substitute the given values into the quadratic formula and then simplify.

1. $a = 1, b = 2, c = -3$

2. $a = 1, b = -3, c = -4$

3. $a = 6, b = 1, c = -2$

4. $a = 9, b = 3, c = -2$

5. $a = 1, b = 3, c = -2$

6. $a = 1, b = -3, c = -1$

7. $a = 2, b = -6, c = 1$

8. $a = 2, b = -7, c = 4$

9. $a = 1, b = -4, c = -2$

10. $a = 1, b = 6, c = 4$

11. $a = 3, b = -8, c = 2$

12. $a = 3, b = 6, c = 1$

Solve each equation using the quadratic formula.

13. $x^2 + 5x + 6 = 0$

14. $x^2 + 5x - 6 = 0$

15. $a^2 - 4a + 1 = 0$

16. $a^2 + 4a + 1 = 0$

17. $x^2 + 6x + 7 = 0$

18. $x^2 - 4x - 1 = 0$

19. $4x^2 + 8x + 1 = 0$

20. $3x^2 + 6x + 2 = 0$

21. $2x^2 - 3x = 5$

22. $3x^2 - 4x = 5$

23. $2x^2 = -6x + 7$

24. $5x^2 = -6x + 3$

25. $x^2 + 3x + 2 = 0$

26. $x^2 - 7x - 8 = 0$

27. $x^2 + 6x + 9 = 0$

28. $x^2 - 10x + 25 = 0$

29. $x^2 - 2x + 1 = 0$

30. $x^2 + 2x - 3 = 0$

31. $x^2 - 5x - 7 = 0$

32. $2x^2 - 6x - 8 = 0$

33. $6x^2 - x = 2$

34. $6x^2 + 5x = 4$

35. $(x - 2)(x + 1) = 3$

36. $(x - 8)(x + 7) = 5$

37. $3x^2 + 4x = 2$

38. $3x^2 - 4x = 2$

39. $2x^2 - 5 = 2x$

40. $5x^2 + 1 = -10x$

41. Solve $3x^2 - 5x = 0$

a. by factoring

b. by the quadratic formula

42. Solve $4x^2 - 9 = 0$

a. by factoring

b. by the quadratic formula

43. Solve the equation $2x^3 + 3x^2 - 4x = 0$ by first factoring out the common factor x and then using the quadratic formula. There are three solutions to this equation.

44. Solve the equation $5y^3 - 10y^2 + 4y = 0$ by first factoring out the common factor y and then using the quadratic formula.

45. Which two of the expressions below are equivalent?

a. $\dfrac{6 + 2\sqrt{3}}{4}$

b. $\dfrac{3 + \sqrt{3}}{2}$

c. $6 + \dfrac{\sqrt{3}}{2}$

46. Which two of the expressions below are equivalent?

a. $\dfrac{8 - 4\sqrt{2}}{4}$

b. $2 - 4\sqrt{3}$

c. $2 - \sqrt{2}$

47. Solve the following equation by first multiplying both sides by the LCD and then applying the quadratic formula to the result.

$$\frac{1}{2}x^2 - \frac{1}{2}x - \frac{1}{6} = 0$$

48. Solve the following equation by first multiplying both sides by the LCD and then applying the quadratic formula to the result.

$$\frac{1}{2}y^2 - y - \frac{3}{2} = 0$$

Applying the Concepts

49. Archery Margaret shoots an arrow into the air. The equation for the height (in feet) of the tip of the arrow is $h = 8 + 64t - 16t^2$. To find the time at which the arrow is 56 feet above the ground, we replace h with 56 to obtain

$$56 = 8 + 64t - 16t^2$$

Solve this equation for t to find the times at which the arrow is 56 feet above the ground.

50. Coin Toss At the beginning of every football game, the referee flips a coin to see who will kick off. The equation that gives the height (in feet) of the coin tossed in the air is $h = 6 + 32t - 16t^2$. To find the times at which the coin is 18 feet above the ground we substitute 18 for h in the equation, giving us

$$18 = 6 + 32t - 16t^2$$

Solve this equation for t to find the times at which the coin is 18 feet above the ground.

Learning Objectives Assessment

The following problems can be used to help assess if you have successfully met the learning objectives for this section.

51. Simplify the quadratic formula if $a = 2$, $b = -5$, and $c = 2$.

a. $-\frac{1}{2}, -2$ **b.** $\frac{5 \pm \sqrt{41}}{4}$ **c.** $\frac{1}{2}, 2$ **d.** $\frac{-5 \pm \sqrt{41}}{4}$

52. Use the quadratic formula to solve $x^2 + 8x + 6 = 0$.

a. $-4 \pm \sqrt{22}$ **b.** $-4 \pm \sqrt{10}$ **c.** $\frac{-8 \pm \sqrt{42}}{2}$ **d.** $4 \pm 2\sqrt{10}$

Getting Ready for the Next Section

Simplify.

53. $7x + 5 + 8x - 6$

54. $3 + x + 2 - 6x$

55. $(2 - 5x) - (3 + 7x)$

56. $2x + 3 - (x + 7)$

Multiply

57. $2x(8x - 7)$

58. $-3x(5x + 2)$

59. $(3x + 6)(3x + 4)$

60. $(x + 2)(x - 4)$

Rationalize the denominator for each expression.

61. $\dfrac{4}{3 + \sqrt{2}}$

62. $\dfrac{2}{1 - \sqrt{3}}$

Complex Numbers 9.4

Learning Objectives

In this section, we will learn how to:

1. Add complex numbers.
2. Subtract complex numbers.
3. Multiply complex numbers.
4. Divide complex numbers.

Introduction

To solve quadratic equations such as $x^2 = -4$, we need to introduce a new set of numbers. If we try to solve $x^2 = -4$ using real numbers, we always get no solution. There is no real number whose square is -4.

The new set of numbers is called the *complex numbers* and is based on the following definition.

DEFINITION ***number i***

The ***number i*** is a number such that $i = \sqrt{-1}$.

The first thing we notice about this definition is that i is not a real number. There are no real numbers that represent the square root of -1. The other observation we make about i is $i^2 = -1$. If $i = \sqrt{-1}$, then, squaring both sides, we must have $i^2 = -1$. The most common power of i is i^2. Whenever we see i^2, we can write it as -1. We are now ready for a definition of complex numbers.

Note The form $a + bi$ is called *standard form* for complex numbers. The definition in the box indicates that if a number can be written in the form $a + bi$, then it is a complex number.

DEFINITION ***complex number***

A ***complex number*** is any number that can be put in the form $a + bi$, where a and b are real numbers and $i = \sqrt{-1}$.

The following are complex numbers:

$$3 + 4i \qquad \frac{1}{2} - 6i \qquad 8 + i\sqrt{2} \qquad \frac{3}{4} - 2i\sqrt{5} \qquad 4i \qquad 8$$

The number $4i$ is a complex number because $4i = 0 + 4i$. The number 8 is a complex number because $8 = 8 + 0i$. From this last case we can see that the real numbers are a subset of the complex numbers, because any real number x can be written as $x + 0i$.

Addition and Subtraction of Complex Numbers

We add and subtract complex numbers according to the same procedure we used to add and subtract polynomials, which is to combine similar terms.

VIDEO EXAMPLES

SECTION 9.4

EXAMPLE 1 Add: $(3 + 4i) + (2 - 6i)$.

SOLUTION

$$\begin{aligned}(3 + 4i) + (2 - 6i) &= (3 + 2) + (4i - 6i) && \text{Commutative and associative properties}\\ &= 5 + (-2i) && \text{Combine similar terms}\\ &= 5 - 2i\end{aligned}$$

EXAMPLE 2 Subtract: $(5 - 4i) - (6 - i)$.

SOLUTION

$$\begin{aligned}(5 - 4i) - (6 - i) &= 5 - 4i - 6 + i && \text{Distributive property}\\ &= (5 - 6) + (-4i + i) && \text{Commutative and associative properties}\\ &= -1 + (-3i) && \text{Combine similar terms}\\ &= -1 - 3i\end{aligned}$$

EXAMPLE 3 Simplify: $(2 - 5i) - (3 + 7i) + (2 - i)$.

SOLUTION

$$\begin{aligned}(2 - 5i) - (3 + 7i) + (2 - i) &= 2 - 5i - 3 - 7i + 2 - i\\ &= (2 - 3 + 2) + (-5i - 7i - i)\\ &= 1 - 13i\end{aligned}$$

Multiplication of Complex Numbers

Multiplication of complex numbers is very similar to multiplication of polynomials. We can simplify many answers by using the fact that $i^2 = -1$.

EXAMPLE 4 Multiply: $4i(3 + 5i)$.

SOLUTION

$$\begin{aligned}4i(3 + 5i) &= 4i(3) + 4i(5i) && \text{Distributive property}\\ &= 12i + 20i^2 && \text{Multiplication}\\ &= 12i + 20(-1) && i^2 = -1\\ &= -20 + 12i\end{aligned}$$

EXAMPLE 5 Multiply: $(3 + 2i)(4 - 3i)$.

SOLUTION

$$\begin{aligned}&(3 + 2i)(4 - 3i)\\ &= 3 \cdot 4 + 3(-3i) + 2i(4) + 2i(-3i) && \text{FOIL method}\\ &= 12 - 9i + 8i - 6i^2\\ &= 12 - 9i + 8i - 6(-1) && i^2 = -1\\ &= (12 + 6) + (-9i + 8i)\\ &= 18 - i\end{aligned}$$

Division of Complex Numbers

We divide complex numbers by applying the same process we used to rationalize denominators. That is, we multiply the numerator and denominator by the conjugate of the denominator (which is referred to as the *complex conjugate*).

EXAMPLE 6 Divide: $\frac{2+i}{5+2i}$.

SOLUTION The conjugate of the denominator is $5 - 2i$:

$$\left(\frac{2+i}{5+2i}\right)\left(\frac{5-2i}{5-2i}\right) = \frac{10 - 4i + 5i - 2i^2}{25 - 4i^2}$$

$$= \frac{10 - 4i + 5i - 2(-1)}{25 - 4(-1)} \qquad i^2 = -1$$

$$= \frac{12+i}{29}$$

When we write our answer in standard form for complex numbers, we get:

$$\frac{12+i}{29} = \frac{12}{29} + \frac{i}{29} = \frac{12}{29} + \frac{1}{29}i$$

Note The conjugate of $a + bi$ is $a - bi$. When we multiply complex conjugates the result is always a real number because:

$$(a+bi)(a-bi) = a^2 - (bi)^2$$
$$= a^2 - b^2i^2$$
$$= a^2 - b^2(-1)$$
$$= a^2 + b^2$$

which is a real number.

Getting Ready for Class

After reading through the preceding section, respond in your own words and in complete sentences.

A. What is the number i?

B. What is a complex number?

C. What kind of number will always result when we multiply complex conjugates?

D. Explain how to divide complex numbers.

Problem Set 9.4

Perform the following additions and subtractions.

1. $(3 - 2i) + 3i$ **2.** $(5 - 4i) - 8i$ **3.** $(6 + 2i) - 10i$

4. $(8 - 10i) + 7i$ **5.** $(11 + 9i) + (4 + i)$ **6.** $(12 + 2i) + (3 - 5i)$

7. $(3 + 2i) + (6 - i)$ **8.** $(4 + 8i) + (7 + i)$ **9.** $(5 + 7i) - (6 + 8i)$

10. $(11 + 6i) - (3 + 6i)$ **11.** $(9 - i) + (2 - i)$ **12.** $(8 + 3i) - (8 - 3i)$

13. $(6 + i) - 4i - (2 - i)$ **14.** $(3 + 2i) - 5i - (5 + 4i)$

15. $(6 - 11i) + 3i + (2 + i)$ **16.** $(3 + 4i) - (5 + 7i) - (6 - i)$

17. $(2 + 3i) - (6 - 2i) + (3 - i)$ **18.** $(8 + 9i) + (5 - 6i) - (4 - 3i)$

Multiply the following complex numbers.

19. $3(2 - i)$ **20.** $4(5 + 3i)$ **21.** $2i(8 - 7i)$

22. $-3i(2 + 5i)$ **23.** $(2 + i)(4 - i)$ **24.** $(6 + 3i)(4 + 3i)$

25. $(2 + i)(3 - 5i)$ **26.** $(4 - i)(2 - i)$ **27.** $(3 + 5i)(3 - 5i)$

28. $(8 + 6i)(8 - 6i)$ **29.** $(2 + i)(2 - i)$ **30.** $(3 + i)(3 - i)$

Divide the following complex numbers.

31. $\frac{2}{3 - 2i}$ **32.** $\frac{3}{5 + 6i}$ **33.** $\frac{-3i}{2 + 3i}$ **34.** $\frac{4i}{3 + i}$

35. $\frac{i}{3 - i}$ **36.** $\frac{-7i}{5 - 4i}$ **37.** $\frac{2 + i}{2 - i}$ **38.** $\frac{3 + 2i}{3 - 2i}$

39. $\frac{4 + 5i}{3 - 6i}$ **40.** $\frac{-2 + i}{5 + 6i}$

41. Use the FOIL method to multiply $(x + 3i)(x - 3i)$.

42. Use the FOIL method to multiply $(x + 5i)(x - 5i)$.

43. The opposite of i is $-i$. The reciprocal of i is $\frac{1}{i}$. Multiply the numerator and denominator of $\frac{1}{i}$ by i and simplify the result to see that the opposite of i and the reciprocal of i are the same number.

44. If $i^2 = -1$, what are i^3 and i^4? (*Hint:* $i^3 = i^2 \cdot i$.)

Learning Objectives Assessment

The following problems can be used to help assess if you have successfully met the learning objectives for this section.

45. Add: $(3 - 5i) + (4 + 2i)$.

a. $4i$ **b.** $22 - 14i$ **c.** 17 **d.** $7 - 3i$

46. Subtract: $(3 - 5i) - (4 + 2i)$

a. $22 - 14i$ **b.** -12 **c.** $-1 - 7i$ **d.** $-1 - 3i$

47. Multiply: $(3 - 5i)(4 + 2i)$.

a. $7 - 3i$ **b.** 12 **c.** 22 **d.** $22 - 14i$

48. Divide: $\dfrac{3 - 5i}{4 + 2i}$.

a. $\dfrac{1}{6} - \dfrac{13}{6}i$ **b.** $\dfrac{1}{10} - \dfrac{13}{10}i$ **c.** $\dfrac{11}{10} - \dfrac{13}{10}i$ **d.** $\dfrac{11}{10} - \dfrac{7}{10}i$

Getting Ready for the Next Section

Simplify.

49. $\sqrt{36}$ **50.** $\sqrt{49}$ **51.** $-\sqrt{75}$ **52.** $-\sqrt{12}$

Solve for x.

53. $(x + 2)^2 = 9$ **54.** $(x - 3)^2 = 9$ **55.** $\dfrac{1}{10}x^2 - \dfrac{1}{5}x = \dfrac{1}{2}$

56. $\dfrac{1}{18}x^2 - \dfrac{2}{9}x = \dfrac{13}{18}$ **57.** $(2x - 3)(2x - 1) = 4$ **58.** $(x - 1)(3x - 3) = 10$

SPOTLIGHT ON SUCCESS *A message from Mr. McKeague*

Dear Student,

Now that you are close to finishing this course, I want to pass on a couple of things that have helped me a great deal with my career. I'll introduce each one with a quote:

Do something for the person you will be 5 years from now.

I have always made sure that I arranged my life so that I was doing something for the person I would be 5 years later. For example, when I was 20 years old, I was in college. I imagined that the person I would be at 25-year-old, would want to have a college degree, so I made sure I stayed in school. That's all there is to this. It is not a hard, rigid philosophy. It is a soft, behind the scenes, foundation. It does not include ideas such as "Five years from now I'm going to graduate at the top of my class from the best college in the country." Instead, you think, "five years from now I will have a college degree, or I will still be in school working towards it."

This philosophy led to a community college teaching job, writing textbooks, doing videos with the textbooks, then to MathTV and the book you are reading right now. Along the way there were many other options and directions that I didn't take, but all the choices I had were due to keeping the person I would be in 5 years in mind.

It's easier to ride a horse in the direction it is going.

I started my college career thinking that I would become a dentist. I enrolled in all the courses that were required for dental school. When I completed the courses, I applied to a number of dental schools, but wasn't accepted. I kept going to school, and applied again the next year, again, without success. My life was not going in the direction of dental school, even though I had worked hard to put it in that direction. So I did a little inventory of the classes I had taken and the grades I earned, and realized that I was doing well in mathematics. My life was actually going in that direction so I decided to see where that would take me. It was a good decision.

It is a good idea to work hard toward your goals, but it is also a good idea to take inventory every now and then to be sure you are headed in the direction that is best for you.

I wish you good luck with the rest of your college years, and with whatever you decide you want to do as a career.

Pat McKeague

Complex Solutions to Quadratic Equations 9.5

Learning Objectives

In this section, we will learn how to:

1. Write square roots of negative numbers in terms of i.
2. Solve quadratic equations having complex solutions.

Introduction

The quadratic formula tells us that the solutions to equations of the form $ax^2 + bx + c = 0$ are always:

$$x = \frac{-b \pm \sqrt{b^2 - 4ac}}{2a}$$

The part of the quadratic formula under the radical sign is called the *discriminant:*

$$\text{Discriminant} = b^2 - 4ac$$

When the discriminant is negative, we have to deal with the square root of a negative number. We handle square roots of negative numbers by using the definition $i = \sqrt{-1}$. To illustrate, suppose we want to simplify an expression that contains $\sqrt{-9}$, which is not a real number. We begin by writing $\sqrt{-9}$ as $\sqrt{9(-1)}$. Then, we write this expression as the product of two separate radicals: $\sqrt{9} \cdot \sqrt{-1}$. Applying the definition $i = \sqrt{-1}$ to this last expression, we have

$$\sqrt{9} \cdot \sqrt{-1} = 3i$$

As you may recall from the previous section, the number $3i$ is called a complex number. Here are some further examples.

VIDEO EXAMPLES

SECTION 9.5

EXAMPLE 1 Write the following radicals as complex numbers:

SOLUTION

a. $\sqrt{-4} = \sqrt{4(-1)} = \sqrt{4} \cdot \sqrt{-1} = 2i$

b. $\sqrt{-36} = \sqrt{36(-1)} = \sqrt{36} \cdot \sqrt{-1} = 6i$

c. $\sqrt{-7} = \sqrt{7(-1)} = \sqrt{7} \cdot \sqrt{-1} = i\sqrt{7}$

d. $\sqrt{-75} = \sqrt{75(-1)} = \sqrt{75} \cdot \sqrt{-1} = 5i\sqrt{3}$

In parts (c) and (d) of Example 1, we wrote i before the radical because it is less confusing that way. If we put i after the radical, it is sometimes mistaken for being under the radical.

Let's see how complex numbers relate to quadratic equations by looking at some examples of quadratic equations whose solutions are complex numbers.

EXAMPLE 2 Solve for x: $(x + 2)^2 = -9$

SOLUTION We can solve this equation by expanding the left side, putting the results into standard form, and then applying the quadratic formula. It is faster, however, simply to apply the square root property:

$$
\begin{aligned}
(x + 2)^2 &= -9 \\
x + 2 &= \pm\sqrt{-9} && \text{Square root property} \\
x + 2 &= \pm 3i && \sqrt{-9} = \sqrt{9}\sqrt{-1} = 3i \\
x &= -2 \pm 3i && \text{Add } -2 \text{ to both sides}
\end{aligned}
$$

The solution set contains two complex solutions. Notice that the two solutions are conjugates.

The solution set is $\{-2 + 3i, -2 - 3i\}$.

EXAMPLE 3 Solve for x: $\frac{1}{10}x^2 - \frac{1}{5}x = -\frac{1}{2}$.

SOLUTION It will be easier to apply the quadratic formula if we clear the equation of fractions. Multiplying both sides of the equation by the LCD of 10 gives us:

$$x^2 - 2x = -5$$

Next, we add 5 to both sides to put the equation into standard form:

$$x^2 - 2x + 5 = 0 \qquad \text{Add 5 to both sides}$$

Applying the quadratic formula with $a = 1$, $b = -2$, and $c = 5$, we have:

$$x = \frac{-(-2) \pm \sqrt{(-2)^2 - 4(1)(5)}}{2(1)} = \frac{2 \pm \sqrt{-16}}{2} = \frac{2 \pm 4i}{2}$$

Dividing the numerator and denominator by 2, we have the two solutions:

$$x = 1 \pm 2i$$

The two solutions are $1 + 2i$ and $1 - 2i$.

EXAMPLE 4 Solve $(2x - 3)(2x - 1) = -4$.

SOLUTION We multiply the binomials on the left side and then add 4 to each side to write the equation in standard form. From there we identify a, b, and c and apply the quadratic formula:

$$
\begin{aligned}
(2x - 3)(2x - 1) &= -4 \\
4x^2 - 8x + 3 &= -4 && \text{Multiply binomials on left side} \\
4x^2 - 8x + 7 &= 0 && \text{Add 4 to each side}
\end{aligned}
$$

Placing $a = 4$, $b = -8$, and $c = 7$ in the quadratic formula we have:

$$
\begin{aligned}
x &= \frac{-(-8) \pm \sqrt{(-8)^2 - 4(4)(7)}}{2(4)} \\
&= \frac{8 \pm \sqrt{64 - 112}}{8} \\
&= \frac{8 \pm \sqrt{-48}}{8} \\
&= \frac{8 \pm 4i\sqrt{3}}{8} && \sqrt{-48} = i\sqrt{48} = i\sqrt{16}\sqrt{3} = 4i\sqrt{3}
\end{aligned}
$$

To reduce this final expression to lowest terms, we factor a 4 from the numerator and denominator and then divide the numerator and denominator by 4:

$$= \frac{4(2 \pm i\sqrt{3})}{4 \cdot 2}$$

$$= \frac{2 \pm i\sqrt{3}}{2}$$

Writing the solutions in standard form, we have

$$\frac{2 \pm i\sqrt{3}}{2} = \frac{2}{2} \pm \frac{i\sqrt{3}}{2} = 1 \pm \frac{\sqrt{3}}{2}i$$

Getting Ready for Class

After reading through the preceding section, respond in your own words and in complete sentences.

A. Describe how you would write $\sqrt{-4}$ in terms of the number i.

B. When would the quadratic formula result in complex solutions?

C. How can we avoid working with fractions when using the quadratic formula?

D. Describe how you would simplify the expression $\frac{8 \pm 4i\sqrt{3}}{8}$.

Problem Set 9.5

Write the following radicals as complex numbers.

1. $\sqrt{-16}$
2. $\sqrt{-25}$
3. $-\sqrt{-49}$
4. $-\sqrt{-81}$
5. $\sqrt{-6}$
6. $\sqrt{-10}$
7. $-\sqrt{-11}$
8. $-\sqrt{-19}$
9. $\sqrt{-32}$
10. $\sqrt{-288}$
11. $\sqrt{-50}$
12. $\sqrt{-45}$
13. $-\sqrt{-8}$
14. $-\sqrt{-24}$
15. $\sqrt{-48}$
16. $\sqrt{-27}$

Solve the following quadratic equations. Use whatever method seems to fit the situation or is convenient for you.

17. $x^2 = 2x - 2$
18. $x^2 = 4x - 5$
19. $x^2 - 4x = -4$
20. $x^2 - 4x = 4$
21. $2x^2 + 5x = 12$
22. $2x^2 + 30 = 16x$
23. $(x - 2)^2 = -4$
24. $(x - 5)^2 = -25$
25. $\left(x + \frac{1}{2}\right)^2 = -\frac{9}{4}$
26. $\left(x - \frac{1}{4}\right)^2 = -\frac{1}{2}$
27. $\left(x - \frac{1}{2}\right)^2 = -\frac{27}{36}$
28. $\left(x + \frac{1}{2}\right)^2 = -\frac{32}{64}$
29. $x^2 + x + 1 = 0$
30. $x^2 - 3x + 4 = 0$
31. $x^2 - 5x + 6 = 0$
32. $x^2 + 2x + 2 = 0$
33. $\frac{1}{2}x^2 + \frac{1}{3}x + \frac{1}{6} = 0$
34. $\frac{1}{5}x^2 + \frac{1}{20}x + \frac{1}{4} = 0$
35. $\frac{1}{3}x^2 = -\frac{1}{2}x + \frac{1}{3}$
36. $\frac{1}{2}x^2 = -\frac{1}{3}x + \frac{1}{6}$
37. $(x + 2)(x - 3) = 5$
38. $(x - 1)(x + 1) = 6$
39. $(x - 5)(x - 3) = -10$
40. $(x - 2)(x - 4) = -5$
41. $(2x - 2)(x - 3) = 9$
42. $(x - 1)(2x + 6) = 9$
43. Is $x = 2 + 2i$ a solution to the equation $x^2 - 4x + 8 = 0$?
44. Is $x = 5 + 3i$ a solution to the equation $x^2 - 10x + 34 = 0$?
45. If one solution to a quadratic equation is $3 + 7i$, what do you think the other solution is?
46. If one solution to a quadratic equation is $4 - 2i$, what do you think the other solution is?

Learning Objectives Assessment

The following problems can be used to help assess if you have successfully met the learning objectives for this section.

47. Write $\sqrt{-20}$ as a complex number.
    a. $2\sqrt{5i}$
    b. $5i\sqrt{2}$
    c. $-2\sqrt{5}$
    d. $2i\sqrt{15}$
48. Solve: $2x^2 + 4x + 5 = 0$.
    a. $-1 \pm \frac{\sqrt{2}}{2}i$
    b. $-1 \pm \frac{\sqrt{6}}{2}i$
    c. $\frac{-2 \pm \sqrt{14}}{2}$
    d. $-1 \pm 2i\sqrt{6}$

Getting Ready for the Next Section

Add the correct term to each binomial so that it becomes a perfect square trinomial. Then factor.

49. $x^2 - 6x$

50. $x^2 - 8x$

Use the equation $y = (x + 1)^2 - 3$ to find the value of y when

51. $x = -4$

52. $x = -3$

53. $x = -2$

54. $x = -1$

55. $x = 1$

56. $x = 2$

Graph the ordered pairs.

57. $(1, 2)$

58. $(2, 1)$

59. $(-2, -2)$

60. $(-1, -1)$

61. $(-4, 1)$

62. $(-1, 2)$

63. $(3, -4)$

64. $(1, -2)$

65. $(5, 0)$

66. $(-3, 0)$

Graphing Parabolas 9.6

Learning Objectives

In this section, we will learn how to:

1. Sketch the graph of a quadratic equation by plotting points.
2. Find the y-intercept for a parabola.
3. Find the x-intercepts (when they exist) for a parabola.
4. Use the intercepts to graph a parabola.

Introduction

In this section we will graph equations of the form $y = ax^2 + bx + c$ and equations that can be put into this form. The graphs of this type of equation all have similar shapes.

We will begin this section by graphing the simplest quadratic equation, $y = x^2$. To get the idea of the shape of this graph, we need to find some ordered pairs that are solutions. We can do this by setting up the following table:

x	$y = x^2$	y

We can choose any convenient numbers for x and then use the equation $y = x^2$ to find the corresponding values for y. Let's use the values -3, -2, -1, 0, 1, 2, and 3 for x and find corresponding values for y. Here is how the table looks when we let x have these values:

x	$y = x^2$	y
-3	$y = (-3)^2 = 9$	9
-2	$y = (-2)^2 = 4$	4
-1	$y = (-1)^2 = 1$	1
0	$y = 0^2 = 0$	0
1	$y = 1^2 = 1$	1
2	$y = 2^2 = 4$	4
3	$y = 3^2 = 9$	9

The table gives us the solutions $(-3, 9)$, $(-2, 4)$, $(-1, 1)$, $(0, 0)$, $(1, 1)$, $(2, 4)$, and $(3, 9)$ for the equation $y = x^2$. We plot each of the points on a rectangular coordinate system and draw a smooth curve between them, as shown in Figure 1.

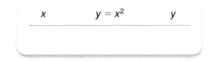

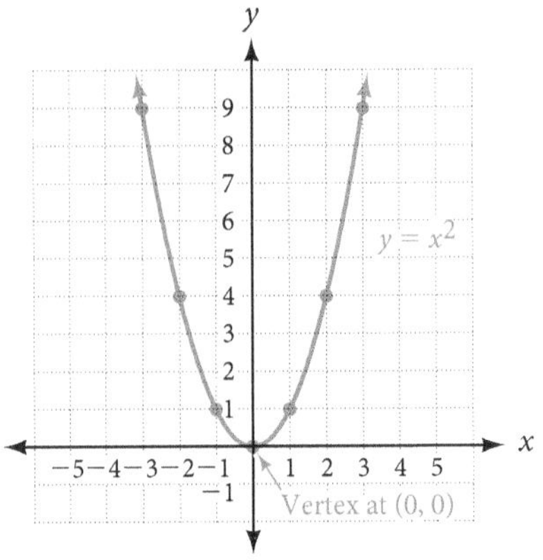

FIGURE 1

This graph is called a *parabola*. All equations of the form $y = ax^2 + bx + c\ (a \neq 0)$ produce parabolas when graphed.

Note that the point (0, 0) is called the *vertex* of the parabola in Figure 1. It is the lowest point on the graph. Although all the parabolas in this section will open up, there are some parabolas that open downward. For those parabolas, the vertex is the highest point on the graph.

VIDEO EXAMPLES

SECTION 9.6

EXAMPLE 1 Graph the equation $y = x^2 - 3$.

SOLUTION We begin by making a table using convenient values for x:

x	$y = x^2 - 3$	y
-2	$y = (-2)^2 - 3 = 4 - 3 = 1$	1
-1	$y = (-1)^2 - 3 = 1 - 3 = -2$	-2
0	$y = 0^2 - 3 = -3$	-3
1	$y = 1^2 - 3 = 1 - 3 = -2$	-2
2	$y = 2^2 - 3 = 4 - 3 = 1$	1

The table gives us the ordered pairs $(-2, 1)$, $(-1, -2)$, $(0, -3)$, $(1, -2)$, and $(2, 1)$ as solutions to $y = x^2 - 3$. The graph is shown in Figure 2.

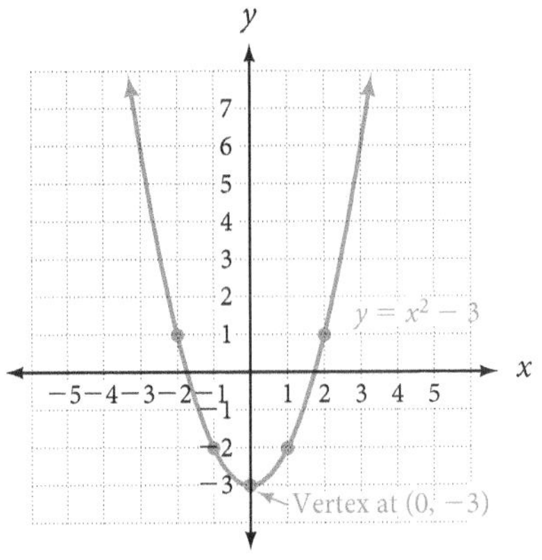

FIGURE 2

EXAMPLE 2 Graph $y = (x - 2)^2$.

SOLUTION Again, we make a table by choosing convenient values of x.

x	$y = (x - 2)^2$	y
-1	$y = (-1 - 2)^2 = (-3)^2 = 9$	9
0	$y = (0 - 2)^2 = (-2)^2 = 4$	4
1	$y = (1 - 2)^2 = (-1)^2 = 1$	1
2	$y = (2 - 2)^2 = 0^2 = 0$	0

We can continue the table if we feel more solutions will make the graph clearer.

3	$y = (3 - 2)^2 = 1^2 = 1$	1
4	$y = (4 - 2)^2 = 2^2 = 4$	4
5	$y = (5 - 2)^2 = 3^2 = 9$	9

Putting the results of the table onto a coordinate system, we have the graph in Figure 3.

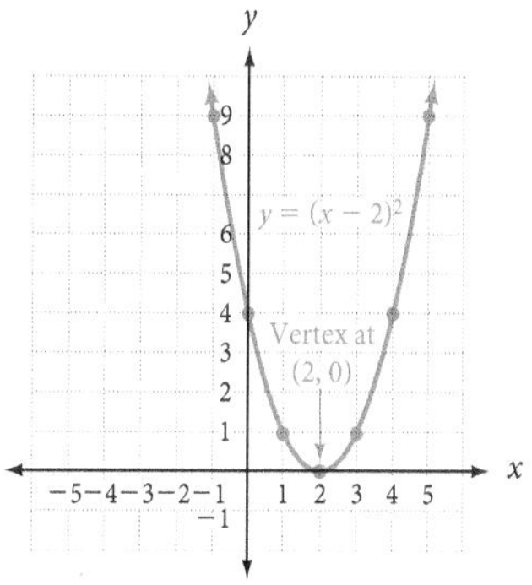

FIGURE 3

Note You may have noticed that the equation in this example (as well as in Example 2 above) is not written in standard form ($y = ax^2 + bx + c$). If you go on to take more math classes, you will learn that this alternate form is called "vertex" form or "graphing" form because the vertex is apparent from the equation itself. The x-coordinate of the vertex is the opposite of the number in parentheses, and the y-coordinate is the number that follows. For Example 3, the vertex is $(-1, -3)$.

EXAMPLE 3 Graph $y = (x + 1)^2 - 3$.

SOLUTION As with the previous two examples, we make a table and then plot points.

x	$y = (x + 1)^2 - 3$	y
-4	$y = (-4 + 1)^2 - 3 = 9 - 3$	6
-3	$y = (-3 + 1)^2 - 3 = 4 - 3$	1
-1	$y = (-1 + 1)^2 - 3 = 0 - 3$	-3
1	$y = (1 + 1)^2 - 3 = 4 - 3$	1
2	$y = (2 + 1)^2 - 3 = 9 - 3$	6

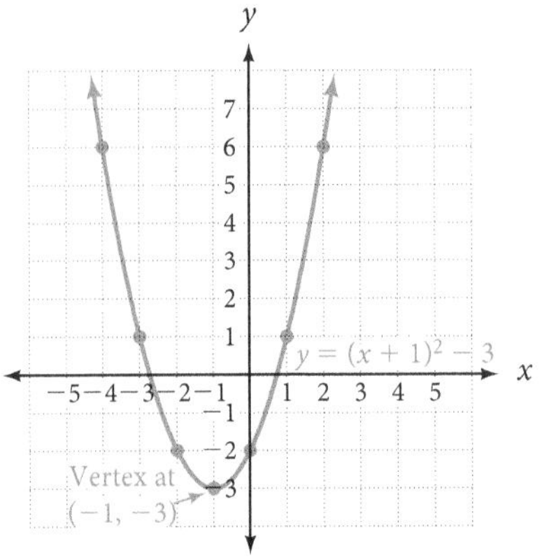

FIGURE 4

Graphing the results of the table, we have Figure 4.

EXAMPLE 4 Graph $y = x^2 - 6x + 5$.

SOLUTION Plotting points as before, we have:

x	$y = x^2 - 6x + 5$	y
0	$y = 0^2 - 6(0) + 5 = 5$	5
1	$y = 1^2 - 6(1) + 5 = 0$	0
2	$y = 2^2 - 6(2) + 5 = -3$	−3
3	$y = 3^2 - 6(3) + 5 = -4$	−4
4	$y = 4^2 - 6(4) + 5 = -3$	−3
5	$y = 5^2 - 6(5) + 5 = 0$	0

The graph is shown in Figure 5.

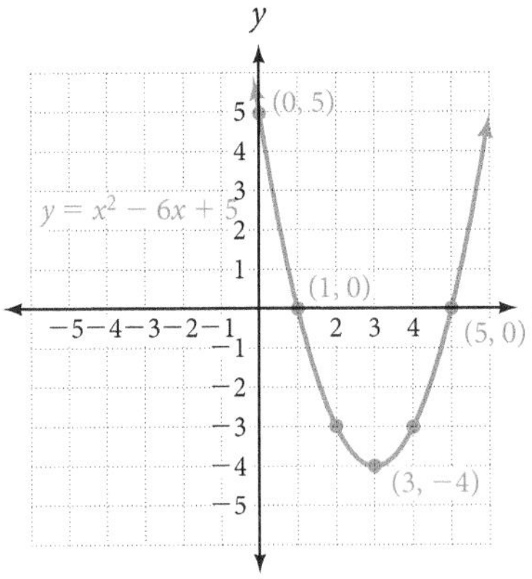

FIGURE 5

Intercepts

Recall that the intercepts for a graph are the points where the graph intersects the x-axis and the y-axis. To find the x-intercepts, which are the points where the graph intersects the x-axis, we substitute $y = 0$ into the equation and solve for x. Likewise, to find the y-intercept, we substitute $x = 0$ into the equation and solve for y to get the point where the graph intersects the y-axis.

For the parabola shown in Figure 5, the x-intercepts are the points (1, 0) and (5, 0). The y-intercept is the point (0, 5). In our next example, we show how to find these points directly from the equation.

EXAMPLE 5 Find the intercepts for $y = x^2 - 6x + 5$.

SOLUTION To find the x-intercepts, we let $y = 0$ and solve for x:

$$0 = x^2 - 6x + 5$$

$$0 = (x - 1)(x - 5) \quad \text{Factor}$$

$$x - 1 = 0 \quad \text{or} \quad x - 5 = 0$$

$$x = 1 \quad \text{or} \quad x = 5$$

The x-intercepts are (1, 0) and (5, 0). To find the y-intercept, we substitute $x = 0$:

$$y = 0^2 - 6(0) + 5$$

$$y = 5$$

The y-intercept is (0, 5).

EXAMPLE 6 Find the intercepts for $y = (x + 1)^2 - 3$.

SOLUTION If $y = 0$, we have:

$$0 = (x + 1)^2 - 3$$

$$3 = (x + 1)^2 \quad \text{Add 3 to both sides}$$

$$\pm\sqrt{3} = x + 1 \quad \text{Square root property}$$

$$-1 \pm \sqrt{3} = x \quad \text{Add } -1 \text{ to both sides}$$

The x-intercepts are $(-1 + \sqrt{3}, 0)$ and $(-1 - \sqrt{3}, 0)$. If we approximate $\sqrt{3} \approx 1.7$, then we have

$$-1 + \sqrt{3} = -1 + 1.7 = 0.7 \quad \text{and} \quad -1 - \sqrt{3} = -1 - 1.7 = -2.7$$

This would give us the points (0.7, 0) and (−2.7, 0).

For the y-intercept, we let $x = 0$ and obtain $y = (0 + 1)^2 - 3 = -2$, giving us the point (0, −2).

We graphed this equation in Example 3. The graph shown in Figure 4 is consistent with these results.

EXAMPLE 7 Find the intercepts for $y = x^2 + 2x - 5$ and use them to sketch the graph of the equation.

SOLUTION We begin by finding the intercepts. If $y = 0$, we have:

$$0 = x^2 + 2x - 5$$

We can solve this equation using the quadratic formula.

$$\begin{aligned} x &= \frac{-2 \pm \sqrt{2^2 - 4(1)(-5)}}{2(1)} \\ &= \frac{-2 \pm \sqrt{24}}{2} \\ &= \frac{-2 \pm 2\sqrt{6}}{2} \\ &= -1 \pm \sqrt{6} \end{aligned}$$

Using $\sqrt{6} \approx 2.4$ gives us

$$x = -1 + 2.4 \quad \text{or} \quad x = -1 - 2.4$$

$$= 1.4 \qquad\qquad = -3.4$$

If $x = 0$, we obtain $y = 0^2 + 2(0) - 5 = -5$. The x-intercepts are $(1.4, 0)$ and $(-3.4, 0)$ and the y-intercept is $(0, -5)$.

To sketch the graph we use a table to obtain a few additional points.

x	$y = x^2 + 2x - 5$	y
-4	$y = (-4)^2 + 2(-4) - 5 = 3$	3
-1	$y = (-1)^2 + 2(-1) - 5 = -6$	-6
2	$y = (2)^2 + 2(2) - 5 = 3$	3

The graph is shown in Figure 6.

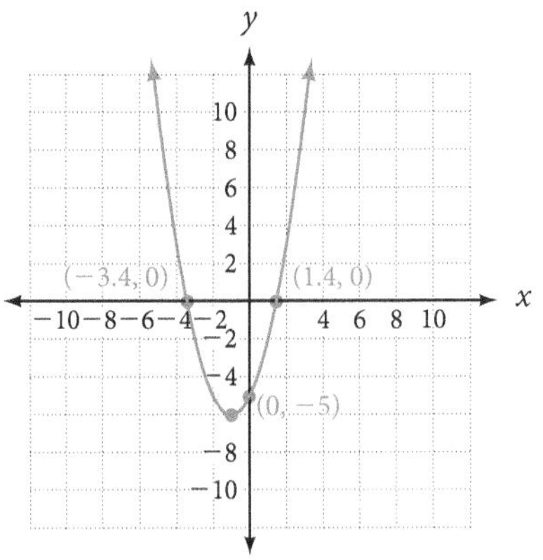

FIGURE 6

Getting Ready for Class

After reading through the preceding section, respond in your own words and in complete sentences.

A. What is a parabola?

B. What is the vertex of a parabola?

C. How do you find the x-intercepts of a parabola?

D. How do you find the y-intercept of a parabola?

Problem Set 9.6

Graph each of the following equations by making a table of values and plotting points.

1. $y = x^2 - 4$
2. $y = x^2 + 2$
3. $y = (x + 2)^2$
4. $y = (x + 5)^2$
5. $y = (x - 3)^2$
6. $y = (x - 2)^2$
7. $y = (x + 1)^2 - 2$
8. $y = (x - 1)^2 + 2$
9. $y = (x + 2)^2 - 3$
10. $y = (x - 2)^2 + 3$
11. $y = x^2 + 6x + 5$
12. $y = x^2 - 8x + 12$

Find the x-intercepts and y-intercept for each equation. Then use them to graph the parabola. Plot additional points as necessary.

13. $y = x^2 - 5$
14. $y = x^2 - 2$
15. $y = (x - 5)^2$
16. $y = (x + 3)^2$
17. $y = (x - 3)^2 - 2$
18. $y = (x + 4)^2 - 1$
19. $y = x^2 - 2x - 3$
20. $y = x^2 + 2x - 3$
21. $y = x^2 - 4x + 2$
22. $y = x^2 + 6x + 4$

The following equations have graphs that are also parabolas. However, the graphs of these equations will open downward; the vertex of each will be the highest point on the graph. Graph each equation by first making a table of ordered pairs using the given values of x.

23. $y = 4 - x^2 \quad x = -3, -2, -1, 0, 1, 2, 3$
24. $y = 3 - x^2 \quad x = -3, -2, -1, 0, 1, 2, 3$
25. $y = -1 - x^2 \quad x = -2, -1, 0, 1, 2$
26. $y = -2 - x^2 \quad x = -2, -1, 0, 1, 2$
27. Graph the line $y = x + 2$ and the parabola $y = x^2$ on the same coordinate system. Name the points where the two graphs intersect.
28. Graph the line $y = x$ and the parabola $y = x^2 - 2$ on the same coordinate system. Name the points where the two graphs intersect.
29. Graph the parabola $y = 2x^2$ and the parabola $y = \frac{1}{2}x^2$ on the same coordinate system.
30. Graph the parabola $y = 3x^2$ and the parabola $y = \frac{1}{3}x^2$ on the same coordinate system.

Learning Objectives Assessment

The following problems can be used to help assess if you have successfully met the learning objectives for this section.

31. Which ordered pair would correspond to a point on the graph of $y = (x + 4)^2 + 1$?

    a. $(6, 5)$ b. $(-6, 5)$ c. $(0, 5)$ d. $(-4, 0)$

32. Find the y-intercept for $y = x^2 - 3x - 4$.

    a. -1 b. 0 c. 4 d. -4

33. Which of the following is an x-intercept for $y = x^2 + 3x - 1$?

    a. $\frac{-3 - \sqrt{13}}{2}$ b. $\frac{3 + \sqrt{13}}{2}$ c. $\frac{-3 + \sqrt{5}}{2}$ d. -1

34. Use the intercepts to graph the equation $y = x^2 - 3x - 2$.

a.

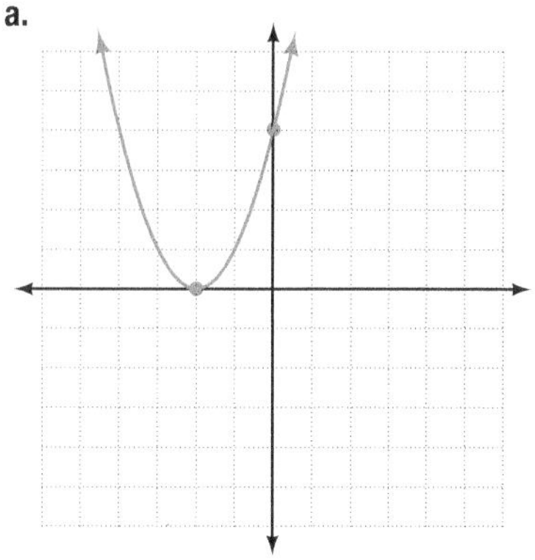

b.

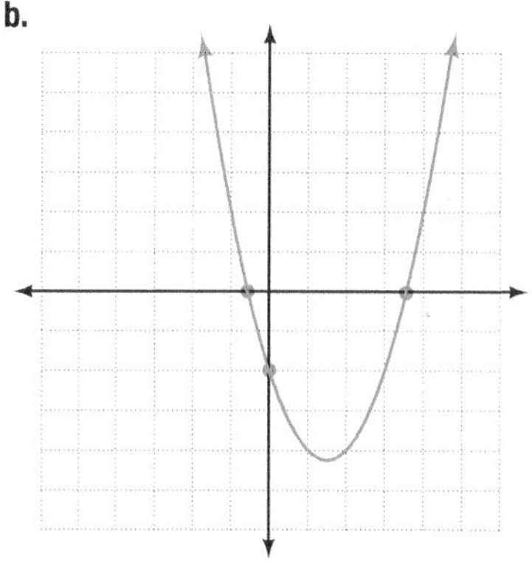

c.

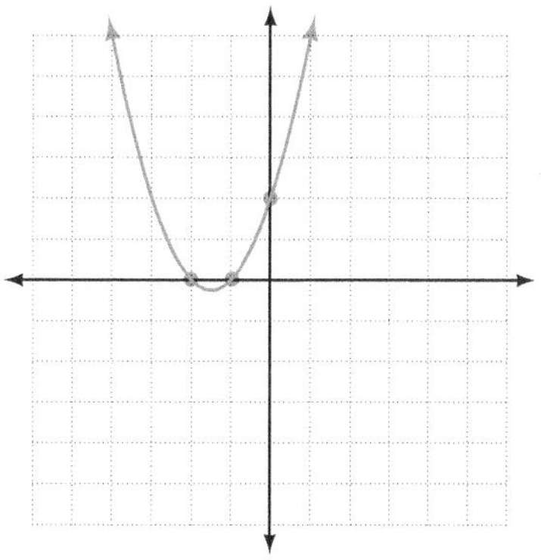

d.

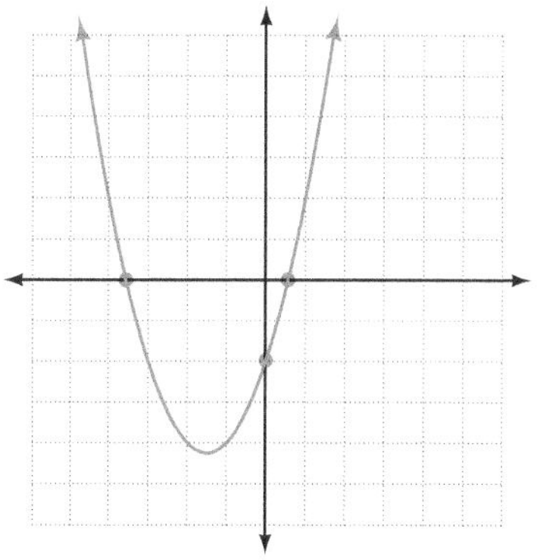

Maintaining Your Skills

Find each root.

35. $\sqrt{49}$

36. $\sqrt[3]{-8}$

Write in simplified form for radicals.

37. $\sqrt{50}$

38. $2\sqrt{18x^2y^3}$

39. $\sqrt{\dfrac{2}{5}}$

40. $\sqrt[3]{\dfrac{1}{2}}$

Perform the indicated operations.

41. $3\sqrt{12} + 5\sqrt{27}$

42. $\sqrt{3}(\sqrt{3} - 4)$

43. $(\sqrt{6} + 2)(\sqrt{6} - 5)$

44. $(\sqrt{x} + 3)^2$

Rationalize the denominator.

45. $\dfrac{8}{\sqrt{5} - \sqrt{3}}$

46. $\dfrac{\sqrt{5} - \sqrt{2}}{\sqrt{5} + \sqrt{2}}$

Solve for x.

47. $\sqrt{2x - 5} = 3$

48. $\sqrt{x + 15} = x + 3$

Chapter 9 Summary

The numbers in brackets refer to the section(s) in which the topic can be found.

EXAMPLES

1. $(x-3)^2 = 25$
$x - 3 = \pm 5$
$x = 3 \pm 5$
$x = -2$ or $x = 8$

Square Root Property for Equations [9.1]

For all positive real numbers b, if $a^2 = b$, then $a = \pm\sqrt{b}$. We can solve equations of the form $(ax+b)^2 = c$ by applying the square root property for equations to write:

$$ax + b = \pm\sqrt{c}$$

2. $x^2 - 6x + 2 = 0$
$x^2 - 6x = -2$
$x^2 - 6x + 9 = -2 + 9$
$(x-3)^2 = 7$
$x - 3 = \pm\sqrt{7}$
$x = 3 \pm \sqrt{7}$

Solving by Completing the Square [9.2]

To complete the square on a quadratic equation as a method of solution, we use the following steps:

Step 1: Move the constant term to one side and the variable terms to the other. Then, divide each side by the coefficient of x^2 if it is other than 1.

Step 2: Take the square of half the coefficient of the linear term and add it to both sides of the equation.

Step 3: Write the left side as a binomial square and then apply the square root property.

Step 4: Solve the resulting equation.

3. If $2x^2 + 3x - 4 = 0$

then $x = \dfrac{-3 \pm \sqrt{9 - 4(2)(-4)}}{2(2)}$

$= \dfrac{-3 \pm \sqrt{41}}{4}$

The Quadratic Formula [9.3, 9.5]

Any equation that is in the form $ax^2 + bx + c = 0$, where $a \neq 0$, has as its solutions:

$$x = \frac{-b \pm \sqrt{b^2 - 4ac}}{2a}$$

The expression under the square root sign, $b^2 - 4ac$, is known as the discriminant. When the discriminant is negative, the solutions are complex numbers.

4. The numbers 5, $3i$, $2 + 4i$, and $7 - i$ are all complex numbers.

Complex Numbers [9.4]

Any number that can be put in the form $a + bi$, where $i = \sqrt{-1}$, is called a complex number.

5. $(3 + 4i) + (6 - 7i)$
$= (3 + 6) + (4i - 7i)$
$= 9 - 3i$

Addition and Subtraction of Complex Numbers [9.4]

We add (or subtract) complex numbers by using the same procedure we used to add (or subtract) polynomials; that is, we combine similar terms.

Multiplication of Complex Numbers [9.4]

We multiply complex numbers in the same way we multiply binomials. The result, however, can be simplified further by substituting -1 for i^2 whenever it appears.

6. $(2 + 3i)(3 - i)$
$= 6 - 2i + 9i - 3i^2$
$= 6 + 7i + 3$
$= 9 + 7i$

Division of Complex Numbers [9.4]

Division with complex numbers is accomplished with the method for rationalizing the denominator that we developed while working with radical expressions. If the denominator has the form $a + bi$, we multiply both the numerator and the denominator by its conjugate, $a - bi$.

7. $\frac{3}{2 + 5i} = \frac{3}{2 + 5i} \cdot \frac{2 - 5i}{2 - 5i}$
$= \frac{6 - 15i}{29}$

Graphing Parabolas [9.6]

The graph of a quadratic equation is a parabola. We can graph the equation by making a table of values and plotting points.

To find the x-intercepts, let $y = 0$ and solve for x. To find the y-intercept, substitute $x = 0$.

8. Graph $y = x^2 - 2$.

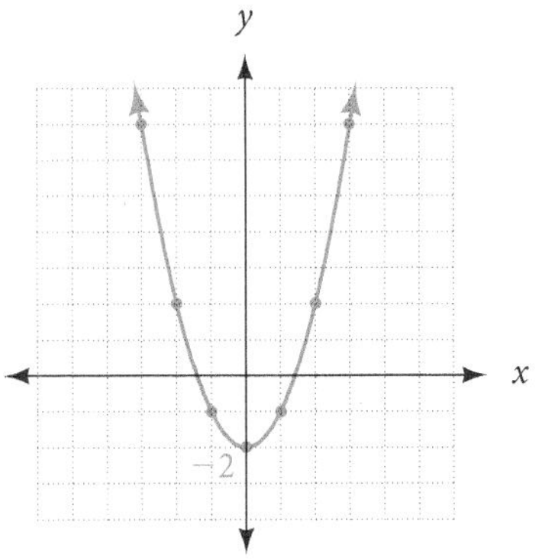

COMMON MISTAKE

1. The most common mistake when working with complex numbers is to say $i = -1$. It does not; i is the *square root* of -1, not -1 itself.
2. The most common mistake when working with the quadratic formula is to try to identify the constants a, b, and c before putting the equation into standard form.

Chapter 9 Test

Solve the following quadratic equations. [9.1, 9.3, 9.5]

1. $x^2 - 8x - 9 = 0$ **2.** $(x - 2)^2 = 18$ **3.** $\left(x - \frac{7}{3}\right)^2 = -\frac{50}{9}$

4. $\frac{1}{3}x^2 = \frac{5}{6}x - \frac{1}{2}$ **5.** $4x^2 + 7x - 2 = 0$ **6.** $(x - 4)(x + 1) = 6$

7. $4x^2 - 20x + 25 = 0$

8. Solve $x^2 - 4x - 9 = 0$ by completing the square. [9.2]

Write as a complex number. [9.5]

9. $\sqrt{-36}$ **10.** $\sqrt{-169}$ **11.** $\sqrt{-45}$ **12.** $\sqrt{-12}$

Perform the indicated operations involving complex numbers. [9.4]

13. $(4i + 3) + (4 + 6i)$ **14.** $(5 - 3i) - (2 - 7i)$ **15.** $(5 - i)(5 + i)$

16. $(2 + 4i)(3 - i)$ **17.** $\frac{i}{3 + i}$ **18.** $\frac{4 - i}{4 + i}$

Graph the following equations. [9.6]

19. $y = x^2 + 3$ **20.** $y = (x + 2)^2$

21. $y = (x - 4)^2 + 2$ **22.** $y = x^2 - 6x + 8$

23. Find the height of the isosceles triangle. [9.1]

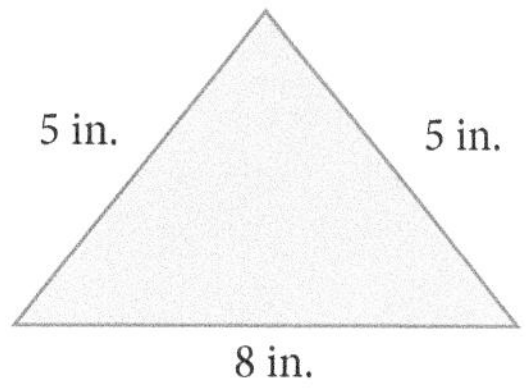

Match each equation with its graph. [9.6]

24. $y = x^2 - 3$ **25.** $y = (x + 3)^2$ **26.** $y = x^2 + 3$ **27.** $y = (x - 3)^2$

A.

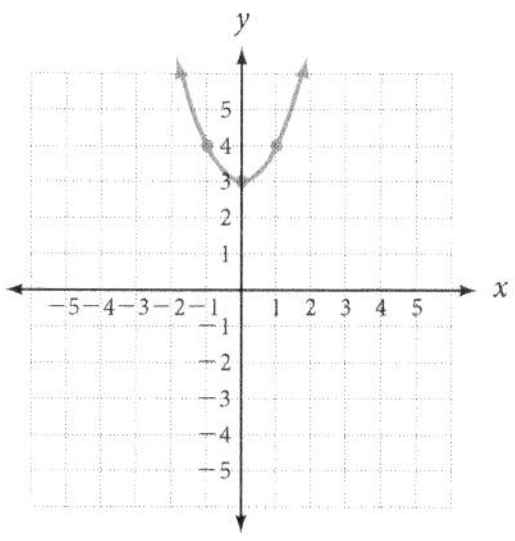

B.

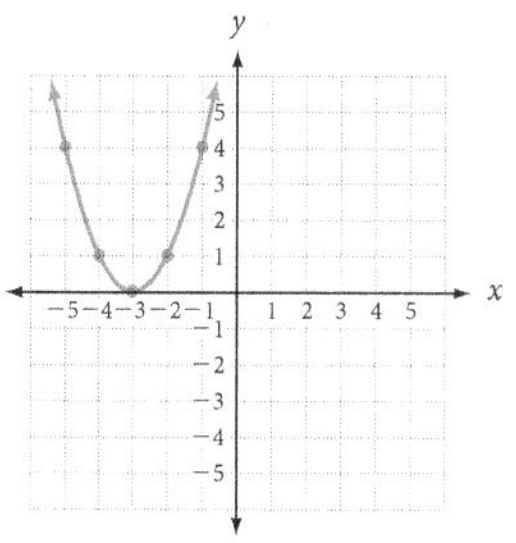

C.

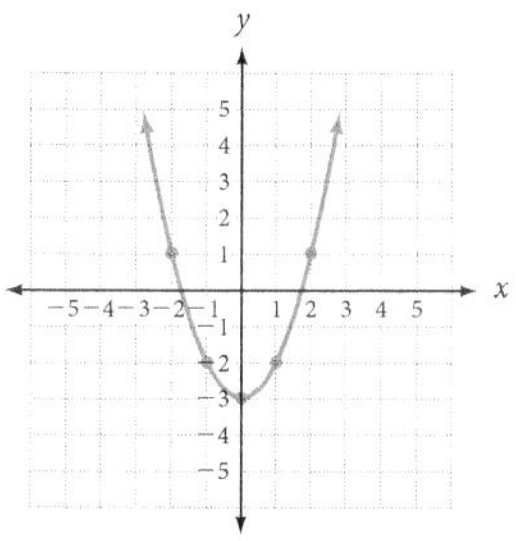

D.

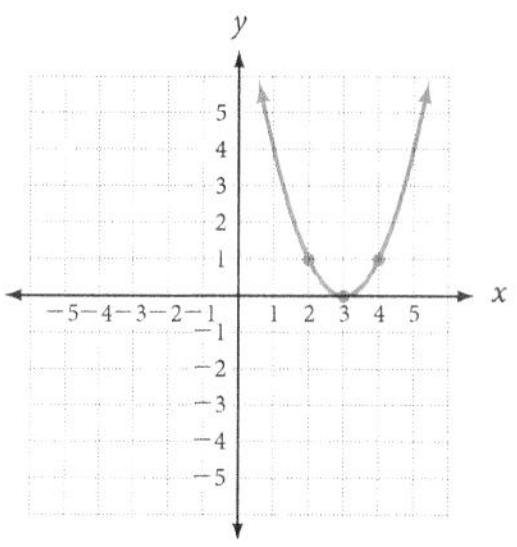

Chapter 1

PROBLEM SET 1.1

1. $x + 5$ **3.** $5y$ **5.** $5(y - 16)$ **7.** $\frac{x}{3}$ **9.** 9
11. 49 **13.** 8 **15.** 64 **17.** 16 **19.** 100
21. 121 **23.** 11 **25.** 16 **27.** 17 **29.** 42
31. 30 **33.** 30 **35.** 24 **37.** 80 **39.** 27
41. 35 **43.** 13 **45.** 4 **47.** 37 **49.** 37 **51.** 16
53. 16 **55.** 81 **57.** 41 **59.** 345 **61.** 2,345
63. 2 **65.** 148 **67.** 36 **69.** 36 **71.** 58
73. 62 **75.** 100 **77.** 9 **79.** 18 **81.** 8
83. 12 **85.** 18 **87.** 42 **89.** 53
91. 4 inches; 1 square inch
93. 4.5 inches; 1.125 square inches
95. 10.25 centimeters; 5 square centimeters
97. 10 **99.** 420 **101.** About 224 chips
103. Less than 95g **105.** **a.** 1,680 mg **b.** 209 mg
107.

Calories Burned by 150-Pound Person

Activity	Calories Burned in 1 Hour
Bicycling	374
Bowling	265
Handball	680
Jogging	680
Skiing	544

109. 93.5 square inches, 39 inches **111.** d **113.** a **115.** b

PROBLEM SET 1.2

1-8.

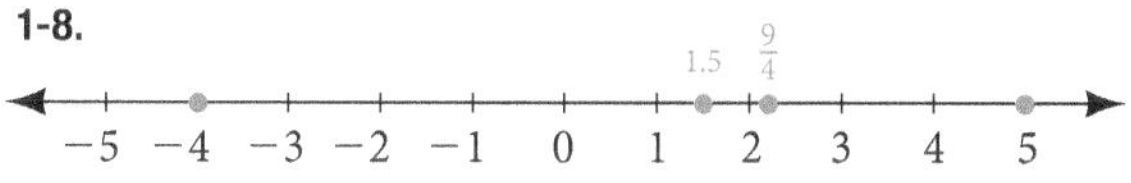

9. 0, 1 **11.** $-3, -2.5, 0, 1, \frac{3}{2}$ **13.** All
15. $-10, -8, -2, 9$ **17.** π **19.** T **21.** F
23. F **25.** T **27.** $-10, 10$ **29.** $-\frac{3}{4}, \frac{3}{4}$
31. $-\frac{11}{2}, \frac{11}{2}$ **33.** 3, 3 **35.** $\frac{2}{5}, \frac{2}{5}$ **37.** $-x, |x|$
39. $<$ **41.** $>$ **43.** $>$ **45.** $>$ **47.** 6
49. 22 **51.** 3 **53.** 7 **55.** 3 **57.** $-8, -2$
59. −64°F; −54°F **61.** −15°F **63.** −100 feet; −105 feet
65. 1,387 calories **67.** 654 more calories
69. **a.** 93 million **b.** False **c.** True **71.** c **73.** a

PROBLEM SET 1.3

1. $3 + 5 = 8, 3 + (-5) = -2, -3 + 5 = 2, -3 + (-5) = -8$
3. $15 + 20 = 35, 15 + (-20) = -5, -15 + 20 = 5, -15 + (-20) = -35$
5. 3 **7.** −7 **9.** −14 **11.** −3 **13.** −25
15. −12 **17.** −19 **19.** −25 **21.** −8 **23.** −4
25. 6 **27.** 6 **29.** 8 **31.** −4 **33.** −14 **35.** −17
37. 4 **39.** 3 **41.** 15 **43.** −8 **45.** 12
47. $5 + 9 = 14$ **49.** $[-7 + (-5)] + 4 = -8$
51. $[-2 + (-3)] + 10 = 5$ **53.** $4 + x$ **55.** $-8 + x$
57. $[x + (-2)] + 3$ **59.** 3 **61.** −3 **63.** 23, 28
65. 30, 35 **67.** 0, −5 **69.** −12, −18 **71.** −4, −8
73. Yes **75.** $-12 + 4$ **77.** $10 + (-6) + (-8) = -\$4$
79. $-30 + 40 = 10$ **81.** \$2,000 **83.** 2006, \$500
85. c

PROBLEM SET 1.4

1. −3 **3.** −6 **5.** 0 **7.** −10 **9.** −16 **11.** −12
13. −7 **15.** 35 **17.** 0 **19.** −4 **21.** 4
23. −24 **25.** −28 **27.** 25 **29.** 4 **31.** 7
33. 17 **35.** 8 **37.** 4 **39.** 18 **41.** 10 **43.** 17
45. 1 **47.** 1 **49.** 27 **51.** −26 **53.** −2
55. 68 **57.** $-7 - 4 = -11$ **59.** $12 - (-8) = 20$
61. $-5 - (-7) = 2$ **63.** $[4 + (-5)] - 17 = -18$
65. $8 - 5 = 3$ **67.** $-8 - 5 = -13$ **69.** $8 - (-5) = 13$
71. $x - 6$ **73.** $-4 - x$ **75.** $(x + 12) - 5$
77. 10 **79.** −2 **81.** $1,500 - 730 = \$770$
83. $-35 + 15 - 20 = -\$40$ **85.** $73 + 10 - 8$, 75°F
87. \$4,500, \$3,950, \$3,400, \$2,850, \$2,300; yes
89.

Day	Plant Height (inches)
0	0
2	1
4	3
6	6
8	13
10	23

b. 12 inches

91. **a.** Yes, the hundreds place **b.** 8,400 **c.** 29,800 **93.** a

PROBLEM SET 1.5

1. −42 **3.** −16 **5.** 3 **7.** 121 **9.** 6 **11.** −60
13. 24 **15.** 49 **17.** −27 **19.** −2 **21.** −3
23. $-\frac{1}{3}$ **25.** 3 **27.** $\frac{1}{7}$ **29.** 0 **31.** 9
33. −15 **35.** −36 **37.** $-\frac{1}{4}$ **39.** $\frac{3}{5}$ **41.** $-\frac{5}{3}$
43. −2 **45.** 6 **47.** 10 **49.** 9 **51.** 45 **53.** 14
55. −2 **57.** −3 **59.** Undefined **61.** Undefined
63. 5 **65.** $-\frac{7}{3}$ **67.** −1 **69.** 216 **71.** −2
73. −18 **75.** 29 **77.** 38 **79.** −7 **81.** $\frac{15}{17}$
83. $-\frac{32}{17}$ **85.** $\frac{1}{3}$ **87.** −5 **89.** 37 **91.** 80
93. 1 **95.** 1 **97.** −2 **99.** $\frac{9}{7}$ **101.** $\frac{16}{11}$
103. −1 **105.** **a.** 25 **b.** −25 **c.** −25 **d.** −25 **e.** 25
107. −25 **109.** −26 **111.** 3 **113.** −10 **115.** −3
117. −8 **119.** $3x - 11$ **121.** 14 **123.** 8
125. −80 **127.** −24 **129.** 1°F **131.** \$350
133. Drops 3.5°F each hour
135. **a.** \$20,000 **b.** \$50,000
c. Yes, the projected revenue for 5,000 email addresses is \$10,000, which is \$5,000 more than the list costs.
137. 465 calories **139.** a **141.** b

PROBLEM SET 1.6

1. Composite, $2^4 \cdot 3$ **3.** Prime
5. Composite, $3 \cdot 11 \cdot 31$ **7.** $2^4 \cdot 3^2$ **9.** $2 \cdot 19$
11. $3 \cdot 5 \cdot 7$ **13.** $2^2 \cdot 3^2 \cdot 5$ **15.** $5 \cdot 7 \cdot 11$ **17.** 11^2
19. $2^2 \cdot 3 \cdot 5 \cdot 7$ **21.** $2^2 \cdot 5 \cdot 31$ **23.** $\frac{7}{11}$ **25.** $\frac{5}{7}$
27. $\frac{11}{13}$ **29.** $\frac{14}{15}$ **31.** $\frac{5}{9}$ **33.** $\frac{5}{8}$

35. $3 \cdot 8 + 3 \cdot 7 + 3 \cdot 5 = 24 + 21 + 15 = 60 = 2^2 \cdot 3 \cdot 5$
37. $\frac{18}{24}$ **39.** $\frac{12}{24}$ **41.** $\frac{15}{24}$ **43.** $\frac{36}{60}$ **45.** $\frac{22}{60}$
47. $<$ **49.** $<$ **51.** $\frac{8}{15}$ **53.** $\frac{3}{2}$ **55.** $\frac{5}{4}$ **57.** 1
59. 1 **61.** 1 **63.** $\frac{9}{16}$ **65.** $\frac{8}{27}$ **67.** $\frac{1}{10{,}000}$
69. $-\frac{10}{21}$ **71.** -4 **73.** 1 **75.** $\frac{9}{16}$ **77.** x
79. $\frac{16}{15}$ **81.** $\frac{4}{3}$ **83.** $-\frac{8}{13}$ **85.** -1 **87.** 1
89. **a.** 10 **b.** 0 **c.** -100 **d.** -20 **93.** $-\frac{1}{4}$
95. $\frac{1}{2}$ **97.** $\frac{x-1}{3}$ **99.** $\frac{3}{2}$ **101.** $\frac{x+6}{2}$
103. $-\frac{3}{5}$ **105.** $\frac{10}{a}$

107.

First Number a	Second Number b	The Sum of a and b $a + b$
$\frac{1}{2}$	$\frac{1}{3}$	$\frac{5}{6}$
$\frac{1}{3}$	$\frac{1}{4}$	$\frac{7}{12}$
$\frac{1}{4}$	$\frac{1}{5}$	$\frac{9}{20}$
$\frac{1}{5}$	$\frac{1}{6}$	$\frac{11}{30}$

109.

First Number a	Second Number b	The Sum of a and b $a + b$
$\frac{1}{12}$	$\frac{1}{2}$	$\frac{7}{12}$
$\frac{1}{12}$	$\frac{1}{3}$	$\frac{5}{12}$
$\frac{1}{12}$	$\frac{1}{4}$	$\frac{1}{3}$
$\frac{1}{12}$	$\frac{1}{6}$	$\frac{1}{4}$

111. $\frac{7}{9}$ **113.** $\frac{7}{3}$ **115.** $\frac{1}{4}$ **117.** $\frac{7}{6}$ **119.** $\frac{19}{24}$ **121.** $\frac{13}{60}$
123. $\frac{29}{35}$ **125.** $\frac{949}{1{,}260}$ **127.** $\frac{13}{420}$ **129.** $\frac{41}{24}$ **131.** $\frac{5}{4}$
133. $-\frac{3}{2}$ **135.** $\frac{3}{2}$ **137.** $\frac{160}{63}$ **139.** $\frac{5}{8}$ **141.** $-\frac{2}{3}$
143. $\frac{7}{3}$ **145.** $\frac{1}{125}$ **147.** $\frac{3}{2}$ ft $= 1\frac{1}{2}$ ft **149.** $\frac{11}{5}$ cm $= 2\frac{1}{5}$ cm
151. 14 blankets **153.** 48 bags **155.** 6 eighth-teaspoons
157. 28 half-pint cartons **159.** $\frac{9}{2}$ pints $= 4\frac{1}{2}$ pints
161. $1,325 **163.** $\frac{2}{5}$

165.

Grade	# of Students	Fraction of Students
A	5	$\frac{1}{8}$
B	8	$\frac{1}{5}$
C	20	$\frac{1}{2}$
below C	7	$\frac{7}{40}$
Total	40	1

167. 10 lots **169.** b **171.** c **173.** a

PROBLEM SET 1.7

1. Commutative **3.** Multiplicative inverse
5. Commutative **7.** Distributive
9. Commutative, associative
11. Commutative, associative
13. Commutative **15.** Commutative, associative
17. Commutative **19.** Additive inverse
21. $3x + 6$ **23.** $9a + 9b$ **25.** 0 **27.** 0
29. 10 **31.** $(4 + 2) + x = 6 + x$
33. $x + (2 + 7) = x + 9$ **35.** $(3 \cdot 5)x = 15x$
37. $(-9 \cdot 6)y = -54y$ **39.** $\left(\frac{1}{2} \cdot 3\right)a = \frac{3}{2}a$
41. $\left(-\frac{1}{3} \cdot 3\right)x = -x$ **43.** $\left(\frac{1}{2} \cdot 2\right)y = y$
45. $\left(-\frac{3}{4} \cdot \frac{4}{3}\right)x = -x$ **47.** $\left[-\frac{6}{5}\left(-\frac{5}{6}\right)\right]a = a$
49. $8x + 16$ **50.** $5x + 15$ **51.** $8x - 16$ **53.** $4y + 4$
55. $18x + 15$ **57.** $-6a - 14$ **59.** $-54y + 72$
61. $x + 2$ **63.** $12x + 18y$ **65.** $12a - 8b$
67. $3x + 2y$ **69.** $-4a - 8$ **71.** $-\frac{3}{2}x + 3$
73. $5x + 6$ **75.** $3x - 20$ **77.** $3 + x$ **79.** $-3x + 7y$
81. 1 **83.** $4a + 25$ **85.** $6x + 12$ **87.** $14x + 38$
89. $-6x + 8$ **91.** $-15x - 30$ **93.** 81 **95.** $\frac{9}{4}$
97. $2x + 1$ **99.** $6x - 3$ **101.** $5x + 10$ **103.** $6x + 5$
105. $5x + 6$ **107.** $6m - 5$ **109.** $7 + 3x$
111. $3x - 2y$ **113.** $\frac{2}{3}x - 2$ **115.** $-2x + y$
117. $0.09x + 180$ **119.** $0.12x + 60$ **121.** $a + 1$
123. $1 - a$ **125.** No
127. Answers may vary. $8 \div 4 \neq 4 \div 8$
129. $4(2 + 3) = 20$
$(4 \cdot 2) + (4 \cdot 3) = 20$ **131.** a

CHAPTER 1 TEST

1. 144 **2.** 64 **3.** 2 **4.** 10 **5.** $-3, 2$
6. $-3, -\frac{1}{2}, 2$ **7.** $6 + (-9) = -3$ **8.** $-5 - (-12) = 7$
9. $6 \cdot (-7) = -42$ **10.** $32 \div (-8) = -4$ **11.** 13
12. 1 **13.** -13 **14.** -10 **15.** -7 **16.** 62
17. 2 **18.** -6 **19.** $2^2 \cdot 3 \cdot 5 \cdot 11$ **20.** $3^3 \cdot 5^2 \cdot 7$
21. $\frac{11}{24}$ **22.** $\frac{5+6}{y} = \frac{11}{y}$ **23.** d **24.** e **25.** a
26. c **27.** $3x + 12$ **28.** $-15y$ **29.** $-10x + 15$
30. $2x + 4$

Chapter 2

PROBLEM SET 2.1

1. $-3x$ **3.** $-a$ **5.** $12x$ **7.** $6a$ **9.** $6x - 3$
11. $7a + 5$ **13.** $5x - 5$ **15.** $4a + 2$ **17.** $-9x - 2$
19. $12a + 3$ **21.** $10x - 1$ **23.** $21y + 6$ **25.** $-6x + 8$
27. $-2a + 3$ **29.** $-4x + 26$ **31.** $4y - 16$
33. $-6x - 1$ **35.** $2x - 12$ **37.** $10a + 33$
39. $4x - 9$ **41.** $7y - 39$ **43.** $-19x - 14$ **45.** 5
47. -9 **49.** 4 **51.** 4 **53.** -37 **55.** -41
57. 64 **59.** 64 **61.** 144 **63.** 144 **65.** 3
67. 0 **69.** 15 **71.** 6

73. **a.**

n	1	2	3	4
$3n$	3	6	9	12

b.

n	1	2	3	4
n^3	1	8	27	64

75. 1, 4, 7, 10, . . . **77.** 0, 1, 4, 9, . . . **79.** $-6y + 4$
81. $0.17x$ **83.** $2x$ **85.** $5x - 4$ **87.** $7x - 5$
89. $-2x - 9$ **91.** $7x + 2$ **93.** $-7x + 6$ **95.** $7x$
97. $-y$ **99.** $10y$ **101.** $0.17x + 180$ **103.** $0.22x + 60$
105. 49 **107.** 40 **109. a.** 42°F **b.** 28°F **c.** −14°F
111. a. $37.50 **b.** $40.00 **c.** $42.50 **113.** c **115.** c
117. 12 **119.** −3 **121.** −9.7 **123.** $-\frac{5}{4}$ **125.** 53
127. $a - 12$ **129.** 7

PROBLEM SET 2.2

1. Yes **3.** No **5. a.** No **b.** Yes **7. a.** Yes **b.** No
9. 11 **11.** 4 **13.** $-\frac{3}{4}$ **15.** −5.8 **17.** −17 **19.** $-\frac{1}{8}$
21. −4 **23.** −3.6 **25.** 1 **27.** $-\frac{7}{45}$ **29.** 3
31. $\frac{11}{8}$ **33.** 21 **35.** 7 **37.** 3.5 **39.** 22
41. −2 **43.** −16 **45.** −3 **47.** 10 **49.** −12
51. 4 **53.** 2 **55.** −5 **57.** −1 **59.** −3 **61.** 8
63. −8 **65.** 2 **67.** 11 **69.** −5.8
71. a. 6% **b.** 5% **c.** 2% **d.** 75% **73.** a **75.** c
77. $-y$ **79.** a **81.** −6 **83.** $\frac{1}{2}$ **85.** 6 **87.** −2
89. −18 **91.** $-\frac{6}{5}$ **93.** $3x$

PROBLEM SET 2.3

1. 2 **3.** 4 **5.** $-\frac{1}{2}$ **7.** −2 **9.** 3 **11.** 4
13. 0 **15.** 0 **17.** 6 **19.** −50 **21.** $\frac{3}{2}$ **23.** 12
25. −3 **27.** 32 **29.** −8 **31.** $\frac{1}{2}$ **33.** 4 **35.** 8
37. −4 **39.** 4 **41.** −15 **43.** $-\frac{1}{2}$ **45.** 3 **47.** 1
49. $\frac{1}{4}$ **51.** −3 **53.** 3 **55.** 2 **57.** $-\frac{3}{2}$ **59.** 1
61. −2 **63.** 3 **65.** −2 **67.** −1 **69.** 2 **71.** −4
73. −2 **75.** 0 **77.** 1 **79.** $\frac{1}{2}$
81. a. $\frac{3}{2}$ **b.** 1 **c.** $-\frac{3}{2}$ **d.** −4 **e.** $\frac{8}{5}$ **83.** 200 tickets
85. $1,390.85 per month **87.** d **89.** 2 **91.** 6
93. 3,000 **95.** $3x - 11$ **97.** $0.09x + 180$
99. $-6y + 4$ **101.** $4x - 11$ **103.** $5x$ **105.** $0.17x$

PROBLEM SET 2.4

1. 1 **3.** 6 **5.** 2 **7.** −1 **9.** $\frac{3}{4}$ **11.** 3
13. $\frac{3}{4}$ **15.** 2 **17.** 2 **19.** 11 **21.** −2 **23.** −6
25. 20 **27.** −2 **29.** $-\frac{1}{3}$ **31.** 8 **33.** 7
35. 0 **37.** $\frac{3}{7}$ **39.** 1 **41.** $-\frac{50}{3}$ **43.** $-\frac{11}{4}$ **45.** $-\frac{3}{2}$
47. 7 **49.** $\frac{3}{4}$ **51.** 6 **53.** 75 **55.** 6
57. 4,000 **59.** 700 **61.** 8 **63.** 8
65. ∅, contradiction **67.** $\frac{5}{3}$
69. All real numbers, identity **71.** ∅, contradiction
73. a. $\frac{5}{4} = 1.25$ **b.** $\frac{15}{2} = 7.5$ **c.** $6x + 20$ **d.** 15
e. $4x - 20$ **f.** $\frac{45}{2} = 22.5$
75. a **77.** c **79.** 14 **81.** −3 **83.** $\frac{1}{4}$ **85.** $\frac{1}{3}$
87. $-\frac{3}{2}x + 3$

PROBLEM SET 2.5

1. 100 feet **3.** 0 **5.** 2 **7.** 15 **9.** 10 **11.** −2
13. 1 **15. a.** 2 **b.** 4 **17. a.** 5 **b.** 18 **19.** $l = \frac{A}{w}$
21. $h = \frac{V}{lw}$ **23.** $a = P - b - c$ **25.** $x = 3y - 1$
27. $y = 3x + 6$ **29.** $y = -\frac{2}{3}x + 2$ **31.** $y = -2x - 5$
33. $y = -\frac{2}{3}x + 1$ **35.** $w = \frac{P - 2l}{2}$ **37.** $v = \frac{h - 16t^2}{t}$
39. $h = \frac{A - \pi r^2}{2\pi r}$
41. a. $y = \frac{3}{5}x + 1$ **b.** $y = \frac{1}{2}x + 2$ **c.** $y = 4x + 3$
43. $y = \frac{3}{7}x - 3$ **45.** $y = 2x + 8$ **47.** 10 **49.** 240
51. 25% **53.** 35% **55.** 64 **57.** 2,000
59. $T = 35$; The temperature is 35°F at an altitude of 10,000 feet.
61. 4,000 **63.** $A = \frac{T - 70}{-0.0035} = \frac{70 - T}{0.0035}$ **65.** 100°C; yes
67. 20°C; yes **69.** $C = \frac{5}{9}(F - 32)$ **71.** 4°F over
73. 44 meters **75.** $\frac{3}{2}$ or 1.5 inches
77. 56.52 cubic centimeters **79.** 132 feet **81.** 60%
83. 26.5% **85.** a **87.** The sum of 4 and 1.
89. The difference of 6 and 2.
91. The difference of a number and 15.
93. Four times the difference of a number and 3
95. $2(6 + 3)$ **97.** $2(5) + 3$ **99.** $x + 5$ **101.** $5(x + 7)$

PROBLEM SET 2.6

1. $x + 5 = 14$ **3.** $\frac{x}{3} = x + 2$ **5.** $2(x - 9) + 5 = 11$
7. $\frac{1}{2}(x + 5) = 3(x - 5)$ **9.** 8 **11.** 5 **13.** −1
15. 3 and 5 **17.** 6 and 14 **19.** Shelly is 39; Michele is 36
21. Evan is 11; Cody is 22 **23.** Barney is 27; Fred is 31
25. Lacy is 16; Jack is 32 **27.** Patrick is 18; Pat is 38
29. $s = 9$ inches **31.** $s = 15$ feet
33. 11 feet, 18 feet, 33 feet **35.** 26 feet, 13 feet, 14 feet
37. $l = 11$ inches; $w = 6$ inches
39. $l = 25$ inches; $w = 9$ inches
41. $l = 15$ feet; $w = 3$ feet **43.** 9 dimes; 14 quarters
45. 12 quarters; 27 nickels **47.** 8 nickels; 17 dimes
49. 7 nickels; 10 dimes; 12 quarters
51. 3 nickels; 9 dimes; 6 quarters
53. c **55.** d **57.** $5x$ **59.** $1.075x$ **61.** $0.09x + 180$
63. 6,000 **65.** 30

PROBLEM SET 2.7

1. 5 and 6 **3.** −5 and −4 **5.** 13 and 15
7. 52 and 54 **9.** −16 and −14 **11.** 17, 19, and 21
13. 42, 44, and 46
15. $4,000 invested at 8%, $6,000 invested at 9%
17. $700 invested at 10%, $1,200 invested at 12%
19. $500 at 8%, $1,000 at 9%, $1,500 at 10%
21. 12 liters of each
23. 15 gallons of 10% solution, 10 gallons of 5% solution
25. 32 pounds of $9.50 beans, 8 pounds of $12.00 beans
27. 45°, 45°, 90° **29.** 22.5°, 45°, 112.5°
31. 80°, 60°, 40° **33.** 16 adult and 22 children's tickets
35. 16 minutes **37.** 39 hours
39. They are in offices 7329 and 7331.
41. Kendra is 8 years old and Marissa is 10 years old.
43. Jeff **45.** $10.38 **47.** $l = 12$ meters; $w = 10$ meters
49. 59°, 60°, 61° **51.** $54.00 **53.** Yes
55. a **57.** a **59. a.** 9 **b.** 3 **c.** −9 **d.** −3
61. a. −8 **b.** 8 **c.** 8 **d.** −8 **63.** −2.3125 **65.** $\frac{10}{3}$

PROBLEM SET 2.8

1. $\{x \mid x < 12\}$

3. $\{a \mid a \le 12\}$

5. $\{x \mid x > 13\}$

7. $\{y \mid y \ge 4\}$

9. $\{x \mid x > 9\}$

11. $\{x \mid x < 2\}$

13. $\{a \mid a \le 5\}$

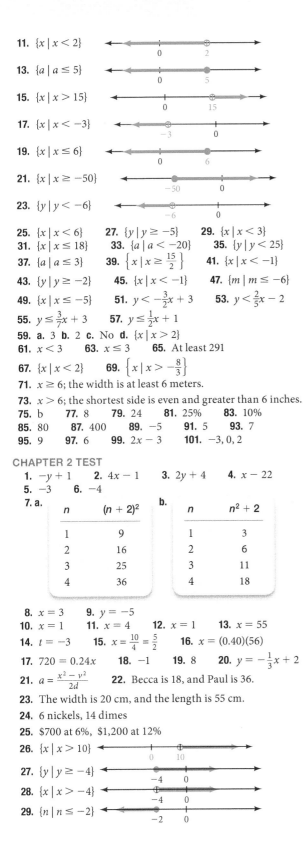

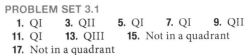

15. $\{x \mid x > 15\}$

17. $\{x \mid x < -3\}$

19. $\{x \mid x \le 6\}$

21. $\{x \mid x \ge -50\}$

23. $\{y \mid y < -6\}$

25. $\{x \mid x < 6\}$ **27.** $\{y \mid y \ge -5\}$ **29.** $\{x \mid x < 3\}$
31. $\{x \mid x \le 18\}$ **33.** $\{a \mid a < -20\}$ **35.** $\{y \mid y < 25\}$
37. $\{a \mid a \le 3\}$ **39.** $\left\{x \mid x \ge \frac{15}{2}\right\}$ **41.** $\{x \mid x < -1\}$
43. $\{y \mid y \ge -2\}$ **45.** $\{x \mid x < -1\}$ **47.** $\{m \mid m \le -6\}$
49. $\{x \mid x \le -5\}$ **51.** $y < -\frac{3}{2}x + 3$ **53.** $y < \frac{2}{5}x - 2$
55. $y \le \frac{3}{7}x + 3$ **57.** $y \le \frac{1}{2}x + 1$
59. a. 3 **b.** 2 **c.** No **d.** $\{x \mid x > 2\}$
61. $x < 3$ **63.** $x \le 3$ **65.** At least 291
67. $\{x \mid x < 2\}$ **69.** $\left\{x \mid x > -\frac{8}{3}\right\}$
71. $x \ge 6$; the width is at least 6 meters.
73. $x > 6$; the shortest side is even and greater than 6 inches.
75. b **77.** 8 **79.** 24 **81.** 25% **83.** 10%
85. 80 **87.** 400 **89.** −5 **91.** 5 **93.** 7
95. 9 **97.** 6 **99.** $2x - 3$ **101.** −3, 0, 2

CHAPTER 2 TEST

1. $-y + 1$ **2.** $4x - 1$ **3.** $2y + 4$ **4.** $x - 22$
5. −3 **6.** −4
7. a.

n	$(n + 2)^2$
1	9
2	16
3	25
4	36

b.

n	$n^2 + 2$
1	3
2	6
3	11
4	18

8. $x = 3$ **9.** $y = -5$
10. $x = 1$ **11.** $x = 4$ **12.** $x = 1$ **13.** $x = 55$
14. $t = -3$ **15.** $x = \frac{10}{4} = \frac{5}{2}$ **16.** $x = (0.40)(56)$
17. $720 = 0.24x$ **18.** −1 **19.** 8 **20.** $y = -\frac{1}{3}x + 2$
21. $a = \frac{x^2 - v^2}{2d}$ **22.** Becca is 18, and Paul is 36.
23. The width is 20 cm, and the length is 55 cm.
24. 6 nickels, 14 dimes
25. \$700 at 6%, \$1,200 at 12%
26. $\{x \mid x > 10\}$

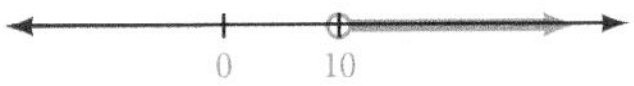

27. $\{y \mid y \ge -4\}$

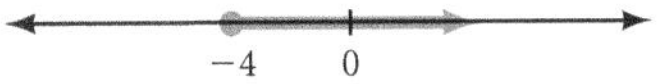

28. $\{x \mid x > -4\}$

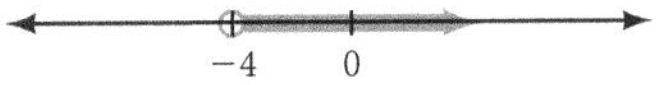

29. $\{n \mid n \le -2\}$

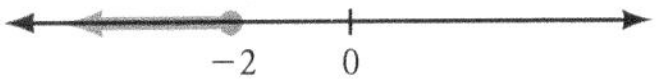

Chapter 3

PROBLEM SET 3.1

1. QI **3.** QII **5.** QI **7.** QI **9.** QII
11. QI **13.** QIII **15.** Not in a quadrant
17. Not in a quadrant

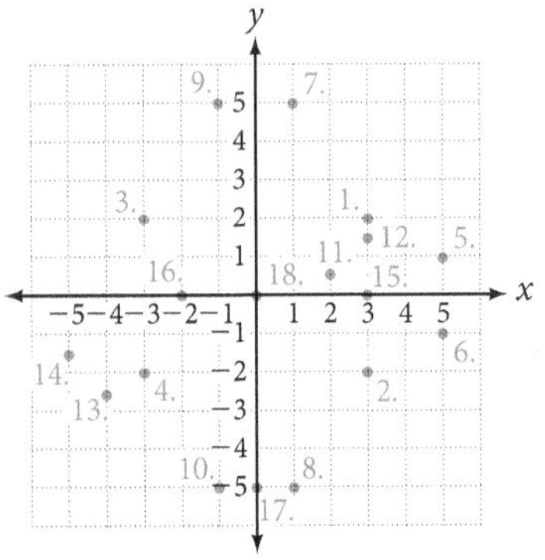

19. (−4, 4) **21.** (−4, 2) **23.** (−3, 0) **25.** (2, −2)
27. (−5, −5) **29.** Yes **31.** No **33.** Yes **35.** No
37. Yes **39.** No **41.** No **43.** No
45.

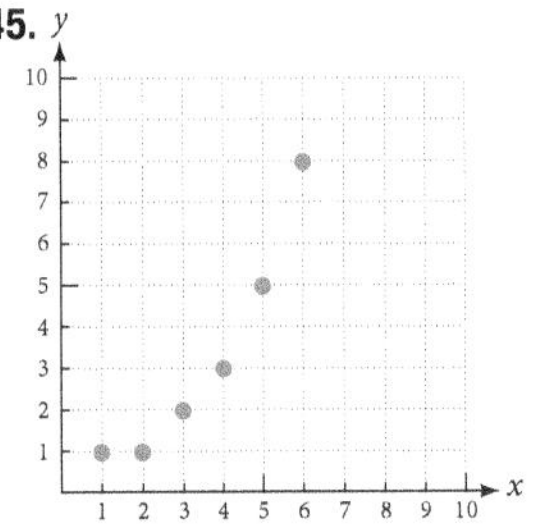

47.

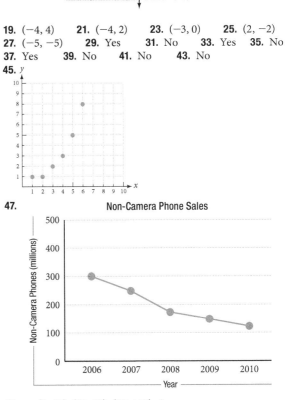

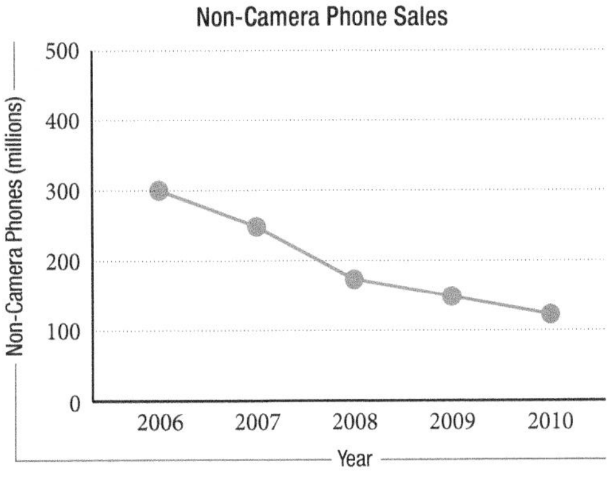

49. a. (5, 40), (10, 80), (20, 160), Answers may vary
b. \$320 **c.** 30 hours
d. No, if she works 35 hours, she should be paid \$280.
51. (1995, 44.8), (2000, 65.4), (2005, 104), (2010, 112.7), (2015, 137.9)
53. $A = (2, 2), B = (2, 5), C = (7, 5)$
55. c **57.** d **59. a.** 10 **b.** −5 **c.** −4 **d.** −2
61. a. 2 **b.** $\frac{5}{3}$ **c.** 4 **d.** −11 **63.** $y = 3x + 5$
65. $y = \frac{2}{3}x - 2$

PROBLEM SET 3.2

1. (0, −2) **3.** (1, 5), (0, −2), and (−2, −16)

5. (2, −2) **7.** (3, 0) and (3, −3) 9. (0, 6), (3,0), (6, −6)

11. (0, 3), (4, 0), (−4, 6) **13.** (1, 1), $\left(\frac{3}{4}, 0\right)$, (5, 17)

15. (2, 13), (1, 6), (0, −1) **17.** (−5, 4), (−5, −3), (−5, 0)

19.

x	*y*
1	3
−3	−9
4	12
6	18

21.

x	*y*
2	3
3	2
5	0
9	−4

23.

x	*y*
2	0
3	2
1	−2
−3	−10

25.

x	*y*
0	−1
−1	−7
−3	−19
$\frac{3}{2}$	8

27. (0, 4), (2, 2), (4, 0)

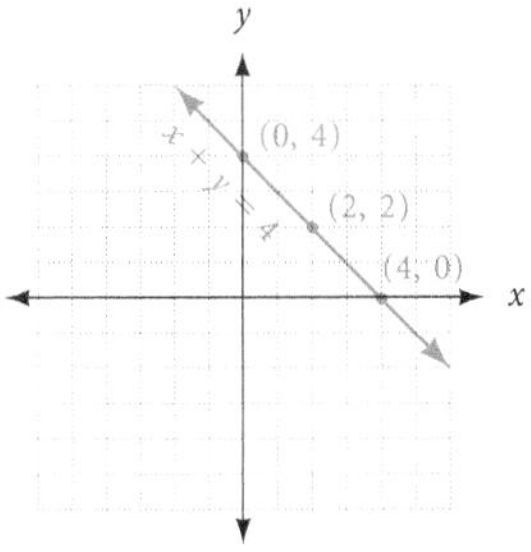

29. (0, 0), (−2, −4), (2, 4)

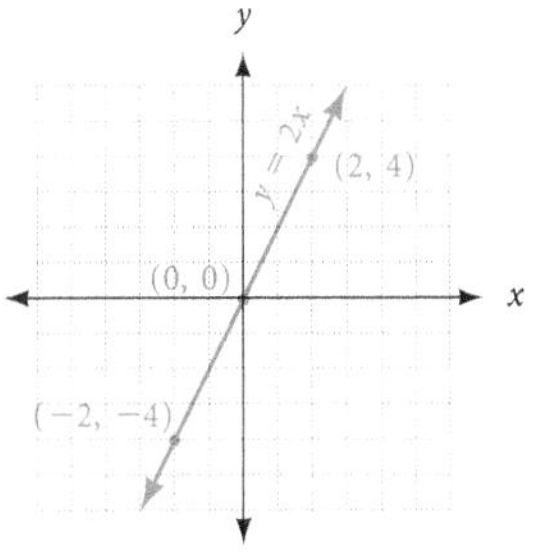

31. (−3, −1), (0, 0), (3, 1)

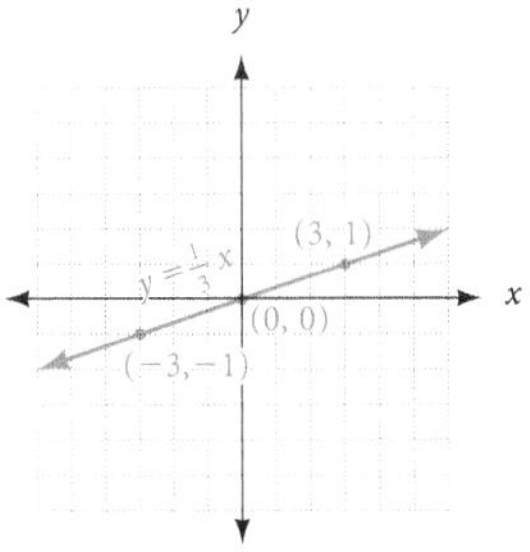

33. (0, 1), (−1, −1), (1, 3)

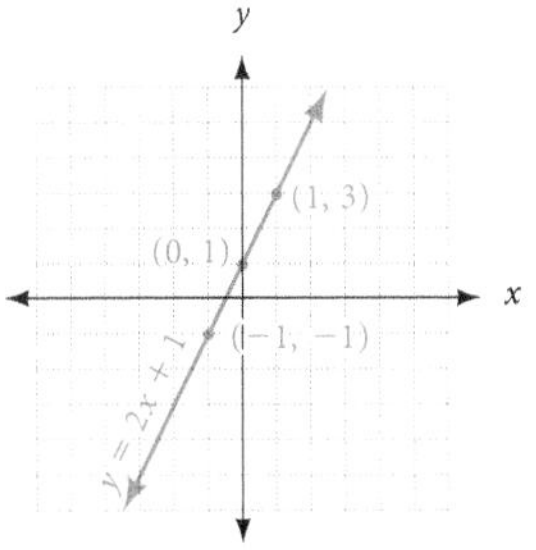

35. (0, 4), (−1, 4), (2, 4)

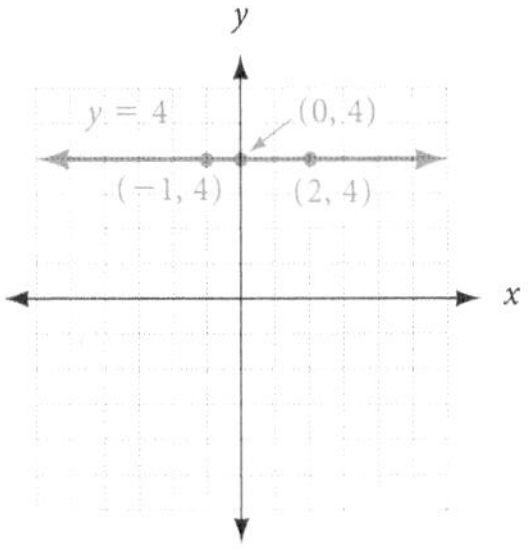

37. (−2, 2), (0, 3), (2, 4)

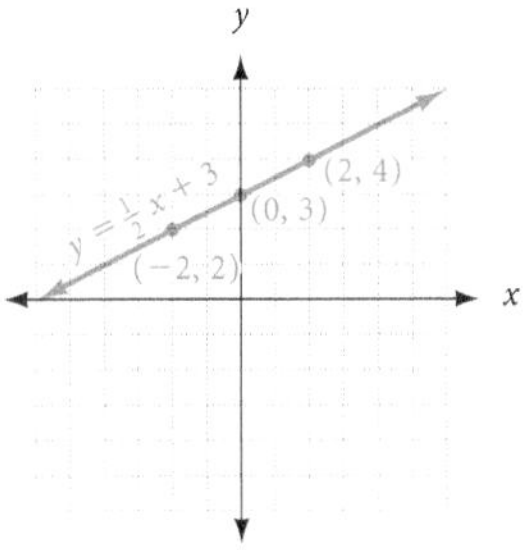

39. (−3, 3), (0, 1), (3, −1)

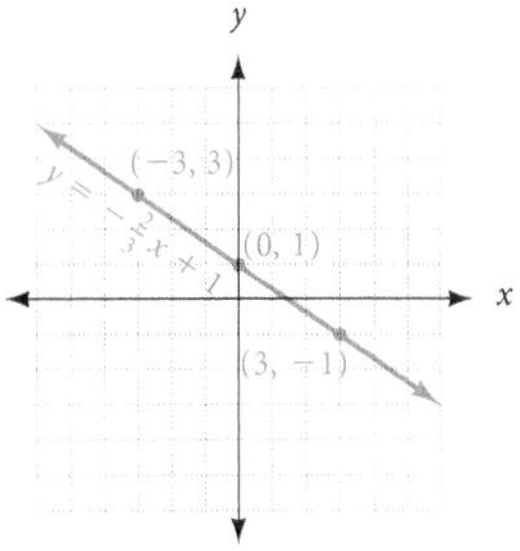

41. (−1, 5), (0, 3), (1, 1)

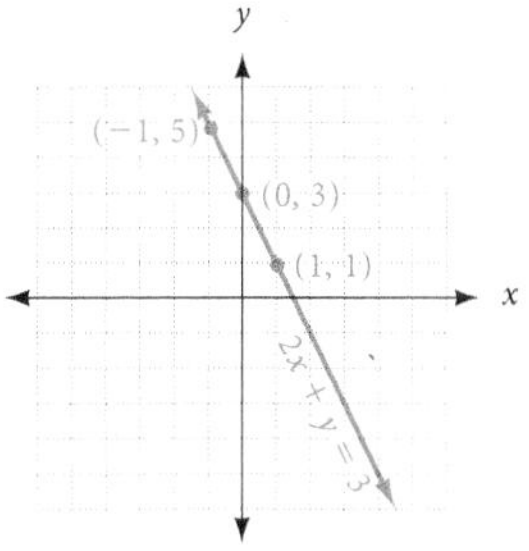

43. (0, 3), (2, 0), (4, −3)

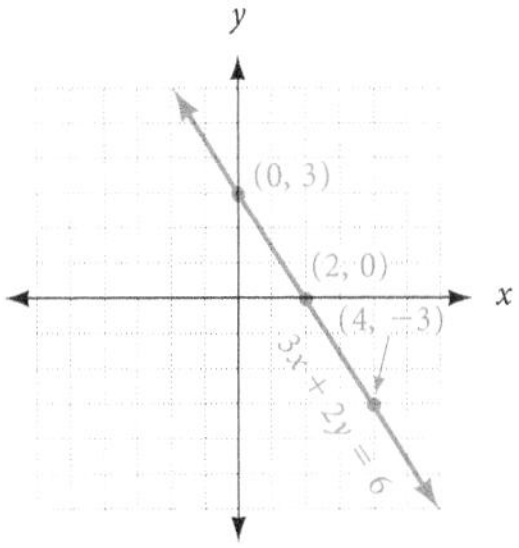

45. (−2, 2), (0, 3), (2, 4)

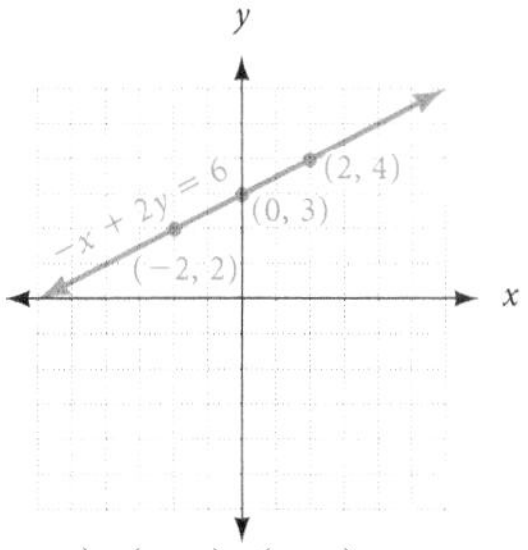

46. (−3, 1), (0, 2), (3, 3)

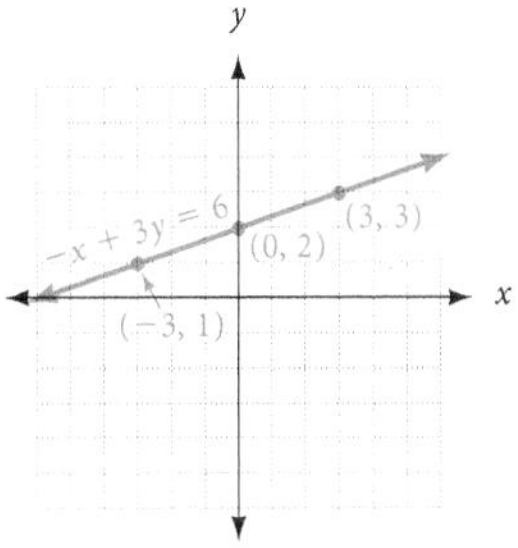

47. $\left(-4, \frac{1}{2}\right), \left(0, \frac{1}{2}\right), \left(4, \frac{1}{2}\right)$

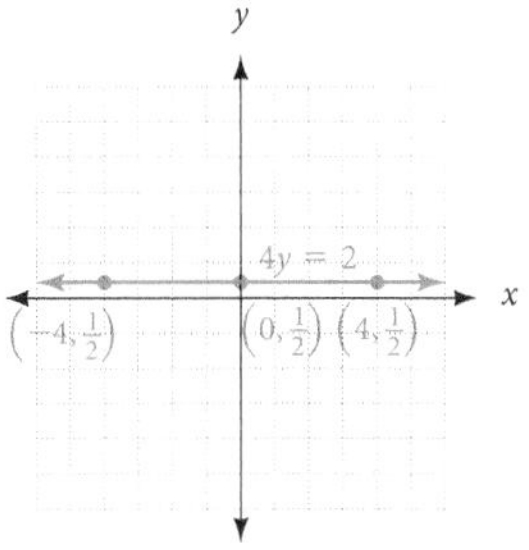

49.

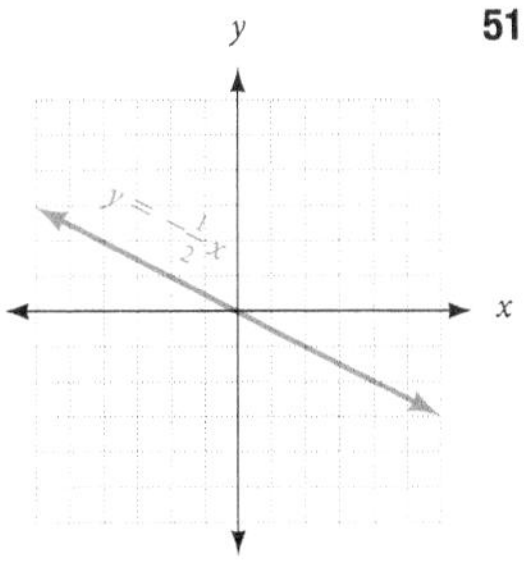

51.

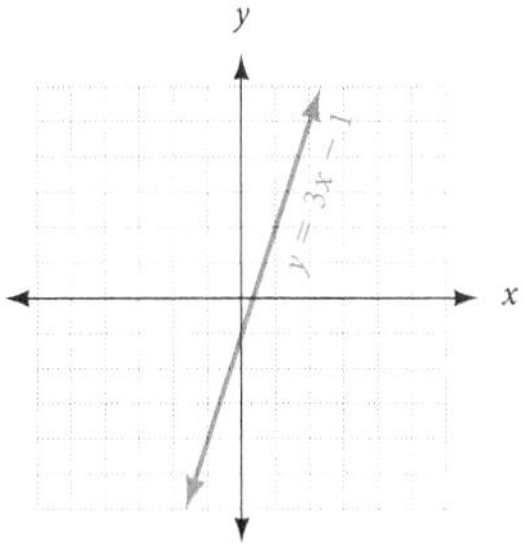

53.

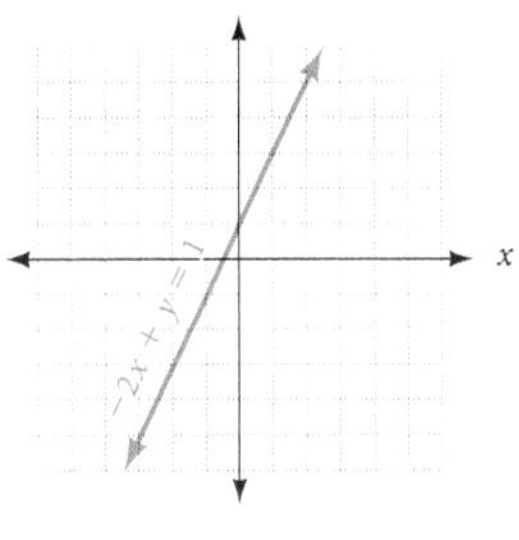

55.

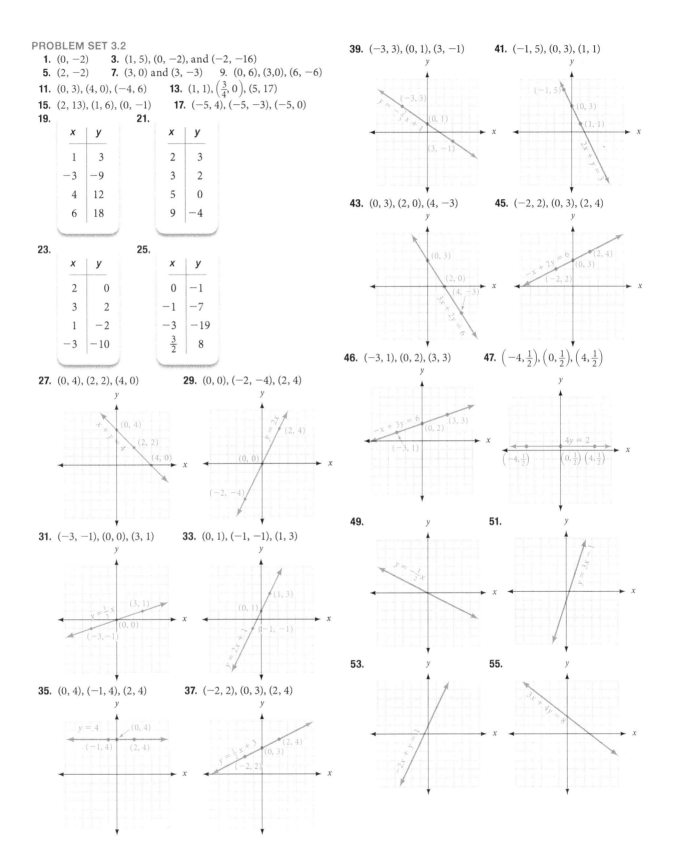

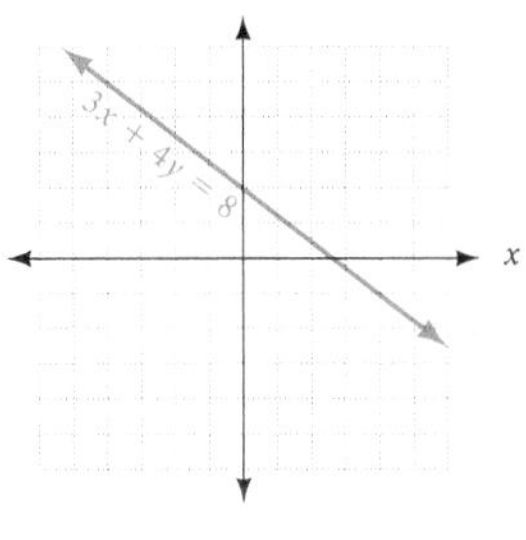

57.

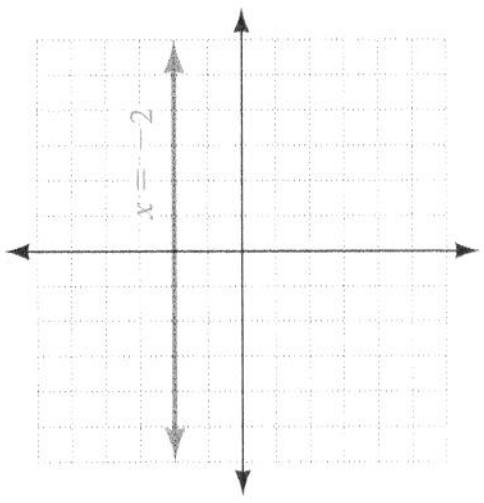

59.

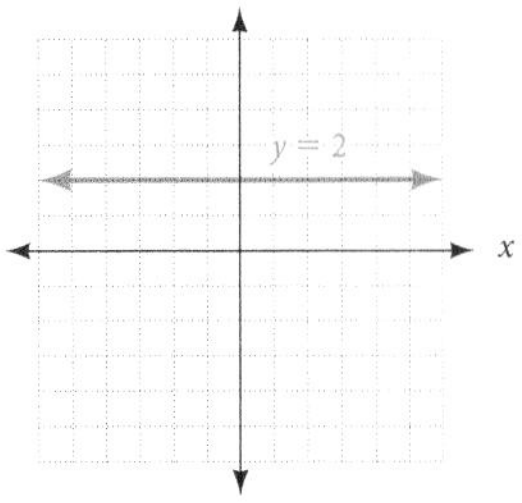

61.

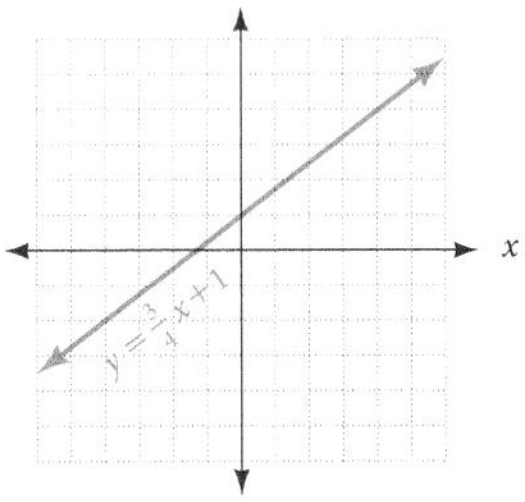

63.

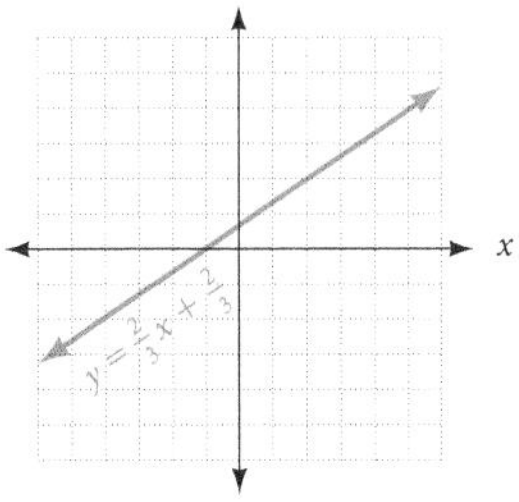

65.

Equation	H, V, and/or O
$x = 3$	V
$y = 3$	H
$y = 3x$	O
$y = 0$	O, H

67.

Equation	H, V, and/or O
$x = -\frac{3}{5}$	V
$y = -\frac{3}{5}$	H
$y = -\frac{3}{5}x$	O
$x = 0$	O, V

69.

x	y
−4	−3
−2	−2
0	−1
2	0
6	2

71. a. $\frac{5}{2}$ **b.** 5 **c.** 2
d.

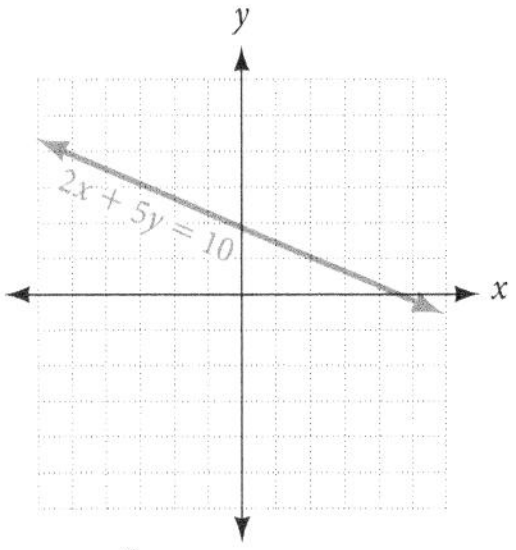

e. $y = -\frac{2}{5}x + 2$

73. 12 inches
75. a. Yes
b. No, she should earn $108 for working 9 hours.
c. No, she should earn $84 for working 7 hours.
d. Yes
77. a. $375,000 **b.** At the end of 6 years.
c. No, the crane will be worth $195,000 after 9 years.
d. $600,000

79. d **81.** c **83. a.** 2 **b.** 3
85. a. −4 **b.** 2 **87. a.** 6 **b.** 2

PROBLEM SET 3.3

1.

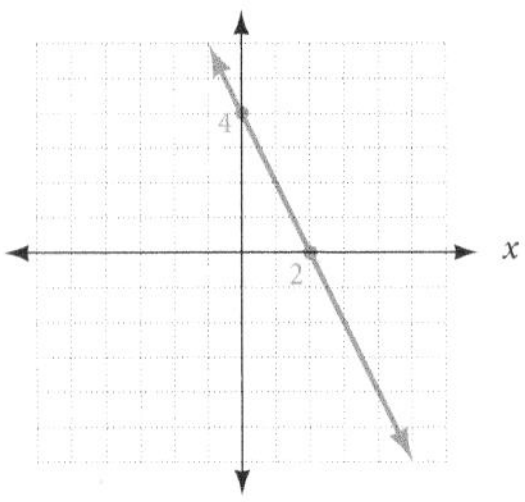

3.

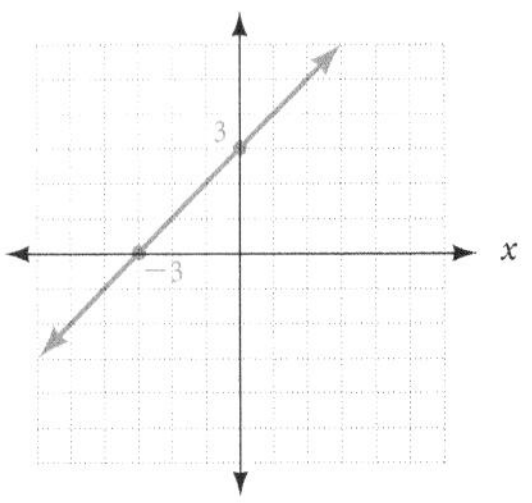

5.

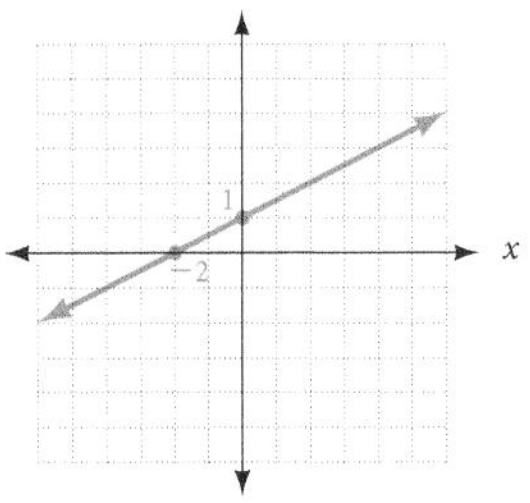

7.

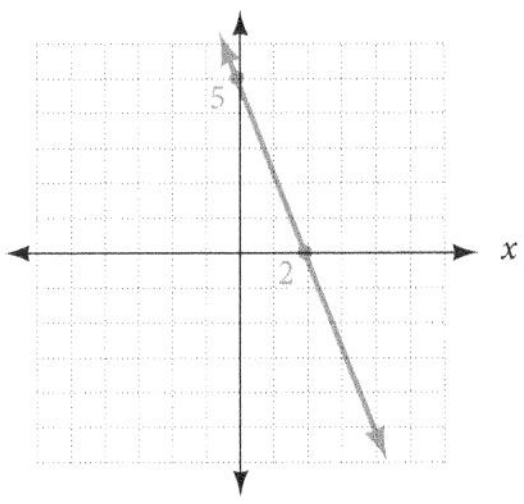

9.

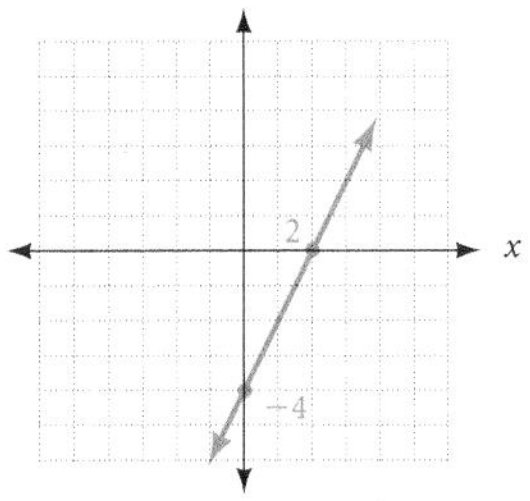

11.

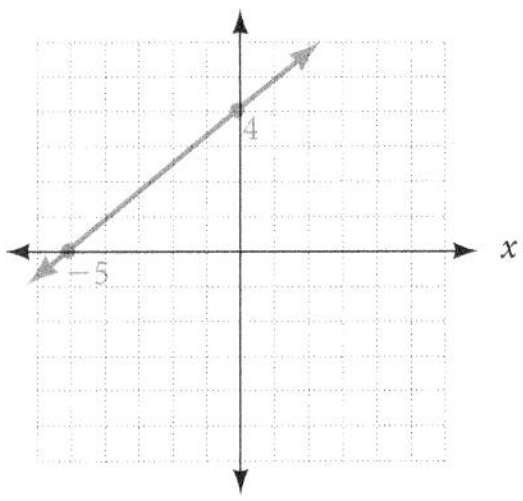

13.

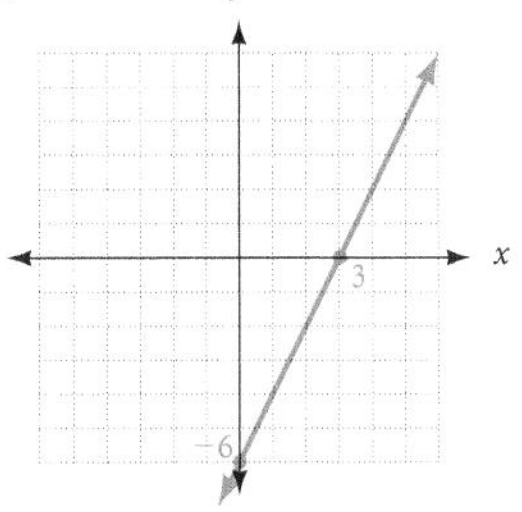

15.

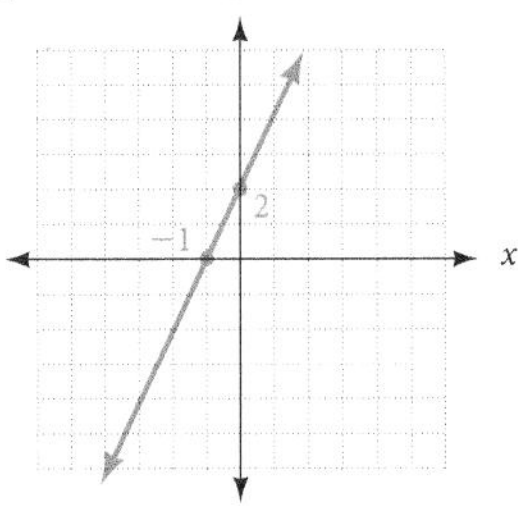

17.

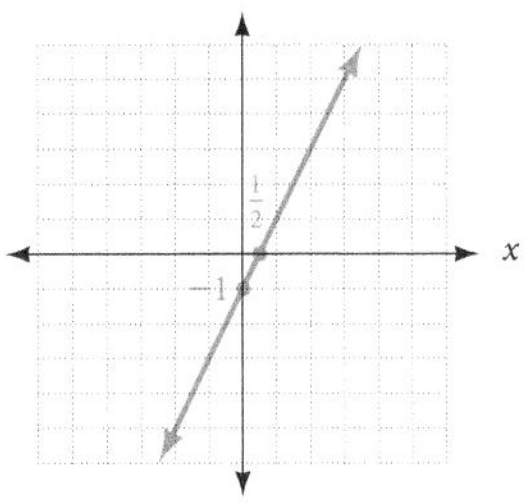

19.

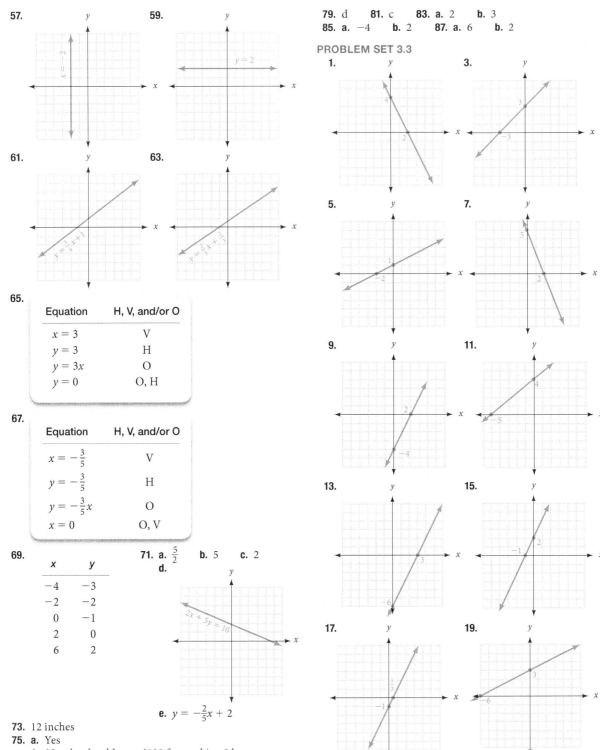

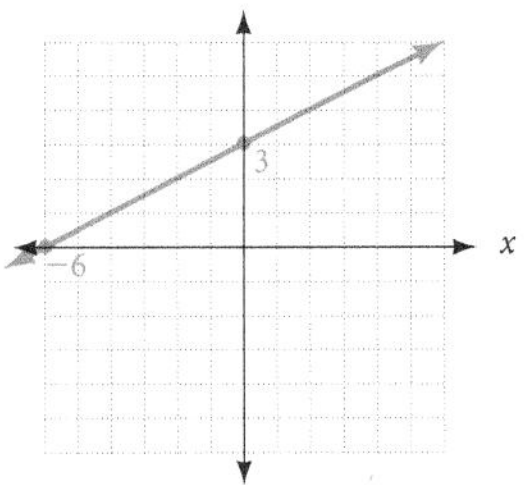

21.

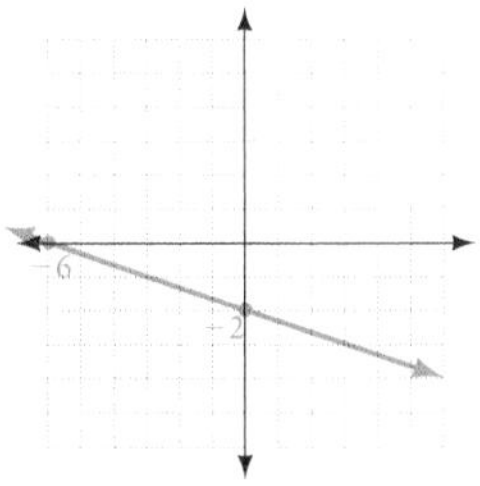

23.

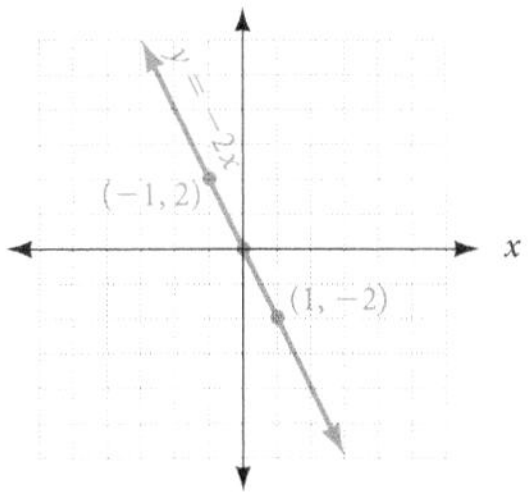

25.

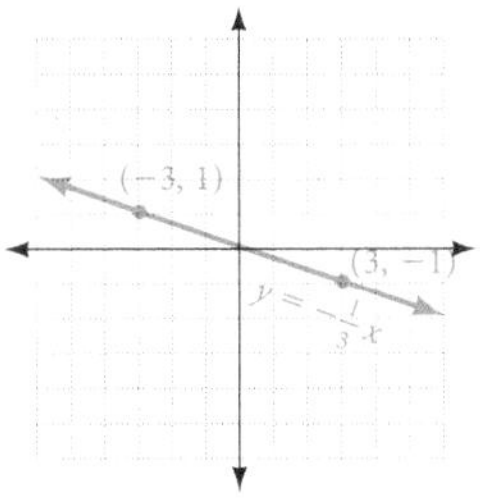

27.

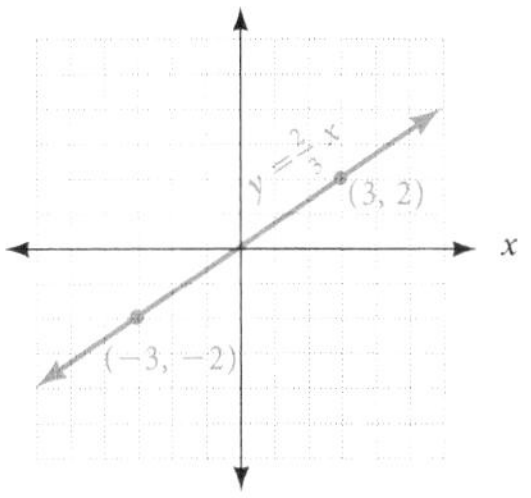

29.

Equation	x-intercept	y-intercept
$3x + 4y = 12$	4	3
$3x + 4y = 4$	$\frac{4}{3}$	1
$3x + 4y = 3$	1	$\frac{3}{4}$
$3x + 4y = 2$	$\frac{2}{3}$	$\frac{1}{2}$

31.

Equation	x-intercept	y-intercept
$x - 3y = 2$	2	$-\frac{2}{3}$
$y = \frac{1}{3}x - \frac{2}{3}$	2	$-\frac{2}{3}$
$x - 3y = 0$	0	0
$y = \frac{1}{3}x$	0	0

33. 5 **35.** $-\frac{2}{3}$ **37.** 4 **39.** $\frac{1}{2}$

41. a. 0 **b.** $-\frac{3}{2}$ **c.** 1

d.

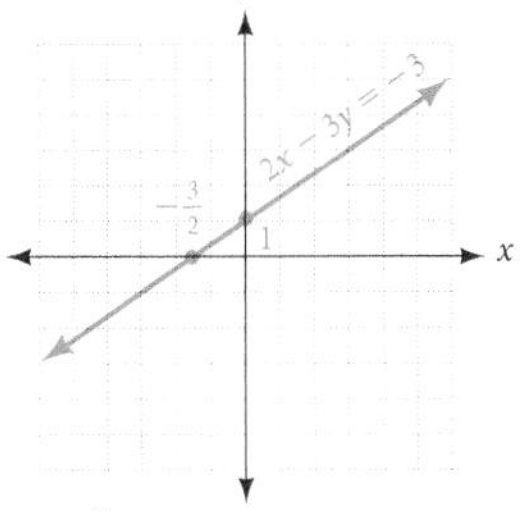

e. $y = \frac{2}{3}x + 1$

43. x-intercept $= 3$; y-intercept $= 5$

44. x-intercept $= -5$; y-intercept $= 2$

45. x-intercept $= -1$; y-intercept $= -3$

47.

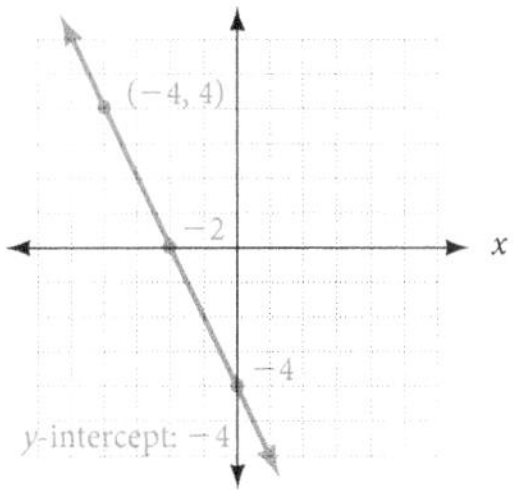

49.

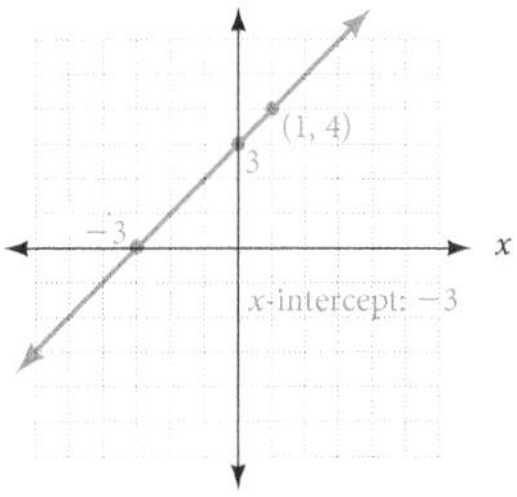

51.

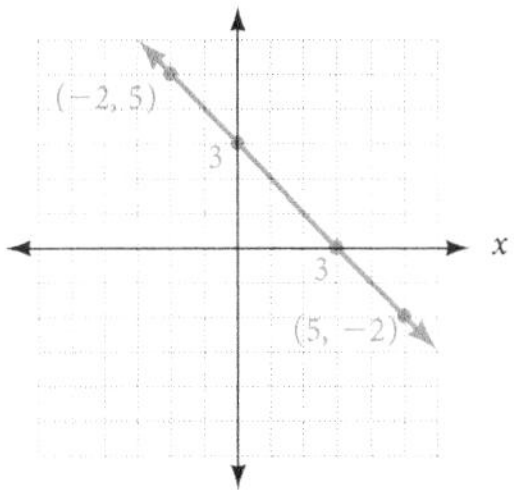

x-intercept $= 3$;
y-intercept $= 3$

53.

x	y
-2	1
0	-1
-1	0
1	-2

55.

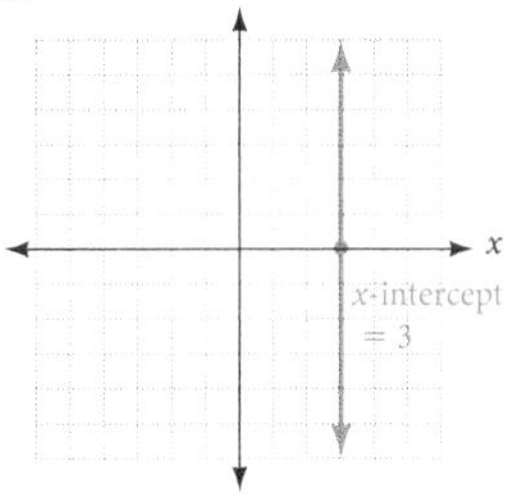

57.

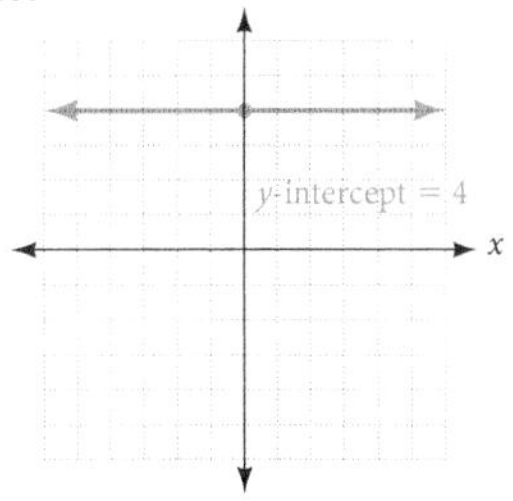

59. θ

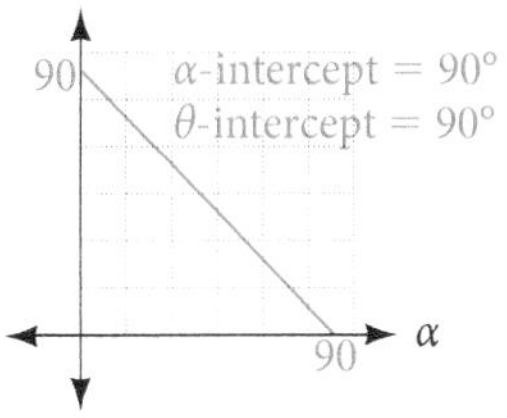

61. d **63.** b **65. a.** $-\frac{5}{7}$ **b.** $-\frac{5}{7}$

67. a. $-\frac{5}{7}$ **b.** $-\frac{5}{7}$

PROBLEM SET 3.4

1. Slope $= \frac{3}{2}$

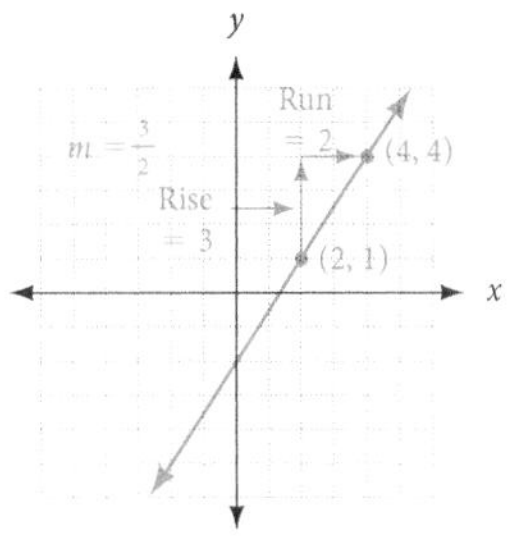

3. Slope $= -\frac{1}{2}$

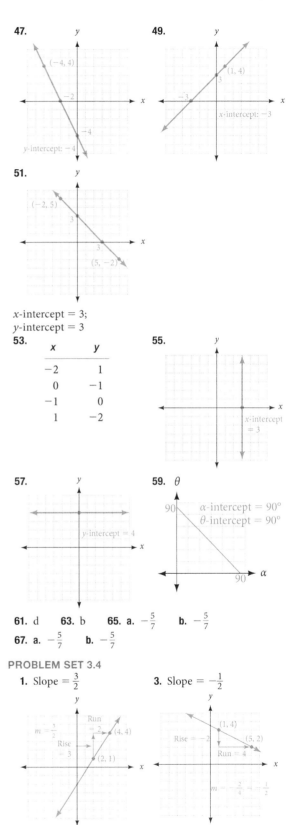

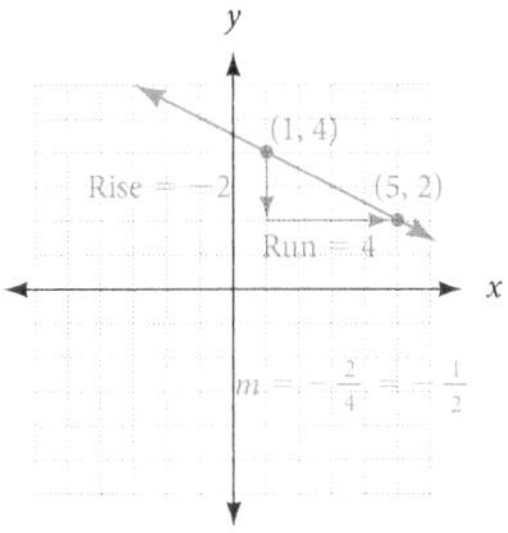

5. Slope $= \frac{5}{3}$

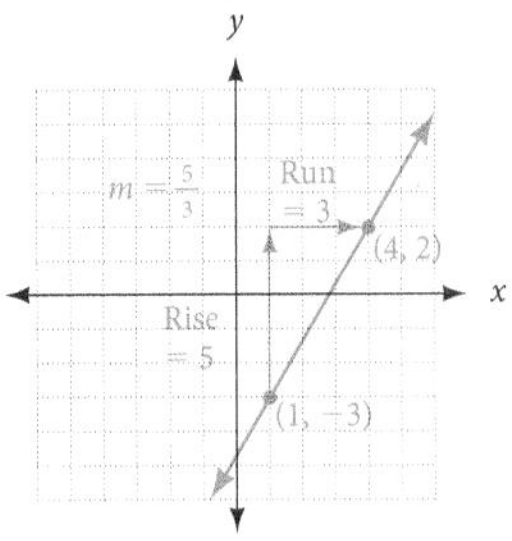

7. Slope $= \frac{5}{4}$

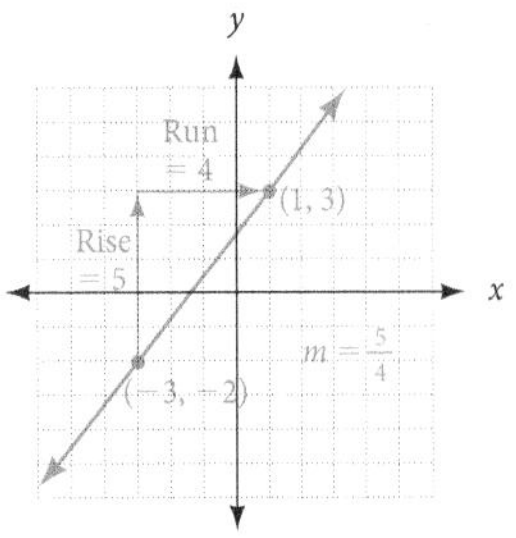

9. Slope $= -\frac{2}{3}$

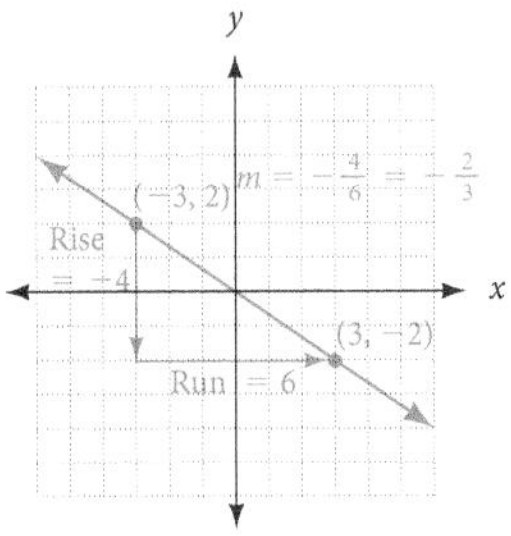

11. Slope = 3

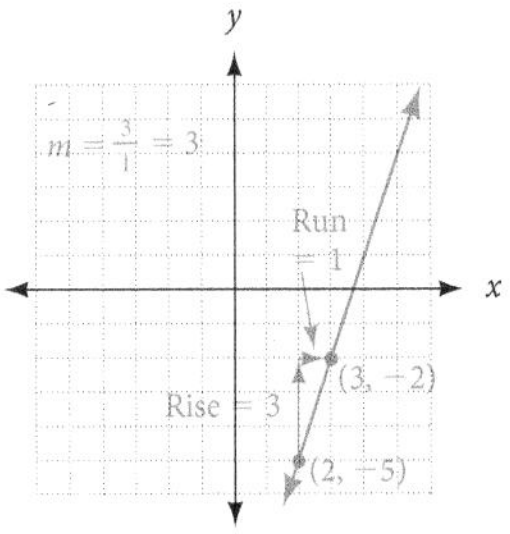

13. No slope

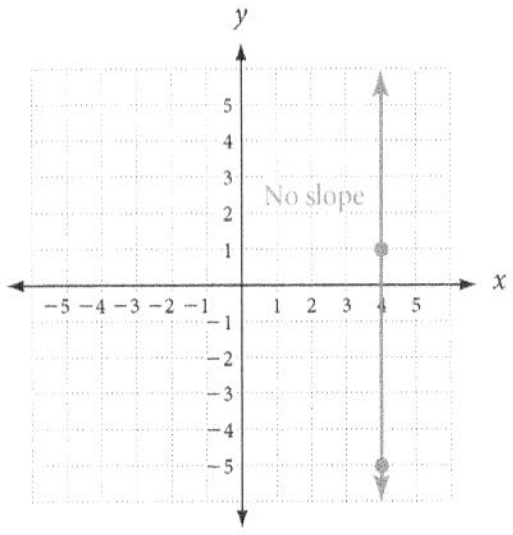

15. Slope = 0

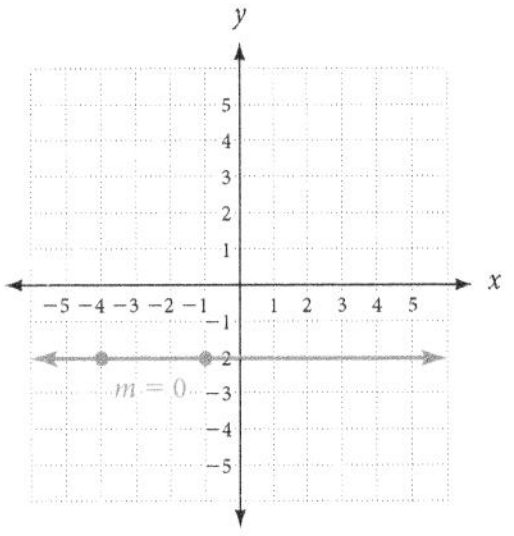

17. Slope $= \frac{1}{3}$

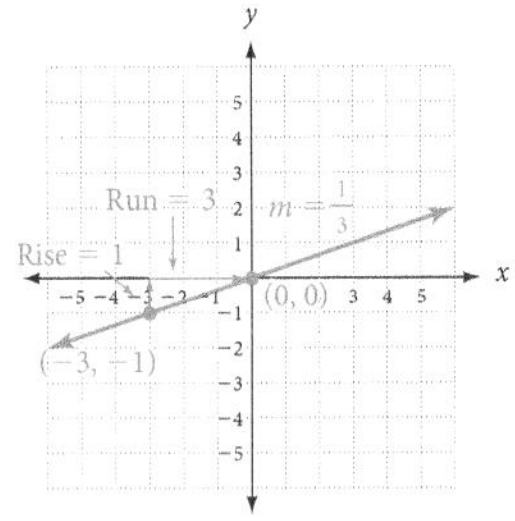

19. Slope = 3 **21.** Slope = 2

23.

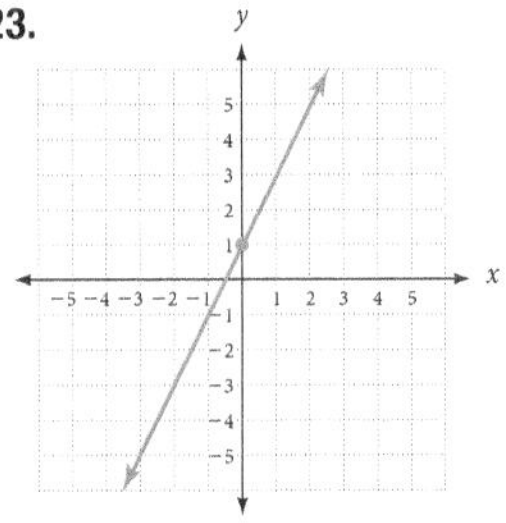

25.

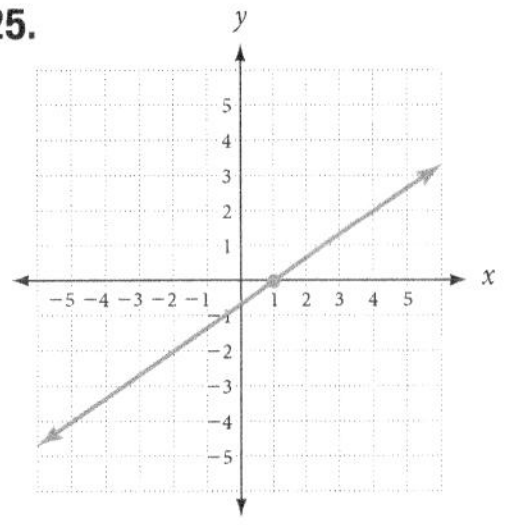

27.

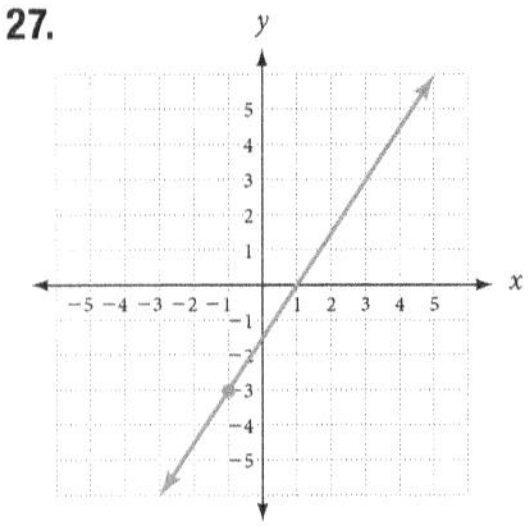

29.

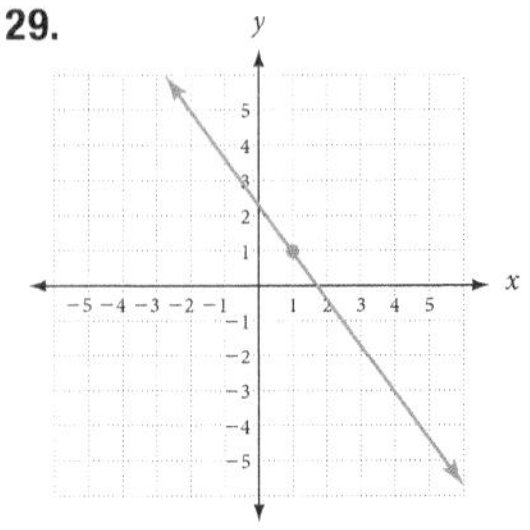

31.

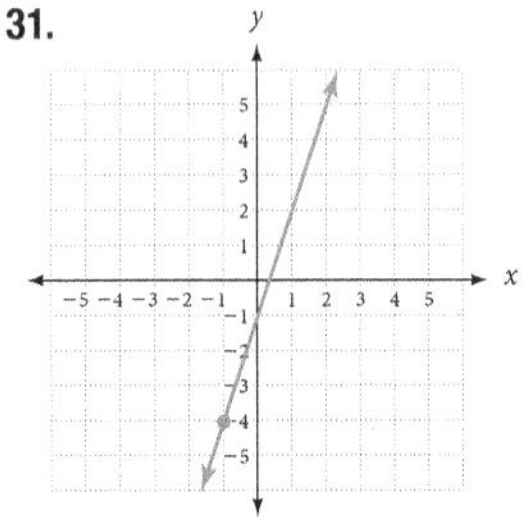

32.

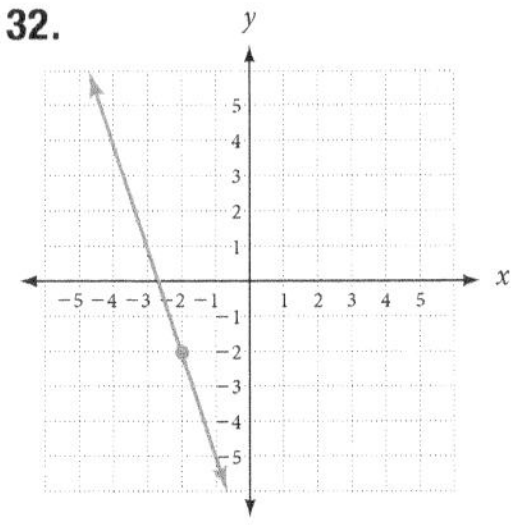

33.

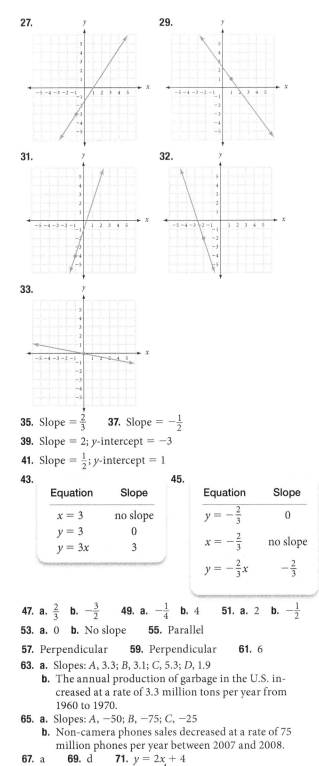

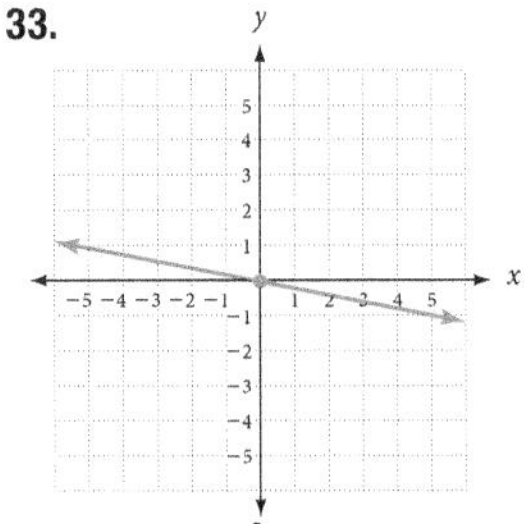

35. Slope $= \frac{2}{3}$ **37.** Slope $= -\frac{1}{2}$

39. Slope = 2; y-intercept = −3

41. Slope $= \frac{1}{2}$; y-intercept = 1

43.

Equation	Slope
$x = 3$	no slope
$y = 3$	0
$y = 3x$	3

45.

Equation	Slope
$y = -\frac{2}{3}$	0
$x = -\frac{2}{3}$	no slope
$y = -\frac{2}{3}x$	$-\frac{2}{3}$

47. a. $\frac{2}{3}$ **b.** $-\frac{3}{2}$ **49. a.** $-\frac{1}{4}$ **b.** 4 **51. a.** 2 **b.** $-\frac{1}{2}$

53. a. 0 **b.** No slope **55.** Parallel

57. Perpendicular **59.** Perpendicular **61.** 6

63. a. Slopes: A, 3.3; B, 3.1; C, 5.3; D, 1.9

b. The annual production of garbage in the U.S. increased at a rate of 3.3 million tons per year from 1960 to 1970.

65. a. Slopes: A, −50; B, −75; C, −25

b. Non-camera phones sales decreased at a rate of 75 million phones per year between 2007 and 2008.

67. a **69.** d **71.** $y = 2x + 4$

73. $y = -2x + 3$ **75.** $y = \frac{4}{5}x - 4$

PROBLEM SET 3.5

1. $m = 5, b = -3$ **3.** $m = -\frac{2}{3}, b = \frac{7}{3}$

5. $m = 1, b = 9$ **7.** $m = \frac{1}{2}, b = -\frac{5}{2}$ **9.** $m = -2, b = \frac{1}{4}$

11. $m = 3, b = 0$ **13.** $m = 0, b = -10$

15.

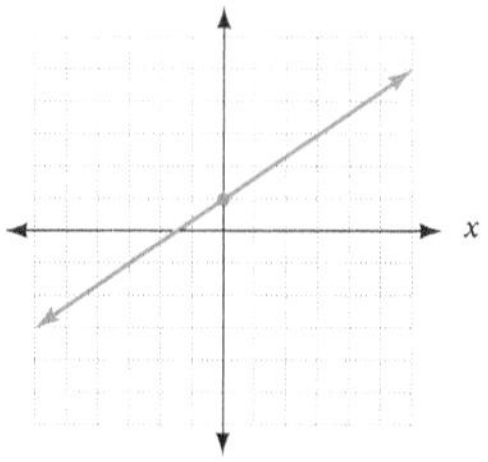

17.

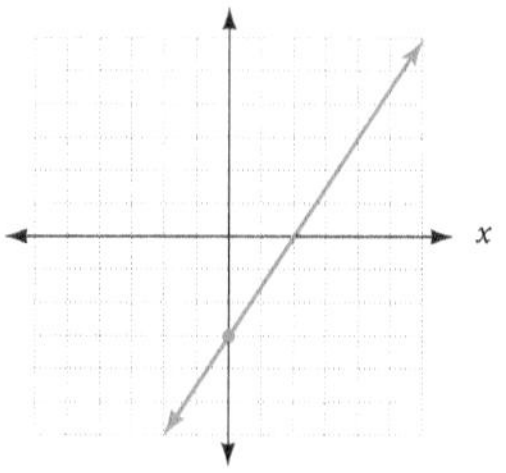

19.

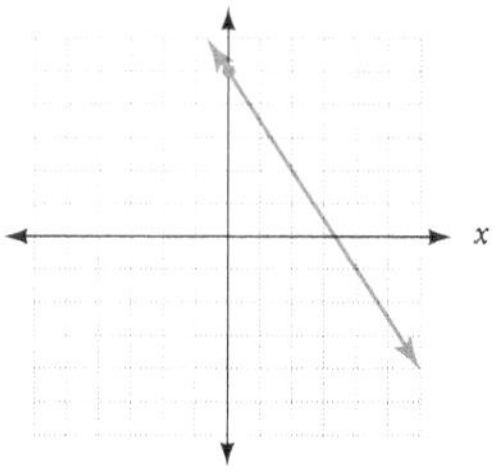

21.

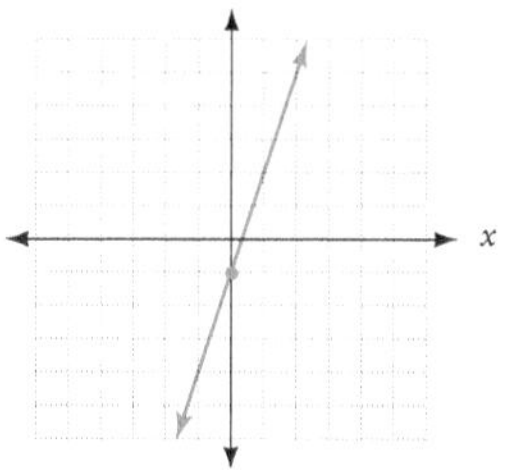

23.

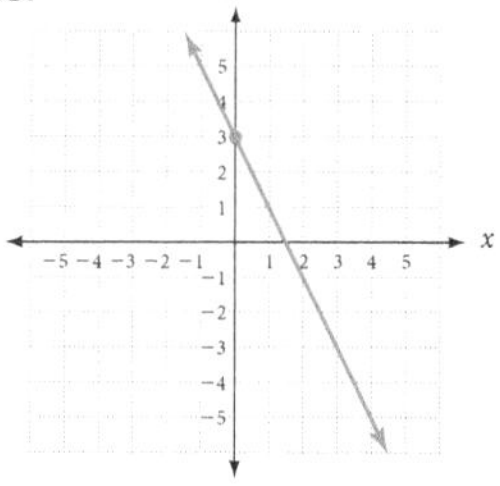

25.

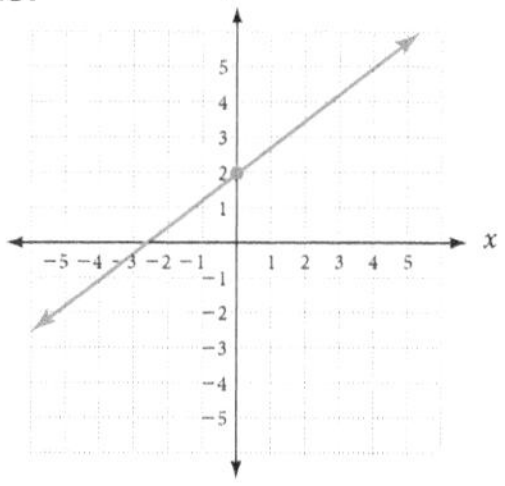

27.

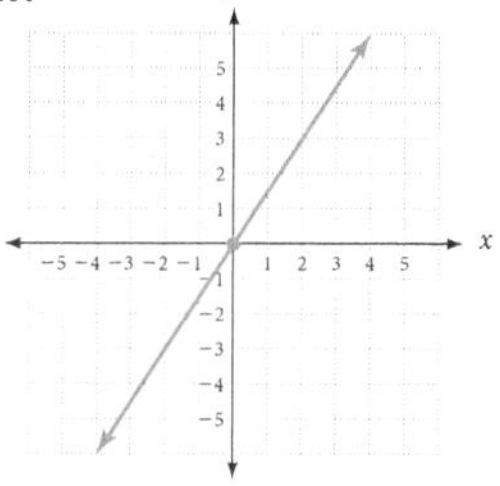

29. $m = 2; b = 4$

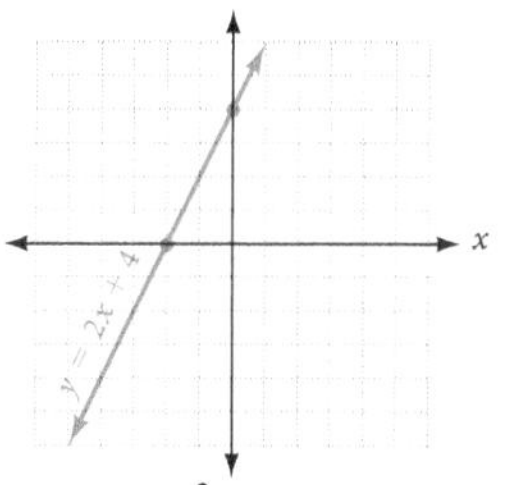

31. $m = -3; b = 3$

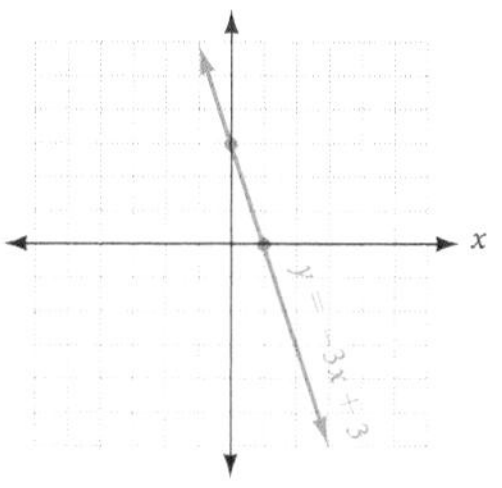

33. $m = -\frac{3}{2}; b = 3$

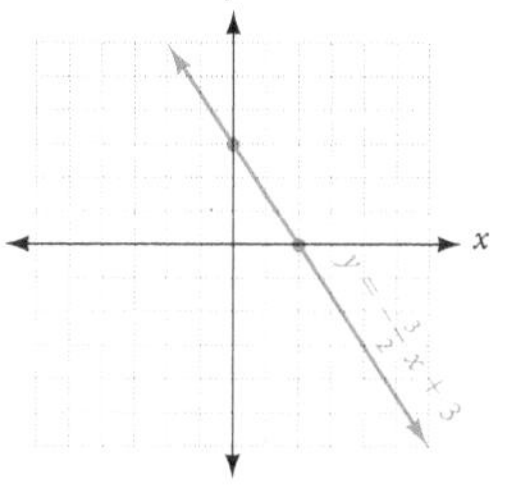

35. $m = \frac{4}{5}; b = -4$

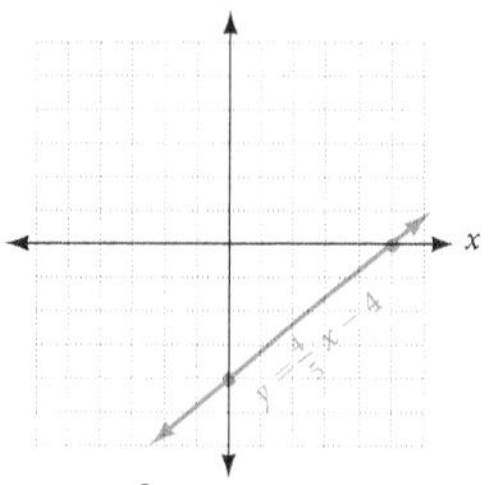

37. $m = -\frac{2}{5}; b = -2$

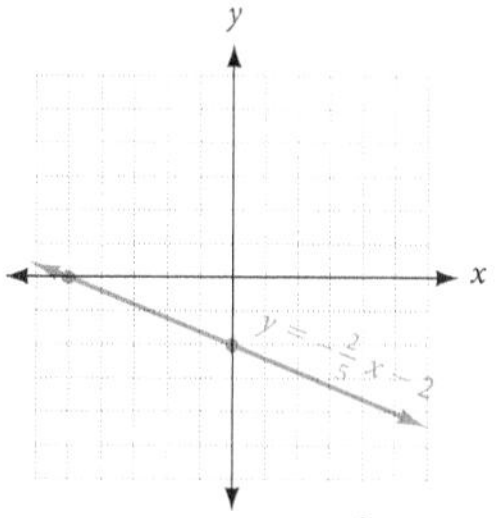

39. $y = \frac{2}{3}x + 1$ **41.** $y = \frac{3}{2}x - 1$ **43.** $y = -\frac{2}{3}x + 3$
45. $y = 2x - 4$ **47.** $m = 3, b = 3, y = 3x + 3$
49. $m = \frac{1}{4}, b = -1, y = \frac{1}{4}x - 1$ **51.** $2x - y = 1$
53. $x + 3y = -11$ **55.** $3x - 5y = -36$
57. $4x + 3y = 16$ **59. a.** $y = 3x - 1$ **b.** $y = -\frac{1}{3}x + \frac{7}{3}$
61. a. $y = -\frac{1}{2}x - 3$ **b.** $y = 2x + 7$
63. a. $y = -\frac{2}{3}x + 6$ **b.** $y = \frac{3}{2}x - \frac{1}{2}$
65. $y = 2$ **67.** $y = 5$
69. a. $-\frac{5}{2}$ **b.** $y = 2x + 6$ **c.** 6 **d.** 2
e.

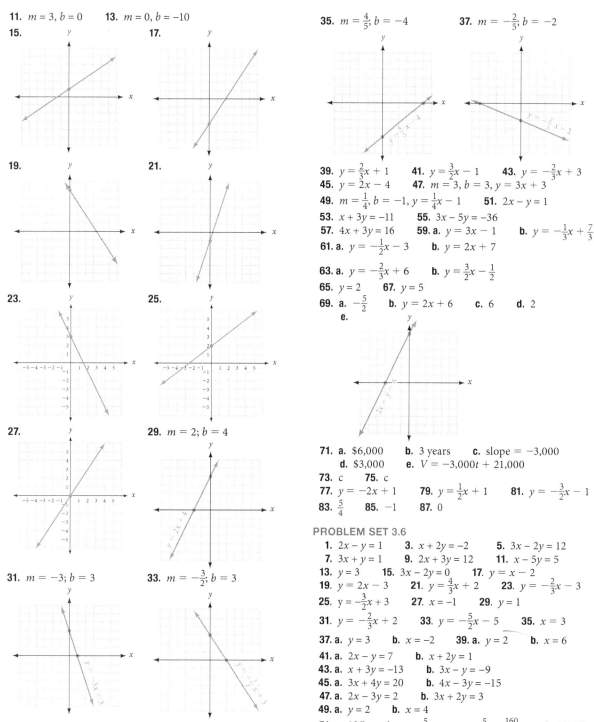

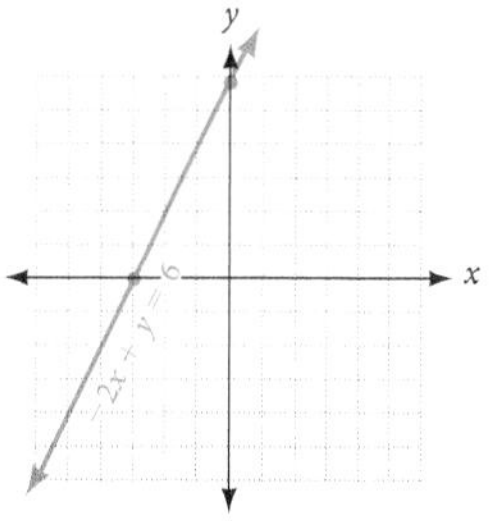

71. a. \$6,000 **b.** 3 years **c.** slope = −3,000
d. \$3,000 **e.** $V = -3{,}000t + 21{,}000$
73. c **75.** c
77. $y = -2x + 1$ **79.** $y = \frac{1}{2}x + 1$ **81.** $y = -\frac{3}{2}x - 1$
83. $\frac{5}{4}$ **85.** −1 **87.** 0

PROBLEM SET 3.6

1. $2x - y = 1$ **3.** $x + 2y = -2$ **5.** $3x - 2y = 12$
7. $3x + y = 1$ **9.** $2x + 3y = 12$ **11.** $x - 5y = 5$
13. $y = 3$ **15.** $3x - 2y = 0$ **17.** $y = x - 2$
19. $y = 2x - 3$ **21.** $y = \frac{4}{3}x + 2$ **23.** $y = -\frac{2}{3}x - 3$
25. $y = -\frac{3}{2}x + 3$ **27.** $x = -1$ **29.** $y = 1$
31. $y = -\frac{2}{3}x + 2$ **33.** $y = -\frac{5}{2}x - 5$ **35.** $x = 3$
37. a. $y = 3$ **b.** $x = -2$ **39. a.** $y = 2$ **b.** $x = 6$
41. a. $2x - y = 7$ **b.** $x + 2y = 1$
43. a. $x + 3y = -13$ **b.** $3x - y = -9$
45. a. $3x + 4y = 20$ **b.** $4x - 3y = -15$
47. a. $2x - 3y = 2$ **b.** $3x + 2y = 3$
49. a. $y = 2$ **b.** $x = 4$
51. a. 0°C **b.** $m = \frac{5}{9}$ **c.** $y = \frac{5}{9}x - \frac{160}{9}$ **d.** 100°C
53. b

55.

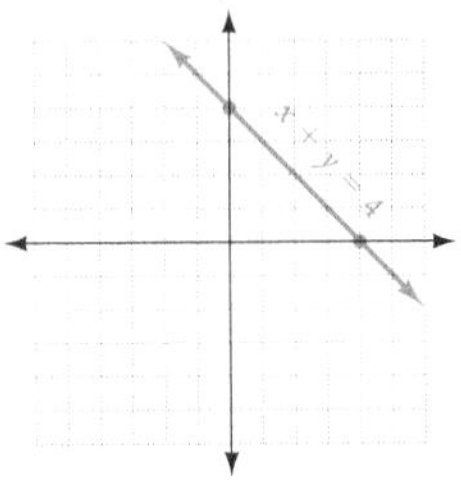

57.

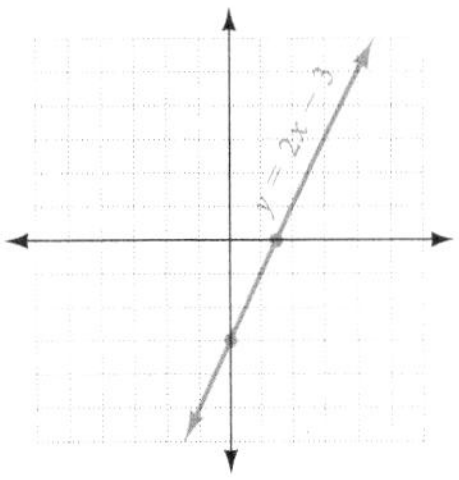

59.

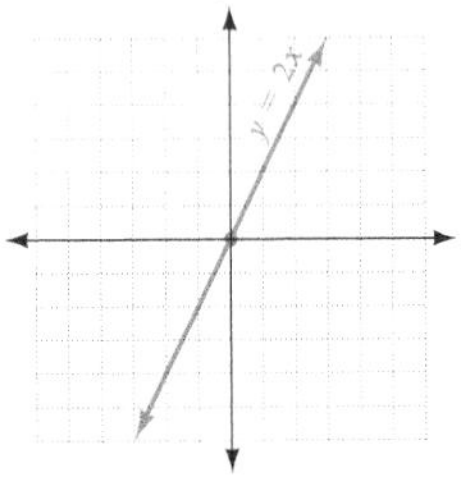

PROBLEM SET 3.7

1.

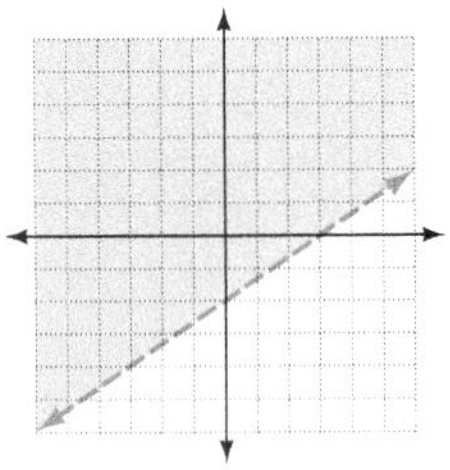

3.

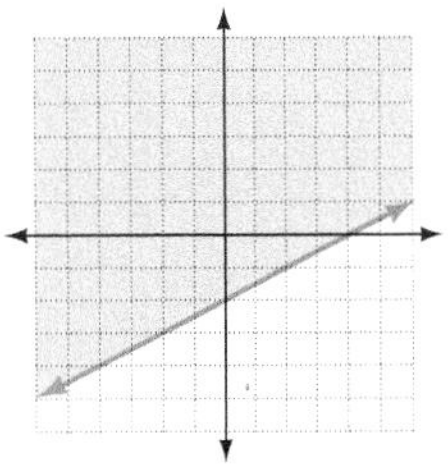

5.

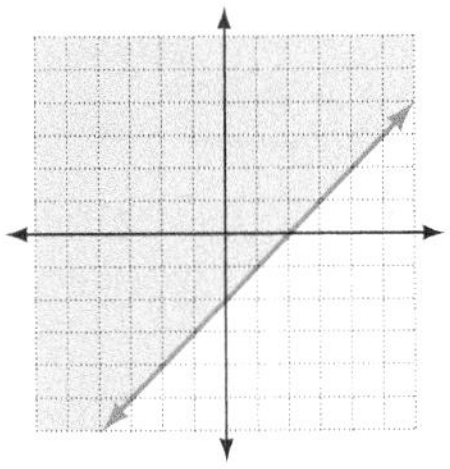

7.

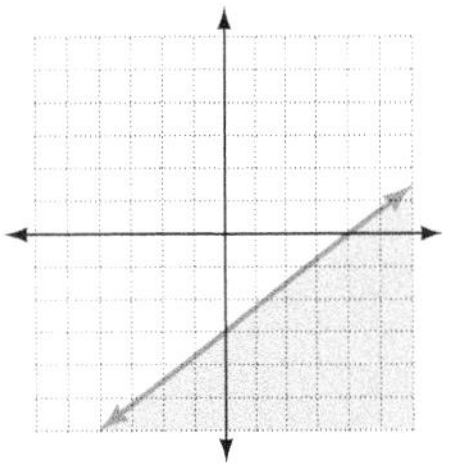

9.

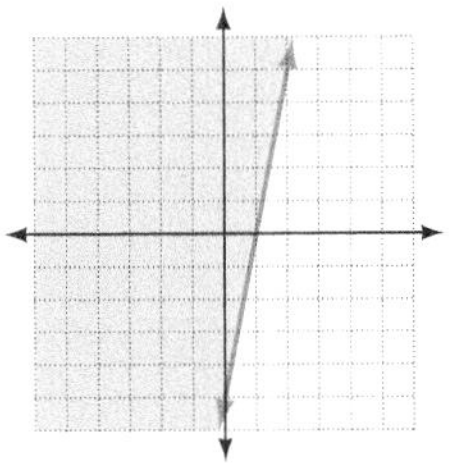

11.

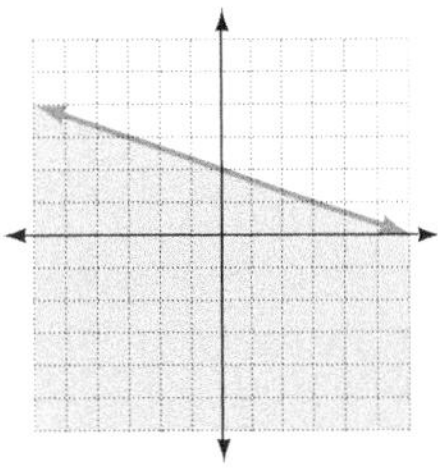

13.

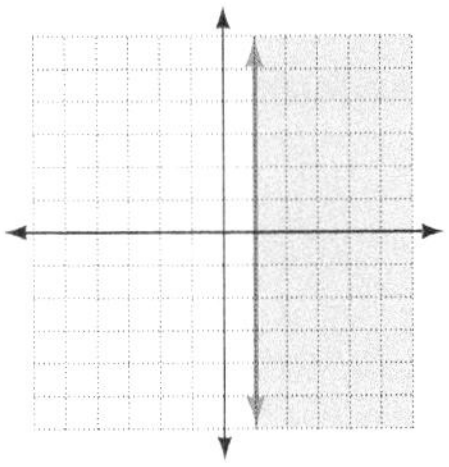

15.

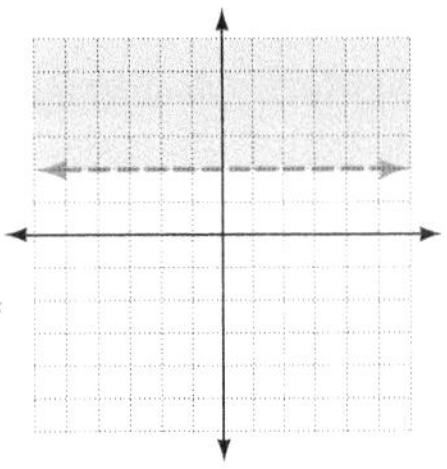

17.

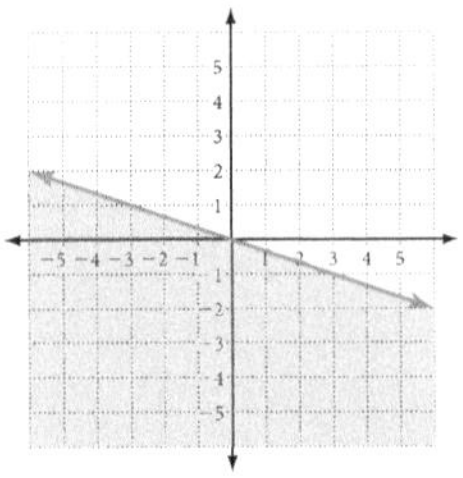

19.

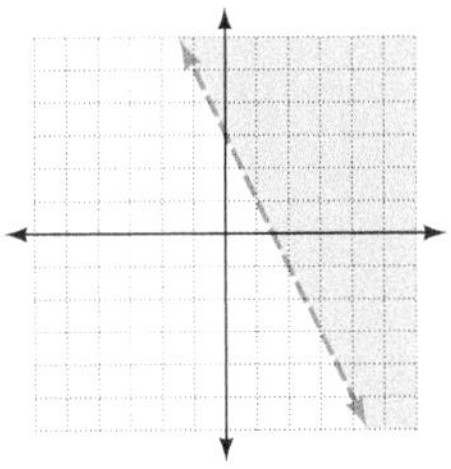

21.

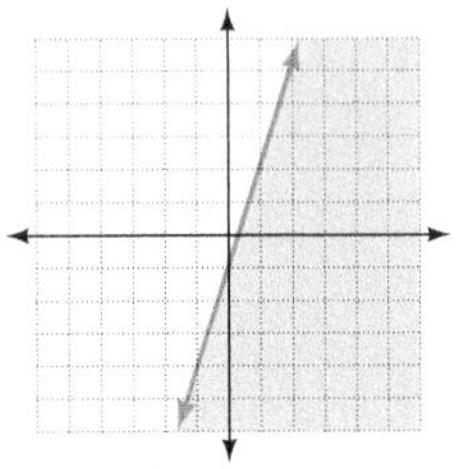

23.

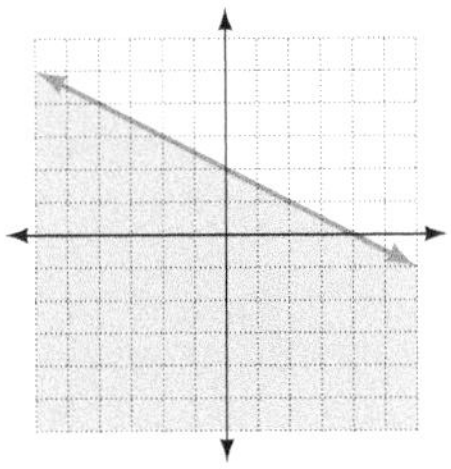

25. a. $y < \frac{8}{3}$ **b.** $y > -\frac{8}{3}$ **c.** $y = -\frac{4}{3}x+4$

d.

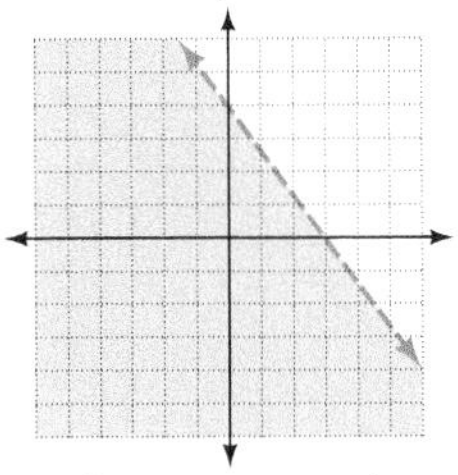

27. a. $y = \frac{2}{5}x + 2$ **b.** $y < \frac{2}{5}x + 2$ **c.** $y > \frac{2}{5}x + 2$

29. b **31.** $-6x + 11$ **33.** -8 **35.** -4

37. $w = \frac{P - 2l}{2}$ **39.** $\{x \mid x < -1\}$

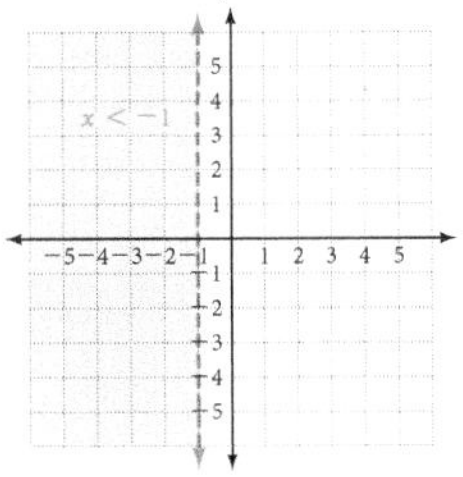

41. $y \geq \frac{3}{2}x - 6$ **43.** Width 2 inches, length 11 inches

CHAPTER 3 TEST

1.-4.

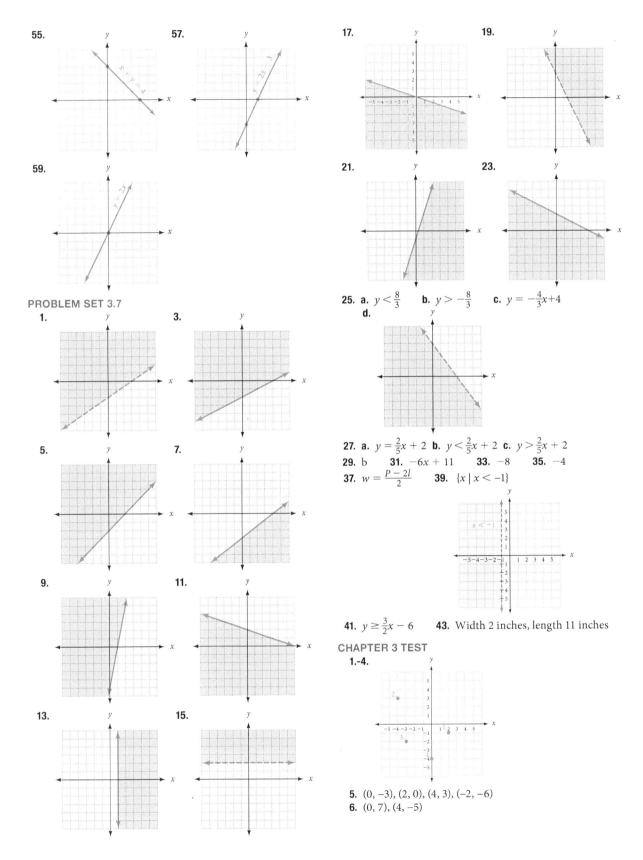

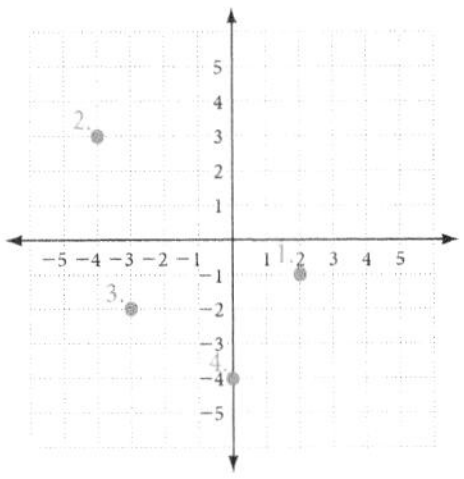

5. (0, −3), (2, 0), (4, 3), (−2, −6)

6. (0, 7), (4, −5)

7.

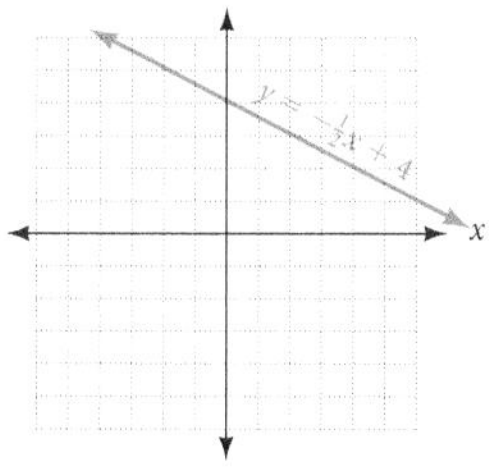

8.

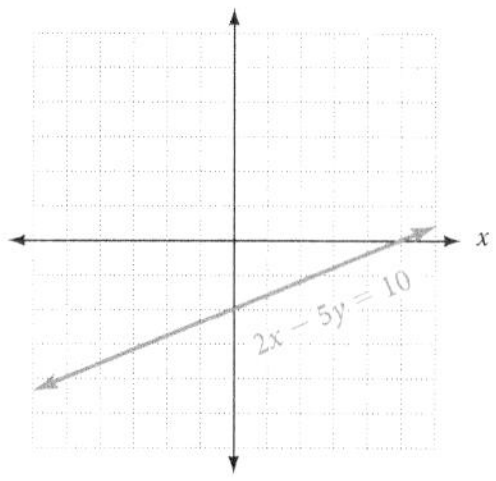

9.

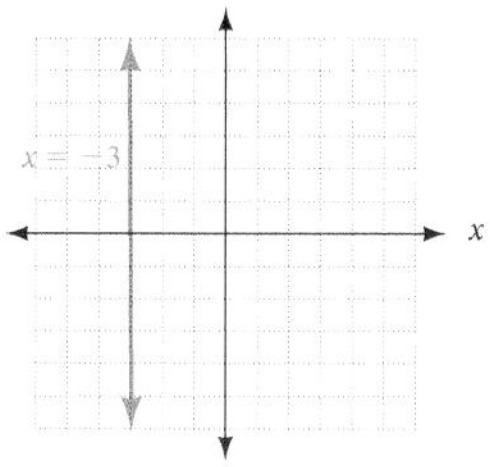

10.

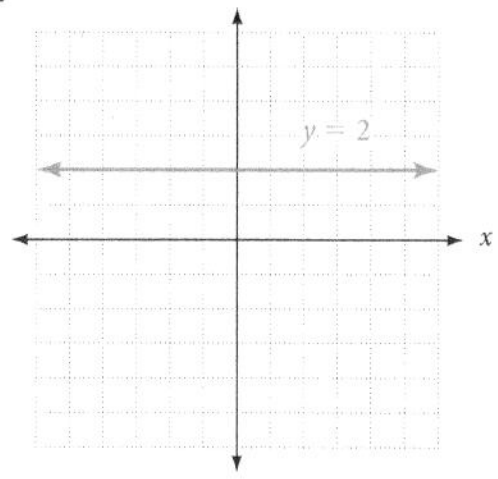

11. x-intercept = 2, y-intercept = −4
12. x-intercept = −4, y-intercept = 6
13. No x-intercept, y-intercept = 3
14. x-intercept = −2, no y-intercept
15. $-\frac{1}{2}$ **16.** $-\frac{8}{7}$ **17.** −3 **18.** No slope **19.** 0
20. $m = -\frac{1}{2}$, y-intercept = 6 **21.** $m = 3$, y-intercept = 0
22. $m = \frac{3}{7}$ **23.** $m = \frac{1}{4}$ **24.** $y = 3x - 5$
25. $y = -\frac{1}{2}x + 3$ **26.** $y = -\frac{2}{3}x - 2$
27. $y = -\frac{6}{5}x + \frac{18}{5}$ **28.** $3x - 2y = 7$
29. $2x - 3y = -17$ **30.** $y = -5$ **31.** $x = -2$

32.

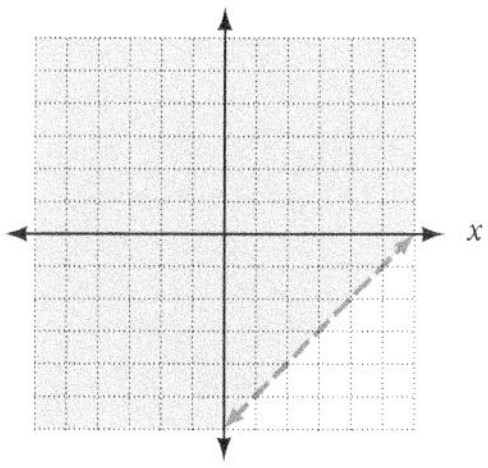

33.

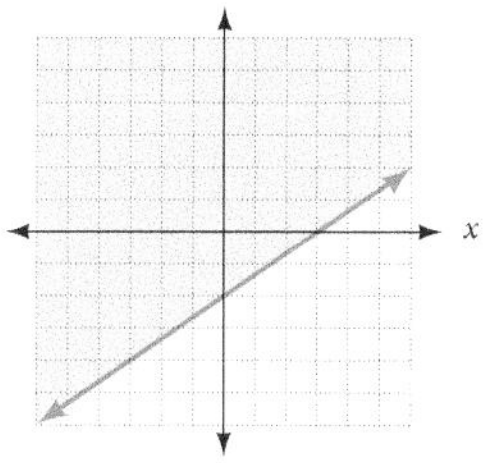

34.

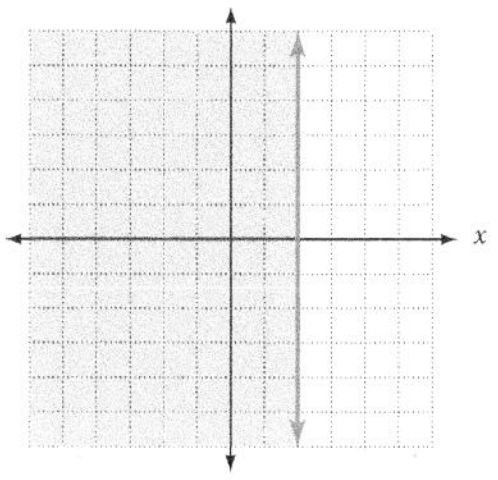

Chapter 4

PROBLEM SET 4.1

1. a. No **b.** Yes **c.** No **3. a.** Yes **b.** No **c.** Yes
4. a. No **b.** Yes **c.** Yes

5. {(2, 1)}

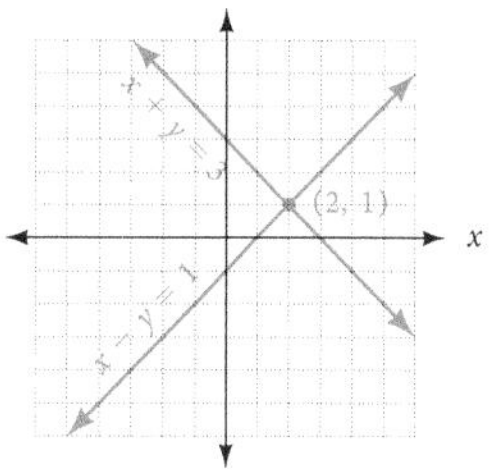

7. {(−1, 2)}

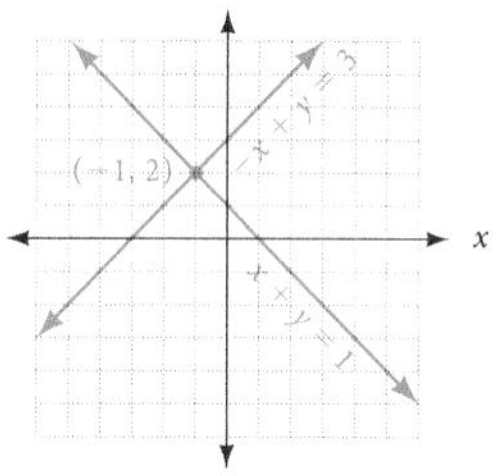

9. {(3, 5)}

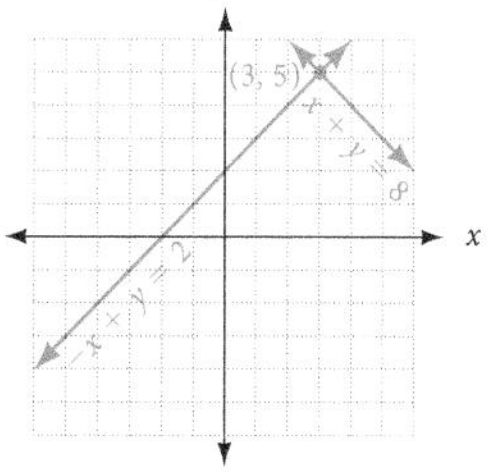

11. {(4, 3)}

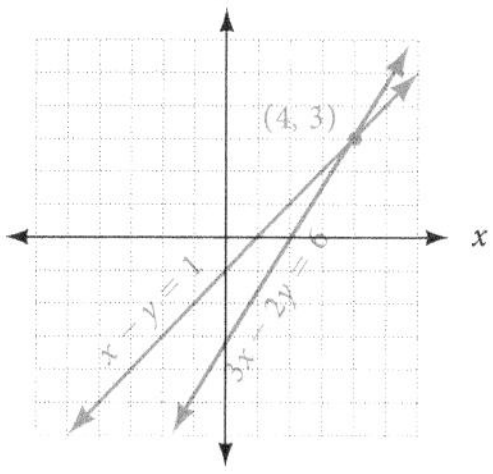

13. {(0, −6)}

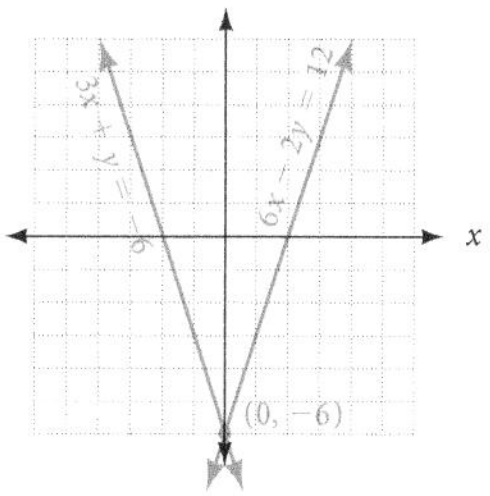

15. {(1, 0)}

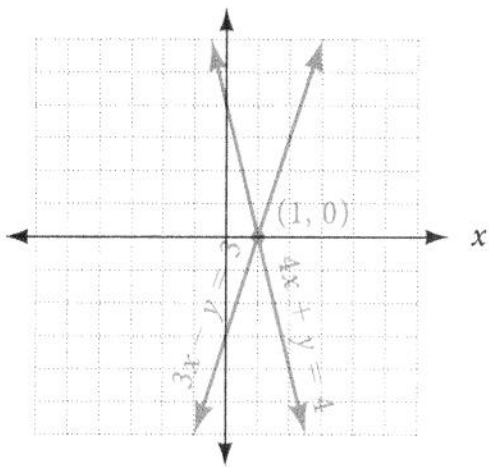

17. {(0, 0)}

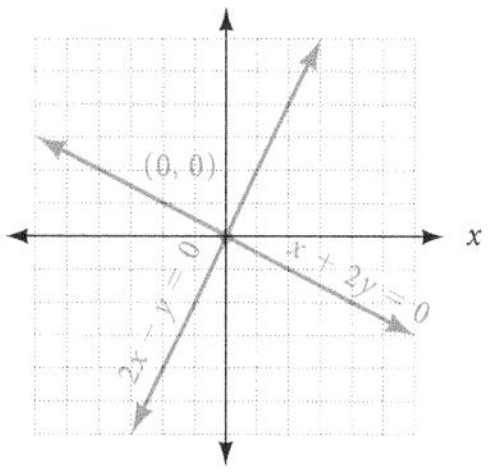

19. {(−5, −6)}

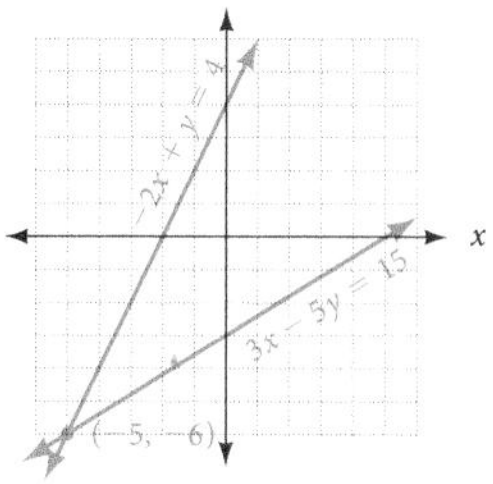

21. {(−1, −1)}

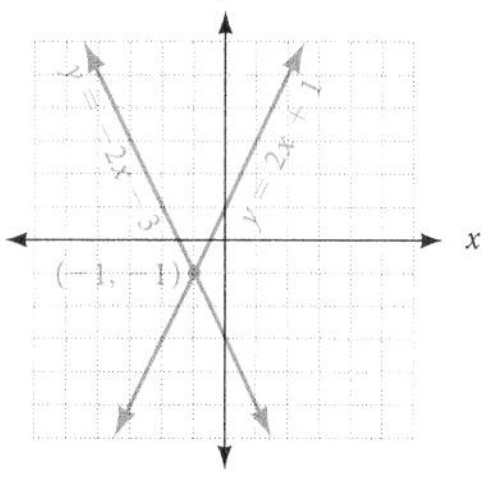

23. {(−3, 2)}

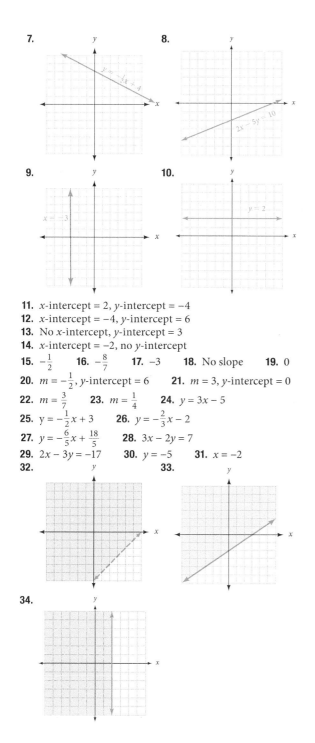

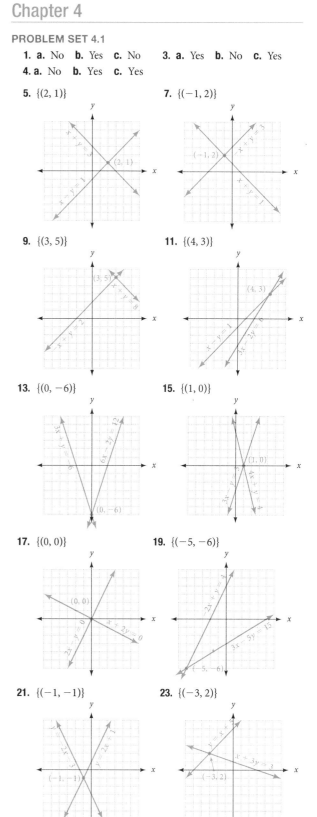

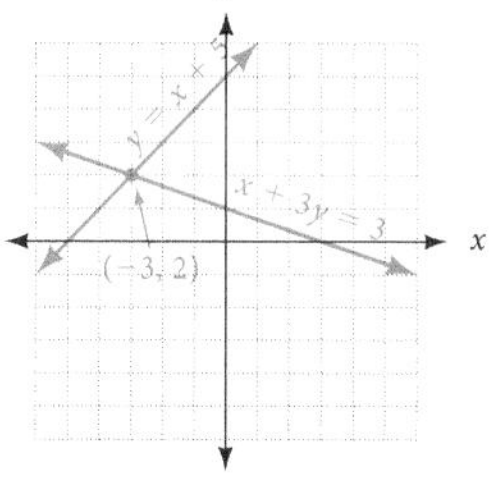

25. $\{(-3, 5)\}$

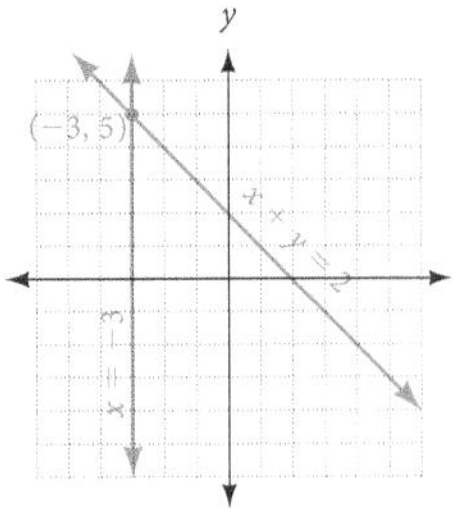

27. $\{(-4, 6)\}$

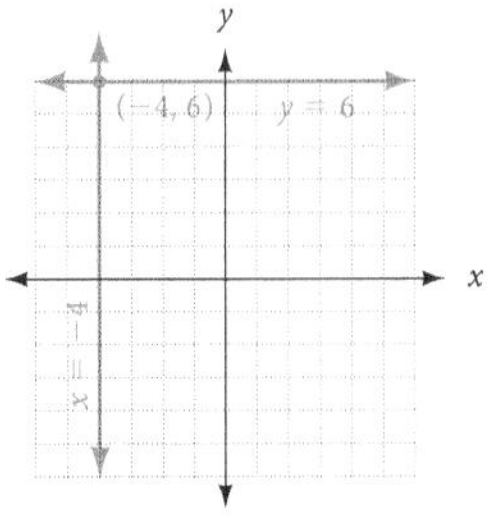

29. $\varnothing$

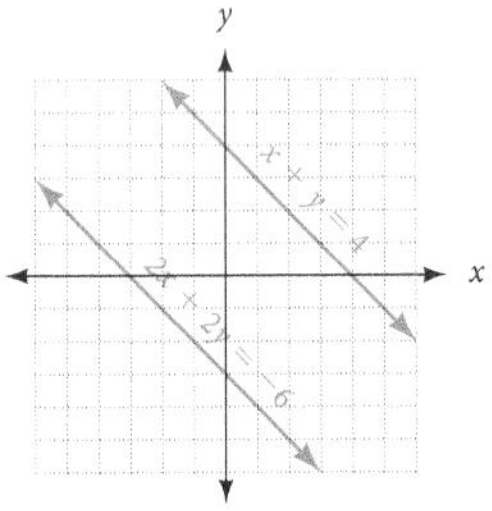

31. $\{(x, y) \mid 2x - y = 4\}$

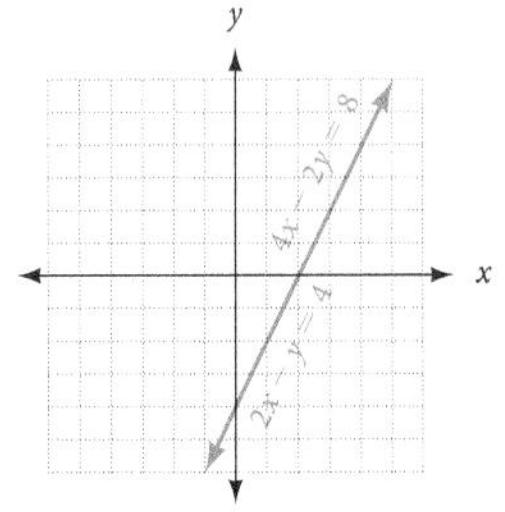

33.

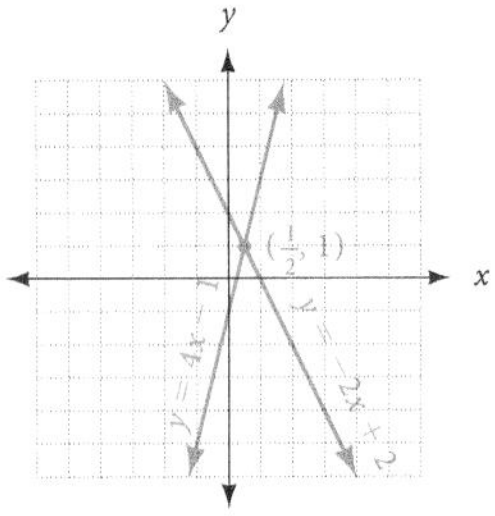

35.

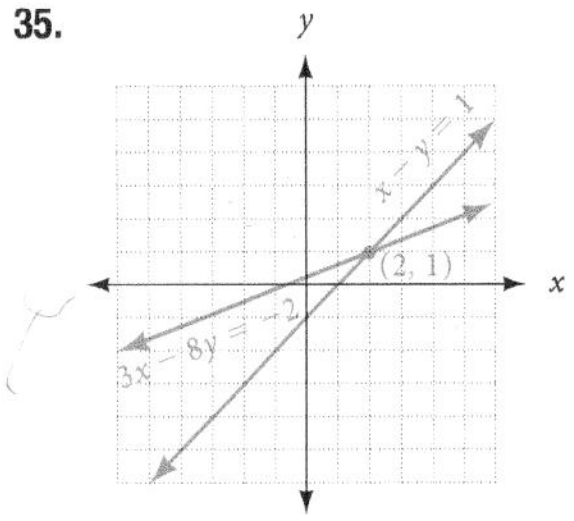

37. $a = 5, b = -5$

39. a. $a = 3, b = 1$ **b.** $a = 3, b \neq 1$ **c.** $x \neq 3$

41. a. 25 hours **b.** Gigi's **c.** Marcy's

43. a **45.** b **47.** $2x$ **49.** $7x$ **51.** $13x$

53. $3x - 2y$ **55.** 1 **57.** 0 **59.** -5

PROBLEM SET 4.2

1. $(2, 1)$ **3.** $(3, 7)$ **5.** $(2, -5)$ **7.** $(-1, 0)$

9. $\{(x, y) \mid 3x + 2y = 1\}$ **11.** $(4, 8)$ **13.** $\left(\frac{1}{5}, 1\right)$

15. $(1, 0)$ **17.** $(-1, -2)$ **19.** $\left(-5, \frac{3}{4}\right)$ **21.** $(-4, 5)$

23. $(-3, -10)$ **25.** $(3, 2)$ **27.** $\left(5, \frac{1}{3}\right)$ **29.** $\left(-2, \frac{2}{3}\right)$

31. $(2, 2)$ **33.** $\varnothing$ **35.** $(1, 1)$**37.** $\varnothing$ **39.** $(10, 12)$

41. c **43.** 1 **45.** 2 **47.** All real numbers

49. $x = 3y - 1$ **51.** 1 **53.** 5 **55.** 34.5 **57.** 33.95

PROBLEM SET 4.3

1. $(4, 7)$ **3.** $(3, 17)$ **5.** $\left(\frac{3}{2}, 2\right)$ **7.** $(2, 4)$ **9.** $(0, 4)$

11. $(-1, 3)$ **13.** $(1, 1)$ **15.** $(2, -3)$ **17.** $\left(-2, \frac{3}{5}\right)$

19. $(-3, 5)$ **21.** $\varnothing$ **23.** $(10, 12)$ **25.** $\left(\frac{2}{7}, -\frac{12}{7}\right)$

27. $(3, 1)$ **29.** $(4, 4)$ **31.** $\left(\frac{1}{3}, \frac{5}{6}\right)$ **33.** $(2, 6)$

35. $(5, -2)$ **37.** $(18, 10)$ **39.** $\{(x, y) \mid x - y = 2\}$

41. $\left(-\frac{1}{2}, \frac{1}{2}\right)$

43. a. 1,000 miles **b.** Car **c.** Truck

d. We are only working with positive numbers.

45. b **47.** 3 and 23 **49.** 15 and 24

51. Length = 23 in.; width = 6 in.

53. 14 nickels and 10 dimes

PROBLEM SET 4.4

1. 10 and 15 **3.** 3 and 12 **5.** 4 and 9 **7.** 6 and 29

9. \$9,000 at 8%, \$11,000 at 6%

11. \$2,000 at 6%, \$8,000 at 5% **13.** 6 nickels, 8 quarters

15. 12 dimes, 9 quarters **17.** 6L of 50%, 12L of 20%

19. 10 gallons of 10% solution, 20 gallons of 7% solution

21. 20 adults, 50 kids **23.** 16 feet wide, 32 feet long

25. 33 \$5 chips, 12 \$25 chips **27.** 50 at \$11, 100 at \$20

29. a **31.** c **33.** 47 **35.** 14 **37.** 70 **39.** 35

41. 5 **43.** 6,540 **45.** 1,760 **47.** 20 **49.** 63

51. 53

CHAPTER 4 TEST

1. $(0, 6)$ **2.**

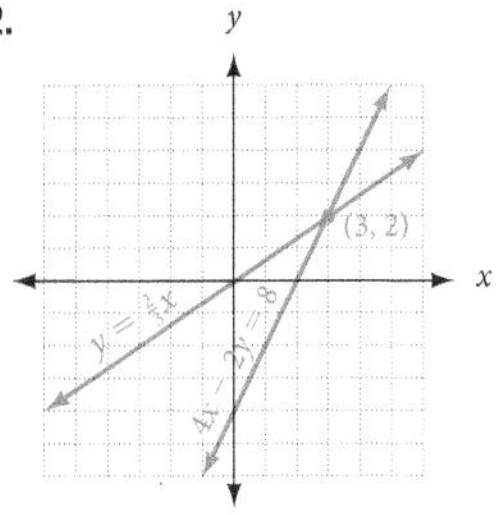

3.

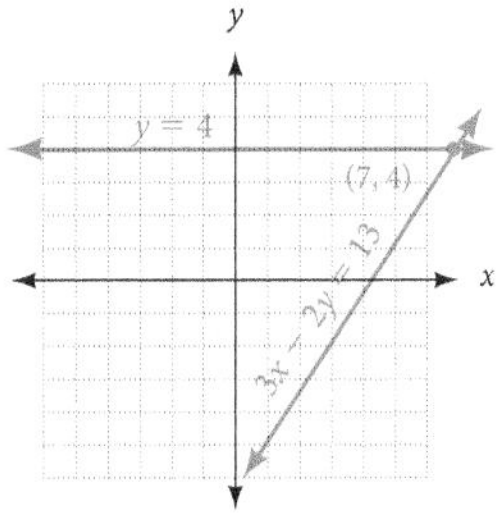

4.

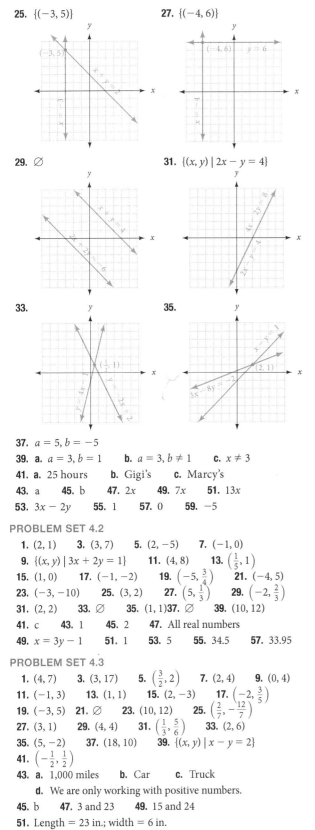

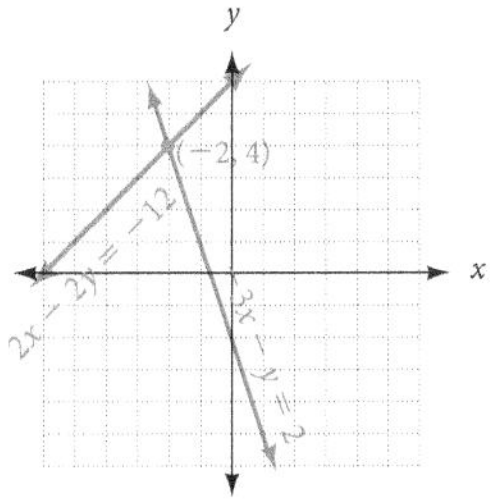

5. $(-4, 5)$ **6.** $(1, 2)$ **7.** $(-6, 3)$

8. Lines coincide, $\{(x, y) \mid 2x + 3y = 4\}$

9. $(4, 0)$ **10.** $(19, 9)$ **11.** $(-3, 4)$ **12.** $(11, 3)$ **13.** 4, 14

14. \$1,520 at 7%, \$480 at 6% **15.** 8 dimes, 11 nickels

16. 20 gallons of 40% and 10 gallons of 70% **17.** 71 ft × 28 ft

Chapter 5

PROBLEM SET 5.1

1. 4, 2, 16 **3.** 0.3, 2, 0.09 **5.** 4, 3, 64 **7.** −5, 2, 25

9. 2, 3, −8 **11.** 3, 4, 81 **13.** $\frac{2}{3}, 2, \frac{4}{9}$ **15.** $\frac{1}{2}, 4, \frac{1}{16}$

17. a.

Number x	1	2	3	4	5	6	7
Square x^2	1	4	9	16	25	36	49

b. larger **19.** x^9 **21.** y^{30} **23.** 2^{12} **25.** x^{28}

27. x^{10} **29.** 5^{12} **31.** y^9 **33.** 2^{50} **35.** a^{3x}

37. b^{xy} **39.** $16x^2$ **41.** $32y^5$ **43.** $81x^4$

45. $0.25a^2b^2$ **47.** $64x^3y^3z^3$ **49.** $8x^{12}$ **51.** $16a^6$
53. x^{14} **55.** a^{11} **57.** $16x^4y^6$ **59.** $\frac{8}{27}a^{12}b^{15}$
61.

Number x	-3	-2	-1	0	1	2	3
Square x^2	9	4	1	0	1	4	9

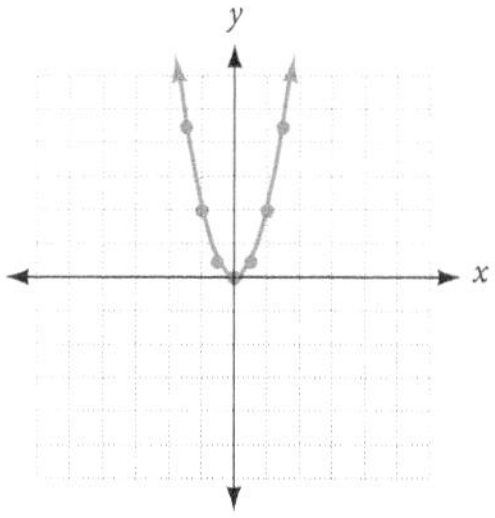

63.

Number x	-2.5	-1.5	-0.5	0	0.5	1.5	2.5
Square x^2	6.25	2.25	0.25	0	0.25	2.25	6.25

65. 4.32×10^4 **67.** -5.7×10^2 **69.** 2.38×10^5
71. 2,490 **73.** -352 **75.** 28,000 **77.** 27 inches^3
79. 15.6 inches^3 **81.** 36 inches^3 **83.** Answers will vary
85. 6.5×10^8 seconds **87.** $740,000 **89.** $180,000
91. 219 inches^3 **93.** 182 inches^3 **95.** a **97.** a
99. -3 **101.** 11 **103.** -5 **105.** 5 **107.** 2
109. 6 **111.** 4 **113.** 3

PROBLEM SET 5.2

1. $\frac{1}{5}$ **3.** $\frac{1}{x}$ **5.** $\frac{1}{9}$ **7.** $\frac{1}{36}$ **9.** $\frac{1}{64}$ **11.** $\frac{1}{125}$
13. $\frac{1}{a^4}$ **15.** $-\frac{1}{16}$ **17.** $-\frac{1}{x}$ **19.** $-\frac{1}{125}$ **21.** $\frac{1}{16}$
23. $\frac{2}{x^3}$ **25.** $\frac{1}{8x^3}$ **27.** $\frac{1}{25y^2}$ **29.** $\frac{1}{100}$
31.

Number x	Square x^2	Power of 2 2^x
-3	9	$\frac{1}{8}$
-2	4	$\frac{1}{4}$
-1	1	$\frac{1}{2}$
0	0	1
1	1	2
2	4	4
3	9	8

33. $\frac{1}{25}$ **35.** x^6 **37.** 64 **39.** $8x^3$ **41.** 6^{10}
43. $\frac{1}{6^{10}}$ **45.** $\frac{1}{2^8} = \frac{1}{256}$ **47.** $2^8 = 256$ **49.** $27x^3$
51. $81x^4y^4$ **53.** 1 **55.** $2a^2b$ **57.** $\frac{1}{49y^6}$

59. $\frac{1}{x^8}$ **61.** $\frac{1}{y^3}$ **63.** x^2 **65.** a^6 **67.** $\frac{1}{y^9}$
69. y^{40} **71.** $\frac{1}{x}$ **73.** x^9 **75.** a^{16} **77.** $\frac{1}{a^4}$
79.

Number x	-3	-2	-1	0	1	2	3
Power of 2 2^x	$\frac{1}{8}$	$\frac{1}{4}$	$\frac{1}{2}$	1	2	4	8

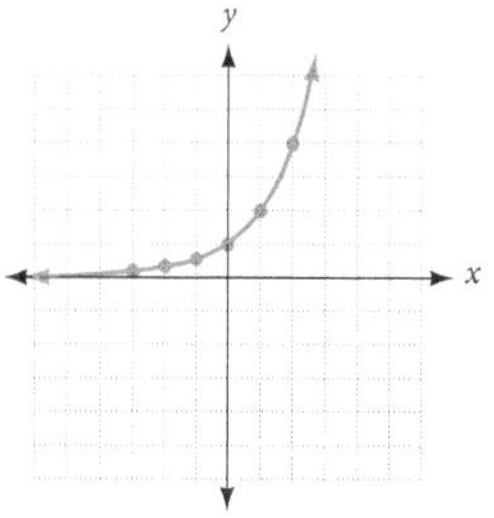

81. 4.8×10^{-3} **83.** 2.5×10^1 **85.** 2.5×10^{-1}
87. 9×10^{-6}
89.

Expanded Form	Scientific Notation $n \times 10^r$
0.000357	3.57×10^{-4}
0.00357	3.57×10^{-3}
0.0357	3.57×10^{-2}
0.357	3.57×10^{-1}
3.57	3.57×10^0
35.7	3.57×10^1
357	3.57×10^2
3,570	3.57×10^3
35,700	3.57×10^4

91. 0.00423 **93.** 0.00008 **95.** 4.2 **97.** 0.24
99. 0.002
101. Craven/Busch 2×10^{-3}; Earnhardt/Irvan 5×10^{-3}; Harvick/Gordon 6×10^{-3}; Kahne/Kenseth 1×10^{-2}; Kenseth/Kahne 1×10^{-2}
103. 2.5×10^4 **105.** 2.35×10^5 **107.** 8.2×10^{-4}
109. 100 inches^2; 400 inches^2; 4 **111.** x^2; $4x^2$; 4
113. 216 inches^3; 1,728 inches^3; 8 **115.** x^3; $8x^3$; 8
117. b **119.** a **121.** d **123.** $\frac{5}{14}$ **125.** 3.2
127. -11 **129.** x^3 **131.** y^3 **133.** x **35.** $\frac{1}{x^3}$
137. 0.0006

PROBLEM SET 5.3

1. Coefficient = 7, degree = 3
3. Coefficient = -1, degree = 1
5. Coefficient = $\frac{1}{2}$, degree = 2
7. Coefficient = -4, degree = 11
9. Coefficient = 8, degree = 0
11. $12x^7$ **13.** $-16y^{11}$ **15.** $32x^2$ **17.** $200a^6$
19. $-24a^3b^3$ **21.** $24x^6y^8$ **23.** $3x$ **25.** $\frac{6}{y^3}$
27. $\frac{1}{2a}$ **29.** $-\frac{3a}{b^2}$ **31.** $\frac{x^2}{9z^2}$

33.

a	b	ab	$\frac{a}{b}$	$\frac{b}{a}$
10	$5x$	$50x$	$\frac{2}{x}$	$\frac{x}{2}$
$20x^3$	$6x^2$	$120x^5$	$\frac{10x}{3}$	$\frac{3}{10x}$
$25x^5$	$5x^4$	$125x^9$	$5x$	$\frac{1}{5x}$
$3x^{-2}$	$3x^2$	9	$\frac{1}{x^4}$	x^4
$-2y^4$	$8y^7$	$-16y^{11}$	$-\frac{1}{4y^3}$	$-4y^3$

35. 6×10^8 **37.** 1.75×10^{-1} **39.** 1.21×10^{-6} **41.** 4.2×10^3 **43.** 3×10^{10} **45.** 5×10^{-3} **47.** $8x^2$ **49.** $-11x^5$ **51.** 0 **53.** $4x^3$ **55.** $31ab^2$

57.

a	b	ab	$a + b$
$5x$	$3x$	$15x^2$	$8x$
$4x^2$	$2x^2$	$8x^4$	$6x^2$
$3x^3$	$6x^3$	$18x^6$	$9x^3$
$2x^4$	$-3x^4$	$-6x^8$	$-x^4$
x^5	$7x^5$	$7x^{10}$	$8x^5$

59. $128x^7$ **61.** $432x^{10}$ **63.** $128x^{17}y^9$ **65.** $4x^2$ **67.** $\frac{27x^2}{16}$ **69.** $\frac{4x^6}{9y^4}$ **71.** $4x^3$ **73.** $\frac{1}{b^2}$ **75.** $\frac{6y^{10}}{x^4}$ **77.** 2×10^6 **79.** 1×10^1 **81.** 4.2×10^{-6} **83.** $9x^3$ **85.** $-20a^2$ **87.** $6x^5y^2$ **89.** b **91.** d **93.** -5 **95.** 6 **97.** 76 **99.** $6x^2$ **101.** $2x$ **103.** $-2x - 9$ **105.** 11

PROBLEM SET 5.4

1. Trinomial, 3 **3.** Trinomial, 3 **5.** Binomial, 1 **7.** Binomial, 2 **9.** Monomial, 2 **11.** Monomial, 0 **13.** $5x^2 + 2$; 2, 5 **15.** $-x^3 + 3x^2 - 6x$; 3, -1 **17.** $6x^2 + x - 1$; 2, 6 **19.** 14 **21.** -8 **23.** 12 **25.** 4 **27.** $5x^2 + 5x + 9$ **29.** $5a^2 - 9a + 7$ **31.** $x^2 + 6x + 8$ **33.** $x^2 - 9$ **35.** $-10x + 5$ **37.** $-5x^2 - x + 2$ **39.** $x^2 - 2x - 3$ **41.** $6x^3 - 13x^2 + 8x - 3$ **43.** $2a^2 - 2a - 2$ **45.** $6x^2 - 13x + 5$ **47.** $3y^2 - 11y + 10$ **49.** $2x^2 - x + 1$ **51.** $-\frac{1}{9}x^3 - \frac{2}{3}x^2 - \frac{5}{2}x + \frac{7}{4}$ **53.** $-4y^2 + 15y - 22$ **55.** $x^2 - 33x + 63$ **57.** $8y^2 + 4y + 26$ **59.** $75x^2 - 150x - 75$ **61.** $12x + 2$ **63.** 56.52 inches^3 **65.** d **67.** b **69.** 5 **71.** -6 **73.** $-20x^2$ **75.** $-21x$ **77.** $2x$ **79.** $6x - 18$

PROBLEM SET 5.5

1. $6x^2 + 2x$ **3.** $6x^4 - 4x^3 + 2x^2$ **5.** $2a^3b - 2a^2b^2 + 2ab$ **7.** $3y^4 + 9y^3 + 12y^2$ **9.** $8x^5y^2 + 12x^4y^3 + 32x^2y^4$ **11.** $x^2 + 7x + 12$ **13.** $x^2 + 7x + 6$ **15.** $x^2 + 2x + \frac{3}{4}$ **17.** $a^2 + 2a - 15$ **19.** $xy + xb - ay - ab$ **21.** $x^2 - 36$ **23.** $y^2 - \frac{25}{36}$ **25.** $2x^2 - 11x + 12$ **27.** $2a^2 + 3a - 2$ **29.** $6x^2 - 19x + 10$ **31.** $2ax + 8x + 3a + 12$ **33.** $25x^2 - 16$ **35.** $2x^2 + \frac{5}{2}x - \frac{3}{4}$ **37.** $3 - 10a + 8a^2$ **39.** $a^3 - 6a^2 + 11a - 6$ **41.** $x^3 + 8$ **43.** $2x^3 + 17x^2 + 26x + 9$ **45.** $5x^4 - 13x^3 + 20x^2 + 7x + 5$ **47.** $6x^4 - 7x^3 - 12x^2 + 3x + 2$ **49.** $a^5 - 3a^4 + 5a^3 - a^2 - 2a + 8$ **51.** $2x^4 + x^2 - 15$ **53.** $6a^6 + 15a^4 + 4a^2 + 10$ **55.** $x^3 + 12x^2 + 47x + 60$ **57.** $x^2 - 5x + 8$ **59.** $8x^2 - 6x - 5$ **61.** $x^2 - x - 30$ **63.** $x^2 + 4x - 6$ **65.** $x^2 + 13x$ **67.** $x^2 + 2x - 3$ **69.** $a^2 - 3a + 6$ **71.** $(x + 2)(x + 3) = x^2 + 2x + 3x + 6 = x^2 + 5x + 6$ **73.** $(x + 1)(2x + 2) = 2x^2 + 4x + 2$ **75. a.** $5x + 1$ **b.** $x - 9$ **c.** $x = 9$ **d.** $6x^2 + 7x - 20$ **77.** $A = x(2x + 5) = 2x^2 + 5x$ **79.** $A = x(x + 1) = x^2 + x$ **81.** $R = 100p - 10p^2$ **83.** d **85.** c **87.** 169 **89.** $-10x$ **91.** 0 **93.** 0 **95.** $-12x + 16$ **97.** $x^2 + x - 2$ **99.** $x^2 + 6x + 9$

PROBLEM SET 5.6

1. $x^2 - 4x + 4$ **3.** $a^2 + 6a + 9$ **5.** $x^2 - 10x + 25$ **7.** $a^2 - a + \frac{1}{4}$ **9.** $x^2 + 20x + 100$ **11.** $a^2 + 1.6a + 0.64$ **13.** $4x^2 - 4x + 1$ **15.** $16a^2 + 40a + 25$ **17.** $9x^2 - 12x + 4$ **19.** $9a^2 + 30ab + 25b^2$ **21.** $16x^2 - 40xy + 25y^2$ **23.** $x^4 + 10x^2 + 25$ **25.** $a^6 + 2a^3 + 1$ **27.** $49m^4 + 28m^2n + 4n^2$ **29.** $36x^4 - 120x^2y^2 + 100y^4$

31.

x	$(x + 3)^2$	$x^2 + 9$	$x^2 + 6x + 9$
1	16	10	16
2	25	13	25
3	36	18	36
4	49	25	49

33.

a	1	3	3	4
b	1	5	4	5
$(a + b)^2$	4	64	49	81
$a^2 + b^2$	2	34	25	41
$a^2 + ab + b^2$	3	49	37	61
$a^2 + 2ab + b^2$	4	64	49	81

35. $a^2 - 25$ **37.** $y^2 - 1$ **39.** $81 - x^2$ **41.** $4x^2 - 25$ **43.** $16x^2 - \frac{1}{9}$ **45.** $4a^2 - 49b^2$ **47.** $36 - 49x^2$ **49.** $x^4 - 9$ **51.** $a^4 - 16b^4$ **53.** $25y^8 - 64$ **55.** $2x^2 - 34$ **57.** $-12x^2 + 20x + 8$ **59.** $a^2 + 4a + 6$ **61.** $8x^3 + 36x^2 + 54x + 27$ **63.** $(50 - 1)(50 + 1) = 2{,}500 - 1 = 2{,}499$ **65.** Both equal 25. **67.** $x^2 + (x + 1)^2 = 2x^2 + 2x + 1$ **69.** $x^2 + (x + 1)^2 + (x + 2)^2 = 3x^2 + 6x + 5$ **71.** $a^2 + ab + ba + b^2 = a^2 + 2ab + b^2$ **73.** b **75.** $2x^2$ **77.** x^2 **79.** $3x$ **81.** $3xy$ **83.** $146\frac{20}{27}$ **85.** $x^2 - 3x$ **87.** $2x^3 - 10x^2$ **89.** $-2x$ **91.** 2

PROBLEM SET 5.7

1. $x - 2$ **3.** $5xy - 2y$ **5.** $7x^4 - 6x^3 + 5x^2$
7. $-4a + 2$ **9.** $-6a^2b + 3ab^2 - 7b^3$
11. $-\frac{a}{2} - b - \frac{b^2}{2a}$ **13.** $3x + 4y$ **15.** $-y + 3$
17. $5y - 4$ **19.** $xy - x^2y^2$ **21.** $-a + 1$
23. $x^2 - 3xy + y^2$ **25.** $2 - 3b + 5b^2$ **27.** $-2xy + 1$
29. $\frac{1}{4x} - \frac{1}{2a} + \frac{3}{4}$ **31.** $\frac{4x^2}{3} + \frac{2}{3x} + \frac{1}{x^2}$ **33.** $3a^{3m} - 9a^m$
35. $2x^{4m} - 5x^{2m} + 7$ **37.** $3x^2 - x + 6$ **39.** 4
41. $x + 5$ **43.** $x - 2$ **45.** $a + 4$ **47.** $x + 3$
49. $x - 3$ **51.** $x^2 - x - 3$ **53.** $1 + \frac{5}{x - 2}$
55. $3 + \frac{-2}{x + 2}$ **57.** $x + 2 + \frac{2}{x + 3}$ **59.** $x + 4 + \frac{9}{x - 2}$
61. $x + 4 + \frac{-10}{x + 1}$ **63.** $x - 3 + \frac{17}{2x + 4}$
65. $3a - 2 + \frac{7}{2a + 3}$ **67.** $2a^2 - a - 3$
69. $1 + \frac{-3x + 11}{x^2 - 3x - 2}$ **71.** $a - 5 + \frac{-3a + 19}{2a^2 + a + 3}$
73. $x^2 - x + 5$ **75.** $x^2 + x + 1$ **77.** $x^2 + 2x + 4$
79. $1 + \frac{3a + 1}{a^2 + 1}$ **81.** $a + 3 + \frac{2a + 7}{a^2 - 2}$
83. $2a^2 + 3 + \frac{1}{2a^2 - 1}$ **85.** Both equal 7.
87. $\frac{3(10) + 8}{2} = 19$; $3(10) + 4 = 34$ **89.** \$491.17
91. \$331.42 **93.** d **95.** $200x^{24}$ **97.** x^7
99. 8×10^1 **101.** $10ab^2$ **103.** $6x^4 + 6x^3 - 2x^2$
105. $9y^2 - 30y + 25$ **107.** $4a^4 - 49$

CHAPTER 5 TEST

1. -32 **2.** -16 **3.** x^{23} **4.** $16x^4y^6$ **5.** $\frac{1}{16}$
6. 1 **7.** x^3 **8.** $\frac{x^3}{27}$ **9.** $\frac{1}{x^3}$ **10.** 4.307×10^{-2}
11. 7,630,000 **12.** $-24a^3b^4$ **13.** $\frac{y^3z^2}{3x^2}$ **14.** $\frac{a^3b^2}{2}$
15. $12x^3$ **16.** 7.5×10^7 **17.** $9x^2 + 5x + 4$
18. $2x^2 + 6x - 2$ **19.** $5x - 4$ **20.** 21
21. $15x^4 - 6x^3 + 12x^2$ **22.** $x^2 - \frac{1}{12}x - \frac{1}{12}$
23. $10x^2 - 3x - 18$ **24.** $x^3 + 64$ **25.** $x^2 - 12x + 36$
26. $4a^2 + 16ab + 16b^2$ **27.** $9x^2 - 36$ **28.** $x^4 - 16$
29. $3x^2 - 6x + 1$ **30.** $3x - 1 + \frac{-5}{3x - 1}$
31. $4x + 3 + \frac{-8x - 5}{x^2 + 2}$ **32.** 32.77 in^3 **33.** $V = w^3$

Chapter 6

PROBLEM SET 6.1

1. 3 **3.** $2x$ **5.** $4a^2b$ **7.** $2x + 1$ **9.** $5(3x + 5)$
11. $3(2a + 3)$ **13.** $4(x - 2y)$ **15.** $3(x^2 - 2x - 3)$
17. $3(a^2 - a - 20)$ **19.** $4(6y^2 - 13y + 6)$
21. $x^2(9 - 8x)$ **23.** $13a^2(1 - 2a)$ **25.** $7xy(3x - 4y)$
27. $11ab^2(2a - 1)$ **29.** $7x(x^2 + 3x - 4)$
31. $11(11y^4 - x^4)$ **33.** $25x^2(4x^2 - 2x + 1)$
35. $8(a^2 + 2b^2 + 4c^2)$ **37.** $4ab(a - 4b + 8ab)$
39. $11a^2b^2(11a - 2b + 3ab)$ **41.** $12x^2y^3(1 - 6x^3 - 3x^2y)$
43. $(x + 3)(y + 5)$ **45.** $(x + 2)(y + 6)$
47. $(a - 3)(b + 7)$ **49.** $(a - b)(x + y)$
51. $(2x + 5)(a - 1)$ **53.** $(9b - 2)(3y + 1)$
55. $(b - 2)(3x - 4)$ **57.** $(x + 2)(x + a)$
59. $(x - b)(x - a)$ **61.** $(x + y)(a + b + c)$
63. $(3x + 2)(2x + 3)$ **65.** $(10x - 1)(2x + 5)$
67. $(4x + 5)(5x + 1)$ **69.** $(x + 2)(x^2 + 3)$
71. $(3x - 2)(2x^2 + 5)$ **73.** 6 **75.** $3(4x^2 + 2x + 1)$
77. $A = 1{,}000(1 + r)$; \$1,120.00
79. **a.** $A = 1{,}000{,}000\,(1 + r)$ **b.** 1,300,000
81. a **83.** c **85.** $x^2 - 9x + 14$ **87.** $x^2 + x - 6$
89. $x^3 - 8$ **91.** $3x^3 - 4x^2 - 16x - 8$
93. $10x^7 + 8x^6 - 6x^5$ **95.** $x^2 + x + \frac{3}{16}$
97. $32a^2 - 52ab + 15b^2$ **99.** $49b^2 + 14b + 1$
101. $x^2 - 16x + 64$ **103.** $x^3 - 27$

PROBLEM SET 6.2

1. $(x + 3)(x + 4)$ **3.** $(x + 1)(x + 2)$
5. $(a + 3)(a + 7)$ **7.** $(x - 2)(x - 5)$
9. $(y - 3)(y - 7)$ **11.** $(x - 4)(x + 3)$
13. $(y + 4)(y - 3)$ **15.** $(x + 7)(x - 2)$
17. $(r - 9)(r + 1)$ **19.** $(x - 6)(x + 5)$
21. $(a + 7)(a + 8)$ **23.** $(y + 6)(y - 7)$
25. $(x + 6)(x + 7)$ **27.** $(x + 2y)(x + 3y)$
29. $(x - 4y)(x - 5y)$ **31.** $(a + 4b)(a - 2b)$
33. $(a - 5b)^2$ **35.** $(a + 5b)^2$ **37.** $(x - 6a)(x + 8a)$
39. $(x + 4b)(x - 9b)$ **41.** $2(x + 1)(x + 2)$
43. $3(a + 4)(a - 5)$ **45.** $100(x - 2)(x - 3)$
47. $100(p - 5)(p - 8)$ **49.** $x^2(x + 3)(x - 4)$
51. $2r(r + 5)(r - 3)$ **53.** $2y^2(y + 1)(y - 4)$
55. $x^3(x + 2)^2$ **57.** $3y^2(y + 1)(y - 5)$
59. $4x^2(x - 4)(x - 9)$ **61.** $-1(a + 5)(a + 6)$
63. $-1(x + 8)(x - 7)$ **65.** $(x^2 - 3)(x^2 - 2)$
67. $(x - 100)(x + 20)$ **69.** $\left(x - \frac{1}{2}\right)^2$
71. $(x + 0.2)(x + 0.4)$ **73.** $x + 16$ **75.** $4x^2 - x - 3$
77. b **79.** c **81.** $6a^2 + 13a + 2$
83. $6a^2 + 7a + 2$ **85.** $6a^2 + 8a + 2$

PROBLEM SET 6.3

1. $(2x + 1)(x + 3)$ **3.** $(2a - 3)(a + 1)$
5. $(3x + 5)(x - 1)$ **7.** $(3y + 1)(y - 5)$
9. $(2x + 3)(3x + 2)$ **11.** $(2x - 3y)^2$
13. $(4y + 1)(y - 3)$ **15.** $(4x - 5)(5x - 4)$
17. $(10a - b)(2a + 5b)$ **19.** $(4x - 5)(5x + 1)$
21. $(6m - 1)(2m + 3)$ **23.** $(4x + 5)(5x + 3)$
25. $(3a - 4b)(4a - 3b)$ **27.** $(3x - 7y)(x + 2y)$
29. $(2x + 5)(7x - 3)$ **31.** $(3x - 5)(2x - 11)$
33. $(5t - 19)(3t - 2)$ **35.** $2(2x + 3)(x - 1)$
37. $2(4a - 3)(3a - 4)$ **39.** $-1(3x + 2)(x - 5)$
41. $-2(2x + 1)(3x - 4)$ **43.** $x(5x - 4)(2x - 3)$
45. $x^2(3x + 2)(2x - 5)$ **47.** $2a(5a + 2)(a - 1)$
49. $3x(5x + 1)(x - 7)$ **51.** $5y(7y + 2)(y - 2)$
53. $a^2(5a + 1)(3a - 1)$ **55.** $3y(2x - 3)(4x + 5)$
57. $2y(2x - y)(3x - 7y)$ **59.** Both equal 25.
61. $4x^2 - 9$ **63.** $x^4 - 81$
65. $h = -2(t - 4)(8t + 1)$

Time t (seconds)	0	1	2	3	4
Height h (feet)	8	54	68	50	0

67. **a.** $V = x(11 - 2x)(9 - 2x)$ **b.** 11 inches × 9 inches
69. d **71.** b **73.** $x^2 - 16$ **75.** $25x^2 - 36y^2$
77. $x^4 - 81$ **79.** $x^2 - 8x + 16$ **81.** $9x^2 - 6xy + y^2$
83. $25x^2 - 60xy + 36y^2$

85. **a.** -1 **b.** -8 **c.** -27 **d.** -64 **e.** -125
87. **a.** $x^3 + 2x^2 + 4x$ **b.** $-2x^2 - 4x - 8$ **c.** $x^3 - 8$
89. **a.** $x^3 + 3x^2 + 9x$ **b.** $-3x^2 - 9x - 27$ **c.** $x^3 - 27$

PROBLEM SET 6.4

1. $(x - 1)^2$ **3.** $(x + 1)^2$ **5.** $(a - 5)^2$ **7.** $(y + 2)^2$
9. $(x - 2)^2$ **11.** $(m - 6)^2$ **13.** $(2a + 3)^2$ **15.** $(7x - 1)^2$
17. $(3y - 5)^2$ **19.** $(x + 5y)^2$ **21.** $(3a + b)^2$
23. $3(a + 3)^2$ **25.** $2(x + 5y)^2$ **27.** $5x(x + 3y)^2$
29. $(x + 3)(x - 3)$ **31.** $(a + 6)(a - 6)$
33. $(x + 7)(x - 7)$ **35.** $4(a + 2)(a - 2)$
37. Cannot be factored. **39.** $(5x + 13)(5x - 13)$
41. $(3a + 4b)(3a - 4b)$ **43.** $(3 + m)(3 - m)$
45. $(5 + 2x)(5 - 2x)$ **47.** $2(x + 3)(x - 3)$
49. $(x - y)(x^2 + xy + y^2)$ **51.** $(a + 2)(a^2 - 2a + 4)$
53. $(3 + x)(9 - 3x + x^2)$ **55.** $(y - 1)(y^2 + y + 1)$
57. $(4 - y)(16 + 4y + y^2)$ **59.** $(5h - t)(25h^2 + 5ht + t^2)$
61. $(x - 6)(x^2 + 6x + 36)$ **63.** $2(y - 3)(y^2 + 3y + 9)$
65. $(4 + 3a)(16 - 12a + 9a^2)$
67. $(2x - 3y)(4x^2 + 6xy + 9y^2)$
69. $32(a + 2)(a - 2)$ **71.** $2y(2x + 3)(2x - 3)$
73. $2(a - 4b)(a^2 + 4ab + 16b^2)$
75. $2(x + 6y)(x^2 - 6xy + 36y^2)$
77. $10(a - 4b)(a^2 + 4ab + 16b^2)$
79. $10(r - 5)(r^2 + 5r + 25)$ **81.** $\left(t + \frac{1}{3}\right)\left(t^2 - \frac{1}{3}t + \frac{1}{9}\right)$
83. $\left(3x - \frac{1}{3}\right)\left(9x^2 + x + \frac{1}{9}\right)$
85. $(4a + 5b)(16a^2 - 20ab + 25b^2)$
87. $\left(\frac{1}{2}x - \frac{1}{3}y\right)\left(\frac{1}{4}x^2 + \frac{1}{6}xy + \frac{1}{9}y^2\right)$
89. $(a^2 + b^2)(a + b)(a - b)$
91. $(4m^2 + 9)(2m + 3)(2m - 3)$
93. $3xy(x + 5y)(x - 5y)$
95. $(a - b)(a^2 + ab + b^2)(a + b)(a^2 - ab + b^2)$
97. $(2x - y)(4x^2 + 2xy + y^2)(2x + y)(4x^2 - 2xy + y^2)$
99. $(x - 5y)(x^2 + 5xy + 25y^2)(x + 5y)(x^2 - 5xy + 25y^2)$
101. $(x + 3 + y)(x + 3 - y)$ **103.** $(x + y + 3)(x + y - 3)$
105. 14 **107.** 25
109. **a.** $x^2 - 16$ **b.** $(x + 4)(x - 4)$
c.

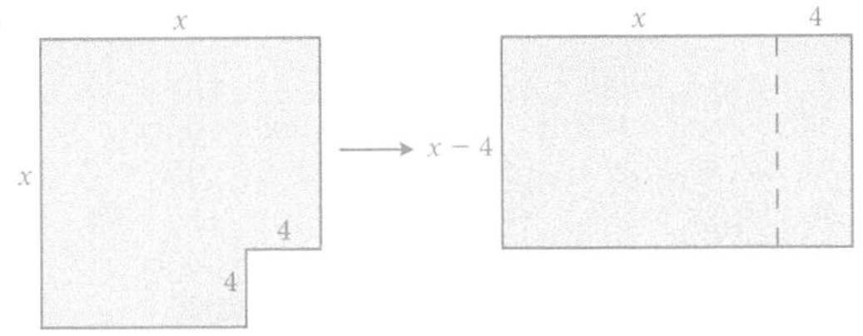

111. $a^2 - b^2 = (a + b)(a - b)$ **113.** c **115.** b
117. $2x^5 - 8x^3$ **119.** $3x^4 - 18x^3 + 27x^2$
121. $y^3 + 25y$ **123.** $15a^2 - a - 2$
125. $4x^4 - 12x^3 - 40x^2$ **127.** $2ab^5 - 8ab^4 + 2ab^3$

PROBLEM SET 6.5

1. $(x + 9)(x - 9)$ **3.** $(x + 5)(x - 3)$ **5.** $(x + 3)^2$
7. $(y - 5)^2$ **9.** $2ab(a^2 + 3a + 1)$
11. Cannot be factored. **13.** $3(2a + 5)(2a - 5)$
15. $(3x - 2y)^2$ **17.** $4x(x^2 + 4y^2)$ **19.** $2y(y + 5)^2$
21. $a^4(a^2 + 4b^2)$ **23.** $(x + 4)(y + 3)$
25. $(x^2 + 4)(x + 2)(x - 2)$ **27.** $(x + 2)(y - 5)$
29. $5(a + b)^2$ **31.** $(4 + x)(16 - 4x + x^2)$
33. $3(x + 2y)(x + 3y)$ **35.** $(2x + 19)(x - 2)$
37. $100(x - 2)(x - 1)$ **39.** $(x + 8)(x - 8)$
41. $(x + a)(x + 3)$ **43.** $a^5(7a + 3)(7a - 3)$
45. Cannot be factored. **47.** $a(5a + 1)(5a + 3)$
49. $(x + y)(a - b)$ **51.** $3a^2b(4a + 1)(4a - 1)$
53. $5x^2(x - 2)(x^2 + 2x + 4)$ **55.** $(3x + 41y)(x - 2y)$
57. $2x^3(2x - 3)(4x - 5)$ **59.** $(2x + 3)(x + a)$
61. $(y^2 + 1)(y + 1)(y - 1)$ **63.** $3x^2y^2(2x + 3y)^2$
65. c **67.** a **69.** 5 **71.** $-\frac{3}{2}$ **73.** $-\frac{3}{4}$

PROBLEM SET 6.6

1. $-2, 1$ **3.** $4, 5$ **5.** $0, -1, 3$ **7.** $-\frac{2}{3}, -\frac{3}{2}$
9. $0, -\frac{4}{3}, \frac{4}{3}$ **11.** $0, -\frac{1}{3}, -\frac{3}{5}$ **13.** $-1, -2$ **15.** $4, 5$
17. $6, -4$ **19.** $2, 3$ **21.** -3 **23.** $4, -4$ **25.** $\frac{3}{2}, -4$
27. $-\frac{2}{3}$ **29.** 5 **31.** $4, -\frac{5}{2}$ **33.** $\frac{5}{3}, -4$ **35.** $\frac{7}{2}, -\frac{7}{2}$
37. $0, -6$ **39.** $0, 3$ **41.** $0, 4$ **43.** $0, 5$ **45.** $2, 5$
47. $\frac{1}{2}, -\frac{4}{3}$ **49.** $4, -\frac{5}{2}$ **51.** $8, -10$ **53.** $5, 8$
55. $6, 8$ **57.** -4 **59.** $5, 8$ **61.** $6, -8$
63. $0, -\frac{3}{2}, -4$ **65.** $0, 3, -\frac{5}{2}$ **67.** $0, \frac{1}{2}, -\frac{5}{2}$
69. $0, \frac{3}{5}, -\frac{3}{2}$ **71.** $\frac{1}{2}, \frac{3}{2}$ **73.** $-5, 4$ **75.** $-7, -6$
77. $-3, -1$ **79.** $2, 3$ **81.** $-15, 10$ **83.** $-5, 3$
85. $-3, -2, 2$ **87.** $-4, -1, 4$
89. **a.** $-4, \frac{1}{2}$ **b.** $(x + 4)(2x - 1)$ **c.** $-3, -\frac{1}{2}$ **d.** $\frac{4}{9}$
91. b **93.** $x(x + 1) = 72$ **95.** $x(x + 2) = 99$
97. $x(x + 2) = 5[x + (x + 2)] - 10$
99. Bicycle \$75, suit \$15 **101.** House \$2,400, lot \$600

PROBLEM SET 6.7

1. 8, 10 and $-10, -8$ **3.** 9, 11 and $-11, -9$
5. 8, 10 and 0, 2 **9.** 2, 12 and $-\frac{12}{5}, -10$
11. 5, 20 and 0, 0 **13.** Width 3 inches, length 4 inches
15. Base 3 inches **17.** 6 inches and 8 inches
19. 12 meters **21.** 2 hundred items or 5 hundred items
23. \$7 or \$10 **25.** **a.** 5 feet **b.** 12 feet
27. **a.** 25 seconds later
b.

t	h
0	100
5	1680
10	2460
15	2440
20	1620
25	0

29. a **31.** d **33.** $(-2, 2)$, $(0, 3)$, $(2, 4)$
35.

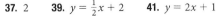

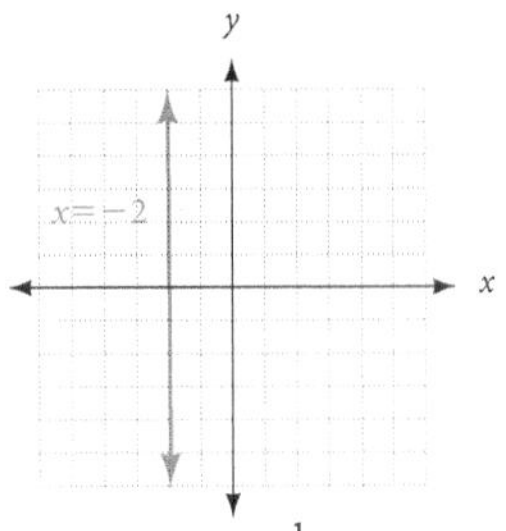

37. 2 **39.** $y = \frac{1}{2}x + 2$ **41.** $y = 2x + 1$

CHAPTER 6 TEST

1. $6(x+3)$ **2.** $4ab(3a-6+2b)$
3. $(x-2b)(x+3a)$ **4.** $(5y-4)(3-x)$
5. $(x+4)(x-3)$ **6.** $(x-7)(x+3)$
7. $(x+5)(x-5)$ **8.** $(x^2+4)(x+2)(x-2)$
9. Cannot be factored. **10.** $2(3x+4y)(3x-4y)$
11. $(x^2-3)(x+4)$ **12.** $(x-3)(x+b)$
13. $2(2x-5)(x+1)$ **14.** $(4n-3)(n+4)$
15. $(3c-2)(4c+3)$ **16.** $3x(2x-1)(2x+3)$
17. $(x+5y)(x^2-5xy+25y^2)$
18. $2(3b-4)(9b^2+12b+16)$ **19.** $5, -3$ **20.** $3, 4$
21. $5, -5$ **22.** $-2, 7$ **23.** $5, -6$ **24.** $0, 3, -3$
25. $\frac{3}{2}, -4$ **26.** $0, -\frac{5}{3}, 6$ **27.** $6, 12$
28. 4, 6 or −4, −2 **29.** 4 ft, 13 ft **30.** 6 ft, 8 ft
31. 200 items, 300 items **32.** \$3, \$5

Chapter 7

PROBLEM SET 7.1

1. $-\frac{1}{6}$, 0, undefined **3.** $-\frac{5}{4}, -\frac{1}{2}$, undefined **5.** 0
7. $-2, 3$ **9.** $-\frac{1}{3}, 1$ **11.** Defined for all real numbers
13. $\frac{1}{x-2}, x \neq 2$ **15.** $\frac{1}{a+3}, a \neq -3, 3$
17. $\frac{1}{x-5}, x \neq -5, 5$ **19.** $\frac{(x+2)(x-2)}{2}$
21. $\frac{2(x-5)}{3(x-2)}, x \neq 2$ **23.** $2, a \neq -2$ **25.** $\frac{5(x-1)}{4}$
27. $\frac{1}{x-3}$ **29.** $\frac{1}{x+3}$ **31.** $\frac{1}{a-5}$ **33.** $\frac{1}{3x+2}$
35. $\frac{x+5}{x+2}$ **37.** $\frac{2m(m+2)}{m-2}$ **39.** $\frac{x-1}{x-4}$
41. $\frac{2(2x-3)}{2x+3}$ **43.** $\frac{2x-3}{2x+3}$ **45.** $\frac{1}{(x^2+9)(x-3)}$
47. $\frac{3x-5}{(x^2+4)(x-2)}$ **49.** $\frac{2x(7x+6)}{2x+1}$ **51.** $\frac{x^2+xy+y^2}{x+y}$
53. $\frac{x^2-2x+4}{x-2}$ **55.** $\frac{x^2-2x+4}{x-1}$ **57.** $\frac{x+2}{x+5}$
59. $\frac{x+a}{x+b}$ **61. a.** x^2-16 **b.** $x^2-8x+16$
c. $4x^3-32x^2+64x$ **d.** $\frac{x}{4}$ **63.** $\frac{4}{3}$ **65.** $\frac{4}{5}$ **67.** $\frac{8}{1}$
69.

Checks Written x	Total Cost $2.00 + 0.15x$	Cost per Check $\frac{2.00 + 0.15x}{x}$
0	2.00	undefined
5	2.75	0.55
10	3.50	0.35
15	4.25	0.28
20	5	0.25

71. 40.7 miles/hour
73. Sale, 1.31; Kershaw, 1.29; Scherzer, 1.21; Archer, 1.19
75. 0.125 miles/minute **77.** d **79.** b
81. $\frac{5}{14}$ **83.** $\frac{9}{10}$ **85.** $(x+3)(x-3)$ **87.** $3(x-3)$
89. $(x-5)(x+4)$ **91.** $a(a+5)$ **93.** $(a-5)(a+4)$
95. 11.8

PROBLEM SET 7.2

1. 2 **3.** $\frac{x}{2}$ **5.** $\frac{3}{2}$ **7.** $\frac{1}{2(x-3)}$ **9.** $\frac{4a(a+5)}{7(a+4)}$
11. $\frac{y-2}{4}$ **13.** $\frac{2(x+4)}{x-2}$ **15.** $-\frac{x+3}{(x-3)(x+1)}$
17. 1 **19.** $\frac{y-5}{(y+2)(y-2)}$ **21.** $-\frac{x+5}{x-5}$ **23.** 1
25. $\frac{a+3}{a+4}$ **27.** $-\frac{2y-3}{y-6}$ **29.** $\frac{(x-1)(x+5)}{(x+1)(x-2)}$
31. $\frac{3}{2}$ **33. a.** $\frac{4}{13}$ **b.** $\frac{x+1}{x^2+x+1}$ **c.** $\frac{x-2}{x^2+x+1}$
d. $\frac{x+1}{x-1}$ **35.** $2(x-3)$ **37.** $(x+2)(x+1)$
39. $-2x(x-5)$ **41.** $\frac{2(x+5)}{x(y+5)}$ **43.** $\frac{2x}{x-y}$
45. $\frac{1}{x-2}$ **47.** $\frac{1}{5}$ **49.** $\frac{1}{100}$ **51.** 2.7 miles
53. 742 miles per hour **55.** 0.45 miles per hour
57. 8.8 miles per hour **59.** b **61.** b **63.** $\frac{6}{7}$
65. $\frac{11}{35}$ **67.** $-\frac{23}{105}$ **69.** x^2+2x **71.** x^2-x-12
73. $\frac{1}{x-7}$ **75.** $-\frac{x-5}{2(x-4)}$ **77.** x^2-x-20

PROBLEM SET 7.3

1. $\frac{7}{x}$ **3.** $\frac{4}{a}$ **5.** 1 **7.** $y+1$ **9.** $x+2$
11. $x-2$ **13.** $\frac{6}{x+6}$ **15.** $5x$ **17.** $x(x-3)$
19. $2y^2(y+4)$ **21.** $(a-2)(a-4)(a+3)$
23. $(x+2)(x+3)^2$ **25.** $\frac{(y+2)(y-2)}{2y}$ **27.** $\frac{2a+3}{6}$
29. $\frac{7x+3}{4(x+1)}$ **31.** $\frac{1}{5}$ **33.** $\frac{1}{3}$ **35.** $\frac{3}{x-2}$
37. $\frac{4}{a+3}$ **39.** $\frac{2(x-10)}{(x+5)(x-5)}$ **41.** $\frac{x+2}{x+3}$ **43.** $\frac{a+1}{a+2}$
45. $\frac{1}{(x+3)(x+4)}$ **47.** $\frac{y}{(y+5)(y+4)}$ **49.** $\frac{3(x-1)}{(x+4)(x+1)}$
51. $-\frac{4(2x-1)}{(x-2)(x+2)^2}$ **53.** $\frac{1}{3}$ **55. a.** $\frac{2}{27}$ **b.** $\frac{8}{3}$
c. $\frac{11}{18}$ **d.** $\frac{3x+10}{(x-2)^2}$ **e.** $\frac{(x+2)^2}{3x+10}$ **f.** $\frac{x+3}{x+2}$
57.

Number x	Reciprocal $\frac{1}{x}$	Sum $1 + \frac{1}{x}$	Quotient $\frac{x+1}{x}$
1	1	2	2
2	$\frac{1}{2}$	$\frac{3}{2}$	$\frac{3}{2}$
3	$\frac{1}{3}$	$\frac{4}{3}$	$\frac{4}{3}$
4	$\frac{1}{4}$	$\frac{5}{4}$	$\frac{5}{4}$

59. $\frac{x+3}{x+2}$ **61.** $\frac{x+2}{x+3}$ **63.** $\frac{x-1}{x+2}$ **65.** $\frac{x+4}{x+1}$
67. $x + \frac{2}{x} = \frac{x^2+2}{x}$ **69.** $\frac{1}{x} + \frac{1}{2x} = \frac{3}{2x}$ **71.** c
73. c **75.** $\frac{3}{4}$ **77.** $\frac{3}{2}$ **79.** $2x^3y^3$ **81.** $\frac{x^2y^3}{2}$
83. $x(xy+1)$ **85.** $\frac{x^2y^2}{2}$ **87.** $\frac{x-2}{x-3}$

PROBLEM SET 7.4

1. 6 **3.** $\frac{1}{6}$ **5.** xy^2 **7.** $\frac{xy}{2}$ **9.** $\frac{y}{x}$ **11.** $\frac{a+1}{a-1}$
13. $\frac{x+1}{2(x-3)}$ **15.** $a-3$ **17.** $\frac{y+3}{y+2}$ **19.** $x+y$
21. $\frac{a+1}{a}$ **23.** $\frac{1}{x}$ **25.** $\frac{2a+3}{3a+4}$ **27.** $\frac{7}{3}, \frac{17}{7}, \frac{41}{17}$
29.

Number x	Reciprocal $\frac{1}{x}$	Quotient $\frac{x}{\frac{1}{x}}$	Square x^2
1	1	1	1
2	$\frac{1}{2}$	4	4
3	$\frac{1}{3}$	9	9
4	$\frac{1}{4}$	16	16

31.

Number x	Reciprocal $\frac{1}{x}$	Sum $1+\frac{1}{x}$	Quotient $\frac{1+\frac{1}{x}}{\frac{1}{x}}$
1	1	2	2
2	$\frac{1}{2}$	$\frac{3}{2}$	3
3	$\frac{1}{3}$	$\frac{4}{3}$	4
4	$\frac{1}{4}$	$\frac{5}{4}$	5

33. d **35.** 3 **37.** 0 **39.** Undefined **41.** $-\frac{3}{2}$
43. $2x + 15$ **45.** $x^2 - 5x$ **47.** -6 **49.** -2

PROBLEM SET 7.5

1. -3 **3.** 20 **5.** -1 **7.** 5 **9.** -2 **11.** 4
13. 3, 5 **15.** $-8, 1$ **17.** 2 **19.** 1 **21.** 8 **23.** 5
25. $\varnothing$; 2 does not check **27.** 3
29. $\varnothing$; -3 does not check **31.** 0
33. 2; 3 does not check **35.** -4 **37.** -1
39. $-6, -7$ **41.** -1; -3 does not check
43. **a.** $\frac{1}{5}$ **b.** 5 **c.** $\frac{25}{3}$ **d.** 3 **e.** $\frac{1}{5}, 5$
45. **a.** $\frac{7}{(a-6)(a+2)}$ **b.** $\frac{a-5}{a-6}$ **c.** 7; -1 does not check
47. a **49.** $\frac{7}{2}$ **51.** $-3, 2$

PROBLEM SET 7.6

1. 1 **3.** 10 **5.** 40 **7.** $\frac{5}{4}$ **9.** $\frac{7}{3}$ **11.** 3, -4
13. 4, -4 **15.** 2, -4 **17.** 6, -1 **19.** 16
21. 15 hits **23.** 21 milliliters **25.** 45.5 grams
27. 14.7 inches **29.** 343 miles **31.** d **33.** $\frac{1}{3}$
35. 30 **37.** 1 **39.** -3

PROBLEM SET 7.7

1. $\frac{1}{4}, \frac{3}{4}$ **3.** $\frac{2}{3}$ and $\frac{3}{2}$ **5.** -2 **7.** 4, 6
9. 16 miles per hour **11.** 300 miles per hour
13. 170 miles per hour; 190 miles per hour
15. 9 miles per hour **17.** 8 miles per hour
19. 36 minutes **21.** 90 minutes **23.** 60 hours
25. $5\frac{5}{11}$ minutes **27.** 12 minutes **29.** c **31.** d
33. $y = 15$ **35.** $y = 4$ **37.** $x = \pm 6$ **39.** $K = 18$
41. $K = 5$

PROBLEM SET 7.8

1. Direct, $K = 10$ **3.** Inverse, $K = 40$
5. Direct, $K = \frac{4}{3}\pi$ **7.** Inverse, $K = \frac{1}{9}$ **9.** $C = Kr^2$
11. $P = \frac{K}{V}$ **13.** $R = K\sqrt{n}$ **15.** 8 **17.** -4
19. 2 **21.** 1 **23.** 3 **25.** $\frac{5}{9}$ **27.** 60
29. 84 pounds **31.** $\frac{735}{2}$ or 367.5 **33.** \$277.50
35. 96 lbs **37.** 12 amperes **39.** c **41.** a
43. $\frac{x+2}{x+3}$ **45.** $\frac{2(x+5)}{x-4}$ **47.** $\frac{1}{x-4}$
49. $\frac{x+5}{x-3}$ **51.** -2 **53.** 24 hours **55.** 24

CHAPTER 7 TEST

1. $\frac{1}{15}$ **2.** 0 **3.** 5 **4.** $-4, 3$ **5.** $\frac{x+3}{x-3}$
6. $\frac{3}{a-4}$ **7.** $\frac{3}{2}$ **8.** $(x-3)(x+4)$ **9.** $\frac{2x-3}{x+4}$
10. $(x+4)(x-1)$ **11.** $\frac{-2}{x-8}$ **12.** $\frac{2(x+2)}{(x+4)(x-4)}$
13. $\frac{2}{(x+3)(x-1)}$ **14.** $\frac{x+2}{x-2}$ **15.** $\frac{x+3}{x+2}$ **16.** $\frac{6}{5}$
17. $-\frac{7}{6}$ **18.** -14 **19.** $\frac{1}{3}, \frac{1}{4}$ **20.** 102 **21.** 15 mph
22. 24 hours **23.** 54 **24.** 2

Chapter 8

PROBLEM SET 8.1

1. 2 **3.** -2 **5.** Not a real number **7.** -12
9. 25 **11.** Not a real number **13.** -8 **15.** -10
17. $\frac{1}{2}$ **19.** $-\frac{1}{10}$ **21.** $\frac{3}{7}$ **23.** $-\frac{2}{11}$ **25.** -35
27. 1 **29.** -2 **31.** -5 **33.** -1 **35.** -3
37. $\frac{1}{4}$ **39.** -2 **41.** 3 **43.** 2 **45.** y **47.** $5x$
49. ab **51.** $a + b$ **53.** $9xy$ **55.** x **57.** $2x$
59. x^2 **61.** $6a^3$ **63.** $5a^4b^2$ **65.** 1.414
67. -4.123 **69.** **a.** 4.243 **b.** 4.243
71. **a.** 7.071 **b.** 7.071 **73.** **a.** $2(4 + 3x)$ **b.** $2(4 + 3\sqrt{3})$
75. 7 **77.** 5 **79.** 17 **81.** 13
83. **a.** 1.618 **b.** -0.618 **c.** 1 **85.** **a.** 3 **b.** 30 **c.** 0.3
87. 6, -1 **89.** **a.** $\frac{\sqrt{3}}{5}$ **b.** $\frac{\sqrt{2}}{3}$ **c.** $\frac{5+2\sqrt{3}}{4}$
d. $1 + 2\sqrt{6}$ **91.** 5 **93.** $\sqrt{125} \approx 11.2$ **95.** 30 feet
97.

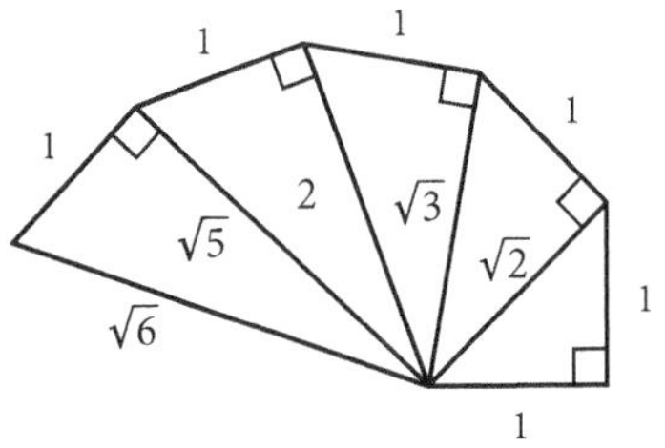

99. $\sqrt{2}, \sqrt{3}, \sqrt{4} = 2$ **101.** b **103.** a **105.** d
107. 12 **109.** $4 \cdot 3$ **111.** $4 \cdot 5$ **113.** $9 \cdot 2$
115. $x^2 \cdot x$ **117.** $4x^2 \cdot 5x$ **119.** $25x^2y^2 \cdot 3y$

PROBLEM SET 8.2

1. $2\sqrt{2}$ **3.** $2\sqrt{3}$ **5.** $2\sqrt[3]{3}$ **7.** $5x\sqrt{2}$ **9.** $3ab\sqrt{5}$
11. $3x\sqrt[3]{2}$ **13.** $4x^2\sqrt{2}$ **15.** $20\sqrt{5}$ **17.** $x\sqrt{7x}$
19. $x^2\sqrt[3]{2x}$ **21.** $6a^2\sqrt[3]{a^2}$ **23.** $4a\sqrt{5a}$ **25.** $15y\sqrt{2x}$
27. $14x\sqrt{3y}$ **29.** $\frac{4}{5}$ **31.** $\frac{2}{3}$ **33.** $\frac{2}{3}$ **35.** $\frac{2}{3}$
37. $2x$ **39.** $3ab$ **41.** $\frac{3x}{2y}$ **43.** $\frac{5\sqrt[3]{2}}{3}$ **45.** $\sqrt{3}$
47. $\frac{8\sqrt{2}}{7}$ **49.** $\frac{12\sqrt{2x}}{5}$ **51.** $\frac{3a\sqrt{6}}{5}$ **53.** $\frac{15\sqrt{2}}{2}$
55. $\frac{14y\sqrt{7}}{3}$ **57.** $\frac{4xy\sqrt{5y}}{3}$ **59.** $\frac{18ab\sqrt{6b}}{5}$
61. $5ab\sqrt{2}$ **63.** $\frac{\sqrt{5}}{2}$ **65.** $2 + \sqrt{3}$ **67.** $\frac{\sqrt{3}}{2}$
69. $3 + \sqrt{3}$ **71.** $\frac{4+\sqrt{7}}{3}$ **73.** $2 + \sqrt{3}$ **75.** $\frac{-2-3\sqrt{3}}{6}$
77. $-2 - \sqrt{2}$ **79.** **a.** $2\sqrt{10}$ **b.** 5 **c.** $3\sqrt{5}$ **d.** $2\sqrt{3}$
81. **a.** $4x^5y^2\sqrt{2y}$ **b.** $2x^3y\sqrt[3]{4xy^2}$ **c.** $2x^2y\sqrt[4]{2x^2y}$ **d.** $2x^2y$
83. **a.** 2 **b.** 0.2 **c.** 20 **d.** 0.02
85.

x	$\sqrt{x}$	$2\sqrt{x}$	$\sqrt{4x}$
1	1	2	2
2	1.414	2.828	2.828
3	1.732	3.464	3.464
4	2	4	4

87.

x	$\sqrt{x}$	$3\sqrt{x}$	$\sqrt{9x}$
1	1	3	3
2	1.414	4.243	4.243
3	1.732	5.196	5.196
4	2	6	6

89. $\frac{5}{4}$ seconds **91.** 1.618 **93.** d **95.** b
97. $2xy\sqrt{x}$ **99.** $2\sqrt{5}$ **101.** $10\sqrt{3}$ **103.** $3\sqrt{6}$
105. $2x$ **107.** $-6x + 3y$

PROBLEM SET 8.3

1. $7\sqrt{2}$ **3.** $2\sqrt{5}$ **5.** $7\sqrt{3}$ **7.** $\frac{11}{56}\sqrt{5}$
9. $13\sqrt[3]{13}$ **11.** $6\sqrt[3]{10}$ **13.** $6\sqrt{5}$ **15.** $4\sqrt{2}$
17. 0 **19.** $-30\sqrt{3}$ **21.** $-4\sqrt[3]{2}$ **23.** 0
25. $\frac{3}{2}\sqrt{2} + \frac{3}{2}\sqrt{3}$ **27.** 0 **29.** $\frac{23}{4}\sqrt{2}$ **31.** $-3\sqrt{7}$
33. $4\sqrt[3]{2}$ **35.** $12\sqrt{6} + 11\sqrt{3}$ **37.** $-6\sqrt{2} - 6\sqrt{3}$
39. $2x\sqrt{x}$ **41.** $4a\sqrt{3}$ **43.** $15x\sqrt{2x}$ **45.** $13x\sqrt{3xy}$
47. $-b\sqrt{5a}$ **49.** $27x\sqrt{2x} - 8x\sqrt{3x}$ **51.** $4x\sqrt[3]{3}$
53. $14ab\sqrt[3]{a^2}$ **55.** $23x\sqrt{2y}$ **57.** $\frac{4-\sqrt{6}}{3}$ **58.** $\frac{2+\sqrt{3}}{2}$
59. $3 + \sqrt{2}$ **61.** $\frac{-2+\sqrt{2}}{2}$ **63. a.** $7x$ **b.** $7y$ **c.** $7\sqrt{5}$
65. a. $7x$ **b.** $7t$ **c.** $7\sqrt{x}$
67.

x	$\sqrt{x^2+9}$	$x+3$
1	3.162	4
2	3.606	5
3	4.243	6
4	5	7
5	5.831	8
6	6.708	9

69.

x	$\sqrt{x+3}$	$\sqrt{x}+\sqrt{9}$
1	2	4
2	2.236	4.414
3	2.449	4.732
4	2.646	5
5	2.828	5.236
6	3	5.449

71. $9\sqrt{3}$ **73.** c **75.** $2\sqrt{3}$ **77.** $2\sqrt{5}$ **79.** $2\sqrt[3]{2}$
81. $2x^2 + 5x$ **83.** $x^2 + 2x + 1$ **85.** $x^2 + 6x - 16$
87. $x^2 - 36$ **89.** $6x^2 + x - 12$ **91.** $19 + 9\sqrt{5}$
93. $x - 4\sqrt{x} - 21$

PROBLEM SET 8.4

1. 2 **3.** 7 **5.** x **7.** 4 **9.** -10 **11.** a
13. $\sqrt{6}$ **15.** $2\sqrt{3}$ **17.** $10\sqrt{21}$ **19.** $24\sqrt{2}$
21. $\sqrt[3]{12}$ **23.** $3\sqrt[3]{2}$ **25.** $36\sqrt[3]{3}$ **27.** $\sqrt{6} - \sqrt{2}$
29. $\sqrt{6} + 2$ **31.** $2\sqrt{6} + 3$ **33.** $6 - \sqrt{15}$
35. $2\sqrt{6} + 2\sqrt{15}$ **37.** $10 + 15\sqrt[3]{20}$ **39.** $3 + 2\sqrt{2}$
41. $x + 6\sqrt{x} + 9$ **43.** $27 - 10\sqrt{2}$ **45.** $a - \sqrt{a} + \frac{1}{4}$
47. $10 + 2\sqrt{21}$ **49.** $12 - 4\sqrt{5}$ **51.** $11 + 5\sqrt{5}$
53. $-28 + \sqrt{2}$ **55.** $30 + \sqrt{7}$ **57.** $6x + 16\sqrt{x} + 8$
59. $\sqrt{6} + \frac{1}{3}\sqrt{3} + \frac{1}{2}\sqrt{2} + \frac{1}{6}$ **61.** $a + \sqrt{a} + \frac{2}{9}$
63. 1 **65.** 23 **67.** 4 **69.** $x - 36$ **71.** $49a - 4b$
73. a. $2\sqrt{3}$ **b.** 4 **c.** $10 - 2\sqrt{21}$ **d.** 10
75. a. $2\sqrt{x}$ **b.** $x - 4$ **c.** $x + 4\sqrt{x} + 4$ **d.** $x + 4$
77. $6\sqrt{5}$ **79.** $52 + 14\sqrt{3}$ **81.** b **83.** c
85. $4\sqrt{3}$ **87.** $5a^2$ **89.** $3xy\sqrt{3x}$ **91.** 2
93. $\sqrt{21}$ **95.** 3 **97.** $6ab$ **99.** -1 **101.** 2

PROBLEM SET 8.5

1. $\sqrt{7}$ **3.** $\sqrt{5}$ **5.** $2\sqrt{5}$ **7.** $2\sqrt{3}$ **9.** $\frac{\sqrt{7}}{2}$
11. $\sqrt[3]{12}$ **13.** $\frac{\sqrt[3]{3}}{3}$ **15.** 2 **17.** $\sqrt{10}$ **19.** $\frac{\sqrt{2}}{2}$
21. $\frac{\sqrt{3}}{3}$ **23.** $\frac{\sqrt{10}}{5}$ **25.** $\frac{\sqrt{6}}{2}$ **27.** $\frac{\sqrt{30}}{2}$
29. $\frac{\sqrt[3]{75}}{5}$ **31.** $\frac{\sqrt[3]{30}}{2}$ 33. $\frac{\sqrt[3]{28}}{2}$ **35.** $xy\sqrt{2}$ **37.** $\frac{x\sqrt{15y}}{3}$
39. $\frac{4a^2\sqrt{5}}{5}$ **41.** $\frac{6a^2\sqrt{10ab}}{5b}$ **43.** $\frac{2xy\sqrt{15yz}}{3z}$
45. $xy\sqrt[3]{4y}$ **47.** $\frac{y\sqrt[3]{6x^2}}{2x}$ **49.** $6 + 2\sqrt{5}$
51. $\frac{4\sqrt{3}-\sqrt{6}}{14}$ **53.** $\frac{x+4\sqrt{x}+4}{x-4}$ **55.** $\frac{\sqrt{15}+\sqrt{6}}{3}$
57. $\frac{5-\sqrt{10}}{3}$ **59.** $5 + 2\sqrt{6}$ **61.** $\frac{5-\sqrt{21}}{2}$
63. $\frac{5-\sqrt{15}-\sqrt{10}+\sqrt{6}}{2}$

65.

x	$\sqrt{x}$	$\frac{1}{\sqrt{x}}$	$\frac{\sqrt{x}}{x}$
1	1	1	1
2	1.414	.707	.707
3	1.732	.577	.577
4	2	.5	.5
5	2.236	.447	.447
6	2.449	.408	.408

67.

x	$\sqrt{x^2}$	$\sqrt{x^3}$	$x\sqrt{x}$
1	1	1	1
2	2	2.828	2.828
3	3	5.196	5.196
4	4	8	8
5	5	11.180	11.180
6	6	14.697	14.697

69. a. 10 **b.** 23 **c.** $27 + 10\sqrt{2}$ **d.** $\frac{27+10\sqrt{2}}{23}$
71. 6 miles **73.** a **75.** a **77.** 49 **79.** 81
81. $x + 1$ **83.** $2x - 3$ **85.** $x^2 + 6x + 9$ **87.** 2
89. $-6, 1$ **91. a.** no **b.** yes
93. undefined, undefined, 0, 3, 6, 9, 12

PROBLEM SET 8.6

1. 3 **3.** 44 **5.** $\varnothing$ **7.** $\varnothing$ **9.** 8 **11.** 4
13. $\varnothing$ **15.** $-\frac{2}{3}$ **17.** 25 **19.** 4 **21.** 7 **23.** 8
25. $\varnothing$ **27.** Possible solutions 1 and 6; only 6 checks

29. $-1, -2$ **31.** -4
33. Possible solutions 5 and 8; only 8 checks
35. 4 **37.** $\frac{13}{4}$ **39.** Possible solutions 2, 9; only 2 checks
41. -12 **43.** 71 **45.** -61 **47.** **a.** 100 **b.** 40 **c.** $\varnothing$
d. Possible solutions 5, 8; only 8 checks
49. **a.** 3 **b.** 9 **c.** 3 **d.** $\varnothing$ **e.** 4 **f.** $\varnothing$
g. Possible solutions 1, 4; only 4 checks
51.

Length L (feet)	1	2	3	4	5	6
Time T (seconds)	1.11	1.57	1.92	2.22	2.48	2.72

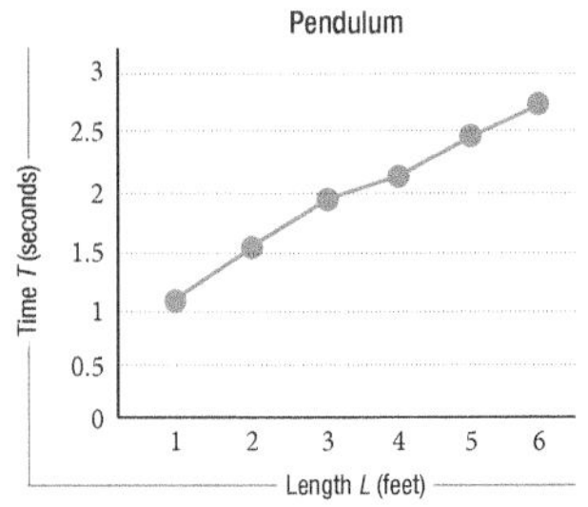

53. $x + 2 = \sqrt{8x}$; $x = 2$
55. $x - 3 = 2\sqrt{x}$; possible solutions 1 and 9; only 9 checks
57. $\frac{392}{121} \approx 3.2$ feet **59.** a **61.** d
63. $x + 2 + \frac{x+2}{x+3}$ **65.** $\frac{3(x+3)}{(x-3)(x-1)}$ **67.** $\frac{-3}{(x+1)(x+4)}$
69. $x = -5, 2$ **71.** 20 mph **73.** 3 liters

CHAPTER 8 TEST

1. 6 **2.** -9 **3.** Not a real number **4.** 4 **5.** -3
6. -5 **7.** $6\sqrt{2}$ **8.** $4\sqrt{6}$ **9.** $8y\sqrt{3}$
10. $xy^2\sqrt{6xy}$ **11.** $\frac{8\sqrt{2}}{5}$ **12.** $2x^2y^2\sqrt[3]{5xy^2}$ **13.** $\frac{4+\sqrt{5}}{2}$
14. $\frac{-1+\sqrt{6}}{3}$ **15.** $3\sqrt{5}$ **16.** $8m\sqrt{3}$ **17.** $-3ab\sqrt[3]{3}$
18. $\sqrt{6} - 3\sqrt{2}$ **19.** $-7 - 3\sqrt{3}$ **20.** $x - 81$
21. $9 - 2\sqrt{14}$ **22.** $\sqrt{6}$ **23.** $4\sqrt[3]{2}$ **24.** $\frac{\sqrt{14}}{7}$
25. $\frac{\sqrt[3]{12x}}{2x}$ **26.** $\frac{5\sqrt{3}+5}{2}$ **27.** $-6 + \sqrt{35}$
28. $\frac{x + 3\sqrt{x}}{x - 9}$ **29.** $x = 13$ **30.** $x = -4, -2$
31. $x = -3$ **32.** $x = -4$ **33.** The number is 9. **34.** $\sqrt{11}$

Chapter 9

PROBLEM SET 9.1

1. ± 5 **3.** $\pm\frac{3}{2}$ **5.** $1, -\frac{5}{2}$ **7.** $-2, 3$ **9.** ± 5
11. $\pm\frac{3}{2}$ **13.** $\pm 2\sqrt{2}$ **15.** $\pm 2\sqrt{3}$ **17.** $0, -4$
19. $3, -5$ **21.** $5 \pm 5\sqrt{3}$ **23.** $-1 \pm 2\sqrt{3}$ **25.** $2, -3$
27. $\frac{11}{4}, -\frac{1}{4}$ **29.** $\frac{1 \pm 2\sqrt{5}}{3}$ **31.** $-2 \pm \sqrt{3}$
33. $3 \pm \sqrt{3}$ **35.** $\frac{1 \pm \sqrt{7}}{2}$ **37.** $\frac{-4 \pm 2\sqrt{3}}{5}$
39. $-5 \pm \sqrt{7}$ **41.** $4, -2$ **43.** $-6 \pm 2\sqrt{2}$
45. **a.** $\frac{1}{2}$ **b.** $\frac{5}{2}$ **c.** $-\frac{1}{2}, \frac{3}{2}$ **d.** 3
47. $(-1 + 5\sqrt{2} + 1)^2 = (5\sqrt{2})^2 = 50$ **49.** $-7, 1$
51. 4,070 feet **53.** 3 feet **55.** c **57.** 81
59. 1 **61.** $\frac{9}{4}$ **63.** $x^2 + 8$ **65.** $(x + 3)^2$
67. $\left(y - \frac{3}{2}\right)^2$

PROBLEM SET 9.2

1. 36, 6 **3.** 36, 6 **5.** 4, 2 **7.** 4, 2 **9.** 25, 5
11. $\frac{25}{4}, \frac{5}{2}$ **13.** $\frac{1}{4}, \frac{1}{2}$ **15.** $\frac{1}{16}, \frac{1}{4}$ **17.** $\frac{1}{9}, \frac{1}{3}$
19. $-6, 2$ **21.** $-3, -9$ **23.** $1 \pm \sqrt{6}$
25. $4 \pm \sqrt{15}$ **27.** $\frac{5 \pm \sqrt{37}}{2}$ **29.** $1 \pm \sqrt{5}$
31. 0, 10 **33.** $-5, 3$ **35.** $-5, -1$ **37.** 1, 2
39. $-1 \pm \sqrt{2}$ **41.** $-2, 1$ **43.** **a.** 0, 6 **b.** 0, 6
45. **a.** $-7, 5$ **b.** $-7, 5$
47. Width = 11.2 in., height = 8.4 in.
49. $-0.4, 2.4$ **51.** **a.** -4 **b.** -3 **53.** 9 **55.** 1
57. c **59.** $2x^2 + 4x - 3 = 0$ **61.** $x^2 + x - 11 = 0$
63. x^2: 1, x: -5, constant: -6
65. x^2: 2, x: 4, constant: -3 **67.** 49 **69.** 40 **71.** 6
73. $\frac{-2 - \sqrt{10}}{2}$

PROBLEM SET 9.3

1. $-3, 1$ **3.** $-\frac{2}{3}, \frac{1}{2}$ **5.** $\frac{-3 \pm \sqrt{17}}{2}$ **7.** $\frac{3 \pm \sqrt{7}}{2}$
9. $2 \pm \sqrt{6}$ **11.** $\frac{4 \pm \sqrt{10}}{3}$ **13.** $-3, -2$ **15.** $2 \pm \sqrt{3}$
17. $-3 \pm \sqrt{2}$ **19.** $\frac{-2 \pm \sqrt{3}}{2}$ **21.** $\frac{5}{2}, -1$
23. $\frac{-3 \pm \sqrt{23}}{2}$ **25.** $-1, -2$ **27.** -3 **29.** 1
31. $\frac{5 \pm \sqrt{53}}{2}$ **33.** $\frac{2}{3}, -\frac{1}{2}$ **35.** $\frac{1 \pm \sqrt{21}}{2}$
37. $\frac{-2 \pm \sqrt{10}}{3}$ **39.** $\frac{1 \pm \sqrt{11}}{2}$ **41.** $0, \frac{5}{3}$
43. $0, \frac{-3 \pm \sqrt{41}}{4}$ **45.** a and b **47.** $\frac{3 \pm \sqrt{21}}{6}$
49. 1 second and 3 seconds **51.** c **53.** $15x - 1$
55. $-12x - 1$ **57.** $16x^2 - 14x$ **59.** $9x^2 + 30x + 24$
61. $\frac{12 - 4\sqrt{2}}{7}$

PROBLEM SET 9.4

1. $3 + i$ **3.** $6 - 8i$ **5.** $15 + 10i$ **7.** $9 + i$
9. $-1 - i$ **11.** $11 - 2i$ **13.** $4 - 2i$ **15.** $8 - 7i$
17. $-1 + 4i$ **19.** $6 - 3i$ **21.** $14 + 16i$
23. $9 + 2i$ **25.** $11 - 7i$ **27.** 34 **29.** 5
31. $\frac{6 + 4i}{13} = \frac{6}{13} + \frac{4}{13}i$ **33.** $\frac{-9 - 6i}{13} = -\frac{9}{13} - \frac{6}{13}i$
35. $\frac{-1 + 3i}{10} = -\frac{1}{10} + \frac{3}{10}i$ **37.** $\frac{3 + 4i}{5} = \frac{3}{5} + \frac{4}{5}i$
39. $\frac{-6 + 13i}{15} = -\frac{6}{15} + \frac{13}{15}i$ **41.** $x^2 + 9$
43. $\frac{1}{i} \cdot \frac{i}{i} = \frac{i}{i^2} = \frac{i}{-1} = -i$ **45.** d **47.** d **49.** 6
51. $-5\sqrt{3}$ **53.** $-5, 1$ **55.** $1 \pm \sqrt{6}$ **57.** $\frac{2 \pm \sqrt{5}}{2}$

PROBLEM SET 9.5

1. $4i$ **3.** $-7i$ **5.** $i\sqrt{6}$ **7.** $-i\sqrt{11}$ **9.** $4i\sqrt{2}$
11. $5i\sqrt{2}$ **13.** $-2i\sqrt{2}$ **15.** $4i\sqrt{3}$ **17.** $1 \pm i$
19. 2 **21.** $\frac{3}{2}, -4$ **23.** $2 \pm 2i$ **25.** $-\frac{1}{2} \pm \frac{3}{2}i$
27. $\frac{1}{2} \pm \frac{\sqrt{3}}{2}i$ **29.** $-\frac{1}{2} \pm \frac{\sqrt{3}}{2}i$ **31.** 2, 3
33. $-\frac{1}{3} \pm \frac{\sqrt{2}}{3}i$ **35.** $\frac{1}{2}, -2$ **37.** $\frac{1 \pm 3\sqrt{5}}{2}$
39. $4 \pm 3i$ **41.** $\frac{4 \pm \sqrt{22}}{2}$ **43.** Yes **45.** $3 - 7i$
47. d **49.** 9; $(x - 3)^2$ **51.** 6 **53.** -2 **55.** 1

57-65.

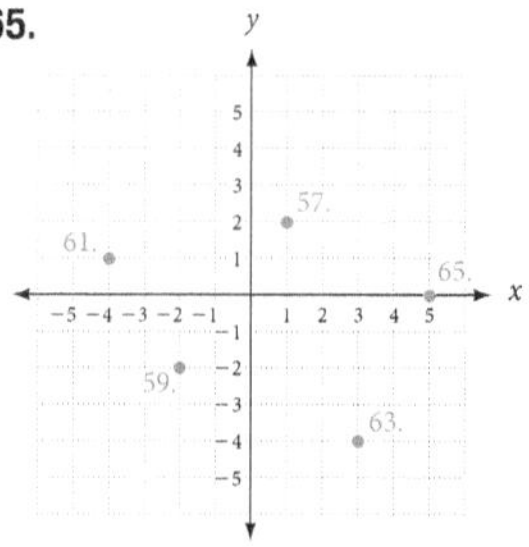

PROBLEM SET 9.6

1.

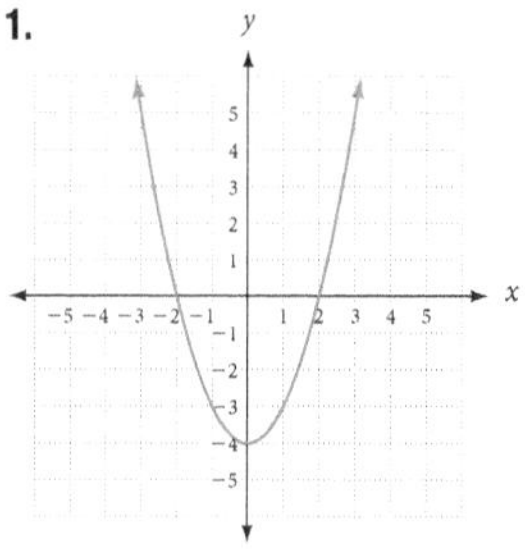

3.

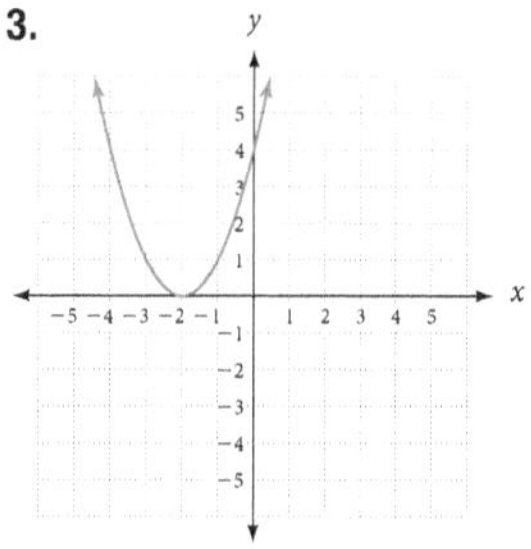

5.

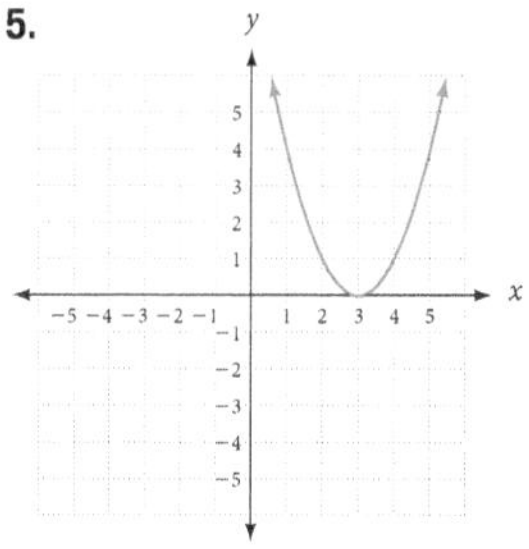

7.

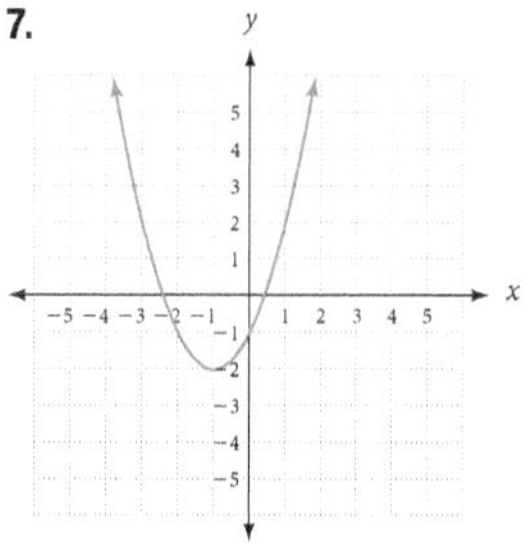

9

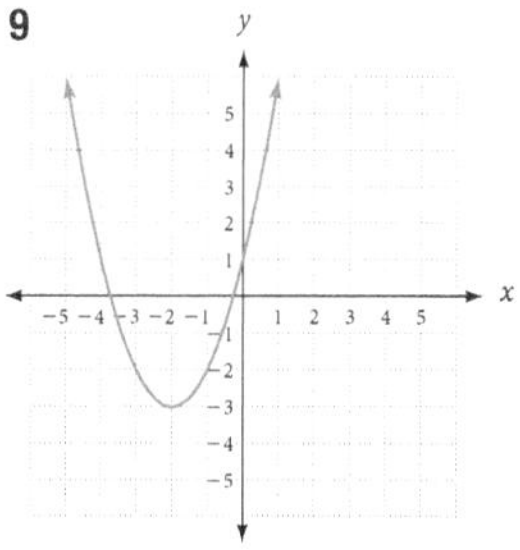

11.

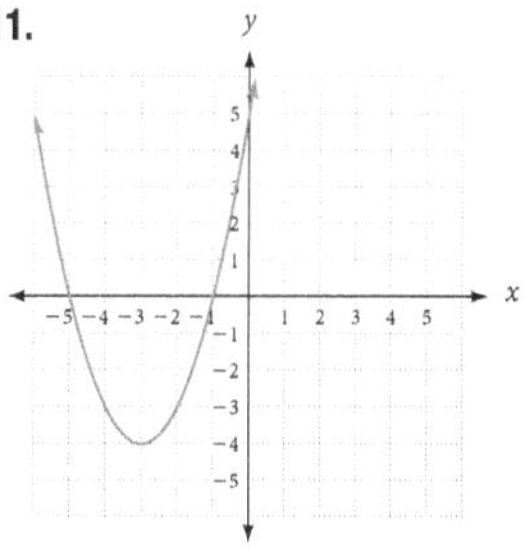

13. $(-\sqrt{5}, 0), (\sqrt{5}, 0), (0, -5)$

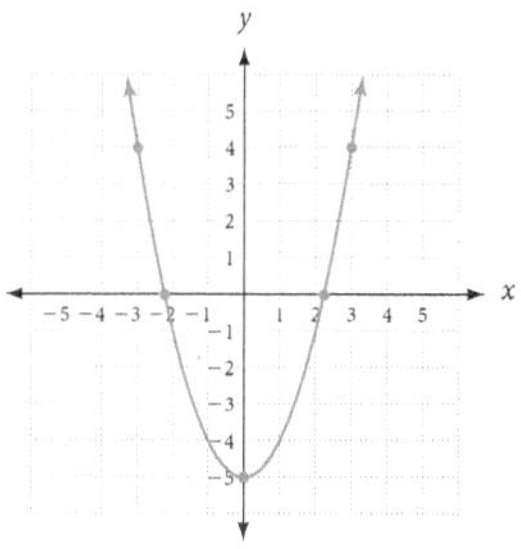

15. $(5, 0), (0, 25)$

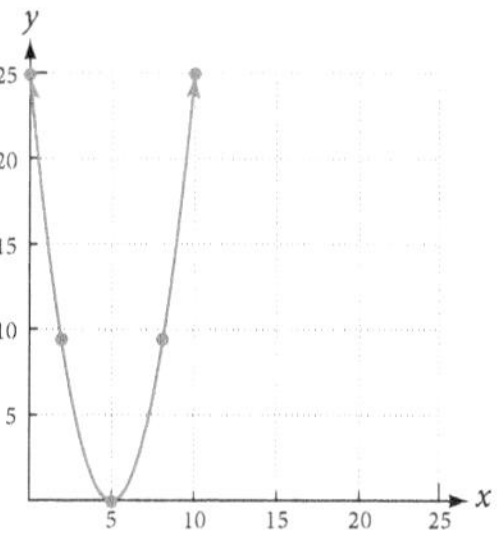

17. $(3 - \sqrt{2}, 0), (3 + \sqrt{2}, 0), (0, 7)$

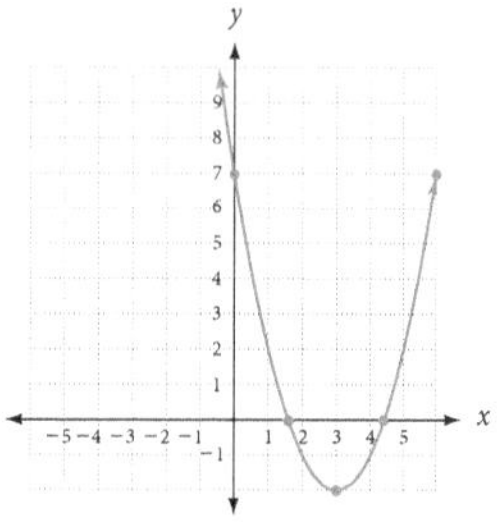

19. $(-1, 0), (3, 0), (0, -3)$

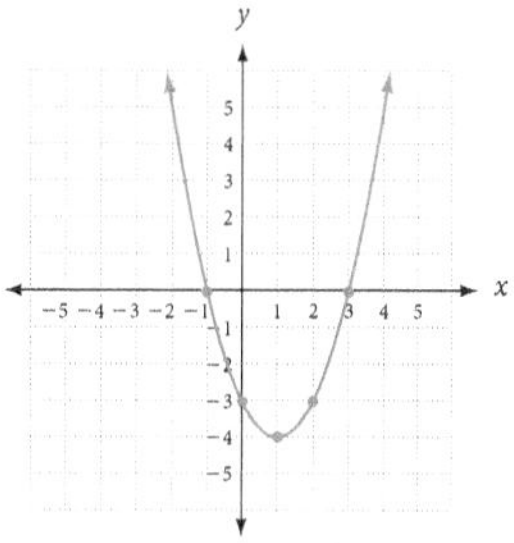

21. $(2 - \sqrt{2}, 0), (2 + \sqrt{2}, 0), (0, 2)$

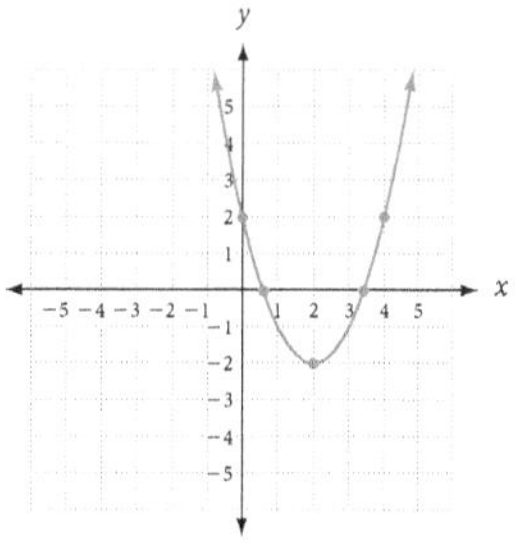

23.

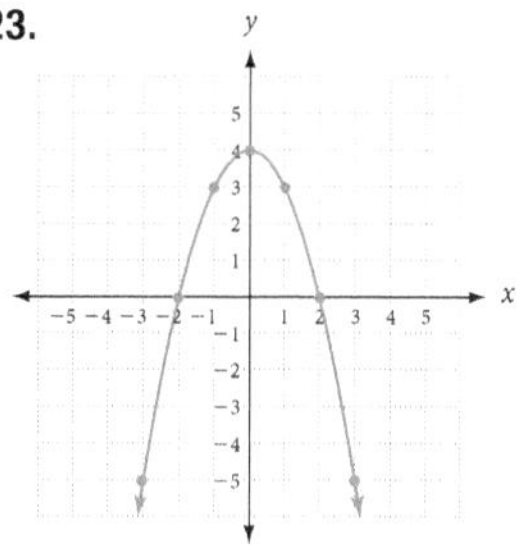

25.

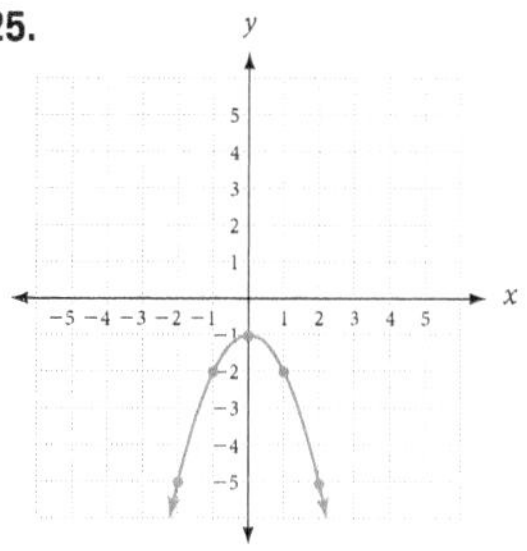

27.

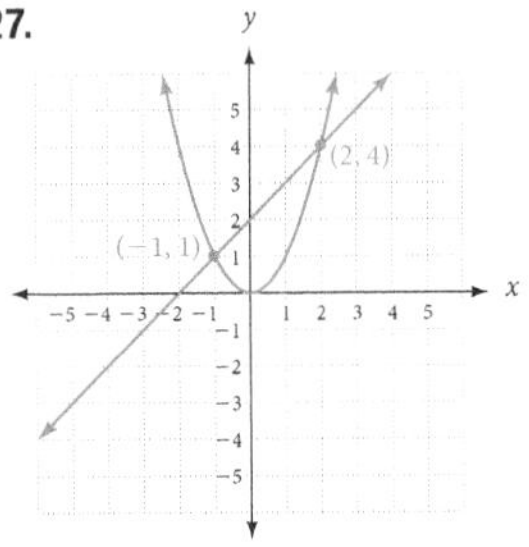

29.

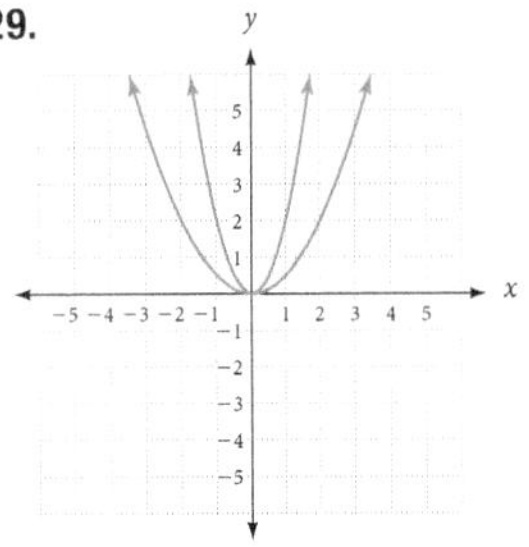

31. b **33.** a **35.** 7 **37.** $5\sqrt{2}$ **39.** $\frac{\sqrt{10}}{5}$

41. $21\sqrt{3}$ **43.** $-4 - 3\sqrt{6}$ **45.** $4\sqrt{5} + 4\sqrt{3}$

47. 7

CHAPTER 9 TEST

1. $-1, 9$ **2.** $2 - 3\sqrt{2}, 2 + 3\sqrt{2}$

3. $\frac{7}{3} - \frac{5\sqrt{2}}{3}i, \frac{7}{3} + \frac{5\sqrt{2}}{3}i$ **4.** $1, \frac{3}{2}$ **5.** $-2, \frac{1}{4}$

6. $-2, 5$ **7.** $\frac{5}{2}$ **8.** $2 - \sqrt{13}, 2 + \sqrt{13}$ **9.** $6i$

10. $13i$ **11.** $3i\sqrt{5}$ **12.** $2i\sqrt{3}$ **13.** $7 + 10i$

14. $3 + 4i$ **15.** 26 **16.** $10 + 10i$ **17.** $\frac{1}{10} + \frac{3}{10}i$

18. $\frac{15}{17} - \frac{8}{17}i$

19.

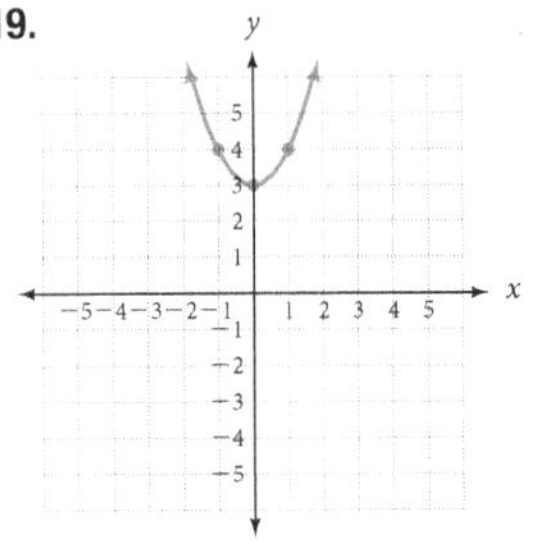

20.

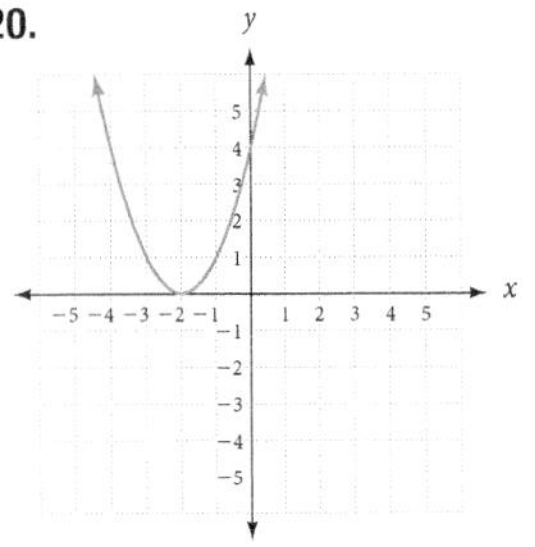

21.

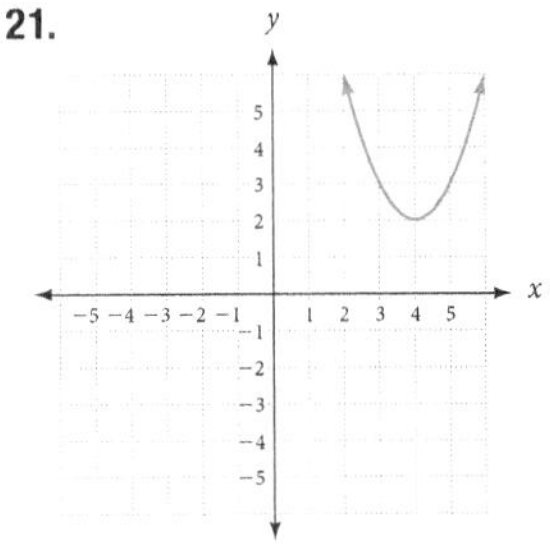

22.

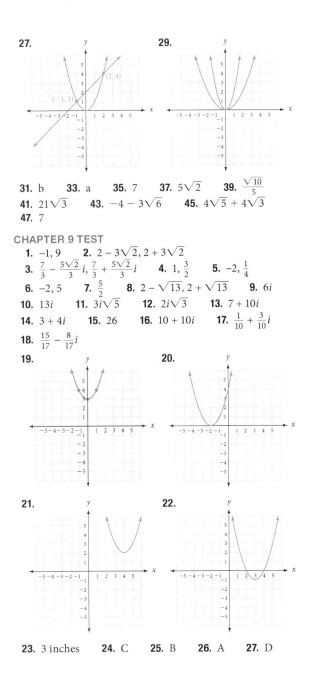

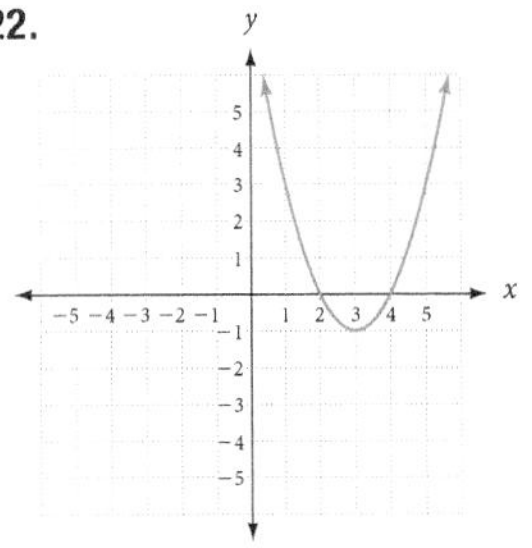

23. 3 inches **24.** C **25.** B **26.** A **27.** D

Index

A

Absolute value 20
Addition 29, 548
  of complex numbers 614
  of radical expressions 548
  of rational expressions 469
  with fractions 68
  with polynomials 346
Addition Property
  for inequalities 168
  of equality 104
Additive identity property 81
Additive inverse property 82
al-Khwarizmi 585
Analytic geometry 198
Area 7
Arithmetic sequence 33
Associative property of addition 78
Associative property of multiplication 78
Average speed 452

B

Bar chart 6
Base 4, 87, 311
Berlinski, David 141
Binomial squares 361
Boundary line 251

C

Cartesian coordinate system 198
Coefficient 93, 335
Common denominator 81
Commutative property of addition 77
Commutative property of multiplication 77
Comparison symbols 16
Completing the square 598
Complex conjugate 615
Complex fraction 479
Complex number 613
Composite 60
Conjugates 557
Consecutive integers 155
  even 155
  odd 155
Constant of variation 509
Constant term 93
Contradiction 124
Conversion factor 462
Coordinate 16
Cost 436
Counting numbers 17, 97

D

Degree 345
Denominator 18
Dependent 271
Descartes, René 198
Difference 4, 43
Difference of cubes 411
Difference of squares 410
Direct variation 509
Discriminant 619
Distributive property 79
Division 49, 112
  of complex numbers 615
  of radicals 563
  of rational expressions 459
  with exponents 322
  with fractions 63
  with polynomials 369
  with the number 0 53

E

Elimination method 277
Equality 16
  addition property of 105
  multiplication property of 111
Equation 103
  containing decimals 123
  containing fractions 122
  containing grouping symbols 120
  equivalent 104
  linear 119
Equivalent fractions 20
Exponent 4, 311
  one as an 324
  zero as an 324
Extraction of roots 589
Extraneous solution 489, 573
Extremes 493

F

Factor 59
Factoring 417
  by grouping 388, 403
  by trial and error 401
  the GCF 386
FOIL method 354
Formula 129
Fraction 18, 61

G

Geometric sequence 54
Golden ratio 541
Graph 195
Graphing 195
  solve linear system by 268
Greatest common factor 385
Grouping symbols 4

H

Horizontal axis 183
Horizontal line 203, 224

I

Identity 124
Inconsistent 271
Index 525, 581
Inductive reasoning 33
Inequality 16, 167
  addition property for 168
  multiplication property for 169
Infinite solutions 272
Integers 17
Interest 156
  rate 156
Inverse variation 510
Irrational numbers 18

L

Largest monomial 385
Least common denominator 65, 89, 470
Like terms 93, 547
Linear equation 119
Linear equation in two variables 199
Linear inequality 167
Linear inequality in two variables 253
Linear term 597
Line graph 183
Lowest terms 20

M

Mathematical model 141
Means 493, 518
Means-extremes property 493
Middle term 597
Monomial 335
  degree of a 335
Multiplication 47
  of complex numbers 614
  of radicals 555
  of rational expressions 459

with exponents 311
with fractions 61
with polynomials 353
Multiplication property
for inequalities 169
of equality 111
Multiplicative identity property 82
Multiplicative inverse property 82

N

Negative square root 524
Notation
inequality 167
set-builder 167
nth power property of equality 577
nth root 525
Number i 613
Numerator 18
Numerical coefficient 335

O

Opposites 21
Ordered pair 195
Order of operations 6
Origin 186

P

Paired data 183
Parabola 626
Parallel lines 225
Percent 132
Perfect square trinomial 409
Perimeter 7, 129
Perpendicular lines 225
Point-slope form 243
Polynomial 345
degree of a 345
leading term of a 345
opposite of a 347
Positive square root 523, 524
Power property for radicals 556
Prime factors 59
Prime number 59
Principal 156
Product 4
Product property for radicals 537, 555
Product property for square roots 536
Pythagorean theorem 435

Q

Quadrant 186
Quadratic equation 423
Quadratic term 597
Quotient 4
Quotient property for radicals 538, 563

R

Radical 523, 581
Radical sign 523, 581
Radicand 523, 581
Rate equation 452
Ratio 451
Rational expression 447
Rationalizing the denominator 565
Rational numbers 17
Real number line 15
Real numbers 16, 18
Reciprocal 63, 89
Rectangle 7, 129
Rectangular coordinate system 185
Revenue 357, 436
Rise 220
Run 220

S

Scatter diagram 183
Scientific notation 314
Sequence 97
arithmetic 33
geometric 54
Set-builder notation 167
Set notation 17
Similar terms 93, 338, 547
Simplified form 535
Slope 219
Slope-Intercept form 234
Solution set 103
Solving a formular for a variable 130
Solving a system of linear equations
by elimination method 283
by graphing 269
by substitution method 290
Solving equations containing square roots 575
Special triangles 592
Spiral of roots 529
Square 7
Square root property 589
Squaring property of equality 573
Standard form 199, 423
Subset 17
Substitution method 287
Subtraction 39, 107
of complex numbers 614
of radical expressions 548
of rational expressions 469
with fractions 68
with polynomials 347
Sum 4
Sum of cubes 411
Systems of linear equations 267
dependent 271
inconsistent 271
solution set 267, 272

T

Term 93, 335, 493
constant 93
like 94
similar 93
unlike 94
Triangle 7
angles of a 159

U

Unit analysis 462

V

Value of an expression 96
Variable 3
Vertex 626
Vertical axis 183
Vertical line 203, 224

W

Whole numbers 17

X

x-axis 186
x-coordinate 195
x-intercept 209

Y

y-axis 186
y-coordinate 195
y-intercept 209

Z

Zero-factor property 423